eyser of hot gas, known as Herbig-Haro 110, from a newborn star ricochets from the core of a cloud of hydrogen molecules; a combination of images taken with the Hubble Space Telescope.

THE COSMOS

Fourth Edition

An exciting introduction to astronomy, the fourth edition of this book uses recent discoveries and stunning photography to inspire non-science majors about the Universe and science. Written by two highly experienced and engaging instructors, each chapter has been fully updated, with more than 200 new images throughout, including recent images from space missions and the world's best observatories. Redesigned, stream-lined pages highlight the breathtaking imagery.

The text is organized as a series of stories, each presenting the history of the field, the observations made and how they fit within the process of science, our current understanding, and what future observations are planned. Math is provided in boxes and easily read around, making the book suitable for courses taking either mathematical or qualitative approaches. New discussion questions encourage students to think widely about astronomy and the role science plays in our everyday lives, and podcasts for each chapter aid studying and comprehension. Tables, Appendices, Selected Readings, and the Glossary have all been updated.

Jay M. Pasachoff is Field Memorial Professor of Astronomy at Williams College, where he teaches the astronomy survey course and works with undergraduate students on a variety of astronomical research projects. He is also Director of the Hopkins Observatory there. He received his undergraduate and graduate degrees from Harvard and was at the California Institute of Technology, where he has also had recent sabbatical leaves, before going to Williams College.

Pasachoff pioneered the emphasis in textbooks on contemporary astronomy alongside the traditional bases. He has taken advantage of his broad experience with a wide variety of ground-based telescopes and spacecraft in writing his texts. Also, "For his eloquent and informative writing of textbooks…, For his devotion to teaching generations of students, For sharing with the world the joys of observing eclipses, For his many popular books and articles on astronomy, For his intense advocacy on behalf of science education in various forums, For his willingness to go into educational nooks where no astronomer has gone before," he received the American Astronomical Society's 2003 Education Prize.

Pasachoff's expedition with students to the 2012 total solar eclipse was his 56th solar eclipse. He has also been carrying out research on transits of Venus and of Mercury. His research has recently been sponsored by the National Science Foundation, NASA, and the National Geographic Society. He is Chair of the Working Group on Eclipses of the solar division of the International Astronomical Union and Chair of the American Astronomical Society's Historical Astronomy Division. He is collaborating with colleagues to observe occultations of stars by Pluto, its largest moon (Charon), and other objects in the outer parts of the Solar System. He also works in radio astronomy of the solar atmosphere and of the interstellar medium, concentrating on deuterium and its cosmological consequences.

Pasachoff has been President of the Commission on Education and Development of the International Astronomical Union and twice Chair of the Astronomy Division of the American Association for the Advancement of Science. He is co-editor of *Teaching and Learning Astronomy: Effective Strategies for Educators Worldwide* (2005) and of *Innovation in Astronomy Education* (2008). Asteroid (5100) Pasachoff is named for him.

Alex Filippenko is a Professor of Astronomy and the Richard & Rhoda Goldman Distinguished Professor in the Physical Sciences at the University of California, Berkeley. His teaching of an astronomy survey course is very popular on campus; he has won the most coveted teaching awards at Berkeley and has nine times been voted "Best Professor" on campus. In 2006, he was named the U.S. National Professor of the Year. In 2010, he won the Emmons Award of the Astronomical Society of the Pacific for excellence in college astronomy teaching. He received his undergraduate degree from the University of California, Santa Barbara, and his doctorate from the California Institute of Technology.

Filippenko has produced five video courses on college-level astronomy through *The Great Courses*. The recipient of the 2004 Carl Sagan Prize for Science Popularization, he lectures widely, and he has appeared frequently on science newscasts and television documentaries, especially *The Universe* series on The History Channel and H2 (about 40 episodes spanning 6 seasons).

Filippenko's primary areas of observational research are exploding stars (supernovae), gamma-ray bursts, active galaxies, black holes, and observational cosmology; he frequently uses the Hubble Space Telescope, the Keck 10-meter telescopes, and other facilities. He and his collaborators have obtained some of the best evidence for the existence of stellar-mass black holes in our Milky Way Galaxy. His robotic telescope at Lick Observatory, together with a large team that includes many undergraduate students, is conducting one of the world's most successful searches for exploding stars in relatively nearby galaxies, having found more than 1,000 of them. He made major contributions to both of the teams that discovered the accelerating expansion of the Universe, and the team leaders received the 2011 Nobel Prize in Physics for their work. One of the world's most highly cited astronomers, his research has been recognized with several prestigious awards, including election to the U.S. National Academy of Sciences.

He has served as a Councilor of the American Astronomical Society and has been Vice President and President of the Astronomical Society of the Pacific. He is an active member of the International Astronomical Union.

"This is an extraordinarily attractive, captivating, and easy-to-follow textbook of modern astronomy. I will happily continue to use it in teaching my course."
Professor Alex Wolszczan, Penn State University

"... the work of two great astronomers and top teachers ... clear and concise. This is the text I would use for my introductory astronomy course."
Professor Arun Venkatachar, Ohio University

"The authors convey a clear and enthusiastic pedagogic presentation of an exciting field. As a textbook, it will be of great benefit to students, providing a valuable starting point to learn about the subject ... *The Cosmos* stands out for its continued excellence over time."
Dr. Roger Kadala, Hawaii Pacific University

"... long considered one of the best descriptive astronomy books, the new Fourth Edition is even better ... not only a beautifully illustrated text, but has many of the latest developments in astronomy."
Professor Victor Kriss, Lewis-Clark State College

"... my first choice for a textbook for introductory astronomy. Hundreds of students in my introductory astronomy class have used *The Cosmos* as the primary text, and I am very happy with the coverage and presentation of the material."
Professor Steinn Sigurðsson, Penn State University

"... a splendid new addition to *The Cosmos* series, distinguished by its crisp, interesting, and informative style and made all the more accessible through the generous use of photographs, charts, and diagrams that greatly facilitate one's understanding."
Mark Rader, Notre Dame High School

THE COSMOS

Astronomy in the New Millennium

Fourth Edition

Jay M. Pasachoff

Williams College, Massachusetts

Alex Filippenko

University of California, Berkeley

CAMBRIDGE
UNIVERSITY PRESS

CAMBRIDGE
UNIVERSITY PRESS

32 Avenue of the Americas, New York, NY 10013-2473, USA

Cambridge University Press is part of the University of Cambridge.

It furthers the University's mission by disseminating knowledge in the pursuit of education, learning, and research at the highest international levels of excellence.

www.cambridge.org
Information on this title: www.cambridge.org/9781107687561

© 2001, 2004, 2007, 2014 Jay M. Pasachoff and Alex Filippenko

First edition published in 2001 by Harcourt College Publishers
Second edition published in 2004 by Brooks/Cole, a division of Thomson Learning, Inc.
Third edition published in 2007 by Brooks/Cole, Cengage Learning
Fourth edition published in 2014 by Cambridge University Press
Reprinted 2015

Printed in the United States of America by Sheridan Books, Inc.

A catalog record for this publication is available from the British Library.

Library of Congress Cataloging in Publication Data
Pasachoff, Jay M.
The cosmos : astronomy in the new millennium / Jay M. Pasachoff, Williams College, Massachusetts, Alex Filippenko, University of California, Berkeley. – 4th edition.
    pages   cm
Includes bibliographical references and index.
ISBN 978-1-107-68756-1 (pbk.)
1. Astronomy.   I. Filippenko, Alexei V.   II. Title.
QB43.3.P37   2012
520–dc23   2012043113

ISBN 978-1-107-68756-1 Paperback

Additional resources for this publication at www.cambridge.org/cosmos

Jay M. Pasachoff collection

The constellation Taurus, the Bull, from the 1603 star atlas by Johann Bayer.

NASA, ESA, and E. Sabbi (ESA/STScI)/R. O'Connell (University of Virginia) and the Wide Field Camera 3 Science Oversight Committee

Merging star clusters in 30 Doradus, in the Large Magellanic Cloud, imaged with the Hubble Space Telescope.

ATST/NSO/NSF, image by L. Phelps

An artist's rendering of the Advanced Technology Solar Telescope with its 4-m-diameter mirror, now being erected on Haleakala, Maui, Hawaii.

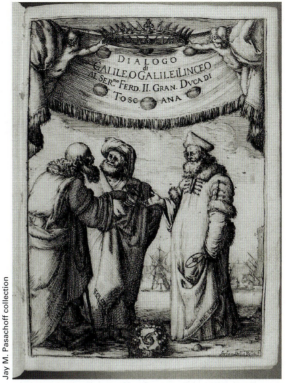

Jay M. Pasachoff collection

The frontispiece of Galileo's *Dialogo*, published in 1632.

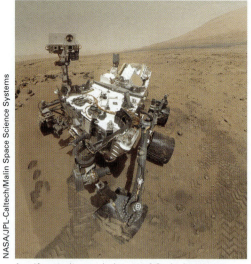

NASA/JPL-Caltech/Malin Space Science Systems

A self-portrait mosaic image of Curiosity, the lander of NASA's Mars Science Laboratory, with Gale Crater's Mount Sharp at background right.

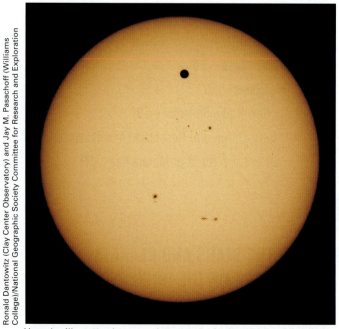

Ronald Dantowitz (Clay Center Observatory) and Jay M. Pasachoff (Williams College)/National Geographic Society Committee for Research and Exploration

Venus's silhouette shows, as do sunspots, in the middle of the six-hour transit of Venus across the face of the Sun on June 5/6, 2012.

Saturn's moon Titan in front of the planet and its rings, from NASA's Cassini spacecraft in 2012.

Jupiter with the Hubble Space Telescope, a composite of an ultraviolet and an infrared image made as part of one of the authors' observations made in 2012 while a transit of Venus as seen from Jupiter was dimming that giant planet by 0.01%.

Jay M. Pasachoff/Williams College
Eclipse Expedition/NSF

The shadow of the Moon surrounds the eclipsed Sun in this view from a helicopter, above an Australian cloud-deck, of the 2012 total solar eclipse; we see the solar corona surrounding the dark silhouette of the Moon.

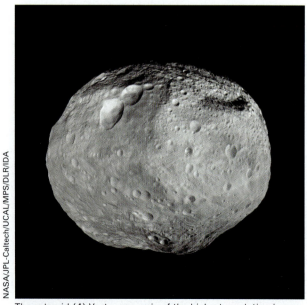

NASA/JPL-Caltech/UCAL/MPS/DLR/IDA

The asteroid (4) Vesta, a mosaic of the highest-resolution images from NASA's Dawn spacecraft, which is now en route to (1) Ceres.

8

PLUTO, COMETS, AND SPACE DEBRIS   197

9

OUR SOLAR SYSTEM AND OTHERS   233

Jay M. Pasachoff/NSF

The diamond ring effect along with reddish chromosphere and prominences mark the end of the 2012 total solar eclipse observed from Australia.

OUR STAR: THE SUN   255

STARS: DISTANT SUNS   279

Solar Dynamics Observatory's Atmospheric Imaging Assembly/LMSAL/NASA

SDO/AIA 171   2012-11-28 23:46:00 UT

A view of the Sun in the extreme ultraviolet part of the spectrum, showing million-degree gas held in place by the solar magnetic field.

12

HOW THE STARS SHINE: COSMIC FURNACES   311

An optical and x-ray composite image of a supernova remnant, incorporating Hubble Space Telescope images showing the pink optical shell surrounding the x-ray images, shown in blue and green, from the Chandra X-ray Observatory. The supernova came from a supergiant star that exploded (Type Ia) 400 years ago in the Large Magellanic Cloud. The bubble is 23 light-years across.

NASA, ESA, CXC, SAO, the Hubble Heritage Team (STScI/AURA), and J. Hughes (Rutgers University)

13

THE DEATH OF STARS: RECYCLING   331

14

BLACK HOLES: THE END OF SPACE AND TIME   361

15

THE MILKY WAY: OUR HOME IN THE UNIVERSE   383

16

A UNIVERSE OF GALAXIES   413

NASA, ESA, the Hubble Heritage Team (STScI/AURA)-ESA/Hubble Collaboration, and W. Keel (University of Alabama)

NGC 3314, a chance alignment of two distant galaxies, imaged with the Hubble Space Telescope. The galaxies are actually separated by a distance ten times that of our galaxy from the Andromeda galaxy and are about 140 million light-years from us.

17

QUASARS AND ACTIVE GALAXIES   451

18

COSMOLOGY: THE BIRTH AND LIFE OF THE COSMOS   477

NASA, ESA, M. Postman (STScI), and the CLASH Team

A cluster of galaxies 4.5 billion light-years away, in a Hubble Space Telescope survey that is mapping dark matter.

Saturn's moons Titan and, smaller, Dione through its haze, with Saturn's disk and rings in the background from NASA's Cassini spacecraft.

We provide, as well, a sampling of the many significant findings from the Hubble Space Telescope. Numerous images from the Chandra X-ray Observatory, the Spitzer Space Telescope, and the Herschel Space Observatory also appear, as do results from the Swift spacecraft that is monitoring gamma-ray bursts, among the most powerful and violent phenomena in the Universe. Of particular interest is the recent advance in our understanding of the age and expansion of the Universe, both through better measurements of the current expansion rate and through the discovery that the Universe's expansion is currently speeding up with time. We also present the recent exciting measurements that allow astronomers to determine the overall geometry of the Universe and the early appearance of large-scale structure through detailed observations of the cosmic background radiation.

COSMIC ORIGINS

The study of our origins, whether it be ourselves as humans, our Earth as a planet, our Sun as a star, or our Galaxy as a whole, is as interesting to many of us as it is to look at our own baby pictures. Most of us have gazed at the stars and wondered how they came to be and what their relationship to us is. NASA Astrophysics has chosen *Cosmic Origins* as one of its major themes for the organization of its missions and has several spacecraft planned in the Cosmic Origins program. We emphasize the study of origins in this text, dealing with a variety of relevant material. NASA Astrophysics also has two additional themes, which we also cover extensively in this text: *Physics of the Cosmos* and *Exoplanet Exploration*.

PEDAGOGY

Within the book (or, in one case, on its website) we have five kinds of features:

1. *Star Parties.* An occasional feature that shows students how to find things in the sky. These include observing exercises and links to the star maps that appear on the inside covers of the book.
2. *Figure It Out.* In some astronomy courses, it may be appropriate to elaborate on equations. Because we wrote *The Cosmos: Astronomy in the New Millennium* to be a descriptive presentation of modern astronomy for liberal-arts students, we kept the use of mathematics to a minimum. However, we recognize that some instructors wish to introduce their students to more of the mathematics associated with astronomical phenomena. Consequently, we provide mathematical features, numbered so they can be assigned or not, at the instructor's option.
3. *Lives in Science.* These boxes provide biographies of important historical figures like Copernicus and Galileo.
4. *A Closer Look.* Using these boxes, students can further explore interesting topics, such as size scales in the Universe, observing with large telescopes, various celestial phenomena, mythology, and naming systems. We are pleased to supply some exciting close-ups of Mars, Titan, and a comet, for example.
5. *People in Astronomy.* Each of these interviews (available on the website) presents a notable contemporary astronomer engaged in conversation about a variety of topics: current and future work in astronomy, what led that person to study and pursue astronomy as a career, and why learning about astronomy is an important scientific *and* human endeavor. We hope that you enjoy reading their comments as much as we enjoyed speaking with them and learning about their varied interests and backgrounds.

We have provided aids to make the book easy to read and to study from. New vocabulary is boldfaced in the text and in the expanded summaries that are given in context at the end of each chapter, and defined in the glossary. The index provides further aid in finding explanations. An expanded set of end-of-chapter questions covers a range of material and includes some that are straightforward to answer from the text and others that require more thought. Appendices provide some information on planets, stars, constellations, and nonstellar objects. The exceptionally beautiful sky maps by Wil Tirion (inside the covers of the book) will help you find your way around the sky when you go outside to observe the stars. Be sure you do.

THE COSMOS WEBSITE

Adopters of *The Cosmos*, 4th edition, have access to a rich array of teaching and learning resources at the book's web page. Students will find multiple-choice self-tests and other useful information.

Recent educational research has shown that students often need to unlearn incorrect ideas in order to understand the correct ones. We have thus placed a list of such "Misconceptions," including common incorrect ideas and the correct alternatives, on the book's website. The book's website also has a list of many relevant URLs from the World Wide Web.

The website is updated occasionally with links to information and photographs from a wide variety of sources. Updates also appear at this book's other website at http://www.thecosmos4.com. Each professor and each student should look at these updates and links, which are organized in chapter order.

STAR CHARTS

Pasachoff's *Peterson Field Guide to the Stars and Planets* is a suggested accompaniment for those wanting monthly star maps, star charts, and other detailed observing aids; see http://www.solarcorona.com. Quarterly star maps, improved and redrawn for this edition by the foremost celestial cartographer, Wil Tirion of The Netherlands, appear on the inside covers in *The Cosmos*, 4th edition.

ACKNOWLEDGMENTS

The publishers join us in placing a heavy premium on accuracy, and we have made certain that the manuscript and past editions have been carefully read and considered. As a result, you will find that the statements in this book, brief as they are, are authoritative.

This fourth edition of *The Cosmos* benefited from the advice and special images provided by Geoff Marcy (University of California, Berkeley), Lynn Cominsky (Sonoma State University), Leon Golub (Harvard-Smithsonian Center for Astrophysics), Joel Weisberg (Carlton College), Daniel Seaton (Royal Observatory of Belgium), John Johnson (Caltech), Robert Simmon (NASA's Goddard Space Flight Center), Richard McCray (University of Colorado), Richard Manchester (CSIRO, Australia), Daniel Stern (JPL), Daniel Mortlock (Imperial College, London), Matthew Templeton, Sarah Beck, and Arne Henden (AAVSO), Stefan Fichtel (Ixtract), Augustin Sanchez Lavega (Valencia), Federica Bianco (Las Cumbres Observatory Global Telescope Network and the University of California, Santa Barbara), Gareth Williams (Harvard-Smithsonian Center for Astrophysics), Lorne McKee (National Resources Canada), Jens Vellev (Aarhus University), Gerard van Belle (Lowell Observatory), Wayne Hammond (Chapin Library of Williams College), Richard Green (Large Binocular Telescope Observatory), Alan Marscher (Boston University), and Axel Mellinger (Central Michigan University).

For comments, corrections, and updates for this fourth edition, we also thank Katherine Stack, Caltech, Mars; Thomas Widemann, Observatoire de Paris, Venus; Imke de Pater, University of California, Berkeley, Neptune; Heidi Hammel, AURA, Uranus; Geoff Marcy, University of California, Berkeley, exoplanets; Paul Kalas, University of California, Berkeley, exoplanet imaging; Alex Wolszczan, Penn State, pulsar planets; Glenn Schneider, Steward Observatory, University of

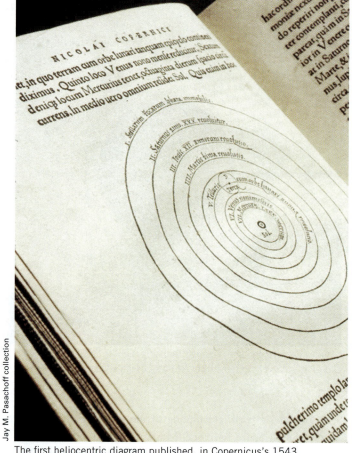

Jay M. Pasachoff collection

The first heliocentric diagram published, in Copernicus's 1543 *De Revolutionibus*.

Arizona, exoplanet imaging; Leon Golub, Harvard-Smithsonian CfA, Sun; Daniel B. Seaton, Royal Observatory of Belgium, Sun; Edw. Ginsberg, UMass, Boston, general and the end-of-chapter questions; Michael Blanton, NYU, Sloan Digital Sky Survey; Gerard Williger, University of Louisville, general; Art McDonald, Sudbury Neutrino Observatory, solar neutrinos; Huajie Cao, Princeton, dark-matter experiments; Caty Pilachowski, University of Indiana, globular clusters; Howard Bond, Space Telescope Science Institute, planetary nebulae; Karen Kwitter, Williams College, planetary nebulae; J. J. Hermes and Don Winget, University of Texas, relativistic white-dwarf system; Laura Brenneman, Harvard-Smithsonian CfA, black holes; Holland Ford, STScI, supermassive black holes; Roeland van der Marel, STScI, supermassive black holes; Seth Shostak, SETI Institute, current SETI; Heather Knutson, Caltech, habitable zones for exoplanets; and Lynn Cominsky, Sonoma State, gamma-ray bursts.

For expert work in the revision and updating of the planetary chapters, we thank Deborah Pasachoff in Pasadena. For her extensive editorial consulting, we thank Madeline Kennedy in Williamstown. For additional assistance, we thank Michele Rech in Williamstown.

At Cambridge University Press, we thank our editor, Vince Higgs, and his assistant, Louis Gulino; James Dunn, Helen Wheeler, Kerry Cahill, and Danielle Poupore in New York; and Catherine Flack and her team in the U.K.

J.M.P. thanks various members of his family, who have provided vital and valuable editorial services in addition to their general support. His wife, Naomi, has always been very helpful both personally and editorially. He is grateful for the family support of Deborah Pasachoff/Ian Kezsbom/Lily Kezsbom/Jacob Kezsbom and of Eloise Pasachoff/Tom Glaisyer/Samuel Glaisyer/Jessica Glaisyer.

A.F. thanks his children, Zoe, Simon, Caprielle, and Orion Filippenko, for the long hours they endured while he was busy working on this book rather than playing with them; they mean the Universe to him. A.F. also thanks his wife, Noelle Filippenko, for her incredible patience, advice, and support; he could not have succeeded without her, and he looks forward to sharing more of the wonders of the cosmos with her.

We are extremely grateful to all the individuals named above for their assistance. Of course, it is we who have put this all together, and we alone are responsible for the content. We would appreciate hearing from readers with suggestions for improved presentation of topics, with comments about specific points that need clarification, with typographical or other errors, or just to tell us how you like your astronomy course. We invite readers to write to us, respectively, at Williams College, Hopkins Observatory, 33 Lab Campus Drive, Williamstown, MA 01267–2630, eclipse@williams.edu, and Department of Astronomy, University of California, Berkeley, CA 94720–3411, alex@astro.berkeley.edu, or through the book's websites. We promise a personal response to each writer.

A Hubble Space Telescope view of the spiral galaxy NGC 5584 with Hubble's new Wide Field Camera 3. Younger stars appear bluer and are in the spiral arms; the stars in the core are mostly older. Reddish, more distant galaxies, can be seen through the nearer galaxy, which is 72 million light years from us in the constellation Virgo. This galaxy had a Type Ia supernova, a supernova caused by the incineration of a white dwarf near its maximum mass limit. The peak apparent brightness of the supernova was measured, and its true peak power was calculated by combining this with the galaxy's distance, obtained through studies of 250 Cepheid variable stars in the galaxy. Type Ia supernovae and Cepheid variables were studied in many additional galaxies, and the calibrated supernovae were subsequently used to measure distances of large numbers of galaxies and determine the expansion rate of the Universe. This project was a main reason for the launch of the Hubble Space Telescope and for naming it after Edwin Hubble.

NASA, ESA, A. Riess (STScI/JHU), L. Macri (Texas A&M University), and the Hubble Heritage Team (STScI/ AURA)

1 A Grand Tour of the Heavens

ORIGINS

We comment on "Origins" and "Structure and Evolution" as organizing themes.

Astronomy is in a golden age, filled with the excitement of new discoveries and a deeper understanding of the Universe, our home – and what an enthralling universe it is!

We have explored most of the planets in the Solar System, revealing an astonishingly wide variety of terrains and moons. We have discovered planets orbiting other stars, increasing our confidence that life exists elsewhere. We have solved many of the mysteries surrounding stellar birth and death, revealing among other things how the chemical elements inside our bodies, like calcium and oxygen, formed inside stars. With the Hubble Space Telescope, the Chandra X-ray Observatory, and the Spitzer Space Telescope we have examined galaxies shortly after their birth, deducing important clues to the origin and evolution of our own Milky Way Galaxy. (When we refer to our Milky Way Galaxy, we say "our Galaxy" or "the Galaxy" with an uppercase "G." When we refer to other galaxies, we use a lowercase "g.")

We have witnessed explosions of stars halfway across the visible Universe whose power is so tremendous that it rivals a galaxy containing ten billion normal stars. Indeed, we are lucky that none have recently occurred too close to Earth, for we wouldn't survive. We have detected black holes – strange objects whose gravitational pull is so strong that nothing, not even light, can escape. And, most recently, we have found strong evidence that what we thought was "empty space" actually contains a dark, gravitationally repulsive kind of energy that is causing the Universe to expand faster and faster with time. The origin of this "dark energy" is a complete mystery, but an understanding of it may revolutionize physics.

AIMS

1. Survey the Universe and the methods astronomers use to study it (Sections 1.1, 1.2, and 1.4).

2. Learn the measurement units used by astronomers (Section 1.1).

3. See how the sky looks in different seasons (Section 1.3).

4. Understand the value of astronomy to humans (Section 1.5).

5. Assess the scientific method and show how pseudoscience fails scientific tests (Sections 1.6 and 1.7).

The study of astronomy enriches our view of the Universe and fills us with awe, increasing our appreciation of the Universe's sheer grandeur and beauty. We hope that your studies will inspire you to ask questions about the Universe all around us and show you how to use your detective skills to search for the answers. Get ready for a thrilling voyage unlike any that you've ever experienced!

1.1 PEERING THROUGH THE UNIVERSE: A TIME MACHINE

Astronomers have deduced that the Universe began almost 14 billion years ago. Let us consider that the time between the origin of the Universe and the year 2013, or 14 billion years, is compressed into one day. If the Universe began at midnight, then it wasn't until slightly after 4 p.m. that the Earth formed; the first fossils date from 6 p.m. The first humans appeared only 2 seconds ago, and it is only 1/300 second since Columbus landed in America. Still, the Sun should shine for another 9 hours; an astronomical timescale is much greater than the timescale of our daily lives (■ Fig. 1–1).

One fundamental fact allows astronomers to observe what happened in the Universe long ago: Light travels at a finite speed, 300,000 km/s

(equal to 186,000 miles per second), or nearly 10 trillion km per year. As a result, if something happens far away, we can't know about it immediately. Light from the Moon takes about a second to reach us (1.3 seconds, more precisely), so we see the Moon as it was roughly one second ago. Light from the Sun takes about eight minutes to reach us. Light from the nearest of the other stars takes over four years to reach us; we say that it is over four "light-years" away. Once we look beyond the nearest stars, we are seeing much farther back in time (see *Figure It Out 1.1: Keeping Track of Space and Time*).

The Universe is so vast that when we receive light or radio waves from objects across our home, the Milky Way Galaxy (a collection of hundreds of billions of stars bound together by gravity), we are seeing back tens of thousands of years. Even for the most nearby other galaxies, light has taken hundreds of thousands, or even millions, of years to reach us. And for the farthest known galaxies, the light has been travelling to us for billions of years. New telescopes on high mountains and in orbit around the Earth enable us to study these distant objects much better than we could previously. When we observe these farthest objects, we see them as they were billions of years ago. How have they changed in the billions of years since? Are they still there? What have they evolved into? The observations of distant objects that we make today show us how the Universe was a long, long time ago – peering into space, we watch a movie of the history of the Universe, allowing us to explore and eventually understand it.

Building on such observations, astronomers use a wide range of technology to gather information and construct theories to learn about the Universe, to discover what is in it, and to predict what its future will be. This book will show you how we look, what we have found, and how we interpret and evaluate the results.

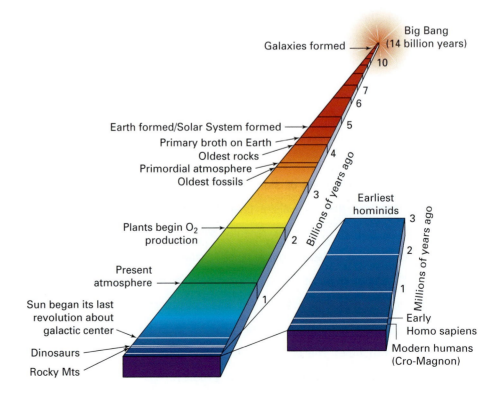

■ **FIGURE 1–1** A sense of time.

FIGURE IT OUT 1.1   Keeping Track of Space and Time

Throughout this book, we shall generally use the metric system, which is commonly used by scientists. The basic unit of length, for example, is the meter, which is equivalent to 39.37 inches, slightly more than a yard. Prefixes are used (Appendix 1) in conjunction with the word "meter," whose symbol is "m," to define new units. The most frequently used prefixes are "milli-," meaning $1/1,000$, "centi-," meaning $1/100$, "kilo-," meaning 1,000 times, and "mega-," meaning one million times. Thus 1 millimeter is $1/1,000$ of a meter, or about 0.04 of an inch, and a kilometer is 1,000 meters, or about $5/8$ of a mile.

As we describe in *Figure It Out 1.2: Scientific Notation*, we keep track of the powers of 10 by which we multiply 1 m by writing the number of tens we multiply together as an exponent; 1,000 m (1,000 meters), for example, is 10^3 m, since 1,000 has 3 zeros following the 1 and thus represents three tens multiplying each other.

The standard symbol for "second" is "s," so km/s is kilometers per second. Astronomers measure mass in kilograms (kg), where each kilogram is 10^3 grams (1,000 g).

We can keep track of distance not only in the metric system but also in units that are based on the length of time that it takes light to travel. The speed of light is, according to Einstein's special theory of relativity, the greatest speed that is physically attainable for objects travelling through space. Light travels at 300,000 km/s (186,000 miles/s), so fast that if we could bend it enough, it would circle the Earth over 7 times in a single second. (Such bending takes place only near a black hole or cluster of galaxies, though; for all intents and purposes, light near the Earth goes straight.)

Even at that fantastic speed, we shall see that it would take years for us to reach the stars. Similarly, it has taken years for the light we see from stars to reach us, so we are really seeing the stars as they were years ago. Thus, we are looking backward in time as we view the Universe, with distant objects being viewed farther in the past than those nearby.

The distance that light travels in a year is called a **light-year**; note that the light-year is a unit of *length* rather than a unit of time even though the term "year" appears in it. It is equal to 9.53×10^{12} km (nearly 10 trillion km) – an extremely large distance by human standards.

A "month" is an astronomical time unit, based on the Moon's orbit, and a "year" is an astronomical time unit, based on the Earth's orbit around the Sun. The measurement of time itself is now usually based on processes in atoms, which are used to define the second. So the second from atomic timekeeping is slightly different from the second that we think of as a sixtieth of a minute, which is a sixtieth of an hour, which is a twenty-fourth of a day, which is based on the rotation of the Earth on its axis. For some purposes, weird stars called pulsars keep the most accurate time in the Universe. In this book, when we are talking about objects billions of years old, it won't matter precisely how we define the second or the year.

1.2 HOW DO WE STUDY THINGS WE CAN'T TOUCH?

The Universe is a place of great variety – after all, it has everything in it! At times, astronomers study things of a size and scale that humans can easily comprehend: the planets, for instance. Most astronomical objects, however, are so large and so far away that we have trouble grasping their sizes and distances. Many of these distant objects are fascinating and bizarre – ultra-dense pulsars that spin on their axes hundreds of times per second, exploding stars that light up the sky and incinerate any planets around them, giant black holes with a billion times the Sun's mass.

In addition to taking photographs of celestial objects, astronomers break down an object's light into its component colors to make a **spectrum** (see Chapter 2), much like a rainbow (■ Fig. 1–2). Today's astronomers, thanks to advances in telescopes and in devices to detect the incoming radiation, study not only the visible part of the spectrum, but also its gamma rays, x-rays, ultraviolet, infrared, and radio waves. We use telescopes on the ground and in space to observe astronomical objects in almost all parts of the spectrum. Combining views in the visible part of the spectrum with studies of invisible radiation gives us a more complete idea of the astronomical object we are studying than we could otherwise have (■ Fig. 1–3). Regardless of whether we are looking at nearby or very distant objects, the techniques of studying in various parts of the spectrum are largely the same.

The tools that astronomers use are bigger and better than ever. Giant telescopes on mountaintops collect visible light with mirrors as large as 10 meters across (■ Fig. 1–4). Up in space, above Earth's atmosphere, the Hubble Space Telescope sends back very clear images (■ Fig. 1–5). Many faraway objects are seen as clearly with Hubble as those closer to us appear with most ground-based telescopes. This accomplishment enables us to study a larger number of distant objects in detail. (But ground-based astronomers have developed methods of seeing very clearly, too, over limited areas of the sky in certain parts of the spectrum.) The Chandra X-ray Observatory produces clear images of a wide variety of objects using the x-rays they emit (■ Fig. 1–6).

FIGURE IT OUT 1.2   Scientific Notation

In astronomy, we often find ourselves writing numbers that have strings of zeros attached, so we use what is called either scientific notation or exponential notation to simplify our writing chores. Scientific notation helps prevent making mistakes when copying long strings of numbers, and so aids astronomers (and students) in making calculations.

In scientific notation, which we use in *A Closer Look 1.1: A Sense of Scale*, included in this chapter, we merely count the number of zeros, and write the result as a superscript to the number 10. Thus the number 100,000,000, a 1 followed by 8 zeros, is written 10^8. The superscript is called the exponent. (In spreadsheets, such as Microsoft Excel, the exponent is written after a caret, ^, as in 10^8.) We also say that "ten is raised to the eighth power."

When a number is not an integer power of 10, we divide it into two parts: a number between 1 and 10, and an integer power of 10. Thus the number 3645 is written as 3.645×10^3. The exponent shows how many places the decimal point was moved to the left.

We can represent numbers between zero and one by using negative exponents. A minus sign in the exponent of a number means that the number is actually one divided by what the quantity would be if the exponent were positive. Thus $10^{-2} = 1/10^2 = 1/100 = 0.01$. As a further example, instead of working with 0.00256, we would move the decimal point three places to the right and write 2.56×10^{-3}.

Powers of 1,000 beyond kilo- (a thousand) are mega- (a million), giga- (a billion), tera- (a trillion), peta-, exa-, zetta-, and yotta-. It has been suggested, not entirely seriously, to use groucho- and harpo-, after two of the Marx Brothers, for the next prefixes.

1.3 FINDING CONSTELLATIONS IN THE SKY

When we look outward into space, we see stars that are at different distances from us. But our eyes don't reveal that some stars are much farther away than others of roughly the same brightness. People have long made up stories about groups of stars that appear in one part of the sky or another. The major star groups are called **constellations**.

These constellations were given names, occasionally because they resembled something (for example, Scorpius, the Scorpion), but mostly to honor a hero or other subject of a story.

The International Astronomical Union put the scheme of constellations on a definite system in 1930. The sky was officially divided into 88 constellations (see Appendix 6 with definite boundaries, and every star is now associated with one and only one constellation. But the constellations give only the directions to the stars, and not the stars' distances. Individual stars in a given constellation generally have

(a)

(b)

■ **FIGURE 1–2** (a) The visible spectrum, light from the Sun spread out in a band. The dark lines represent missing colors, which tell us about specific elements in space or in the Sun absorbing those colors. (b) Spectra of all the stars in a cluster of extremely hot stars, including even a so-called Wolf-Rayet star whose spectrum shows some bands of emission as opposed to the usual absorption; a drawing.

(a) National Optical Astronomy Observatories/Kitt Peak;
(b) Harry Roberts, Australia

(a) (b) (c)

Spitzer infrared data: NASA/JPL-Caltech/I. Bourke (Harvard-Smithsonian CfA) and c2d Legacy Team; VLT visible light data: J. Alves (ESO), E. Tolstoy (Groningen), R. Fosbury (ST-ECF), and R. Hook (ST-ECF)

■ **FIGURE 1–3** Objects often look very different in different parts of the spectrum. At left we see a visible-light image of the nebula BHR 71, obtained with the Very Large Telescope in Chile; a large amount of dust makes it opaque. The center panel shows the same view at infrared wavelengths, obtained with the Spitzer Space Telescope and displayed in false color; a jet from a protostar deep in the nebula appears, and many other stars can be seen as well. At right the images have been combined, again using false color.

(a)

Richard Wainscoat, Institute for Astronomy, University of Hawaii

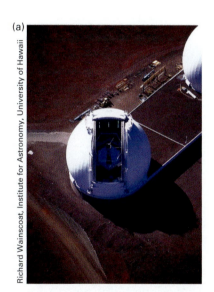

(b)

Peter Michaud, Gemini Observatory

■ **FIGURE 1–4** Near the twin 10-m Keck telescopes (a), we find the 8.2-m-diameter Gillett Gemini North telescope (b) on Mauna Kea volcano in Hawaii. The Giant Magellan Telescope is to be erected in Chile out of several 8-m mirrors in order to have the equivalent of a 21-m-diameter telescope. A Thirty-Meter Telescope is also planned, and the European Southern Observatory is building its 39-m European Extremely Large Telescope (E-ELT); both are based on the Keck design consisting of many relatively small hexagonal segments.

ESA/NASA, ESO, and Danny LaCru

■ **FIGURE 1–5** An image of part of a small, neighboring galaxy to our own, the Tarantula Nebula in the Large Magellanic Cloud. It was taken with the Hubble Space Telescope and a European Southern Observatory ground-based telescope. It was mosaicked and processed by a 23-year-old amateur astronomer using publicly available software. Shock waves from exploding stars have compressed the gas into the visible filaments and sheets.

(a)

European Southern Observatory

(b)

NASA Chandra X-ray Observatory Center/SAO

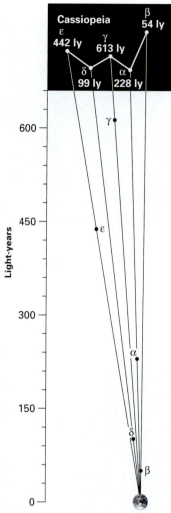

■ **FIGURE 1–6** (left) The center of the Crab Nebula, the remnant of a star that exploded almost 1,000 years ago, in visible light. This highly detailed image was taken with the Very Large Telescope in Chile. (right) An x-ray image of the heart of the Crab Nebula taken with the Chandra X-ray Observatory. The image reveals tilted 1-light-year rings of material lighted by electrons flowing from the central pulsar, plus jets of gas extending perpendicular to the rings.

■ **FIGURE 1–7** The stars we see as a constellation are actually at different distances from us. In this case, we see the true relative distances of the stars in the "W" of Cassiopeia, as determined from the Hipparcos spacecraft. The stars' appearance projected on the sky is shown in the upper part.

quite different distances from us (■ Fig. 1–7); these stars aren't physically associated with each other, and they were born at different times and locations.

Thus, the constellations show where things appear to be in the sky, but not what they are like or what they are made of. In most of this book, you will study the "how and why" of astronomy, not merely where things are located or what they are called. Still, it can be fun to look up at night and recognize the patterns in the sky.

Some groupings of stars are very familiar to many people but are not actually constellations. These configurations, made of parts of one or more constellations, are known as **asterisms**. The Big Dipper, for example, is an asterism but isn't a constellation, since it is but part of the constellation Ursa Major (the Big Bear). As we will see further in Chapter 4, some asterisms and constellations are sufficiently close to the celestial north pole in the sky that they are visible at all times of year, as seen from the United States. The Big Dipper is an example. But other asterisms and constellations, farther from celestial north, are visible at night for only part of the year. Let us now survey some

of the prominent asterisms and constellations that you can see in each season; see also *Star Party 1.1: Using the Sky Maps*. (Amateur astronomers often hold viewing sessions informally known as "star parties," during which they observe celestial objects. Likewise, the occasional "Star Party" boxes in this text highlight interesting observations that you can make.)

1.3a The Autumn Sky

As it grows dark on an autumn evening, you will see the Pointers in the Big Dipper – the two end stars – point upward toward Polaris. Known as the "north star," Polaris is not one of the brightest or nearest stars in the sky, but is well known because it is close to the direction of the celestial north pole. As we will see in Chapter 4, that means it uniquely appears almost motionless in the sky throughout the night, and provides a bearing that can help you get safely out of the woods. Almost an equal distance on the other side of Polaris is a

■ **FIGURE 1–8** The constellation Cassiopeia is easily found in the sky from its distinctive "W" shape, marked in red on the photograph.

"W"-shaped constellation named Cassiopeia (■ Fig. 1–8). In Greek mythology, Cassiopeia was married to Cepheus, the king of Ethiopia (and the subject of the constellation that neighbors Cassiopeia to the west). Cassiopeia appears sitting on a chair.

As we continue across the sky away from the Pointers, we come to the constellation Andromeda, named for Cassiopeia's daughter in Greek mythology. In Andromeda, on a very dark night you might see a faint, hazy patch of light; this is actually the center of the nearest large galaxy to our own, and is known as the Andromeda Galaxy. Although at about 2.4 million light-years away it is one of the nearest galaxies to us, it is much farther away than any of the individual stars that we see in the sky, since they are all in our own Milky Way Galaxy.

Southwest in the sky from Andromeda, but still high overhead, are four stars that appear to make a square known as the Great Square of Pegasus. One of the corners of this asterism is actually in the constellation Andromeda.

If it is really dark outside (which probably means that you are far from a city and also that the Moon is not full or almost full), you will see the hazy band of light known as the "Milky Way" crossing the sky high overhead, passing right through Cassiopeia. This dim band with ragged edges, which marks the plane of our disk-shaped galaxy (see Chapter 16), has many dark patches that make rifts in its brightness.

Moving southeast from Cassiopeia, along the Milky Way, we come to the constellation Perseus; he was the Greek hero who slew the Medusa. (He flew off on Pegasus, the winged horse, who is conveniently nearby in the sky, and saw Andromeda, whom he saved.) On the edge of Perseus nearest to Cassiopeia, with a small telescope or binoculars we can see two hazy patches of light that are really clusters of hundreds of stars called "open clusters," a type of grouping we will discuss in Chapter 11. This "double cluster in Perseus," also known as h and χ (the Greek letter "chi") Persei, provides two of the open clusters that are easiest to see with small telescopes. (They already

appeared in Figure 1–3.) In 1603, Johann Bayer assigned Greek letters to the brightest stars and lowercase Latin letters to less-bright stars (■ Fig. 1–9), but in this case the system was applied to name the two clusters as well.

Along the Milky Way in the other direction from Cassiopeia (whose "W" is relatively easy to find), we come to a cross of bright stars directly overhead. This "Northern Cross" is an asterism marking part of the constellation Cygnus, the Swan (■ Fig. 1–10). In this direction, spacecraft detect x-rays whose brightness varies with time, and astronomers have deduced in part from that information that a black hole is located there. Also in Cygnus is a particularly dark region of the Milky Way, called the Northern Coalsack. Dust in space in that direction prevents us from seeing as many stars as we see in other directions of the Milky Way.

Slightly to the west is another bright star, Vega, in the constellation Lyra (the Lyre). And farther westward, we come to the constellation Hercules, named for the mythological Greek hero who performed twelve great labors, of which the most famous was bringing back the golden apples. In Hercules is an older, larger type of star cluster called a "globular cluster," another type of grouping we will discuss in Chapter 11. It is known as M13, the great globular cluster in Hercules. It resembles a fuzzy mothball whether glimpsed with the naked eye or seen with small telescopes; larger telescopes have better clarity and so can reveal the individual stars.

1.3b The Winter Sky

As autumn proceeds and winter approaches, the constellations we have discussed appear closer and closer to the western horizon for the same hour of the night. By early evening on January 1, Cygnus is setting in the western sky, while Cassiopeia and Perseus are overhead.

STAR PARTY 1.1

Using the Sky Maps

Because of Earth's motion around the Sun over the course of a year, the parts of the sky that are "up" after dark change slightly each day. A given star rises (and crosses the meridian, or highest point of its arc across the sky) about 4 minutes earlier each day. By the time a season has gone by, the sky has apparently slipped a quarter of the way around at sunset as the Earth has moved a quarter of the way around the Sun in its yearly orbit. Some constellations are lost in the afternoon and evening glare, while others have become visible just before dawn.

In December of each year, the constellation Orion crosses the meridian at midnight. Three months later, in March, when the Earth has moved through one-quarter of its orbit around the Sun, the constellation Virgo crosses the meridian at midnight, when Orion is setting. Orion crosses the meridian at sunset (that is, 6 hours earlier than in December – consistent with 4 minutes/day × 90 days = 360 minutes = 6 hours). Another three months later, in June, Orion crosses the meridian an additional 6 hours earlier – that is, at noon. Hence, it isn't then visible at night. Instead, the constellation Ophiuchus crosses the meridian at midnight.

Because of this seasonal difference, inside the front and back covers of this book we have included four Sky Maps, one of which is best for the date and time at which you are observing. Suitable combinations of date and time are marked. Note also that if you make your observations later at night, it is equivalent to observing later in the year. Two hours later at night is the same as shifting later by one month.

Hold the map above your head while you are facing north or south, as marked on each map, and notice where your zenith is in the sky and on the map. The horizon for your latitude is also marked. Try to identify a pattern in the brightest stars that you can

see. Finding the Big Dipper, and using it to locate the pole star, often helps you to orient yourself. Don't let any bright planets confuse your search for the bright stars – knowing that planets usually appear to shine steadily instead of twinkling like stars (see Chapter 4) may assist you in locating the planets.

Come back and look at Sections 1.3a–d at the appropriate time of year – even after you have finished with this course.

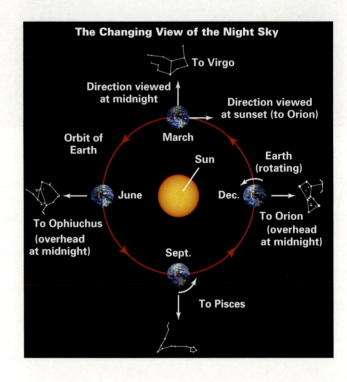

Farther toward the east, rising earlier every evening, is the constellation Orion, the Hunter (■ Fig. 1–12). Orion is perhaps the easiest constellation of all to pick out in the sky, for three bright stars close together in a line make up its belt. Orion is warding off Taurus, the Bull, whose head is marked by a large "V" of stars. A reddish

To the south of the Milky Way, near Perseus, we can now see a group of six stars close together in the sky (■ Fig. 1–11). The tight grouping tends to catch your attention as you scan the sky. It is the Pleiades (pronounced "plee′a-deez"), traditionally the Seven Sisters of Greek mythology, the daughters of Atlas. (We can usually see six stars with the unaided eye now, so either one of the stars has faded over the millennia or it was never visible and the association with the Pleiades myth was loose.) These stars are another example of an open cluster of stars. Binoculars or a small telescope will reveal dozens of stars there, whereas a large telescope will ordinarily show too small a region of sky for you to see the Pleiades well. So a bigger telescope isn't always better.

star, Betelgeuse ("bee′tl-juice" is an acceptable pronunciation, though some say "beh′tl-jouz"), marks Orion's armpit, and symmetrically on the other side of his belt, the bright bluish star Rigel ("rye′jel") marks his heel. Betelgeuse is an example of a red supergiant star; it is hundreds of millions of kilometers across, far bigger itself than the Earth's orbit around the Sun!

Orion's sword extends down from his belt. A telescope, or a photograph, reveals a beautiful region known as the Orion Nebula. Its general shape can be seen in even a smallish telescope; however, only images that have detected much light clearly reveal the vivid colors – though whether it is reddish or greenish in an image depends on what kind of filters were used. It is a site where new stars are forming right now, as you read these words.

Rising after Orion is Sirius, the brightest star in the sky. Orion's belt points directly to it. Sirius appears blue-white, which indicates that its surface is very hot. Sirius is so much brighter than the other stars that it stands out to the naked eye. It is part of the constellation

Jay M. Pasachoff

■ **FIGURE 1–9** Johann Bayer, in 1603, used Greek letters to mark the brightest stars in constellations; he also used lowercase Latin letters. Here we see Orion, the great hunter. The inset shows the red-supergiant star Betelgeuse, α Orionis (that is, alpha of Orion), marking Orion's shoulder.

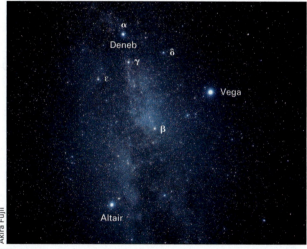

Akira Fujii

■ **FIGURE 1–10** The Northern Cross, composed of the brightest stars in the constellation Cygnus, the Swan. Deneb, also called alpha (α) Cygni, gamma (γ) Cygni, and beta (β) Cygni make the long bar; epsilon, gamma, and delta Cygni make the crossbar. The bright star Vega, alpha in Lyra, is nearby. Also marked is the bright star Altair, alpha in Aquila, the Eagle. These stars lie in the Milky Way, which shows clearly on the image.

NASA, ESA and AURA/Caltech

■ **FIGURE 1–11** The Pleiades, the Seven Sisters, in the constellation Taurus, the Bull. It is a star cluster, and long exposures like this one show dust around the stars reflecting starlight, preferentially the bluish colors. When the Pleiades and the Hyades, another star cluster, rose just before dawn, ancient peoples in some parts of the world knew that the rainy season was about to begin.

Canis Major, the Big Dog. (You can remember that it is near Orion by thinking of it as Orion's dog.)

Back toward the top of the sky, between the Pleiades and Orion's belt, is a group of stars that forms the "V"-shaped head of Taurus. This open cluster is known as the Hyades ("hy´a-deez"). The stars of the Hyades mark the bull's face, while the stars of the Pleiades ride on the bull's shoulder. In a Greek myth, Jupiter turned himself into a bull to carry Europa over the sea to what is now called Europe.

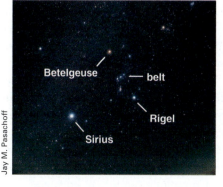

Jay M. Pasachoff

■ **FIGURE 1–12** The constellation Orion, marked by reddish Betelgeuse, bluish Rigel, and a belt of three stars in the middle. The Orion Nebula is the reddish object below the belt. It is M42 on the Winter Sky Map at the end of the book. Sirius, the brightest star in the sky, is to the lower left of Orion in this view.

1.3c The Spring Sky

We can tell that spring is approaching when the Hyades and Orion get closer and closer to the western horizon each evening, and finally are no longer visible shortly after sunset. Now Castor and Pollux, a pair of equally bright stars, are nicely placed for viewing in the western sky. Castor and Pollux were the twins in the Greek pantheon of gods. The constellation is called Gemini, the twins.

On spring evenings, the Big Bear (Ursa Major) is overhead, and anything in the Big Dipper – which is part of the Big Bear – would spill out. Leo, the Lion, is just to the south of the overhead point, called the zenith (follow the Pointers backward). Leo looks like a backward question mark, with the bright star Regulus, the lion's heart, at its base. The rest of Leo, to the east of Regulus, is marked by a bright triangle of stars. Some people visualize a sickle-shaped head and a triangular tail.

If we follow the arc made by the stars in the handle of the Big Dipper, we come to a bright reddish star, Arcturus, another supergiant. It is in the kite-shaped constellation Boötes, the Herdsman.

Sirius sets right after sunset in the spring; however, a prominent but somewhat fainter star, Spica, is rising in the southeast in the constellation Virgo, the Virgin. It is farther along the arc of the Big Dipper through Arcturus. Vega, a star that is between Sirius and Spica in brightness, is rising in the northeast. And the constellation Hercules, with its notable globular cluster M13, is rising in the east in the evening at this time of year.

1.3d The Summer Sky

Summer, of course, is a comfortable time to watch the stars because of the generally warm weather. Spica is over toward the southwest in the evening. A bright reddish star, Antares, is in the constellation Scorpius, the Scorpion, to the south. ("Antares" means "compared with Ares," another name for Mars, because Antares is also reddish.)

Hercules and Cygnus are high overhead, and the star Vega is prominent near the zenith. Cassiopeia is in the northeast. The center of our Galaxy is in the dense part of the Milky Way that we see in the constellation Sagittarius, the Archer, in the south (■ Fig. 1–13).

Around August 12 every summer is a wonderful time to observe the sky, because that is when the Perseid meteor shower occurs. (Meteors, or "shooting stars," are not stars at all, as we will discuss in Chapter 8.) In a clear, dark sky, with not much moonlight, one bright meteor a minute may be visible at the peak of the shower. Just lie back and watch the sky in general – don't look in any specific direction. The rate of meteors tends to be substantially higher after midnight than before midnight, since our part of Earth has then turned so that it is plowing through space, crossing through the paths of pebbles and ice chunks that streak through the sky as they heat up. Although the Perseids is the most observed meteor shower, partly because it occurs at a time of warm weather in North America and Europe, many other meteor showers occur during various parts of the year.

The summer is a good time of year for observing a prime example that shows that stars are not necessarily constant in brightness. This

■ FIGURE 1–13 The Milky Way – the band of stars, gas, and dust that marks the plane of our disk-shaped Galaxy – is easily recognized in a dark sky. It is shown here with Halley's Comet (left, middle, with a tail) passing by.

Akira Fujii

"variable star," δ (delta) Cephei, appears in the constellation Cepheus, which is midway between Cassiopeia and Cygnus. δ Cephei varies in brightness with a 5.4-day period. As we will see in Chapters 11 and 16, studies of its variations have been important for allowing us to measure the distances to galaxies. This fact reminds us of the real importance of studying the sky – which is to learn what things are and how they work, and not just where they are. The study of the sky has led us to better understand the Universe, which makes astronomy beautiful and exciting to so many people.

1.4 HOW DO YOU TAKE A TAPE MEASURE TO THE STARS?

It is easy to see the direction to an object in the sky but much harder to find its distance. Astronomers are always seeking new and better ways to measure distances of objects that are too far away to touch. Our direct ability to reach out to astronomical objects is limited to our Solar System. We can send people to the Moon and spacecraft to the other planets. We can even bounce radio waves off the Moon, most of the other planets, and the Sun, and measure how long the radio waves' round trip takes to find out how far they have travelled. The distance (d) travelled by light or by an object is equal to the constant rate of travel (its speed v) multiplied by the time (t) spent travelling ($d = vt$).

If we were carrying on a conversation by radio with someone at the distance of the Moon, there would be pauses of noticeable length after we finished speaking before we heard an answer. This is because radio waves, even at the speed of light, take over a second to travel each way. Astronauts on the Moon had to get used to these pauses when their messages travelled by radio waves to people on Earth, and spacecraft roaming Mars have to be autonomous to travel safely over rocky fields at any reasonable speed.

Once we go further afield, we must be more ingenious. Repeatedly in this book, we will discuss ways of measuring distance. For the nearest hundreds of thousands of stars, we have recent results from a spacecraft that showed how much their apparent position shifts when we look at them from slightly different angles (see Chapter 11). For more distant stars, we find out how far away they are by comparing how bright they actually (intrinsically) are and how bright they appear to be. We often tell their intrinsic brightness from looking at their spectra (Chapter 11), though we will also see other methods.

Once we get to galaxies other than our own, we will see that our methods are even less precise. For the nearest galaxies, we search for stars whose specific properties we recognize. Some of these stars are thought to be identical in type to the same kinds of stars in our own Galaxy whose intrinsic brightnesses we know. Again, we can then compare intrinsic brightness with apparent brightness to give distance. For the farthest galaxies, as we shall discuss in Chapter 16, we find distances using the 1920s discovery that shifts in color of the spectrum of a galaxy reveal how far away it is. Of course, we continue to test these methods as best we can, and some of the most exciting investigations of modern astronomy are related to the determination of distances. For example, one of the Hubble Space Telescope's Key Projects was devoted to finding distances of galaxies, and made good progress toward resolving a long-term controversy over the expansion rate and age of the Universe. Subsequent studies of a similar nature obtained even more accurate and precise results.

1.5 THE VALUE OF ASTRONOMY

1.5a The Grandest Laboratory of All

Throughout history, observations of the heavens have led to discoveries that have had a major impact on people's ideas about themselves and the world around them. Even the dawn of mathematics may have stemmed from ancient observations of the sky, made in order to keep track of seasons and seasonal floods in the fertile areas of the Earth. Observations of the motions of the Moon and the planets, which are free of such complicating terrestrial forces as friction and which are massive enough so that gravity dominates their motions, led to an understanding of gravity and of the forces that govern all motion.

The regions of space studied by astronomers serve as a cosmic laboratory where we can investigate matter or radiation, often under conditions that we cannot duplicate on Earth (■ Fig. 1–14). These studies allow us to extend our understanding of the laws of physics, which govern the behavior and evolution of the Universe.

A new importance has been given to astronomy by the realization that large asteroids and comets have hit the Earth every few tens of millions of years with enough power to devastate our planet. We shall see in Chapter 8 how the dinosaurs and many other species were extinguished 65 million years ago by an asteroid or comet, according to a theory that has become widely favored. All those movies about

SWAP is a project of the Centre Spatial de Liège and the Royal Observatory of Belgium funded by the Belgian Federal Science Policy Office (BELSPO); image courtesy of Daniel B. Seaton

■ **FIGURE 1–14** The Sun is made of plasma, a strange stage of matter that is gas so hot that it is ionized and subject to magnetic fields. Here we see eruptions on the Sun's edge and active regions on the solar disk, imaged with the Sunwatcher camera in space. A partial solar eclipse, visible up to 87 per cent from Antarctica and 30 per cent from New Zealand (where one of us, J.M.P., saw it), shows at lower right.

asteroids and comets on collision courses with Earth aren't entirely science fiction!

Many of the discoveries of tomorrow – perhaps the control of nuclear fusion or the discovery of new sources of energy, or maybe even something so revolutionary that it cannot now be predicted – will undoubtedly be based on advances made through such basic research as the study of astronomical systems. Considered in this sense, astronomy is an investment in our future.

The impact of astronomy on our conception of the Universe has been strong through the years. Discoveries that the Earth is not at the center of the Universe, or that the Universe has been expanding for billions of years, affect our philosophical conceptions of ourselves and of our relations to space and time.

Yet most of us study astronomy not for its technological and philosophical benefits but rather for its beauty and inherent interest. We must stretch our minds to understand the strange objects and events that take place in the far reaches of space. The effort broadens us and continually enthralls us all. Ultimately, we study astronomy because of its fascination and mystery.

1.5b Origins

It is always fun to look at our own childhood pictures. Similarly, astronomers are increasingly looking at the baby pictures of the Universe and of various types of objects in it. NASA has made its "Origins" program a centerpoint of its strategy for investigation. Under this program, many scientists will use spacecraft, telescopes on Earth, computers, and just plain clear thought to study all kinds of origins: the origin of the Universe, of our Galaxy, of stars

A CLOSER LOOK 1.1

A Sense of Scale: Measuring Distances

Let us try to get a sense of the scale of the Universe, starting with sizes that are part of our experience and then expanding toward the enormously large. Imagine a series of cubes, of which we see one face (a square).

1 mm = 0.1 cm

ACL 1.1–a We begin our journey through space with a view of something 1 millimeter across, an electron-microscope view of a fly. Every step we take will show a region 100 times larger in diameter than that in the previous picture.

ACL 1.1–b A square 100 times larger on each side is 10 centimeters x 10 centimeters. (Since the area of a square is the length of a side squared, the area of a 10-cm square is 10,000 times the area of a 1-mm square.) The area encloses a flower.

ACL 1.1–c Next we move far enough away to see an area 10 meters on a side.

ACL 1.1–d A square 100 times larger on each side is now 1 kilometer square, about 250 acres. An aerial view of New York shows a slightly bigger region, with Central Park at lower left and the United Nations building at lower right, and the future site of Cornell University's science graduate school on the island at the top.

ACL 1.1–e The next square, 100 km on a side, encloses California's Bay Area, including San Francisco, Berkeley, Oakland, and San Jose. Note that though we are still bound to the limited area of the Earth, the area we can see is increasing rapidly.

10,000 km = 10^7 m

ACL 1.1–f A square 10,000 km on a side covers nearly the entire Earth. We see the southwestern United States, through the clouds, and northwestern Mexico, including the Baja California peninsula.

A Sense of Scale: Measuring Distances (*continued*)

$1{,}000{,}000 \text{ km} = 10^9 \text{ m} = 3.3 \text{ lt sec}$

ACL 1.1–g When we have receded 100 times farther, we see a square 100 times larger in diameter: 1 million kilometers across. It encloses the orbit of the Moon around the Earth. We can measure with our wristwatches the amount of time that it takes light to travel this distance.

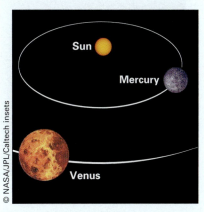

$10^{11} \text{ m} = 5.5 \text{ lt min}$

ACL 1.1–h When we look on from 100 times farther away still, we see an area 100 million kilometers across, $2/3$ the distance from the Earth to the Sun. We can now see the Sun and the two innermost planets in our field of view.

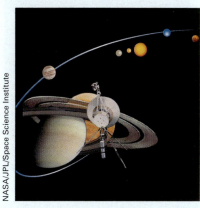

$10^{13} \text{ m} = 9 \text{ lt hr}$

ACL 1.1–i An area 10 billion kilometers across shows us the entire Solar System in good perspective. It takes light about 10 hours to travel across the Solar System. The outer planets have become visible and are receding into the distance as our journey outward continues.

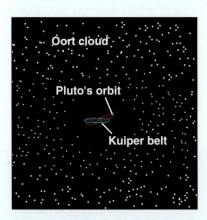

$10^{15} \text{ m} = 38 \text{ lt days}$

ACL 1.1–j From 100 times farther away, 1 trillion km across, we see Pluto and other small bodies in the Kuiper belt. At even larger distances from the Sun, the Oort cloud is a vast collection of comets. We have not yet reached the scale at which another star besides the Sun is in a cube of this size.

$10^{17} \text{ m} = 10 \text{ lt yr}$

ACL 1.1–k As we continue to recede from the Solar System, the nearest stars finally come into view. We are seeing an area 10 light-years across, which contains only a few stars, most of whose names are unfamiliar (see the Appendices). The brightest stars tend to be intrinsically very powerful, but more distant than these dim, nearby stars.

ACL 1.1–l When we are 100 times farther away, we can see a 1000 light-year fragment of our Galaxy, the Milky Way Galaxy. We see not only many individual stars, including in this image the Southern Cross, but also many clusters of stars and many "nebulae" – regions of glowing, reflecting, or opaque gas or dust.

A Sense of Scale: Measuring Distances (*continued*)

10^{21} m $= 10^5$ lt yr

ACL 1.1–m In a field of view 100 times larger in diameter, 100,000 light-years across, we would be able to view our entire Milky Way Galaxy, with its spiral arms, in one go. (The picture shown is a Hubble image of a galaxy similar to ours.)

10^{23} m $= 10^7$ lt yr

ACL 1.1–n Next we move sufficiently far away so that we can see an area 10 million light-years across. There are about as many centimeters across this image as there are grains of sand in all of Earth's beaches. The image shows part of a cluster of galaxies; our Galaxy is in such a grouping, the Local Group.

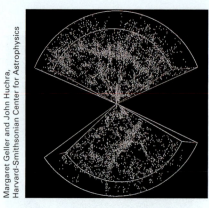

10^{25} m $= 10^9$ lt yr

ACL 1.1–o If we could see a field of view 100 times larger, 1 billion light-years across, our Local Group of galaxies would appear as but part of a supercluster – a cluster of galaxy clusters. The image shows a mapping of a wedge of the Universe centered on Earth and containing hundreds of thousands of galaxies.

© ESA, HFI and LFI consortia/Planck collaboration

ACL 1.1–p This map of the whole sky comes from the Planck spacecraft, which is now in orbit. It is mapping the sky in several microwave wavelengths to show the structure of our Galaxy. Subtracting that structure has allowed it to show a slight difference in the Universe's apparent temperature in opposite directions (Chapter 19), resulting from our Sun's motion. It has provided even more detailed maps than were made by earlier spacecraft, to see the seeds from which today's galaxies and clusters of galaxies grew. We are continuing our study of the Universe's earliest years. A combination of radio, ultraviolet, x-ray, and optical studies, together with theoretical work and experiments with giant atom smashers on Earth, is allowing us to explore the past and predict the future of the Universe.

(■ Fig. 1–15), of the Sun, of our Earth and the elements in it, and of life itself – maybe someday not only life here on Earth but also on one of the planets around other stars that are being discovered at an increasing rate. After all, we now know of more planets outside our Solar System than we do inside it, a big change in our view that occurred in the past decade.

NASA's Origins program has three official major scientific goals: (1) to understand how galaxies formed early in the Universe and to determine the role of galaxies in the emergence of planetary systems and life; (2) to understand how stars and planetary systems form and to determine whether life-sustaining planets exist around other stars; and (3) to understand how life originated on Earth and to determine whether it began and may still exist elsewhere as well. Existing or planned missions relevant to the Origins program include the Hubble Space Telescope, the James Webb Space Telescope, various Mars missions, a mission to the outer planets and their moons, and the airplane known as the Stratospheric Observatory for Infrared Astronomy (SOFIA), which carries a 2.5-m telescope aloft.

■ **FIGURE 1–15** A giant cloud of gas and dust, with thousands of stars forming. The view, with false color to bring out different wavelengths of infrared radiation, was taken with the Spitzer Space Telescope. It shows the Tarantula Nebula in the Large Magellanic Cloud. See Figure 1–5 for an optical view.

Another major theme of NASA's scientific investigations is Structure and Evolution of the Universe. In this book, you will learn about the structure of the Solar System, of our Galaxy, and of space. You will learn the life stories of stars, and see how different kinds of stars evolve. And you will learn about the evolution of the Universe as a whole.

In this book, we will make these themes explicit as the central organizing structure. In particular, we will try to point out the links to Origins – to our Universe's childhood pictures – on a chapter-by-chapter basis.

1.6 WHAT IS SCIENCE?

Science is not merely a body of facts; it is also a process of investigation. The standards that scientists use to assess their ideas and to decide which to accept – in some sense, those that are "true" – are the basis of much of our technological world. One of the guiding principles of science is that results should be reproducible; that is, other scientists should be able to get essentially the same result by repeating the same experiment or observation. Science is thus a self-checking way of carrying out investigations.

Though acceptable scientific investigations are actually carried out in many ways, there is a standard model for the **scientific method**. In this standard model, one first looks at a body of data (singular: datum) and makes educated guesses as to what might explain them. An educated guess is a **hypothesis**. Then one thinks of consequences that would follow if the hypothesis were true and tries to carry out experiments or make observations that test the hypothesis. If, at any time, the results are contrary to the hypothesis, then the hypothesis is discarded or modified. (There may, though, be other assumptions that could be modified or discarded instead, because they are inappropriate. Also, one needs to check for experimental or observational errors.) If the hypothesis passes its tests and is established in some basic framework or set of equations, it can be called a **theory**.

If the hypothesis or theory survives test after test, it is accepted as being "true." Still, at any time a new experiment or observation could show it was false after all. This process of "falsification," being able to find out if a hypothesis or theory is false, is basic to the definition of science formulated by a leading set of philosophers of science.

Sometimes, you see something described as a "law" or a "principle" or a "fact." "Laws of nature" – like Kepler's laws of planetary motion, Newton's laws of motion, or Newton's law of gravitation (see Chapter 5) – are actually *descriptions* of how nature behaves, and they might in fact be incorrect. Newton used the word "law" (in Latin) over three hundred years ago. The words "law" and "principle" are historical usages associated with certain basic theories. Scientists consider something a "fact" when it is so well established that it would be extremely unlikely that it is incorrect.

What is an example of the scientific method? Albert Einstein advanced a theory in 1916 to explain how gravity works. (It was developed so completely that it was far beyond a hypothesis.) His "general theory of relativity" was based on mathematical equations he worked out from some theoretical ideas. His result turned out to explain some

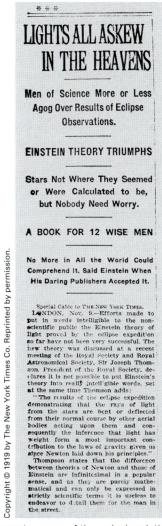

■ **FIGURE 1–16** This report was one of the early descriptions of general relativity's success that captured the public's fancy. From *The New York Times* of November 10, 1919.

observations of the orbit of Mercury (the orbit's elliptical shape slowly rotates with time) that had been puzzling up to that time, though this hadn't been Einstein's motivation. Still, its basic tests lay ahead.

Einstein's theory predicted that starlight would appear to be bent by a certain amount if it could be observed to pass very near the Sun. Such bending would be visible only during a total solar eclipse, when the sky was dark but the Sun was still present in the sky. The theory seemed so important that expeditions were made to total solar eclipses to test it. When the first telegram came back that the strange prediction of Einstein's theory had been verified, the theory was quickly accepted. Successful predictions are usually given more weight than mere explanations of already known facts (■ Fig. 1–16). Einstein immediately gained a worldwide reputation as a great scientist. The theory's nearly complete acceptance by the scientific community was established as viable alternatives could not be found.

Most of the time, though, the scientific method does not work so straightforwardly. We can consider, for example, our understanding of how the stars shine. Until the 1930s, it was thought that the Sun got its energy from its contracting gas, but this idea seemed wrong, because it could not explain how the Sun could be as old as rocks on Earth whose ages we measured. (Presumably the Sun and Earth formed nearly simultaneously.) Then scientists suggested that nuclear fusion – the merging of 4 hydrogen nuclei to make a single helium nucleus, in particular – could provide the energy for the Sun to live 10 billion years (see Chapter 12), and they even worked out the detailed ways in which the hydrogen could fuse. Thus a theory of nuclear energy as a source of power for the Sun and stars exists. But how could it be tested? The observed properties of a wide variety of stars around the Universe fit, in several ways, with the deductions of the theory. Still, a direct test had to wait.

Over the last four decades, measurements have finally been made of individual subatomic particles called "neutrinos" that should be given off as the fusion process takes place. Neutrinos from fusion in the Sun were indeed measured, but only at one-third to one-half the rate expected, as we will describe in Chapter 12. Had our main theory of stellar energy failed? No, because explanations were discovered for why some of the neutrinos may not reach us in a detectable form. No physicists doubt that the Sun and stars get their energy from nuclear fusion, so our theory of the properties of neutrinos themselves is changing. Thus the astronomical study of neutrinos from the Sun has led to important changes in the most basic ideas of physics.

So the "scientific method" isn't cut-and-dried. But it does demand a rigor and honesty in scientific testing. The standards in science are high, especially those of evidence, and we hope that this book will provide enough examples to enable you to form an accurate impression of how the process actually works.

1.7 WHY IS SCIENCE FAR BETTER THAN PSEUDOSCIENCE?

Though science is itself fascinating, too many people have beliefs that may seem related to science but either have no present verification or are false. Such beliefs, such as astrology or the idea that UFOs (unidentified flying objects) are now definitely bringing aliens from other planets, are pseudoscience rather than science. ("Pseudo" means that something is not authentic or sincere, in spite of looking somewhat real.)

Astrology is not at all connected with astronomy, except in a historical context (they had similar origins, and hence the same root), so it does not really deserve a place in a text on astronomy. But since so many people incorrectly associate astrology with astronomy, and since astrologers claim to be using astronomical objects to make their predictions, let us use our knowledge of astronomy and of the scientific method to assess astrology's validity. Millions of Americans – a number that shows no signs of decreasing – believe in astrology as more than just a source of amusement, so the topic is too widespread to ignore.

Astrology is an attempt to predict or explain our actions and personalities on the basis of the positions of the stars and planets now and at the instants of our births. Astrology has been around for a long time, but it has never definitively been shown to work.

Chapin Library/Williams College, image by Jay M. Pasachoff

■ **FIGURE 1–17** A 16th-century zodiac from a book by Regiomontanus.

Believers may cite incidents that reinforce their faith in astrology, but statistical measurements do not compellingly show a real effect. If something happens to you that you had expected because of an astrological prediction, you would more certainly notice that this event occurred than you would notice the hundreds of unpredicted things that also happened to you that day. Yet we do enough things, have sufficiently varied thoughts, and interact with enough people that if we make many predictions, some of them are likely to be at least partially fulfilled by chance. We simply forget the rest.

In fact, even the traditional astrological alignments are not accurately calculated, for the Earth's pole points in different directions in space as time passes over millennia. In truth, stars are overhead at different times of year compared with the case millennia ago, when the astrological tables that are often still in current use were computed. At a given time of year, the Sun is usually in a different sign of the zodiac from its traditional astrological one (■ Fig. 1–17). Furthermore, we know that the constellations are illusions; they don't even exist as physical objects. They are merely projections of the positions of stars that are at very different distances from us.

Studies have shown that superstition actively constricts the progress of science and technology in various countries around the world and is therefore not merely an innocent force. It is not just that some people harmlessly believe in astrology; their lack of understanding of scientific structure may actually impede the training or work of people needed to solve the problems of our age, such as AIDS, pollution, food shortages, and the energy crisis. Polio has been wiped out almost everywhere in the world, but superstitious beliefs have set back the finality of the still-hoped-for triumph over this disease.

In addition to the lack of evidence corroborating astrological predictions, one minor reason why scientists in general and astronomers in particular doubt astrology is that they cannot conceive of a

Jay M. Pasachoff

■ FIGURE 1–18 The symbols for the constellations of the zodiac shown on the historic 15th-century (repainted more recently) astronomical clock in Prague.

way in which it could possibly work. The human brain is so complex that it seems most improbable that any celestial alignment can affect people, including newborns, in an overall way. The celestial forces that are known are not sufficient to set personalities or influence daily events. Any unknown force must have truly remarkable properties to exert any noticeable effect on humans.

A team of psychologists from the California State University at Long Beach arranged for a magician to perform three psychic-like stunts in front of psychology classes. Even when they emphasized to the students that the performer was a magician performing tricks, 50 per cent of the class still believed the magician to be psychic. Evidently, a lot of self-deception is involved with believing in pseudoscience.

Let us mention a specific study showing that astrology doesn't work, one in which a psychologist tested certain values. Do Libras and Aquarians rank "Equality" highly? Do Sagittarians especially value "Honesty"? Do Virgos, Geminis, and Capricorns treasure the quality "Intellectual"? Several astrology books agreed that these and other similar examples are values typical of those signs. Although believers often criticize the objections of skeptics on the grounds that these "group horoscopes" are not as valuable or accurate as individualized charts, surely some general assumptions and rules hold in common, as claimed in newspaper horoscope columns. (Signs of the zodiac appear in ■ Fig. 1–18.)

The subjects, 1,600 psychology graduate students, did not know in advance what was being tested. They gave their birth dates, and the questioners determined their astrological signs. The results: no special correlation with the values they were supposed to hold was apparent for any of the signs. Also, when asked to what extent they shared the qualities of each given sign, as many subjects ranked themselves above average as below, regardless of their astrological signs.

From an astronomer's view, astrology is meaningless, unnecessary, and impossible to explain without contradicting the broad set of physical laws that we have developed and tested over the years and that very well explains what happens on the Earth and in the sky. Astrology snipes at the roots of all pure science. Let's learn from the stars, but let's learn the truth.

Science is more than just a set of facts, since a methodology of investigation and standards of proof are involved, but science is also more than just a methodology. In this course, you should not only learn certain facts about the Universe, but in addition appreciate the way that theories and facts come to be accepted.

We will be discussing such fascinating things as quasars, pulsars, and black holes. We will consider complex molecules that have spontaneously formed in interstellar space, and try to decide whether the Universe will expand forever or collapse. We have sent a rocket into interstellar space bearing a portrait of humans, and have beamed a radio message toward a cluster of stars 24,000 light-years away. These topics and actions are part of modern astronomy: what contemporary, often conservative, scientists are doing and thinking about the Universe. So, though it would be tremendously exciting if celestial alignments could affect individuals, the evidence is so strongly against this idea that it seems fruitless to spend time on astrology instead of the exciting aspects of the real astronomical Universe.

We will have more to say about pseudoscience in the final chapter of this book, where we discuss the notion that UFOs are currently bringing aliens to Earth.

CONCEPT REVIEW

» This is an enormously exciting time to study astronomy. Telescopes and the instruments on them are better than ever, providing new clues to a wide variety of phenomena and deepening our understanding of the cosmos.

» The Universe is about 14 billion years old. Compared with a 24-hour day scaled to match the age of the Universe, the Earth formed about 8 hours ago, and humans appeared only 2 seconds ago (Sec. 1.1).

» Scientists often use a version of the metric system, in which prefixes like "kilo-" for one thousand and "mega-" for one million go with units like meter for distance and second for time, or are found in kilogram for mass. Astronomers also use special units, like the **light-year** for the distance light travels in one year (Sec. 1.1). By viewing objects that are at different distances from Earth, we look back to different times in the past, because light travels at a finite speed. Thus, the observation of very distant objects we see today provides a glimpse of the Universe as it was a long, long time ago.

» Astronomers gain most of their information about the Universe by studying radiation from objects (Sec. 1.2). In addition to taking photographs, we can break down an object's light into its component colors to make a **spectrum**, like a rainbow. We study not only the visible part of the spectrum, but also its gamma rays, x-rays, ultraviolet, visible light, infrared, and radio waves.

» The sky is divided into 88 regions known as **constellations** (Sec. 1.3). The individual stars in a given constellation are generally at quite different distances from Earth and are not physically associated with each other; they are simply apparent groupings of stars. Some familiar groupings, such as the Big Dipper, are not actually constellations but rather asterisms, which are easily recognizable parts of one or more constellations. As the Earth rotates about its axis and revolves around the Sun, different constellations and asterisms appear in the night sky. It can be fun to look at the sky and know some of the constellations, as well as their related mythology (Sec. 1.3a–d).

» Some of the most exciting studies in modern astronomy are related to the determination of distances (Sec. 1.4). We can measure how far away the Moon or a planet is by bouncing radio waves off it and measuring the total amount of time the waves were in flight. It is much more difficult to measure the distances of stars, galaxies, and other objects beyond our Solar System, but later in this book we explain the methods used by astronomers.

» The study of astronomy has had a major impact on the development of science throughout human history (Sec. 1.5). For example, observations of the changing positions of the Moon and the planets led to an understanding of gravity and the laws governing the motions of bodies. Moreover, astronomy allows us to investigate human origins, including the creation of the chemical elements and the formation of the Sun and Earth. The Universe is a cosmic laboratory to test ideas of science in conditions often not available in laboratories on Earth. Observations of the skies may even help humans avoid or delay extinction, if we are able to discover (and subsequently deflect) large asteroids or comets that are on a collision course with Earth.

» One of the main principles of science is that results should be reproducible; other scientists should be able to get essentially the same result by repeating the same experiment or observation (Sec. 1.6). Astronomers and other scientists follow the **scientific method**, which is difficult to define but provides a standard about which scientists agree. In the basic form of the scientific method, a **hypothesis** passes observational tests to become a **theory**.

» Many people have beliefs that may seem related to science but either have no present verification or are false; they are based on pseudoscience rather than on authentic science (Sec. 1.7). Astrology (as opposed to astronomy), for example, passes no scientific tests and so is not a science. Furthermore, astrology has been shown not to work.

QUESTIONS

1. Why might we say that our senses have been expanded in recent years?

†2. The speed of light is 3×10^5 km/s. Express this number in m/s and in cm/s.

3. Geologists study different strata to determine conditions on Earth long in the past. For example, dinosaur bones are found in strata dating from about 245 million to 65 million years ago. How is it that astronomers are able to see parts of the Universe appearing as they were in the past?

4. Distinguish between a hypothesis and a theory. Give a non-astronomical example of each.

5. Discuss an example of science and an example of pseudoscience, and distinguish between their respective values.

†6. (a) Write the following in scientific notation: 4642; 70,000; 34.7. (b) Write the following in scientific notation: 0.254; 0.0046;

0.10243. (c) Write out the following in an ordinary string of digits: 2.543×10^6; 2.0043×10^2; 7.673×10^{-4}.

†7. Suppose you wish to construct a scale model of the Solar System. The distance between the Earth and the Sun (one "Astronomical Unit" – 1 au) is 1.5×10^8 km, and you represent it by 15 km, about the size of a reasonably big city. (a) What is the scaled distance of the dwarf-planet Pluto (39.5 au)? (Hint: Use ratios here and elsewhere when possible!) (b) The Earth is about 12,800 km in diameter. How large is Earth on your scale? Compare its size with that of a common object. (c) The Sun's diameter is about 1.4×10^6 km. How large is it relative to Earth? (d) The nearest star, Proxima Centauri, is 4.2 light-years away (i.e., 4.0×10^{13} km). How far is this in your scale model? (e) Compare your answer in part (d) with the true distance to the Moon, 3.84×10^5 km. Comment on whether this gives you some idea of the enormous distances of even the nearest stars.

†8. Suppose an asteroid is found that appears to be heading directly toward Earth (using Earth's location at the time of discovery). It is 4.2 au away (i.e., 4.2 times the Earth-Sun distance), and travelling through the Solar System with a speed of 20 km/s. (a) How long would it take the asteroid to reach Earth (at Earth's location at the time of discovery), assuming the asteroid's speed is constant? (b) Will the Earth be at approximately the same location in its orbit around the Sun after the time interval computed in part (a)? Discuss what could happen.

9. Look up the horoscopes for a given day in a wide variety of newspapers. (This can most easily be done over the Internet.) How well do you think they agree with each other? Do you think the predictions are sufficiently general or vague that quite a few people might think they came true, regardless of whether they had anything to do with celestial influences?

10. True or false? When we observe very distant stars and galaxies, we see them as they are at the present time (or nearly the present time), since light travels at the maximum possible speed.

11. True or false? Astronomers gain information about celestial objects by studying many forms of radiation, including x-rays and infrared, not just visible light.

12. True or false? Astrology is a pseudoscience, originally related to astronomy, that has been shown to be incorrect.

13. True or false? Constellations are groups of stars that in general are physically close together, all at roughly the same distance from Earth.

14. True or false? When added to a unit of measure, the prefixes "kilo," "mega," "giga," and "nano" mean one thousand, one million, one billion, and one trillion (respectively).

†15. True or false? If the Sun (radius $R \approx 700,000$ km) is scaled down to a basketball ($R \approx 25$ cm), then the Earth ($R \approx 7,000$ km) is roughly like a golf ball ($R \approx 2.5$ cm).

†16. Multiple choice: Suppose the distance from the Sun to Pluto, 40 au = 6×10^9 km, were compressed to the size of a pen (15 cm). On this scale, what would be the distance from the Sun to Aldebaran, a bright star (the Eye of Taurus, the Bull) whose true

distance is roughly 60 light-years? (Note: 1 light-year is about 10^{13} km.) (a) 15 km. (b) 1.5×10^6 km. (c) 1.5×10^5 cm. (d) 6×10^{14} km. (e) 6×10^{14} cm.

†17. Multiple choice: An asteroid is found in space with a constant speed of 4.0×10^3 m/s. How far does it travel in 2 minutes? (a) 33.3 m. (b) 4.8×10^2 m. (c) 2.0×10^3 m. (d) 8.0×10^3 m. (e) 4.8×10^5 m.

†18. Multiple choice: Suppose the first person ever to land on Mars has just touched down, and she needs to send a message to her mother, letting her know she's there safely. If Mars is about 120 million km from Earth at that moment, what is the *fastest* her message can reach her mother? (a) In about 0.4 s. (b) In about 40 s. (c) In about 400 s. (d) In about 4,000 s. (e) It depends on whether she uses FedEx or the U.S. Postal Service.

19. Multiple choice: Scientific results must be (a) hypothetical; (b) reproducible; (c) controversial; (d) believed by at least 50 per cent of all scientists; or (e) believed by 100 per cent of all scientists.

20. Multiple choice: The process of breaking light down into its component colors creates (a) a spectrum; (b) an image; (c) a pulse; (d) a hologram; or (e) a movie.

21. Fill in the blank: The distance travelled by light or by an object moving with a constant speed is equal to speed multiplied by _____.

†22. Fill in the blank: If 14 billion years (the approximate current age of the Universe) were compressed to 1 year (i.e., 12 months), then on the same scale the Earth (whose true age is about 4.6 billion years) formed about ___ months after the Universe was born.

23. Fill in the blank: Distinct groups of stars, called _____, were given names by ancient astronomers; stories were associated with them.

24. Fill in the blank: A light-year is a unit of _____.

†This question requires a numerical solution.

TOPICS FOR DISCUSSION

1. What is the value of astronomy to you? How do you rank National Science Foundation (NSF) and National Aeronautics and Space Administration (NASA) funds for research with respect to other national needs? Re-answer this question when you have completed this course.

2. Weigh the importance of continuing to build the James Webb Space Telescope compared with other uses of money within NASA for other missions.

3. Discuss the role of pseudoscience, like astrology, in our modern society.

The reddish color that shows in this visible-light image of gas, part of the Trifid Nebula (M20), comes from emission by hydrogen, one of the topics covered in this chapter. The reddish emission is crossed by about 3 dark, absorbing lanes of dust, giving the nebula its name ("tri-" means three). At top, separately, a blue reflection nebula surrounds a hot star.
ESO

2

Light, Matter, and Energy: Powering the Universe

ORIGINS

We discuss the basic constituents of atoms and some of the techniques that allow us to explore the Universe nearly as far back as the beginning of time. We discover from studies of light that distant stars and galaxies are made of the same kinds of elements as those found on Earth.

The light that reaches us from the stars and planets is only one type of radiation, a certain way in which energy moves through space. Radiation in this sense results from the continuous changes in electricity and magnetism at each point of space, so it is more formally known as electromagnetic radiation or electromagnetic waves. Gamma rays, x-rays, ultraviolet, ordinary light, infrared, and radio waves are all merely electromagnetic radiation of different wavelengths.

In this chapter, we discuss the properties of radiation in what we call the spectrum and how analysis of this radiation enables scientists to study the Universe. After all, we cannot touch a star! Despite having brought bits of the Moon back to Earth for study, we cannot yet do the same for even the nearest planets, though we have found a few rocks from the Moon and from Mars in the form of meteorites (see Chapter 8). At various points in the book, we will discuss other types of contact that we on Earth have with the Universe beyond, including particles called cosmic rays and extremely elusive particles called neutrinos. Remember that this book's index can take you straight to any given topic, should you choose to learn about it before the text formally addresses it.

Studies of atoms are important for understanding radiation and how it is given off. The simplest atom is hydrogen, and we will see how it gives off or takes up radiation at certain colors, which are known as spectral lines.

The dust lanes that give the Trifid Nebula its name are clearly shown in this Hubble Space Telescope image. Newly formed bright, hot stars at the center illuminate the bright rim seen to their right in this image.
NASA, ESA, and The Hubble Heritage Team (AURA/STScI)/F. Yusef-Zadeh (Northwestern University)

These spectral lines sometimes are visible as bright colors; they are known as emission lines. When seen from certain angles (to be discussed later), a gas's spectral lines can appear as gaps in a continuous band of color; they are then known as absorption lines. (Note that "absorption" is spelled with a "p," not with a second "b.") We will see how studying the spectral lines can even tell you whether an object is moving toward or away from us and how fast.

Finally, we will discuss the measurement scales most commonly used to describe temperature: Fahrenheit, Celsius, and kelvin. We also show how to convert from one temperature scale to another.

FIGURE 2–1 When white light passes through a prism, a full optical spectrum results, with different colors bent by different amounts.

2.1 STUDYING A STAR IS LIKE LOOKING AT A RAINBOW

Over 300 years ago Isaac Newton showed that when ordinary sunlight is passed through a prism, a band of color like the rainbow comes out the other side. Thus "white light" is composed of all the colors of the rainbow (■ Fig. 2–1). A graph of color versus the brightness (we won't make technical distinctions in this book between "brightness" and "intensity") at each color is called a spectrum (plural: spectra), as is the actual display of color spread out. A very dense gas or a solid gives off a continuous spectrum, a spectrum that changes smoothly in brightness from one color to the next.

Technically, each color corresponds to light of a specific wavelength, the distance between two consecutive wave crests or two consecutive wave troughs; see *Figure It Out 2.1: The Nature of Light*. This "wave theory" of light is not the only way we can consider light, but it does lead to very useful and straightforward explanations. Astronomers often measure the wavelength of light in angstroms (abbreviated Å, after A. J. Ångström, a Swedish physicist of the 19th century who mapped the solar spectrum). One angstrom is 10^{-10} m (that is, one ten-billionth of a meter, 1/10,000,000,000 meter).

Expanding the narrow rainbow of light in the center of Figure 2–2, the range of visible light from violet light through blue, green, yellow, orange, and red light extends only from about 4,000 Å to about 6,500 Å in wavelength. The entire visible region of the spectrum is in this narrow range of wavelength, which varies by less than a factor of 2. You can also see the wavelengths of visible light if you look ahead to Figure 2–4. These wavelengths are about half a millionth of a meter long, since 10,000 Å is 1 micrometer. We can remember the colors we perceive from the name of the friendly fellow ROY G BIV: Red Orange Yellow Green Blue Indigo Violet,

going from longer to shorter wavelengths. (The name Indigo is no longer as commonly used.)

Astronomers still tend to use angstroms rather than the unit nanometers that is preferred in the metric system known as Système International (SI). Because 10 Å = 1 nm, divide all the numbers above by 10 to show their values in nanometers; violet light is about 400 nm, for example.

Only certain parts of the electromagnetic spectrum can penetrate the Earth's atmosphere. We say that our atmosphere has "windows" for the parts of the spectrum that can pass through it. The atmosphere is transparent (completely clear) at these windows and opaque (not passing light at all) at other parts of the spectrum. One window passes what we generally call "light," and what astronomers technically call visible light, or "the visible." (In this book, we will use the term "light" to refer to any form of electromagnetic radiation.) Another window falls in the radio part of the spectrum, and modern astronomers can thus base their "radio telescopes" on Earth and still detect that radiation (■ Fig. 2–2).

But we of the Earth are no longer bound to our planet's surface. Balloons, rockets, and satellites carry telescopes high up in or above our atmosphere. They can observe parts of the electromagnetic spectrum that do not reach the Earth's surface. In recent years, we have been able to make observations all across the spectrum. It may seem strange, in view of the long-time identification of astronomy with visible-light observations, to realize that optical studies no longer dominate astronomy.

2.2 "BLACKBODIES" AND THEIR RADIATION

Many things in astronomy seem simple to study. Stars, for example, are gravitationally bound balls of gas, and they give off visible

FIGURE IT OUT 2.1   The Nature of Light

Light can be described as an electromagnetic wave: self-propagating, oscillating (that is, getting larger and smaller) electric and magnetic fields.

The wavelength, denoted by the Greek lowercase λ (lambda), is the distance from one wave crest to the next; see the figure. This has units of length, such as cm.

1 Wavelength

FIGURE FIO 2.1

The frequency, often denoted by the Roman lowercase f [though astronomers and physicists tend to prefer the Greek lowercase ν (nu)], is the number of times per second that a crest passes a fixed point. The unit is 1/s, or hertz (Hz). Hence, the period of the wave, P (in seconds), is simply $1/f$.

In general, the length per wave (λ) multiplied by the number of waves per second (f) gives the length per second traversed by the wave. This is its speed v, so we have

$$\lambda f = v.$$

In our case, $v = c$, the speed of light, so $\lambda f = c$.

(Note that, throughout this book, as is customary in many science books, variables are written in italics and constants in regular, non-italic type. Here, the velocity v is variable while c, the speed of light, is a constant.)

All electromagnetic waves in a vacuum travel with the same speed, c, regardless of their wavelength. The measured speed is independent of the relative speeds of the observer and the light source. This is admittedly counterintuitive, but it has been completely verified; indeed, it is one of the foundations of Einstein's theory of relativity. Electromagnetic waves do slow down in substances such as glass and water, and the speed generally depends on wavelength. This, in fact, is what leads to the spreading out of the colors when light passes through a prism.

Light can also behave as discrete particles known as *photons* (wave or energy "packets"). This is a fundamental aspect of quantum theory. With the right equipment, photons can be detected as discrete lumps of energy.

A photon has no mass, but its energy E is given by the product of Planck's constant h (named after the physicist Max Planck) and its frequency f: $E = hf$. Planck's constant is a very small number. (Its value is listed in Appendix 2 rather than here, to show that you shouldn't be memorizing it.)

Photons of higher energy, therefore, have higher frequency and shorter wavelength:

$$E = hf = hc/\lambda, \text{ since } \lambda f = c.$$

(All photons, regardless of their energy, travel through a vacuum with the same speed, c.)

■ **FIGURE 2–2** In a window of transparency in the terrestrial atmosphere, radiation can penetrate to the Earth's surface. The curve specifies the altitude at which the intensity of arriving radiation is reduced to half its original value. When this level of reduction is reached high in the atmosphere, little or no radiation of that wavelength reaches the ground.

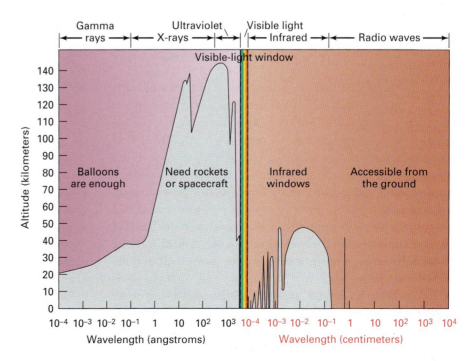

A blackbody is a perfect *absorber*: It absorbs all incident radiation, reflecting and transmitting none. It has a certain temperature, which is a measure of the average speed with which the particles within it jiggle around: the higher the temperature, the greater the average speed. The randomly moving particles, some of which are charged (that is, have electric charge), emit electromagnetic radiation. This emission is called *thermal radiation*, or *blackbody radiation*.

A blackbody is also a perfect *emitter*: The shape of the object's emitted spectrum depends only on its temperature, not on its chemical composition or other properties. This spectral shape is called the Planck curve, in honor of the physicist Max Planck; its derivation was a fundamental problem in quantum physics.

As can be seen in the figure above, at all wavelengths the spectrum (Planck curve) of a hot blackbody is higher (that is, brighter) than that of a colder blackbody having the same surface area.

An important property is that the spectrum of a hot blackbody peaks at a shorter wavelength than that of a colder blackbody. The product of the temperature (T) and the wavelength at which the spectrum peaks (λ_{peak}) is a constant:

$$\lambda_{peak} T = 2.9 \times 10^7 \text{ Å K} = 0.29 \text{ cm K}.$$

This relation is known as *Wien's law*. It is mathematically derivable from the formula for the Planck curve. For example, the spectrum of the Sun, whose surface temperature is about 5,800 K, peaks at a wavelength of

$$\lambda_{peak} = (2.9 \times 10^7 \text{ Å K})/(5{,}800 \text{ K}) = 5{,}000 \text{ Å}.$$

A star that is twice as hot as the Sun has a spectrum that peaks at half this wavelength, at 2,500 Å; the product of temperature and the wavelength of the spectral peak must remain constant.

FIGURE FIO 2.2 The brightness of radiation at different wavelengths for different temperatures. "Blackbodies" – objects that give off (emit) radiation in an ideal manner – have spectra that follow these particular curves, which depend only on the temperature of the object. The spectra of stars are fairly similar to these curves; stars are nearly perfect blackbodies.

light and other electromagnetic radiation that on the whole follow an especially simple rule. If you have two stars of the same size and one is hotter than the other, it is also intrinsically brighter. Furthermore, the hotter star gives off more of its energy at shorter wavelengths.

Let us consider the simplest possible object that gives off radiation. It is called a blackbody, which indicates that it is, in principle, a simple thing – it isn't like a polka-dotted body, for example, that gives off radiation with different properties from different areas. A blackbody is an opaque object that absorbs all radiation that is incident upon it; absolutely no radiation is reflected or transmitted. Atomic motions within that object (because of its nonzero temperature) then cause it to emit (radiate) energy in a manner that depends only on its temperature, not on chemical composition or anything else. This result is called blackbody radiation (or thermal radiation, because one number, a temperature, exists that can be used to generate the Planck curve).

Understanding blackbodies and their radiation is a key to understanding stars. Each bit of gas in a star radiates as a blackbody. However, overall a star is not a perfect blackbody; its spectrum

actually does depend in a minor way on the chemical composition of its outermost layers of gas. Moreover, these outer layers are partly transparent, so we may see gas having a small range of temperatures when we look at a star. Nevertheless, for many purposes we can approximate the spectrum of a star as being that of a blackbody. (There are, in fact, very few objects in nature that are perfect, idealized blackbodies, yet the concept is useful in a wide range of circumstances.)

Spectra of hotter blackbodies peak (that is, have their highest brightness) at shorter (bluer) wavelengths than spectra of colder blackbodies (see Figure It Out 2.2: *Blackbody Radiation and Wien's Law*). Also, a hot blackbody emits much more energy per second than a cold blackbody of equal surface area (see *Figure It Out 2.3: Blackbody Radiation and the Stefan-Boltzmann Law*).

Planets are much cooler than stars, so they give off most of their radiation at long wavelengths, in the infrared. The Sun, on the other hand, gives off most of its radiation in visible light. When we look at the spectrum of a planet, we see that some of the radiation is strongest in the visible, which means it is reflected sunlight, while other radiation is strongest in the infrared and therefore emitted by

Another useful property of blackbody radiation is that, per unit of surface area, a hot blackbody emits much more energy per second than a cold blackbody. In fact, the energy emitted is proportional to the fourth power of the temperature: $T \times T \times T \times T$. This relation is known as the *Stefan-Boltzmann law*, which (like Wien's law) is derivable from the Planck curve. It can be expressed as

$$E = \sigma T^4,$$

where E is the energy emitted per unit area (for example, cm²) per second, T is the surface temperature in kelvins (it is important that the temperature scale start at absolute zero), and σ (sigma) is a constant (known as the Stefan-Boltzmann constant).

For example, if two stars (labeled with subscripts 1 and 2) have the same surface area, but one is twice as hot as the other, the hotter star emits $2^4 = 16$ times as much energy (per second) as the colder star:

$$E_1/E_2 = (\sigma T_1^4) /(\sigma T_2^4) = T_1^4/T_2^4 = (T_1/T_2)^4 = 2^4 = 16.$$

(Note that the constant cancelled out, so you didn't need to know its value and multiply it. Very often, ratios like this one simplify calculations; here, you just take 2 to the fourth power. Use ratios whenever you can!)

However, in the above example, if the hotter star's radius (R) is $1/4$ that of the colder star, then its surface area is $(1/4)^2 = 1/16$ as large as that of the colder star, because the surface area of a sphere is given by $4\pi R^2$. This lesser area exactly balances the greater emission per unit area, making the two stars equally luminous.

In general, the luminosity (L) (that is, the energy emitted per second, also known as power or intrinsic brightness) of a blackbody is given by its surface area (S) multiplied by the energy emitted per unit area per second (E): $L = SE$. For a sphere of radius R and temperature T, we have

$$L = 4\pi R^2 \sigma T^4.$$

Thus, if we know the luminosity and surface temperature of a star, we can derive its radius R.

Note that Fraunhofer labelled lines with letters from A through H and then used I for the end of the spectrum. Though most scientists don't realize it, the K line of ionized calcium so often referred to in astronomy as a Fraunhofer line wasn't given its notation until the mid-nineteenth century.

the planet itself as thermal radiation, largely in the infrared. Thus a graph of the energy given off by a planet shows two peaks – one in the visible and one in the infrared (■ Fig. 2–3). A planet, therefore, is not a perfect blackbody, but by considering each of its spectral peaks and its blackbody approximation, the concept is nonetheless often useful.

Similarly, humans reflect visible light from the Sun or from room lamps, and so are not perfect blackbodies. But we do emit our own thermal (blackbody) radiation, which is most intense at infrared wavelengths that are visible to certain infrared cameras but not to our eyes.

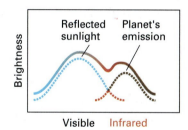

■ **FIGURE 2–3** The spectrum of a planet includes one peak in the visible, produced by reflected sunlight, and another peak in the infrared. The infrared peak is the thermal emission (nearly blackbody radiation) from the planet itself, based on its surface temperature. The observed spectrum is the sum of the two curves; here, it is offset upward a little for clarity.

2.3 WHAT ARE THOSE MISSING COLORS AND WHERE ARE THEY?

In the early 1800s, when Joseph Fraunhofer looked in detail at the spectrum of the Sun, he noticed that the continuous range of colors in the Sun's light was crossed by dark gaps (■ Fig. 2–4). He saw a dozen or so of these "Fraunhofer lines"; astronomers have since mapped millions. The presence of such lines indicates that the Sun

is not a perfect blackbody, but they allow astronomers to learn much more about the Sun (and other stars as well).

The dark Fraunhofer lines turn out to be from relatively cool gas absorbing radiation from behind it. We thus say that they are absorption lines. Atoms of each of the chemical elements in the gas absorb light at a certain set of wavelengths. By seeing what wavelengths are absorbed, we can tell what elements are in the gas, and the proportion of them present. We can also measure the temperature of the gas more accurately than from the peak of the nearly blackbody spectrum.

Imagine the band of color shown in Figure 2–4 without the dark vertical lines crossing it. That continuous band of color would be

6563 Å   5890/96 Å   5270 Å   4861 Å   4308 Å   3968 Å

Red   Orange   Yellow   Green   Blue   Indigo   Violet

Jay M. Pasachoff

■ **FIGURE 2–4** The dark lines that extend from top to bottom, marking the absence of color at a specific wavelength, are the Fraunhofer lines. The underlying drawing is Fraunhofer's original. We still use the capital-letter notation today for the "D" line (sodium) and the "H" line (from ionized calcium, not hydrogen!). We now know that Fraunhofer's "C" and "F" lines are basic lines of hydrogen. The "E" line is from iron and the "G" line is from iron and calcium. Fraunhofer's lowercase "b" is magnesium. The features marked "A" and "B" are bands from oxygen in the Earth's atmosphere.

the "continuous spectrum" we defined in Section 2.1. With the dark lines crossing the spectrum, however, we have an "absorption-line spectrum" or a "Fraunhofer spectrum."

How does the absorption take place? To understand it, we have to study processes inside the atoms themselves. Atoms are the smallest particles of a given chemical element. For example, all hydrogen atoms are alike, all iron atoms are alike, and all uranium atoms are alike. As was discovered by Ernest Rutherford in 1911, atoms contain relatively massive central objects, which we call nuclei (■ Fig. 2–5). Distributed around the nuclei are relatively light particles, which we call electrons. The nucleus contains protons, which have a positive electric charge, and neutrons, which have no electric charge and so are neutral. Thus the nucleus – protons and neutrons together – has a positive electric charge. Electrons, on the other hand, have a negative electric charge. Each of the chemical elements has a different number of protons in its nucleus. (Protons and neutrons are each made of smaller particles called quarks, which we discuss toward the end of this book; they are not needed for discussing the present matters of spectra.)

Why aren't the electrons pulled into the nucleus? In 1913, Niels Bohr (■ Fig. 2–6) made a suggestion that some rule (at that time the rule was arbitrary) kept the electrons at fixed distances, in what are often (inaccurately) thought of as orbits. His suggestion was an early step in the development of the quantum theory, which was elaborated in the following two decades. The theory incorporates the idea that light consists of individual packets – quanta of energy; the utility of considering such individual, non-continuous objects had been developed a decade earlier.

We can think of each quantum of energy as a massless particle of light, which is called a photon; see *Figure It Out 2.1: The Nature of Light*. Photons of blue light have higher energy than photons of red light, but they all travel at the *same* speed through a vacuum.

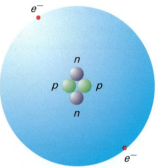

■ **FIGURE 2–5** A helium atom contains two protons and two neutrons in its nucleus and two "orbiting" electrons. The nucleus is drawn much larger than its real scale with respect to the overall atom; the actual radius of the electron cloud is over 10,000 times larger than that of the nucleus.

American Institute of Physics, Emilio Segrè Visual Archives, Uhlenbeck Collection

■ **FIGURE 2–6** Niels Bohr and his wife, Margrethe. They have become widely known in the last few years through Michael Frayn's play *Copenhagen*, about different perceptions of a controversial meeting Bohr had with his former assistant Werner Heisenberg in a fraught time leading up to the invention of the atomic bomb.

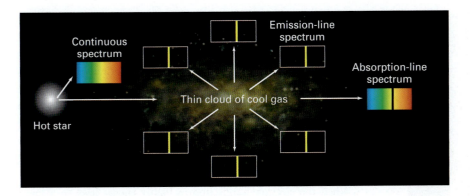

■ **FIGURE 2–7** When we view a hotter source that emits continuous radiation through a cooler gas, we may see absorption lines. These absorption lines appear at the same wavelengths at which the gas gives off emission lines when viewed with no background or against a cooler, darker background. Each of the spectra in the little boxes is the spectrum you would see looking back along the arrow. Note that the view from the right shows an absorption line. Only when you look through one source silhouetted against a hotter source do you see the absorption lines.

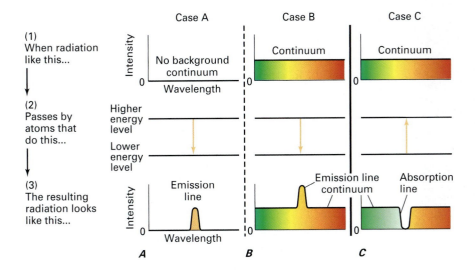

■ **FIGURE 2–8** When photons are emitted, we see an emission line. Continuous radiation may or may not be present (Cases A and B, respectively). An absorption line must absorb radiation from something; hence, an absorption line only appears when there is also continuous radiation (Case C). The top and bottom horizontal rows are graphs of the spectrum before and after the radiation passes by an atom. A schematic diagram of the atom's energy levels is shown in the center row (to simplify the situation, only two levels are given).

For some purposes, it is best to consider light as waves, while for others it is best to consider light as particles (that is, as photons). Though we will not explore quantum theory in detail here, we should mention that it was one of the major intellectual advances of the 20th century.

Let us return to absorption lines. We mentioned that they are formed as light passes through a gas (■ Fig. 2–7). The atoms in the gas take up (absorb) some of the light (photons) at specific wavelengths. The energy of this light goes into giving the electrons in those atoms more energy, putting them in higher levels. Each change in energy corresponds to a fixed wavelength of absorbed light; the greater the energy absorbed, the shorter the absorbed photon's wavelength.

But energy can't pile up in the atoms. If we look at the atoms from the side, so that they are no longer seen in silhouette against a background source of light, we would see the atoms give off (emit) just as much energy as they take up (absorb), in random directions. Essentially, they emit this energy at the same set of wavelengths as the absorbed light, producing emission lines (abrupt spikes in the brightness of light) at these wavelengths; see ■ Figure 2–8.

Almost all stars have absorption lines. They are formed, basically, as light from layers just inside the star's visible surface passes through atoms right at the surface. The surface is cooler, so the atoms there absorb energy, making absorption lines. (Detailed models are much more complicated.) Emission lines occur only in special cases

for stars. We see absorption lines on the Sun's surface, but we can see emission lines by looking just outside the Sun's edge, where only dark sky is the background. Extended regions of gas called "nebulae" (■ Fig. 2–9) give off emission lines, as we will explain later, because they are not silhouetted against background sources of light.

As we will discuss in more detail below, each of the chemical elements has its own distinctive, unique set of spectral lines, like a fingerprint, because of the different numbers of electrons and their allowed energy levels. So if you take the spectrum of a distant star, you may see several overlapping patterns of spectral lines. These patterns allow you to tell what chemical elements are in the star. We know what stars consist of from an analysis of their spectra!

2.4 THE STORY BEHIND THE BOHR ATOM

Bohr, in 1913, presented a model of the simplest atom – hydrogen – and showed how it produces emission and absorption lines. Stars often show the spectrum of hydrogen. Laboratories on Earth can also reproduce the spectrum of hydrogen; the same set of spectral lines appears both in emission and in absorption.

NASA, ESA, and The Hubble Heritage Team (STScI/AURA); P. McCullough (STScI)

■ FIGURE 2–9 An interstellar cloud known as NGC 281 (from the New General Catalogue, a 19th-century list of largely nonstellar objects) glows in the reddish color of hydrogen gas. Dark dust is silhouetted against it.

Hydrogen consists of a single central particle (a proton) in its nucleus with a single electron surrounding it (■ Fig. 2–10). The model for the hydrogen atom explained why hydrogen has only a few spectral lines (■ Fig. 2–11), rather than continuous bands of color. It postulated that a hydrogen atom could give off (emit) or take up (absorb) energy only in one of a fixed set of amounts, just as you can climb up stairs from step to step but cannot float in between. The position of the electron relative to the nucleus determines its specific, discrete energy level, the amount of energy in a hydrogen atom. So the electron can only jump from energy level to energy level, and not hover in between values of the fixed set of energies allowable.

When light hits an electron and makes it jump from a lower energy level to a higher energy level, a photon is absorbed, contributing to an absorption line. (When there are many absorbed photons, we see an absorption line in an otherwise continuous spectrum.) The photon that can be absorbed has an energy exactly equal to the difference between the higher and lower electronic energy levels. An atom cannot absorb a photon having too little energy, and it cannot take just part of the energy of a high-energy photon (releasing the excess energy as a lower-energy photon). Either all or none of the energy is absorbed.

When the electron jumps from a higher energy level to a lower one, a photon is given off (emitted). Such photons can go away from the atom in any direction, regardless of the direction of the original photon that was absorbed by the atom. Many photons together make an emission line.

The Bohr atom is Bohr's model that explains the hydrogen spectrum. In it, electrons can have orbits of different sizes. Each orbit corresponds to an energy level (■ Fig. 2–12). Only certain orbits are allowable. Bohr himself couldn't give a clear reason for this; he said that's just the way it is. (Note, however, that according to modern quantum theory, the electron orbits are not like planetary orbits; instead, the electron behaves as though it were distributed throughout its orbit. Also, we now understand the details of atoms in the context of quantum theory, but these are complicated.)

We use the letter n to label the energy levels. We call the energy level for $n = 1$ the ground level, as it is the lowest possible energy state. Excited levels have $n = 2$ and higher. The hydrogen atom's series of transitions from or to the ground level is called the Lyman series, after the American physicist Theodore Lyman. The lines fall in the ultraviolet, at wavelengths far too short to pass through the Earth's atmosphere. Telescopes above the atmosphere now enable us to observe the Lyman lines of stars.

The series of transitions with $n = 2$ as the lowest level is the Balmer series, which is in the visible part of the spectrum. The bright red emission line in the visible part of the hydrogen spectrum, known as the Hα (H-alpha) line, corresponds to the transition from level 3 to level 2. The transition from $n = 4$ to $n = 2$ causes the Hβ (H-beta) line, and so on. Because the series falls in the visible where it is so well observed, we usually call the lines simply H-alpha, etc., instead of Balmer alpha, etc. So the Lyman series is transitions to or

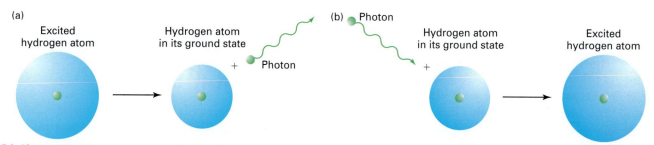

■ FIGURE 2–10 Since the hydrogen atom has only a single electron, it is a particularly simple case to study. The lowest possible energy state of an atom is called its ground state; the electron is closest to the nucleus. All other energy states are called excited states; the electron is farther from the nucleus. (b) The formation of an absorption line, by taking up (absorbing) photons from incoming continuous radiation. Figure 2–12 shows more details for the specific transitions.

■ **FIGURE 2–11** The Balmer series, representing transitions down to (or up from) the second energy state of hydrogen. The strongest line in this series, Hα (H-alpha), is in the red. Note how the lines get closer and closer together as you go to shorter wavelengths. This distinct, converging pattern makes the Balmer series, and thus hydrogen, easy to identify in a spectrum.

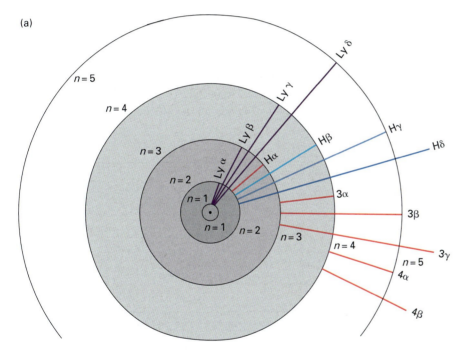

(a)

■ **FIGURE 2–12** The representation of hydrogen energy levels known as the Bohr atom. (a) Each of the circles shows the lone electron in a different energy level. (b) A schematic transition in which an electron is excited from level 1 to level 3 by an incoming photon. It can then jump down to level 2, emitting a red photon, or directly to level 1 (a greater difference in energy), emitting an ultraviolet photon, which has more energy than the red photon.

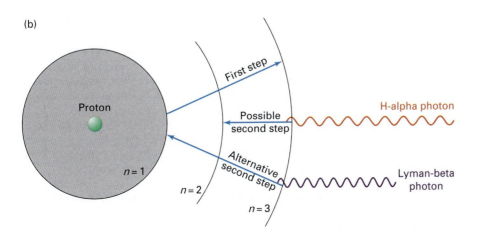

(b)

from lower level 1, and the Balmer series is transitions to or from lower level 2. Other series of hydrogen lines correspond to transitions with still different lower levels.

Since the higher energy levels have greater energy (they are higher above the ground state), a spectral line caused solely by transitions from a higher level to a lower level is an emission line. When, on the other hand, continuous radiation falls on cool hydrogen gas, some of the atoms in the gas can be raised to higher energy levels and absorption lines result.

Each of the chemical elements has its own set of discrete energy levels. Although the details are much more complex than in the simple Bohr model of the hydrogen atom, the general principles are similar. Thus, each of the chemical elements produces a unique pattern of spectral lines, like a fingerprint. The relative strength of the lines actually seen in the spectrum of a particular element depends on the distribution of electrons among the various energy levels, and this in turn depends on temperature and other factors.

Some atoms have lost one or more electrons, through either collisions with other particles or absorption of high-energy photons; they are said to be ionized (■ Fig. 2–13). The spectrum of an ionized element is different from the spectrum of the same element when it is not ionized, because the distribution of electron energy levels is different. Similarly, the spectrum of each molecule (two or more atoms bound to each other) is unique.

2.5 THE DOPPLER EFFECT AND MOTION

Even though we can't reach out and touch a star, we can examine its light in great detail. Above, you saw how studying starlight reveals the temperature of a star (from its color) and what chemical elements it has near its surface (from its spectral lines). Studying starlight also tells us how fast the star is moving toward or away from us. The technique applies to all kinds of objects, including the planets.

The effect that motion has on waves is known as the Doppler effect, and was studied by Christian Doppler about 170 years ago (see ■ Fig. 2–14). It works for any waves – water waves, sound waves, or light waves. We will give more details about it when we study the motion of stars (Chapter 11).

For the moment, we need only realize that when an object is coming toward you, the waves you are getting from it are compressed, so the wavelength becomes shorter. When an object is going away from you, the waves you are getting from it are spread out, and the wavelength becomes longer.

Since the visible part of the spectrum has blue at short wavelengths and red at long wavelengths, the wavelengths are shifted toward the blue when objects are approaching and toward the red

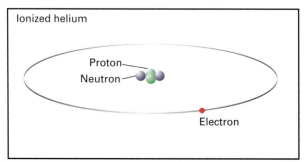

■ **FIGURE 2–13** An atom missing one or more electrons is ionized. Neutral helium is shown at left; it has two electrons to balance the charge of the two protons. Singly ionized helium is shown at right. It has only one electron, so its net charge is +1.

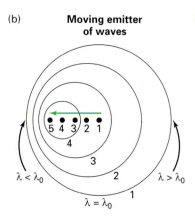

■ **FIGURE 2–14** (a) A stationary source is emitting waves of wavelength λ_0, in a circular pattern around it. (The subscript "0" is pronounced "naught" and indicates that the source is at rest.) Four wave crests, labeled 1 through 4, are shown. The observed wavelength is independent of the observer's position relative to the source. (b) The source is moving to the left (green arrow) with a constant speed. Along the direction of motion, the source partially keeps up with its most recently emitted wave crest before it emits another wave crest, so the crests get bunched closer together. From the left, the observed wavelength λ is shorter than λ_0; this is a "blueshift." Conversely, an observer on the right would see $\lambda > \lambda_0$; this is a "redshift." Perpendicular to the direction of motion, the spacing of the wave crests is unaffected, and an observer would therefore measure a wavelength $\lambda = \lambda_0$.

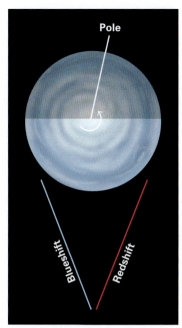

FIGURE 2–15 If a planet is rotating, one side will be approaching us, blueshifting its spectral lines, while the other side will be receding from us, redshifting its spectral lines.

when objects are receding. Astronomers use the term blueshift to describe the shifts in wavelength of objects that are approaching and the term redshift to describe the shifts in wavelength of objects that are receding. Note that the light still travels at c, the speed of light in a vacuum; wavelengths and frequencies change, but the measured speed is independent of the motion of the source or of the observer.

One example of redshifts and blueshifts deals with planets in the Solar System. If you take spectra of several parts of a planet, and see that one side is blueshifted and the other side is redshifted, then you know that the first side is approaching you and the other side is receding (■ Fig. 2–15). You have found out that the planet is rotating! By measuring how much the light is shifted, you can even tell how fast the planet is rotating. The Doppler effect is a powerful tool for studying distant objects.

Perhaps the most exciting use of the Doppler effect has been the discovery of many planetary systems other than our own Solar System. Since 1995, hundreds of planets have been discovered around other stars with this method of looking for small redshifts and blueshifts. The discovery in a few cases of several planets around the same star confirms that at least most of these objects are indeed planets rather

FIGURE IT OUT 2.4   Temperature Conversions

Though most Americans use the Fahrenheit temperature scale, in which water freezes at 32°F and boils at 212°F, most of the rest of the world uses the Celsius scale, in which water freezes at 0°C and boils at 100°C. Note that the difference between freezing and boiling points for each scale is 180°F and 100°C, respectively, so a change of 180°F equals a change of 100°C, or 9°F for every 5°C. There is little, if any, water on the stars or on most planets, so astronomers use a more fundamental scale. Their scale, the kelvin scale (whose unit is kelvins, not degrees kelvin, and whose symbol is K, no longer °K), begins at absolute zero, the coldest temperature that can ever be

approached. Since absolute zero is about –273.16°C, and a rise of one kelvin (1 K) is the same as a rise of 1°C, the freezing point of water (0°C) is about 273 K (see the figure below).

To convert from kelvins to °C, simply subtract 273. To change from °C to °F, we must first multiply by $9/5$ and then add 32. You may (or may not) find it easy to remember: times two, minus point two, plus thirty-two. That is, multiply by 2, subtract two-tenths of the original (which gives you $9/5$), and then add 32, to get °F. For stellar temperatures, the 32° is too small to notice, and a sufficient level of approximation is often simply to multiply °C by 2 (see figure below).

FIGURE FIO 2.4a Temperature scales in the range useful for terrestrial and planetary studies.

FIGURE FIO 2.4b Temperature scales in the range useful for stellar studies.

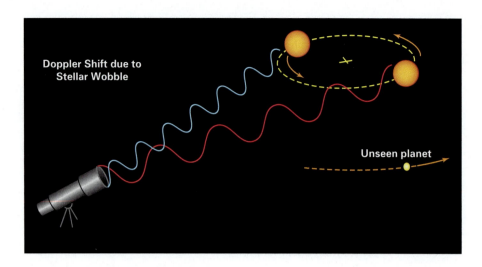

■ **FIGURE 2–16** The technique used by several teams of scientists to detect planets around Sun-like stars (Chapter 9) by searching for periodically changing Doppler shifts in the stars' spectra. The diagram shows a particular star at two positions in its orbit. The successive redshifts and blueshifts of that star's spectrum reveal the otherwise unseen planet. It must always be on the opposite side of the system's "center of mass" (like the fulcrum of a see-saw; Chapter 11 has more details) from the star.

Doppler Shift due to Stellar Wobble

Unseen planet

than some kind of failed companion star; we would not expect there to be so many failed stars in a system.

The technique first used to find these planets relies upon the Doppler effect, since the planets are too faint and too close to their parent stars to be seen directly, even with our best telescopes on Earth or in space. (It is independent of the transit technique that we describe in Chapter 9, fruitfully used by the Kepler spacecraft; our situation is best when we can use both techniques on the same objects.) The Doppler technique depends on the fact that the star isn't entirely steady in space, but rather moves to and fro as the planet around it moves fro and to.

Consider, for example, a pair of dancers waltzing: They usually rotate around a point between them. If one of the dancers were invisible, we could still see the other dancer moving around. Similarly, we can see the visible object, the star, moving slightly toward and away from us, and infer that there must be an invisible object, the planet, moving in step but in the opposite direction at every time. Since the star is so much more massive than the planet, it moves less than the planet (■ Fig. 2–16).

The breakthrough in 1995 that enabled astronomers to detect these planets came when they developed some extremely precise ways of measuring Doppler shifts, more precise than had been possible before. Many more planets are now known outside the Solar System than inside it. We will describe those "extra-solar planets," or "exoplanets," in Chapter 9.

CONCEPT REVIEW

» Electromagnetic radiation is a combination of electricity and magnetism changing together in space; it is also known as electromagnetic waves (introductory section). The wavelength is the distance between two consecutive crests (or troughs) of the wave, and different wavelengths correspond to different colors (Sec. 2.1). The rainbow produced by passing light through a prism, as well as the graph of color (wavelength) versus the brightness at each color, is called a spectrum. Such spectra are extremely valuable to astronomers; they can be used to determine the temperature, chemical composition, and other properties of the object being studied. From the Earth's surface, we view through atmospheric windows that pass only certain types of electromagnetic radiation, chiefly visible and radio.

» A very dense gas or solid gives off a **continuous spectrum** that changes smoothly in brightness from one color to the next (Sec. 2.2). In the case of a **blackbody**, an object that absorbs all radiation that is incident upon it (none is reflected or transmitted), the shape of the emitted continuous spectrum depends only on the object's surface temperature, not on its chemical composition or other properties. This result is called **blackbody radiation** or **thermal radiation**. Spectra of hotter blackbodies are brightest at shorter (bluer) wavelengths while spectra of cooler blackbodies are brightest at longer (redder) wavelengths. Also, a hot blackbody is intrinsically brighter than a cold blackbody of equal surface area.

» **Atoms** are composed of positively charged **nuclei** "orbited" by negatively charged **electrons** (Sec. 2.3). The nuclei are made of positively charged **protons** (which distinguish each chemical element) and neutral **neutrons**.

» **Quantum theory** explains how to describe atoms, and the rules that prevent electrons from spiraling into nuclei. When an atom loses one or more electrons, it is **ionized**. Two or more atoms bound together form a **molecule**.

» Light sometimes acts as though it were made of particles, which are called **photons** (Sec. 2.3). **Absorption lines** are dark gaps in a spectrum caused by atoms (or molecules) absorbing some of the photons that are passing through them; the electrons jump from lower to higher energy levels. Conversely, **emission lines** are locations in a spectrum where extra photons are present. They are caused by electrons jumping from higher to lower energy levels, thereby emitting light.

» The **Bohr atom**, with electrons at different radii centered on the nucleus, explains the spectrum of hydrogen in detail (Sec. 2.4). Transitions to the lowest **energy level**, called the **ground level**, do not cause lines in the visible part of the spectrum, but rather in the ultraviolet. Transitions between the second energy level and higher levels (called **excited levels**) lead to photons being added or subtracted in the visible part of the spectrum. The resulting spectral lines are known as the **Balmer series**.

» The **Doppler effect** is a change in wavelength of light caused by an object's motion along a line toward or away from you (Sec. 2.5). If you and the object are receding from each other, the object's spectrum appears **redshifted**, while **blueshifts** are seen if you and the object are approaching each other. Larger relative speeds produce bigger Doppler shifts.

QUESTIONS

1. What is the difference between a gamma ray and a radio wave?

2. Why is electromagnetic radiation so important to astronomers?

3. Discuss the difference between the general concept of the spectrum and the particular spectrum we see as a rainbow.

4. Does white light contain red light? Green light? Discuss.

5. After Newton separated sunlight into its component colors with a prism, he reassembled the colors. What should he have found and why?

6. Identify Roy G Biv. Identify A. J. Ångström. What relation do they have to each other?

7. How can our atmosphere have a "window"? Can our air leak out through it? Explain.

8. Describe the relation of Fraunhofer lines to emission lines and to absorption lines.

9. Describe how the same atoms can sometimes cause emission lines and at other times cause absorption lines.

10. Sketch an atom, showing its nucleus and its electrons.

11. Sketch the energy levels of a hydrogen atom, showing how the Balmer series arises.

12. On a sketch of the energy levels of a hydrogen atom, show with an arrow the transition that matches Hα in absorption. Also show the transition that matches Hα in emission.

13. If you were to take an emission nebula and put it in front of a very hot, bright source of continuous radiation, what change would you see in the nebula's spectrum? Explain.

14. If the surface of a star were increasing in temperature as you went outward, would you see absorption lines or emission lines in its spectrum? Explain how this case differs from what we see in normal stars, which decrease in temperature as you go outward in the region from which they give off most of their light.

15. Does the Bohr atom explain iron atoms? Describe from this chapter what type of atom the Bohr atom explains.

16. If someone were to say that we cannot know the composition of distant stars, since there is no way to perform experiments on them in terrestrial laboratories, how would you respond?

17. A spectral line from the left side of Saturn's rings, as you see them, is at a wavelength slightly shorter than the same spectral line measured from the right side of Saturn's rings. Which way (direction) is Saturn rotating, and why?

†18. Give the temperature in degrees Celsius for a 60°F day.

†19. What Fahrenheit temperature corresponds to 30°C?

†**20.** The Earth's average temperature is about 27°C. What is its average temperature in kelvins?

†**21.** The Sun's surface is about 5,800 K. What is its temperature in °F?

22. An iron rod is heated by a welder's torch. Initially it glows a dull red, then a brighter orange, and finally very bright white. Discuss this sequence in terms of blackbody radiation.

23. Does the Doppler effect depend on the distance between the source of light and the observer? (Assume the direction of motion is purely radial: directly toward or away from the observer.) Explain.

†**24.** The Hβ line of hydrogen has a wavelength of 4,861 Å. (**a**) What is its frequency? (**b**) What is the frequency of a line with twice the wavelength of Hβ?

†**25.** Announcers at a certain radio station say that they are at "95 FM on your dial," meaning that they transmit at a frequency of 95 MHz (95 megahertz, or 95 million cycles per second). What is the wavelength of the radio waves from this station?

†**26.** (**a**) If one photon has 10 times the frequency of another photon, which photon is the more energetic, and by what factor? (**b**) Answer the same question for the case where the first photon has twice the wavelength of the second photon.

†**27.** Consider a blackbody whose temperature is 3 K. At what wavelength does its spectrum peak? (We will see the importance of this radiation in the discussion of cosmology in Chapter 19.)

†**28.** Suppose the peak of a particular star's spectrum occurs at about 6,000 Å. (**a**) Use Wien's law to calculate the star's surface temperature. (**b**) If this star were a factor of four hotter, at what wavelength would its spectrum peak? In what part of the electromagnetic spectrum is this peak?

†**29.** (**a**) Compare the luminosity (amount of energy given off per second) of the Sun with that of a star the same size but three times hotter at its surface. (**b**) Answer the same question, but now assume the star also has twice the Sun's radius.

†**30.** How many times hotter than the Sun's surface is the surface of a star the same size, but that gives off twice the Sun's energy per second (that is, is twice as luminous)?

31. **True or false?** When we see other people, our eyes are detecting the visible light that each of us radiates as an approximate "blackbody" (thermal emitter).

†**32.** **True or false?** If the surface temperature of Star Sam is 3 times that of Star Lily, then the wavelength at which Star Sam's spectrum peaks is ⅓ the wavelength at which Star Lily's spectrum peaks.

33. **True or false?** The Doppler effect has been used to measure the rotation rates of planets in our Solar System, as well as to detect the presence of planets around some other stars.

34. **True or false?** If the spectrum of Star Jessica peaks at optical wavelengths, and Star Jacob's surface is twice as hot as Jessica's surface, then Jacob's spectrum peaks in the infrared.

35. **True or false?** The type of spectral feature usually observed from a hot gas with no star behind it along the line of sight is an absorption line.

36. **True or false?** The formula $c = \lambda f$ for light in a vacuum is correct even if the source of light is moving toward the observer.

37. **True or false?** Per photon, ultraviolet photons have more energy than gamma-ray photons.

38. **Multiple choice:** Which one of the following statements about atoms is *false*? (**a**) Electrons have discrete energy levels. (**b**) Each element produces a unique pattern of spectral lines, like a fingerprint. (**c**) Photon emission occurs randomly, in any direction. (**d**) An electron in an atom may absorb either part or all of the energy of a photon. (**e**) Absorption occurs when an electron in an atom jumps from a lower energy level to a higher energy level.

39. **Multiple choice:** In a vacuum, photons of higher energy (**a**) move faster than lower-energy photons; (**b**) have higher frequencies and shorter wavelengths than lower-energy photons; (**c**) have more mass than lower-energy photons; (**d**) are not as likely to become redshifted as lower-energy photons; or (**e**) travel less distance between their source and the observer than lower-energy photons.

†**40.** **Multiple choice:** Suppose atom Orion has an electron with permitted levels of energy $E = 1$, $E = 2$, $E = 5$, and $E = 10$ units. If the electron is sitting in the level with energy $E = 2$, which one of the following photons could Orion absorb? (**a**) A photon having an energy of 8. (**b**) A photon having an energy of 5. (**c**) A photon having an energy of π. (**d**) A photon having an energy of 1. (**e**) A photon having an energy of –1.

†**41.** **Multiple choice:** A local radio station broadcasts at 100.3 megahertz (megahertz = 10^6 cycles per second; symbol MHz). What is the approximate wavelength of the signal? (**a**) 3,000 angstroms. (**b**) 30 meters. (**c**) 1.86 miles. (**d**) 30 centimeters. (**e**) 3 meters.

42. **Multiple choice:** Which one of the following statements about electromagnetic waves is *false*? (**a**) Human eyes are able to detect only a tiny fraction of all possible magnetic waves. (**b**) If electromagnetic wave Zoe has twice the wavelength of electromagnetic wave Simon, then Zoe also has twice the frequency of Simon. (c) "White light" such as sunlight actually consists of many electromagnetic waves, having different wavelengths, mixed together. (d) An electromagnetic wave consists of oscillating electric and magnetic fields that are perpendicular to each other and perpendicular to the direction of motion. (e) The measured speed of an electromagnetic wave is independent of the speed of its source relative to the observer.

†43. **Multiple choice:** Suppose Star Caprielle, having the same size and surface temperature as our Sun, suddenly triples in temperature and shrinks to a third of its original diameter. The amount of light emitted from Caprielle will (a) increase by a factor of 27; (b) increase by a factor of 9; (c) increase by a factor of 3; (d) remain unchanged; or (e) decrease by a factor of 9.

44. **Fill in the blank:** If a neutral atom loses one or more electrons, the atom is said to be _____.

45. **Fill in the blank:** By observing the _____ of a star or planet, we can determine what kinds of atoms or molecules are present and their relative abundances.

46. **Fill in the blank:** Electromagnetic radiation behaves as though it has properties of both _____ and _____.

†47. **Fill in the blank:** According to the Stefan-Boltzmann law, the ratio of the energy emitted per second by two blackbodies having the same surface area is proportional to the _____ power of the ratio of their temperatures.

†This question requires a numerical solution.

TOPICS FOR DISCUSSION

1. What are some examples in which you know that magnetic or electric fields play a prominent role? Is there evidence that one type of field produces or interacts with the other type?

2. Why is spectroscopy useful in astronomy? Describe and explain three possible applications.

3. Is any object truly a perfect "blackbody," absorbing absolutely all of the energy that hits it? If there are few such objects, why is the blackbody concept useful?

4. What do you think of the dual nature of light, being both a wave and a particle? Is this contrary to your intuition?

Top: The Hubble Space Telescope as seen in 2009 from the space shuttle that had brought astronauts to service it for the last time, with the photograph taken as the astronauts departed.

NASA's Johnson Space Center

Bottom: Light pollution is a problem for ground-based observatories, a major reason why the major ones are located at isolated sites such as Mauna Kea in Hawaii and a ridge near the Andes in Chile. Here we see an International Space Station view showing the eastern United States, with lights from such cities as Washington, Baltimore, Philadelphia, and New York, framed by two docked Russian spacecraft. Though major cities will never be good venues for optical telescopes, light-pollution-fighting ordinances, such as those preventing upward spotlights, can help make the sky visible from other locations and can protect observatory sites.

International Space Station Expedition 30, NASA

3

Light and Telescopes: Extending Our Senses

ORIGINS

We discuss the telescopes used by astronomers to see the farthest and faintest objects, including not only the current generation of huge telescopes on the ground and major visible, x-ray, and infrared space telescopes aloft but also even better new telescopes under construction. These instruments allow us to learn about the earliest epochs of the Universe, to study how stars form, and to search for other planetary systems, among other things.

AIMS

1. Discuss the historical importance of the development of telescopes (Section 3.1).

2. See how telescopes work, and understand the main uses of telescopes (Sections 3.2 and 3.3).

3. Learn about a variety of telescopes on the ground and in space for studying radiation inside and outside the visible spectrum (Sections 3.4 to 3.8).

Everybody knows that astronomers use telescopes, but not everybody realizes that the telescopes astronomers use are of very different types. Moreover, very few modern telescopes are used directly with the eye. In this chapter, we will first discuss the telescopes that astronomers use to collect visible light, as they have for hundreds of years. Then we will see how astronomers now also use telescopes to study gamma rays, x-rays, ultraviolet, infrared, and radio waves.

3.1 THE FIRST TELESCOPES FOR ASTRONOMY

Over four hundred years ago, a Dutch optician put two eyeglass lenses together, and noticed that distant objects appeared closer (that is, they looked magnified). The next year, in 1609, the English scientist Thomas Harriot built one of these devices and looked at the Moon. But all he saw was a blotchy surface, and he didn't make anything of it.

Credit for first using a telescope to make serious astronomical studies goes to Galileo Galilei. In 1609, Galileo heard that a telescope had been made in Holland, so in Venice he made one of his own and used it to look at the Moon. Perhaps as a result of his training

in interpreting light and shadow in drawings (he was surrounded by the Renaissance and its developments in visual perspective), Galileo realized that the light and dark patterns on the Moon meant that there were craters there (■ Fig. 3–1). With his tiny telescopes – using lenses only a few centimeters across and providing, with an eyepiece, a magnification of only 20 or 30, not much more powerful than a modern pair of binoculars and showing a smaller part of the sky – he went on to revolutionize our view of the cosmos, as will be further discussed in Chapter 5.

Whenever Galileo looked at Jupiter through his telescope, he saw that it was not just a point of light, but appeared as a small disk. He also spotted four points of light that moved from one side of Jupiter to another (■ Fig. 3–2). He eventually realized that the points of

■ **FIGURE 3–1** (a) An engraving of Galileo's observations of the Moon from his book *Sidereus Nuncius* (*The Starry Messenger*), published in 1610. (b) A modern photo appears for comparison. Galileo was the first to report that the Moon has craters. It seems reasonable that Galileo had been sensitized to interpreting surfaces and shadows by the Italian Renaissance and by his related training in drawing. Aristotle and Ptolemy had held that the Earth was imperfect but that everything above it was perfect, so Galileo's observation contradicted them.

(a)

Jay M. Pasachoff

(b)

Jay M. Pasachoff, after Ewan Whitaker, Lunar and Planetary Laboratory, University Arizona

(a)

Mars Orbiter Camera on Mars Global Surveyor, NASA/JPL-Caltech/Malin Space Science Systems; colorized with a Cassini image, NASA/JPL-Caltech/Space Science Institute

■ **FIGURE 3–2** (a) Jupiter and three of its four Galilean satellites; from left to right: Callisto, Ganymede, Jupiter, Europa. Io was hidden behind Jupiter. The relative brightness of the satellites was enhanced. (b) Part of Galileo's printed book that described his first observations of Jupiter's moons. Simon Marius also observed the moons at about the same time; we now use the names Marius proposed for them: Io, Europa, Ganymede, and Callisto, though we call them the Galilean satellites. (c) An image of Jupiter (circled in purple) and four of its moons made in 2002 with a replica of one of Galileo's 2.5-cm (1-in) telescopes.

(b)

Jay M. Pasachoff

(c)

Tom Pope and Jim Mosher

■ **FIGURE 3–3** Venus goes through a full set of phases, from new to full (crescent to nearly full shown here).

light were moons orbiting Jupiter, the first proof that not all bodies in the Solar System orbited the Earth.

The existence of Jupiter's moons contradicted the ancient Greek philosophers – chiefly Aristotle and Ptolemy – who had held that the Earth is at the center of all orbits (see Chapter 5). Further, the ancient ideas that the Earth could not be in motion because the Moon (and other objects) would be "left behind" was also wrong. Galileo's discovery of the moons thus backed the newer theory of Copernicus, who had said in 1543 that the Sun and not the Earth is at the center of the Universe. And Galileo's lunar discovery – that the Moon's surface had craters – had also endorsed Copernicus's ideas, since the Greek philosophers had held that celestial bodies were all "perfect." Galileo published these discoveries in 1610 in his book *Sidereus Nuncius* (*The Starry Messenger*).

Seldom has a book been as influential as Galileo's slim volume. He also reported in it that his telescope revealed that the Milky Way was made up of a myriad of individual stars. He drew many individual stars in the Pleiades, which we now know to be a star cluster. And he reported some stars in the middle of what is now known as the Orion Nebula. But one attempt made by Galileo in his *Sidereus Nuncius* didn't stick: he proposed to use the name "Medicean stars," after his financial backers, for the moons of Jupiter. Nowadays, recognizing Galileo's intellectual breakthroughs rather than the Medicis' financial contributions, we call them the "Galilean moons."

Galileo went on to discover that Venus went through a complete set of phases, from crescent to nearly full (■ Fig. 3–3), as it changed dramatically in size (■ Fig. 3–4). These variations were contrary to the prediction of the Earth-centered (geocentric) theory of Ptolemy and Aristotle that only a crescent phase would be seen (see Chapter 5). The Venus observations were thus the fatal blow to the standard geocentric hypothesis. He also found that the Sun had spots on it (which we now call "sunspots"), among many other exciting things.

3.2 HOW DO TELESCOPES WORK?

The basic principles of telescopes are easy to understand. In astronomy we normally deal with light rays that are parallel to each other, which is the case for light from the stars and planets, since they are

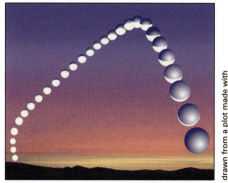

■ **FIGURE 3–4** The position in the sky of Venus at different phases, in an artist's rendition of a computer plot. Venus is a crescent only when it is in a part of its orbit that is relatively close to the Earth, and so looks larger at those times.

extremely far away (■ Fig. 3–5). Certain curved lenses and mirrors can bring starlight to a single point, called the focus (■ Fig. 3–6). The many different points of light coming from an extended object (like a planet) together form an image of the object in the focal plane. If an eyepiece lens is also included, then the image becomes magnified and can be viewed easily with the human eye. For example, each "monocular" in a pair of binoculars is a simple telescope of this kind, much like the ones made by Galileo.

But Galileo's telescopes had deficiencies, among them that white-light images were tinged with color, and somewhat out of focus. This effect, known as chromatic aberration (■ Fig. 3–7), is caused by the fact that different colors of light are bent by different amounts as the light passes through a lens, similar to what happens to light in a prism (as discussed in Chapter 2). Each color ends up having a different focus.

Toward the end of the 17th century, Isaac Newton, in England, had the idea of using mirrors instead of lenses to make a telescope. Mirrors do not suffer from chromatic aberration. When your focusing mirror is only a few centimeters across, however, your head would block the incoming light if you tried to put your eye to this "prime" focus. Newton had the bright idea of putting a small, flat "secondary mirror" just in front of the focus to reflect the light out to the side, bringing the focus point outside the telescope tube. This Newtonian telescope (■ Fig. 3–8) is a design still in use by many amateur astronomers. But many telescopes instead use the Cassegrain

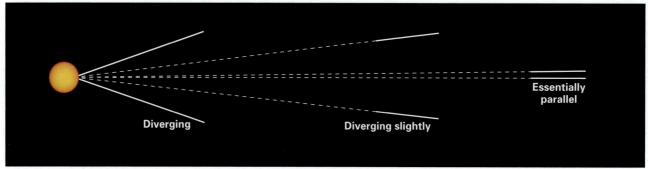

■ **FIGURE 3–5** The light we see from a star is almost parallel, diverging imperceptibly, because stars are so distant.

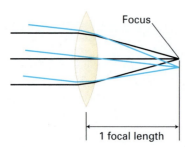

■ **FIGURE 3–6** The focal length is the distance behind a lens to the point at which a very distant object is focused. [The focal length of the human eye is about 2.5 cm (1 inch)]. An object in a slightly different part of the sky (indicated by parallel light rays tilted with respect to the horizontal ones) is imaged at a slightly different location in the focal plane, as shown.

that are in the shape of a paraboloid (a parabola rotated around its axis, forming a curved surface) since paraboloids bring parallel light near the mirror's axis to a focus (■ Fig. 3–11). However, light that comes in from a direction substantially tilted relative to the mirror's axis is still out of focus; thus, simple reflecting telescopes generally have a narrow field of view.

Through the 19th century, telescopes using lenses (refracting telescopes, or refractors) and telescopes using mirrors (reflecting telescopes, or reflectors) were made larger and larger. The pinnacle of refracting telescopes was reached in the 1890s with the construction of a telescope with a lens 40 inches (1 m) across for the Yerkes Observatory in Wisconsin, now part of the University of Chicago (■ Fig. 3–12).

It was difficult to make a lens of clear glass thick enough to support its large diameter; moreover, such a thick lens may sag from its weight, absorbs light, and also suffers from chromatic aberration. And the telescope tube had to be tremendously long. Because of these difficulties, no larger telescope lens has ever been put into long-term service. (A 1.25-m refractor, mounted horizontally to point at a mirror that tracked the stars, was set up for a few months at an exposition in Paris in 1900.) In 2002, over 100 years later, a lens also 40 inches across was put into use at the Swedish 1-m Solar Telescope on La Palma, Canary Islands.

design, in which a secondary mirror bounces the light back through a small hole in the middle of the primary mirror (■ Fig. 3–9).

Note that the hole, or an obstruction of part of the incoming light by the secondary mirror, only decreases the apparent brightness of the object, without altering its shape or producing a hole in it. Every part of the mirror forms a complete image of the object.

Spherical mirrors reflect light from their centers back onto the same point, but do not bring parallel light to a good focus (■ Fig. 3–10). This effect is called spherical aberration. We now often use mirrors

■ **FIGURE 3–7** Chromatic aberration occurs when white light passes through a simple lens. The various colors of light come to a focus at different distances from the lens; there is no single "best focus," and objects end up showing a tinge of color.

■ **FIGURE 3–8** The path of light in a Newtonian telescope; note the diagonal mirror that brings the focus out to the side. Many amateur telescopes are of this type.

Cassegrain

Eyepiece

Objective mirror

Convex secondary mirror

Prime focus

Cassegrain focus

■ **FIGURE 3–9** A Cassegrain telescope, in which a curved secondary mirror bounces light through a hole in the center of the primary mirror.

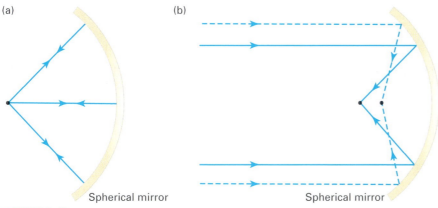

(a)

(b)

Spherical mirror

Spherical mirror

■ **FIGURE 3–10** (a) A spherical mirror (that is, part of the interior of a sphere) focuses light that originates at its center of curvature back on itself, as does the circular arc shown here. (b) A spherical mirror (or a circular arc, as shown here) suffers from spherical aberration in that it does not perfectly focus light from very large distances, for which the incoming rays are essentially parallel (recall Fig. 3–5). Lenses that don't have the proper shape can also show spherical aberration.

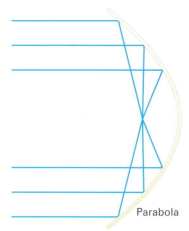

Parabola

■ **FIGURE 3–11** A paraboloid is a three-dimensional curve (two-dimensional surface) created by spinning a parabola on its axis of symmetry. As shown here, a parabola focuses parallel light to a single point, as does a paraboloid.

The size of a telescope's primary lens or mirror is particularly important because the main job of most telescopes is to collect light – to act as a "light bucket." All the light is brought to a common focus, where it is viewed or recorded. (See *Figure It Out 3.1: Light-Gathering Power of a Telescope.*) The larger the telescope's lens or mirror, the fainter the objects that can be viewed or the more quickly observations can be made. A larger telescope would also provide better angular resolution (often called simply "resolution") – clarity, the ability to detect fine detail – if it weren't for the shimmering (turbulence) of the Earth's atmosphere, which limits all large telescopes to about the same angular resolution. Only if you can improve the resolution is it worthwhile magnifying images. For the most part, then, the fact that telescopes magnify is secondary to their ability to gather light.

Yerkes Observatory

■ **FIGURE 3–12** The Yerkes refractor, still one of the two largest in the world, has a 1-m-diameter lens.

3.3 MODERN TELESCOPES

From the mid-19th century onward, progressively larger reflecting telescopes were constructed. But the mirrors, then made of shiny metal, tended to tarnish. This problem was avoided by evaporating a thin coat of silver onto a mirror made of glass. More recently, a thin coating of aluminum turned out to be longer lasting, though silver with a thin transparent overcoat of tough material is now coming back into style. The 100-inch (2.5-m; see *Figure It Out 3.2: Changing Units*) "Hooker" reflector at the Mt. Wilson Observatory in California became the largest telescope in the world in 1917. Its use

FIGURE IT OUT 3.1 Light-Gathering Power of a Telescope

A telescope's main purpose is to collect light quickly and in large quantities. The mirror or lens of a telescope acts as a gigantic eye pupil that intercepts the light rays from a distant object. The larger its area, the more light it will collect in a given time, allowing fainter objects to be seen. (The length of the telescope tube is irrelevant to this "light-gathering power.")

The area of a circle is proportional to the square of its radius r (or diameter $D = 2r$):
$$A = \pi r^2 = \pi(D/2)^2 = \pi D^2/4.$$
Therefore, the ratio of the areas of two telescopes with circular mirrors of diameter D_1 and D_2 is given by
$$A_2/A_1 = D_2^2/D_1^2 = (D_2/D_1)^2.$$

Let's say that $D_2 = 8$ m (a typical size for a large optical telescope), and $D_1 = 8$ mm $= 0.008$ m (as for the pupil opening of a dilated eye). The ratio of areas is
$$(D_2/D_1)^2 = (8 \text{ m } /0.008 \text{ m})^2 = 1,000^2 = 10^6.$$
Thus, looking through the eyepiece of such a telescope, one could see stars a million times fainter than with the unaided eye!

By attaching a detector to the telescope, the exposure time can be made very long, making even fainter stars visible. Some electronic detectors, such as charge-coupled devices (CCDs), are far more sensitive than eyes, and detect most of the photons that hit them. With large telescopes, long exposures, and high-quality CCDs, objects over 10^9 (a billion) times fainter than the limit of the unaided eye have been detected.

led to discoveries about distant galaxies that transformed our view of what the Universe is like and what will happen to it and us in the far future (Chapters 16 and 18).

In 1948, the 200-inch (5-m) "Hale" reflecting telescope opened at the Palomar Observatory, also in California, and was for many years the largest in the world. Current electronic imaging detectors, specifically charge-coupled devices (CCDs) similar to those in camcorders and digital cameras, have made this and other large telescopes many times more powerful than they were when they recorded images on film.

3.3a Current Large Telescopes Around the World

There are now many large telescopes at various locations on Earth; see Table 3–1. For example, the National Optical Astronomy Observatories, supported by the National Science Foundation, have 63 per cent in two telescopes, each with 8-m mirrors, the northern-hemisphere one in Hawaii and the southern-hemisphere one in Chile. The project is called Gemini, since "Gemini" are the twins in Greek mythology (and the name of a constellation) and these are twin telescopes. The other share in the project is divided among Canada, Chile, Australia, Argentina, and Brazil. By sharing, the United States has not only most of the time on a telescope in the northern hemisphere, but also most of the time on a telescope in the southern hemisphere, a better case than having a full telescope in the north and little or no share in the south. (Some of the most interesting astronomical objects are in the southern sky. For example, the nearest galaxies to our own – known as

the Magellanic Clouds – are not observable from the continental United States.)

The observatory with the greatest number of large telescopes is now on top of the dormant volcano Mauna Kea in Hawaii, partly because its latitude is as far south as +20°, allowing much of the southern sky to be seen, and partly because the site is so high that it is above 40 per cent of the Earth's atmosphere. To detect the infrared part of the spectrum, telescopes must be above as much of the water vapor in the Earth's atmosphere as possible, and Mauna Kea (altitude 4,200 m, or 13,800 feet) is above 90 per cent of it. In addition, the peak is usually above the atmospheric inversion layer that keeps the clouds from rising, typically giving about 300 nights each year of clear skies with steady images. Consequently, several of the world's dozen largest telescopes are there (■ Fig. 3–13).

In particular, the California Institute of Technology (Caltech) and the University of California have built the two Keck 10-m telescopes (■ Fig. 3–14), whose mirrors are each twice the diameter and four times the surface area of Palomar's largest reflector. Hence, each one is able to gather light four times faster. When it was built, a single 10-m mirror would have been prohibitively expensive, so University of California scientists worked out a plan to use a mirror made of 36 smaller hexagons (■ Fig. 3–15). The first telescope worked so well that a twin (Keck II) was quickly built beside it. Mauna Kea is also the site of the Gillett Gemini North 8-m telescope and a Japanese 8-m telescope named Subaru (for the star cluster known in English as the Pleiades).

The University of Texas and Pennsylvania State University have built a 9.2-m telescope in Texas. It is on an inexpensive mount and has limited mobility, but it is very useful for gathering a lot of light for spectroscopy. A clone – the South Africa Large Telescope (SALT) – has been built in South Africa by many international partners.

Mirror diameter		Observatory	Year completed
m	**in.**		
10.4	410	Gran Telescopio Canarias (GTC), La Palma, Canary Islands	2009
10.0 × 2	400 × 2	Keck Telescopes I and II, Mauna Kea, Hawaii	1993 (I), 1996 (II)
9.2	362	South African Large Telescope (SALT), Sutherland, South Africa	2004
9.2	362	Hobby-Eberly Telescope, McDonald Observatory, Davis Mountains, Texas	1997
11.9 (8.4 × 2)	468 (330 × 2)	Large Binocular Telescope (LBT), Mount Graham, Arizona (two mirrors on a common mount, acting as a single telescope of aperture 11.9 m; effectively the largest telescope)	2005
8.2	323	Subaru Telescope, National Astronomical Observatory, Japan; Mauna Kea, Hawaii	1999
8.2 × 4	323 × 4	Very Large Telescope (VLT) (Antu, Kueyen, Melipal, and Yepun), ESO Paranal Observatory, Cerro Paranal, Chile	1998–2001
8.1	318	Gemini North (Gillett), National Optical Astronomy Observatory, Mauna Kea, Hawaii	1999
8.1	318	Gemini South, National Optical Astronomy Observatory, Cerro Pachón, Chile	2001
6.5	256	MMT, Mount Hopkins, Arizona	2000
6.5 × 2	256 × 2	Magellan Telescopes (Baade and Clay), Cerro Las Campanas, Chile	2000, 2002
6.0	238	Special Astrophysical Observatory, Zelenchukskaya, Caucasus, Russia	1975
5.08	200	California Institute of Technology/Palomar Observatory, Palomar Mountain, California	1948
4.25	167	Discovery Channel Telescope, Happy Jack (Mogolo Rim), Arizona	2012

Table 3–1 Largest reflecting telescopes in the world

■ **FIGURE 3–13** Mauna Kea, with its many huge telescopes. The Keck twin domes with the Subaru dome near them are at the middle left; the NASA Infrared Telescope Facility is above them; and the Canada-France-Hawaii, Gemini North, University of Hawaii 88-inch, United Kingdom Infrared, and University of Hawaii 24-inch telescopes are on a ridge at top. In the "submillimeter valley" at lower right, we see the Caltech 10-m and James Clerk Maxwell 15-m submillimeter telescopes (Maxwell, one of the greatest physicists of all time along with Newton and Einstein, unified electricity and magnetism theoretically), as well as seven of eight 6-m telescopes and the assembly building of the Smithsonian Astrophysical Observatory's submillimeter array.

■ **FIGURE 3–14** The dome of the Keck I 10-m telescope silhouetted against the Milky Way.

Jay M. Pasachoff

■ **FIGURE 3–15** Each 10-m mirror of the Keck telescopes is made of 36 contiguous hexagonal segments, which are continuously adjustable.

Another major project is the European Southern Observatory's Very Large Telescope, an array of four 8-m telescopes (■ Fig. 3–16) in Chile. Most of the time they are used individually, but technology has advanced to allow them to sometimes be used in combination to give still more finely detailed images. (The Keck pair has also been used occasionally in that mode.) The images are superb (■ Fig. 3–17). A pair of 6.5-m telescopes on Las Campanas, another Chilean peak, is the Magellan project, a collaboration among the Carnegie Observatories, the University of Arizona, Harvard, the University of Michigan, and MIT.

The Large Binocular Telescope in Arizona has two 8.4-m mirrors on a common mount, observing the same object; thus, this can legitimately be considered the world's largest optical telescope, equivalent to a single 12-m telescope. Partners (half American and half European) include the University of Arizona (for the Arizona university system), Research Corporation for Science Advancement (for Notre Dame, University of Minnesota, and University of Virginia), the Italian national astrophysical observatory, and a consortium of German research institutes. The Astrophysical Institute of the Canary Islands, with European membership, and with additional participation from Mexico and the University of Florida, has a 10.4-m telescope of the Keck design on La Palma, Canary Islands.

As we mentioned, the angular size of the finest details you can see (the angular resolution) is basically limited by turbulence in the Earth's atmosphere, but a technique called adaptive optics is improving the resolution of more and more telescopes. In adaptive

When changing from one system of units to another, it is help-ful to keep careful track of the conversion factors, to make sure you aren't getting them upside down. For example, 1 inch = 2.54 cm. You can always divide both sides of an equation by the same number (except zero) without destroying the equality. So dividing both sides of the previous equation by 2.54 cm, we get (1 inch)/(2.54 cm) = 1.

We can also always multiply anything by 1 without changing its value. So if we are given, say, a telescope size that is 8 m = 800 cm, as for some in the current generation of large telescopes, we can convert that unit to inches by multiplying by the conversion factor just derived and making sure that units at the top and the bottom of the equation cancel properly. Thus

$$800 \text{ cm} \times (1 \text{ inch})/(2.54 \text{ cm}) = (800) \text{ inches}/(2.54) = 300 \text{ inches}$$
$$\text{(approximating } 800/2.54 = 300\text{)}.$$

European Southern Observatory

■ **FIGURE 3–16** Domes of the Very Large Telescope, in Chile. The entire project consists of four individual 8-m telescopes plus four smaller telescopes (a few of which were still under construction when this photograph was taken). One of these movable 1.8-m "auxiliary telescopes" is visible here. From left to right in this 2004 image, we see Antu, Kueyen, Melipal, an auxiliary telescope, the 2.6-m VLT Survey Telescope (a wide-field imaging telescope), and Yepun.

optics, the light from the main mirror eventually hits a deformable mirror whose shape can be slightly distorted many times a second to compensate for the atmosphere's distortions. With this technol-ogy and other advances, the resolution from ground-based telescopes has been improving recently after a long hiatus. However, the images show fine detail only over very small areas of the sky, such as the disk of Jupiter.

3.3b The Next Generation of Optical and Infrared Ground-Based Telescopes

The compound Giant Magellan Telescope (GMT), with seven 8.4-m mirrors, each already huge, making a surface equivalent to that of a 22-m telescope, is now under construction, in a project led by

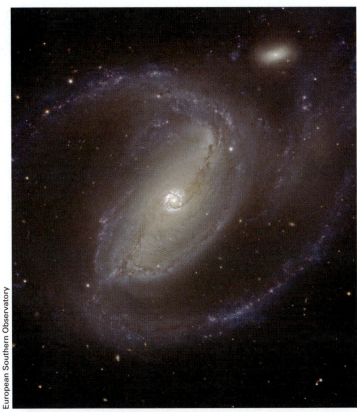

European Southern Observatory

■ **FIGURE 3–17** The spiral galaxy NGC 1097, imaged with one of the telescopes of the Very Large Telescope in Chile. The president of Chile, on an official visit, helped operate the telescope for this image. This galaxy has a central black hole whose mass is tens of millions of times that of the Sun.

parts of the Magellan collaboration (■ Fig. 3–18a). It is to go on Las Campanas, the mountain in Chile where the Pasadena-based Observatories of the Carnegie Institution for Science already has telescopes, including the pair of telescopes known as Magellan. The consortium developing it includes the Carnegie Observatories, University of Chicago, University of Arizona, University of Texas at Austin, Harvard University, Texas A&M University, Smithsonian Astrophysical Observatory, and universities and institutes in Australia and Korea.

A 30-m Keck-style telescope spearheaded by Caltech and the University of California is known, not astonishingly, as the Thirty Meter Telescope (■ Fig. 3–18b); in its initial stages, the concept was referred to as the California Extremely Large Telescope (C-ELT). With 492 mirror segments, the TMT is to be built on Mauna Kea in Hawaii, by an international collaboration that also includes Canada, Japan, China, and India.

The European Southern Observatory had considered an OWL, the Overwhelmingly Large Telescope, at 100-m diameter. But that jump in size seemed too much, and in recent years they have instead been planning the E-ELT, the European Extremely Large Telescope, with a mirror 42 m across. In 2011, they reduced that anticipated size to 39.3 m, in 798 hexagonal segments, to move the project ahead faster (■ Fig. 3–18c). The superb site of Cerro Armazones in Chile has been selected.

All these telescopes are planned for completion late in this decade or early in the next, but funding is still a limitation and additional partners are being sought. The U.S. National Science Foundation had been advised to select between one of the two American projects (GMT or TMT), though it wasn't clear that the budget would soon allow substantial support so the GMT decided not to apply. Let us hope that within 10 years from when you read this, at least one of these giant telescopes is collecting photons that can compose images of planets around other stars, or images of the most distant supernovae that will help us explain why the expansion of the Universe is accelerating rather than slowing down. (To keep up with the latest on the giant telescope projects, see gmto.org for the GMT, tmt.org for the TMT, and eso.org/public/teles-instr/e-elt.html for the E-ELT.)

3.4 THE BIG PICTURE: MAPPING THE SKY

As mentioned above, ordinary optical telescopes see a fairly narrow field of view – that is, a small part of the sky is in focus. Even the most modern have images of less than about 1° × 1° (for comparison, the full moon appears roughly half a degree in diameter), which means it would take decades to make images of the entire sky (over 40,000 square degrees). The German optician Bernhard Schmidt, in the 1930s, invented a way of using a thin lens ground into a complicated shape together with a spherical mirror to image a wide field of sky (■ Fig. 3–18).

The largest Schmidt telescopes, except for one of interchangeable design, are at the Palomar Observatory in California and at the United Kingdom Schmidt site in Australia (■ Fig. 3–19); these telescopes have front lenses 1.25 m (49 inches) in diameter and mirrors half again as large. (The back mirror is larger than the front lens in order to allow study of objects off to the side.) They can observe a field of view some 7° × 7° – almost the size of your fist held at the end of your outstretched arm, compared with only the size of a grain of sand at that distance in the case of the Hubble Space Telescope!

The Palomar camera, now named the "Oschin" Schmidt telescope, was used in the 1950s to survey the whole sky visible from southern California with photographic film and filters that made pairs of images in red and blue light. This Palomar Observatory Sky Survey is a basic reference for astronomers. Hundreds of thousands of galaxies, quasars, nebulae, and other objects have been discovered on them. The Schmidt telescopes in Australia and Chile have extended this survey to the southern hemisphere. In the 1990s, Palomar carried out a newer survey with improved films and more overlap between adjacent regions. Among other things, it is being compared with the first survey to see which objects have changed or moved. Both surveys have been digitized, to improve their scientific utility. The Palomar Schmidt has now been converted to digital detection with CCDs. For a time it was largely being used to search for Earth-approaching asteroids.

(a)

(b)

(c)

European Southern Observatory

(d)

Jay M. Pasachoff

■ **FIGURE 3–18** Artist's renditions of huge ground-based telescopes now under construction: (a) the Giant Magellan Telescope (GMT), (b) the Thirty Meter Telescope (TMT), and (c) the European Extremely Large Telescope (E-ELT). (d) The size of the mirrors of the GMT laid out in the parking lot of the Carnegie Observatories in Pasadena, California.

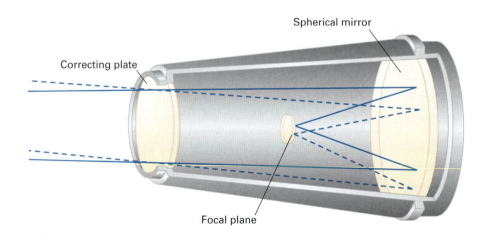

Spherical mirror

Correcting plate

Focal plane

■ **FIGURE 3–19** By having a nonspherical thin lens called a correcting plate, a Schmidt camera is able to focus a wide angle of sky onto a curved piece of film. Since the image falls at a location where you cannot put your eye, the image is always recorded on film. Accordingly, this device is often called a Schmidt camera rather than a Schmidt telescope.

The Oschin Schmidt is now the workhorse of the Palomar Transient Factory (PTF); it scans wide regions of the sky, looking for objects that move with respect to the stars or that change brightness. New discoveries found with the computers examining the Oschin Schmidt's output can be funneled over to the Palomar 1.5-m telescope for a more detailed view, and even over to the 5-m Hale telescope. One triumph for PTF was the discovery in August 2011 of Supernova PTF11kly (also known as SN 2011fe, the designation of the International Astronomical Union) in the galaxy M101; it was found just 11 hours after the stellar explosion, much earlier than in the case of most supernovae. We discuss its significance in Chapters 13 (on stellar evolution) and 18 (on cosmology). The notation "kly" shows that PTF discovered a very large number of objects in 2011, running through the alphabet many times (starting with 2011a through 2011z, then 2011aa through 2011az, and so on).

An interesting method of surveying wide regions of sky was developed by the Sloan Digital Sky Survey team. They use a telescope (not of Schmidt design) at Apache Point, New Mexico, to record digital CCD images of the sky as the Earth turns, observing simultaneously through several different filters (■ Fig. 3–20). So many data were collected that developing new methods of data handling was an important part of the project. They have mapped hundreds of millions of galaxies (Chapter 16) and half a million "quasars" (extremely distant, powerful objects in the centers of galaxies; see Chapter 17). The telescope took spectra of more than a million galaxies and a hundred thousand quasars that it mapped.

A major plan for mapping the sky over and over is the University of Hawaii's Pan-STARRS, the Panoramic Survey Telescope & Rapid Response System. Ultimately the hope is to use four telescopes on Mauna Kea, but at present it is being tested with one prototype on Haleakala on the Hawaiian island of Maui at 3,000 m altitude (10,000 feet). A major goal of Pan-STARRS is to discover near-Earth asteroids, allowing astronomers to assess how likely one might hit Earth and even potentially giving us enough notice that we could divert it. These days, it is important to note that the Pan-STARRS system relies not only on the telescope and its camera (which has 1.4 gigapixels, 100 times more pixels than CCDs on the fanciest ordinary cameras), but also on the computers that analyze the massive amounts of data, identifying steady and moving objects. By being sensitive to objects up to 100 times fainter than current near-Earth objects being tracked, the final system is expected to catalog up to 10 million regular asteroids and tens of thousands of objects that either pass near Earth (Near-Earth Objects) or are in the outer Solar System beyond Neptune (trans-Neptunian objects). We discuss the science involved in Chapter 8.

A major national project is the Large Synoptic Survey Telescope, LSST (lsst.org), which will use a 3 gigapixel camera on a single 8.4-m telescope, whose mirror system is specially designed to image a wide field some 3.5° across, to scan the sky rapidly to look for motion or changes in brightness. (Astronomers like to use the word "synoptic" to indicate studies over time. Day-after-day studies of the Sun and its sunspots are also synoptic, for example.) Its wide-field capability will allow it to image the sky over and over, repeating almost

University of Chicago News Office and the SDSS Collaboration

■ **FIGURE 3–20** The filters of the Sloan Digital Sky Survey; the charge-coupled devices (CCDs) are behind them. The color bands include ultraviolet, blue-green, red, far red, and near-infrared.

the whole accessible sky from its location every 3 days or so. It is to be erected in Chile on Cerro Pachón, already the site of Gemini South. The 2010 Astronomy and Astrophysics Decadal Survey, a scientific-community-wide effort, ranked LSST first in its category of large ground-based projects. A wide variety of universities and institutions across the U.S. are already members of the consortium that will run the effort, which is directed by a professor at the University of California at Davis.

3.5 AMATEURS ARE PARTICIPATING

It is fortunate for astronomy as a science that so many people are interested in looking at the sky. Many are just casual observers, who may look through a telescope occasionally as part of a course or on an "open night," when people are invited to view through telescopes at a professional observatory, but others are quite devoted "amateur astronomers" for whom astronomy is a serious hobby. Some amateur astronomers make their own equipment, ranging up to quite large telescopes perhaps 60 cm in diameter. But most amateur astronomers use one of several commercial brands of telescopes.

Computer power and the techniques for CCD image processing have advanced so much that these days, amateur astronomers are producing pictures that professionals using the largest telescopes would have been proud of a decade ago. One interesting technique is to take thousands of photos very quickly, use a computer to throw out the blurriest ones, and combine the rest to form a sharp image. Some amateurs are even contributing significantly to joint work with professional research astronomers, obtaining high-quality complementary data.

Some of the amateur telescopes are Newtonian reflectors, with mirrors 15 cm in diameter being the most popular size (■ Fig. 3–21a). It is quite possible to shape your own mirror for such a telescope. The Dobsonian telescope is a variant of this type, made with very inexpensive mirrors and construction methods to provide affordable systems with relatively large mirrors. Ordinary Dobsonians

(a)

Jay M. Pasachoff

(b)

Eyepiece

Diagonal mirror

(c)

Secondary mirror

Eyepiece

Correcting plate

Objective mirror

Flat reflector for convenience

■ **FIGURE 3–21** (a) An amateur astronomer's Newtonian telescope. Note that the eyepiece is perpendicular to the tube near the tube's top. (b) A cutaway drawing of this Newtonian telescope. (c) A cutaway drawing of a compound Schmidt-Cassegrain design, now widely used by amateurs because of its light-gathering power and portability.

don't track the stars as the sky revolves overhead, but they are easy to turn by hand in the up-down direction (called "altitude") and in the left-right direction (called "azimuth") on cheap Teflon bearings, so they are mainly used for visual observing. (Computerized tracking can be added.)

Compound telescopes that combine some features of both reflectors and Schmidt telescopes are very popular. This Schmidt–Cassegrain design (■ Fig. 3–21b) bounces the light, so that the telescope is relatively short, making it easier to transport and set up. Many of the current versions of these telescopes are now being provided with a computer-based "Go-To" function, where you press a button and the telescope "goes to" point at the object you have selected. Some have a Global Positioning System (GPS) installed, so that the telescope knows where it is located. Then you point to at first one and then another bright star whose names you know, and the telescope's computer calculates the pointing for the rest of the sky. There are even amateur-grade telescopes now that can set themselves up and align themselves, thus making humans unnecessary for those tasks. They photograph the sky, figure out from the patterns where they are pointing, and make the necessary adjustments and calibrations.

3.6 GLORIOUS HUBBLE AFTER INITIAL TROUBLE

The first moderately large telescope to be launched above the Earth's atmosphere was the Hubble Space Telescope (HST or Hubble, for short; ■ Fig. 3–22), built by NASA with major contributions from the European Space Agency. The set of instruments on board Hubble is sensitive not only to visible light but also to ultraviolet and infrared radiation that are impeded by Earth's atmosphere or don't pass through it at all. (These regions of the electromagnetic spectrum will be discussed in more detail below.)

NASA's Johnson Space Center

■ **FIGURE 3–22** Astronaut John Grunsfeld works on the Hubble Space Telescope during the 2009 servicing mission when the Wide Field Camera 3 was installed and the Advanced Camera for Surveys was restored to working order. The telescope is named after Edwin Hubble, who discovered the true nature of galaxies and the redshift–distance relation for galaxies that led to our understanding of the expansion of the Universe (see Chapter 16).

Goodrich Corp., Optical and Space Systems Division

■ **FIGURE 3–23** The 2.4-m (94-inch) mirror of the Hubble Space Telescope, before launch but after it was ground, polished, and covered with a reflective coating. This single telescope has made numerous discoveries but is raising as many questions as it answers. So we need still more ground-based telescopes as partners in the enterprise, in addition to more telescopes in space.

Table 3–2   Hubble Space Telescope Servicing Missions
Servicing Mission 1, December 1993
Installed Wide Field and Planetary Camera 2 with corrective optics and a *corrective-optics device for the other instruments*
Servicing Mission 2, February 1997
Installed Space Telescope Imaging Spectrograph and NICMOS
Servicing Mission 3A, December 1999
Replaced all six gyroscopes, a Fine Guidance Sensor, and the computer
Servicing Mission 3B, March 2002
Revived NICMOS; installed Advanced Camera for Surveys
Servicing Mission 4, May 2009
Installed Wide Field Camera 3 and Cosmic Origins Spectrograph

The Hubble saga is a dramatic one. When it was launched in 1990, the telescope, with its 94-inch (2.4-m) mirror, was supposed to provide images with about 10 times better angular resolution than ground-based images, because of Hubble's location outside of Earth's turbulent atmosphere. (But the latest advances in adaptive optics now eliminate, for at least some types of infrared observations involving narrow fields of view, the advantage Hubble formerly had over ground-based telescopes.) Unfortunately, the main mirror (■ Fig. 3–23) turned out to be made with slightly the wrong shape. Apparently, an optical system used to test it was made slightly the wrong size, and it indicated that the mirror was in the right shape when it actually wasn't. The result is some amount of spherical aberration, which blurred the images and caused great disappointment when it was discovered soon after launch.

Fortunately, the telescope was designed so that space-shuttle astronauts could visit it every few years to make repairs; see Table 3–2. A mission launched in 1993 carried a replacement for the main camera and correcting mirrors for other instruments, and brought the telescope to full operation. A second generation of equipment, installed in 1997, included a camera, the Near Infrared Camera and Multi-Object Spectrometer (NICMOS), that is sensitive to the infrared; however, it ran out of coolant sooner than expected. A mission in December 1999 was used to replace the gyroscopes that hold the telescope steady and made certain other repairs and improvements, such as installing better computers. Astronauts on a mission in 2002 installed the Advanced Camera for Surveys (ACS) and a cooler (refrigerator) that made the infrared camera work again. With increased sensitivity

and a wider field of view than the best camera then on board, ACS is about 10 times more efficient in getting images, so it took Hubble to a new level before it failed electronically. A final space-shuttle mission to repair and upgrade Hubble instruments took place in 2009, after it was cancelled and subsequently restored. Astronauts on the mission fixed ACS and installed the Wide Field Camera 3, with enhanced optical and infrared capability. It also installed the Cosmic Origins Spectrograph.

But the space shuttle program ended in 2011, so we do not see how another mission to Hubble can take place, though some plans were developed for a possible robotic repair mission. It is likely that Hubble will work a few more years and eventually fail in some known (gyroscopes, for example) or unknown and perhaps abrupt way. The plan is then to attach a rocket to it and direct it into the ocean, so that people won't worry that they will be hit by debris as it reenters the lower parts of Earth's atmosphere. (Even at its altitude of hundreds of kilometers overhead, it experiences atmospheric friction and will eventually reenter; each past repair mission boosted it into a higher orbit.)

Hubble's high resolution is able to concentrate the light of a star into an extremely small region of the sky – the star or galaxy isn't blurred out. This, plus the very dark background sky far from Earth's surface, even at optical wavelengths but especially in the infrared (where Earth's warm atmosphere glows brightly as seen from the ground), allows us to examine much fainter objects than we could formerly. The combination of high angular resolution and high sensitivity has led to great advances toward solving several basic problems of astronomy. As we shall see later (Chapter 16), we were able to pin down our whole notion of galaxy distances and the age of the Universe much more accurately than before. Thus, Hubble fills a unique niche, and was well worth its much larger cost compared with a ground-based telescope of similar size.

A Next Generation Space Telescope, named the James Webb Space Telescope (JWST) after the Administrator of NASA in its glory, moon-landing days, is under construction with a nominal launch date that at present has been postponed until 2018. With a 6.5-m mirror (■ Fig. 3–24), the telescope will be optimized to work at

■ **FIGURE 3–24** The first six mirrors out of the eighteen, all coated in gold for its infrared reflectivity, that will unfold to make the 6.5-m-diameter collecting area of the James Webb Space Telescope, which is scheduled for launch in 2018. These mirrors are shown in a test chamber.

NASA/MSFC/David Higginbotham

infrared wavelengths, which will make it especially useful in studying the origins of new stars and planets, and in looking for extremely distant objects in the Universe. But it will not have high-resolution visible-light capabilities, so it won't be a direct replacement for Hubble. At this writing, JWST has passed various milestones, but it is so late and over budget that many people despair over its effect on NASA's other programs; thus, JWST has become controversial among scientists, administrators, and funders. Also, it is to be so far out in space that no repair mission could reach it, so a Hubble-style repair or servicing would not be possible. Still, if it is successfully built and launched, and if all systems operate as planned, its scientific return would be enormous.

3.7 YOU CAN'T LOOK AT THE SUN AT NIGHT

Telescopes that work at night usually have to collect a lot of light. Solar telescopes work during the day, and often have far too much light to deal with. They have to get steady images of the Sun in spite of viewing through air turbulence caused by solar heating. So solar telescopes are usually designed differently from nighttime telescopes. The Swedish 1-m Solar Telescope on La Palma, with its 1-m mirror, uses adaptive optics to get fantastically detailed solar observations. Planning is under way for a 4-m (13-ft) Advanced Technology Solar Telescope, much larger than any previous solar telescope, with the 10,000-foot (3,000-m) altitude of Haleakala on Maui, Hawaii, selected as the site.

Solar observatories in space have taken special advantage of their position above the Earth's atmosphere to make x-ray and ultraviolet observations. The Japanese Yohkoh (Sunbeam) spacecraft, launched in 1991, carried x-ray and ultraviolet telescopes for study of solar activity. Yohkoh was killed by a solar eclipse in 2001; its trackers, which were used to find the edge of the round Sun, got confused, allowing the telescope to drift away and to start spinning. In 2006, the Japanese Space Agency launched a successor with improved imaging detail: Hinode (Sunrise). It carries not only an x-ray telescope but also a high-resolution optical telescope, both American built.

The Solar and Heliospheric Observatory (SOHO), a joint European Space Agency/ NASA mission, is located a million kilometers upward toward the Sun, for constant viewing of all parts of the Sun. It is still working well even at its great age, given its launch way back in 1995.

The Transition Region and Coronal Explorer (TRACE) gave even higher-resolution images of loops of gas at the edge of the Sun (■ Fig. 3–25), during 1998–2010. All these telescopes gave especially interesting data during the year 2000–2001 maximum of sunspot and other solar activity.

The major new high-resolution solar spacecraft in place for the most recent sunspot maximum (2013/14), is NASA's Solar Dynamics Observatory (SDO), launched in 2010. One of its instruments, the Atmospheric Imaging Assembly (AIA), has the same high resolution (half-arc-second pixels) as TRACE had, but it includes the whole Sun and takes a full set of images through 8 filters every 12 seconds. SDO is in an orbit that is continually in view of a ground station in New Mexico to handle this high data rate. A second SDO instrument, the Helioseismic and Magnetic Imager (HMI), measures the solar magnetic field with a resolution even slightly

Jay M. Pasachoff and Kamen Kozarev, Williams College-Hopkins Observatory

■ **FIGURE 3–25** Spikes of hot chromospheric gas from the Transition Region and Coronal Explorer (TRACE) spacecraft imaged in the ultraviolet as part of one of the authors' (J.M.P.) research.

better than AIA's. And a third, the Extreme-ultraviolet Variability Experiment (EVE), measures the Sun's changing ultraviolet spectrum. AIA is controlled by Lockheed Martin Solar and Astrophysics Laboratory, with a telescope built by the Smithsonian Astrophysical Observatory scientists and engineers; HMI is controlled by Stanford scientists; and EVE is controlled by scientists at the University of Colorado at Boulder. We will show various images taken with SDO instruments in Chapter 10.

■ **FIGURE 3–26** Waves of electric and magnetic fields travelling across space are called "radiation." The wavelength is the length over which a wave repeats.

3.8 HOW CAN YOU SEE THE INVISIBLE?

As discussed more fully in Chapter 2, we can describe a light wave by its wavelength (■ Fig. 3–26). But a large range of wavelengths is possible, and visible light makes up only a small part of this broader spectrum (■ Fig. 3–27). Gamma rays, x-rays, and ultraviolet light have shorter wavelengths than visible light, and infrared and radio waves have longer wavelengths.

3.8a X-ray and Gamma-ray Telescopes

The shortest wavelengths would pass right through the glass or even the reflective coatings of ordinary telescopes, so special imaging devices have to be made to study them. And x-rays and gamma rays do not pass through the Earth's atmosphere, so they can be observed only from satellites in space. NASA's series of three High-Energy Astronomy Observatories (HEAOs) was tremendously successful in the late 1970s. One of them even made detailed x-ray observations of individual objects with resolution approaching that of ground-based telescopes working with ordinary light.

NASA's current and best x-ray telescope is called the Chandra X-ray Observatory, named after a scientist (S. Chandrasekhar; see Chapter 13) who made important studies of white dwarfs and black holes. It was launched in 1999. It makes its high-resolution images with a set of nested mirrors made on cylinders (■ Figs. 3–28). Ordinary mirrors could not be used because x-rays would pass right through them. However, x-rays bounce off mirrors at low angles, just as stones can be skipped across a lake at low angles (■ Fig. 3–29). Chandra joins Hubble as one of NASA's series called the "Great Observatories." The European Space Agency's XMM-Newton mission has more telescope area and so is more sensitive to faint sources than Chandra, but it doesn't have Chandra's high resolution. (An earlier European x-ray satellite, Roentgensatellite, or Rosat for short – named after Wilhelm Roentgen, a German who was the discoverer of x-rays over 100 years ago – was long defunct and reentered the atmosphere in 2011, crashing into the Indian Ocean as expected and not injuring anyone, but nonetheless causing a bit of worldwide panic over the possibility of being hit.) The Japanese Suzaku x-ray satellite has taken x-ray images since its 2005 launch.

NASA's Swift spacecraft, named in part for the swiftness (within about a minute) with which it can turn its ultraviolet/visible-light and x-ray telescopes to point at "gamma-ray burst" positions, started its observations in 2005. We will discuss gamma-ray bursts, a major topic of current interest, in Chapter 14.

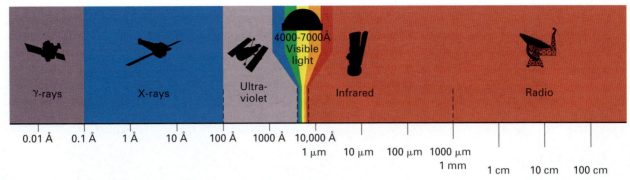

■ **FIGURE 3–27** The spectrum of electromagnetic waves. The silhouettes represent telescopes or spacecraft used for observing that part of the spectrum: the Compton Gamma Ray Observatory, the Chandra X-ray Observatory, the Hubble Space Telescope, the dome of a ground-based telescope, the Herschel Space Observatory, and a ground-based radio telescope called the Green Bank Telescope.

■ **FIGURE 3–28** (a) An outside view of the nested Chandra X-ray Observatory mirrors. (b) One of the cylindrical mirrors from the Chandra X-ray Observatory. Its inside is polished to reflect x-rays by grazing incidence.

NASA's NuSTAR, The Nuclear Spectroscopic Telescope Array, is mapping the sky in x-rays of especially high energies. This pair of grazing-incidence telescopes was launched in June 2012.

Unfortunately, as successful as Chandra has been, NASA's major Constellation-X quartet of x-ray spacecraft, which had been on the drawing board for 2019, is no longer on the list. There is no foreseeable new American x-ray mission.

In the gamma-ray part of the spectrum, the Compton Gamma Ray Observatory was launched in 1991, also as part of NASA's Great Observatories program. NASA destroyed it in 2000, since the loss of one more of its gyros would have made it more difficult to control where its debris would land on Earth. It made important contributions to the study of exotic gamma-ray bursts (Chapter 14) and other high-energy objects.

The European Space Agency launched its "Integral" (International Gamma-Ray Astrophysics Laboratory) telescope in 2002. Besides observing gamma rays, it can make simultaneous x-ray and visible-light observations.

The biggest and best gamma-ray telescope in space is the Fermi Gamma-ray Space Telescope, named after the atomic scientist Enrico Fermi. It was launched in 2008. Its observations show a variety of

■ **FIGURE 3–29** Follow the arrows on the light rays from the left to see how each bounces off first a paraboloid and then a hyperboloid (the surface formed by spinning a hyperbola on its axis of symmetry) to come to a common focus. As the three-dimensional drawing in the lower part of the diagram shows, four similar paraboloid/hyperboloid mirror arrangements, used at low angles to incoming x-rays and all sharing the same focus, are nested within each other to increase the area of telescope surface that intercepts x-rays in NASA's Chandra X-ray Observatory. A photo of one segment was shown in the preceding figure (Fig. 3–28).

NASA/JPL-Caltech/R. Hurt (SSC)

■ **FIGURE 3–30** An artist's conception of how NASA's Galaxy Evolution Explorer (GALEX) spacecraft uses the concept of the brightness of a "standard candle" at different distances to support the idea that dark energy is accelerating the Universe's expansion. GALEX's survey of 200,000 galaxies as far away as 7 billion light-years, together with ground-based observations from Australia, led to a three-dimensional map on which astronomers could study the patterns of distances between the galaxies. From the characteristic distances between pairs of galaxies, the astronomers could analyze the stretching of space as we go farther and farther from us. We discuss these concepts in Chapter 19.

violent events in the Universe, such as gamma-ray bursts from mergers of neutron stars and black holes, and high-energy jets of gas emitted from the vicinity of black holes. Its Large Area Telescope is more sensitive than its predecessor on Compton by a factor of 30. A second instrument monitors the sky for gamma-ray and x-ray bursts. Fermi's international collaborators include space agencies in France, Germany, Japan, Italy, and Sweden.

Some huge "light buckets" are giant ground-based telescopes, bigger than the largest optical telescopes, but focusing only well enough to pick up (but not accurately locate) flashes of visible light in the sky (known as "Cherenkov radiation") that are caused by gamma rays hitting particles in our atmosphere. The MAGIC (Major Atmospheric Gamma-ray Imaging Cherenkov) telescopes on La Palma in the Canary Islands are such devices. MAGIC 1, built in 2001–2002, was joined in 2009 by a second MAGIC. They have detected cosmic rays from quasars and other activity in galaxies. The MAGIC telescopes are run by an international consortium of 17 institutes.

A huge High Energy Stereoscopic System (H.E.S.S.) II telescope opened in the Namibian desert with a segmented mirror 24 m wide and 32 m tall. (Its acronym honors Victor Hess, who discovered cosmic rays, for which he won the 1936 Nobel Prize in Physics.) It is also measuring flashes of light from atmospheric air showers caused

by gamma rays. H.E.S.S. II joins four large but smaller cosmic-ray telescopes on that African site that opened a decade earlier.

Other devices for detecting such optical effects in the Earth's upper atmosphere, to observe cosmic-ray showers, are part of the Pierre Auger Cosmic Ray Observatory, an international set of devices located in Argentina, as we discuss in Chapter 13.

Both MAGICs, all five H.E.S.S.s, and Pierre Auger are part of what is now called astroparticle physics, merging astronomical observations with the study of elementary particles formerly carried out exclusively in atom smashers on Earth.

3.8b Telescopes for Ultraviolet Wavelengths

Ultraviolet wavelengths are longer than x-rays but still shorter than visible light. All but the longest wavelengths of ultraviolet light do not pass through the Earth's atmosphere, so they must be observed from space. For about two decades, the 20-cm telescope on the International Ultraviolet Explorer spacecraft sent back valuable ultraviolet observations. Overlapping it in time, the 2.4-m Hubble Space

Telescope was launched, as discussed above; it has a much larger mirror and so is much more sensitive to ultraviolet radiation.

Several ultraviolet telescopes have been carried aloft for brief periods aboard space shuttles, and brought back to Earth at the end of the shuttle mission. NASA's Far Ultraviolet Spectrographic Explorer (FUSE) took high-resolution spectra largely in order to study the origin of the elements in the Universe. Starting in 2003, NASA's Galaxy Evolution Explorer (GALEX) sent back ultraviolet views of distant and nearby galaxies to find out how galaxies form and change. Among its accomplishments is contributing to a verification of the idea that dark energy is accelerating the Universe's expansion (■ Fig. 3–30).

3.8c Infrared Telescopes

From high-altitude sites such as Mauna Kea, parts of the infrared can be observed from the Earth's surface. From high aircraft altitudes, even more can be observed, and NASA has refitted a Boeing 747 airplane with a 2.5-m telescope to operate in the infrared. Cool objects such as planets and dust around stars in formation emit most of their radiation in the infrared, so studies of planets and of how stars form have especially benefited from infrared observations. After long delays, this Stratospheric Observatory for Infrared Astronomy (SOFIA) telescope had its first scientific flights in 2011. It isn't easy to be able to certify a plane as airworthy when you cut a hole into one of its sides big enough for a 2.5-m telescope to see through. On one of the first science flights, we can report that it successfully maneuvered near to the path from which an occultation of a star by Pluto was aligned, allowing observations of Pluto's atmosphere as it focused the starlight around the dwarf planet and through that atmosphere (as we discuss in Chapter 8).

An international observatory, the Infrared Astronomical Satellite (IRAS), mapped the sky in the 1980s, and then was followed by the European Infrared Space Observatory (ISO) in the mid-1990s. Since the telescopes and detectors themselves, because of their warmth, emit enough infrared radiation to overwhelm the faint signals from space, the telescopes had to be cooled way below normal temperatures using liquid helium. These telescopes mapped the whole sky, and discovered a half-dozen comets, hundreds of asteroids, hundreds of thousands of galaxies, and many other objects. ISO took infrared spectra of many of these objects.

Since infrared penetrates the haze in space, its whole-sky view reveals our own Milky Way Galaxy. During 1990–1994, the Cosmic Background Explorer (COBE) spacecraft mapped the sky in a variety of infrared (■ Fig. 3–31) and radio wavelengths, primarily to make cosmological studies. NASA's Wilkinson Microwave Anisotropy Probe (WMAP) was launched in 2001 to make higher-resolution observations. We now have the European Space Agency's Planck spacecraft making even higher-resolution observations of the whole sky in these infrared and short-wavelength radio regions of the spectrum. We shall discuss their cosmological discoveries in Chapter 19.

NASA's COBE Science Team

■ **FIGURE 3–31** The sky, mapped by the Cosmic Background Explorer (COBE) spacecraft in 1990, reveals mainly the Milky Way and thus the shape of our Galaxy. The image here is false color, a translation into visible wavelengths of three infrared wavelengths that penetrate interstellar dust.

The whole sky was mapped in the infrared in the 1990s by 2MASS, a joint ground-based observation by the University of Massachusetts at Amherst and the Imaging Processing and Analysis Center at Caltech. (The "2" in the acronym is for its 2-micron wavelength and the initials of 2 Micron All Sky Survey are "MASS.")

A sensitive infrared mission, the Spitzer Space Telescope, was launched as NASA's infrared Great Observatory in 2003. It worked very well until its coolant ran out, and even then some of its wavelength bands remained usable in its "warm mode." It was superseded in the space-infrared-telescope honor roll by the European Space Agency's Herschel infrared telescope (■ Fig. 3–32a), launched in 2009 on the same rocket as Planck. (Sir William Herschel, having discovered the planet Uranus, discovered infrared radiation in about 1800.) It produced a wide variety of phenomenal images and other measurements (for example, ■ Fig. 3–32b), before running out of coolant in 2013.

3.8d Radio Telescopes

Since Karl Jansky's 1930s discovery (■ Fig. 3–33) that astronomical objects give off radio waves, radio astronomy has advanced greatly. Huge metal "dishes" are giant reflectors that concentrate radio waves onto antennae that enable us to detect faint signals from objects in outer space.

A still larger dish in Arecibo, Puerto Rico, is 1,000 feet (330 m) across, but points only more or less overhead. Still, all the planets and many other interesting objects pass through its field of view. This telescope and one discussed below starred in the movie *Contact*.

Instead of converting incoming radio signals to sound with amplifiers and loudspeakers, astronomers use computers to convert them to graphs or digital intensity values.

(a)

(b)

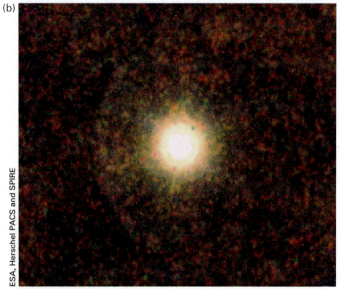

■ **FIGURE 3–32** (a) The European Space Agency's Herschel Space Observatory before launch. With its 3.5-m-diameter mirror, it is the largest telescope in space, and will remain so until the launch of the James Webb Space Telescope. (b) One of thousands of Herschel images, this image of the variable star CW Leonis shows that it is plowing through interstellar space; we see a bow shock ("bow" rhymes with "wow") to the left of the star produced by the interaction of gas leaving the star (the star's "stellar wind") with the gas in interstellar space around it. The false color reproduces observations at wavelengths of 160 μm (blue), 250 μm (green), and 350 μm (red) from two separate Herschel instruments that receive light from the main 3.5-m mirror.

Radio telescopes were originally limited by their very poor angular resolution. The resolution of a telescope depends only on the telescope's diameter, but we have to measure the diameter relative to the wavelength of the radiation we are studying. For a radio telescope studying waves 10 cm long, even a 100-m telescope is only 1,000 wavelengths across. A 10-cm optical telescope studying ordinary light is 200,000 wavelengths across, so it is effectively much larger and gives much finer images (■ Fig. 3–34); see *Figure It Out 3.3: Angular Resolution of a Telescope*. For the human eye, it is the lens's optical quality rather than its diameter that limits the resolution that we can perceive to about 1 minute of arc, about the size of Jupiter in the sky.

A radio telescope at the National Radio Astronomy Observatory's site at Green Bank, West Virginia, replaced one that collapsed. This Byrd Green Bank Telescope has a reflecting surface 100 m in diameter in an unusual design (■ Fig. 3–35). New technology has made it possible to observe radio waves well at relatively short wavelengths, those measured to be a few millimeters. Molecules in space are especially well studied at these wavelengths. The Byrd Green Bank Telescope is useful for studying such molecules.

A breakthrough in providing higher resolution has been the development of arrays of radio telescopes that operate together and give the resolution of a single telescope spanning kilometers or even continents. The Very Large Array (VLA) is a set of 27 radio telescopes, each 26 m in diameter, and was also seen in the movie *Contact* (■ Fig. 3–36). All the telescopes operate together, and powerful computers analyze the joint output to make detailed pictures of objects in space. These telescopes are linked to allow the use of "interferometry" to mix the signals; analysis later on gives images of very high resolution. The VLA's telescopes are spread out over hundreds of square kilometers on a plain in New Mexico. After the VLA's abilities were expanded by 2011 with improved electronics, fiber-optics links, and other enhancements, transforming it into a much more sensitive and versatile device, it was renamed the Karl G. Jansky Very Large Array, after the founder of radio astronomy.

The Europeans also have networks of radio telescopes, including Merlin. LOFAR is a network of low-frequency (long wavelength) radio telescopes based in the Netherlands with telescopes also in Germany, England, France, and Sweden. The system started working in 2011.

To get even higher resolution, astronomers have built the Very Long Baseline Array (VLBA), spanning the whole United States. Its images are many times higher in resolution than even those of the VLA. Astronomers often use the technique of "very-long-baseline interferometry" to link telescopes at such distances, or even distances spanning continents, but the VLBA dedicates telescopes full-time to such high-resolution work.

3.8e The Major New Radio Projects

The major new worldwide project in radio astronomy is ALMA, the Atacama Large Millimeter/submillimeter Array. It is under

(a)

Debra Elmegreen, Vassar College

(b)

Robert W. Wilson

David B. Jansky, courtesy of Woodruff Sullivan and the Astronomical Society of the Pacific

■ **FIGURE 3–33** (a) Karl Jansky (inset) and the full-scale model of the rotating antenna with which he discovered radio waves from space. (b) A sculpture honoring the original Jansky telescope, erected at the original site in Holmdel, New Jersey, in 1998.

FIGURE IT OUT 3.3

Angular Resolution of a Telescope

Telescopes improve the clarity with which objects are seen: they have higher (greater, better) angular resolution (often called simply "resolution," the ability to see fine detail) than the human eye.

Angular measure is important in astronomy. The full circle is divided into 360 degrees (360°). The Moon and the Sun each subtend (cover) about $1/2°$. Each degree consists of 60 arc minutes (60'). Each minute of arc consists of 60 arc seconds (60"). A second of arc is very small – approximately the angle subtended by a dime viewed from a distance of 3.7 km. One can also use radians for angular measure. There are 2π radians in 360°, so 1" = 1/206,265 radian.

If two point-like objects are closer together than about 1 arc minute (1'), the unaided eye will perceive them as only one object because their individual "blur circles" merge together. With a telescope, the size of the individual blur circles decreases, and the objects become resolved (see the figure). The angular diameter of the blur circle is proportional to λ/D, where λ is the observation wavelength and D is the diameter of the lens or mirror. Hence, in principle, large telescopes are able to resolve finer details than small telescopes (at a given wavelength); large telescopes have smaller blur circles (i.e., higher angular resolution), so they don't merge together as much.

A rule of thumb found by the amateur astronomer Dawes in the 19th century is that the theoretical angular resolution (in arc seconds) equals 0.002 λ/D, if λ is given in angstroms (Å) and D is given in centimeters. This resolution is the angular separation at which a double star is detectable as having two separate components (if they have equal brightness).

As an example, the theoretical angular resolution of a telescope 1 m (i.e., 100 cm) in diameter for green light of approximately 5,000 Å is given by 0.002(5000/100) = 0.1 arc second. The theoretical resolution of a 10-m (1,000-cm) telescope is a factor of 10 higher, or 0.01 arc second. [Note that the *smaller* the numerical value (in arc seconds) for the angular resolution, the *greater* the clarity of the image. "High resolution" means two stars that appear very close together can be seen individually.]

Unfortunately, in practice, Earth's atmosphere blurs starlight: Layers of air with different densities move in a turbulent way relative to each other, and the rays of light bend in different directions. This is related to the twinkling of starlight. The angular resolution of telescopes larger than 20–30 cm is limited by the blurring effects of the atmosphere (typically 1 arc second, and rarely smaller than $1/3$ arc second), not by the size of the mirror or lens. Thus, even bigger ground-based telescopes do not give clearer images, unless advanced techniques such as adaptive optics are used.

Radio wave

Light wave

$1\frac{3}{4}\lambda$

8λ

Radio telescope dish

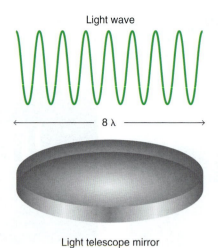

Light telescope mirror

■ **FIGURE 3–34** For familiarity's sake, we often refer to telescopes by their physical size, though this is not the most significant measure. It is more meaningful to measure the diameter of telescope mirrors in terms of the wavelength of the radiation that is being observed than it is to measure them in terms of units like centimeters that have no particularly relevant significance. The radio dish at left is only $1\frac{3}{4}$ wavelengths across, whereas the mirror at right is 8 wavelengths across, making it effectively much bigger. The wavelengths are greatly exaggerated in this diagram relative to the size of any actual reflectors. For the Bonn radio telescope used to observe 10-cm waves, the 100-m diameter divided by 0.1 m per wave = 1,000 wavelengths.

Ric Noyle, W. M. Keck Observatory

■ **FIGURE 3–35** The 100-m Robert C. Byrd Green Bank Telescope, which went into use in 2001.

Riccardo Giovanelli and Martha Haynes, Cornell University

■ **FIGURE 3–36** The Karl G. Jansky Very Large Array (VLA) is laid out in a giant "Y" with a diameter of 27 km, and sprawls across a plain in New Mexico. It contains 27 telescopes (plus a spare). This photo shows many of the dishes in the VLA's most compact configuration, but the dishes move on railroad tracks so they can be much more widely separated. The VLA is so large that even the weather is sometimes different from one side to another. Its imaging capability has been important for understanding radio sources. In the years leading up through 2011, the electronics and computers were replaced and modernized, and data links between individual dishes on fiber optics were provided, to make the array much more sensitive at each frequency and also sensitive over a wider range of frequencies than its predecessor.

construction by American, European, and Japanese astronomers on the very high Atacama desert, at an exceptionally high altitude of 5,000 m, over 16,500 feet in altitude. Though working conditions are extremely difficult, much of the observing will be done remotely from lower altitudes, and having the telescopes be above so much of the water vapor in the Earth's atmosphere will greatly extend the array's scientific capability. ALMA is to consist of 66 individual dishes (■ Fig. 3–37), but in 2012 it was already working in a partial configuration (■ Fig. 3–38).

A further worldwide project in radio astronomy is the Square Kilometre Array (SKA), in which a collecting area totaling a square kilometer of Earth is to be covered with thousands of radio telescopes.

With that huge collecting area, the array should be 50 times more sensitive than our current best radio telescopes and work thousands of times more rapidly. The seven governmental and research organizations that formed the SKA Organisation in late 2011 include representatives from Australia, China, Italy, the Netherlands, New Zealand, South Africa, and the U.K. (the last of which will host the headquarters). The prime sites fighting it out for the main part of the telescope were in barren plains in western Australia and in South Africa, and in 2012 the planners compromised by arranging to put some dishes in each of those locations: Australia for the lowest frequencies and South Africa for the middle and highest frequencies. Among the thousands of individual component telescopes, some will be thousands

A CLOSER LOOK 3.1

A Night at Mauna Kea

An "observing run" with one of the world's largest telescopes is a highlight of the work of many astronomers. The construction of several huge telescopes at the top of Mauna Kea ("White Mountain") in Hawaii has made the mountain the site of one of the world's major observatories. It was chosen because of its clear and steady air. These conditions stem from the fact that the summit is so high – 4,200 m (13,800 ft) above sea level – and the mountain is surrounded by ocean. Though there are other tall mountains in the world, few have such favorable conditions for astronomy.

Let us imagine you have applied for observing time and have been chosen. During the months before your observing run, you make lists of objects to observe, prepare charts of the objects' positions in the sky based on existing star maps and photographs, and plan the details of your observing procedure. The air is thin at the telescope; you may not think as clearly or rapidly as you do at lower altitudes. Therefore, it is wise and necessary to plan each detail in advance.

The time for your observing run comes, and you fly off to the island of Hawaii, the largest of the islands in the state of Hawaii. Increasingly, observations are done remotely, from the telescope's headquarters near the island's west coast at low altitude (where there are good data links to the summit) or even from the mainland (using satellite connections). But let us describe what it is like to observe from the top.

You drive up the mountain, as the scenery changes from tropical to relatively barren, crossing dark lava flows. First stop is the mid-level facility (Hale Pohaku) at an altitude of 2,750 m (9,000 ft). The Onizuka Center for International Astronomy at Hale Pohaku is named after Ellison Onizuka, the Hawaiian astronaut who died in the space-shuttle Challenger explosion in 1986.

The next night is yours. During the afternoon, you test the instrument with which the data will be obtained and determine the exact configuration in which it is to be used. For example, if you will be obtaining optical spectra of objects, the spectrograph needs to be focused properly and you must choose specific wavelength ranges to observe. An afternoon trip to the summit might be necessary.

After dinner you return to the summit to start your observing. You dress warmly, since the nighttime temperature approaches freezing at this altitude, even in the summertime. Although most telescope domes are equipped with warm rooms in which astronomers sit

when collecting data, occasionally you might step outside to absorb the magnificence of the dark sky – or to monitor the motion of encroaching clouds.

During the night you are assisted by a Telescope Operator who knows the telescope and its systems very well, and who is responsible for the telescope, its operation, and its safety. The operator points the telescope to the object you have chosen to observe and makes sure the telescope is properly compensating for the Earth's rotation. A television camera mounted on the telescope shows objects within the telescope's field of view. You examine that region of the sky, carefully comparing it with your chart to identify the right object among many random, unwanted stars. That object is then centered at the proper position to take data.

Nearly everything else is computer-controlled. You measure the brightness of the object at the wavelength you are observing, or begin an exposure to get a spectrum, by starting a computer program on a console provided for the observer. The data are stored in digital form on computer disks and tapes, and you might even transfer them over the Internet to your university computer.

Suppose you are interested in what kinds of chemical elements are produced and ejected by supernovae (exploding stars) or how they reveal the overall structure of the Universe and its expansion. One of us (A.F.) is a regular user of the Keck Telescopes on various supernova projects. Or, suppose you are studying Pluto's atmosphere by watching it pass in front of a star. One of us (J.M.P.) made such

Ric Noyle, W. M. Keck Observatory

FIGURE ACL 3.1 The view from one of the ridges of telescopes across Mauna Kea, with the submillimeter valley at left, the Subaru and twin Keck telescopes in the middle, and the NASA Infrared Telescope Facility at right.

an observation at Mauna Kea and continues collaborating through remote observing with one of the telescopes there. In either case, you labor through the night, occasionally seeing an obviously exciting result, but often just getting tantalizing hints that demand much more detailed and time-consuming analysis.

At some time during the night, you may go into the telescope dome to ponder the telescope at work. The telescope all but blocks your view of the sky. Hardly anyone ever actually looks through the telescope; indeed, there is usually no good way to do so. Still, you may sense a deep feeling for how the telescope is looking out into space.

You drive down the mountain to Hale Pohaku in the early morning sun. Over the next few days and nights, you repeat your new routine. When you leave Mauna Kea, it is a shock to leave the pristine air above the clouds. But you take home with you data about the objects you have observed. It may take you months or years to fully study the data that you gathered in a few brief days and nights at the top of the world. But you hope that your data will allow a more complete understanding of some aspect of the Universe. Sometimes major conceptual breakthroughs are made.

(a)

Carlos Padilla, NRAO/AUI/NSF

(b)

ESO/S. Rossi

(c)

J. Stoke, NRAO/AUI/NSF

■ **FIGURE 3–37** Antennas being installed on the 5,000-m Chajnantor plain of Chile's Atacama Desert, the driest desert in the world, to make up ALMA, the Atacama Large Millimeter/submillimeter Array.

of kilometers away from the central cores, in New Zealand and on islands in the Indian Ocean. It is hoped that construction could start in 2016 and be completed by 2024, but the funding is not yet fully secure. The SKA will take data at such a rate that a single day's raw data could fill 5 million 160 GB iPods!

The radio arrays can study the Sun, but time is rarely allotted. One of us (J.M.P.), with colleagues and students, used the Jansky VLA to observe the May 20, 2012, annular solar eclipse, taking observations second by second to see what changed over those brief intervals to provide higher angular resolution on the Sun than had previously been available in the radio part of the spectrum. There is a radio array in China devoted to studies of the Sun (■ Fig. 3–39).

■ **FIGURE 3–38** One of the first observations with a partial ALMA array is of a pair of colliding galaxies known as The Antennae. The inset shows the ALMA observations with detail better than at any other wavelength. Gravitational interactions of these galaxies as they passed near each other drew out the long tidal tails shown by their radio radiation in blue. Recent starbursts show better in optical radiation (whites and pinks). Some millimeter and submillimeter regions of current star formation are also visible (orange and yellows). See also Figures 16–14 and 16–40.

NRAO/AUI/NSF; ALMA (ESO/NAOJ/NRAO); HST (NASA, ESA, AND B. Whitmore [STScI]; J. Hibbard (NRAO/AUI/NSF); NOAO/AURA/NSF

Yihua Yan, Chinese Spectral Radioheliograph Team

■ **FIGURE 3–39** The Chinese Spectral Radioheliograph, constructed in Inner Mongolia about 4 hours drive north of Beijing, is used to make high-resolution images of the Sun. It consists of forty 4.5 m antennas and sixty 2 m antennas spreading out with a maximum baseline of 3 km in both the north-south and east-west directions.

CONCEPT REVIEW

» About 400 years ago, Galileo was the first to use telescopes for extensive astronomical studies (Sec. 3.1). He discovered lunar craters, the disk and four main moons of Jupiter, countless stars in the Milky Way, the phases of Venus, sunspots, and other phenomena. These findings contributed to the Copernican revolution, to be further discussed in Chapter 5.

» **Refracting telescopes** use lenses and **reflecting telescopes** use mirrors to collect light (Sec. 3.2). **Spherical aberration** occurs when parallel light is not reflected or refracted to a good **focus**, the place where the image is formed; it is solved for on-axis light with a mirror in the shape of a **paraboloid**. **Chromatic aberration**, a property only of lenses, occurs when light of different colors does not reach the same focus. **Newtonian telescopes** reflect the light from the main mirror diagonally out to the side; **Cassegrain telescopes** use a secondary mirror to reflect the light from the main mirror through a hole in the primary.

» The larger the area of the telescope's main lens or mirror, the better its light-gathering power, the ability to detect faint objects (Sec. 3.2). Also, the larger the diameter, the higher the theoretical angular **resolution** – the ability to distinguish detail. In practice, turbulence in the Earth's atmosphere limits the resolution of large telescopes, but a new technique called **adaptive optics** allows astronomers to achieve very clear images over small fields of view (Sec. 3.3). **Charge-coupled devices** (**CCDs**) are a common type of electronic detector that is extremely sensitive to light, allowing very faint objects to be studied.

» **Schmidt telescopes** have a very wide field of view (Sec. 3.4). One of the most important astronomical surveys ever conducted was the Palomar Observatory Sky Survey; in the 1950s, it mapped the whole sky visible from southern California. Recently, the Sloan Digital Sky Survey carried out a CCD survey of a large portion of the sky. Pan-STARRS and the future Large Synoptic Survey Telescope are new projects for mapping the sky every few days with high sensitivity.

» Amateur astronomy has long been a popular hobby. Improvements in telescope design and computer technology have increased the portability and ease of operation of telescopes (Sec. 3.5). Equipped with CCDs and fancy image-processing computer programs, some amateur astronomers have even been contributing high-quality data to research projects conducted by professional astronomers.

» The 2.4-m **Hubble Space Telescope** has been in the limelight since 1990 (Sec. 3.6). Although initially disappointing because its primary mirror suffered from spherical aberration, corrective optics were subsequently installed by space-shuttle astronauts, and the telescope began delivering phenomenal results. For example, Hubble provides optical observations with resolution about ten times better than that of most ground-based telescopes. It has been critical for ultraviolet observations, and many of its studies in the infrared have been superior as well.

» Ground-based solar telescopes, which collect enormous amounts of light and must deal with the effects of solar heating, have technical differences from nighttime telescopes (Sec. 3.7). Solar telescopes in space provide access to wavelengths not visible from Earth's surface and also obtain optical images with very high resolution. Telescopes to observe **x-rays**, **gamma rays**, and most **ultraviolet** light must be above the Earth's atmosphere.

» For nearly a decade, NASA's now-defunct Compton Gamma Ray Observatory provided data at the highest energies (Sec. 3.8a). The **Chandra X-ray Observatory**, one of NASA's "Great Observatories," has been an outstanding success. The Hubble Space Telescope has been NASA's premier ultraviolet telescope, but there have been other important ultraviolet telescopes as well (Sec. 3.8b). While some parts of the **infrared** can be observed with telescopes on the ground, telescopes in airplanes or in orbit are also necessary (Sec. 3.8c). NASA's Spitzer Space Telescope ran out of coolant and has been superseded by the European Space Agency's **Herschel Space Observatory**, whose 3.5-m-diameter mirror is the largest in space.

» Radio telescopes on the ground observe a wide range of wavelengths covered by **radio waves** (Sec. 3.8d). Arrays of radio telescopes, such as the **Jansky Very Large Array** (**VLA**) in New Mexico and the continent-spanning **Very Long Baseline Array** (**VLBA**), give high resolution through a process known as interferometry, linking the data from the different telescopes.

» **ALMA**, the **Atacama Large Millimeter/submillimeter Array**, is being built on a very high plain in Chile by an international consortium. It will provide extremely high angular resolution in the millimeter and submillimeter, at and below the shortest radio wavelengths. A future major project is the **Square Kilometre Array**.

QUESTIONS

1. What are three discoveries that immediately followed the first use of the telescope for astronomy?

2. What advantage does a reflecting telescope have over a refracting telescope?

3. What limits a large ground-based optical telescope's ability to see detail?

4. List the important criteria in choosing a site for an optical observatory meant to study stars and galaxies.

5. Describe a method that allows us to make large optical telescopes more cheaply than simply scaling up designs of previous large telescopes.

6. Name some of the world's largest telescopes and their locations.

7. What are the similarities and differences between making radio observations and using a reflector for optical observations? Compare the path of the radiation, the detection of signals, and limiting factors.

8. Why is it sometimes better to use a small telescope in orbit around the Earth than it is to use a large telescope on a mountaintop?

9. Why is it better for some purposes to use a medium-size telescope on a mountain instead of a telescope in space?

10. Describe the problem that initially occurred with the Hubble Space Telescope, and how it was corrected.

11. Describe and compare the Hubble Space Telescope, the Herschel Space Observatory, and the Chandra X-ray Observatory. Mention their uses, designs, and other relevant factors.

12. What are two reasons why the Hubble Space Telescope can observe fainter objects than we can now study from the ground?

†13. What is the light-gathering power of a telescope that is 3 meters in diameter, relative to a 1-m telescope? What is it relative to a dilated human eye (pupil 6 mm in diameter)?

†14. How many times more light is gathered by a single Keck telescope, whose mirror is equivalent in area to that of a circle 10 m in diameter, than by the Palomar Hale telescope, whose mirror is 5 m in diameter, in any given interval of time?

†15. Assume for simplicity that you have only one eye. Let's say that on a dark, clear night you can barely see Star Moe with your fully dilated (pupil 6 mm in diameter) naked eye. (a) How big an eye would you need to barely see Star Curly, which is 100 times fainter than Star Moe? Although human eyes aren't that large, one can buy small, inexpensive telescopes that have mirrors or lenses this size. (b) However, suppose you instead want to see Star Shemp, a million times fainter than Star Moe. How big would your telescope have to be? (c) Compare the telescope in (b) with existing telescopes at major observatories. Do you think you could afford to buy one?

16. Why might some stars appear double in blue light, though they could not be resolved (i.e., seen clearly as individuals) in red light with the same telescope?

†17. What mirror diameter gives 0.1 arc second resolution for infrared radiation of wavelength 2 micrometers?

†18. What mirror diameter gives 1 arc second resolution for radio radiation of wavelength 1 m? Compare this with the size of existing optical telescopes.

19. Why is light from stars and planets considered to be "parallel light"?

20. Why are optical and radio telescopes often built in groups, or arrays?

21. How is it possible to focus x-rays into images if x-rays pass right through most materials?

†22. **True or false?** All other things being equal, the faintest star one can see through a 10-m diameter telescope is 10 times fainter than the faintest star one can see through a 1-m diameter telescope.

23. **True or false?** In some cases, radio telescopes at many different locations have been used together, combining the light, to produce images with much higher resolution than that of the individual telescopes.

24. **True or false?** The primary purpose of an astronomical telescope is to magnify images of stars and other objects.

25. **True or false?** The first optical telescopes ever made were refractors, but for most or all of the past century the world's largest optical telescopes have been reflectors.

26. **True or false?** The resolution (clarity) of images obtained with large ground-based optical telescopes is generally degraded by turbulence in Earth's atmosphere, unless special correction techniques are used.

27. **True or false?** Although Galileo was not the first to use two lenses in the form of a telescope, he was the first to conduct systematic astronomical studies with a telescope.

28. **True or false?** Space telescopes will still see stars twinkle because the light from the stars does not consist of parallel rays.

29. **True or false?** Adaptive optics reduces the blurring of images by using smoother (better polished) mirrors.

30. **Multiple choice:** Which one of the following is *not* an advantage of the Hubble Space Telescope (Hubble) over ground-based telescopes in gathering information about the Universe? **(a)** Hubble is better able to collect ultraviolet light. **(b)** Hubble detects fainter objects at optical wavelengths because it has a darker sky. **(c)** Hubble detects fainter objects at infrared wavelengths because it has a darker sky. **(d)** Hubble produces clearer images of celestial objects at optical wavelengths. **(e)** Hubble was cheaper to build than ground-based telescopes of comparable diameter.

†31. **Multiple choice:** Sam's telescope has a mirror 10 m in diameter, while Lily's telescope has a mirror 2 m in diameter. The tube of Sam's telescope is twice the length of the tube of Lily's telescope. When both telescopes are being used from the same location and in the same manner, the faintest stars Sam can see are _____ times fainter than the faintest stars Lily can see. **(a)** 5. **(b)** 10. **(c)** 12.5. **(d)** 25. **(e)** 50.

32. **Multiple choice:** Which one of the following statements about telescopes is *false*? **(a)** The light-gathering power is proportional to the square of the diameter. **(b)** Atmospheric turbulence can limit the angular resolution (clarity) of a telescope. **(c)** Reflecting telescopes can suffer from chromatic aberration. **(d)** The angular resolution (clarity) of a telescope of fixed diameter improves as you observe at shorter wavelengths. **(e)** The Hubble Space Telescope initially suffered from a problem known as spherical aberration.

33. **Multiple choice:** A ground-based telescope to observe x-rays with x-ray detectors would **(a)** give astronomers a chance to study the hot interiors of stars and planets; **(b)** be worthless because x-rays cannot get through the Earth's atmosphere; **(c)** be worthless because astronomers have not yet devised detectors sensitive to x-rays; **(d)** be worthless because no astronomical objects emit x-rays; or **(e)** be useful only to Superman, who has x-ray vision.

†34. **Multiple choice:** Suppose you own an optical telescope, 1 m in diameter, which detects light at a wavelength of 5,000 Å (note: 1 Å $= 10^{-8}$ m). If you wish to have a radio telescope with the same angular resolution (clarity), how big must its diameter be, if it detects radio signals at a wavelength of 5 m? **(a)** 10,000 km. **(b)** 100 km. **(c)** 1 km. **(d)** 1 m. **(e)** 1 cm.

35. **Multiple choice:** All other things being equal, and ignoring atmospheric effects, a reflecting telescope with a large primary mirror will have, compared with a telescope with a smaller primary mirror: **(a)** better light-gathering power only; **(b)** better angular resolution only; **(c)** better light-gathering power and better angular resolution; **(d)** better light-gathering power but poorer angular resolution; or **(e)** no clear advantage unless the secondary mirror is also larger.

36. **Multiple choice:** Chromatic aberration is a problem in _____ telescopes and refers to _____. **(a)** refracting: differing focal lengths for different wavelengths of light. **(b)** reflecting: differing focal points from different parts of the mirror. **(c)** refracting: the smearing of light due to atmospheric turbulence. **(d)** reflecting: the smearing of light due to atmospheric turbulence. **(e)** space: the deformation of the glass due to the absence of gravity.

37. **Fill in the blank:** About 300 years ago, _____ put a small diagonal mirror in front of the focal point of the primary mirror to produce the first workable reflecting telescope.

38. **Fill in the blank:** Electronic gadgets called _____, rather than photographic film, are now used almost exclusively to record optical images and spectra of astronomical objects.

39. **Fill in the blank:** The ability to distinguish between details or to distinguish two adjacent objects as separate is known as _____.

†40. **Fill in the blank:** A radio telescope 100 m in diameter, when used to measure 1-cm waves, is _____ wavelengths across.

†41. **Fill in the blank:** The theoretical angular resolution (neglecting Earth's atmosphere) of a telescope whose primary mirror diameter is 30 cm is _____ arc seconds at a wavelength of 6,000 Å (note: 1 Å $= 10^{-8}$ m).

42. **Fill in the blank:** Substantial improvements have recently been made to narrow-field images made with ground-based telescopes by using the technique of _____, which corrects for distortions in the waves of incoming light.

†This question requires a numerical solution.

TOPICS FOR DISCUSSION

1. Is it a coincidence that major improvements in astronomical detectors were made at nearly the same time as the electronics and computer revolution that began in the late 1970s?

2. Was the Hubble Space Telescope worth the roughly $2 billion that it cost? Reconsider this question after completing this textbook or your astronomy course.

3. Will the James Webb Space Telescope likely be worth the billions of dollars that it will ultimately cost? Reconsider this question after completing this textbook or your astronomy course.

4. Does it make sense that an obstruction inside of a telescope (such as the secondary mirror), or the hole in the center of the primary mirror of a Cassegrain telescope, doesn't produce a hole in the object that is being viewed?

Top: Star trails beyond the Gemini Observatory on Mauna Kea, from which a laser beam forms a guide star on the Earth's ionosphere that enables the use of adaptive optics. In this long exposure, the laser beam tracks across the sky, making the visible arc. (The object in the sky at which the laser was pointed changed several times during the night.) Note that stars rise set at an angle (not perpendicular) relative to the horizon, because Hawaii is about 20° north of Earth's equator. Also, the farther a star is from the pole, the more its path looks straight over a short distance. Telescope domes from left to right are Subaru, Keck (twins), the NASA Infrared Telescope Facility, and, behind and to the left of Gemini, the Canada-France-Hawaii Telescope. Dusk twilight is visible at right, and dawn twilight and the setting moon are visible at left.

Gemini Observatory/image by Joy Pollard

Bottom: A panoramic view showing star trails around the north celestial pole at left and around the south celestial pole (which is below the horizon at right). The view is from the Teide Observatory on Tenerife in the Canary Islands.

Image by J. C. Casado – Teide Observatory (IAC) – starryearth.com/TWAN (The World At Night)

4

Observing the Stars and Planets: Clockwork of the Universe

ORIGINS

Throughout their existence, humans have used the apparent positions of celestial objects to define the day, month, year, and seasons, as well as for navigation.

The Sun, the Moon, and the stars rise every day in the eastern half of the sky and set in the western half. If you leave your camera on a tripod with the lens open for a few minutes or hours in a dark place at night, you will photograph the "star trails" – the trails across the photograph left by the individual stars. In this chapter, we will discuss the phases of the Moon and planets, and how to find stars and planets in the sky. Stars twinkle; planets don't twinkle as much (■ Fig. 4–1). We will also discuss the motions of the Sun, Moon, and planets, as well as of the stars in the sky.

Pekka Parviainen

■ FIGURE 4–1 A camera was moved steadily from left to right, with the bright star Sirius (bottom trail) in view. The star trail shows twinkling. Also, the camera was moved from left to right (top trail) when pointed at Jupiter, which left a more solid, less twinkling trail. Notice the contrast between the less-twinkling planet trail and the more-twinkling star trail. The fact that the star trail breaks up into bits of different colors was caused by part of the twinkling effect of the Earth's atmosphere, given that Sirius was low in the sky when the picture was taken.

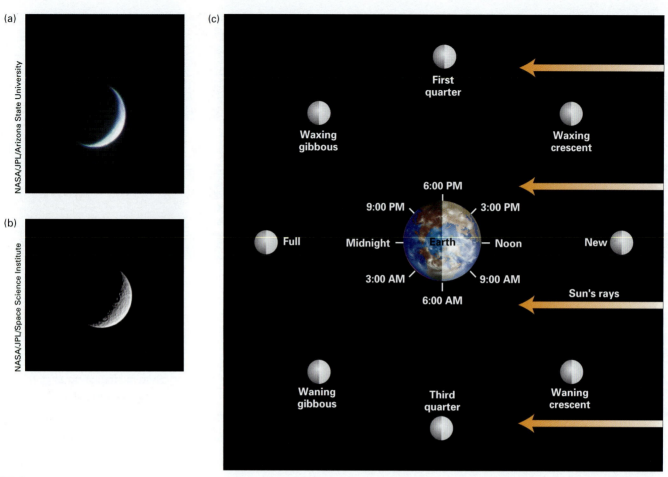

FIGURE 4–2 (a) As the 2001 Mars Odyssey spacecraft left Earth, its visible-light camera recorded Earth as a crescent of reflected sunlight, since at that time the spacecraft was behind most of the hemisphere facing the Sun. (b) Though Saturn's disk always looks full when seen from Earth, the views backward toward the Sun from the Cassini spacecraft show Saturn and its moons as crescents. Here we see a 2005 view of Rhea. (c) The phases of the Moon depend on the Moon's position in orbit around the Earth. Here we visualize the situation as if we were looking down from high above the Earth's orbit. Each of the Moon images shows how the Moon is actually lighted. Note that the Sun always lights one hemisphere of the Moon – the one directly facing the Sun, which is generally not the one facing the Earth.

4.1 THE PHASES OF THE MOON AND PLANETS

From the simple observation that the apparent shapes of the Moon and planets change, we can draw conclusions that are important for our understanding of the mechanics of the Solar System. In this section, we shall see how the positions of the Sun, Earth, and other Solar-System objects determine the appearance of these objects.

The phases of moons or planets are the shapes of the sunlit areas as seen from a given vantage point. The fact that the Moon goes through a set of such phases approximately once every month is perhaps the most familiar astronomical observation, aside from the day/night cycle and the fact that the stars come out at night (■ Fig. 4–2c, 4–3). In fact, the name "month" comes from the word "moon." The actual period of the phases, the interval between a particular phase of the Moon and its next repetition, is approximately 29½ Earth days. This type of period that depends on the orientation with respect to a third body (in this case, the Sun) is known as a **synodic period**.

The Moon is not the only object in the Solar System that goes through phases. Mercury and Venus both orbit inside the Earth's orbit, and so sometimes we see the side that faces away from the Sun and sometimes we see the side that faces toward the Sun; see Chapter 5 for more details. Thus at times Mercury and Venus are seen as crescents, though it takes a telescope to observe their shapes. Spacecraft to the outer planets have looked back and seen the Earth as a crescent (■ Fig. 4–2a) and the other planets and moons as crescents (■ Fig. 4–2b) as well.

The explanation of the phases is quite simple: The Moon is a sphere that shines by reflecting sunlight, and at all times the side that faces the Sun is lighted and the side that faces away from the Sun is dark. The phase of the Moon that we see from the Earth, as the Moon revolves (orbits) around us, depends on the relative orientation of the three bodies: Sun, Moon, and Earth. The situation is simplified by the fact that the plane of the Moon's revolution around the Earth is nearly, although not quite, the same as the plane of the Earth's revolution around the Sun; they are inclined relative to each other by 5°.

When the Moon is almost exactly between the Earth and the Sun, the dark side of the Moon faces us. We call this a "new moon." A few

(a)

Williams College – Hopkins Observatory by Kevin Reardon, now at National Solar Observatory, Sunspot, NM

(b)

Jay M. Pasachoff

■ **FIGURE 4–3** (a) The phases of the Moon, as seen from Earth. Each phase is labeled to match the corresponding Sun-Moon-Earth relative position that Fig. 4–2c showed from a point of view high above the Solar System. (b) The crescent Moon, shortly after sunset, with the planet Mercury slightly above and to the left of it. Note the "earthshine" that faintly illuminates the dark side of the Moon. Earthshine can also be seen in the first image of the waxing crescent moon in sequence (a), but not in the last image because of the high contrast.

■ **FIGURE 4–4** Because the phase of the Moon depends on its position in the sky with respect to the Sun, a full moon always rises at sunset. A crescent moon is either setting shortly after sunset, as shown here, or rising shortly before sunrise. (These statements are true when viewing the Moon from latitudes relatively near Earth's equator. Close to either pole, the relationship between lunar phase and time of rising or setting is more complicated.)

East   West   East   West

days earlier or later we see a sliver of the lighted side of the Moon, and call this a "crescent." As the month progresses, the crescent gets bigger (a "waxing crescent"), and about seven days after a new moon, half the face of the Moon that is visible to us is lighted. We are one-quarter of the way through the cycle of phases, so we have the "first-quarter moon."

When over half the Moon's disk is visible, it is called a "gibbous moon" (pronounced "gibb'-us"). As the sunlit portion visible to us grows, the gibbous moon is said to be "waxing." One week after the first-quarter moon, the Moon is on the opposite side of the Earth from the Sun, and the entire face visible to us is lighted. This is called a "full moon." Thereafter we have a "waning gibbous moon." One week after full moon, when again half the Moon's disk that we see appears lighted, we have a "third-quarter moon." This phase is followed by a "waning crescent moon," and finally by a "new moon" again. The cycle of phases then repeats.

Note that since the phase of the Moon is related to the position of the Moon with respect to the Sun, if you know the phase, you can tell approximately when the Moon will rise. For example, since the Moon is 180° across the sky from the Sun when it is full, a full moon rises just as the Sun sets (■ Fig. 4–4). Each day thereafter, the Moon rises an average of about 50 minutes later. The third-quarter moon, then, rises near midnight, and is high in the sky at sunrise. The new moon rises with the Sun in the east at dawn, and sets with the Sun in the west at dusk. The first-quarter moon rises near noon and is high in the sky at sunset. Contrary to popular belief, the Moon is often visible during the day; it simply isn't noticed as much as at night, because the contrast with the bright daytime sky is lower.

An interesting phenomenon occurs when the Moon is in its crescent phase: The "dark" side of the Moon is faintly visible, as can be seen in Figure 4–3b. This glow is known as "earthshine"; we are seeing sunlight that was first reflected by Earth toward the Moon, and then from the Moon back to Earth. After all, when the Moon is in its crescent phase, the Earth would appear gibbous (and thus very bright) as seen from the Moon, so a lot of light would reach the Moon and reflect back toward Earth. But when the Moon's phase is closer to quarter or gibbous, the amount of light reflected by the Earth diminishes; also, it becomes more difficult to detect the fainter earthshine given the greater glare from the directly illuminated part of the Moon.

It is natural to ask why the Earth's shadow doesn't generally hide the Moon during full moon, and why the Moon doesn't often block the Sun during new moon. These phenomena are rare because the Moon's orbit is tilted by about 5° relative to the Earth–Sun plane, making it difficult for the Sun, Earth, and Moon to become exactly

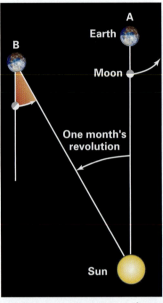

■ **FIGURE 4–5** After the Moon has completed its 27⅓-day revolution around the Earth with respect to the stars (**a sidereal month**), it has moved from point A to B. But the Earth has moved one month's worth around the Sun. It takes about an extra two days for the Moon to complete its revolution around the Earth as measured with respect to the Sun (**a synodic month**).

aligned. However, when they do reach the right configuration, eclipses occur, as we will soon discuss.

The Moon revolves around the Earth every 27⅓ days with respect to the stars, a **sidereal period**. But during that time, the Earth has moved partway around the Sun, so it takes a little more time for the Moon to complete a revolution with respect to the Sun (■ Fig. 4–5). Thus the cycle of phases that we see from Earth repeats with this 29½-day period.

We noted above that the Moon rises an average of about 50 (actually 48.8) minutes later each day. Multiplying 48.8 minutes/day by 29½ days (the lunar synodic month) gives 1440 minutes, which is 24 hours. Thus, at a given phase, the calculation verifies that the Moon rises at the same time each month.

4.2 CELESTIAL SPECTACLES: ECLIPSES

Because the Moon's orbit around the Earth and the Earth's orbit around the Sun are not precisely in the same plane (■ Fig. 4–6),

(a)

(b)

NASA/JPL/Arizona State University

■ **FIGURE 4–6** (a) The plane of the Moon's orbit is tipped by 5° with respect to the plane of the Earth's orbit, so the Moon usually passes above or below the Earth's shadow. Therefore, we don't have lunar eclipses most months. (The diagram is not to scale.) (b) An actual image of the Earth and Moon, illustrating their true separation: about 30 Earth diameters. The image was obtained with the infrared camera on board the 2001 Mars Odyssey spacecraft. Taken at a wavelength of 9 µm, the image shows colder regions, such as the South Pole near the Earth's bottom, as dark, and Australia, which is warmer, as bright. Each pixel corresponds to 900 km across, about the size of Texas.

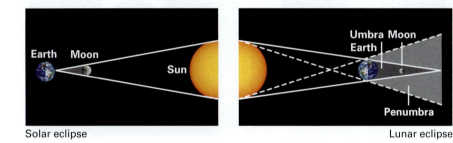

Solar eclipse   Lunar eclipse

■ **FIGURE 4–7** When the Moon is between the Earth and the Sun (as at left), we observe an eclipse of the Sun. When the Moon is on the far side of the Earth from the Sun (as at right), we see a lunar eclipse. The part of the Earth's shadow that is only partially shielded from the Sun's view is called the "penumbra"; the part of the Earth's shadow that is entirely shielded from the Sun is called the "umbra." (The diagram is not to scale.)

the Moon usually passes slightly above or below the Earth's shadow at full moon, and the Earth usually passes slightly above or below the Moon's shadow at new moon. But up to seven times a year, full moons or new moons occur when the Moon is at the part of its orbit that crosses the Earth's orbital plane. At those times, we have a **lunar eclipse** or a **solar eclipse** (■ Fig. 4–7). Thus up to seven eclipses (mostly partial) can occur in a given year.

Many more people see a total lunar eclipse than a total solar eclipse when one occurs. At a total lunar eclipse, the Moon lies entirely in the Earth's full shadow and sunlight is entirely cut off from it (neglecting the atmospheric effects we will discuss below). So for anyone on the entire hemisphere of Earth for which the Moon is up, the eclipse is visible (weather permitting!). In a total solar eclipse, on the other hand, the alignment of the Moon between the Sun and the Earth must be precise, and only those people in a narrow band on the surface of the Earth see the eclipse. Therefore, it is much rarer for a typical person on Earth to see a total solar eclipse – when the Moon covers the whole surface of the Sun – than a total lunar **eclipse**.

We will discuss the science of the Sun in Chapter 10. Historically, many important things about the Sun, such as the hot "corona" (see below) and its spectrum, were discovered during eclipses. Now, satellites in space are able to observe the middle and outer corona on a daily basis, and to study the Sun in a variety of important ways. A few observatories on high mountains study some aspects of the inner corona, but what they can observe is limited. There are still gaps in what spacecraft and ground-based observatories can do to observe the corona, and eclipses remain scientifically useful for studies that fill in those gaps. For example, coronagraphs in space have to hide not only the Sun's everyday surface but also a region around it, for technological reasons (■ Fig. 4–8). So images taken on the days of eclipses cover those physical gaps. In addition, eclipse images can be obtained over shorter intervals than images with current spacecraft.

STAR PARTY 4.1

Observing Total Solar Eclipses

The next total solar eclipse visible from the United States isn't until 2017, and from Canada and again from the U.S. until 2024. In general, one must travel far to witness totality: to Africa in 2013, to the Arctic (Svalbard, for example) in 2015, and to Indonesia in 2016. At a given location on Earth, the average interval between total solar eclipses is roughly 380 years!

Should you go out of your way to observe a total solar eclipse? The entire event is indescribably beautiful and moving. Photographs and words, and even videos, simply do not convey the drama, beauty, and thrill of a total solar eclipse: It must be witnessed in person. In our opinion, everyone should see at least one!

Only during the total part of a total eclipse can one look at the Sun directly without special filters. (During the Baily's beads and diamond ring that mark the last few seconds before or after totality, it is also safe to observe the eclipse directly with the naked eye, though not with binoculars or a telescope.) During other phases and types of eclipses, you can observe the Sun safely only through special filters or by projecting the image so that you are not looking directly at the Sun. Appropriate filters to use include shade 14 welder's glass, which is available at welding supply shops, or one of the special solar-viewing filters available from certain vendors (www. eclipses.info).

Alternatively, to make a pinhole camera (Fig. 4–11b) that projects the image of the Sun (so that you don't look directly at the Sun), punch a small hole (about 3 mm across) in a piece of thin cardboard. Face the cardboard toward the Sun and project the resulting image of the Sun onto a flat surface, such as another piece of cardboard, which you look at while you are facing away from the Sun. This method works best if the flat surface is in the shade, directly behind the cardboard with the hole in it. If you punch multiple holes, you will get an image of the Sun from each one; be creative by spelling your name, drawing a simple object, etc.

(a)

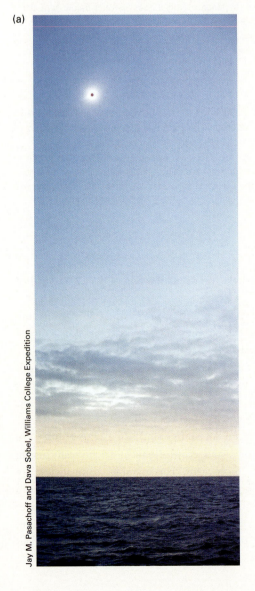

Jay M. Pasachoff and Dava Sobel, Williams College Expedition

(b)

Jay M. Pasachoff and Shelby B. Kimmel, Williams College Expedition

(c)

Jay M. Pasachoff and Shelby B. Kimmel, Williams College Expedition

(d)

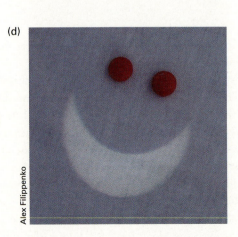

Alex Filippenko

FIGURE SP 4.1 The total part of the hybrid solar eclipse of April 8, 2005. (a) A view during totality showing the eclipsed Sun with darker sky below. At the horizon, we see in the image part of a 360° view of the sky glowing yellow/orange from sunlight outside the Moon's shadow reflected into the shadow, where the camera was located. (b) The diamond-ring effect just before totality began. (c) The solar corona surrounding the dark silhouette of the Moon in the middle of the 30-second period of totality. (d) Image of the partially eclipsed Sun projected by a monocular (half of a pair of binoculars), next to two round pieces of candy, showing the festive mood of successful eclipse viewers after totality.

■ FIGURE 4–8 A compound image of a solar eclipse. The disk of the Sun is from the Extreme-Ultraviolet Imaging Telescope on the Solar and Heliospheric Observatory (SOHO) spacecraft, taken from a position in space from which the spacecraft was not in the path of totality; the middle zone is from the Williams College Eclipse Expedition headed by one of the authors (J.M.P.), and the outer part is from the Large-Angle Spectrographic Coronagraph on SOHO. Note how features can be traced from lower levels in the corona to higher levels because of the way the images are used together.

4.2a Eerie Lunar Eclipses

A total lunar eclipse is a much more leisurely event to watch than a total solar eclipse. The partial phase, when the Earth's shadow gradually covers the Moon, usually lasts over an hour, similar to the duration of the partial phase of a total solar eclipse. But then the total phase of a lunar eclipse, when the Moon is entirely within the Earth's shadow, can also last for over an hour, in dramatic contrast with the few minutes of a total solar eclipse (see Section 4.2b).

During a total lunar eclipse, the sunlight is not entirely shut off from the Moon: A small amount is refracted (bent) around the edge of the Earth by our atmosphere. Much of the short-wavelength light (violet, blue, green) is scattered out by air molecules, and scattered or absorbed by dust and other particles, during the sunlight's passage through our atmosphere. The remaining light is reddish-orange, and this is the light that falls on the Moon during a total lunar eclipse. (For the same reason, the light from the Sun that we see near sunset or sunrise is orange or reddish; the blue light has been scattered away by molecules and scattered or absorbed by particles. See *A Closer Look 4.1: Colors in the Sky*.) Thus, the eclipsed Moon, though relatively dark, looks orange or reddish (■ Fig. 4–9); its overall appearance, in fact, is rather eerie and three-dimensional.

(a)

(b)

■ FIGURE 4–9 (a) A multiple exposure sequence of a total lunar eclipse, showing both the partial and total phases. The totally eclipsed Moon appears faintly in the center of the shadow and looks reddish. (b) Another composite image, this time of the total lunar eclipse of December 2011, observed from the Himalayas in India. Note that the Earth's umbral shadow is round, as is always the case during lunar eclipses, and thus the Earth must also be round. The diameter of Earth's umbral shadow is nearly three times the size of the Moon, at the Moon's distance from Earth.

A CLOSER LOOK 4.1 — Colors in the Sky

Why is the sky blue? Air molecules in the Earth's atmosphere scatter (reflect) short-wavelength photons of light (violet, blue, and green) more effectively than they scatter longer-wavelength photons (orange or red), preferentially sending the short-wavelength light around in all directions, including toward our eyes. So, away from the Sun, the sky looks some shade of blue; see the figure. (The Sun doesn't emit much violet light compared with blue, and the eye's color cones are not as sensitive to violet, so the sky is blue instead of violet.)

Why are sunsets and sunrises red? When the Sun is high in the sky, it appears white (not yellow, a very common misconception). As the Sun moves progressively lower in the sky, we have to look more and more diagonally through the Earth's layer of air. The path of light through the atmosphere increases, so more of the short-wavelength photons get scattered away by air molecules before they reach us; the Sun then looks yellow, sometimes followed by orange (see the figure).

Also, various particles in the atmosphere, such as dust and smoke, tend to scatter and absorb blue light more than red light, further contributing to the Sun's changing color. As it approaches the horizon, it looks progressively more orange and red. Just before sunset

(or just after sunrise), the Sun is sometimes very red if there is much dust aloft. Reflection of these orange and red rays of sunlight by air, dust, and clouds gives the sky its vivid sunset/sunrise colors.

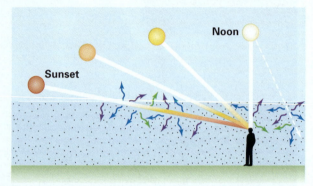

FIGURE ACL 4.1 Air molecules preferentially scatter violet, blue, and green light, causing the sky away from the Sun to appear some shade of blue. The longer the path through the air, the more short-wavelength light is lost, so the setting Sun appears orange or red.

The next total lunar eclipses will be on April 15, 2014, October 8, 2014, April 4, 2015, and September 28, 2015, and then not until January 21, 2018.

4.2b Glorious Solar Eclipses

The hot, tenuous outer layer of the Sun, known as the **corona**, is fainter than the blue sky. To study it, we need a way to remove the blue sky while the Sun is up. A total solar eclipse does just that for us.

Solar eclipses arise because of a happy circumstance: Though the Moon is about 400 times smaller in diameter than the solar **photosphere** (the disk of the Sun we see every day, also known as the Sun's "surface"), it is also about 400 times closer to the Earth. Because of

this coincidence, the Sun and the Moon cover almost exactly the same angle in the sky – about $1/2°$ (■ Fig. 4–10).

Occasionally the Moon passes close enough to the Earth–Sun line that the Moon's shadow falls upon the surface of the Earth. The central part of the lunar shadow barely reaches the Earth's surface. This lunar shadow sweeps across the Earth's surface in a band up to 300 km wide. Only observers stationed within this narrow band can see the total eclipse.

From anywhere outside the band of totality, one sees only a partial eclipse. Sometimes the Moon, Sun, and Earth are not precisely aligned and the darkest part of the shadow – called the **umbra** – never hits the Earth. We are then in the intermediate part of the shadow, which is called the **penumbra**. Only a partial eclipse is visible on Earth under these circumstances (■ Fig. 4–11a).

■ **FIGURE 4–10** An apple, the Empire State Building, the Moon, and the Sun are very different from each other in physical size, but here they cover the same angle because they are at different distances from us.

(a)

Jay M. Pasachoff

(b)
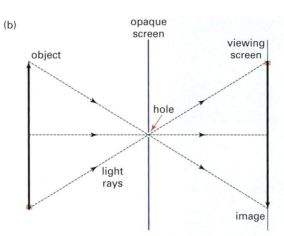

(c)

Alex Filippenko

■ **FIGURE 4–11** (a) The partially eclipsed Sun, as seen through a safe filter that eliminates all but one part in about 100,000 of the incident sunlight. This view, partly through clouds, was at the November 25, 2011, partial solar eclipse viewed from New Zealand. (b) Diagram of a pinhole camera, essentially just a hole in an opaque screen such as cardboard. Two of the rays of light, in this case from the extreme ends of an arrow, show the formation of an inverted image of the arrow. (c) The multiple images of a partially eclipsed Sun, produced by the pinhole-like gaps between leaves in a tree.

As long as the slightest bit of photosphere is visible, even as little as 1 per cent, one cannot see the important eclipse phenomena – the faint outer layers of the Sun – which are both beautiful and the subject of scientific study. Thus partial eclipses are of little value for most scientific purposes. After all, the photosphere is 1,000,000 times brighter than the outermost layer, the corona; if even only 1 per cent of the photosphere is showing, then we still have 10,000 times more light from the photosphere than from the corona, which is enough to ruin our opportunity to see the corona.

To see a partial eclipse or the partial phase of a total eclipse, you should not look at the Sun except through a special filter, just as you shouldn't look at the Sun without such a special filter on a non-eclipse day (see *Star Party 4.1: Observing Total Solar Eclipses*). You need the filter to protect your eyes because the photosphere is visible throughout the partial phases before and after totality. Its direct image on your retina for an extended time could cause burning, heating, and blindness. Alternatively, you can project the image of the Sun with a telescope or binoculars onto a surface, being careful not to look up through the eyepiece. A homemade pinhole camera (■ Fig. 4–11b) is the simplest and safest device to use, forming an image of the Sun. Multiple images of the Sun appear with cameras having many pinholes, and are sometimes produced by natural phenomena such as holes between leaves in trees (■ Fig. 4–11c).

You still need the special filter or pinhole camera to watch the final minute of the partial phases. As the partial phase ends, the bright light of the solar photosphere passing through valleys on the edge of the Moon glistens like a series of bright beads, which are called Baily's beads. At that time, the eclipse becomes safe to watch unfiltered, but only with the naked eye, not with binoculars or telescopes. The last bit of the uneclipsed photosphere seems so bright that it looks like the diamond on a ring – the "diamond-ring effect" (■ Fig. 4–12a).

The phenomenon of the darkening of the sky around you as totality approaches is dramatic. Watching on television loses most of the glory. For a few seconds, the chromosphere is visible as a pinkish band around the leading edge of the Moon. Then, as totality begins, the corona comes into view in all its glory and magnificence (■ Fig. 4–12b), often having streamers of gas near the Sun's equator and finer plumes near its poles. You can also see prominences (■ Fig. 4–12c), glowing pockets of hydrogen gas that appear pink. The total phase may last just a few seconds, more typically a couple of minutes, or (extremely rarely) as long as about 7 minutes. Astronomers use their spectrographs and make images through special filters. Tons of equipment are used to study the corona during this brief time of totality.

At the end of the eclipse, the diamond ring appears on the other side of the Sun, followed by Baily's beads and then the more mundane partial phases.

Somewhere in the world, a total solar eclipse occurs about every 18 months, on average (■ Fig. 4–13). On August 21, 2017, a total eclipse will cross the continental United States, the first time since 1979. Thereafter, one will next cross the U.S. in 2024.

Sometimes the Moon covers a slightly smaller angle in the sky than the Sun, because the Moon is in the part of its elliptical orbit that is relatively far from the Earth. When a well-aligned eclipse occurs in such a circumstance, the Moon doesn't quite cover the Sun. An annulus – a ring – of the photosphere remains visible, so we call this special type of partial eclipse an **annular eclipse**. In the continental United States, on May 20, 2012, we witnessed the end of an annular eclipse that started in Asia (■ Fig. 4–14). The next annular eclipse crossing the United States won't be until 2023; Canada will have one in 2021.

Jay M. Pasachoff, Muzhou Lu, and Craig Malamut, Williams College Expedition

■ **FIGURE 4–12** The total solar eclipse of 2010, viewed from Easter Island in the middle of the Pacific Ocean, since totality was not visible from land. The diamond-ring effect, with just a little bit of the bright underlying solar disk not covered by the edge of the Moon, signals the beginning and the end (in this particular case) of the total phase of a solar eclipse. Here, sunlight shining through valleys or between mountains on the edge of the Moon led to the effect looking like a double diamond ring. The totally eclipsed Sun (upper left and lower left) is safe to view with the naked eye, binoculars, or a telescope; the corona is about the same brightness as the full moon and equally safe to look at. Some prominences (generally pink, or red to some viewers), a thin layer known as the chromosphere (pinkish as well), and the inner corona during totality can also be seen around the time of the diamond rings (lower right).

Rarely, a solar eclipse is annular over some parts of the eclipse path on Earth, and total over other parts; such a "hybrid" case will occur in 2013 (Fig. 4–14), though the eclipse will be total over most of its path, including the entire part of its path over land (in Africa). In Figure 4–12 and in *Star Party 4.1: Observing Total Solar Eclipses*, we show photographs taken during the totality part of the hybrid eclipse of April 8, 2005. Totality was visible only in the middle of the Pacific Ocean, and was observed by just 1,000 people who were on the three ships that sailed to observe it. Fortunately, both of the authors were included in that group.

We will discuss the scientific value of eclipses in the chapter on the Sun, Chapter 10.

4.3 TWINKLE, TWINKLE, LITTLE STAR...

If you look up at night in a place far from city lights, you may see hundreds of stars with the naked eye. They will seem to change in brightness from moment to moment – that is, to **twinkle**. This twinkling comes from the moving regions of air in the Earth's atmosphere. The air bends starlight, just as a glass lens bends light. As the

Michael Zeiler

■ **FIGURE 4–13** Paths of total solar eclipses from 2001 to 2025. In 2013, totality will be visible only in the middle of a band whose ends show just an annular eclipse, so its path is marked in purple. A similar hybrid eclipse previously occurred in 2005, and there will also be one in 2023 (purple paths).

(a)

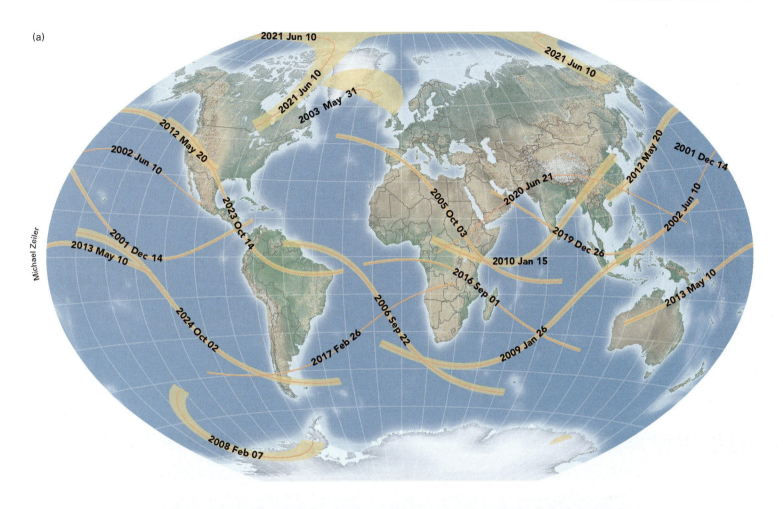

Michael Zeiler

2021 Jun 10
2021 Jun 10
2003 May 31
2021 Jun 10
2012 May 20
2002 Jun 10
2023 Oct 14
2001 Dec 14
2013 May 10
2024 Oct 02
2017 Feb 26
2006 Sep 22
2008 Feb 07
2005 Oct 03
2020 Jun 21
2019 Dec 26
2010 Jan 15
2016 Sep 01
2009 Jan 26
2012 May 20
2001 Dec 14
2002 Jun 10
2013 May 10

(b)

Jay M. Pasachoff and the Williams College Eclipse Expedition; montage by Muzhou Lu '13

■ **FIGURE 4–14** (a) Annular eclipses of the Sun from 2001 to 2025. (b) A montage of annularity at the May 20, 2012, eclipse, with some images crossed by clouds.

air moves, the starlight is bent by different amounts and the strength of the radiation hitting your eye varies, making the distant, point-like stars seem to twinkle (as we saw in Figure 4–1).

Unlike stars, planets are close enough to us that they appear as tiny disks when viewed with telescopes, though we can't quite see these disks with the naked eye. As the air moves around, even though the planets' images move slightly, there are enough points on the image to make the average amount of light we receive keep relatively steady. (At any given time, some points are brighter than average and some are fainter.) So planets, on the whole, don't twinkle as much as stars. Generally, a bright object in the sky that isn't twinkling is a planet.

But when a planet is low enough over the horizon, as Venus often is when we see it, it too can twinkle because there is so much turbulent air along the line of sight. Since the bending of light by air is different for different colors, we sometimes even see Venus or a bright star turning alternately reddish and greenish. On those occasions, professional astronomers sometimes get calls that UFOs have been sighted – especially when the crescent moon is nearby, drawing attention to the unusual configuration. See Chapter 20 for a discussion of why astronomers are not impressed with UFO reports.

4.4 THE CONCEPT OF APPARENT MAGNITUDE

To describe the apparent brightness of stars in the sky, astronomers – professionals and amateurs alike – usually use a scale that stems from the ancient Greeks. Over two millennia ago, Hipparchus described the typical brightest stars in the sky as "of the first magnitude," the next brightest as "of the second magnitude," and so on. The faintest stars were "of the sixth magnitude."

The **magnitude scale** – in particular, **apparent magnitude**, since it is how bright the stars appear – is fixed by comparison with the historical scale (■ Fig. 4–15). The higher the number, the fainter the star; see *Figure It Out 4.1: Using the Magnitude Scale*. If you hear about a star of 2nd or 4th magnitude (■ Fig. 4–16), you should know that it is relatively bright and that it would be visible to your naked eye. If you read about a 13th magnitude quasar, on the other hand, it is much too faint to see with the naked eye. If you read about a telescope on the ground or in space observing a 30th-magnitude galaxy, it is among the faintest objects we can currently study.

FIGURE IT OUT 4.1

Using the Magnitude Scale

Thousands of years since the terminology was first used by Hipparchus, we still use a similar scale, the **magnitude scale**, though now it is on a mathematical basis. Its basic structure – characterizing the brightness of the naked-eye stars with easy-to-visualize numbers between about 0 and 6 – is simple to use.

For example, if you are outdoors looking at the sky, you can tell in a moment that a 1st-magnitude star is brighter than a 3rd-magnitude star. Whether you use more complicated aspects of the magnitude scale (below) is optional, and later in this book we give brightnesses not only using magnitudes but also using the actual values for brightness or relative brightness that some people prefer.

Technically, each *difference* of 5 magnitudes is a *factor* of 100 times in brightness. A 1st-magnitude star is exactly 100 times brighter than a 6th-magnitude star – that is, we receive exactly 100 times more energy in the form of light. Sixth magnitude is still typically the faintest that we can see with the naked eye (seventh magnitude, in truly exceptional conditions), though because of urban sprawl the dark skies necessary to see such faint stars are harder to find these days than they were long ago. Objects too faint to see with the naked eye in quite dark skies have magnitudes greater than 6 (that is, they are fainter than 6th magnitude).

A few stars and planets are brighter than 1st magnitude, so the scale has also been extended in the opposite direction into negative numbers. For example, Sirius (the brightest star in the sky) has a magnitude of −1.5, and Venus can reach magnitude −4.4.

Since each difference of 5 magnitudes is a factor of 100 times, each one magnitude is a number that, when five of them are multiplied together, equals 100. No integer has this property. After all, $1 \times 1 \times 1 \times 1 \times 1 = 1$, $2 \times 2 \times 2 \times 2 \times 2 = 32$, which is less than 100, and $3 \times 3 \times 3 \times 3 \times 3 = 243$, which is greater than 100, so the number we want is somewhere between 2 and 3. A number with this property is simply called "the fifth root of 100." So, each difference of 1 magnitude is a factor of the fifth root of 100 (which is approximately equal to 2.512, or even just 2.5) in brightness. This choice is necessary to make a *difference* of 5 magnitudes ($1 + 1 + 1 + 1 + 1$, an additive process) equal to a *factor* of 100 ($2.5 \times 2.5 \times 2.5 \times 2.5 \times 2.5$, a multiplicative process).

For those who would like the formula linking magnitudes m (which we add and subtract) with brightnesses b (which we multiply or divide), it is

$$b_A = 2.512(m_B - m_A) \times b_B$$

Examples: If Star B is 1 magnitude fainter than Star A (say 3rd magnitude instead of 2nd magnitude), $m_B - m_A = 3 - 2 = 1$, $b_A / b_B = 2.512^1 = 2.512$, and $b_A = 2.512 b_B$, making Star A approximately 2.512 times brighter than Star B.

If Star B is 2 magnitudes fainter than Star A, $b_A / b_B = 2.512^2$, which is approximately 6, so Star A is about 6 times brighter than Star B.

A magnitude difference of 5 corresponds to $(2.512...)^5 = 100$ times (exactly) in brightness.

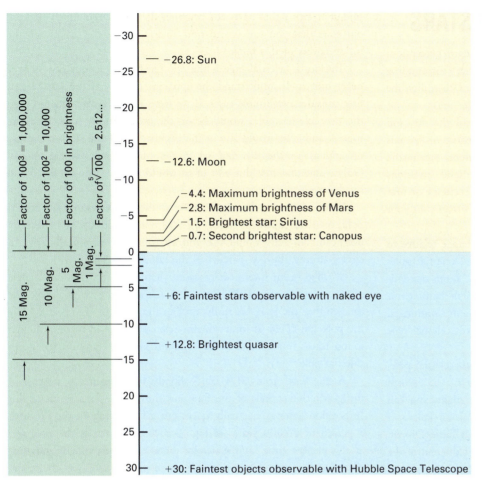

■ **FIGURE 4–15** The apparent magnitude scale is shown on the vertical axis. At the left, sample intervals of 1, 5, 10, and 15 magnitudes are marked and translated into multiplicative factors of 2.512 ..., 100, 10,000, and 1,000,000, respectively. For example, a difference of 1 magnitude is a factor of about 2.5 times in brightness; a difference of 5 magnitudes is a factor of exactly 100 times in brightness.

■ **FIGURE 4–16** The constellation Taurus in the star atlas by Johann Bayer from 1603. In this *Uranographia*, Bayer assigned the Greek letters to the stars that we still use. α Tauri, Aldebaran, is visual magnitude 0.85, and β Tauri, Alnath, is visual magnitude 1.65. These two stars are colorized for this reproduction.

4.5 RISING AND SETTING STARS

Stars and planets are so distant that we have no depth perception: they all seem to be glued to an enormously large sphere surrounding the Earth, the celestial sphere. This sphere is imaginary; in reality, the stars and planets are at different distances from Earth.

Though stars seem to rise in the east, move across the sky, and set in the west each night, the Earth is actually turning on its axis and the celestial sphere is holding steady. If you extend the Earth's axis beyond the north pole and the south pole, these extensions point to the celestial poles. Stars appear to traverse circles or arcs around the celestial poles. (See *A Closer Look 4.2: Photographing the Stars*.)

Since the orientation of the Earth's axis doesn't change relative to the distant stars (at least on timescales of years), the celestial poles don't appear to move during the course of the night. From our latitudes (the United States ranges from about 20° north latitude for Hawaii, to about 49° north latitude for the northern continental United States, to 65° for Alaska), we can see the north celestial pole but not the south celestial pole.

A star named Polaris happens to be near the north celestial pole, only about 1° away, so we call Polaris the **pole star** or the "North Star." If you are navigating at sea or in a desert at night, you can always go due north by heading straight toward Polaris. Polaris is conveniently located at the end of the handle of the Little Dipper, and can easily be found by following the "Pointers" at the end of the bowl of the Big Dipper. Polaris isn't especially bright, but you can find it if city lights have not brightened the sky too much. (Check out stars that you can use to find Polaris on the Sky Maps, on the inside covers of this book.)

The north celestial pole is the one fixed point in our sky, since it never moves. To understand the apparent motion of the stars, let us first consider two simple cases. If we were at the Earth's equator, then the two celestial poles would be on the horizon (due north and due south), and stars would rise in the eastern half of the sky, go straight up and across the sky, and set in the western half (■ Fig. 4–17). Only a star that rose due east of us would pass directly overhead and set due west.

If, on the other hand, we were at the Earth's north (or south) pole, then the north (or south) celestial pole would always be directly overhead. No stars would rise and set, but they would all move in circles around the sky, parallel to the horizon (■ Fig. 4–18).

We live in an intermediate case, where the stars rise at an angle relative to the horizon (■ Fig. 4–19). Close to the celestial pole, we can see that the stars are really circling the pole (■ Fig. 4–20). Only the pole star (Polaris) itself remains relatively fixed in place, although it too traces out a small circle of radius about 1° around the north celestial pole.

As the Earth spins, it wobbles slightly, like a giant top, because of the gravitational pulls of the Sun and the Moon. As a result of this "precession," the axis actually traces out a large circle in the sky with a period of 26,000 years (■ Fig. 4–21). So Polaris is the pole star at the present time, and generally there isn't a prominent star near either celestial pole when the orientation differs.

A CLOSER LOOK 4.2 — Photographing the Stars

It is easy to photograph the stars if you have a dark sky far from city lights. If you can, you should use a camera that allows the option of taking a long exposure; most "point-and-shoot" cameras won't take such long exposures, at least not with your choice of an exposure time of 30 seconds or 5 minutes. Place your camera on a tripod, turn off your flash (or put black tape over the flash), set the lens as wide open (the lowest "*f*-number," where *f* is the "focal ratio" marked on many lenses and cameras; *f*/2 is a good choice, though many cameras don't go past *f*/4, a less-wide lens setting), and set the focus to infinity.

You also have a choice of sensitivity, called "speed," of the sensor in your camera, rated by the International Standards Organization (ISO). Choosing a traditional standard ISO such as 400 or 800 would be a place to start. Cameras often allow you to choose faster ISOs such as 1,600 or 3,200, which would be more sensitive, but a consequence may be color spots or graininess on the resulting image, caused by the way the sensitivity is increased electronically.

Use a cable release to open the shutter for minutes; don't take exposures of several minutes without first testing shorter times to make sure that background skylight doesn't overwhelm ("fog") the image. You will have a picture of star trails. If the North Star (Polaris) is roughly centered in your field of view, and you expose for many minutes, you will see the circles that the stars make as a result of the Earth spinning on its axis. To show star trails, you might have to use a low ISO, to keep the background sky dark in the image.

With the new fast electronic sensors, you can record the stars in a constellation with exposures of a few seconds. Try a series: 1, 2, 4, 8, and 16 seconds, if you have control; otherwise, hope that the camera's automatic setting will make the exposure long enough to show stars. The stars will not noticeably trail on the shorter exposures. The constellation Orion, visible during the winter, is a particularly interesting constellation to photograph, since your image will show the greenish (or sometimes reddish, depending on the sensitivity of your detector) Orion Nebula in addition to the stars. Your eyes are not sensitive to color when viewing such faint objects, but your camera will record them.

FIGURE IT OUT 4.2 Sidereal Time

Since a whole circle is divided into 360°, and the sky appears to turn completely around the Earth once every 24 hours, the sky appears to turn 15 degrees per hour (15°/hr). Astronomers therefore divide right ascension (RA) into hours, minutes, and seconds (of time) instead of degrees. At the equator, one hour of RA = 15°. Dividing by 60, 1 minute of RA = $1/4° = 15$ minutes of arc (since there are 60 minutes of arc in a degree). Dividing by 60 again, 1 second of RA = $1/4$ minute of arc = 15 arc seconds (since there are 60 seconds of arc in a minute of arc).

We can keep "star time" by noticing the right ascension of a star that is "crossing the meridian," where the meridian is the line that extends from the north celestial pole through the point right over our heads (the zenith) to the horizon due south of us. This star time is called "sidereal time"; astronomers keep special clocks that run on sidereal time to show them which celestial objects are crossing the meridian – that is, crossing the north-zenith-south line. A sidereal day (a day by the stars) is about 4 minutes shorter than a solar day (a day by the Sun), which is the one we usually keep track of on our watches.

■ **FIGURE 4–17** Viewed from the equator, the stars rise straight up, pass across the sky, and set straight down.

■ **FIGURE 4–18** Viewed from the pole, the stars move around the sky in circles parallel to the horizon, never rising or setting.

4.6 CELESTIAL COORDINATES TO LABEL THE SKY

Geographers divide the surface of the Earth into a grid, so that we can describe locations. The equator is the line halfway between the poles (■ Fig. 4–22). Lines of constant longitude run from pole to pole, crossing the equator perpendicularly. Lines of constant latitude circle the Earth, parallel to the equator.

Astronomers have a similar coordinate system in the sky. Imagine that we are at the center of the celestial sphere, looking out at the stars. The **zenith** is the point directly over our heads. The **celestial equator** circles the sky on the celestial sphere, halfway between the

■ **FIGURE 4–19** (a) This long exposure shows stars circling the celestial north pole; the Hawaiian volcano Mauna Kea, where many telescopes are located, is in the foreground. The shortest bright arc is Polaris, the "North Star," only about 1° from the pole. (Though only the 49th-brightest star in the sky, it is famous because of its location.) (b) When we look toward the north celestial pole, stars appear to move counterclockwise in giant circles about the pole. Here, we are looking past one of the telescopes of the Kitt Peak National Observatory in Arizona.

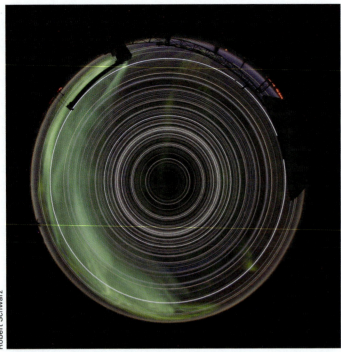

■ **FIGURE 4–20** A 24-hour star-trail image taken in 2012 by a camera pointing straight up from its location at the south pole. There is no "south star" to act as an analog to Polaris, but an aurora is visible. The brightest track, near the rim, is Sirius, the brightest star in the nighttime sky.

celestial poles. It lies right above the Earth's equator. Lines of constant **right ascension** run between the celestial poles, crossing the celestial equator perpendicularly. (See *Figure It Out 4.2: Sidereal Time.*) They are similar to terrestrial longitude. Lines of constant **declination** circle the celestial sphere, parallel to the celestial equator, similarly to the way that terrestrial latitude circles our globe (■ Fig. 4–23). The right ascension and declination of a star are essentially unchanging, just as each city on Earth has a fixed longitude and latitude. (Precession actually causes the celestial coordinates to change very slowly over time.)

From any point on Earth that has a clear horizon, one can see only half of the celestial sphere at any given time. This can be visualized by extending a plane that skims the Earth's surface so that it intersects the very distant celestial sphere (■ Fig. 4–24).

4.7 THE REASON FOR THE SEASONS

Though the stars appear to turn above the Earth at a steady rate, the Sun, the Moon, and the planets slowly drift among the stars in the sky, as discussed in more detail in Chapter 5. The planets were long ago noticed to be "wanderers" among the stars, slightly deviating from the daily apparent motion of the entire sky overhead.

The path that the Sun follows among the stars in the sky is known as the **ecliptic**. We can't notice this path readily because the Sun is so bright that we don't see the stars when it is up, and because the Earth's rotation causes another more rapid daily motion (the rising and setting of the Sun). The ecliptic is marked with a dotted line on the Sky Maps, on the inside covers of this book.

The Earth and the other planets revolve around the Sun in more or less a flat plane. So from Earth, the apparent paths of the other planets are all close to the ecliptic. (See *Star Party 4.2: The Paths of the Moon and Planets.*) But the Earth's axis is not perpendicular to the ecliptic; rather, it is tipped from perpendicular by 23½°.

The ecliptic is therefore tipped with respect to the celestial equator. The two points of intersection are known as the **vernal equinox** and the **autumnal equinox**. The Sun is at those points at the beginning of our northern hemisphere spring and autumn, respectively. On the equinoxes, the Sun's declination is zero.

Three months after the vernal equinox, the Sun is on the part of the ecliptic that is farthest north of the celestial equator. The Sun's

■ **FIGURE 4–21** The Earth's axis precesses with a period of 26,000 years. (a) The two positions shown are separated by 13,000 years. As the Earth's pole precesses, the equator moves with it (since the Earth is a rigid body). The celestial equator and the ecliptic – the Sun's apparent path through the stars – will always maintain the 23½° angle between them (with some slight variations due to subtle effects), but the points of intersection (the equinoxes) will change. Thus, over the 26,000-year cycle, the vernal equinox will move through all signs of the zodiac. (b) The vernal equinox is now in the constellation Pisces and approaching Aquarius (and thus the celebration, in the musical *Hair*, of the "Age of Aquarius").

Jay M. Pasachoff

■ **FIGURE 4–22** In Ecuador (which means "Equator" in Spanish), a monument marks a 19th-century French expedition to measure the length of a section of the equator. The monument, supposedly on the equator itself, is often visited. (Our group's GPS devices showed that the monument is actually many yards from the true equator.) Though charlatans there claim to show that water swirls down a drain oppositely in the two hemispheres, water does not actually do this unless other factors have forced it. (The "coriolis force" that causes hurricanes and tornados to rotate in opposite directions is much too small to affect water going down a drain.)

declination is then $+23\frac{1}{2}°$, and we say it is at the summer solstice, the first day of summer in the northern hemisphere. (Conversely, when the Sun's declination is $-23\frac{1}{2}°$, we say it is at the winter solstice, the first day of winter in the northern hemisphere.) The summer is hot because the Sun is above our horizon for a longer time and because it reaches a higher angle above the horizon when it is at high declinations (■ Fig. 4–25). A consequence of the latter effect is that a given beam of light intercepts a smaller area than it does when the light strikes at a glancing angle, so the heating *per unit area* is greater (■ Fig. 4–26). Moreover, there is less absorption, or attenuation, of the sunlight by Earth's atmosphere when the Sun is high than when it's low. (For the same two reasons, on a given day the temperature usually begins to decrease in the late afternoon, when the Sun is low above the horizon.)

The seasons (■ Fig. 4–27), thus, are caused by the variation in the declination of the Sun. This variation, in turn, is caused by the

---

STAR PARTY 4.2   The Paths of the Moon and Planets

Watch the Moon from night to night and note its position among the stars on each occasion. The Moon goes through its full range of right ascension and of declination each month as it circles the Earth.

The motion of the planets is much more difficult to categorize. For whatever planets are visible, plot their paths among the stars once a night every 3 days or so. Mercury and Venus circle the Sun more quickly than does Earth, and have smaller orbits than the Earth. They wiggle in the sky around the Sun over a period of weeks or months, and are never seen in the middle of the night. They are the "evening stars" or the "morning stars," depending on which side of the Sun they are on.

Mars, Jupiter, and Saturn (as well as the other planets too faint to see readily) circle the Sun more slowly than does Earth. So they drift across the sky with respect to the stars, and don't change their right ascension or declination rapidly. (On a given night, they seem to rise and set at the same rate as the stars; it takes observations over several nights to notice their motion.)

The U.S. and U.K. governments jointly put out a volume of tables each year, *The Astronomical Almanac*. It is an "ephemeris," a list of changing things; the word comes from the same root as "ephemeral." The book includes tables of the positions of the Sun, Moon, and planets. One of us (J.M.P.) has written the *Field Guide to the Stars and Planets*, which includes graphs and less detailed tables.

Near the end of 2011, Samoa (100 km from American Samoa) changed from the American side of the international date line to the Asian side, to reflect their increased business relations there. (On Fridays in Samoa, businesses were closed in Sydney, for example, since it was Saturday there; while on Sundays in Samoa, while businesses there were closed, businesses in Sydney with whom they might have worked were open.) 11:59 p.m. on Thursday night, December 29, 2011, was followed two minutes later by 12:01 a.m. on Saturday, December 31, 2011. Arrangements had been made for people to be paid for the Friday, but people with Friday birthdays that week no doubt felt a small loss.

■ **FIGURE 4–23** The celestial equator is the projection of the Earth's equator onto the sky, and the ecliptic is the Sun's apparent path among the stars in the course of a year. (The ecliptic, again, is a path in the sky; it lies in a plane called the ecliptic plane; though the two are related, do not confuse them.) The vernal equinox is one of the intersections of the ecliptic and the celestial equator, and is the zero-point of right ascension. From a given location at the latitude of the United States, the stars nearest the north celestial pole never set and the stars nearest the south celestial pole never rise above the horizon. Right ascension is measured along the celestial equator. Each hour of right ascension equals 15°. Declination is measured perpendicularly (+10°, +20°, etc.).

■ **FIGURE 4–24** The zenith is the point directly over your head; the nadir is the point straight down below your feet. The horizon marks as far as you can see. It is clear that if the celestial sphere were nearby (innermost circle), much less than half of it would be above the horizon. We can see progressively more of the celestial sphere at any given time if it is farther away, as shown (middle and outer circles). In fact, it is so far away that we see essentially half of it at a time.

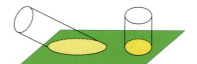

■ **FIGURE 4–26** There is more heating per unit area of the ground when the Sun is high than when it is low, because when the Sun is high a given beam of light intercepts a smaller area. In addition (though the effect is not shown here), the beam of light is attenuated more when it is low; there is more scattering from molecules and particles when the path through Earth's atmosphere is longer.

■ **FIGURE 4–25** The blue arcs from the equator up and over the top of the sphere shown are the path of the Sun at different times of the year. Around the summer solstice, June 21, the Sun (shown in yellow) is at its highest declination, rises highest in the sky, stays up longer (because, as shown, more of its path is above the horizon), and rises and sets farthest to the north. The opposite is true near the winter solstice, December 21. The diagram is drawn for latitude 40°.

Summer in northern hemisphere
Winter in southern hemisphere

Winter in northern hemisphere
Summer in southern hemisphere

■ **FIGURE 4–27** A schematic view from space of the Sun and the Earth (not to scale). The seasons occur because the Earth's axis is tipped with respect to the plane of the orbit in which it revolves around the Sun. The dotted line is drawn perpendicular to the plane of the Earth's orbit. When the northern hemisphere is tilted toward the Sun, it has its summer; at the same time, the southern hemisphere is having its winter. At both locations of the Earth shown here, the Earth rotates through many 24-hour day-night cycles before its motion around the Sun moves it appreciably.

fact that the Earth's axis of spin is tipped by $23^1/2°$. Many, if not most, people misunderstand the cause of the seasons. Note that the seasons are *not* a consequence of the changing distance between the Earth and the Sun. If they were, then seasons would not be opposite in the northern and southern hemispheres, and there would be no seasonal changes in the number of daytime hours. In fact, the Earth is closest to the Sun each year around January 4, which falls in the northern hemisphere winter.

The word "equinox" means "equal night," implying that in theory the length of day and night is equal on those two occasions each year. The equinoxes mark the dates at which the center of the Sun crosses the celestial equator. But the Sun is not just a theoretical point, which is what is used to calculate the equinoxes. Since the top of the Sun obviously rises before its middle, the daytime is actually a little longer than the nighttime on the day of the equinox. Also, bending (refraction) of sunlight by the Earth's atmosphere allows us to see the Sun when it is really a little below our horizon, thereby lengthening the daytime. So the days of equal daytime and nighttime are displaced by a few days from the equinoxes.

Because of its apparent motion with respect to the stars, the Sun goes through the complete range of right ascension and between $-23^1/2°$ and $+23^1/2°$ in declination each year. As a result, its height above the horizon varies from day to day. If we were to take a photograph of the Sun at the same hour each day, over the year the Sun would sometimes be relatively low and sometimes relatively high.

If we were at or close to the north pole, we would be able to see the Sun whenever it was at a declination sufficiently above the celestial equator. This phenomenon is known as the midnight sun (■ Fig. 4–28). Indeed, from the north or south poles, the Sun is continuously visible for about six months of the year, followed by about six months of darkness.

4.8 TIME AND THE INTERNATIONAL DATE LINE

Every city and town on Earth used to have its own time system, based on the Sun, until widespread railroad travel made this inconvenient. In 1884, an international conference agreed on a series of longitudinal time zones. Now all localities in the same zone have a standard time (■ Fig. 4–29).

Since there are twenty-four hours in a day, the 360° of longitude around the Earth are divided into 24 standard time zones, each modified to some degree from a basic north-south swath that is 15° wide. Each of the standard time zones is centered, in principle, on a meridian of longitude exactly divisible by 15, though the actual zones are modified for geopolitical reasons. Because the time is the same throughout each zone, the Sun is not directly overhead at noon at each point in a given time zone, but in principle is less than about a half-hour off. (There are some exceptions; for example, note China, or the west coast of Canada, in Figure 4–29.) Standard time is based on the average length of a solar day. As the Sun seems to move in the sky from east to west, the time in any one place gets later.

We can visualize noon, and each hour, moving around the world from east to west, minute by minute. We get a particular time back 24 hours later, but if the hours circled the world continuously the date would not be able to change at midnight. So we specify a north-south line and have the date change there. We call it the **international date line** (■ Fig. 4–30). With this line present on the globe, as we go eastward the hours get later and we go into the next day. At some time in our eastward trip, which you can visualize on Figure 4–30, we cross the international date line, and we go back one day. Thus we have only 24 hours on Earth at any one time.

(a)

Emil Schulthess, Black Star

(b)

Babak Tafreshi, TWAN (The World At Night)

(c)

■ **FIGURE 4–28** (a) In this series taken in June from northern Norway, above the Arctic Circle, one photograph was obtained each hour for an entire day. The Sun never set, a phenomenon known as the "midnight sun." Since the site was not quite at the north pole, the Sun and stars move somewhat higher and lower in the sky during the course of a day. (b) Hourly images showing the 2012 transit of Venus observed from Norway with midnight-sun conditions, with the silhouette of a mountain obscuring part of the Sun in some frames. (c) Midnight sun on Mars, a mosaic made from NASA's Phoenix Mars Lander, which was near Mars's north pole, over 11 "sols" (martian days).

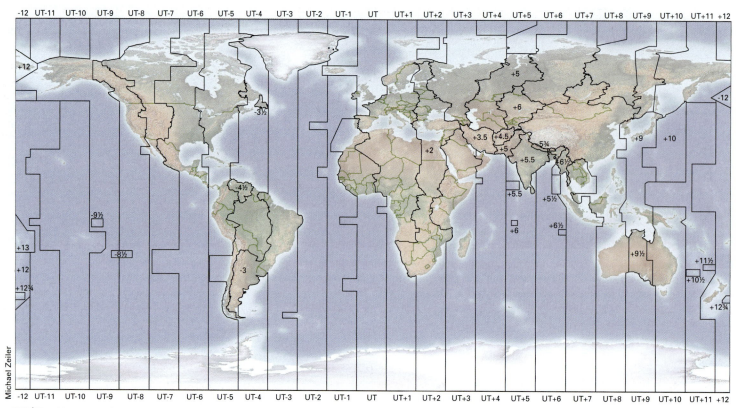

Michael Zeiler

■ **FIGURE 4–29** International time zones for standard time. In some places they are irregular; China, for example, should span 5 time zones, but the Chinese government prefers to have only a single time zone. At a few locations (e.g., India, central Australia), time zones are shifted by half an hour. Note the international date line, which is crooked in several places to avoid crossing land.

■ **FIGURE 4–30** This top view of the Earth, looking down from above the north pole, shows how days are born and die at the international date line. When you cross the international date line, your calendar changes by one day. To be among the first locations to see the new millennium, countries even changed their time zones, which shows how timekeeping can be arbitrary.

Jay M. Pasachoff

■ **FIGURE 4–31** The international zero circle of longitude at the Royal Observatory Greenwich is marked by a telescope that looked, historically, up and down along the north-south line. This site, in Greenwich, England, is a museum on the former site of the Royal Observatory there.

A hundred years ago, England won the distinction of having the basic line of longitude, 0° (■ Fig. 4–31), for Greenwich, then the site of the Royal Observatory. Realizing that the international date line would disrupt the calendars of those who crossed it, that line was put as far away from the populated areas of Europe as possible – near or along the 180° longitude line. The international date line passes from north to south through the Pacific Ocean and actually bends to avoid cutting through continents or groups of islands, thus providing them with the same date as their nearest neighbor.

In the summer, in order to make the daylight last into later hours, many countries have adopted "daylight saving time." Clocks are set ahead 1 hour on a certain date in the spring. Thus if darkness falls at 7 p.m. Eastern Standard Time (E.S.T.), that time is called 8 p.m. Eastern Daylight Time (E.D.T.), and most people have an extra hour of daylight after work (but, naturally, one less hour of daylight in the morning). In most places, clocks are set back one hour in the fall, though some places have adopted daylight saving time all year. The phrase to remember to help you set your clocks is "fall back, spring forward." Of course, daylight saving time is just a bookkeeping change in how we name the hours, and doesn't result from any astronomical changes.

It is interesting to note that daylight saving time is of political interest. In 2005, the energy bill passed by Congress and signed by the president extended daylight saving time for an extra month, to save energy by providing more light in the evening. (Traditionally, this desire has been counterbalanced by the farmers' lobby, which didn't want more morning darkness, but their influence is waning. Children waiting for schoolbuses in the morning darkness, though,

remain a counterargument for extending the period of daylight saving time.) A 2008 attempt to extend daylight saving time in Scotland failed for political reasons: the Parliament at Westminster in England didn't want to legislate for the newly empowered Scottish Parliament, and then people began to worry that England and Scotland would wind up in different time zones. Would Wales and Northern Ireland demand their own time zones, too, leaving the U.K., small in area, in four time zones? They dropped the effort. As of 2011, some parliamentarians were still working on the issue.

4.9 CALENDARS

The period of time that the Earth takes to revolve once around the Sun is called, of course, a year. This period is about 365¼ average solar days.

Roman calendars had, at different times, different numbers of days in a year, so the dates rapidly drifted out of synchronization with the seasons (which follow solar years). Julius Caesar decreed that 46 B.C. would be a 445-day year in order to catch up, and defined a calendar, the **Julian calendar**, that would be more accurate. This calendar had years that were normally 365 days in length, with an extra day inserted every fourth year in order to bring the average year to 365¼ days in length. The fourth years were, and still are, called leap years.

The name of the fifth month, formerly Quintillis, was changed to honor Julius Caesar; in English we call it July. The year then began in March; the last four months of our year still bear names from this system of numbering. (For example, October was the eighth month, and "oct" is the Latin root for "eighth.") Augustus Caesar, who carried out subsequent calendar reforms, renamed August after himself. He also transferred a day from February in order to make August last as long as July.

The Julian calendar was much more accurate than its predecessors, but the actual **solar year** is a few minutes shorter than 365¼ days. By 1582, the calendar was about 11 days out of phase with the date at which Easter had occurred at the time of a religious council 1250 years earlier, and Pope Gregory XIII issued a proclamation to correct the situation: he simply dropped 11 days from 1582. Many citizens of that time objected to the supposed loss of the time from their lives and to the commercial complications. Does one pay a full month's rent for the month in which the days were omitted, for example? "Give us back our 11 days," they supposedly cried (though this is now thought to probably be a myth based in part on a painting by William Hogarth).

In the **Gregorian calendar**, years that are evenly divisible by four are leap years, except that three out of every four century years – the ones *not* evenly divisible by 400 – have only 365 days. Thus 1600 was a leap year; 1700, 1800, and 1900 were not; and 2000 was again a leap year. Although many countries immediately adopted the Gregorian calendar, Great Britain (and its American colonies) did not adopt it until 1752, when 11 days were skipped. (Current

■ **FIGURE 4–32** In George Washington's family Bible, his date of birth is given as the "11th day of February 1731/2." Some contemporaries would have said 1731; we now say February 22, 1732.

U.S. states that were then Spanish were already using the Gregorian calendar.) As a result, we celebrate George Washington's birthday on February 22, even though he was born on February 11 (■ Fig. 4–32). Actually, since the year had begun in March instead of January, the year 1752 was cut short; it began in March and ended the next January. Washington was born on February 11, 1731, then often written February 11, 1731/2, but since 1752, people have referred to his date of birth as February 22, 1732. The Gregorian calendar is the one in current use. It will be over 3,000 years before this calendar is as much as one day out of step.

When did the new millennium start? It is tempting to say that it was on "January 1, 2000." That marks the beginning of the thousand years whose dates start with a "2." But, if you are trying to count millennia, since there was no "year zero," two thousand years after the beginning of the year 1 would be January 1, 2001. (Nobody called it "year one" then, however; the dating came several hundred years later.) If the first century began in the year 1, then the 21st century began in the year 2001! We, the authors, had two new millennium parties – a preliminary one on New Year's Eve 2000 and then the real one on New Year's Eve 2001.

A Johns Hopkins University astronomy professor keeps trying to reform the calendar. His 2011/12 proposal has the same set of dates matching days of the week each year, with leap year replaced by a leap week every five or six years. He and an economist colleague also suggest that we all keep Universal Time – time that matches the time zone of Greenwich, England – dropping time zones. We think it's unlikely that many (or even any) governments will adopt his plan.

4.10 KEEPING TIME

The accuracy of timekeeping has improved drastically over the years. Hundreds of years ago, daylight was divided into twelve hours, and as the length of the day changed over the seasons, the length of the hour changed, too.

Galileo's 16th-century discovery that a swinging chandelier kept a steady beat led to the development of pendulum clocks. They had

■ **FIGURE 4–33** An atomic clock based on transitions of a strontium atom. We see the beam of a blue laser shining onto ultracold strontium atoms to measure how well a previous burst from a red laser has raised the strontium atoms to an excited state. This clock is at the Joint Institute of Laboratory Astrophysics at the University of Colorado, in proximity to the Boulder headquarters of the National Institute of Standards and Technology.

their own inaccuracies, including the effect of temperature changes affecting the length of the pendulum, making the clocks run a bit fast or slow. Furthermore, pendulum clocks don't work on ships. That lack led to a famous 18th-century prize being offered in England for whoever could find a way to determine longitude accurately from a ship, clearly of practical importance for navigation.

The prize was won, finally, not by astronomers who could keep time with the positions of the moons of Jupiter, for example, but by the clockmaker John Harrison. He invented what we would call a pocket watch that ran from a mainspring. A book about the subject by the author Dava Sobel, *Longitude*, became an improbable best seller.

Nowadays we keep very accurate time with atomic clocks, counting vibrations of a cesium atom to define the second of time officially. Other atoms have been used to make even more accurate beats (■ Fig. 4–33). These clocks vie with the steady beat of distant pulsars (Chapter 13) in keeping accurate time over millennia.

But the Earth's rotation isn't as regular as the time we keep with atomic clocks; for example, Earth's rotation is gradually slowing down because of tidal effects (see Chapter 6). We still keep time by the Earth's rotation in step with atomic time by inserting "leap seconds" whenever time by the Earth's rotation differs from atomic time by more than 1 second. Such a leap second was inserted in 2008 and again in the middle of 2012. If we didn't do that, then eventually, over many thousands of millennia, noon by the Sun would slide earlier in the morning, and the seasons would drift with respect to the calendar. The difference wouldn't be noticeable in our lifetimes, though.

Nonetheless, a debate has been going on in the International Telecommunications Union as to whether leap seconds should be dropped, letting official time drift from the astronomical time. Some people say that computer networks could get confused by leap seconds, while others say that maintaining the link with astronomical time is more important. A decision had been expected early in 2012, but at that time, the decision was put off until 2015.

CONCEPT REVIEW

» Planets and moons have detectable disks that show **phases**, depending on how much of the sunlit side we see (Sec. 4.1). In particular, the phase of the Moon changes over the course of a month because of the changing angle formed by the Moon-Earth-Sun system.

» During a **lunar eclipse**, the Moon is in the shadow cast by the Earth (Sec. 4.2a). Even when a lunar eclipse is total, however, the Moon can be faintly seen because Earth's atmosphere bends some sunlight toward it. The Moon then has a reddish-orange color for the same reason that sunsets and sunrises are reddish-orange: air molecules preferentially scatter blue light in random directions (causing the daytime sky to appear blue), and particles such as dust in the atmosphere scatter and absorb blue light more than red light.

» At a total **solar eclipse**, the **umbra**, or darkest part, of the Moon's shadow hits the Earth (Sec. 4.2b). People in the **penumbra** see only a partial eclipse, which must be viewed through special filters or by using a projection technique like a pinhole camera. Just before an eclipse is total, the last bits of the Sun's surface, or photosphere, shine through lunar valleys and appear as **Baily's beads**. The last bit of the uneclipsed photosphere seems so bright that it looks like the diamond on a ring – the **diamond-ring effect**. During totality, you can see the Sun's hot, tenuous atmosphere, the **corona**; a total eclipse is perfectly safe to view with the naked eye, binoculars, or a telescope. When the Moon's disk appears too small to fully cover the Sun's photosphere, we have an **annular eclipse**. Because the Moon's orbit around the Earth and the Earth's orbit around the Sun are not precisely in the same plane, lunar and solar eclipses each occur only a few times a year, not every month.

» The point-like images of stars **twinkle** because we see them through the Earth's turbulent atmosphere (Sec. 4.3). Planets don't twinkle as much as stars, because the disk-like images of planets consist of many independently twinkling points that tend to average out.

» Astronomers use a simple scale of brightness, where the brightest visible stars are of the first magnitude, the next brightest of the second magnitude, and so on (Sec. 4.4). Long ago this was set on a mathematical scale, the **magnitude scale**, in which each factor of 100 in brightness corresponds to a difference of exactly 5 magnitudes. A star's **apparent magnitude** is how bright it looks to us.

» Stars appear to rise and set each day because the Earth turns (Sec. 4.5). We see the stars as though they were on a giant **celestial sphere**, with **celestial poles** over the Earth's north and south poles. Polaris, now the **pole star**, marks the north celestial pole, but **precession**, the very slowly changing orientation of the Earth's axis, will change that in the future (Sec. 4.5).

» Astronomical coordinate systems include one based on the Earth (Sec. 4.6). The **celestial equator** is the line on the celestial sphere above the Earth's equator. **Right ascension** corresponds to terrestrial longitude and **declination** corresponds to terrestrial latitude. The **meridian** is the north-south line extending through the **zenith**, the point overhead. Sidereal time, time by the stars, differs from solar time by about 4 minutes per day.

» The planets "wander" among the stars as their coordinates change (Sec. 4.7). They travel close to the path of the Sun across the sky, the **ecliptic**. The ecliptic meets the celestial equator at the **vernal equinox** and at the **autumnal equinox**. The Sun is at those points at the beginning of our northern hemisphere spring and autumn, respectively. The Sun's highest and lowest declinations correspond to the summer solstice (the first day of summer in the northern hemisphere) and the winter solstice (the first day of winter), respectively. Thus, the seasons are caused by the tilt of the Earth's axis. When you are far enough north or south on the Earth so that the Sun never sets at the current season, you see the **midnight sun**.

» Standard time is based on the average length of a solar day (Sec. 4.8). There are 24 time zones, each nominally 15° wide in longitude (though there are large variations throughout the Earth for geopolitical reasons). Days change at the **international date line**.

» The **Julian calendar** included leap years, but the seasons drifted because the actual solar year is not exactly 365¼ days (Sec. 4.9). We now use the **Gregorian calendar**, in which leap years are sometimes omitted to adjust for the fact that the year is not composed of exactly 365¼ days.

QUESTIONS

†1. On the picture opening the chapter, estimate the angle covered by the star trails and deduce how long the exposure lasted.

2. Describe the phases of the Earth you would see over a month if you were on the Moon.

†3. If the Moon were placed at twice its current distance from the Earth, how large (physically) would it have to be so that total solar eclipses could still occur?

†4. Compare the average frequencies of total and annular eclipses over a 25-year interval, say, 2001–2025.

5. In what phases of a total eclipse must one use eye protection and in what phase do you look directly? Is it ever safe to look directly at an annular eclipse? Explain.

6. If you look toward the horizon, are the stars you see likely to be twinkling more or less than the stars overhead? Explain.

7. Is the planet Saturn, which is farther away and physically smaller than Jupiter, likely to twinkle more or less than Jupiter?

8. Explain how it is that some stars never rise in our sky, and others never set.

9. Using the Appendices and the Sky Maps, comment on whether Polaris is one of the twenty brightest stars in the sky.

†10. Compare a 6th-magnitude star and an 11th-magnitude star in brightness. Which is brighter, and by how many times?

†11. (a) Compare a 16th-magnitude star with an 11th-magnitude star in brightness, as in Question 10. (b) Now compare the 16th-magnitude with the 6th-magnitude star, specifying which is brighter and by how many times.

†12. Since the sky revolves once a day, how many degrees does it appear to revolve in 1 hour?

13. Use the suitable Sky Map to list the bright stars that are near the zenith during the summer in the early evening, at midnight, and approaching dawn.

14. Explain why Christmas comes in the summer in Australia.

15. Are the lengths of day and night exactly equal at the vernal equinox? Explain.

16. Does the Sun pass due south of you at exactly noon on your watch each day? Explain why or why not.

17. Explain why for an observer at the north pole the Sun changes its altitude in the sky over the year, but the stars do not.

†18. When it is 5 p.m. on November 1 in New York City, what time of day and what date is it in Tokyo, Japan?

†19. Describe how to use Figure 4–29 to find the date and time in England when it is 9 a.m. on February 21 in California. Carry out the calculation both by going eastward and by going westward, and describe why you get the same answer.

†20. List the leap years between 1800 and 2200. How many will there be?

21. **True or false?** Suppose one Ewok is standing on Earth and another on the Moon while in the waxing crescent phase (as seen from Earth), and the Sun is directly overhead as seen by both Ewoks. If we assume the Ewoks have good telescopes, then each one can see the location where the other is standing at this time.

22. **True or false?** Earth's atmosphere scatters (reflects) blue light more easily than red light, causing the daytime sky to appear blue and contributing to the reddish-orange appearance of sunsets.

23. **True or false?** As observed from Earth, Jupiter sometimes appears to be in the crescent phase.

24. **True or false?** If Earth's axis of rotation were perpendicular to Earth's orbital plane (i.e., if Earth's axis weren't tilted), there would be no major seasonal variations at a given location on Earth throughout the year.

25. **True or false?** At mid-northern latitudes, the full moon passes higher in the sky in winter than in summer.

26. **True or false?** Venus and Mercury are sometimes visible at a position in the celestial sphere diametrically opposite the Sun.

27. **True or false?** More "circumpolar stars" (i.e., stars that are always above the horizon) are visible from Alaska than from Hawaii.

28. **True or false?** Some parts of the celestial sphere that are visible from New York at night in December are also visible from the South Pole at night in June.

29. **True or false?** As seen from Earth's surface, stars appear to twinkle because our eyes don't see photons outside visible wavelengths, which are emitted when the star appears to be dimming slightly.

30. **Multiple choice:** Which one of the following statements about lunar eclipses is *false*? (a) Lunar eclipses don't occur monthly because of the inclination of the Moon's orbit around the Earth relative to the Earth's orbit around the Sun. (b) Total lunar

eclipses last longer than total solar eclipses. (c) Lunar eclipses are predictable. (d) At a given time, a total lunar eclipse is visible only from a small part of the Earth's surface. (e) The Moon is still visible during a total lunar eclipse because of light going through the Earth's atmosphere.

31. **Multiple choice:** If the Moon as seen from the Earth is in the waning crescent phase, what phase of the Earth would be seen from the near side of the Moon? (a) Full. (b) Waxing gibbous. (c) Waning crescent. (d) The Earth does not go through phases as seen from the Moon. (e) The Earth's phases as seen from the Moon are unrelated to the Moon's phases as seen from the Earth.

32. **Multiple choice:** Which one of the following is the main reason why Earth has seasons? (a) Earth's orbit is elliptical. We have summer when we are closer to the Sun and winter when we are farther from the Sun. (b) The tilt of Earth's axis of rotation causes one hemisphere of the planet to be substantially closer to the Sun during the day than the other hemisphere. Because it is closer to the Sun, it receives much more solar energy per hour. (c) Over the course of the year, the tilt of Earth's axis of rotation varies from 23.5° to 0° in such a way as to bring more heating per hour in the summer than in the winter. (d) The tilt of Earth's axis of rotation causes the Sun to pass higher in the sky during the day in one hemisphere than in the other, thereby giving more daylight hours and more heating per hour. (e) It has seasons so that sunbathers will know when to go to the beach, and skiers will know when to go skiing.

33. **Multiple choice:** You are in California at 3 p.m. and you want to look at the full Moon. The best idea would be to (a) look toward the southern horizon; (b) look toward the western horizon; (c) look toward the eastern horizon; (d) look overhead, or nearly overhead; or (e) give up and try again later.

34. **Multiple choice:** If the tilt of Earth's rotation axis were 40° instead of 23.5°, which one of the following statements regarding what an observer at latitude 40° would experience would be *false*? (a) The summer would have even longer days and even shorter nights. (b) There would be a longer combined spring and summer period. (c) Summer days would be even hotter, on average. (d) The Sun would pass essentially overhead (the zenith) around June 22, the summer solstice. (e) The days and nights would each remain roughly 12 hours long around March 22 and September 22, the equinoxes.

35. **Multiple choice:** If the Moon were half its current distance from the Earth, then (a) we would no longer see the Sun's photosphere fully blocked by the Moon during solar eclipses; (b) we would no longer see the entire corona at a given time during total solar eclipses; (c) solar prominences would be easier to see during solar eclipses; (d) partial solar eclipses would not be visible from far-northern and far-southern latitudes on Earth; or (e) there would never be partial lunar eclipses.

36. **Multiple choice:** Which one of the following statements is *true* during totality for *both* total solar and total lunar eclipses? (a) They can be observed from an entire hemisphere on Earth. (b) They look reddish primarily due to the refraction and scattering of light passing through the Earth's atmosphere. (c) They last about the same amount of time. (d) They can be viewed safely with the naked eye. (e) They occur when the Sun or Moon are in Earth's shadow.

†37. **Fill in the blank:** A given star near the celestial equator rises about _____ minutes earlier each night because of Earth's revolution around the Sun.

38. **Fill in the blank:** When the Sun reaches the point in the sky known as the _____, it is the first day of spring in the northern hemisphere.

39. **Fill in the blank:** The celestial analogues of longitude and latitude are _____ and _____, respectively.

†40. **Fill in the blank:** A magnitude 3 star is _____ times brighter than a magnitude 10 star.

41. **Fill in the blank:** As viewed from the Earth's equator, the North Star (Polaris) appears to be located _____.

42. **Fill in the blank:** In order to account for the fact that the Earth's orbital period around the Sun is about 365¼ days rather than 365 days, we introduce _____.

†This question requires a numerical solution.

TOPICS FOR DISCUSSION

1. Physically, how is it possible that a given solar eclipse can sometimes be both total and annular along different parts of the eclipse path?

2. During a total lunar eclipse, what would someone on the Moon see when looking toward the Sun?

3. When the Sun is on either of the equinoxes, observers at both the North Pole and the South Pole of the Earth can see it completely above the horizon. How is this possible?

Kepler's diagram from *Cosmographium Mysterium* (*Mystery of the Cosmos*, 1596), in which he drew a stacking of the 5 "Platonic solids" – pyramid (tetrahedron), cube, octahedron, dodecahedron, and icosahedron – to explain the orbital spheres of the planets. The idea was entirely wrong, but it helped make him attractive as a scholar when he was considered by Tycho for an assistantship.

Jay M. Pasachoff

5

Gravitation and Motion: The Early History of Astronomy

ORIGINS

We explore the overall structure of the Solar System and how ideas about it were developed over many centuries, providing basic information for theories of its formation.

AIMS

1. Summarize theories of the Universe developed two millennia ago (Sections 5.1 and 5.2).

2. Follow how modern astronomy progressed through the ideas of Copernicus, Kepler, Galileo, and Newton (Sections 5.3 to 5.7).

3. Introduce the orbital motions of planets and how they provide clues to the formation of the Solar System (Section 5.8).

Ancient peoples knew of five planets – Mercury, Venus, Mars, Jupiter, and Saturn. When we observe these planets in the sky, we join the people of long ago in noticing that the positions of the planets vary from night to night with respect to each other and with respect to the stars. In this chapter, we will discuss the motions of the planets, and you will learn how major figures in the history of astronomy explained these motions. These explanations have led to today's conceptions of the Universe and of our place in it.

5.1 A BRIEF SURVEY OF THE SOLAR SYSTEM

Our ideas of the Solar System took shape before even the telescope was invented, but nowadays we have visited almost all of the Solar System with spacecraft (■ Fig. 5–1). Let us first give an overview of the Solar System before discussing the historical figures who started us on our current path. Discussion of the individual Solar-System objects occurs in later chapters.

Since 1957, when the first spacecraft were launched into orbit around the Earth, we have made giant strides. Satellites routinely study the Earth and even the Moon, mapping them in detail and watching how they change (■ Fig. 5–2). The series of crewed landings on the Moon during 1969–1972 revolutionized our understanding of that relatively nearby object.

The planet that comes closest to Earth is Venus, which is only 70 per cent our distance from the Sun. Venus is covered with clouds, but radar has been used to map its surface. The European Space Agency's Venus Express is now orbiting that planet, sending back images and other data. Beyond Venus, Mercury orbits so close to the Sun that any possible atmosphere should be hot, making the gas particles move too quickly to be permanently retained by the planet's relatively weak gravitational pull. It was therefore a surprise that a thin atmosphere containing many kinds of atoms has been detected, first with telescopes on Earth and now from NASA's MESSENGER spacecraft, which is in orbit around Mercury. Even early in its mission, the images and other observations from MESSENGER are changing our understanding of Mercury.

NASA/JPL/Caltech

■ **FIGURE 5–1** A composite of spacecraft images of the Solar System's 8 planets: Mercury, Venus, Earth, Mars, Jupiter, Saturn, Uranus, and Neptune.

Farther outward than the Earth's orbit is Mars. Its seasonal changes have long made it an object of interest, though we now know that they are from blowing dust rather than from vegetation. The signs (photographed from orbiting spacecraft and from landers on Mars's surface) that water once flowed on its surface make it all the more interesting, especially since liquid water is thought to be necessary for life to arise.

The four planets beyond Mars are much larger than the Earth or Mars. Giant Jupiter is second only to the Sun in ruling the Solar System by its gravity. NASA's *Galileo* spacecraft sent back detailed images of Jupiter and most of its many moons. Saturn, with its extensive, beautiful rings, is always a pretty object to see in a telescope. Its rings are made of material broken up by the same kinds of factors that cause the tides on Earth, as we shall learn in the next chapters. NASA's *Cassini* mission continues to orbit among

Naval Research Laboratory

■ **FIGURE 5–2** The Earth, with a lunar crater in the foreground, imaged by the Clementine spacecraft in orbit around the Moon. The angular distance between the images of the Earth and Moon was electronically reduced in this image. (This spacecraft, named after the prospector's daughter in the old song, was the first spacecraft launched that was named after a female – and not a real one, at that. The Juno spacecraft to Jupiter wasn't named after a real woman, either.)

STAR PARTY 5.1

Prograde and Retrograde Motions

If Mars or Jupiter is visible during times when you are awake, you can monitor the apparent position of either one in the sky to see prograde or even retrograde motion. Both planets, but especially Mars, move fairly quickly relative to the stars. Note that the planets move almost imperceptibly on a given night with respect to the stars; the stars and planets rise together in the east and set together in the west at almost the same rate. (To locate the current positions of planets, consult magazines such as *Sky and Telescope* or *Astronomy*, consult the *Field Guide to the Stars and Planets*, or look them up with planetarium software.)

With your naked eye (or through binoculars), note the position of Mars or Jupiter relative to stars near them in the sky. Carefully sketch or photograph what you observe. (See the hints in Chapter 4 on photographing the sky at night.) Repeat this exercise every 6 to 9 days, or more often if you wish. The best results are obtained with at least 10 observations over the course of 2 to 3 months.

Examining your sequence of observations made over several days, can you see the planet in its prograde motion? (Prograde is from west to east among the stars.) Were you viewing at a time when it went through retrograde motion?

Saturn and its moons, providing a treasure trove of images and other measurements.

Uranus and Neptune were visited by one of NASA's pair of Voyager spacecraft after it passed Jupiter and Saturn, providing our only close-up views of these large planets. The Hubble Space Telescope is useful for monitoring changes in their atmospheres, as are large ground-based telescopes equipped with adaptive optics (recall the discussion in Chapter 3). These objects can also be studied in other ways, such as when they pass in front of stars, providing "occultations."

Beyond the orbit of Neptune lies Pluto, which was considered the ninth planet for 75 years. In 2006, Pluto was reclassified as the first of a new class of objects known as "dwarf planets," and specifically the first "plutoid," so in some ways it is more accurate to say that Pluto was promoted rather than demoted. We will discuss dwarf planets in Chapter 8.

Along with the planets and dwarf planets, we find comets and asteroids, as well as interplanetary dust and, beyond Neptune, what are called Kuiper-belt objects or trans-Neptunian objects (KBOs or TNOs). The planets, dwarf planets, asteroids, and Kuiper-belt objects, but not most of the comets, orbit the Sun in nearly the same plane rather than having their orbits oriented every which way. Before we describe all these objects in more detail, let us tell an interesting chronological story.

5.2 THE EARTH-CENTERED ASTRONOMY OF ANCIENT GREECE

Most of the time, planets appear to drift slowly in one direction (west to east) with respect to the background stars. This forward motion over the course of weeks and months, moving slightly slower in the sky than the stars as they rise and set, is called **prograde motion**. But sometimes a planet drifts in the opposite direction (that is, east to west) with respect to the stars. We call this backward drift **retrograde motion** (■ Fig. 5–3). (See *Star Party 5.1: Prograde and Retrograde*

Motions.) Note that the slow drifting of planets ("the wanderers" in Greek) among the stars should *not* be confused with the much more rapid, daily rising in the east and setting in the west of stars and planets caused by Earth's rotation.

When we make a mental picture of how something works, we call it a **model**, just as sometimes we make mechanical models. The ancient Greeks made theoretical models of the Solar System in order to explain the motion of the planets. For example, by comparing the lengths of the periods of retrograde motion of the different planets, they were able to discover the order of distances of the planets.

One of the earliest and greatest philosophers, Aristotle (■ Fig. 5–4), lived in Greece in about 350 BCE. Aristotle thought, and actually believed that he definitely knew, that the Earth was at the center of the Universe. Furthermore, he thought that he knew that the planets, the Sun, and the stars revolved around it (■ Fig. 5–5).

In Aristotle's model, the Universe was made up of a set of 55 celestial spheres that fit around each other. Each sphere's natural motion was rotation. The planets were carried around by some of the spheres, and the motions of the spheres affected the other spheres. Some retrograde motion could be accounted for in this way. The outermost sphere was that of the *fixed stars*. Outside this sphere was the "prime mover" that caused the rotation of the stars.

Aristotle's theory dominated scientific thinking for about 1,800 years, until the Renaissance. Unfortunately, his theories were accepted so completely that they may have impeded scientific work that might have led to new theories.

In about 140 CE, almost 500 years after Aristotle, the Greek scientist Claudius Ptolemy (■ Fig. 5–6) flourished in Alexandria. He presented a detailed theory of the Universe that explained the retrograde motion. Ptolemy's model was **geocentric** (Earth-centered), as was Aristotle's. To account for the planets' retrograde motion, Ptolemy's model has the planets travelling along small circles that move on the larger circles of the planets' overall orbits. The small circles are called **epicycles**, and the larger circles are called **deferents** (■ Fig. 5–7). (See *A Closer Look 5.1: Ptolemaic Terms*.) The center of an epicycle moved with a constant angular speed relative to a point called the **equant** (Fig. 5–7). Since circles were thought to be "perfect" shapes, it seemed natural that planets should follow circles in their motion.

Simulation with the Zeiss ZKP3/B planetarium projector of Williams College's Hopkins Observatory by Jay M. Pasachoff, Yariv Pierce, and Megan Bruck Syal

■ **FIGURE 5–3** A planetarium simulation of the path of Mars over a nine-month period. In the middle we see a period of retrograde motion. Mars begins at the right (west) and proceeds toward the left (east) on this image. After it passes above the V-shaped Hyades star cluster (whose brightest star is Aldebaran), it goes into retrograde motion. Then it goes into prograde motion again. Because Mars and the Earth orbit in different planes, our perspective changes over time and the backward motion doesn't retrace the forward motion exactly; rather, it makes a loop in the sky.

■ **FIGURE 5–4** Aristotle (pointing down) and Plato (pointing up) in *The School of Athens* by Raphael (1483–1520), a fresco in the Vatican.

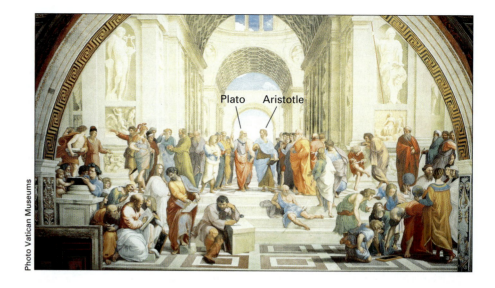

Photo Vatican Museums

Ptolemy's views were very influential in the study of astronomy for a long time. His tables of planetary motions, which were reasonably accurate considering how long ago he developed them, were accepted for nearly 15 centuries. His major work became known as the *Almagest (Greatest)*. It contained not only his ideas but also a summary of the ideas of his predecessors. Most of our knowledge of Greek astronomy comes from Ptolemy's *Almagest*, which in medieval times reappeared in Latin from its Arabic version, with a first printed version in 1515.

5.3 A HERETICAL IDEA: THE SUN-CENTERED UNIVERSE

We credit Nicolaus Copernicus, a 16th-century Polish astronomer, with our modern view of the Solar System and of the Universe beyond. (See *Lives in Science 5.1: Copernicus*.) Copernicus suggested a **heliocentric** (Sun-centered) theory (■ Figs. 5–8, 5–9, and 5–10). He explained the retrograde motion with the Sun rather than the Earth at the center of the Solar System.

Aristarchus of Samos, a Greek scientist, had suggested a heliocentric theory 18 centuries earlier. We do not know, though, how

from Dopplemyer (1742) by Jay M. Pasachoff

■ **FIGURE 5–5** Aristotle's cosmological system, with water and Earth at the center, surrounded by air, fire, the Moon, Mercury, Venus, the Sun, Mars, Jupiter, Saturn, and the firmament of fixed stars.

A CLOSER LOOK 5.1   Ptolemaic Terms

deferent – A large circle centered approximately on the Earth, actually centered midway between the Earth and the equant (see below).

epicycle – A small circle whose center moves along a deferent. The planets move on the epicycles.

equant – The point around which the epicycle moves at a uniform angular rate.

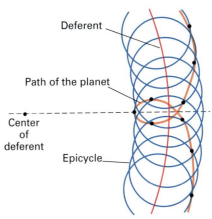

FIGURE 5–6 In the Ptolemaic system, a planet would move along an epicycle. The epicycle, in turn, moved on a deferent. The observed changing speed of planets in the sky would mean that the center of the epicycle did not move at a constant speed around the center of the deferent. Rather, it moved at a constant angular speed (that is, an angle changing with time at a constant rate) as seen from a point called the equant, which was not quite at the center of the deferent.

FIGURE 5–7 In the Ptolemaic system, the projected path in the sky of a planet in retrograde motion is shown.

detailed a picture he presented. His ancient heliocentric suggestion required the counterintuitive notion that the Earth itself moved. This idea seems contrary to our senses, and it contradicts the theories of Aristotle. If the Earth is either rotating or moving through space, for example, why aren't birds and clouds "left behind"? Only the 17th-century discovery by Isaac Newton of laws of motion and gravitation that differed substantially from Aristotle's solved this dilemma, as we shall see later in this chapter.

Copernicus's heliocentric theory, first conceived around 1510 CE though we don't know why or how, still assumed that the planets moved in circles, though the circles were not quite centered

FIGURE 5–8 The pages of Copernicus's original manuscript in which he drew his heliocentric system.

FIGURE 5–9 Copernicus's heliocentric diagram as printed in his book *De Revolutionibus* (1543). The printed version was not an exact copy of the hand-drawn one in the manuscript.

FIGURE 5–10 The first English diagram of the Copernican system, which appeared in 1576.

LIVES IN SCIENCE 5.1

Copernicus

Nicolaus Copernicus was born in 1473. Around 1505–1510, he formulated his heliocentric ideas. He wrote a lengthy manuscript, and continued to develop his model even while carrying out official duties in Frauenberg (now Frombork, Poland).

He published his book *De Revolutionibus* (*Concerning the Revolutions*) in 1543. Though it is not certain why he did not publish his book earlier, his reasons for withholding the manuscript from publication probably included a general desire to perfect the manuscript and his realization that to publish the book would involve him in controversy. (The 2011 book by Dava Sobel, *A More Perfect Heaven*, describes in prose and in a play how the younger scholar Rheticus drew the book out of Copernicus.)

When the book appeared, an unsigned preface by a theologian who had been drafted by the printer to proofread the book incorrectly implied that Copernicus disavowed any true belief in the physical reality of his hypothesis. The preface implied that Copernicus thought it was only a means for computation. Copernicus did not receive a full copy of his book until the day of his death.

Although some people, including Martin Luther, noticed a contradiction between the Copernican theory and the Bible, little controversy occurred at that time. In particular, the Pope – to whom the book had been dedicated – did not object.

The Harvard-Smithsonian astronomer and historian of science Owen Gingerich has, by valiant searching, located over 300 extant copies of the first edition of Copernicus's masterpiece and over 300 copies of the second edition. By studying which were censored and what notes were handwritten into the book by readers of the time, he has been able to study the spread of belief in Copernicanism.

Copernicus was known to have been buried in the cathedral at Frauenberg, and excavations starting in 2004 turned up bones likely to be his. In 2010, he was reburied with ceremony and with a new plaque.

FIGURE LIS 5.1 Copernicus, in a painting hanging in a museum in Torun, Poland.

on the Sun. The notion that celestial bodies had to follow such "perfect" shapes (circles) shows that Copernicus had not broken entirely away from the old ideas; the observations that would later force scientists to do so were still over 100 years in the future. Copernicus's manuscript had the mechanisms that drove the planets housed in zones. The exact notion of "orbits" did not yet exist.

Copernicus still invoked the presence of some minor epicycles in order to have his predictions agree better with observations. Copernicus was proud, though, that by using circular orbits not quite centered on the Sun, he had eliminated the equant. In any case, the detailed predictions that Copernicus himself computed were not in much better agreement with the existing observations than tables based on Ptolemy's model. The heliocentric theory appealed to Copernicus mostly on philosophical rather than on observational grounds. Scientists did not then have the standards we now have for how to compare observations with theory.

Copernicus published his theory in 1543 in the book he called *De Revolutionibus* (*Concerning the Revolutions*). The model explained the retrograde motion of the planets in terms of a projection effect, as follows (■ Fig. 5–11).

Let us consider, first, a planet that is farther from the Sun than the Earth is. Such planets travel more slowly than the Earth; orbital speed decreases with increasing distance from the Sun. Using Mars as an example, Figure 5–11 shows that when Earth is roughly approaching Mars, the projection of the Earth-Mars line outward to the distant, apparently motionless stars moves in the forward, prograde direction. But notice what happens when the Earth gets close to Mars and starts passing it; the projection of the Earth-Mars line now moves backward (retrograde) compared with the way it had previously been moving. Then, as the Earth and Mars continue around their orbits, the projection of the Earth-Mars line appears to go forward (prograde) again. A similar analysis holds for Mercury and Venus, which orbit the Sun at smaller distances than does the Earth.

The idea that the Sun was at the approximate center of the Solar System led Copernicus to two additional important results. First, he was able to work out the relative distances to the planets. Second, he was able to derive the length of time the planets take to orbit the Sun from observations of how long they take between appearances in the same place in our sky. His ability to derive these results played a large part in persuading Copernicus that his heliocentric system was better than the geocentric, Ptolemaic model.

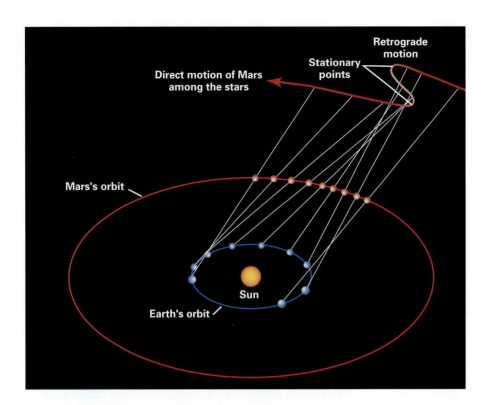

■ **FIGURE 5–11** The Copernican theory explains retrograde motion as an effect of projection. For each of the nine positions of Mars shown from right to left on the red line, follow the white line from Earth's position through Mars's position to see the projection of Mars against the sky (relative to distant background stars). Mars's forward motion appears to slow down as the Earth overtakes it. Between the two "stationary points," Mars appears in retrograde motion; that is, it appears to move backward with respect to the stars. A similar argument works for planets closer to the Sun than the Earth.

5.4 THE KEEN EYES OF TYCHO BRAHE

In the last part of the 16th century, not long after Copernicus's death, Tycho Brahe (see *Lives in Science 5.2: Tycho Brahe*) began to observe Mars and the other planets to improve quantitative predictions of the positions of planets. Tycho, a Danish nobleman, set up an observatory on an island off the mainland of Denmark (■ Fig. 5–12). The first building there was called Uraniborg (after Urania, the muse of astronomy). You can still see its foundations when visiting this now-Swedish island.

Though the telescope had not yet been invented, Tycho used giant instruments to make observations that were unprecedented in their accuracy. In 1597, Tycho lost his financial support in Denmark, and he arrived in Prague two years later. There, Johannes Kepler came to work with him as a young assistant. (See *Lives in Science 5.3: Johannes Kepler.*) At Tycho's death in 1601, after some battles for access, Kepler was left to analyze all the observations that Tycho and his assistants had made (■ Fig. 5–13).

5.5 JOHANNES KEPLER AND HIS LAWS OF ORBITS

Johannes Kepler studied with one of the first professors to believe in the Copernican view of the Universe. Kepler came to agree, and

■ **FIGURE 5–12** Tycho's observatory at Uraniborg, Denmark. In this image from one of Tycho's books, *Astronomiae Instauratae Mechanica* (1602), a portrait of Tycho was drawn into the mural quadrant (a marked quarter-circle on a wall) that he used to measure the altitudes at which stars and planets crossed the meridian (their highest point in the sky, for all but the most northerly stars).

LIVES IN SCIENCE 5.2

Tycho Brahe

Tycho Brahe was born in 1546 to a Danish noble family. As a child he was taken away and raised by a wealthy uncle. In 1560, a total solar eclipse was visible in Portugal, and the young Tycho witnessed the partial phases in Denmark. Though the event itself was not visually spectacular – partial eclipses rarely are – Tycho, at the age of 14, was so struck by the ability of astronomers to predict the event that he devoted his life from then on to making an accurate body of observations.

When Tycho was 20, he dueled with swords with a fellow student over which of them was the better mathematician. During the duel, part of his nose was cut off. For the rest of his life, he wore a metal replacement and was frequently rubbing the remainder with ointment. His bones were exhumed in 2010 by Danish and Czech archaeologists and anthropologists, and a surprise finding was that his artificial nose was made of brass rather than of gold or silver, as had long been thought. Portraits made during his life and the relief on his tomb (see figure) show a line across his nose, though just how much was actually cut off is not now definitely known.

In 1572, Tycho was astounded to discover an apparently new star in the sky, so bright that it outshone Venus. It was what we now call a "supernova"; indeed, we call it "Tycho's supernova." It was the explosion of a star, and remained visible in the sky for 18 months. As Europe was emerging from the Dark Ages, Tycho's observations provided important evidence that the heavens were not immutable. He published a book about the supernova, and his fame spread. In 1576, the king of Denmark offered to set up Tycho on the island of Hveen with funds to build a major observatory, as well as various other grants.

Unfortunately for Tycho, a new king came into power in Denmark in 1588, and Tycho's influence waned. Tycho had always been an argumentative and egotistical fellow, and he fell out of favor in the countryside and in the court. Finally, in 1597, his financial support cut, he left Denmark. Two years later he settled in Prague, at the invitation of the Holy Roman Emperor, Rudolph II.

In 1601, Tycho attended a dinner where etiquette prevented anyone from leaving the table before the baron (or at least so Tycho thought). The guests drank freely and Tycho wound up with a urinary tract infection that probably contributed to his death 11 days later. Though it had been suggested that he might have died of an overdose of mercury, maybe even administered by Kepler, the 2010 analysis of his beard showed that he had consumed only low levels of mercury over the 5 years or so before his death, ruling out mercury poisoning.

(a)

Jay M. Pasachoff

(b)

Jay M. Pasachoff

(c)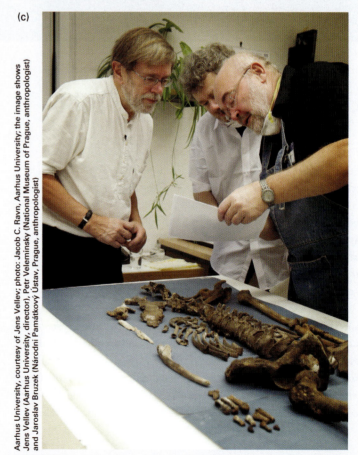

Aarhus University, courtesy of Jens Vellev; photo: Jacob C. Ravn, Aarhus University; the image shows Jens Vellev (Aarhus University, director), Petr Veleminsky (National Museum of Prague, anthropologist) and Jaroslav Bruzek (Národní Památkový Ústav, Prague, anthropologist)

FIGURE LIS 5.2 (a) Tombstone of Tycho Brahe. (b) Tycho, in 1596. (c) Tycho, in 2010, as his bones were inspected by Danish and Czech archaeologists and anthropologists, in part to see if he was poisoned and to use DNA and other modern medical methods. He has since been reburied in Prague.

Jay M. Pasachoff

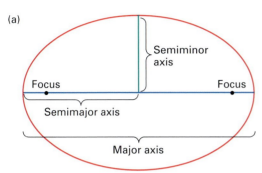

FIGURE 5–13 The plaque over Johannes Kepler's house in Prague.

(a)

(b)

Jay M. Pasachoff

FIGURE 5–14 (a) The parts of an ellipse, a closed curve. (right) The ellipse shown has the same perihelion distance (closest approach to the Sun) as does the circle. Its eccentricity, the distance between the foci divided by the major axis, is 0.8. If the perihelion distance remains constant but the eccentricity reaches 1, then the curve is a parabola, whose ends do not join. Curves for eccentricities greater than 1 are hyperbolas (which, like parabolas, are open curves). (b) It is easy to draw an ellipse. Put two thumbtacks in a piece of cardboard, and link them with a loop of string that has some slack in it. If you pull a pen around while the pen keeps the string taut, the pen will necessarily trace out an ellipse. It does so since the string doesn't change in length; the sum of the lengths of the lines from the point to the two foci remains constant, and this is the defining property of an ellipse.

made some mathematical calculations involving geometrical shapes (■ see Chapter Opener). His ideas were generally wrong, but the quality of his mathematical skill had impressed Tycho before Kepler arrived in Prague.

Tycho's new, more thorough and precise observational data showed that the tables of the positions of the planets then in use were not very accurate. When Kepler joined Tycho, he carried out detailed calculations to explain the planetary positions. In the years after Tycho's death, Kepler succeeded in explaining, first, the orbit of Mars. But he could do so only by dropping the idea that the planets orbited in circles, the perfect shape. (See *A Closer Look 5.2: Kepler's Laws.*)

5.5a Kepler's First Law

Kepler's first law, published in 1609, says that *the planets orbit the Sun in ellipses, with the Sun at one focus*. An ellipse is defined in the following way: Choose any two points on a plane; these points are called the **foci** (each is a **focus**). From any point on the ellipse, we can draw two lines, one to each focus. The sum of the lengths of these two lines is the same for each point on the ellipse.

The **major axis** of an ellipse is the line within the ellipse that passes through the two foci, or the length of that line (■ Fig. 5–14a). The **minor axis** is the line perpendicular to the center of the major axis, or its length. The **semimajor axis** is half the length of the major axis, and the **semiminor axis** is half the length of the minor axis. A circle is the special case of an ellipse where the two foci are in the same place (the center of the circle).

It is easy to draw an ellipse (■ Fig. 5–14b). A given spacing of the foci and a given length of string define each ellipse. The shape of the ellipse will change if you change the length of the string or the distance between the foci.

According to Kepler's first law, the Sun is at one focus of the elliptical orbit of each planet. What is at the other focus? Nothing special; we say that it is "empty."

5.5b Kepler's Second Law

Kepler's second law describes the speed with which the planets travel in their orbits. It states that *the line joining the Sun and a planet sweeps through equal areas in equal times*. It is thus also known as the **law of equal areas**.

A CLOSER LOOK 5.2

To this day, we consider Kepler's laws to be the basic description of the motions of objects in our Solar System and others, and even for stars orbiting in galaxies such as one in a ten-year orbit around our galaxy's central supermassive black hole (Chapter 15):

1. The planets orbit the Sun in ellipses, with the Sun at one focus.

2. The line joining the Sun and a planet sweeps through equal areas in equal times.

3. The square of the orbital period of a planet is proportional to the cube of the semimajor axis of its orbit – half the longest dimension of the ellipse. (Loosely: the squares of the periods are proportional to the cubes of the planets' average distances from the Sun.)

When a planet is at its greatest distance from the Sun as it follows an elliptical orbit, the line joining it with the Sun sweeps out a long, skinny sector. This sector has two straight lines starting at a focus and extending to the ellipse. The curved part of the ellipse is relatively short for a planet at its great distance from the Sun. On the other hand, according to Kepler's second law, the area of this long, skinny sector must be the same as that of any short, fat sector formed for the planet nearer to **perihelion**, the closest point to the Sun in a planet's orbit (■ Fig. 5–15). Thus, the planet moves most quickly when it is near the Sun. (The farthest point from the Sun in a planet's orbit is called "aphelion," in contrast to perihelion.)

Since the time of the Greeks, the idea that the planets travelled at a constant rate had been thought to be important. Kepler's second law replaced this old notion with the idea that the total area swept out changes at a constant rate.

Kepler's second law is especially noticeable for comets, which have very eccentric (that is, flattened) elliptical orbits. For example,

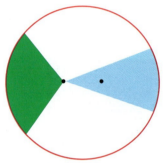

■ **FIGURE 5–15** Kepler's second law states that the two shaded sectors are equal in area. They represent the areas covered by a line drawn from the Sun, which is at a focus of the ellipse, to an orbiting planet in a given length of time.

LIVES IN SCIENCE 5.3

Shortly after Tycho Brahe arrived in Prague, he was joined by a mathematically inclined assistant, Johannes Kepler. Kepler was born in 1571 in Weil der Stadt, near what is now Stuttgart, Germany. He was from a poor family and had gone to school on scholarships.

In 1604, Kepler observed a supernova and soon wrote a book about it. It was the second exploding star to appear in 32 years, after one seen and written about by Tycho. It was also seen by Galileo, but Kepler's study has led us to call it "Kepler's supernova." We have not seen another bright supernova in our Milky Way Galaxy since then.

By the time of the supernova, Kepler had discovered what we now call his second law through analysis of Tycho's data. He was soon to find his first law. It took four additional years, though, to arrange and finance publication. Much later on, he developed his third law.

The abdication of Kepler's patron, the Emperor, led Kepler to leave Prague and move to Linz. Much of his time had to be devoted to financing the printing of the planetary tables and to his other expenses – and to defending his mother against the charge that she was a witch. He died in 1630, while travelling to collect an old debt.

FIGURE LIS 5.3 Prague statue of Tycho Brahe and Johannes Kepler; sculptor: Josef Vajce, 1983.

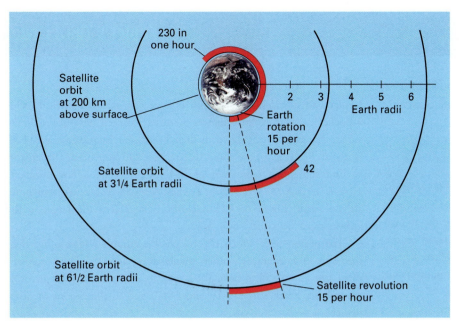

■ **FIGURE 5–16** Most satellites, including the recently decommissioned space shuttles, orbit only a few hundred kilometers above the Earth's surface. They orbit Earth in about 90 minutes. From Kepler's third law, we can see that the speed in orbit of a satellite decreases as the satellite gets higher; see *Figure It Out 5.3: Orbital Speed of Planets*. But distances are measured from the center of the Earth, so even a doubling of the orbital height above the Earth's surface is only a slight increase in the orbit's radius. When a satellite is as high as 6 1/2 Earth radii from Earth's center, the satellite orbits in 24 hours. This 24-hour period is the same as the period with which a point on the Earth's surface below rotates, so the satellite appears from the ground to hover in one place. It is then called a "geostationary satellite."

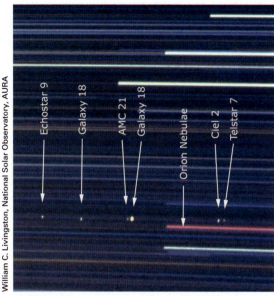

■ **FIGURE 5–17** In this time exposure, the stars trail but the geostationary satellites that hover over the Earth's equator, many delivering TV or telephone signals, appear as points. In 2013, an asteroid came within the orbits of the geostationary satellites and passed safely by (see *A Closer Look 8.3: February 15, 2013 – An Exploding Meteor; A Nearby Asteroid*).

Kepler's second law demonstrates why Halley's Comet sweeps so quickly through the inner part of the Solar System. It moves much more slowly when it is farther from the Sun, since the sector swept out by the line joining it to the Sun is skinny but long.

5.5c Kepler's Third Law

Kepler's third law deals with the length of time a planet takes to orbit the Sun, which is its **period** of revolution. Kepler's third law relates the period to some measure of the planet's distance from the Sun. Specifically, it states that *the square of the period of revolution is proportional to the cube of the semimajor axis of the ellipse.* (Mathematically, $P^2 = kR^3$, where k is a constant; or, $P^2 \propto R^3$, where "$\propto$" means "proportional to.") That is, if the cube of the semimajor axis of the ellipse goes up, the square of the period goes up by the same factor (see *Figure It Out 5.1: Kepler's Third Law*).

A terrestrial application of Kepler's third law is in "geostationary satellites," which are so high that they orbit Earth once a day as Earth rotates at the same rate underneath (■ Fig. 5–16). They thus appear to hover over the equator (■ Fig. 5–17), and are used for relaying television and telephone signals.

5.6 THE DEMISE OF THE PTOLEMAIC MODEL: GALILEO GALILEI

Galileo Galilei flourished in what is now Italy at the same time that Kepler was working farther north. Galileo began to believe in the Copernican heliocentric system in the 1590s, and identified some of our basic laws of physics in the following years. In late 1609 or early 1610, Galileo was the first to use a telescope for systematic astronomical studies (■ Fig. 5–18).

In 1610, he reported in a book that with his telescope he could see many more stars than he could detect with his unaided eye. Indeed, the Milky Way and certain other hazy-appearing regions of the sky actually contained numerous individual stars. He described views of the Moon, including the discovery of mountains, craters, and the relatively dark lunar "seas"; see *Star Party 5.2: Galileo's Observations*.

Galileo's discovery that small bodies revolved around Jupiter (■ Fig. 5–19) was very important because it proved that not all celestial objects revolved around the Earth. Moreover, the moons were obviously not "left behind" while Jupiter moved across the sky, suggesting that the Earth might move as well without leaving objects behind.

FIGURE IT OUT 5.1 — Kepler's Third Law

It is often easiest to express Kepler's third law, $P^2 = kR^3$ (where P is the period of revolution, R is the semimajor axis, and k is a constant), by relating values for a planet to values for the Earth:

$$\frac{P_{\text{Planet}}^2}{P_{\text{Earth}}^2} = \frac{kR_{\text{Planet}}^3}{kR_{\text{Earth}}^3} = \frac{R_{\text{Planet}}^3}{R_{\text{Earth}}^3}$$

We can choose to work in units that are convenient for us here on Earth. We use the term **1 Astronomical Unit (1 au)** to mean the semimajor axis of the Earth's orbit. This semimajor axis can be shown to be the average distance from the Sun to the Earth. We use a unit of time based on the length of time the Earth takes to orbit the Sun: **1 year**. When we use these values, Kepler's law appears in a simple form, since the numbers on the bottom of the equation are just 1, and we have

$$P_{\text{Planet}}^2 = R_{\text{Planet}}^3 \text{ (distance in au; period in Earth years)}.$$

Example: We know from observation that Jupiter takes 11.86 years to revolve around the Sun. What is Jupiter's average distance from the Sun?

Answer: $(11.86)^2 = R_{\text{Jupiter}}^3$, so $R_{\text{Jupiter}} = 5.2$ au. Kepler had to calculate by hand, but we can calculate R easily with a pocket calculator. Astronomers are often content with approximate values that can be calculated in their heads. For example, since 11.86 is about 12, $(11.86)^2$ is about $12^2 = 144$. Do you need a calculator to find the rough cube root of 144? No. Just try a few numbers. $1^3 = 1$, $2^3 = 8$, $3^3 = 9 \times 3 = 27$, and $4^3 = 16 \times 4 = 64$, which are all too small. But $5^3 = 125$ is closer. $6^3 = 216$ is too large, so the answer must be a little over 5 au.

The period with which objects revolve around bodies other than the Sun (for example, other planetary systems) follows Kepler's third law as well. The laws also apply to artificial satellites in orbit around the Earth (Figs. 5–16 and 5–17) and to the moons of other planets.

■ **FIGURE 5–18** Two of Galileo's telescopes.

By displaying facts that Aristotle and Ptolemy obviously had not known, Galileo showed that Aristotle and Ptolemy had not been omniscient, and opened the idea that more remained to be discovered. But not all of Galileo's contemporaries accepted that what was seen through the newfangled telescope was real, or were even willing to look through this contraption at all.

Critical to the support of the heliocentric model was Galileo's discovery that Venus went through an *entire series of phases* (■ Fig. 5–20). Note that it is the fact that it is an entire series of phases (especially the gibbous and full phases), and not just phases

On the 7th of January Jupiter is seen thus

On the 8th thus   it was therefore direct and not retrograde

On the 12th day it is seen in this arrangement

The 13th are seen very close to Jupiter 4 stars   or better so

On the 14th it is cloudy

The 15th   the nearest to Jupiter was smallest the 4th was distance from the 3rd about double

The spacing of the 3 to the west was no greater than the diameter of Jupiter and they were in a straight line.

long. 7138' lat. 113'

■ **FIGURE 5–19** A translation (*right*) of Galileo's original notes (*left*) summarizing his first observations of Jupiter's moons in 1610. The highlighted areas were probably added later. It had not yet occurred to Galileo that the objects were moons in revolution around Jupiter.

Galileo Galilei was born in Pisa, in northern Italy, in 1564. In the 1590s, he became a professor at Padua, in the Venetian Republic. At some point in this decade, he adopted Copernicus's heliocentric theory. After hearing that a device to magnify far-off objects had been made, he figured out the principles himself and constructed his own version. In late 1609 or early 1610, Galileo was among the first to turn a telescope toward the sky.

In 1613, Galileo published his acceptance of the Copernican theory. The book in which he did so was received and acknowledged with thanks by the Cardinal who, significantly, was Pope at the time of Galileo's trial. Just how Galileo got out of favor with the Church has been the subject of much study. His bitter quarrel with a Jesuit astronomer, Christopher Scheiner, over who deserved the priority for the discovery of sunspots did not put him in good stead with the Jesuits. Still, the disagreement on Copernicanism was more fundamental than personal, and the Church's evolving views against the Copernican system were undoubtedly more important than personal feelings. Also, the Church apparently objected to his belief that matter consists of atoms.

In 1615, Galileo was denounced to the Inquisition, and was warned against teaching the Copernican theory. There has been much discussion of how serious the warning was and how seriously Galileo took it.

Over the following years, Galileo was certainly not outspoken in his Copernican beliefs, though he also did not cease to hold them. In 1629, he completed his major manuscript, *Dialogue on the Two Great World Systems*, these systems being the Ptolemaic and Copernican. With some difficulty, clearance from the censors was obtained, and the book was published in 1632. It was written in contemporary Italian, instead of the traditional Latin, and could thus reach a wide audience. The book was condemned, and Galileo was again called before the Inquisition.

Galileo's trial provided high drama, and has indeed been transformed into drama on the stage. He was convicted, and sentenced to house arrest. He lived nine more years, until 1642, his eyesight and his health failing. He devoted his time to writing *Two New Sciences*, an excellent book that describes the experimental studies he had conducted about four decades earlier on the motions of falling bodies and on the strength of materials. Indeed, most of the principles in Newton's subsequent laws of motion are based on experimental facts determined by Galileo and documented in this book. In 1992, Pope John Paul II acknowledged that Galileo had been correct to favor the Copernican theory, an implicit though not an explicit pardon.

from Galileo's *Macchie Solare*, 1613; collection of Jay M. Pasachoff

FIGURE LIS 5.4 Galileo.

in general, that is significant: The non-crescent phases could not be explained with the Ptolemaic system (■ Fig. 5–21). After all, if Venus travelled on an epicycle located between the Earth and the Sun, it should always appear as a crescent (or "new"). Thus, the complete set of phases of Venus provided the fatal blow to the Ptolemaic model. (Tycho proposed a hybrid model where all the other planets orbit the Sun, which in turn orbits the Earth, thus preserving some aspects of the geocentric hypothesis. Though consistent with the observed phases of Venus, and accepted at the time by quite a few people, such a model seems cumbersome, and it still violates the main ideas of Aristotle and Ptolemy.)

Moreover, in 1612, Galileo used his telescope to project an image of the Sun. On the images, he discovered sunspots, additional evidence that celestial objects were not perfect (■ Fig. 5–22). (A couple of others vie for the credit for discovering sunspots, a battle still not resolved.) In his book on sunspots, Galileo drew a series of images showing how sunspots looked from day to day as they rotated with the Sun's surface.

In *Lives in Science 5.4: Galileo Galilei*, we discuss Galileo's controversy with the Roman Catholic Church, following the publication in Italian (rather than Latin) of his *Dialogo* (*Dialogue between the Two World Systems*), after which he was sentenced to house arrest (■ Fig. 5–23). The controversy echoes down to today, and some of the truth about the real reasons behind the fight may still be hidden in the Vatican archives. But by now, 400 years since Galileo made his discoveries and more than 400 years since his near-contemporary Giordano Bruno was burned at the stake at least in part because of his view that there were worlds beyond our Solar System, peace generally reigns between the Church and scientists. Indeed, the Vatican supports a modern observatory in Arizona as well as research facilities in and near Rome, and it has several respected contemporary astronomers on its staff.

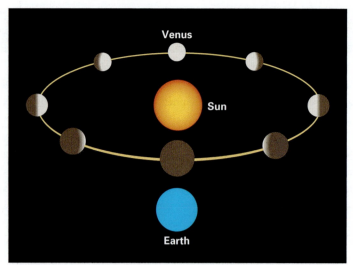

■ **FIGURE 5–20** When Venus is on the far side of the Sun as seen from the Earth, it is more than half illuminated. We thus see it go through a whole cycle of phases, from thin crescent (or even "new") through full. Venus thus matches the predictions of the Copernican system.

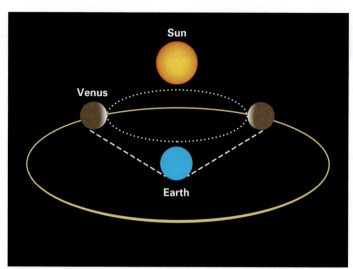

■ **FIGURE 5–21** In the Ptolemaic theory, Venus and the Sun both orbit the Earth. However, because Venus is never seen far from the Sun in the sky, Venus's epicycle (which is shown dotted) must be restricted to always fall on the line joining the center of the deferent (which is near the Earth) and the Sun. In this diagram, Venus could never get farther from the Sun than the dashed lines. It would therefore always appear as a crescent.

from Galileo's *Macchie Solare*, 1613; collection of Jay M. Pasachoff

■ **FIGURE 5–22** Sunspots observed by Galileo in 1612, as engraved in his book of the following year.

5.7 ON THE SHOULDERS OF GIANTS: ISAAC NEWTON

Kepler discovered his laws of planetary orbits by trial and error; they were purely empirical, and he had no physical understanding of them. Only with the work of Isaac Newton 60 years later did we find out why the laws existed.

Newton was born in England in 1642, the year of Galileo's death. He became the greatest scientist of his time and perhaps of all time. (Albert Einstein and James Clerk Maxwell are the two physical scientists often ranked with Newton.) His work on optics, his invention of the reflecting telescope, and his discovery that visible sunlight can be broken down into a spectrum of colors (see Chapter 2) and again

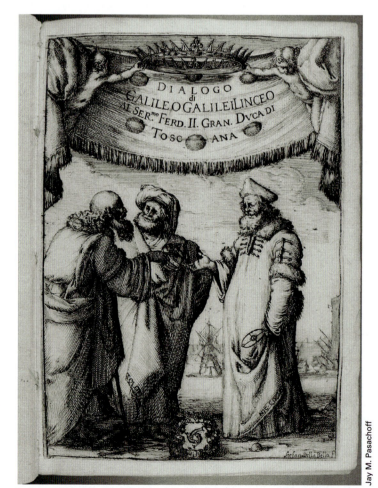

Jay M. Pasachoff

■ **FIGURE 5–23** The frontispiece of Galileo's *Dialogue on the Two Great World Systems*, published in 1632. According to the labels, Copernicus is to the right, with Aristotle and Ptolemy to the left; Copernicus was drawn with Galileo's face, however.

With a small telescope, it is easy to recreate many of Galileo's observations. Try these with the telescope on your campus.

1. Observe Venus over a period of several months. This is most easily done when it is visible in the evening, but at some times of the year the morning is more suitable. You should be able to see it go through different phases, as the Moon does.
2. Observe Jupiter over the course of several nights. You should be able to see its four bright moons, the "Galilean satellites." Their relative positions change daily, or even over the course of a few hours. If you draw a picture each night for a few weeks, you'll even be able to determine their orbital periods and at least roughly verify that they obey Kepler's third law.
3. Observe the Moon. You should see craters and maria ("seas"), as did Galileo. The views will be best during the crescent, quarter, or somewhat gibbous phases, especially near the "terminator" (the border between the sunlit and dark parts of the Moon). During a

full moon, the features tend to appear washed out because of the lack of shadows.

4. Point the telescope to a bright region of the Milky Way and notice that many stars are visible. Use the eyepiece that gives the widest field of view for this observation. Do you think Galileo was surprised to find that the Milky Way consists of many faint stars?
5. A small telescope can be used to view sunspots, as did Galileo. But do *not* look directly at the Sun through the telescope; it is dangerous to do so and may cause blindness. A better technique is to project the Sun's image onto a sheet of cardboard or some other clean surface. (Galileo also used this projection technique.) You can look directly through the telescope *only if* you are certain that you have the right kind of filter at the top end of the telescope, and that it is securely fastened. Never use a filter in the eyepiece of the telescope, since it will get very hot and can break, leading to possible blindness.

recombined into "white light" would merit him a place in astronomy texts. But still more important was his work on motion and on gravity.

Newton set modern physics on its feet by deriving laws showing how objects move on Earth and in space, and by finding the law that describes the force of gravity. In order to work out the law of gravity, Newton had to invent calculus, a new branch of mathematics! Newton long withheld publishing his results, until Edmond Halley (whose name we now associate with the famous comet) convinced him to publish his work. Newton's great book, *The Principia* (pronounced "Prin-ki´pee-a"), short for the Latin form of *Mathematical Principles of Natural Philosophy*, appeared in 1687.

The Principia contains Newton's three laws of motion. The first is that *bodies in motion tend to remain in motion, in a straight line with constant speed, unless acted upon by an external (that is, an outside) force.* It is a law of inertia, which was really discovered by Galileo.

Newton's second law relates a force with its effect on accelerating (speeding up) a mass. *A larger force will make the same mass accelerate faster (F = ma, where F is the force, m is the mass, and a is the acceleration).*

Newton's third law is often stated, *For every action, there is an equal and opposite reaction.* The flying of jet planes and rockets is only one of the many processes explained by this law.

The Principia also contains the law of gravity; see *A Closer Look 5.3: Newton's Law of Universal Gravitation.* One application of Newton's law of gravity is "weighing" the planets. (See *Figure It Out 5.2: Newton's Version of Kepler's Third Law.*)

One of the most popular tales of science is that an apple fell on Newton's head, leading to his discovery of the concept of gravity. Though no apple actually hit Newton's head, the story that Newton himself told, years later, is that he saw an apple fall and realized that just as the apple fell to Earth, the Moon is falling to Earth, though its forward motion keeps it far from us. (In any short time interval, the distance the Moon travels toward Earth gets compensated by the distance the Moon also travels forward; the result over many such time intervals is a stable orbit, rather than a collision course with Earth.)

When Newton calculated that the acceleration of the Moon fit the same formula as the acceleration of the apple, he knew he had the right method. Newton was the first to realize that gravity was a force that acted in the same way throughout the Universe. Many pieces of the original apple tree have been enshrined (■ Fig. 5–24). There is no knowing which – if any – of the apple trees now growing in front of Newton's house, still standing at Woolsthorpe, actually descended from the famous original.

Newton's most famous literary quotation is, "If I have seen so far, it is because I have stood on the shoulders of giants." In fact, a whole book has been devoted to all the other usages of this phrase, which turns out to have been in wide use before Newton and thereafter. (Actually, Newton's statement was not merely praising his predecessors; one of his scientific rivals, Robert Hooke, was very short.) In the past half century, the pace of science has been so fast that it has even been remarked, "Nowadays we are privileged to sit beside the giants on whose shoulders we stand."

A CLOSER LOOK 5.3

Newton's Law of Universal Gravitation

When Isaac Newton famously saw an apple fall, he masterfully deduced that the same force that makes the apple fall toward Earth similarly attracts the Moon toward Earth. He published the result in his *Philosophiae Naturalis Principia Mathematica* in 1687.

Newton realized that gravitation follows an inverse-square law – that is, the strength of the force is directly related to the inverse of ("one over") the square of the distance. The force is also directly proportional to each of the two masses involved, with no preference for one over the other. Thus the formula fits Newton's third law of motion, that for every action there is an equal and opposite reaction. Gravitational tugs are mutual – the Earth is pulling on your mass (giving you weight) just as strongly as your mass is pulling on the Earth.

Newton's law of gravitation shows that for two masses m_1 and m_2

$$F \propto \frac{m_1 m_2}{d^2}.$$

For Kepler's third law, the constant is different for each different central object; see *Figure It Out 5.2: Newton's Version of Kepler's Third Law*. But in Newton's law of gravitation, the constant of proportionality (G) is the same in all circumstances throughout the Universe: It is universal. Newton's law of gravitation, then, is

$$F = \frac{G m_1 m_2}{d^2}.$$

From this law, we see that if the same body were twice as far away, the force of gravitation on it from another body would be only one-fourth (one-half squared) as great. Furthermore, if the central body doubles in mass, its force of gravitation at a given distance is twice as great (not four times as great).

See *Lives in Science 5.5: Isaac Newton* for a brief summary of Newton's life.

FIGURE IT OUT 5.2

Newton's Version of Kepler's Third Law

Modern astronomers often make rough calculations to test whether physical processes under consideration could conceivably be valid. Astronomy has also had a long tradition of exceedingly accurate calculations. Pushing accuracy to one more decimal place sometimes leads to important results.

For example, Kepler's third law, in its original form – the orbital period of a planet squared is proportional to its distance from the Sun cubed ($P^2 = \text{constant} \times R^3$) – holds to a reasonably high degree of accuracy and seemed completely accurate when Kepler did his work. But now we have more accurate observations, and they verify Newton's version.

Newton derived the formula for Kepler's third law, taking account of his own formula for the law of universal gravitation. He found that

$$P^2 = [4\pi^2/G(m_1 + m_2)] \, R^3,$$

where m_1 and m_2 are the masses of the two bodies. For planetary orbits, if m_1 is the mass of the Sun, then m_2 is the mass of the planet.

Newton's formulation of Kepler's third law shows that the constant of proportionality (the constant number by which you multiply the variable on the right-hand side of the equation to find the result given as the variable on the left-hand side) is determined by the mass of the Sun. Thus, we can use the formula to find out the Sun's mass.

Newton's version of Kepler's third law, as applied to any planet's orbit around the Sun, also shows that the planet's mass contributes to the value of the proportionality constant, but its effect is very small because the Sun is so much more massive. (The effect can be detected even today only for the most massive planets.) Newton therefore actually predicted that the constant in Kepler's third law depends slightly on the mass of the planet, being different for different planets. The observational confirmation of Newton's prediction was a great triumph.

Kepler's third law, and its subsequent generalization by Newton, applies not only to planets orbiting the Sun but also to any bodies orbiting other bodies under the control of gravity. Thus it also applies to satellites orbiting planets. We determine the mass of the Earth by studying the orbit of our Moon, and we determine the mass of Jupiter by studying the orbits of its moons.

For many decades, we were unable to reliably determine the mass of Pluto because we could not observe a moon in orbit around it. The discovery of a moon of Pluto in 1978 finally allowed us to determine Pluto's mass, and we found that our previous best estimates (based on Pluto's now-discredited gravitational effects on Uranus) were way off. The same formula can be applied to binary stars to find their masses.

In recent years, adaptive-optics techniques have allowed infrared observations of stars orbiting within light-days of the center of our Galaxy. Using Newton's version of Kepler's third law, two groups of scientists have measured that our Galaxy's center contains about 4 million times the mass of our Sun, as we shall discuss in Chapter 15.

Jay M. Pasachoff

■ **FIGURE 5–24** A piece of Newton's apple tree, now in the Royal Astronomical Society, London. The donor's father was present in about 1818 when some logs were sawed from this famous apple tree, which had blown over.

5.8 CLUES TO THE FORMATION OF OUR SOLAR SYSTEM

The scientists just discussed provided the basis of our current understanding of our Solar System, which in their time was essentially equivalent to the Universe. Studying the orbits of the planets shows that the Solar System has some basic properties. All the planets, for example, move in the same direction around the Sun and are more or less in the same plane.

Considering these regularities gives important clues about how the Solar System may have formed, a topic we will discuss more thoroughly in Chapter 9. In studying the other planetary systems now being discovered at a rapid pace, we are eager to determine how many things about our own Solar System are truly fundamental, and how many may be specific to our case. As we will learn, some of the ideas we thought were basic turned out not to apply to many of the other planetary systems we are finding.

As we have seen, the motion of a planet around the Sun in its orbit is its **revolution**. The spinning of a planet on its own axis is its **rotation**. The Earth, for example, revolves around the Sun in one year and rotates on its axis in one day. (These motions define the terms "year" and "day.") According to Kepler's third law, planets far from the Sun have longer periods of revolution (orbital periods) than planets close to the Sun. The distant planets also move more slowly (see *Figure It Out 5.3: Orbital Speeds of Planets*).

The orbits of the planets take up a disk rather than a full sphere (■ Fig. 5–25). The **inclination** of a planet's orbit is the angle that the plane of its orbit makes with the plane of the Earth's orbit. None of the planets' orbits is inclined more than about 7°.

The fact that the planets all orbit the Sun in roughly the same plane is one of the most important generalities about the Solar System. Spinning or rotating objects have a property called angular momentum. This angular momentum is larger if an object is more massive, if it is spinning faster, or if its mass distribution is farther from the axis of spin.

LIVES IN SCIENCE 5.5   Isaac Newton

Isaac Newton was born in Woolsthorpe, Lincolnshire, England, on December 25, 1642. Even now, over 370 years later, scientists still refer regularly to "Newton's laws of motion" and to "Newton's law of gravitation."

Newton's most intellectually fertile years were those right after his graduation from college, when he returned home to the country because fear of the plague shut down many cities. There, he developed his ideas about the nature of motion and about gravitation. In order to derive them mathematically, he invented calculus. Newton long withheld publishing his results, possibly out of shyness. Finally Edmond Halley – who became known by applying Newton's law of gravity to show that a series of reports of some bright comets all really referred to this single comet that we now know as Halley's Comet – persuaded Newton to publish his results. A few years later, in 1687, the *Philosophiae Naturalis Principia Mathematica* (*Mathematical Principles of Natural Philosophy*), known as *The Principia*, was published with Halley's practical and financial assistance. In it, Newton showed that the motions of the planets and comets could all be explained by the same law of gravitation that governed bodies on Earth. In fact, he derived and generalized Kepler's laws on theoretical grounds.

Newton was professor of mathematics at Cambridge University and later in life went into government service as Master of the Mint

in London. He was knighted in 1705 (primarily for his governmental rather than his scientific contributions) and lived until 1727. His tomb in Westminster Abbey bears the epitaph: "Mortals, congratulate yourselves that so great a man has lived for the honor of the human race."

From Newton's *Principia Mathematica*, 3rd ed., 1726; collection of David Park

FIGURE LIS 5.5 Isaac Newton in 1702. He was knighted three years later, although mostly for his public work rather than his science.

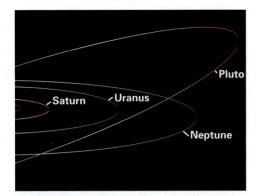

■ **FIGURE 5–25** The orbits of the planets have only small inclinations to the plane of the Earth's orbit. The orbits are nearly circular, but look more extended and obviously elliptical here because of the inclined perspective. The fact that Pluto's orbit is much more inclined than those of the other planets was an early hint that Pluto is not a genuine planet; in 2006 it was placed in a separate category, "dwarf planet" (a "plutoid," most recently and specifically).

■ **FIGURE 5–26** Ptolemy, Copernicus, Kepler, and Tycho, on the frontispiece of the 1742 book by Dopplemyer that included many astronomical diagrams.

As will be discussed more extensively in Chapter 9, one important property of angular momentum is that the total angular momentum of a system doesn't change (we say that it is "conserved"), unless the system interacts with something from outside that brings in or takes up **angular momentum**. (For example, as a spinning ice skater pulls her arms closer to her torso, her rate of spin increases because of the conservation of angular momentum.) Thus, if one body were to spin more and more slowly, or were to begin spinning in the opposite direction, the angular momentum would have to be taken up by another, external body.

The planets undoubtedly all revolve around the Sun in the same direction because of how they were formed out of a large spinning disk of gas and dust. They have angular momentum associated with them, and this must have also been the case for the material from which they coalesced. These concepts will be useful when we consider the hundreds of systems of planets newly discovered around other stars (Chapter 9). After all, it is always better to have more than one example of some concept.

FIGURE IT OUT 5.3 — Orbital Speed of Planets

The more distant a planet is from the Sun, the slower is its orbital speed. We can derive this from Kepler's third law. Since most of the planets have nearly circular orbits, a reasonable approximation for our purposes here is to assume that the distance travelled by a planet in one full orbit is $2\pi R$, the circumference of a circle of radius R.

If the planet travels at a constant speed (which it would if the orbit were exactly circular), the distance travelled would equal its speed multiplied by time: $d = vt$. Applying this to one full orbit, we find that $2\pi R = vP$, where P is the planet's orbital period. Thus,
$$P = 2\pi R/v, \text{ so } P \propto R/v,$$
where the symbol "$\propto$" means "proportional to."

But Kepler's third law states that $P^2 \propto R^3$. When we substitute the expression $P \propto R/v$ for P into Kepler's law, we get $(R/v)^2 \propto R^3$. Rearranging gives $v^2 \propto 1/R$, and hence

$$v \propto (1/R)^{1/2}.$$

In words, the speed of a planet is inversely proportional to the square root of its orbital radius; distant planets move more slowly than those near the Sun. For example, a planet whose orbital radius is 4 times that of the Earth has an orbital speed that is $(1/4)^{1/2} = 1/2$ as fast as the Earth's orbital speed, which is itself about 30 km/s. If the planet is 9 times as distant as the Earth, its orbital speed is $(1/9)^{1/2} = 1/3$ that of the Earth.

This "inverse-square-root" relationship is characteristic of systems in which a large central mass dominates over the masses of orbiting particles. Later, we will see that a different result is found for rotating galaxies, where the material is spread out over a large volume.

CONCEPT REVIEW

» The Solar System has been largely explored with spacecraft over the last five decades (Sec. 5.1). Long before that, and even before the telescope was invented, it was known that planets somehow undergo **retrograde motion** (slowly drifting east to west among the stars) besides their generally **prograde motion** (drifting west to east among the stars). Aristotle's cosmological **model**, a description of reality, had the Sun, Moon, and planets orbiting the Earth (Sec. 5.2). Ptolemy enlarged on Aristotle's **geocentric** (Earth-centered) theory and explained retrograde motion by having the planets orbit the Earth on **epicycles**, small circles whose centers move along larger circles called **deferents** at a constant angular rate relative to a point called the **equant** (Sec. 5.2). As a calculational tool, Ptolemy's model could be used to predict the positions of planets with reasonable accuracy.

» Copernicus, in 1543, advanced a **heliocentric** (Sun-centered) theory that explained retrograde motion as a projection effect (Sec. 5.3). Copernicus still used circular orbits and a few epicycles, and so had not completely broken with the past, but his idea was nevertheless revolutionary and had lasting consequences. However, Copernicus did not have clear observational evidence favoring his model over that of Ptolemy; instead, the appeal was largely philosophical. See Fig. 5–26.

» Tycho Brahe made unprecedentedly accurate observations of planetary positions (Sec. 5.4), which Johannes Kepler interpreted quantitatively in terms of **ellipses** (Sec. 5.5a). An ellipse is a closed curve, all of whose points have the same summed distance from two interior points, each of which is called a **focus** (plural **foci**). The **major axis** of an ellipse is the longest line that can be drawn within the ellipse, or the length of that line; it passes through the two foci. The **semimajor axis** is half this length. The **minor axis** is perpendicular to the center of the major axis, or the length of this line; the **semiminor axis** is half of the minor axis. The **eccentricity** of an ellipse, a measure of how far it is out of round, is equal to the distance between the foci divided by the major axis.

» Kepler formulated three empirical laws of planetary motion (Sec. 5.5a–c): (1) The orbits are ellipses with the Sun at one focus.

(2) (**The law of equal areas**.) The line joining the Sun and a planet sweeps out equal areas in equal times. Thus, a planet moves fastest when it is at **perihelion**, the point on the orbit closest to the Sun. (3) The orbital **period** squared is proportional to the semimajor axis cubed. When measured in **Astronomical Units** (au; the semimajor axis of the Earth's orbit, or Earth's average distance from the Sun) and Earth **years**, Kepler's third law has a particularly simple form.

» Galileo first systematically turned a telescope on the sky and discovered features on the Moon, moons of Jupiter, and a full set of phases of Venus (Sec. 5.6). His observational discoveries strongly endorsed the Copernican heliocentric theory and provided evidence against the Ptolemaic geocentric model. Near the end of his life, Galileo published his much earlier studies of the motions of falling bodies, providing the foundation for Newton's subsequent work.

» Newton developed three basic laws of motion (Sec. 5.7): (1) If no external forces act on an object, its speed and direction of motion stay constant. (2) The force on an object is equal to the object's mass times its acceleration ($F = ma$). (3) When two bodies interact, they exert equal and opposite forces on each other. Newton also discovered the law of gravity, namely, that the gravitational force between two objects is proportional to the product of their masses and inversely proportional to the square of the distance between them (Sec. 5.7). He derived Kepler's laws mathematically and invented calculus, which he needed for these studies.

» A planet **revolves** around the Sun and **rotates** on its axis (Sec. 5.8). Its orbit is generally **inclined** by only a few degrees to the plane of the Earth's orbit. Spinning or rotating objects have **angular momentum**, roughly a product of an object's mass, physical size, and rotation rate. The total angular momentum of a system can change only if the system interacts with something from the outside. The angular momentum associated with planetary orbits probably came from the material that formed the planets.

QUESTIONS

1. Describe how you can tell in the sky whether a planet is in prograde or retrograde motion.

2. Describe the difference between the Ptolemaic and the Copernican systems in explaining retrograde motion.

3. Can planets interior to the Earth's orbit around the Sun appear to undergo retrograde motion? Explain with a diagram.

4. If Copernicus's heliocentric model did not give significantly more accurate predictions than Ptolemy's geocentric model, why do we now prefer Copernicus's model?

5. Draw the ellipses that represent Neptune's and Pluto's orbits. Show how one can cross the other, even though they are both ellipses with the Sun at one of their foci.

6. The Earth is closest to the Sun in January each year. Use Kepler's second law to describe the Earth's relative speeds in January and July.

†7. Pluto orbits the Sun in about 250 years. From Kepler's third law, calculate its semimajor axis. Show your work. The accuracy you can easily get by estimating the roots by hand rather than with a calculator is sufficient.

†8. Mars's orbit has a semimajor axis of 1.5 au. From Kepler's third law, calculate Mars's period. Show your work. The accuracy you can easily get by estimating the roots by hand rather than with a calculator is sufficient.

9. Explain in your own words why the observation of a full set of phases for Venus backs the Copernican system instead of the standard Ptolemaic (geocentric) model.

10. Do you expect Venus to have a larger angular size in its crescent phase or gibbous phase? Explain.

11. Discuss Galileo's main scientific contributions.

12. Discuss Newton's main scientific contributions.

†13. Explicitly show from Newton's version of Kepler's third law that Kepler's "constant" actually depends on the planet under consideration. (Substitute values for the mass of the Sun and of various planets into the constant.) Does it vary much for different planets?

†14. The discovery of a star orbiting the center of our Galaxy with a 15-year period has led to the finding that the mass of the giant black hole at the center of our Galaxy is about 4 million times that of the Sun (as described in Chapter 15). Using ratios, calculate the semimajor axis of the star's orbit. Convert the answer to light-years.

†15. If a hypothetical planet is 4 times farther from the Sun than Earth is, what is its orbital *speed* (not period) relative to Earth's?

†16. What is the period of a planet orbiting 3 au from a star that is 3 times as massive as the Sun? (*Hint*: Consider Newton's version of Kepler's third law. Use ratios!)

17. Which is greater: The Earth's gravitational force on the Sun or the Sun's gravitational force on the Earth? Explain.

†18. Compare the periods of rotation and revolution of our Moon.

19. If you were standing on a spinning turntable holding barbells close to your chest and you extended your arms, how would your spin rate be affected? Why?

†20. If you were standing on a turntable and a friend pushed you to give you angular momentum, compare how fast you would spin if you were holding 10-kg vs. 30-kg barbells.

21. **True or false?** An external force is needed to keep an object in motion at a constant speed and a constant direction.

22. **True or false?** In Ptolemy's geocentric theory of the Solar System, Venus could not go through the complete set of phases observed by Galileo.

23. **True or false?** Although the values of the constants may differ, Newton showed that Kepler's laws apply equally well to the moons orbiting Jupiter, if "moon" and "Jupiter" are substituted for the words "planet" and "Sun," respectively.

24. **True or false?** The orbital period and distance of a planet from the Sun can be used, with Newton's version of Kepler's third law, to estimate the mass of the Sun.

25. **True or false?** Copernicus's greatest contribution to astronomy was that he put forth the notion that retrograde motion can be explained by having the planets move along epicycles.

26. **True or false?** The outer planets orbit the Sun with a smaller physical speed (km/s) than the inner planets.

27. **True or false?** Suppose Astronaut Tamara is in space, outside and at rest relative to her spaceship (which she wanted to fix with a wrench), but a few feet away, just out of reach… and her tether is broken. Then she can move back toward her spaceship by throwing the wrench in the direction away from the spaceship.

28. **True or false?** The planets all orbit around the Sun in the same direction because the planets gravitationally tug on each other, causing them to move in the same direction.

29. **Multiple choice:** Which one of the following statements is *false*? **(a)** If you lived on Mercury, you would notice that Earth exhibits retrograde motion for a while every year. **(b)** As seen from Earth's surface, planets rise in the east and set in the west, even when they undergo retrograde motion. **(c)** Copernicus' model of the Solar System was accepted shortly after being proposed because it provided significantly more accurate positions of planets than did the Ptolemaic system. **(d)** Our observations of a full or gibbous Venus provide evidence against the Ptolemaic, Earth-centered model of the Solar System. **(e)** According to Kepler's first law, the orbits of planets are ellipses with the Sun at one focus; there is no object at the second focus.

30. **Multiple choice:** Which one of the following observational discoveries was *not* made by Galileo? **(a)** A given planet's speed among the background stars varies with time. **(b)** Jupiter's four large moons. **(c)** Large numbers of stars in the Milky Way. **(d)** The full set of phases of Venus. **(e)** Craters and mountains on the Moon.

†31. **Multiple choice:** Two planets orbit a star that has the same mass as the Sun. Planet Alex orbits at a distance of 2 au while Planet Noelle orbits at a distance of 1 au. What is Planet Alex's orbital period compared to Planet Noelle's orbital period? **(a)** 8 times longer. **(b)** $\sqrt{8}$ times longer. **(c)** 4 times longer. **(d)** 2 times longer. **(e)** The two periods are equal, since the mass of the star is the same in both cases.

†32. **Multiple choice:** Suppose you discover a comet whose orbital period around the Sun is 10^6 years. What is the semimajor axis of the comet's orbit, in au? **(a)** 10^2. **(b)** 10^3. **(c)** 10^4. **(d)** 10^6. **(e)** 10^9.

33. **Multiple choice:** The retrograde motion of Venus across the sky **(a)** is caused by the "backward" rotation of Venus about its own axis (it spins in a direction opposite that of most planets); **(b)** is caused by the gravitational tug of other planets on Venus; **(c)** is caused by the change in perspective as Venus catches up with, and passes, Earth while both planets orbit the Sun; **(d)** is caused by the motion of Venus along an epicycle whose center orbits the Sun (and hence Earth as well, from our perspective); or **(e)** was used by Galileo to explain the complete set of phases of Venus that he observed through his telescope.

†34. **Multiple choice:** Which one of the following statements is *true*? **(a)** A given force accelerates a massive object more than a lower-mass object. **(b)** If the distance between two objects is doubled, the gravitational force between them decreases to half its previous value. **(c)** The force on an object due to gravity does not depend on the object's mass. **(d)** Neglecting air resistance, a light tennis ball falls just as fast as a heavy lead ball dropped from the same height above the ground. **(e)** If the Sun's gravitational attraction were suddenly cut off, at that instant the Earth would start flying directly away from the Sun – that is, along a path radially away from the Sun.

35. **Multiple choice:** Which one of the following statements is *not* a consequence of Newton's laws? **(a)** If two satellites are orbiting the Earth at the same distance, the satellite with greater mass will feel a greater acceleration from the gravitational force than the one with less mass. **(b)** Kepler's "constant" is actually not a constant; it is a bit smaller for Jupiter's orbit around the Sun than for Earth's orbit around the Sun. **(c)** The gravitational force that the Earth exerts on you is equal in strength to the gravitational force that you exert on the Earth. **(d)** A force is required to change the direction of an object moving with constant speed. **(e)** The Moon is continually falling toward the Earth, but it never hits Earth because of its tangential motion.

36. **Fill in the blank:** Although Johannes Kepler developed three laws of planetary motion, _____ developed a physical understanding of these laws.

37. **Fill in the blank:** According to Kepler's first law, the orbits of planets are ellipses with the Sun at one _____.

38. **Fill in the blank:** The line that extends from one edge of an ellipse to the other, passing through both foci of the eclipse, is called the _____.

39. **Fill in the blank:** In size, the gravitational force pulling an apple toward the Earth is _____ the gravitational force pulling the Earth toward the apple.

40. **Fill in the blank:** If a spinning body expands, its rate of spin decreases, providing an example of the conservation of _____.

41. **Fill in the blank:** In the model of the Universe developed by Aristotle, the outermost sphere was that of the _____.

42. **Fill in the blank:** A satellite that orbits Earth at the same rate that Earth rotates about its axis is called a _____.

†This question requires a numerical solution.

TOPICS FOR DISCUSSION

1. Had you been alive 2,000 years ago, do you think you would have believed in a geocentric model of the Universe or a heliocentric model, as did Aristarchus of Samos? Explain.

2. Why does a satellite speed up as it spirals toward the Earth because of friction with the outer atmosphere? Naively, it seems that the friction would cause it to slow down.

3. Referring back to the scientific method discussed in Chapter 1, why was it important to verify Newton's prediction that the "constant" in Kepler's third law actually depends on the mass of the planet under consideration?

Painting by Jean-Leon Huens; used with permission of National Geographic Stock.

Nicolaus Copernicus, in a painting commissioned for *National Geographic Magazine*

Jay M. Pasachoff

The frontispiece of Johannes Kepler's Rudolphine Tables (1627).

A montage, in the correct relative scale, of NASA space views of the inner planets, plus the Earth's Moon, with the Sun's edge at extreme left. We see an approach image of Mercury from NASA's MESSENGER spacecraft, which has been in orbit around Mercury since 2011; Venus in a reconstruction of Mariner 10 data; a new "blue marble" image of Earth from NASA-Canada-Japan's Terra MODIS (**Mod**erate Resolution **I**maging **S**pectroradiometer), which maps the world every day or two in 36 spectral bands); the Moon as mapped by NASA's Lunar Reconnaissance Observatory; and Mars from the Mars Global Surveyor orbiter. At left, we see detail on the Sun from NASA's Solar Dynamics Observatory.

NSSDC at NASA's Goddard Space Flight Center, courtesy of Jay Friedlander

6

The Terrestrial Planets: Earth, Moon, and Their Relatives

ORIGINS

The small, rocky, terrestrial planets and the Moon show many similarities and differences among their interiors, surfaces, and atmospheres. By studying these objects, we can begin to understand the formation and evolution of the Earth, whose properties seem uniquely suitable for the development of complex life.

AIMS

1. Describe our Earth, our Moon, the Moon's origin, and its tidal influence on Earth (Sections 6.1 and 6.2).

2. Learn about the inner planets, four rocky worlds including the Earth (Sections 6.3 to 6.5).

3. Examine the similarities and differences among the terrestrial planets (Sections 6.3 to 6.5).

Mercury, Venus, Earth, and Mars share many similar features. Small compared with the huge planets beyond them, these inner planets also have rocky surfaces surrounded by relatively thin and transparent atmospheres, in contrast with the larger, gaseous/liquid planets. Together, we call these four the **terrestrial planets** (from the Latin "terra," meaning earth), which indicates their significance to us in our attempts to understand our own Earth. In this chapter, we discuss each of these rocky bodies, as well as their moons. (See *A Closer Look 6.1: Comparative Data for the Terrestrial Planets and Their Moons.*)

Venus and the Earth are often thought of as "sister planets," in that their sizes, masses, and densities are about the same (■ Fig. 6–1). But in many respects they are as different from each other as the wicked stepsisters were from Cinderella. The Earth is lush; it has oceans of water, an atmosphere containing oxygen, and life. On the other hand, Venus is a hot, foreboding planet with a uniform temperature of 735 K (865°F), a planet on which life seems unlikely to develop. Why is Venus like that? How did these harsh conditions come about? Can it happen to us here on Earth?

The one Solar-System body other than Earth that humans have visited is our Moon. It is so large relative to Earth that it joins us almost as a type of a "double-planet system." We will see how space exploration has revealed many of its secrets.

Table 6.1 Comparative planetology: atmospheres

Venus	= 93 × Earth
Mars	= Earth/160
Mercury	= exosphere of very low density

Mars is only 53 per cent the diameter of Earth and has 10 per cent of Earth's mass. Its atmosphere is much thinner than Earth's, too thin for visitors from Earth to rely on to breathe. But Mars has long been attractive as a site for exploration. We remain interested in Mars as a place where we may yet find signs of life or, indeed, where we might encourage life to grow. Mars's two tiny moons are but chunks of rock in orbit.

Mercury, the innermost planet, is so close to the Sun and has such a relatively weak gravitation pull that it was a surprise that NASA's MESSENGER spacecraft, which went into orbit around it in 2011, has detected a variety of particles surrounding it. Our understanding of Mercury is changing rapidly thanks to findings from MESSENGER, even early in its mission.

NASA's Ames Research Center

■ **FIGURE 6–1** Surface topography on Earth and Venus at the same resolution, imaged with radar.

6.1 EARTH: THERE'S NO PLACE LIKE HOME

On the first trip that astronauts ever took to the Moon, they looked back and saw for the first time the Earth floating in space. Nowadays we see that space view every day from weather satellites, so the views from the Jupiter-bound Galileo spacecraft and the Saturn-bound Cassini spacecraft as they passed allowed us to test the instruments on known objects, the Earth and Moon (■ Fig. 6–2).

A CLOSER LOOK 6.1   Comparative Data for the Terrestrial Planets and Their Moons

	Semimajor Axis	Orbital Period	Equatorial Radius (Earth = 100%)	Mass (Earth = 100%)
PLANETS				
Mercury	0.39 au	88 days	38%	5.5%
Venus	0.72 au	225 days	95%	82%
Earth	1.00 au	365.25 days	100%	100%
Mars	1.52 au	687 days	53%	11%
EARTH'S SATELLITE				
The Moon	384,000 km	27$\frac{1}{3}$ days	27% (1738 km)	1.2%
MARS'S SATELLITES				
Phobos	9,378 km	7 hr 39 min	14 × 11 × 9 km³	1 millionth %
Deimos	23,459 km	1 d 6 hr 18 min	8 × 6 × 6 km³	0.3 millionth %

■ **FIGURE 6–2** A composite of the Earth and Moon, viewed from the Galileo spacecraft as it passed by our region of the Solar System on its mission to Jupiter. Structure on the Earth's surface, especially in South America and Central America, shows clearly. The relative sizes are to scale.

The realization that Earth is an oasis in space helped inspire our present concern for our environment. Until fairly recently, we studied the Earth only in geology courses and the other planets only in astronomy courses, but now the lines are very blurred. Not only have we learned more about the interior, surface, and atmosphere of the Earth but we have also seen the planets in enough detail to be able to make meaningful comparisons with Earth. The study of **comparative planetology** is helping us to understand weather, earthquakes, and other topics. This expanded knowledge will help us improve life on our own planet.

6.1a The Earth's Interior

The study of the Earth's interior and surface (■ Fig. 6–3a) is called **geology**. Geologists study, among other things, how the Earth vibrates as a result of large shocks, such as earthquakes. Much of our knowledge of the structure of the Earth's interior comes from **seismology**, the study of these vibrations. The vibrations travel through different types of materials at different speeds.

From seismology and other studies, geologists have been able to develop a picture of the Earth's interior. The Earth's innermost region, the **core,** consists primarily of iron and nickel. Outside the core is the **mantle**, and on top of the mantle is the thin outer layer called

the **crust**. The upper mantle and crust are rigid and contain a lot of silicates, while the lower mantle is partially melted. (Silicates, which contain molecules with silicon combined with two or four oxygen atoms, make up about 90 per cent of the rock-forming minerals on the Earth's surface.)

Such a layered structure must have developed when the Earth was young and molten; the denser materials (like iron) sank deeper than the less-dense ones, as discussed below. But from where did Earth get sufficient heat to become molten?

The Earth, along with the Sun and the other planets, was probably formed from a cloud of gas and dust. Some of the original energy, though not enough to melt the Earth's interior, came from gravitational energy released as particles came together to form the Earth; such energy is released from gravity between objects when the objects move closer together and collide. The water at the base of a waterfall, for example, is slightly (unnoticeably) hotter than the water at the top, if we ignore heat losses to the air while the water is falling; part of the water's energy of motion, gained by the pull of gravity, is converted to heat by the collision on the rocks at the waterfall's base.

Also, the young Earth was subject to constant bombardment from the remaining debris (dust and rocks), which carried much energy of motion. This bombardment heated the surface to the point where it began to melt, producing lava.

However, scientists have concluded that the major source of energy in the interior, both at early times and now, is the natural radioactivity within the Earth. Certain forms of atoms are unstable – that is, they spontaneously change into more stable forms. In the process, they give off energetic particles that collide with the atoms in the rock and give some of their energy to these atoms. The rock heats up.

The Earth's interior became so hot that the iron melted and sank to the center since it was denser, forming the core. Eventually other materials also melted. As the Earth cooled, various materials, because

(a)

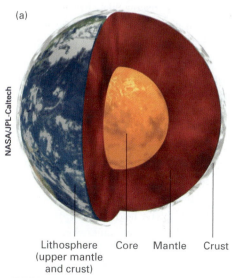

Lithosphere   Core   Mantle   Crust
(upper mantle
and crust)

■ **FIGURE 6–3 (a)** The interior structure of the Earth.

A CLOSER LOOK 6.2

Density

The density of an object is its mass divided by its volume. The scale of mass in the metric system was set up so that 1 cubic centimeter of water would have a mass of 1 gram (symbol: g). Though the density of water varies slightly with the water's temperature and pressure, the density of water is always about 1 g/cm³. Density is always expressed in units of mass (grams, kilograms, and so on) per volume (cubic centimeters, and so on), where the slash means "per." In the units in general use in the United States (but not elsewhere in the world), the density of water is about 62 pounds/cubic foot (though, technically, pounds is a measure of weight – or gravitational force – rather than of mass). Always be careful to note if you are using metric or pound/foot-type units. A spacecraft was lost just as it reached Mars in 1999 because the software on board used metric units while the ground-control software used pound/foot-type units.

To measure the density of any amount of matter, astronomers must independently measure its mass and its size. They are often able to identify the materials present in an object on the basis of density. For example, since the density of iron is about 8 g/cm³, Saturn's

density of only 0.7 g/cm³ shows that Saturn is not made largely of iron. On the other hand, the measurement that the Earth's density is about 5 g/cm³ shows that the Earth has a substantial content of dense elements, of which iron is a likely candidate.

Astronomical objects cover the extremes of density. In the space between the stars, the density is incredibly tiny fractions of a gram in each cubic centimeter (about 10^{-24} g/cm³), corresponding to just one hydrogen atom. And in neutron stars, the density exceeds a billion tons per cubic centimeter, comparable to that of an atomic nucleus.

Some sample values of the densities of materials are as follows:

water	1 g/cm³
aluminum	3 g/cm³
typical rock	3 g/cm³
iron	8 g/cm³
lead	11 g/cm³

of their different densities (density is mass divided by volume; see *A Closer Look 6.2: Density*) and freezing points (the temperature at which they change from liquid to solid), solidified at different distances from the center. This process, called "differentiation," is responsible for the present layered structure of the Earth.

Geologists have known for decades that the Earth's iron center consists of a solid inner core surrounded by a liquid outer core. The inner core is solid, in spite of its high (5,000°C) temperature, because of the great pressure on it. A study of 30 years' worth of earthquake waves that passed through the Earth's core revealed that the inner core has a different inner region, like the pit in a peach. This inner region is less than 10 per cent of the diameter of the "inner core," to continue to use that technical term.

Why does this "peach-pit" innermost core make earthquake waves act differently than they do in the surrounding solid inner core? It could be because this innermost core is a remnant of the original ball of material from which the Earth formed 4.6 billion years ago. Less exciting is the possibility that iron crystals deposited on it had a different orientation after the innermost core reached a critical size. Perhaps the temperature and pressure in that innermost region pack iron crystals differently.

The rotation of the Earth's metallic core helps generate a magnetic field on Earth. The magnetic field has a north magnetic pole and a south magnetic pole that are not quite where the regular north and south geographic poles are. The Earth's magnetic north pole is now in the Arctic Ocean north of Canada. The location of the magnetic poles wanders across the Earth's surface over time. The north magnetic

pole is currently moving northward at an average speed of 50 km/year (■ Fig. 6–3b,c).

6.1b Continental Drift

Some geologically active areas exist in which heat flows from beneath the surface at a rate much higher than average (■ Fig. 6–4). The outflowing **geothermal energy**, sometimes tapped as an energy source, signals what is below. The Earth's rigid outer layer is segmented into **plates**, each thousands of kilometers in extent but only about 50 km thick. Because of the internal heating, the top layers float on an underlying hot layer (the mantle) where the rock is soft, though it is not hot enough to melt completely.

The mantle beneath the rigid plates of the surface churns very slowly, thereby carrying the plates around. This theory, called **plate tectonics**, explains the observed **continental drift** – the drifting of the continents, over eons, from their original positions – at the rate of a few centimeters per year, about the speed your fingernails grow. ("Tectonics" comes from the Greek word meaning "to build.")

Although the notion of continental drift originally seemed unreasonable, it is now generally accepted. The continents were once connected as two super-continents, which may themselves have separated from a single super-continent called Pangaea ("all lands"). Over the past two hundred million years or so, the continents have moved apart as plates have separated. We can see from their shapes how they once fit together (■ Fig. 6–5). We even find similar fossils and rock

■ **FIGURE 6–3 (b)** (1) The motion of the Earth's magnetic north pole, which has accelerated from about 15 km/year a century ago to the current 50 km/year, moving north through the Northwest Passage of Canada. (2) The 2007 ground-based survey of the pole's position is likely to be the last; magnetic-mapping satellites will now take over the task, providing modeled positions for the north magnetic pole. Three ESA/Canadian satellites called Swarm, with a 2013 launch, will measure precisely the magnetic signals that stem from Earth's core, mantle, crust, and oceans, as well as its ionosphere and magnetosphere.

■ **FIGURE 6–4** A geyser and its surrounding steam, at Rotorua, New Zealand.

types along two opposite coastlines that were once adjacent but are now widely separated. Remnants of the magnetic field as measured in rocks laid down when the Earth's magnetic poles had flipped, north and south magnetic poles interchanging (as they do occasionally), are also among the strongest evidence.

Pangaea itself probably formed from the collision of previous generations of continents. The coming together and breaking apart of continents may have had many cycles in Earth's history. In the future, we expect part of California to separate from the rest of the United States, Australia to be linked to Asia, and the Italian "boot" to disappear.

The boundaries between the plates are geologically active areas (■ Fig. 6–6). Therefore, these boundaries are traced out by the regions where earthquakes and most of the volcanoes occur. The boundaries where two plates are moving apart mark regions where molten

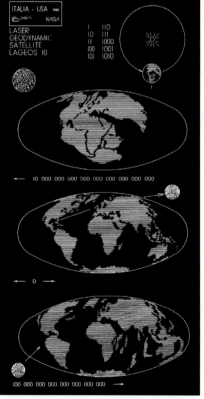

■ **FIGURE 6–5** LAGEOS I and LAGEOS II (**La**ser **Geo**dynamic **S**atellites) bear plaques showing the continents in their positions 270 million years in the past (*top*), at present (*middle*), and 8 million years in the future (*bottom*), according to our knowledge of continental drift. (The dates are given in the binary system.) Data on continental drift are gathered by reflecting laser beams from telescopes on Earth off the 426 retroreflectors that cover the satellites' surfaces. (Each retroreflector is a cube whose interior reflects incident light back in the direction from which it came.) Measuring the time until the beams return can be interpreted to show the accurate position of the ground station. We can now measure the position to an accuracy of about 1 cm.

■ **FIGURE 6–6** The San Andreas fault on Earth marks the boundary between the North American and Pacific plates. It is a strike-slip fault, in which two blocks of Earth move past each other. This view is paired with a Galileo-spacecraft image of a similar strike-slip fault on Jupiter's moon Europa, shown at the same scale and resolution.

NASA/JPL/Caltech and University of Arizona

material is being pushed up from the hotter interior to the surface, such as the mid-Atlantic ridge (■ Fig. 6–7). Molten material is being forced up through the center of the ridge and is being deposited as lava flows on either side, producing new seafloor.

The motion of the plates relative to each other is also responsible for the formation of most great mountain ranges. When two plates come together, one may be forced under the other and the other rises. The great Himalayan mountain chain, for example, was produced by the collision of India with the rest of Asia. The "ring of fire" volcanoes around the Pacific Ocean (including Mt. St. Helens in Washington) were formed when molten material made its way through gaps or weak points between plates.

6.1c Tides

It has long been accepted that **tides** are most directly associated with the Moon and to a lesser extent with the Sun. We know of their association with the Moon because the tides – like the Moon's passage across your meridian (the imaginary line in the sky passing from north to south through the point overhead) – occur about an hour later each day.

Tides result from the fact that the force of gravity exerted by the Moon (or any other body) gets weaker as you get farther away from it. Tides depend on the *difference* between the gravitational attraction of a massive body at different points on another body.

To explain the tides in Earth's oceans, suppose, for simplicity, that the Earth is completely covered with water. We might first say that the water closest to the Moon is attracted toward the Moon with the greatest force and so is the location of high tide as the Earth rotates. If this were the whole story, high tides would occur about once a day. However, *two* high tides occur daily, separated by roughly 12¹/₂ hours.

To see why we get two high tides a day, consider three points, *A, B,* and *C,* where *B* represents the solid Earth (which moves all together as a single object and is marked by a point at its center), *A* is the ocean nearest the Moon, and *C* is the ocean farthest from the Moon (■ Fig. 6–8). Since the Moon's gravity weakens with distance, it is greater at point *A* than at *B,* and greater at *B* than at *C.* If the

Earth and Moon were not in orbit around each other, all these points would fall toward the Moon, moving apart as they fell because of the difference in force. Thus the high tide on the side of the Earth that is near the Moon is a result of the water being pulled away from the Earth. The high tide on the opposite side of the Earth results from the Earth being pulled away from the water. In between the locations of the high tides, the water has rushed elsewhere (to the regions of high tides), so we have low tides.

Since the Moon is moving in its orbit around the Earth, a point on the Earth's surface has to rotate longer than 24 hours to return to a spot nearest to the Moon. Thus a pair of tides repeats about every 25 hours, making 12¹/₂ hours between high tides.

The Sun's effect on the Earth's tides is only about half as much as the Moon's effect. Though the Sun exerts a greater gravitational force on the Earth than does the Moon, the Sun is so far away that its force does not change very much from one side of the Earth to the other. And it is only the *difference* in force from one place to another that counts for tides.

Nonetheless, the Sun does matter. We tend to have very high and very low tides when the Sun, Earth, and Moon are aligned (as is the case near the time of full moon or of new moon), because their effects reinforce each other. Conversely, tides are less extreme when the Sun, Earth, and Moon form a right angle (as near the time of a first-quarter or third-quarter moon).

The effect of the tides on the Earth–Moon system slows down the Earth's rotation slightly, currently by about 2 seconds per 100,000 years, as we can verify from the timing of solar eclipses that took place thousands of years ago. Also, the interaction is leading to a gradual spiraling away of the Moon from the Earth, though the rate is only 4 centimeters per year. Thus, far in the future (about a billion years from now), it will not be possible to witness a total solar eclipse from Earth; the Moon's angular diameter will be less than that of the Sun's photosphere.

6.1d The Earth's Atmosphere

We name layers of our atmosphere (■ Fig. 6–9) according to the composition and the physical processes that determine their

(a)

L. A. Lawver, I. W. D. Dalziel, I. O. Norton, and L. M. Gahagan, PLATES project, University of Texas Institute for Geophysics; Plate boundaries are from the PLATES Project digital data compilation; Plate boundaries are from the PLATES Project digital data compilation. Topography is from Smith and Sandwell 1997 (*Science*, **277**, 1956–1962) plotted using GMT mapping software (Wessel and Smith, 1991, *EOS, Transactions AGU*, **72**, 441, 445–444). Map produced at the University of Texas Institute for Geophysics, 2012.

(b)

Stefan Fichtel www.ixtract.de and Johanes Bouman DGFI

■ **FIGURE 6–7** (a) The Earth's ocean floor, mapped with radar altimetry bounced off the ocean's surface from spacecraft as well as with soundings from ships. Features on the seafloor with excess gravity attract the water enough to cause structure on the ocean's surface that can be mapped, though a 2,000-meter-high seamount produces a bump on the surface only 2 meters high. High gravity is shown in yellow, orange, and red; low gravity is shown in blue and violet. The feature running from north to south in the middle of the Atlantic Ocean is the *mid-Atlantic ridge*. It marks the boundary between plates that are moving apart. Continents drift at about the rate that fingernails grow. In 2012, film director James Cameron (*Titanic*, *Avatar*) dived in a new type of submarine nearly 7 miles (35,756 feet, 10.9 km), to the Mariana Trench, the deepest known point in the Earth's oceans, about 200 miles (320 km) southwest of Guam. (b) A gravity map of the Earth also shows the continental plates. ESA's GOCE (the Gravity field and steady-state Ocean Circulation Explorer) satellite, launched in 2009, has mapped Earth's gravity with unrivalled precision, exaggerated 10,000 times in this false-color image to make the variations stand out. The satellite's data have provided the most accurate model of the "geoid," the Earth's gravitational shape, ever available. The gravity data from GOCE are helping to develop a deeper knowledge of the processes that cause earthquakes, such as the event that devastated part of Japan in 2011.

temperatures. The atmosphere contains about 20 per cent oxygen, the gas that our bodies use when we breathe; almost all of the rest is nitrogen. Later, we will see how the small amounts of carbon monoxide, carbon dioxide, methane, and other gases are affecting our climate. When we find a planet around another star (Chapter 9) whose spectrum shows such a high percentage of oxygen, we might infer the presence there of life-forms making the oxygen.

The Earth's weather is confined to the very thin **troposphere**. The ground is a major source of heat for the troposphere, so the temperature of the troposphere decreases as altitude increases. The rest of the Earth's atmosphere, as well as the Earth's surface, is heated mainly by solar energy from above.

A higher layer of the Earth's atmosphere is the "thermosphere." It is also known as the **ionosphere**, since many of the atoms there are ionized – that is, stripped of some of the electrons they normally contain. Most of the ionization is caused by x-ray and ultraviolet radiation from the Sun, as well as from solar particles. Thus the ionosphere forms during the daytime and diminishes at night. The free electrons in the ionosphere reflect very-long-wavelength radio signals. When the conditions are right, radio waves bounce off the ionosphere, which allows us to tune in distant radio stations.

Observations from high-altitude balloons and satellites have greatly enhanced our knowledge of Earth's atmosphere. Scientists carry out calculations using the most powerful supercomputers to interpret the

Low tide (as at the surface between point B and us)

High tide (as at points A and C)

■ **FIGURE 6–8** (a) A schematic representation of the tidal effects caused by the Moon (not to scale). The arrows represent the gravitational pull produced by the Moon (exaggerated in the drawing). The water at point *A* is pulled toward the Moon more than the Earth's center (point *B*) is; since the Earth is solid, the whole Earth moves with its center. (Tides in the solid Earth exist, but are much smaller than tides in the oceans.) Similarly, the solid Earth is pulled away from the water at point *C*. Water tends to collect at points *A* and *C* (high tide) and to be depleted in regions halfway between them on the Earth's surface, such as on the surface between point *B* and us (low tide).

Note that although the tidal force results from gravity, it is the *difference* between the gravitational forces at two places and is therefore not the same as the "gravitational force." Tidal forces are important in many astrophysical situations, including interactions between closely spaced galaxies, and rings around planets.

(b) The tidal range at Kennebunkport, Maine, where the shape of the ocean floor leads to an especially high tidal range.

Note that the Moon and the Earth are drawn to a different scale from their separation.

■ **FIGURE 6–9** (a) Temperature in the Earth's atmosphere as a function of altitude. In the troposphere, the energy source is the ground, so the temperature decreases as you go up in altitude. Higher temperatures in other layers result from ultraviolet radiation and x-rays from the Sun being absorbed. (b) The location of the ozone layer, which is depleted over Antarctica in its springtime.

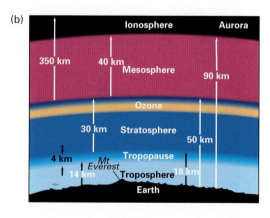

global data and to predict how the atmosphere will behave. The equations are essentially the same as those for the internal temperature and structure of stars, except that the sources of energy are different, with stars heated from below and the Earth's atmosphere mostly heated from above.

Winds are caused partly by uneven heating of different regions of Earth. The rotation of the Earth also has a very important effect in determining how the winds blow. Comparison of the circulation of winds on the Earth (which rotates in 1 Earth day), on slowly rotating Venus (which rotates in 243 Earth days), and on rapidly rotating Jupiter and Saturn (each of which rotates in about 10 Earth hours) helps us understand the weather on Earth. Our improved understanding allows forecasters of weather (day to day) and climate (long term) to be more accurate.

Comparison with the planet Venus (Section 6.4d) led to our realization that the Earth's atmosphere traps some of the radiation from the Sun, and that we are steadily increasing the amount of trapping. (The word "trapping" is used here in a figurative rather than a literal sense; the energy is actually transformed from one kind to another when air molecules absorb it, and no particular light photons are physically "trapped.") The process by which light is trapped, resulting in the extra heating of Earth's atmosphere and surface, is similar to the process that is generally (though incorrectly) thought to occur in terrestrial greenhouses; it is thus called the **greenhouse effect**. It is caused largely by the carbon dioxide in Earth's atmosphere (■ Fig. 6–10), the amount of which is growing each year mostly because of our use of fossil fuels.

Section 6.4d (below) more fully describes the greenhouse effect, which greatly affects the temperature of Venus, which has a much

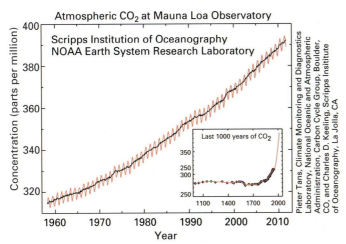

FIGURE 6–10 The concentration of atmospheric carbon dioxide in parts per million (ppm) of dry air as it changed with time, observed at the Mauna Loa Observatory in Hawaii. In addition to the yearly cycle, there is an overall increase. Carbon dioxide is an important contributor to the greenhouse effect. In 2013, the daily average surpassed 400 ppm for the first time.

higher carbon dioxide content. In brief, Earth's atmosphere is warmed both from above by solar radiation, and from below by radiation from the Earth's surface, which is itself warmed by solar radiation. Most solar radiation is in the visible, corresponding to the peak of the Sun's blackbody radiation (recall the discussion of this radiation in Chapter 2). The radiation from the Earth's surface is in the form of infrared, which corresponds to the peak wavelength of the blackbody curve at Earth's temperature. Some of this infrared radiation is absorbed in the atmosphere by carbon dioxide, water, and, to a lesser extent, by other "greenhouse gases" such as methane.

The greenhouse effect itself is good for us; it warms us by about 33°C, bringing Earth's atmospheric temperature to the livable range it is now. The important question is whether the amount of greenhouse warming is increasing, progressively raising the Earth's temperature – a phenomenon known as "global warming" regardless of its origin.

Quite a separate problem is a discovery about the ozone (O_3) in our upper atmosphere. This ozone becomes thinner over Antarctica each Antarctic spring, a phenomenon known as the "ozone hole." The ozone hole started forming only around 1980, and its maximum size grew nearly each year for about a decade, though international agreements have now stabilized it (■ Fig. 6–11). Worries about the formation of an Arctic ozone hole are still present.

The ozone hole is caused by the interaction in the cold upper atmosphere of sunlight with certain gases we give off near the ground, such as chlorofluorocarbons used in air conditioners and refrigerators. International governmental meetings have arranged cutbacks in the use of these harmful gases. Some success has been achieved, but we have far to go to protect our future atmosphere.

Don't confuse the greenhouse effect (global warming from gases like carbon dioxide trapping sunlight) with the ozone hole (a decrease of atmospheric ozone over Antarctica and surrounding areas each southern-hemisphere springtime, caused by chlorofluorocarbons). It was, however, reported in 2012 that global warming may have led to

an increase in violent storms that inject water vapor to relatively high atmospheric levels, where the water vapor interacts with chlorofluorocarbons already there, leading to a diminution of the ozone layer. So there might be a link between the two effects after all.

6.1e The Van Allen Belts

In 1958, the first American space satellite carried aloft, among other things, a device to search for particles carrying electric charge that might be orbiting the Earth. This device, under the direction of James A. Van Allen of the University of Iowa, detected a region filled with charged particles having high energies. Two such regions – the **Van Allen belts** – were found to surround the Earth, like a small and a large doughnut, containing mostly protons and electrons (■ Fig. 6–12). They start a few hundred kilometers above the Earth's surface and extend outward to about 8 times the Earth's radius. A temporary belt, formed after a solar flare, was observed by NASA's Van Allen Probes mission soon after its 2012 launch. After 4 weeks, the new belt was destroyed by an interplanetary shock wave from the Sun.

The particles in the Van Allen belts are trapped by the Earth's magnetic field. Charged particles preferentially move in the direction of magnetic-field lines, and not across the field lines. These particles, often from solar magnetic storms, are guided by the Earth's magnetic field toward the Earth's magnetic poles. When they interact with air molecules, they cause our atmosphere to glow, which we see as the beautiful northern and southern lights – the **aurora borealis** and **aurora australis**, respectively (■ Fig. 6–13).

Every planet with a magnetic field – most notably Jupiter (as we will see in the next chapter) – has regions filled with charged particles. Earth's Van Allen belts were the first to be studied.

6.2 THE MOON

The Earth's nearest celestial neighbor – the Moon – is only 380,000 km (238,000 miles) away from us, on the average. At this distance, it appears sufficiently large and bright to dominate our nighttime sky. The Moon's stark beauty has captured our attention since the beginning of history. Now we can study the Moon not only as an individual object but also as an example of a small planet or a large planetary satellite, since spacecraft observations have told us that there may be little difference between small planets and large moons.

6.2a The Moon's Appearance

Even binoculars reveal that the Moon's surface is pockmarked with craters. Other areas, called **maria** (pronounced mar´ee-a; singular **mare,** pronounced mar´eyh), are relatively smooth and dark. Indeed, the name comes from the Latin word for sea (■ Fig. 6–14). But

Total Ozone Mapping Spectrometer (TOMS), NASA's Goddard Space Flight Center

1981   1986   1991   1996

Total Ozone Mapping Spectrometer (TOMS), NASA's Goddard Space Flight Center

2001   2005   2011

http://earthobservatory.nasa.gov/IOTD/view.php?id=49874

2010   2011

■ **FIGURE 6–11** The amount of ozone (O_3) over Antarctica in the spring (September and early October for the Earth's southern hemisphere) is much lower (*blue and purple*) than the amount in other seasons. Note the outline of the Antarctic continent (*also shown in blue*) in this display of data from NASA's TOMS (**T**otal **O**zone **M**onitoring **S**pectrometer). (a) We see over two decades largely a decline in the total amount of ozone, shown as the spreading of the area in deep blue, the "ozone hole." Fortunately, through international agreements and changes in various manufactured objects and gases, the decline has been slowed or reversed. (b) An ozone hole occasionally opens in the Arctic, as we see in the contrast between this 2010 view (left), of a relatively strong amount of ozone even at the crucial time of year, Arctic spring, and the 2011 view (right), of a strong depletion of ozone. We hope the ozone depletion is a rare occurrence in the Arctic, quite possible given the unusually cold circumstances that led to the 2011 situation.

there are no ships sailing on the lunar seas and no water in them; the Moon is a dry, airless, barren place.

The gravity at the Moon's surface is only one-sixth that of the Earth. Typically you would weigh only 20 or 30 pounds there if you stepped on a scale! The gravity is so weak that any atmosphere and any water that may once have been present would long since have escaped into space.

The Moon rotates on its axis at the same rate that it revolves around the Earth, thereby always keeping essentially the same face in our direction. To understand this idea, put a quarter on your desk

and then slide a dime around it, keeping both flat on the desk and keeping the top of the head on the dime always on the side that is away from you. Notice that though the dime isn't rotating as seen from above, a viewer on the quarter would see the dime at different angles. Then move the dime around the quarter so that the same point on the dime always faces the quarter. Notice that as seen from above, the dime rotates as it revolves around the quarter.

The Moon used to rotate about its axis more rapidly, but tides in the solid Moon raised by the Earth's gravity caused a gradual loss of rotation energy caused by friction. Eventually, Earth's gravity locked

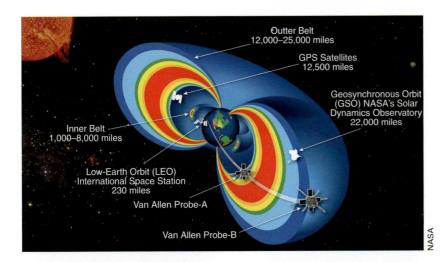

■ **FIGURE 6–12** The outer Van Allen belt contains mainly electrons; the traditional "inner belt" contains mainly protons. A temporary belt that lasted only days inside the so-called inner belt contained mainly ions of oxygen, nitrogen, and neon, for which the distribution of isotopes indicated that they were from interstellar space. 2013's Van Allen Probes mission found a transient belt that lasted for about 4 weeks.

(a)

(b)

(c)

■ **FIGURE 6–13** (a) The aurora borealis and the aurora australis (northern and southern lights) looking down from NASA's Polar spacecraft. These auroras followed a coronal mass ejection (see Chapter 10) from the Sun, which provided particles that travelled for a day and a half until they were trapped in Earth's magnetic field. (b) The aurora looking down from NASA's IMAGE (**I**mager for **M**agnetopause to **A**urora **G**lobal **E**xploration) spacecraft. NASA's Interstellar Boundary Explorer (IBEX), launched in October 2008 into a high-eccentricity 8-day orbit around the Earth, is observing the interaction of the solar wind with the Earth's magnetosphere in addition to watching the solar wind interact farther out in the Solar System. (c) A ground-based view of an aurora along with a meteor from the Perseid shower.

■ **FIGURE 6–14** A waning moon (just past third quarter) showing dark lunar seas (maria) and lighter highlands.

the Moon into "synchronous rotation," pulling on a bulge in the distribution of the lunar mass to prevent the Moon from rotating freely. But the tidal locking isn't perfect, because of several effects such as the Moon's elliptical orbit around the Earth (the Moon's orbital speed isn't constant, unlike its rotation rate), so although we always see almost the same face of the Moon from our vantage point on Earth, over time we can actually observe close to 60 per cent of the Moon's surface.

When the Moon is full, it is bright enough to cast shadows or even to read by. But a full moon is a bad time to try to observe lunar surface structure, for any shadows we see are short, and lunar features appear washed out. When the Moon is a crescent or even a quarter moon, however, the sunlit part of the Moon facing us is covered with long shadows. The lunar features then stand out in bold relief. Shadows are longest near the **terminator**, the line that separates day from night. Note that nature photographers on Earth, concluding that views with shadows are more dramatic, generally take their best photos when the Sun is low.

Six teams of astronauts in NASA's Apollo program landed on the Moon in 1969–1972. (See *A Closer Look 6.3: The First People on the Moon.*) In some sense, before this period of exploration, we knew more about bright stars than we did about the Moon. As a relatively cold, solid body, the Moon reflects the spectrum of sunlight rather than emitting its own optical spectrum, so we were hard pressed to determine even the composition or the physical properties of the Moon's surface (such as whether you would sink into it!).

6.2b The Lunar Surface

The kilometers of film exposed by the astronauts, the 382 kg of rock brought back to Earth, the lunar seismograph data recorded on tape, meteorites from the Moon that have been found on Earth, and other sources of data have been studied by hundreds of scientists from all over the world. (Meteorites are rocks from space that have landed on Earth; see Chapter 8 for more details.) The data have led to new views of several basic questions, and have raised many new questions about the Moon and the Solar System.

The rocks that were encountered on the Moon are types that are familiar to terrestrial geologists (■ Fig. 6–15). Almost all the rocks are the kind that were formed by the cooling of lava, known as igneous rocks. Basalts are one example.

The Moon and the Earth seem to be similar chemically, though significant differences in overall composition do exist. Some elements that are rare on Earth – such as uranium and thorium – are found in greater abundances on the Moon. (Will we be mining on the Moon

one day?) None of the lunar rocks contain any trace of water bound inside their minerals.

Meteoroids, interplanetary rocks that we will discuss in Chapter 8, hit the Moon with such high speeds that huge amounts of energy are released upon impact. The effect is that of an explosion, as though TNT or an H-bomb had exploded. As a result of the Apollo missions, we know that almost all the craters on the Moon come from such impacts.

One way of dating the surface of a moon or planet (that is, figuring out when it was last molten, or when it formed) is to count the number of craters in a given area, a method that was used before Apollo. Surely those locations with the greatest number of craters must be the oldest. Relatively smooth areas – like maria – must have been covered over with molten volcanic material at some relatively recent time (which is still billions of years ago, though). The superposition of one crater on another – when one crater lies wholly or partially over the other – indicates that the underlying crater formed first and is therefore older.

Obvious rays of lighter-colored matter splattered outward during the impacts that formed a few of the craters. Since these rays extend over other craters, the craters with rays must have formed more recently. The youngest rayed craters may be very young indeed – perhaps only a few hundred million years. The rays darken with time, so rays that may have once existed near other craters are now indistinguishable from the rest of the surface.

Crater counts and the superposition of one crater on another give only *relative* ages. We found the *absolute* ages only when rocks from the Moon were physically returned to Earth. Scientists worked out the dates by comparing the current ratio of radioactive forms of atoms to nonradioactive forms present in the rocks with the ratio that they would have had when they were formed. (Varieties of the chemical elements having different numbers of neutrons are known as "isotopes," and radioactive isotopes are those that decay spontaneously; that is, they change into other isotopes even when left alone. Stable isotopes remain unchanged. For certain pairs of isotopes – one

A CLOSER LOOK 6.3 The First People on the Moon

On July 20, 1969, Neil Armstrong and Buzz Aldrin left Michael Collins orbiting the Moon and descended to the lunar surface. After Armstrong stepped out of the spacecraft first and took that first step on the Moon, he, the first person on the Moon, made the famous statement, "That's one small step for man, one giant leap for mankind."

At the time of the 30th anniversary of the Moon landing, looking back, Aldrin said, "I was full of goose bumps when I stepped down onto the surface. I felt a bit disoriented because of the nearness of the horizon. On Earth when you look at the horizon it appears flat, but on the Moon, so much smaller and without hills, the horizon in all directions curved visibly down away from us.

"There was about every type of rock imaginable, all covered with a very fine powder. The rocks themselves actually had no color until you looked closely at the crystals. The thought briefly occurred to me that these rocks had been sitting there for hundreds of millions or billions of years, and that we were the first living beings to see them. But we were too busy to be philosophical for long or to study them closely, and we just grabbed what looked like an interesting assortment. I felt them crunch beneath my feet as we walked around." (*Griffith Observer* interview)

■ **FIGURE 6–15** Different types of lunar rocks brought back to Earth by the Apollo 15 astronauts. A 1-cm block appears for scale. (a) A rock that is a compound of other types of rock. It weighs 1.5 kg. (b) The dark part is dust that covered another rock to which this rock was attached. (c) A 4.5-kg compound rock.

Table 6.2 Astronauts who landed on the Moon and their orbiting colleagues

Apollo 11 July 20, 1969; landed: Neil Armstrong and Buzz Aldrin; orbited: Michael Collins

Apollo 12 November 19, 1969; landed: Pete Conrad and Alan Bean; orbited: Dick Gordon

Apollo 14 February 5, 1971; landed: Alan Shepard and Ed Mitchell; orbited: Stu Roosa

Apollo 15 July 30, 1971; landed: Dave Scott and Jim Irwin; orbited: Al Worden

Apollo 16 April 20, 1972; landed: John Young and Charlie Duke; orbited: Ken Mattingly

Apollo 17 December 11, 1972; landed: Gene Cernan and Harrison Schmitt; orbited: Ron Evans

radioactive and one stable – we know the proportion of the two when the rock was formed. Since we know the rate at which the radioactive one is decaying, we can calculate how long it has been decaying from a measurement of what fraction is left.) The oldest rocks that were found on the Moon solidified 4.4 billion years ago. The youngest rocks ever found on the Moon solidified 3.1 billion years ago.

This "radioactive dating" is widely used, not only for dating rocks from the Moon but also for dating rocks on Earth and for finding the ages of archaeological sites, of ancient pottery, and so on.

The lunar observations can be explained on the basis of the following general sequence (■ Fig. 6–16). The Moon formed about 4.6 billion years ago. From the oldest rocks, we know that at least the surface of the Moon was molten about 200 million years later. Then the surface cooled. Studies released in 2011 of a rock collected by Apollo 16 astronauts in 1972 used new methods to date it to 4.360 billion years, with an uncertainty of only 3 million years. This value is significantly younger than the 4.43 billion years that has been considered up to now to be the age of the lunar crust. Perhaps the lunar surface cooled more slowly than had been thought, or, in a more radical idea, a no-longer-surviving second early moon of Earth smashed into the surviving Moon.

From 4.2 to 3.9 billion years ago, bombardment by interplanetary rocks caused most of the craters we see today. About 3.8 billion years ago, the interior of the Moon heated up sufficiently (from radioactive elements inside) that volcanism began. Lava flowed onto the lunar surface and filled the largest basins that resulted from the earlier bombardment, thus forming the maria (■ Fig. 6–17). By 3.1 billion years ago, the era of volcanism was over. The Moon has been geologically pretty quiet since then.

Up to this time, the Earth and the Moon shared similar histories. But active lunar history stopped about 3 billion years ago, while the Earth continued to be geologically active. Almost all the rocks on the Earth are younger than 3 billion years of age; the oldest single rock ever discovered on Earth has an age of 4.5 billion years, but few such old rocks have been found. Erosion and the remolding of the continents as they move slowly over the Earth's surface have taken their toll. So we must look to extraterrestrial bodies – the Moon or meteorites – that have not suffered the effects of plate tectonics or erosion (which occurs in the presence of water or an atmosphere) to study the first billion years of the Solar System.

After the manned missions to the Moon in 1969–1972, it was not until the 1990s that spacecraft revisited the Moon. The Clementine spacecraft (named after the prospector's daughter in the old song, since the spacecraft was looking for minerals) took photographs and other measurements. Photographs of the far side of the Moon (■ Fig. 6–18) have shown us that the near and far hemispheres are quite different in overall appearance. The maria, which are so conspicuous on the near side, are almost absent from the far side, which is cratered all over. We shall see in the next section that the difference probably results from the different thicknesses of the lunar crust on the sides of the Moon nearest Earth and farthest from Earth. The difference was first seen in the fuzzy photographs of the far side that were taken by the Soviet Lunik 3 spacecraft in 1959.

In the 1990s, NASA's Lunar Prospector and Clementine spacecraft mapped the Moon with a variety of instruments (■ Fig. 6–19). Lunar Prospector confirmed indications from the Clementine spacecraft that there is likely to be water ice on the Moon, by detecting more neutrons coming from the Moon's polar regions than elsewhere. Clementine and Lunar Prospector scientists think that these neutrons are given off in interactions of particles coming from the Sun with hydrogen in water ice in craters near the lunar poles, where they are shaded from the Sun's rays.

But the detection is not of water directly, and Apollo 17 astronaut Harrison Schmitt, the only geologist ever to have walked on the Moon, told one of the authors (J.M.P.) in 1999 that the neutrons

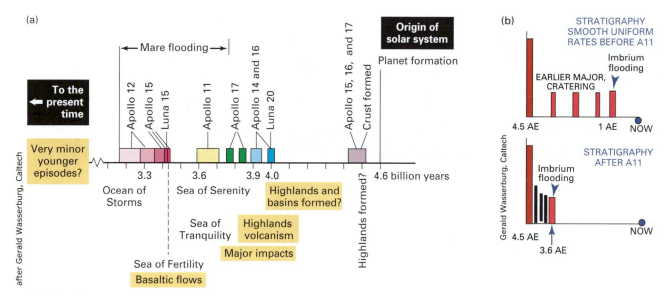

■ **FIGURE 6–16** (a) The chronology of the formation of the lunar surface based on the work of Gerald Wasserburg and collaborators at Caltech, the so-called Lunatic Asylum. The ages of rocks found in 8 missions are shown. Note how many crewed American Apollo missions and uncrewed Soviet Luna missions it took to get a sampling of many different ages on the lunar surface. A "late heavy bombardment" took place about 4 billion years ago, which is now explained by the migration of Jupiter and Saturn (a conclusion based on exoplanet observations; see Chapter 9). (b) Changes in our idea of the chronology based on discoveries from Apollo 11. Before Apollo 11, it was believed that the whole cratering history was spread over about 3.5 billion years, and this belief governed all of the U.S. Geological Survey's mission-planning approach. After Apollo 11, we realized that most of the cratering must have occurred within the first 0.9 billion years.

■ **FIGURE 6–17** An artist's view of the formation of the Mare Imbrium region of the lunar surface. (a) An asteroid impact on the Moon, sometime between 3.85 and 4.0 billion years ago. (b) The shock of the asteroid impact began the Imbrium crater. (c) As the dust and heat subsided, the 1,300-km Imbrium crater was left. (d) The molten rock (lava) flowed over outlying craters and cooled, leaving lunar mountains. (e) Lava welled up from inside the Moon 3.8 billion years ago. It filled the basin. (f) By 3.3 billion years ago, the lava flooding was nearly complete. (g) The final flow of thick lava came 2.5–3.0 billion years ago. (h) Subsequent cratering has left the Mare Imbrium of today's Moon.

NASA's Ames Research Center

■ **FIGURE 6–18** The far side of the Moon from Lunar Reconnaissance Observatory, a mosaic of over 15,000 images acquired in 2009–2011.

(a) U.S. Naval Research Laboratory and NASA

(b) NASA's Ames Research Center

(c) U.S. Naval Research Laboratory and NASA

■ **FIGURE 6–19** (a) An oblique view of part of the lunar far side, from Clementine, shows how rough it looks. (b) A Lunar Prospector map of the distribution of gravity on the Moon, showing hemispheres of the near side and the far side. (c) A Clementine map of the distribution of iron on the Moon's surface.

may instead have come from the solar wind (■ Fig. 6–20), though as of this writing most scientists do not agree. Schmitt would love to find water there, because it would increase the chance that a crewed Moon base could be supplied on the Moon itself, much easier than bringing everything from Earth. In his estimation, it would take about 10 years to set up such a base, once basic funding is available on Earth.

The European Space Agency (ESA) sent a test satellite in their Small Mission for Advanced Research and Technology (SMART) to the Moon in 2003, using propulsion by the ejection of ions instead of with the chemical rockets used by all the other lunar missions. The ESA mission was successful, sending back many images and other data. It was crashed into the Moon in 2006.

In 2007, the newly formed Japan Aerospace Exploration Agency (JAXA) sent its lunar **Sel**enographic and **En**gineering **E**xplorer (SELENE) mission, nicknamed Kaguya. (Selene was the Greek goddess of the Moon; the name Kaguya was that of a lunar princess in a traditional Japanese story and was selected by the public.) It was in lunar orbit and crashed into the lunar surface in 2009, sending dramatic videos as it did so.

The Chinese National Space Administration has launched two lunar orbiters: Chang'e 1 in 2007 and Chang'e 2 in 2010. Chang'e 3, with a lander carrying a rover, is scheduled for launch in 2013, with Chang'e 4 as a backup. They hope for a sample-return mission, Chang'e 5, sometime after 2017. (The spacecraft are named after a Chinese goddess of the Moon.)

The Indian Space Research Organization launched its Chandrayaan-1 (Moon-Traveler, in Sanskrit) in 2008. It sent a probe to impact near the south pole, as had been done by several previous

spacecraft; a NASA instrument on board found water ice in the plume that was thrown up. The orbiter stopped sending radio signals in 2009. It had carried not only Indian equipment but also experiments from the U.S., the U.K., ESA, Germany, Poland, Norway, and Bulgaria. Chandrayaan-2, a joint mission between India and Russia that is to consist of an orbiter and a rover, is scheduled for not earlier than 2014.

NASA sent its Lunar Reconnaissance Orbiter (LRO) to the Moon in 2009. It remains in polar orbit around the Moon, only 50 km high, on a mapping mission (■ Fig. 6–21). The launch also carried the **L**unar **Cr**ater **O**bservation and **S**ensing **S**atellite (LCROSS), again to search for water ice in a shadowed crater near a lunar pole; it was sent to crash into such a crater on October 9, 2009. The rocket's upper stage crashed first, and the following probe sent back images. Analysis seems to show some amount of water was thrown up in the plume, but whether it is at a useful density for eventual astronauts to make use of is under debate.

The American **G**ravity **R**ecovery **a**nd **I**nterior **L**aboratory (GRAIL) was a pair of spacecraft that orbited the Moon during most of 2012 to map the Moon's gravitational field sufficiently carefully that its

■ **FIGURE 6–20** Scientist–astronaut Harrison Schmitt, the only geologist to have stood on the Moon, during the Apollo 17 mission. The presence of the giant rock indicates that the later Apollo missions went to less-smooth sites than the earlier ones.

NASA's Johnson Space Center

(a)

(b)

NASA's Ames Research Center

NASA's Ames Research Center

■ **FIGURE 6–21** (a) Tycho crater's central peak, about 15 km wide, imaged with NASA's Lunar Reconnaissance Observatory. (b) A closeup of Tycho's summit.

interior structure could be analyzed. Monitoring the distance between the two spacecraft revealed the lunar gravitational field in detail. Depending on how far apart the spacecraft were on a large scale determined how deep in the lunar interior the data are sensitive. The results show that, once the gravitational effects of surface features and the mascons under lunar basins are removed from consideration, the layer of the Moon under the surface is highly fractured throughout. Most of the upper part of the lunar crust is apparently highly porous and pulverized, without regions of dense rock. It also turns out that the Moon's crust averages between 21 and 27 miles thick (34 to 43 km), about 25% thinner than previously thought.

When the GRAIL spacecraft entered lunar orbit, it was announced that the winner of the naming contest was a fourth-grade class from Montana. The names, Ebb and Flow; the prize: choosing the pointing of the first image of a camera on board for remote use by school students. MIT professor Maria Zuber, the Principal Investigator of the project, explains that despite more than 100 spacecraft having visited the Moon, we don't even know why the front side and the back side are so different; the answer probably lies in the interior, which Ebb and Flow studied. They were crashed into the Moon on purpose in December 2012 after monitoring exactly how much fuel they had left, to check the accuracy of NASA's fuel-monitoring methods.

NASA plans a **L**unar **A**tmosphere and **D**ust **E**nvironment **E**xplorer (LADEE) for 2013. In 2017, NASA's/Canada's **R**egolith and **E**nvironment **S**cience and **O**xygen and **L**unar **V**olatile **E**xtraction (RESOLVE) mission will carry a lunar rover to find and analyze water frozen near the lunar south pole and to liquefy it there. The Russian and Indian space agencies' joint 2015 lunar mission includes a Luna-Glob (lunar sphere) Russian lander and Chandrayaan-2 Indian orbiter and rover. Chinese and Indian scientists separately announced plans to eventually land people on the Moon, perhaps within a decade. Colloquially, the Chinese equivalent to our astronaut is taikonaut; the Indian equivalent is vyomanaut.

6.2c The Lunar Interior

Before the Moon landings, it was widely thought that the Moon was a simple body, with the same composition throughout. But we now know it to be differentiated (■ Fig. 6–22), like the planets. Most experts believe that the Moon's core is molten, but the evidence is not conclusive. GRAIL measured the lunar crust to be about 25 km thick on the near side and twice as thick on the far side. This asymmetry may explain the different appearances of the sides, because lava would be less likely to flow through the far side's thicker crust.

The Apollo astronauts brought seismic equipment to the Moon (■ Fig. 6–23). One type of earthquake wave moves material to the side. Only if the Moon's core were solid would the material move and then return to where it started, continuing the wave. Since that type of wave doesn't come through the Moon, scientists deduce that the Moon's core is probably molten. When new computer analysis methods were applied in 2004 to the old data, thousands of additional moonquakes were discovered.

Even before GRAIL's results, tracking the orbits of the Apollo Command Modules and more recently the Clementine and Lunar Prospector spacecraft that orbited the Moon told us about the lunar interior. If the Moon were a perfect, uniform sphere, the spacecraft orbits would have been perfect ellipses. But they weren't. One of the major surprises of the lunar missions of the 1960s, refined more recently, was the discovery in this way of **mascons**, regions of **mass** **con**centrations near and under most maria. The mascons may be lava that is denser than the surrounding matter, providing a stronger gravitational force on satellites passing overhead.

We also find out about the lunar interior by bouncing powerful laser beams off the Moon's surface. This "laser ranging" (where "ranging" means finding distances) uses several sets of reflectors left on the Moon by the Apollo astronauts. The laser-ranging programs find the distance to the Moon to within a few centimeters – pretty good for an object about 400,000 km away! Variations in the distance result in part from conditions in the lunar interior.

NASA's GRAIL satellites, Ebb and Flow, orbiting in formation and monitoring the distance between them to an incredible precision, improved the accuracy of our understanding of the distribution of density in the lunar interior by a factor of at least 100. (A similar pair of satellites is mapping the density of our Earth's interior.)

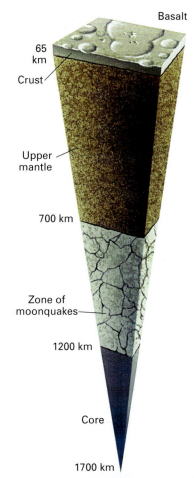

■ **FIGURE 6–22** The Moon's interior. The depth of solidified lava is greater under maria, which are largely on the side of the Moon nearest the Earth.

■ **FIGURE 6–23** Buzz Aldrin and the experiments he deployed on Apollo 11, in an image by Neil Armstrong. In the foreground, we see the seismic experiment. The laser-ranging retroreflector is behind it.

6.2d The Origin of the Moon

The leading models for the origin of the Moon that were considered at the time of the Apollo missions were as follows:

1. *Fission.* The Moon was separated from the material that formed the Earth; the Earth spun up and the Moon somehow spun off.

2. *Capture.* The Moon was formed far from the Earth in another part of the Solar System, and was later captured by the Earth's gravity.

3. *Condensation.* The Moon was formed near to (and simultaneously with) the Earth in the Solar System.

But work over the last four decades has all but ruled out the first two of these and has made the third seem less likely. The model now strongly favored, especially because of computer simulations, is

4. *Ejection of a Gaseous Ring.* A planet-like body perhaps twice the size of Mars hit the young Earth, ejecting matter in gaseous (and perhaps some in liquid or solid) form (■ Fig. 6–24). Most of the projectile fell back onto Earth within 24 hours, but two or so lunar masses formed a disk of liquid and vapor in orbit around Earth. Over time, about half of the disk spiraled back to Earth, and a small part escaped entirely. The remaining orbiting material eventually coalesced to form the Moon in perhaps a hundred to a thousand years.

Comparing the chemical composition of the lunar surface with the composition of the Earth's surface has been important in narrowing down the possibilities. The mean lunar density of 3.3 grams/cm^3 is close to the average density of the Earth's major upper region (the mantle), and the Moon seems especially deficient in iron. This fact favors the fission hypothesis; had the Moon condensed from the same

material as Earth, as in the condensation scenario, it would contain much more iron.

Geochemical evidence, especially isotopic evidence, increasingly supports the view that the Moon differs little from Earth's mantle, except for some loss of volatiles: potassium, for example. Although this evidence is compatible with the fission hypothesis, calculations considering angular momentum (recall the discussion of this concept in Chapter 5) strongly suggest that the fission mechanism doesn't work. Plate-tectonic theory now explains the formation of the Pacific Ocean basin. Before the ejection-of-a-ring theory was considered most probable, it seemed possible that the Pacific Ocean could be the scar left behind when the Moon was ripped from the Earth according to the fission hypothesis. It is possible that the angular momentum of the Earth and Moon were initially larger than at present (perhaps by as much as a factor of two) and this would allow a larger fraction of the disk-forming material to come from Earth rather than the projectile, thus leading to an outcome that is in some respects not unlike a fission story, except that very severe heating is required to explain the moon, and this must be supplied by the impact event (and is not possible by fission alone). The excess angular momentum must then be extracted by the Sun, perhaps by a resonance, but many aspects of this story remain uncertain.

We are obtaining additional evidence about the capture model by studying the moons of Jupiter and Saturn. The outermost moon of Saturn, for example, is apparently a captured asteroid (small bodies orbiting the Sun, mostly between Mars and Jupiter; see Chapter 8). However, the similarities in composition between the Moon's and the Earth's mantles argue against the capture hypothesis for the Earth–Moon system. Thus as of now, ejection of a gaseous ring is the most accepted model.

6.2e Rocks from the Moon

A handful of meteorites found in Antarctica, Australia, and Africa have been identified by their chemical composition as having come from the Moon (■ Fig. 6–25). They presumably were ejected from the Moon when craters formed. So we are still getting new moon rocks to study and, as we saw above, restudying the ones we had with improved methods! A few other meteorites have even been found to come from Mars, as we shall discuss later in this chapter.

Robin Canup, Southwest Research Institute

■ **FIGURE 6–24** A computer simulation of a collision between the young Earth and an impactor. Each is composed of an iron core and a rocky mantle. The colors indicate temperature. In the first images, the impactor hits the young Earth, separates, and then falls in again, pulling out a tail of matter. This sequence takes about 24 Earth hours; the large clump probably disperses. But in about a hundred Earth years, the Moon forms out of the material in the disk.

NASA's Johnson Space Center

■ **FIGURE 6–25** This rock is one of many meteorites found in the Earth's continent of Antarctica. Under a microscope, it seems like a sample from the lunar highlands and quite unlike any terrestrial rock. Eighteen of the known meteorites are thought to have come from the Moon.

FIGURE 6–26 Since Mercury's orbit is much closer to the Sun than ours (a), Mercury is never seen against a really dark sky. As a result, even Copernicus apparently never saw it. A simulated view from the Earth appears in (b), showing Mercury and Venus at their greatest respective distances from the Sun, looking west soon after sunset. (c) A very rare case where four bright planets are visible in the same region of the sky shortly after sunset (May 4, 2002). Venus is the brightest object, Mars is to its upper left, Saturn is slightly below and to the left of Mars, and Mercury is the relatively bright object (comparable to Mars and Saturn) to the lower right of Venus.

FIGURE 6–27 Mercury silhouetted against the Sun during its 1999 transit, observed from the solar spacecraft known as TRACE. (Other dark spots are sunspots; see Chapter 10.) One of the authors (J.M.P.) participated in analysis of these observations in order to prepare for the much rarer transits of Venus that occurred in 2004 and in 2012 (Section 6.4a). (b) A high-resolution view of Mercury during the 2006 transit and fine-scale solar features imaged by our team at the Sacramento Peak Observatory in New Mexico, part of the National Solar Observatory.

6.3 MERCURY

Mercury is the innermost of our Sun's eight planets. Its average distance from the Sun is $^4/_{10}$ of the Earth's average distance, or 0.4 au. Its elliptical orbit around the Sun is the most elongated (eccentric) of the planets.

Since we on the Earth are outside Mercury's orbit looking in at it, Mercury always appears close to the Sun in the sky (■ Fig. 6–26). At times Mercury rises just before sunrise, and at times it sets just after sunset, but it is never up when the sky is really dark. The Sun always rises or sets within an hour or so of Mercury's rising or setting. As a result, whenever Mercury is visible, its light has to pass obliquely through the Earth's atmosphere. This long path through turbulent air leads to blurred images. Thus astronomers have never had a really clear view of Mercury from the Earth, even with the largest telescopes. Even the best photographs taken from the Earth show Mercury as only a fuzzy ball with faint, indistinct markings. Most people have never seen it at all.

About 15 times per century, Mercury goes into **transit** across the Sun; that is, we see it as a black dot crossing the Sun. Transits of Mercury (■ Fig. 6–27) occurred in 1999, 2003, and 2006. The next transit of Mercury will occur on May 9, 2016. Understanding what we see as Mercury transits helps us understand the much rarer transits of Venus, which we will discuss later in this chapter, and potentially transits of exoplanets across the faces of their parent stars (Chapter 9).

6.3a The Rotation of Mercury

When the only data available were provided by ground-based drawings and photographs, astronomers did as well as they could to describe Mercury's surface. A few features seemed to be barely distinguished, and the astronomers watched to see how long those features took to rotate around the planet. From these observations they decided that Mercury rotates in the same amount of time that it takes to revolve

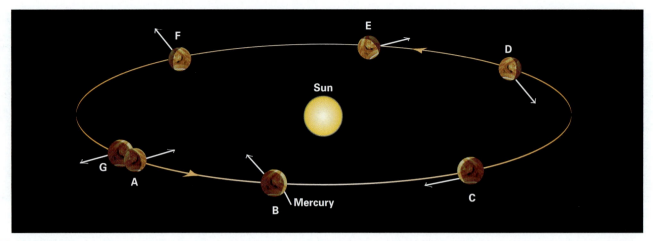

■ **FIGURE 6–28** Follow, from A to G. the arrow that starts facing to the right, toward the Sun in A. Mercury rotates once with respect to the stars in 59 days, when Mercury has moved only ²/₃ of the way around the Sun (E). Note that after one full revolution of Mercury around the Sun (G), the arrow faces away from the Sun. It takes another full revolution, a second 88 days, for the arrow to again face the Sun. Thus Mercury's rotation period with respect to the Sun is twice 88, or 176, days.

around the Sun in its orbit, 88 Earth days. Thus they thought that one side always faces the Sun and the other side always faces away from the Sun. (Recall that one side of the Moon always faces the Earth, a similar phenomenon.) This discovery led to the fascinating conclusion that Mercury could be both the hottest planet (on the Sun-facing side) and the coldest planet (on the other side) in the Solar System.

But when the first measurements were made of Mercury's radio radiation, the planet turned out to be giving off more energy than had been expected. This meant that it was hotter than expected. The dark side of Mercury was too hot for a surface that was always in the shade. (The visible light we see is merely sunlight reflected by Mercury's surface and doesn't tell us the surface's temperature. The radio waves are actually being emitted by the surface, as part of its "thermal" or nearly blackbody radiation; see the discussion in Chapter 2.)

Later, we became able to transmit radar from Earth to Mercury. (Radar – **ra**dio **d**etection **a**nd **r**anging – is sending out radio waves so that they bounce off another object, allowing you to study their reflection.) Since one edge of the visible face of Mercury is rotating toward Earth, while the other edge is rotating away from Earth, the reflected radio waves were slightly smeared in wavelength according to the Doppler effect (recall the description of this effect in Chapter 2). This measurement allowed astronomers to determine Mercury's rotation speed, similarly to the way that radar is used by the police to tell if a car is breaking the speed limit. Knowing the rotation speed and Mercury's radius, we could determine the rotation period.

The results were a surprise: it actually rotates every 59 days, not 88 days. Mercury's 59-day period of rotation with respect to the stars is exactly ²/₃ of the 88-day orbital period, so the planet rotates three times for every two times it revolves around the Sun.

Mercury's rotation and revolution combine to give a value for the rotation of Mercury relative to the Sun (that is, a mercurian solar day) that is neither 59 nor 88 days long (■ Fig. 6–28). If we lived on Mercury we would measure each "day" (that is, each day/night cycle) to be 176 Earth days long. We would alternately be fried for 88 Earth days and then frozen for 88 Earth days. Since each point on Mercury faces the Sun at some time, the heat doesn't build up forever at the place under the Sun, nor does the coldest point cool down as much as it would if it never received sunlight. The hottest temperature is about 700 K (800°F). The minimum temperature is about 100 K (–280°F).

No harm was done by the scientists' original misconception of Mercury's rotational period, but the story teaches all of us a lesson: we should not be too sure of "facts." Scientists are currently learning new things about Mercury, and we realize that even some of the information in this book may be out of date by the time it is finished printing. It should be fun for you to look back in twenty years and see how much of what we now think we know about astronomy actually turned out to be wrong. After all, science is a dynamic process.

6.3b Mercury's History

Mercury may be the fragment of a giant early collision that nearly stripped it to its core. The stripping would account for the large proportion of iron in Mercury relative to Earth. Still early on, the core heated up and the planet's crust expanded. The expansion opened paths for molten rock to flow outward from the interior, producing the intercrater plains. As the core cooled, the crust contracted, producing cliffs and wrinkles. Solar tides (tides induced by the Sun) slowed Mercury's original rotation until its rotational period became ²/₃ of the orbital period.

A CLOSER LOOK 6.4

Naming the Features of Mercury

Seeing features on the surface of Mercury for the first time led to an immediate need for names. The scarps were named for historical ships of discovery and exploration, such as Endeavour (Captain Cook's ship, with which he went to Tahiti to see the transit of Venus and wound up also exploring what we now call Australia and New Zealand), Santa Maria (Columbus's ship), and Victoria (the first ship

to sail around the world, which it did from 1519 to 1522 under Magellan and his successors). Some plains were given the names of the gods equivalent to the Greeks' Mercury, such as Tir (in ancient Persian), Odin (an ancient Norse god), and Suisei (Japanese). Craters were named for nonscientific authors, composers, and artists, in order to complement the lunar naming system, which honors scientists.

Light rays

High albedo   Medium albedo   Low albedo

■ **FIGURE 6–29** Albedo is the fraction of radiation reflected. A surface of low albedo looks dark.

6.3c Mercury Observed from the Earth

Even though the details of the surface of Mercury can't be seen very well from the Earth, other properties of the planet can be better studied. For example, we can measure Mercury's **albedo**, the fraction of the sunlight hitting Mercury that is reflected (■ Fig. 6–29). We can measure the albedo because we know how much sunlight hits Mercury (we know the brightness of the Sun and the distance of Mercury from the Sun). Then we can easily calculate at any given time how much light Mercury reflects, knowing how bright Mercury looks to us and its distance from the Earth. Comparison with the albedos of materials on the Earth and on the Moon can teach us something of what the surface of Mercury is like.

Let us consider some examples of albedo. An ideal mirror reflects all the light that hits it; its albedo is thus 100 per cent. (The very best real mirrors have albedos of as much as 96 per cent.) A black cloth reflects essentially none of the light that hits it; its albedo (in the visible part of the spectrum, anyway) is almost 0 per cent. Mercury's overall albedo is only about 10 per cent. Its surface, therefore, must be made of a dark (that is, poorly reflecting) material (though a few regions are very reflective, with albedos of up to 45 per cent). The albedos of the Moon's maria are similarly low, 6 to 10 per cent. In fact, Mercury (or the Moon) appears bright to us only because it is contrasted against a relatively dark sky; if it were silhouetted against a white bedsheet, it would look relatively dark.

6.3d Spacecraft Views of Mercury

In 1974, an uncrewed NASA spacecraft flew right by Mercury. (See *A Closer Look 6.4: Naming the Features of Mercury.*) The tenth in the series of Mariner spacecraft launched by the United States went to Mercury with a variety of instruments on board. First the 475-kg spacecraft passed by Venus and had its orbit changed by Venus's gravity to direct it to Mercury. Tracking its orbit improved our measurements of the gravity of these planets and thus of their masses.

NASA has finally sent a spacecraft back to Mercury for the first time since Mariner 10. The **ME**rcury **S**urface, **S**pace **EN**vironment, **GE**ochemistry and **R**anging (or MESSENGER) spacecraft was launched in 2004 and, after conducting two flybys of Mercury in 2008, entered into orbit around Mercury in 2011. Within six months of entering its orbit, MESSENGER had confirmed certain theories and in other ways changed astronomers' understanding of Mercury.

The most striking overall impression, first from Mariner 10 and now from MESSENGER, is that Mercury is heavily cratered (■ Fig. 6–30). At first glance, it looks like the Moon! But there were several basic differences visible between the features on the surface of Mercury and those on the lunar surface. Mercury's craters seem flatter than those on the Moon, and they have thinner rims (■ Fig. 6–31). Scientists theorized that Mercury's higher gravity at its surface may have caused the rims to slump more. Also, Mercury's surface may have been softer, more plastic-like, when most of the cratering occurred. The craters may have been eroded by any of a number of methods, such as the impacts of meteorites or micrometeorites (large or small bits of interplanetary rock). Alternatively, erosion may have

■ **FIGURE 6–30** (a) One side of Mercury, imaged from NASA's MESSENGER spacecraft. At left, a mosaic of thousands of individual images, each taken with a resolution of 1 km/pixel. At right, color differences are brought out by coding wide-angle camera filters at 1,000, 750, and 430 nm as red, green, and blue, respectively. (b) A global mosaic made from 22,000 images.

■ **FIGURE 6–31** Comparative planetology: A comparison of (a) a crater on Mercury, with hollows (also called horcruxes from the Harry Potter books) on its floor, observed from MESSENGER, (b) lunar craters, (c) craters on Mars, and (d) craters on Venus. Though it is not noticeable on these images, material has been thrown out less far on Mercury than on the Moon because of Mercury's higher gravity. Craters on Mars and Venus show flows across their surfaces, possibly resulting from the melting of a buried ice-rich layer and lava flows, respectively. The photographs are dominated by Copernicus on the Moon, 93 km in diameter; a similarly sized flat-bottomed crater on Mercury; Cerulli on Mars, 115 km in diameter; and craters on Venus, 37 to 50 km in diameter.

occurred during a much earlier period when Mercury may have had an atmosphere, undergone internal activity, or been flooded by lava.

Most of the craters seem to have been formed by impacts of meteorites. The Caloris Basin, in particular, is the site of a major impact. The secondary craters, caused by material ejected as primary craters were formed, are closer to the primaries than on the Moon, presumably because of Mercury's higher surface gravity. In many areas, the craters appear superimposed on relatively smooth volcanic plains. Their age is estimated to be 4.2 billion years, making them the oldest features on Mercury.

Smaller, brighter craters are sometimes, in turn, superimposed on the larger craters and thus must have been made afterward. Some craters have rays of higher albedo emanating from them (■ Fig. 6–32), just as some lunar craters do. The ray material represents relatively recent crater formation (that is, within the last hundred million years). The ray material must have been tossed out in the impact that formed the crater.

Lines of cliffs hundreds of miles long are visible on Mercury; on Mercury, as on Earth, such lines of cliffs are called **scarps**. The scarps are particularly apparent in the region of Mercury's south pole (■ Fig. 6–33). Unlike fault lines on the Earth, such as California's

famous San Andreas fault (responsible for the 1906 San Francisco earthquake), on Mercury there are no signs of geologic tensions like rifts or fissures nearby. These scarps are global in scale, not just isolated. The scarps may actually be wrinkles in Mercury's crust. Perhaps Mercury was once molten, and shrank by 1 or 2 km as it cooled. This shrinking would have caused the crust to buckle, creating the scarps in the quantity that we now observe.

Data from Mariner's infrared radiometer indicate that the surface of Mercury is covered with fine dust, as is the surface of the Moon, to a depth of at least several centimeters. Astronauts sent to Mercury, whenever they go, will leave footprints behind. Part of Mercury's surface is very jumbled, probably from the energy released by an impact.

The biggest surprise of the Mariner 10 mission was the detection of a magnetic field in space near Mercury. The field is weak, only about 1 per cent of the Earth's surface field. It had been thought that magnetic fields were generated by the rapid rotation of molten iron cores in planets, but Mercury is so small that its core should have quickly solidified after forming. So the magnetic field is probably not now being generated. Perhaps the magnetic field has been frozen into Mercury since the time when its core was molten.

FIGURE 6–32 (a) The crater Degas, with the area shown in the image from Mariner 10 at left, displayed with higher resolution at right in an image from MESSENGER. Note how the MESSENGER image revealed cracks in the floor of the crater, which is 52 km across. (b) Dominici, the bright crater at the top, shows bright rays as well as the newly discovered feature on Mercury known as halos. (c) Rays extending hundreds of kilometers from Enwonwuu crater, which is 38 km across.

FIGURE 6–33 (a) Yeats, named for the poet and dramatist, is crossed by a scarp from top to bottom in this image from MESSENGER. The crater is 92 km across. (b) A scarp across the crater Ts'ai Wen-chai, named for a Han Dynasty poet. (c) The scarp Beagle Rupes, one of the tallest and longest scarps on Mercury.

6.3e Mercury Research Rejuvenated

After Mariner 10, decades passed with no additional images of Mercury, but in 2000 two teams of scientists released their composite ground-based images. Each was better than any previous ground-based image of Mercury (■ Fig. 6–34a).

A dozen years after Mariner 10, an important new discovery about Mercury was made with a telescope on Earth: Mercury has an atmosphere! The atmosphere is very thin, but is still easily detectable in spectra. It was a surprise that Mercury has an atmosphere because it is so hot, since it is relatively close to the Sun. "Hot" means that the individual atoms or molecules are moving rapidly in random directions, and to keep an atmosphere a planet must have enough gravity to hold in the constituent particles. Since Mercury has relatively little mass but is hot, the atmospheric particles escape easily (■ Fig. 6–34b) and few are left – none from long ago. They must be replenished by some process.

Mercury's surface features can be mapped from Earth with radar. The radar observations provide altitudes and show how rough the surface is. The radar features show, in part, the half of Mercury not imaged by Mariner 10. It, too, is dominated by intercrater plains, though its overall appearance is different. The craters, with their floors flatter than the Moon's craters, show clearly on the radar maps. The scarps are obvious as well. The highest-resolution images (■ Fig. 6–35) reveal probable water-ice deposits near Mercury's poles. They have been shielded from sunlight for so long that ice (deposited by comets) could have persisted.

6.3f Mercury from MESSENGER

The MESSENGER spacecraft has found that Mercury's "exosphere," a term used for the wispy atmosphere, consists primarily of sodium and oxygen particles kicked up by the solar wind. As the only planet in

(a)

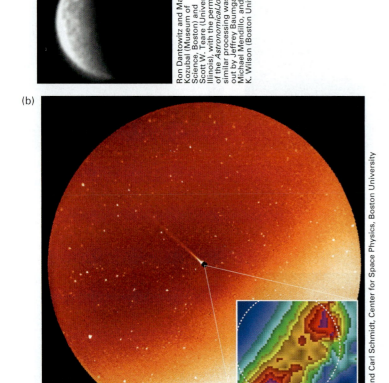

(b)

Ron Dantowitz and Marek Kozubal (Museum of Science, Boston) and Scott W. Teare (University of Illinois), with the permission of the *AstronomicalJournal*; similar processing was carried out by Jeffrey Baumgardner, Michael Mendillo, and Jody K. Wilson (Boston University)

J. K. Wilson and Carl Schmidt, Center for Space Physics, Boston University

■ **FIGURE 6–34** (a) The best visible-light image of Mercury ever taken from Earth. It is a combination of several dozen very short exposures, the result of an effort to find moments when the Earth's atmosphere was the most steady as we looked through it (that is, when the "seeing" was best). (b) Mercury's sodium tail, 2° long (four times the Moon's angular diameter), imaged from the ground. The brightness of Mercury itself is hidden behind an occulting disk at the center of the image. The image was taken by Boston University astronomers from the McDonald Observatory in Texas. The inset image, of Mercury behind the occulting disk, was taken with adaptive optics on a telescope on Maui. With this clarity, we can see that the sodium's source regions are mainly at high latitudes.

(a)

John Harmon, National Astronomy and Ionosphere Center

(b)

NASA/Johns Hopkins University Applied Physics Laboratory/ Carnegie Institute of Washington

■ **FIGURE 6–35** (a) Mercury's north pole, imaged with the radar capabilities of the giant radio telescope at Arecibo, Puerto Rico. The image measures 450 km on a side and shows detail down to 1.5 km. The bright features are most likely water-ice deposits located in the permanently shaded floors of craters. (b) An composite image centered on Mercury's south pole, taken from MESSENGER.

the Solar System besides Earth with magnetized poles, Mercury exhibits a greater concentration of these particles at its poles, evidence that the planet's weak magnetosphere is providing little protection from the solar wind. Helium ions, believed to be generated by the solar wind interacting with the Mercury's surface, were found throughout the magnetosphere.

High levels of sulfur and potassium were also detected. Since these elements vaporize at relatively low temperatures, this finding is inconsistent with previous theories that Mercury experienced extremely high temperatures early in its formation.

Huge expanses of volcanic plains surrounding the north pole of Mercury and covering more than 6 per cent of the planet's surface were discovered. Vents of up to 25 km (15.5 miles) in length also have been discovered. These are likely the source of a large amount of hot lava that flowed over the planet and created the current landscape.

"Hollows" has been adopted as a term to describe small, shallow, irregularly shaped depressions often found in clusters in many areas of the planet (■ Fig. 6–36). Many of these hollows have bright interiors and halos, and appear not to have accumulated craters from small impacts, indicating that they are relatively young features. They may have formed by the loss of volatile materials.

Scientists analyzing MESSENGER's orbit around Mercury have mapped Mercury's topography and gravitational field. From the latter, they conclude that Mercury has an exceptionally large iron core, which is covered with a layer of iron sulfide. That solid layer, in

25 km

■ **FIGURE 6–36** Hollows in a crater, shown in enhanced color.

turn, is topped by a thin mantle and crust of silicates. The scientists were further surprised by the discovery that a large number of Mercury's craters are filled in, which suggests that processes deforming Mercury's surface were active longer than expected.

Mercury was once thought to be similar to our Moon. However, MESSENGER has already proven to scientists that Mercury has a much more interesting past than anyone had predicted.

6.3g Continuing Exploration of Mercury

ESA and Japan are planning to launch BepiColombo to Mercury in 2015 for arrival in 2022. (Bepi – a nickname – Colombo was a scientist who studied Mercury and spacecraft trajectories.) It will carry two orbiters, one to map the planet in a variety of wavelengths and the other to study Mercury's magnetosphere.

6.4 VENUS

Venus orbits the Sun at a distance of 0.7 au. Although it comes closer to us than any other planet, we did not know much about it until recently because it is always shrouded in highly reflective clouds (■ Fig. 6–37).

6.4a Transits of Venus

Venus was in transit (crossing the Sun's disk) in 2004 and again in 2012, for the first times since the pair of transits in 1874 and 1882. Previous transits of Venus were important for determining the scale of the Solar System, once called "the noblest problem in

■ **FIGURE 6–37** A crescent Venus, observed with a large telescope on Earth. In visible-light images like this one, we see only a layer of clouds.

astronomy," which we can now measure better by other means. Still, the matter is of historical interest and can provide some scientific insights into properties of Venus's atmosphere. In addition, transits of Venus allow us to make relatively close-up and detailed observations of what a transit of a star by a planet looks like in detail, providing an analogue to similar observations of exoplanets around their parent stars, which have led to the discovery of literally thousands of probable exoplanets from NASA's Kepler spacecraft over the last few years (Chapter 9).

The accuracy of observations of transits of Venus in 1761, 1769, 1874, and 1882 (the only ones ever scientifically observed until this century) was impeded by the dreaded "black-drop effect," in which a dark apparent connection between the silhouette of Venus and the edge of the Sun remained for about a minute before Venus was clearly distinctly separated from the edge. (Captain Cook went to Tahiti to observe the 1769 transit; his subsequent exploration of the coast of Australia and of New Zealand can be considered a bonus of this astronomical expedition.) Timing the transit accurately determined how well the distances between the planets could be known, since Kepler's third law gave only relative distances.

One of the authors (J.M.P.) led a collaboration that used modern instrumentation on the ground and in space to observe the June 8, 2004, transit and the June 5, 2012, transit (■ Fig. 6–38). It was striking to see the black silhouette of Venus against the Sun, something that could be seen looking through a suitable filter even without a telescope. It looked as though a hole had been drilled through the Sun. With spacecraft in orbit and telescopes from the

(a)   (b)

■ **FIGURE 6–38** (a) The transit of Venus of June 8, 2004. (b) A frame from the middle of the six-hour transit of Venus of June 5, 2012, at a time of higher solar activity than that of the previous transit.

ground, we observed the black drop, a small effect compared with its reports from old records.

Data we obtained from the **T**ransition **R**egion **a**nd **C**oronal **E**xplorer (TRACE) spacecraft confirmed our earlier work based on observations of the 1999 transit of Mercury (whose atmosphere is negligible, yet a black-drop effect was still seen to some extent) that the black-drop effect has nothing to do with Venus's atmosphere. The black-drop effect as seen from space is caused by a combination of two things: (1) the inherent blurring produced by the optics of a telescope, and also, importantly, (2) solar limb-darkening, the fact that the brightness of the Sun's disk diminishes very rapidly just before its apparent edge. (In ground-based observations, the black-drop effect can be enhanced by turbulence in our Earth's atmosphere.) Though Venus has a substantial atmosphere, it is much too small in height to make the black-drop effect observed during Venus transits.

We were even able to use two different spacecraft to detect how the amount of light that the Earth received from the Sun dimmed by the 0.1 per cent that corresponded to the fraction of the Sun that was covered by Venus. Even more interesting, the measured amount of light was not constant, as Venus covered parts of the Sun that had different brightnesses. By seeing up close the causes of transit effects that astronomers are now studying with exoplanets (Chapter 9), the 21st-century transits of Venus provided links between historical astronomy and one of the most exciting current and future areas of astronomical investigation.

Though there won't be another transit of Venus visible from Earth until 2117 and 2125, if you are elsewhere in the Solar System, you might see a transit. One of us (J.M.P.) and colleagues used the newest camera on the Hubble Space Telescope to try to detect a transit of Venus in reflection off Jupiter's clouds on September 20, 2012. And we also arranged to point NASA's Cassini mission (in orbit in the Saturn system) toward the Sun to try to directly detect the dimming of sunlight on December 21, 2012, from a transit of Venus in the Sun-Venus-Saturn line.

6.4b The Atmosphere of Venus

The clouds on Venus are primarily composed of droplets of sulfuric acid, H_2SO_4, which are in acid solution in water. Sulfuric acid may seem like a peculiar constituent of a cloud, but the Earth, too, has a significant layer of sulfuric acid droplets in its stratosphere, a higher layer of the atmosphere. Sulfuric acid, also present in the lower levels of the Earth's atmosphere, is chemically processed and transported because of sunlight and weather.

Observations from Earth show a high concentration of carbon dioxide in the thick atmosphere of Venus. In fact, carbon dioxide makes up over 96 per cent of the mass of Venus's atmosphere (■ Fig. 6–39). The Earth's atmosphere, for comparison, is mainly nitrogen (79 per cent), with a fair amount of oxygen (20 per cent) as well. Carbon dioxide makes up less than 1 per cent of the terrestrial atmosphere.

96% CO_2

<4% N_2 / <0.1% H_2O / <1% O_2

■ **FIGURE 6–39** The composition of Venus's atmosphere.

Because of the large amount of carbon dioxide in its atmosphere, which leads to the atmosphere's massive nature, Venus's surface pressure is about 90 times higher than the pressure of Earth's atmosphere. Carbon dioxide on Earth dissolved in seawater and rain, eventually forming some types of terrestrial rocks, often with the help of life forms. (Limestone, for example, has formed from deceased marine life.) If this carbon dioxide were released from the Earth's rocks, along with other carbon dioxide trapped in seawater, our atmosphere would become as dense and have as high a pressure as that of Venus. Venus, slightly closer to the Sun than Earth and thus hotter, has no oceans or rain in which the carbon dioxide can dissolve to help take it up from the atmosphere. Thus the carbon dioxide remains in Venus's atmosphere.

ESA's Venus Express spacecraft has orbited Venus since 2006. One of its primary objectives is studying the planet's atmosphere. It identified hydroxyl, the first detection of this radical in the atmosphere of a planet other than Earth; it also detected nitric oxide nightglow, the first time this has been found in the infrared for any planet. We will continue to learn more about Venus's atmosphere as this satellite is scheduled to send back data until at least 2014.

6.4c The Rotation of Venus

In 1961, the radio waves used in radar astronomy penetrated Venus's clouds, allowing us to determine accurately how fast Venus rotates. Venus rotates in 243 days with respect to the stars, in the direction opposite that of the other planets. Venus revolves around the Sun in 225 Earth days. Venus's periods of rotation and revolution combine so that a solar day/night cycle on Venus corresponds to 117 Earth days; that is, the Sun returns to the same position in the sky every 117 days.

The notion that Venus rotates backward used to seem very strange to astronomers, since all the planets revolve around the Sun in the forward direction, and most of the other planets (except Uranus, which is on its side; dwarf-planet Pluto rotates on its side as well) and most satellites also rotate in that forward direction. Because the laws of physics do not allow the amount of spin (angular momentum; recall the discussion in Chapter 5) to change on its own, and since the original material from which the planets coalesced was undoubtedly rotating, we had expected all the planets to revolve and rotate in the same sense.

Nobody knows definitely why Venus rotates "the wrong way." One possibility is that in the chaos that reigned in the early Solar System, when Venus was forming, a large clump of material struck it at an angle that caused the merged resulting planet to rotate backward. Recent studies of chaotic changes in Venus's angle of rotation show alternative ways that Venus could have wound up rotating backward, perhaps depending on the fact that Venus's atmosphere is so dense. The Sun's gravity raises tidal bulges in Venus's dense atmosphere. Gravitational tugs from other planets on these tidal bulges contribute to making the system chaotic, and eventually the axis could flip over.

The slow rotation of Venus's solid surface contrasts with the rapid rotation of its clouds. The tops of the clouds rotate in the same sense as the surface rotates but about 60 times more rapidly, once every 4 Earth days. Lower parts of the atmosphere, however, rotate very slowly.

6.4d Why Is Venus So Incredibly Hot?

We can determine the temperature of Venus's surface by studying its radio emission, since radio waves emitted by the surface penetrate the clouds. The surface is very hot, about 735 K (865°F), even on the night side. In addition to measuring the temperature on Venus, scientists theoretically calculate what the temperature would be if Venus's atmosphere were transparent to radiation of all wavelengths, both coming in and going out. This value – less than 375 K (215°F) – is much lower than the measured values. The high temperatures derived from radio measurements indicate that Venus traps much of the solar energy that hits it.

The process by which the surface of Venus is heated so much is called the greenhouse effect (■ Fig. 6–40), already discussed briefly in Section 6.1d. The fraction of sunlight not reflected outward by the high clouds and molecules passes through the venusian atmosphere in the form of radiation in the visible part of the spectrum. The sunlight is absorbed by the surface of Venus, which heats up. At the resulting temperatures, the thermal (nearly blackbody) radiation that the surface gives off is mostly in the infrared. But the carbon dioxide and other constituents of Venus's atmosphere are largely

Table 6.3 Comparative planetology: greenhouse effect	
Earth:	33°C extra
Venus:	375°C extra
Mercury:	Negligible
Mars:	3°C extra

(but not entirely) opaque to infrared radiation; thus, much of the energy is absorbed. Since it doesn't get out, we commonly say that it is "trapped," though not literally of course (the radiation no longer exists after being absorbed). The atmosphere becomes hotter, thereby heating the planet's surface as well.

Thus Venus heats up far above the temperature it would reach if the atmosphere were transparent to infrared radiation. The surface radiates more and more energy as the planet heats up. Finally, a balance is struck between the rate at which energy flows in from the Sun and the rate at which it trickles out (as infrared) through the atmosphere. The situation is so extreme on Venus that we say a "runaway greenhouse effect" has taken place there. Understanding such processes involving the transfer of energy is but one of the practical results of the study of astronomy.

Greenhouses on Earth don't work quite this way. In actual greenhouses, closed glass on Earth prevents the mixing of air inside them (which is mainly heated by conduction from the warmed ground and by radiation from the ground and the greenhouse glass) with cooler outside air. The trapping of solar energy by the "greenhouse effect" – the inability of infrared radiation, once formed, to get out as readily, since it is absorbed by atmospheric carbon dioxide and water – is a less important process in an actual greenhouse or in a car when it is left in the sunlight.

As mentioned in Section 6.1d, the greenhouse effect can be beneficial; indeed, it keeps Earth's temperature at a comfortable level. Earth would be quite chilly without it – about 33°C (60°F) colder, on average (Mars's greenhouse effect is only 3°C).

Earth manages to achieve a comfortable greenhouse effect by recycling its greenhouse gases, primarily CO_2 and H_2O (water). As the oceans evaporate, water vapor builds up in the atmosphere, but then the rain comes down. The rain dissolves atmospheric CO_2, producing carbonic acid. This mild "acid rain" dissolves rocks and forms carbonates, and there are other forms of weathering as well. The carbonate solution eventually reaches the ocean. Various sea creatures then use the carbonates in their shells. When these organisms die, their carbonate shells form sediments on ocean bottoms. Because of plate tectonics, the sediments can later dive down under other plates, where they become part of the magma inside the Earth. The CO_2 is subsequently outgassed by volcanism, and the process repeats.

How did Venus reach its hellish state? It may have begun its existence with a pleasant climate, and possibly even with oceans. Perhaps rain and other processes were not quick enough in removing the atmospheric CO_2 outgassed by volcanoes. As CO_2 accumulated, the greenhouse effect increased, and temperatures rose; oceans evaporated faster, and a runaway greenhouse effect was in progress.

■ FIGURE 6–40 Though most sunlight is reflected from Venus's clouds, some sunlight penetrates them, so the surface is illuminated with radiation in the visible part of the spectrum. Venus's own radiation is mostly in the infrared, because it is much cooler than the Sun, pushing the peak of its blackbody curve to longer wavelengths. This infrared radiation does not escape as readily, leading to a warming of the atmosphere and hence of the surface, a phenomenon known as the greenhouse effect.

Table 6.4 Comparative planetology: temperature at surface
Earth: about 0°C to 40°C (just right)
Venus: about 462°C (much too hot)
Mercury: about –175°C to 425°C (much too cold to much too hot)
Mars: about –125°C to 25°C (too cold to just right)

The water vapor was gradually broken apart by ultraviolet radiation from the Sun, and the hydrogen (being so light) escaped from Venus. Eventually there were no oceans and no rain – but continued outgassing of CO_2 led to more greenhouse warming and extremely high temperatures.

It is also possible that Venus was essentially born in a hot state, or reached it very quickly. Without oceans, it is difficult for atmospheric CO_2 to become incorporated into rocks as it did on Earth. Extensive greenhouse heating is therefore maintained.

Studies of Venus are important for an overall understanding of Earth's climate and atmosphere. It is essentially undeniable that Earth's average temperature has increased during the past century, and there is a scientific consensus (though not complete agreement) that at least part of this global warming is caused by humans. The politics and what to do about it, however, remain controversial. As shown in Figure 6–10, the terrestrial atmospheric CO_2 concentration is increasing. Most scientists in the field have concluded that the human-induced increase in CO_2 (from the burning of fossil fuels) is one of the main causes of the current episode of global warming.

Questions have been raised whether a "positive feedback" mechanism that leads to sustained global warming will necessarily occur; there are many variables, and their effects are not yet completely understood. For example, an increase in global temperatures leads to more evaporation of the oceans, but this can increase the cloud cover and hence Earth's reflectivity (albedo), thereby producing a decrease in the amount of visible sunlight reaching Earth's surface. This is a "negative feedback" mechanism that tends to stabilize temperatures. On the other hand, recent computer models suggest that the higher water-vapor content of the atmosphere actually increases greenhouse heating by an amount that more than compensates for the greater reflectivity of the Earth. Earth's atmosphere is a very complicated system, but scientists are gaining confidence in the accuracy of their computer models.

It is very unlikely that the Earth will experience a *runaway* greenhouse effect that makes the planet unfit for *all* forms of life. In the past, there have been times when Earth was much warmer than now, yet a runaway did not occur. Moreover, life on Earth recovered from collisions with large asteroids and comets, some of which dumped enormous quantities of greenhouse gases into the atmosphere.

On the other hand, even a rise of a few degrees in Earth's global average temperature over a timescale of a few decades, which is now generally predicted, would have terrible consequences for humans, including flooding of coastal cities and destruction of many current agricultural areas. Also, since we cannot accurately predict what will happen under various circumstances, it is dangerous to act in ways that might disrupt the balance in our atmosphere and adversely affect life on Earth. Venus has shown us the importance of being very careful about how we treat the only home we have.

We need a major push in research, even on a 50-year timescale, to develop large-scale energy supplies from solar power, nuclear power (improved fission plants now and more research into possible future fusion plants), wind power (about which various problems have turned out to be more serious than originally thought), and other sources that do not give off greenhouse gases and so do not lead to an increased greenhouse effect. Whether such research funds will be available is a different question. Some people are now putting substantial emphasis on starting to *adapt* to the effects of greenhouse warming in place of (or in addition to) limiting the greenhouse warming itself, especially given the likelihood of using coal for economic growth in the developing world.

The "Kyoto Protocol," a multination agreement signed in Japan in 1997 to limit the emission of greenhouse-causing gases by industrialized nations, went into effect in 2005, when enough countries ratified it. The United States had signed it originally but never ratified it, so the U.S. was not part of the agreement, though over 190 countries were. At the Climate Gateway meeting in 2012 in Doha (in Qatar), the Kyoto Protocol was extended until 2020. A stronger treaty with more worldwide effect is being negotiated to start then.

6.4e Spacecraft Observations of Venus's Atmosphere

Venus was an early target of both American and Soviet space missions. During the 1960s, American spacecraft flew by Venus, and Soviet spacecraft dropped through its atmosphere. In 1970, the Soviet Venera 7 spacecraft radioed 23 minutes of data back from the surface of Venus before it succumbed to the high temperature and pressure. Two years later, the lander from Venera 8 survived on the surface of Venus for 50 minutes. Both landers confirmed the Earth-based results of high temperatures, high pressures, and high carbon dioxide content.

Several United States spacecraft have observed Venus's clouds and followed the changes in them. Some views of changes in Venus's clouds came from the Galileo spacecraft (■ Fig. 6–41) and from the Hubble Space Telescope. The most recent, ongoing views are from ESA's Venus Express, in orbit since its 2006 arrival. Structure in the clouds shows only when viewed in ultraviolet light. At mid-latitudes, the clouds appear as long, delicate streaks, looking like terrestrial cirrus clouds.

Such studies of Venus have practical value. The principles that govern weather on Venus are similar to those that govern weather on Earth, though there are major specific differences, such as the absence of oceans on Venus. The better we understand the interaction of solar heating, planetary rotation, and chemical composition in setting up an atmospheric circulation, the better we will understand

(a)

NASA/JPL/Caltech

(b)

ESA/CNR-IASF, Rome, Italy, and Observatoire de Paris, France

■ **FIGURE 6–41** (a) Changes in Venus's clouds over several hours, observed from the Galileo spacecraft. (b) Venus's day side on the left and night side on the right, in false color from Venus Express. At left, we see mainly sunlight reflected from the tops of clouds, down to about 65 km above the surface. At right, we see a near-infrared view at 1.7 μm that shows dynamic spiral cloud structures around 55 km above the surface. Some regions are brighter because the clouds are thinner, allowing more of the radiation from hotter, lower levels to leak through.

our Earth's atmosphere. We then may be better able to predict the weather. The potential financial return from such knowledge is enormous: it would be many times the investment we have made in planetary exploration.

Not one of the spacecraft to Venus has detected an internal magnetic field. The absence of a magnetic field may indicate either that Venus does not have a liquid core or that a possible liquid core does not rotate fast enough.

Spacecraft have provided evidence that volcanoes may be active on Venus. The abundance of sulfur dioxide they found varied at different times by a factor of 10. The effect could come from eruptions at least 10 times greater than that of even the largest recent

terrestrial volcanic eruptions. Lightning that was detected is further evidence that those regions are sites of active volcanoes. On Earth, it is common for the dust and ash ejected by volcanoes to rub together, generating static electricity that produces lightning.

Probes that penetrated the atmosphere and went down to the surface found that high-speed winds at the upper levels are coupled to other high-speed winds at lower altitudes. The lowest part of Venus's atmosphere, however, is relatively stagnant. The probes detected three distinct layers of venusian clouds, separated from each other by regions of relatively high transparency.

One of the probes measured that only about 2 per cent of the sunlight reaching the outer atmosphere of Venus filters down to the surface, making it like a dim terrestrial twilight. Thus most of the Sun's energy is absorbed in or reflected by the clouds, unlike the situation with Earth, and Venus's clouds are more important than Earth's for controlling weather. Most of the light that reaches the surface is orange, so photographs taken on the surface have an orange cast.

6.4f Radar Observations of Venus's Surface

From Venus's size and from the fact that its mean density is similar to that of the Earth, we conclude that its interior is also probably similar to that of the Earth. This means that we expect to find volcanoes and mountains on Venus.

We can study the surface of Venus by using radar to penetrate Venus's clouds. Radars using huge Earth-based radio telescopes, such as the giant 1,000-ft (305-m) dish at Arecibo, Puerto Rico, have mapped a small amount of Venus's surface with a resolution of up to about 20 km. (The **resolution** is the size of the finest details that could be detected, so we mean that only features larger than about 20 km can be detected.) Regions that reflect a large percentage of the radar beam back at Earth show up as bright, and other regions as dark.

NASA's Pioneer Venus, the Soviet Venera 15/16, and NASA's Magellan spacecraft were in orbit around Venus, allowing them to make observations over a lengthy time period. They carried radars to study the topography of Venus's surface, mapping a much wider area than could be observed from the Earth.

■ **FIGURE 6–42** The Arecibo Observatory's high-resolution radar image of Maxwell Montes, with a resolution of about 2 km. Brighter areas (red and yellow false color) show regions where more power bounces back to us from Venus, which usually corresponds to rougher terrain than darker (blue and green false color) areas.

Courtesy of D. B. Campbell

From the radar maps (look back at Figure 6–1), we now know that 60 per cent of Venus's surface is covered by a rolling plain, flat to within plus or minus 1 km. Only about 16 per cent of Venus's surface lies below this plain, a much smaller fraction than the two-thirds of the Earth covered by ocean floor.

Two large features, the size of small Earth continents, extend several km above the mean elevation. A northern "continent," Terra Ishtar, is about the size of the continental United States. The giant chain of mountains on it known as Maxwell Montes (■ Fig. 6–42) is 11 km high, which is 2 km taller than Earth's Mt. Everest

stands above terrestrial sea level. It had formerly been known only as a bright spot on Earth-based radar images. Ishtar's western part is a broad plateau, about as high as the highest plateau on Earth (the Tibetan plateau) but twice as large. An equatorial "continent," Aphrodite Terra, is about twice as large as the northern one and is much rougher.

From the radar maps, it appears that Venus, unlike Earth, is made of only one continental plate. We observe nothing on Venus equivalent to Earth's mid-ocean ridges, at which new crust is carried upward. In particular, the high-resolution Magellan observations (■ Fig. 6–43) show no signs of crust spreading laterally on Venus. Venus may well have such a thick crust that any plate tectonics that existed in the distant past was choked off. Thus "venusquakes" are probably much less common than earthquakes.

Perhaps giant hot spots, like those that created the Hawaiian Islands on Earth, force mountains to form on Venus in addition to causing volcanic eruptions. The rising lava may also make the broad, circular domes that are seen. Venus got rid of its internal energy through widespread volcanism. Apparently, virtually the whole planet was resurfaced with lava about 500 million years ago. Thus the surface we now study is relatively young, and the history of Venus's surface has been obliterated.

In the 1970s and 1980s, a series of Soviet spacecraft landed on Venus and sent back photographs (■ Fig. 6–44). They also found that the soil resembles basalt in chemical composition and density, in common with the Earth, the Moon, and Mars. They measured temperatures of about 735 K and pressures over 90 times that of the Earth's atmosphere, confirming earlier, ground-based measurements.

Our recent space results, coupled with our ground-based knowledge, show us that Venus is even more different from the Earth than had previously been imagined. Among the differences are Venus's slow rotation, its one-plate surface, the complete resurfacing that occurred relatively recently, the absence of a satellite, the extreme weakness or

■ **FIGURE 6–43** Maat Mons in a simulated perspective view based on radar data from Magellan and arbitrarily given the orange color that comes through Venus's clouds. Our viewpoint is 560 km north of Maat Mons at an elevation of 1.7 km. We see lava flows extending hundreds of kilometers across the foreground fractured plains. Maat Mons is 8 km high; its height here is exaggerated by a factor of about 20. The absence of plate tectonics on Venus may result in part because its high temperature baked carbon dioxide and water out of the rocks.

NASA/JPL/Caltech

Courtesy of Valeriy Barsukov and Yuri Surkov

absence of a magnetic field, the lack of water in its atmosphere, and its high surface temperature and pressure.

6.4g Venus Exploration in the 21st Century

As mentioned previously, ESA's Venus Express spacecraft has been in orbit around Venus since 2006, studying the planet's atmosphere and creating an infrared map of its surface from a 24-hour elliptical, quasi-polar orbit ((■ Fig. 6–45). Among its major breakthroughs, continuous radio-occultation measurements by the VeRa instrument have provided the most comprehensive dataset to date of the atmospheric density and temperature. Stellar and solar occultations have provided very high spatial resolution vertical profiles of the abundance of both major and trace chemical species in the mesosphere. Observations of the surface in the near-infrared (1.18 μm) have evidenced several areas of recent crustal activity and thermal anomalies, corresponding to suspected "hot spots" previously identified in the Magellan radar and gravity-field maps.

In coordination with ground-based Doppler measurements in several wavelength ranges, Venus Express has characterized atmospheric wind circulation regimes by cloud tracking at several altitudes, while new airglow species (hydroxyl, nitric oxide) have been

described along with O_2 luminescence at 1.27 μm. Escape rates for the major escaping ions (H, He, O) have yielded new constraints on how an intrinsic magnetic field protects a planetary atmosphere. We will continue to learn more about Venus's atmosphere, as this satellite is scheduled to send back data until at least 2014.

NASA's MESSENGER mission to Mercury (Section 6.3f) conducted flybys of Venus in 2006 and in 2007. In addition, the Japanese Aerospace Exploration Agency (JAXA) launched the Akatsuki probe, also known as Venus Climate Orbiter, in May 2010. However, though it reached Venus, the spacecraft failed to enter orbit around the planet. ESA will try again in 2015–16.

6.5 MARS

Mars has long been the planet of greatest interest to scientists and nonscientists alike. Its unusual appearance as a reddish object in the night sky with some color changes with its seasons, coupled with some past scientific studies, have made Mars the first place we have looked for extraterrestrial life.

In 1877, the Italian astronomer Giovanni Schiaparelli published the results of a long series of telescopic observations he had made of Mars. He reported that he had seen "canali" on the surface. When this Italian word for "channels" was improperly translated into "canals," which seemed to connote that they were dug by intelligent life, public interest in Mars increased. Percival Lowell, in particular, was fascinated by the prospect of an ancient civilization on Mars. In the late 19th century, he drew detailed maps of Mars that showed an elaborate system of canals, presumably built to bring water to arid regions.

Over the next decades, there were endless debates over just what had been seen. We now know that the canals Schiaparelli and other observers reported are not present on Mars. They were an illusion. In fact, positions of the "canali" do not even always overlap the spots and markings that are actually on the martian surface (■ Fig. 6–46). But hope of finding life elsewhere in the Solar System springs eternal, and the latest studies have indicated the presence of considerable quantities of liquid water in Mars's past, a fact that leads many astronomers to suggest that life could have formed during those periods.

ESA/NASA/Caltech-JPL/Sue Smrekar

■ **FIGURE 6–45** A Venus Express image of heat patterns (showing minerals recently changed by flowing lava) derived from surface brightness, superimposed on contours from Magellan's radar; the vertical scale is exaggerated by a factor of about 30.

(a)

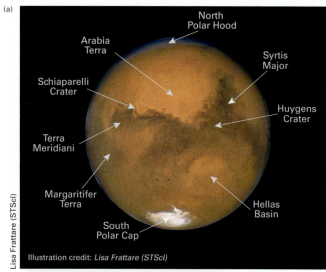

North
Polar Hood

Arabia
Terra

Schiaparelli
Crater

Syrtis
Major

Huygens
Crater

Terra
Meridiani

Margaritifer
Terra

South
Polar Cap

Hellas
Basin

Illustration credit: *Lisa Frattare (STScI)*

Lisa Frattare (STScI)

(b)

NASA/JPL/Malin Space Science Systems

(c)

Andrew Chaikin

■ **FIGURE 6–46** (a) Mars viewed from the Hubble Space Telescope. (b) A composite view from Mars Global Surveyor in 2005, showing Mars's Tharsis plain and set of giant volcanoes. The images in the composite were taken at the same Mars longitude daily over a year, showing northern-hemisphere autumn and southern-hemisphere spring. (c) Views from Earth of Mars going close to, and being partly occulted by, the Moon.

6.5a Characteristics of Mars

Mars is a small planet, 6792 km across, which is only about half the diameter and one-eighth the volume of Earth or Venus, although somewhat larger than Mercury. It has two tiny (about 20 km in diameter), irregularly shaped moons named Phobos and Deimos (■ Fig. 6–47). Mars rotates on its axis in about 24$\frac{1}{2}$ Earth hours, meaning that its day/night cycle is similar to that of our own planet.

Mars's atmosphere is very thin, only 1 per cent of Earth's, but it might be sufficient for certain kinds of life. Unlike the orbits of Mercury or Venus, the orbit of Mars is outside the Earth's, so we can observe Mars in the late-night sky.

Mars revolves around the Sun in 1.9 Earth years. The axis of its rotation is tipped at a 25° angle from perpendicular to the plane of its orbit, nearly the same as the Earth's 23$\frac{1}{2}$° tilt. Because the tilt of the axis causes the seasons, we know that Mars goes through its year with four seasons just as the Earth does.

NASA/JPL/Caltech

■ **FIGURE 6–47** Mars's moons Deimos (*lower left*) and Phobos (*lower right*), with the asteroid Gaspra (*top*) for comparison, shown to the same scale. Phobos is about 27 km across. Deimos measures about 15 km × 12 km × 11 km. Its surface is heavily cratered and thus presumably very old. The surface seems smoother than that of Phobos, though, because soil 10 m thick covers Deimos. Spectra of Phobos and Deimos are similar to those of one type of dark asteroid. The Phobos and Deimos images are from Viking; the Gaspra image is from Galileo. See Figures 6–59 and 6–60 for more recent images.

We have watched the effects of the seasons on Mars over the last century. In the martian winter, in a given hemisphere, there is a polar cap. As the martian spring comes to the northern hemisphere, the north polar cap shrinks and material at more temperate zones darkens.

The surface of Mars appears mainly reddish-orange when seen from Earth against dark space, probably due to the presence of iron oxide (rust), with darker gray areas that appear blue-green for physiological reasons, as the human eye and brain misjudge the color contrast. The changes in color over time apparently result from dust that is either covering surface rock or is blown off by winds that can have speeds of hundreds of kilometers per hour. Sometimes, global dust storms occur, and we can later watch the color change as the dust blows off the places where it first lands, exposing the dark areas underneath.

From Mars's mass and radius we can easily calculate that it has an average density about that of the Moon, substantially less than the densities of Mercury, Venus, and Earth. The difference indicates that Mars's overall composition must be fundamentally different from that of these other planets. Mars probably has a smaller iron-rich core and a thicker crust than does Earth.

6.5b Mars's Surface

Mars has been the target of several series of spacecraft, most notably some U.S. spacecraft, including Mariner 9 in 1971; a pair of Vikings that started orbiting in 1976 and dropped landers onto the martian surface; and Mars Pathfinder, which landed on July 4, 1997, and placed the Sojourner rover on Mars's surface. NASA's Mars Global Surveyor orbited the planet from 1997 through 2006, when it lost contact with Earth during its third extended mission phase. NASA's Spirit and Opportunity rovers, which arrived in 2004 (and, in the case of Spirit, ended communication in 2010), and Mars Phoenix in 2008, have sent back detailed views (■ Fig. 6–48) and have found many signs that water flowed on Mars in the past. NASA's Mars Odyssey, NASA's Mars Reconnaissance Orbiter, and ESA's Mars Express are currently orbiting the planet. NASA's Curiosity, a much larger and more capable rover than the Spirit and Opportunity rovers, landed in Gale Crater on Mars in August 2012. (See *A Closer Look 6.5: Mars Exploration Rovers, Mars Phoenix, and Mars Science Lab with Curiosity.*)

Surface temperatures measured on Mars from orbiters and landers range from a low of about 150 K (–190°F) to over 300 K (80°F). The temperature can vary each day by 35 to 50°C (60 to 90°F).

The surface of Mars can be divided into four major types: volcanic regions, canyon areas, expanses of craters, and terraced areas near the poles. A chief surprise was the discovery of extensive areas of volcanism. The largest volcano – which corresponds in position to the surface marking long known as Nix Olympica, "the Snow of Olympus" – is named Olympus Mons, "Mount Olympus." It is a huge volcano, 600 km at its base and about 21 km high (■ Fig. 6–49). It is crowned with a crater 65 km wide; Manhattan Island could be easily dropped inside. (The tallest volcano on Earth is Mauna Kea

in the Hawaiian Islands, if we measure its height from its base deep below the ocean. Mauna Kea is slightly taller than Mount Everest, though still only 9 km high.)

Perhaps the volcanic features on Mars can get so huge because continental drift is absent there, as on Venus. If molten rock flowing upward causes volcanoes to form, then on Mars the features just get bigger and bigger for hundreds of millions of years, since the volcanoes stay over the sources and do not drift away. (Each of the Hawaiian Islands was formed over a single "hot spot," but gradually drifted away from it. The Big Island of Hawaii is still over the hot spot, which explains why it has active volcanoes and why it is still growing.) Mars's surface gravity is lower than Earth's, adding to the ability of Olympus Mons to get so big.

Another surprise on Mars was the discovery of canyons (■ Fig. 6–50). One tremendous system of canyons – about 5,000 km long – is as big as the continental United States and comparable in size to the Rift Valley in Africa, the longest geological fault on Earth. Known as Valles Marineris, it does not appear to have been produced by water, unlike some of the other features discussed below.

Perhaps the most amazing discovery on Mars was the presence of sinuous channels. These are on a smaller scale than the "canali" that Schiaparelli and Lowell believed they had seen, and are entirely different phenomena. Some of the channels show the same characteristic features as streambeds on Earth. Even though liquid water cannot exist on the surface of Mars under today's conditions, it is difficult to think of ways to explain the channels satisfactorily other than to say that they were cut in the past by running water. Why water? Calculations indicate that conditions on Mars's early surface would have been suitable for liquid water, but not for liquid carbon dioxide, methane, or other substances.

The orbiting spacecraft show fans and deltas that no doubt arose from flowing water. The high-resolution Mars Global Surveyor images show what look like sedimentary layers of rock (■ Fig. 6–51). Mars Global Surveyor and the ESA's Mars Express, which is also now in orbit, found additional signs, such as terracing and gullies, showing that water flowed, perhaps fairly recently in geological times (■ Fig. 6–52). A leading model for the gullies suggests that the source of this water is melting snow on the undersides of snowpacks.

This indication that water most likely flowed on Mars was particularly exciting because biologists feel that water is necessary for the formation and evolution of life as we know it. The presence of water on Mars, even in the past, is an indication that life could have formed and may even have survived.

Scientists theorized that much of the water is probably in a permafrost layer – permanently frozen subsoil – beneath middle latitudes and polar regions. Crustal hydrated minerals and the mantle may also contain much of the water. Also, a substantial fraction may have escaped from Mars in gaseous form. Some of the water is bound in the polar caps (■ Fig. 6–53). The large polar caps that extend to latitude 50° during the winter consist of carbon dioxide. But when a cap shrinks during its hemisphere's summer, a residual polar cap of water ice remains in the north, while the south has a residual polar cap of carbon dioxide ice, also with water ice below.

■ **FIGURE 6–48** A 360° panorama of an arctic plane from the Phoenix Mars Lander in 2008, with the solar arrays and some of its scientific instruments, including the wind gauge that is part of its meteorological instrumentation, in the foreground. The polygonal patterns on the ground are similar to those in terrestrial permafrost. At right, we see some of the trenches the spacecraft dug to expose subsurface layers.

(a)

(b)

(c)

■ **FIGURE 6–49** (a) Olympus Mons, from Mars Global Surveyor. (b) A high-resolution view of a strip across a tiny piece of Olympus Mons, from 2001 Mars Odyssey. This enlargement of a scarp shows lava that flowed in one direction, down the volcano's flank. The low contrast indicates that the lava flows are covered with dust. Few craters show in the lava flows, making them younger than about 500 million years. (c) The caldera of Olympus Mons, from the European Space Agency's Mars Express.

ESA's Mars Express joined NASA's Mars Global Surveyor in orbit around Mars in late 2003. Since entering its long-term orbit on January 28, 2004, it has been performing studies and global mapping of Mars's atmosphere and surface, analyzing their chemical composition, and sending back images of martian landscapes.

NASA's Mars Reconnaissance Orbiter arrived in 2006, carrying a 50-cm-diameter telescope, cameras, a spectrometer for the visible and near-infrared (which has found hydrated minerals and thus contributed to our understanding of water on Mars), and a radar. It carries a 50-cm-diameter telescope mirror, the largest ever sent to another planet, so its HiRISE camera (**H**igh **R**esolution **I**maging **S**cience **E**xperiment) resolves surface features as small as a desktop. It is also surveying Mars's weather and climate as well as acting as a high-speed relay for communications from the other spacecraft in orbit around Mars or on Mars's surface.

NASA's Phoenix Mars Mission arrived on Mars, landing in 2008 to search for organic chemicals and research the history of water on the planet. In the six months it survived, Phoenix completed the experiments necessary to confirm the presence of water ice. Curiosity's capabilities are even greater, as we discuss in Section 6.5e.

6.5c Mars's Atmosphere

Scientists have found that the martian atmosphere consists of 95 per cent carbon dioxide with small amounts of carbon monoxide, oxygen, and water. The surface pressure is less than 1 per cent of that near Earth's surface. The current atmosphere is too thin to significantly affect the surface temperature, in contrast to the huge effects that the atmospheres of Venus and the Earth have on climate.

But long ago, up to about a billion years after it formed, Mars had a thicker atmosphere and a hospitable climate. When the iron

NASA/JPL–Caltech/University of Arizona/Texas A&M University

NASA/JPL/University of Arizona/ Los Alamos National Laboratories

■ **FIGURE 6–50** Mars, an image made from earlier spacecraft mapping with colors superimposed from observations of mid-energy neutrons detected by the 2001 Mars Odyssey spacecraft. The valleys Valles Marineris (which were not carved by water, but rather are gashes in Mars's crust) appear horizontally across the center; together, they are as long as the United States is wide. At left we see the giant volcanoes. Colors indicate the relative abundances of hydrogen (and thus of water), with decreasing amounts shown ranging from deep blue through green to yellow to red. The hydrogen in the far north is hidden beneath a layer of carbon-dioxide frost (dry ice).

(a)

NASA/JPL/Caltech and Malin Space Science Systems

(b)

NASA/JPL/Caltech and Malin Space Science Systems

■ **FIGURE 6–51** (a) Layered rock, apparently sedimentary, imaged with Mars Global Surveyor, an enlargement of the box shown in (b).

(a)   (b)

■ **FIGURE 6–52** The past presence of water on Mars is revealed by features like the ones shown here, imaged with the European Space Agency's Mars Express. (a) Erosion in Ares Vallis, no doubt caused by flowing water. (b) This 2011 Mars Express view of Oralbi crater, which is 32 km across, shows that the crater was filled with sediment and that its southern rim (at left) has been eroded by flowing water.

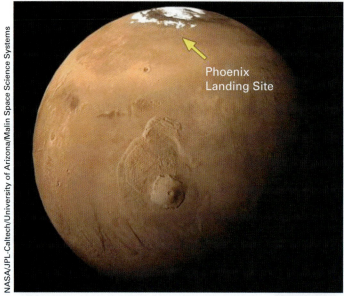

■ **FIGURE 6–53** The martian north polar cap, observed from Mars Global Surveyor just before the start of southern spring on Mars, with Mars Phoenix's landing site marked.

Phoenix Landing Site

■ **FIGURE 6–54** Terra Sirenum, in a mosaic of several 2001 Mars Odyssey strips, using images through three infrared filters translated into visible-light colors. The false-color differences correspond to differences in the composition of the surface materials. The field of view is about 30 km by 30 km.

core cooled sufficiently and solidified, Mars lost its magnetic field, and the solar wind partially stripped the atmosphere into space. This led to colder overall temperatures and the condensation of water from the remaining atmosphere. Carbon dioxide then dissolved in the water, leading to less greenhouse heating and hence even colder temperatures. This mechanism may have led to essentially the reverse of what happened on Venus; it is sometimes called a "runaway refrigerator" as opposed to Venus's runaway greenhouse.

Observations from the Viking orbiters showed Mars's atmosphere in some detail. The lengthy period of observation led to the discovery of weather patterns on Mars. The current spacecraft orbiting Mars, NASA's Mars Odyssey, NASA's Mars Reconnaissance Observer, and ESA's Mars Express, map the weather there even better. The 2001 Mars Odyssey records narrow strips of Mars's surface at exceedingly high resolution (■ Fig. 6–54). Additional information continues to come from the Hubble Space Telescope. With Mars's rotation period similar to that of Earth, some features of its weather are similar to our own. Studies of Mars's weather have already helped us better

understand windstorms in Africa that affect weather as far away as North America.

Studies of the effect of martian dust storms on the planet led to the idea that the explosion of many nuclear bombs on Earth might lead to a "nuclear winter." Dust thrown into the air would shield the Earth's surface from sunlight for a lengthy period, with dire consequences for life on Earth. Improvement of computer models for the circulation of atmospheres is contributing to the investigations. The effect, at present, seems smaller than first feared, so it is perhaps better described as "nuclear fall," which would still be something to avoid. A dust-caused winter would have resulted from the collision of one or more asteroids with Earth, as at least one did 65 million years ago; we shall discuss such collisions in Chapter 8 when commenting on the cause of the extinction of the dinosaurs at that time.

These models, and observations of Mars, also help us to understand the effects of smaller amounts of matter we are putting into the atmosphere from factories and fossil-fuel power plants and by the burning of forests. Of course, the absence of oceans on Mars (and their associated effects) complicates direct comparisons with Earth.

A half-dozen years after NASA's tiny Pathfinder rover won the hearts of Internet downloaders around the world, NASA landed a pair of Mars Exploration Rovers in early 2004. Named by a 9-year-old schoolgirl in a contest run by The Planetary Society and LEGO, Spirit and Opportunity roamed across Mars's surface for far longer than the few months of the original plan. Spirit last sent back data in 2010; as of this writing, Opportunity continues to explore Mars.

Mars Phoenix landed on Mars and sent back images and other data in 2008. (The name Phoenix stems from the project's birth as a resurrection of an unused spare twin spacecraft to the Mars Polar Lander that had crashed during landing.)

Opportunity entered a shallow crater and examined its walls, sending back images that reveal the past presence of water. Small, round pebbles, called "blueberries" by the scientists, are universally considered to have formed in water. Ripples on the surface at Opportunity's landing site was yet another sign that the material from which the rocks formed was laid down under water. A small amount of water may have flowed as recently as half a million years ago, though most disappeared within Mars's first billion years.

The rovers' wheels have dug trenches in the dirt to expose the soil and what is just under it. Mars Phoenix found some frost slightly under the surface, as shown here in a trench it dug.

Astonishingly, a NASA spacecraft orbiting Mars imaged the Mars Science Laboratory's parachute (which was in the orange-and-white colors of Caltech, which runs JPL) as the spacecraft went successfully through what was known as 7 minutes of terror, as a series of over 70 sequenced occurrences, including dozens of small explosive bolt releases, had to occur perfectly. (Some of the scientists at Cornell who are operating the spacecraft kept Mars time in their personal lives; with days that are $24^1/_2$ Earth days long, for the first 90 sols (martian days), they had unusual sleeping habits from the point of view of more ordinary earthlings.)

NASA/JPL/Caltech

FIGURE ACL 6.5–1 A mosaic of images from the Mars Exploration Rover named Opportunity made this wide-angle (180°) view of the face of Burns Cliff.

NASA/JPL/Cornell

FIGURE ACL 6.5–2 The fine, parallel laminations in part of this rock, named El Capitan after a peak in a national park in Texas, indicate that it formed in water. Small spherical objects, nicknamed "blueberries," were found in the same region as the laminations. A spectrometer revealed the presence of jarosite, a mineral that contains water. The image is from Opportunity.

NASA/JPL/Cornell/USGS

FIGURE ACL 6.5–3 These spherical beads of hematite, 1 to 6 mm across, were remarked to resemble blueberries embedded in a muffin of martian soil. The name stuck. This microscopic image was taken with Opportunity. On Earth, hematite usually forms in the presence of liquid water.

NASA/JPL/Cornell

FIGURE ACL 6.5–4 Sand dunes on Endurance Crater, imaged by Opportunity and displayed in false color. The chance of getting stuck in the dunes was too high to risk the rover's trying to traverse them.

NASA/JPL/Cornell

FIGURE ACL 6.5–5 Spirit's tracks show in this view taken en route to Mars's Columbia Hills.

FIGURE ACL 6.5–6 Curiosity carries a drill that makes holes 1.6 cm across and 6.4 cm deep (at left), with a test hole also showing (at right), to measure rocks' chemical compositions.

FIGURE ACL 6.5–7 The rovers' wheels have dug trenches in the dirt to expose the soil and what is just under it. Mars Phoenix found some frost slightly under the surface, as shown here in a trench it dug.

FIGURE ACL 6.5–8 A view of Mars Science Laboratory at the end of its parachute as it landed in Gale Crater.

FIGURE ACL 6.5–9 A mosaic of images from Curiosity showing 5-km-high Mount Sharp (also called Aeolis Mons) in the middle of Gale Crater. The color has been balanced to show the terrain as though it was lighted up on Earth, though that makes the martian sky, which is actually pinkish, look too blue. Nonetheless, examining martian scenes lighted this way helps scientists recognize similar rocks to those on Earth.

FIGURE ACL 6.5–10 An image by Curiosity of layered, apparently sedimentary rock (that is, formed in water) in the middle of Gale Crater.

NASA's **M**ars **A**tmosphere and **V**olatile **E**volutio**N** (MAVEN) spacecraft will arrive in 2014 to study Mars's upper atmosphere. It will measure the rate at which Mars's atmosphere, now only about 1 per cent the density of Earth's, is currently escaping to space, which will allow scientists to extrapolate back to periods when Mars's atmosphere was denser. At the 2016 and 2018 launch opportunities, an ESA/Roscosmos (Russian) ExoMars collaboration plan to launch spacecraft to study the martian atmosphere and, with landers, its surface, with some NASA participation. NASA plans the InSight (**In**terior Exploration using **S**eismic **I**nvestigations, **G**eodesy and **H**eat **T**ransport) lander for 2016 and a rover for 2020.

6.5d Mars's Satellites

Two small moons orbiting Mars were discovered in the 19th century using the U.S. Naval Observatory's telescope in Washington, D.C. In the 20th century, close-up views from spacecraft allowed scientists to learn more about these satellites. Named after the Roman companions of Mars, Phobos (which translates to Fear) and Deimos (which translates to Terror) might be captured asteroids.

Phobos, the larger of the two moons, is only 27 km in its longest dimension and 19 km in its shortest. It is covered with craters and revolves around Mars in 7 hours 40 minutes. Deimos is only 15 km at its longest and 11 km at its shortest. Dust about 30 feet deep seems to have partially filled its craters, making them less visible.

Given that one moon orbits more slowly than the planet rotates (in 25 hours) and the other moon more quickly, an observer on Mars's surface would see one moon moving forward and the other moving backward.

In 2011, Russia sent its Fobos-Grunt (where "grunt" is Russian for "soil") mission to Phobos. Most unfortunately, the spacecraft didn't leave Earth orbit, and fell into the Pacific Ocean early in 2012. The Chinese orbiter Yinghuo-1, planned for studying especially Mars's magnetic field and ionosphere, was on the same rocket, and so it also failed.

Fobos-Grunt also carried, controversially, a set of microorganisms to see if they survived the round-trip, including the irradiation accumulated during the flight to Phobos and back. Their capsule was supposedly hardened, though some people worried that some of these microbes could be set free in a crash and then survive, contaminating Phobos before we could adequately study it. The experiment was arranged by The Planetary Society (planetary.org), a public-interest group. Anyway, the controversy is now moot.

Table 6.5 Comparative planetology: moons
Mercury: none
Venus: none
Mars: 2, only about 20 km diameter each
Earth: 1, of 3476 km diameter

6.5e The Search for Life on Mars

On July 20, 1976, exactly seven years after the first crewed landing on the Moon, Viking 1's lander descended safely onto a martian plain called Chryse. The views showed rocks of several kinds, covered with yellowish-brown material that is probably an iron oxide (rust) compound. (Many of the initial photographs incorrectly showed an orange-red color, but subsequent careful color balancing of the images revealed the true yellowish-brown.) Some sand dunes were visible. The sky on Mars also turned out to be yellowish-brown, almost pinkish (■ Fig. 6–55); the color is formed as sunlight is scattered by dust suspended in the air as a result of Mars's frequent dust storms.

A series of experiments aboard the lander was designed to search for signs of life. A long arm was deployed and a shovel at its end dug up a bit of the martian surface. The soil was dumped into three experiments that searched for such signs of life as respiration and metabolism. The results were astonishing at first: The experiments sent back signals that seemed similar to those that would be caused on Earth by biological processes. But later results were less spectacular, and nonbiological chemical explanations seem more likely in all cases.

One important experiment gave much more negative results for the chance that there is life on Mars. It analyzed the soil and looked for traces of organic compounds. On Earth, many organic compounds left over from dead forms of life remain in the soil; living organisms themselves are only a tiny fraction of the organic material. Yet these experiments found no trace of organic material.

Unfortunately, any ancient organic material on the surface would have been destroyed by solar ultraviolet radiation, so its absence isn't proof that Mars never had any life. Still, it is a strong argument against the presence of recent life on the surface of Mars. But even if the life signs detected by Viking come from chemical rather than biological processes, as seems likely, we have still learned of fascinating new chemistry going on.

NASA/JPL/Caltech

■ **FIGURE 6–55** A Viking 2 view of Mars. The spacecraft landed on a rock and so was at an angle. The boom that supports Viking's weather station cuts through the center of the picture. ("Chance of precipitation," the local newscaster would say, "is 0 per cent.")

■ **FIGURE 6–56** A microscopic view of a small (2.3 mm across), thin section from one of the Mars meteorites. It shows globules of carbonate materials.

■ **FIGURE 6–57** Structures in the "Mars meteorite," seen with a special kind of microscope, are thought by a few scientists to be remnants of primitive life-forms but by most other scientists to be natural formations, too small to be fossils of life-forms.

In 1996, study of a martian meteorite found in the Earth's Antarctic revealed possible evidence for ancient primitive (microbial) organisms. The chunk of rock, estimated to be 4.1 billion years old, was blasted from the surface of Mars about 15 million years ago by a collision. There is little doubt that the rock is from Mars, based on analysis of gases trapped within it; also, calculations show the likelihood that rocks blasted off Mars can reach Earth. The meteorite landed in Antarctica about 13,000 years ago. It was the first meteorite from the Allan Hills analyzed in 1984 at the Johnson Space Center, and so is called ALH 84001.

The meteorite contained carbonate globules (■ Fig. 6–56), which generally form in liquid water. It is thought that these globules were produced when water flowed through the rock, in the first billion years after the formation of Mars. Within the globules, certain types of organic compounds were found. Although these types of compounds can be formed in a number of ways other than by life, and are sometimes seen in normal meteorites, these particular ones were relatively unusual. A few other substances typical of life (such as magnetite, a mineral produced by some types of bacteria) were also seen. But photographs of tube-like structures resembling the tiniest Earth bacteria received the most publicity (■ Fig. 6–57).

Although none of these findings definitively implied the presence of life, taken together they were certainly intriguing (yet still not compelling). Several researchers challenged the conclusions on a number of grounds. For example, they cited evidence for contamination of the meteorite by substances on Earth. The original research team responded to these criticisms with reasonable counterarguments, and stand by their original announcement. Moreover, they published a report stating that they had found signs of magnetism of a type formed by life. Still, the results are highly controversial, and additional tests are being carried out; most scientists remain skeptical.

Regardless of whether the discovery of ancient microbial life is correct, it raises an interesting possibility: Earth life may have originated on Mars, and subsequently migrated to Earth in a meteorite. If so, we are the descendants of Martians!

In 2003, ESA's Mars Express reached Mars. It carried a last-minute, poorly funded add-on search-for-life laboratory named Beagle 2. (The original Beagle was Charles Darwin's ship in the Galápagos Islands.) The last communication from Beagle 2 occurred when it was released from the mother ship, but unfortunately it was never heard from again. Its rocket may have misfired, or something might have broken as it bounced around off giant balloons as it landed on Christmas Day.

In 2008, NASA's Phoenix Mission, the first of its smaller, lower-cost series of spacecraft known as Scouts, landed on Mars's surface and carried out experiments for about six months on an arctic plain. In trenches that it dug, it found water ice, as well as chemistry and minerals in the soil that showed that the climate there was wetter and warmer within the last few million years, a relatively recent time for Mars. It found signs of perchlorate, a chemical that could mix with water to make a brine that stays liquid at the surface temperatures of Mars and is compatible with some terrestrial microbes. Perchlorate is in the news now since it is a rocket-fuel component that has contaminated many terrestrial water supplies. Astronauts could potentially use it as rocket fuel or to generate oxygen for breathing.

The most ambitious set of instruments to search for life (or conditions potentially suitable for life) is on NASA's Curiosity, the nickname for the Mars Science Laboratory's rover, which landed on Mars on August 5, 2012. While Sojourner was the size of your arm, and Spirit and Opportunity are each the size of a wheelbarrow, Curiosity is the size of a small car (■ Fig. 6–58a). It landed close to a mound of layered rocks in Gale Crater (■ Fig. 6–58b,c), a crater the size of Connecticut that is just south of Mars's equator, near the dividing line between low-lying northern-hemisphere material and southern highlands that have many craters.

The scientists chose this site because it contains a 5-km-high (3-mile-high) stack of presumably sedimentary material, Mt. Sharp,

(a)

NASA/JPL

(b)

NASA/JPL-Caltech/ASU/UA

(c)

Choosing a landing site

Map by Emily S. Lakdawalla and Loren A. Roberts atop Viking color MDIM, courtesy of The Planetary Society

Viking 2
SUCCESSFUL
Landed: September 3, 1976
Duration: September 3, 1976 to July 25, 1978

Phoenix
SUCCESSFUL
Landed: May 25, 2008
Duration: May 25, 2008 to November 2, 2008

Viking 1
SUCCESSFUL
Landed: July 20, 1976
Duration: July 20, 1976 to November 13, 1982

Pathfinder
SUCCESSFUL
Landed: July 4, 1997
Duration: July 4, 1997 to September 27, 1997

Beagle 2
UNSUCCESSFUL
Landed: December 19, 2003

Curiosity
Projected landing: between August 6 and August 20, 2012
Duration: to be determined

Spirit
SUCCESSFUL
Landed: January 4, 2004
Duration: January 4, 2004 to March 22, 2010

Opportunity
SUCCESSFUL
Landed: January 25, 2004
Duration: January 25, 2004 to present

Mars 3
UNSUCCESSFUL
Landed: December 2, 1971

Mars 6
UNSUCCESSFUL
Landed: March 12, 1974

Mars 2
UNSUCCESSFUL
Landed: November 27, 1971

Mars Polar Lander
UNSUCCESSFUL
Landed: December 3, 1999

(d)

■ **FIGURE 6–58** (a) We see a test lander of Curiosity, formally the Mars Science Laboratory, before Curiosity's 2011 launch to Mars. For comparison, we also see a test sibling to Spirit and Opportunity, which landed on Mars in 2004, and, at bottom left, a flight spare ("Marie Curie") for the first Mars rover, Sojourner, which landed on Mars in 1997 as part of the Mars Pathfinder Project. Curiosity is 3 m (10 feet) long; Spirit and Opportunity are 1.6 m (5.2 feet) long; and Sojourner was only 0.65 m (2 feet) long. (b) Gale Crater, Curiosity's landing site, in the ellipse in front of the mound of layered rocks. The image is a combination of imaging from NASA's Mars Odyssey orbiter and the Mars Orbiter Laser Altimeter on the Mars Global Surveyor orbiter, with color imaging added from the Mars Reconnaissance Orbiter. Unlike many images from spacecraft, the vertical dimension is not exaggerated. (c) The locations on Mars of the various landers. (d) A view of layered rock on the central peak, informally known as Mount Sharp, in the middle of Gale Crater. Curiosity will eventually roll right up to these apparently sedimentary layers and make direct measurements that trace the history of water on Mars back for billions of years in time.

NASA/JPL-Caltech/University of Arizona

■ **FIGURE 6–59** Phobos, about 21 km (13 miles) across by 9 km (5.6 miles). The crater Stickney, whose color difference with the rest of Phobos is emphasized in this false-color image, is slightly bluer than the rest of Phobos, probably meaning that its surface is relatively fresh and therefore young. Each pixel in this image, from the HiRISE camera on NASA's Mars Reconnaissance Orbiter, shows detail as small as 7 m (22 feet) per pixel.

NASA/JPL-Caltech/University of Arizona

■ **FIGURE 6–60** Deimos, only 12 km (7.5 miles) in diameter. The resolution of this image with the HiRISE camera is 20 m (66 feet) per pixel. Deimos's surface is much smoother than Phobos's, perhaps from a blanket of broken-up rock. Deimos, much farther out than Phobos, is slowly spiraling out, while Phobos is slowly spiraling in because of tidal forces.

Though earlier Mars landers were powered by solar panels, which eventually get covered with martian dust and lose power, Curiosity is powered by the decay of plutonium, so it should be able to send back data for many years, including its chemical and mineralogical analyses of dozens of samples it is extracting with its drill from rock interiors.

containing intriguing evidence for hydrated materials (clays and sulfates). At the base of the mound, orbiters have detected signatures of both clay minerals and sulfate salts, minerals that, as far as we know, depend on water to form. These minerals are prime locations for potential microbial life. When Curiosity rolls over to nearby canyons that are cut through the layers of rock, they expect to see a geologic history of Mars's surface, as well as investigate sediment carried by the water that no doubt cut the canyons, another prime location to look for signs of life. Its sampling mechanism should be able to handle the clumpy martian soil that foiled some of Phoenix's investigations, and its results should be very exciting as it traverses over 20 km.

The discovery by Mars Phoenix of calcium carbonate and perchlorate on Mars has led some of the Curiosity scientists to question the conclusion of most Viking scientists that no signs of life were found. The perchlorate could be an energy source for microbes on Mars. Its presence could explain a detection of chloromethanes by Viking that had been attributed to a pre-launch contamination. The new idea is that the heating of Mars soil by an oven aboard Viking may have led to the emission of chloromethanes from a perchlorate-bearing soil along with an output of oxygen as the organics burned and thus went undetected. Curiosity will help clear up this question.

6.5f Crewed Missions to Mars

Because of the tremendous interest in Mars and the possibility of finding out about the origin of life by studying it, space agencies in the United States, Europe, and Russia are planning a set of spacecraft to Mars over the rest of this decade. Eventually, martian rock and soil will be brought to Earth, though the original goal of doing so by about 2015 has been postponed for budgetary reasons. Some scientists feel that only with this material to study using sophisticated Earth-based microscopes and other instruments can we really understand what Mars is like.

The American government has made it a long-term goal for NASA to send astronauts to Mars, and they reconsidered many of NASA's priorities toward that aim, though it is not a priority in the current budget. Note that though astronauts reached the Moon in only four days, Mars is so much farther away that a trip there would take a couple of years. Because of the cost and other problems, no such astronaut mission is likely for decades. Some private efforts are under way to get there sooner.

In the meantime, we watch Mars and its diminutive moons (■ Figs. 6–59, 6–60) from a variety of spacecraft on and around it. From Earth, Mars is only a tiny – though intriguingly reddish – spot in the sky (■ Fig. 6–61).

The Moon

Venus Mars Jupiter Saturn Uranus Neptune

Pillars in the
Eagle Nebula

Tadpole Galaxy

Galaxy Cluster
CL 0024 +1654

Hubble Deep Field

■ **FIGURE 6–61** Sizes of celestial objects as seen from Earth. Shown here are the Moon, six planets, and some images from the Hubble Space Telescope at the correct relative size as they would appear in the sky. Note how small a field of view Hubble has when compared to the full moon. Hubble observes less than one-millionth of the sky at a time. In these images, the pillars in the Eagle Nebula and the Hubble Deep Field were observed with the Wide Field and Planetary Camera 2 (note its irregular shape), whereas the Tadpole Galaxy and the cluster of galaxies were observed with the Advanced Camera for Surveys.

Moon image: T. A. Rector, I. P. Dell'Antonio NOAO/AURA/NSF; All other images: Hubble Space Telescope STScI/AURA/NASA; Designed graphics: J. R. Gott (Princeton) and F. J. Summers (STScI)

CONCEPT REVIEW

» The **terrestrial planets**, the four inner planets, are rocky worlds. Venus and the Earth have similar sizes, masses, and densities, but are otherwise very different. Mercury and Mars are about a third and a half of Earth's diameter, respectively. Compared with the giant planets we will discuss in the next chapter, the terrestrial planets are smaller, rocky, and dense, and they have fewer moons and no rings. **Comparative planetology** has applications for understanding weather, earthquakes, and other topics of use to us on Earth (Sec. 6.1).

» **Geology** is the study of the Earth's interior and surface, and often uses **seismology**, the study of waves that pass through the Earth, to study the Earth's interior (Sec. 6.1a). The Earth has a layered structure, with a **core**, **mantle**, and **crust**. Radioactivity heats the interior, providing **geothermal energy** (Sec. 6.1b). Continental **plates** move around the surface, as explained by **plate tectonics**, the current theory that elaborates on the earlier notion of **continental drift**.

» **Tides** are a differential force caused by the fact that the near side of the Earth is closer to the Moon than the Earth's center, and so is subject to higher gravity, while the far side is farther than the center and is subject to lower gravity (Sec. 6.1c). The Earth's weather is confined in its atmosphere to the **troposphere** (Sec. 6.1d). The atmosphere and surface of the Earth are substantially warmer than they would have been without the **greenhouse effect**. In the **ionosphere**, atoms are ionized and the temperature is higher. The **Van Allen belts** of charged particles were a space-age discovery (Sec. 6.1e). Such particles lead to the **aurora borealis** and **aurora australis**, the northern and southern lights.

» Our own Moon is midway in size between Pluto and Mercury. Its volcanic surface shows smooth **maria** (singular: **mare**) and cratered highlands (Sec. 6.2a). Shadows reveal relief best along the **terminator**, the day-night line. Most of the craters are from meteoritic impacts (Sec. 6.2b). The oldest rocks are from 4.4 billion years ago; on the Earth, erosion and plate tectonics have erased most of the oldest rocks though a few equally old terrestrial rocks have been found. **Mascons** are mass concentrations of high density under the Moon's surface (Sec. 6.2c). The current leading model for the Moon's formation is that a body perhaps twice the size of Mars hit the proto-Earth, ejecting matter into a ring that coalesced (Sec. 6.2d). A handful of meteorites found on Earth have been identified by their chemical composition as having come from the Moon (Sec. 6.2e).

» Mercury always appears close to the Sun in the sky; rarely it even goes into **transit** across the Sun's disk (Sec. 6.3). The correct rotation period of Mercury was first measured using radar, and proved to be linked to its orbital period (Sec. 6.3a). Like the Moon, Mercury has a dark (low **albedo**) surface (Sec. 6.3b). Mercury is heavily cratered, and shows **scarps** that resulted from a planet-wide shrinkage (Sec. 6.3c).

» Mercury has a weak magnetic field and is the only planet besides Earth to have magnetized poles. It is too close to the Sun to retain a thick atmosphere, but a very thin atmosphere does exist (Sec. 6.3d). NASA's MESSENGER spacecraft entered orbit around Mercury in 2011 and is redefining our understanding of the planet (Sec. 6.3f).

» In 2004 and 2012, Venus transited the Sun, for the first pair of transits in over 100 years (Sec. 6.4a); the next pair of transits of Venus won't occur until 2117 and 2124. The highly reflective clouds that shroud Venus, giving it a high albedo, are mainly droplets of sulfuric acid, though the atmosphere consists primarily of carbon dioxide (Sec. 6.4b). Venus's atmospheric surface pressure is about 90 times higher than Earth's. Radar penetrated Venus's clouds to show that Venus is rotating slowly backward and to map its surface (Sec. 6.4c).

» Venus's surface temperature is very high, about 735 K, heated by the greenhouse effect (Sec. 6.4d). Earth escaped such a fate, but we should be careful about disrupting the balance in our atmosphere, lest conditions change too quickly. The atmosphere of Venus has been studied with various spacecraft (Sec. 6.4e). Radar is used to map the surface of Venus (Sec.6.4f). Higher-**resolution** radar reveals smaller surface features. ESA's Venus Express spacecraft has been in orbit around Venus since 2006 (Sec. 6.4g).

» Stories about life on Mars have long inspired study of this planet. We now know the seasonal surface changes to be the direct result of winds that arise as the sunlight hits the planet at different angles over a martian year (Sec. 6.5a). Mars's weather and dust storms help us understand our own weather. Mars's surface appears reddish-orange because of rusty dust; careful calibration of photos taken on Mars shows that it is actually yellowish-brown.

» Mars boasts of giant volcanoes and a canyon longer than the width of the continental United States (Sec. 6.5b). There is strong evidence that Mars had liquid water flowing on its surface long ago. Mars's current atmosphere is very thin, with a surface pressure less than 1 per cent of that near Earth's surface (Sec. 6.5c). But long ago, up to about a billion years after it formed, Mars had a thicker atmosphere and a hospitable climate.

» In the 1970s, the Viking landers found no conclusive evidence for life on Mars, despite initially promising signs (Sec. 6.5d). Studies of a martian meteorite provided suggestive, but not convincing, evidence for the presence of ancient, primitive life on Mars. Mars Science Laboratory, nicknamed Curiosity, landed on Mars in 2012 with the most ambitious set of instruments yet to search for the conditions needed for life (Sec. 6.5e).

QUESTIONS

1. How did the layers of the Earth arise? Where did the energy that is flowing as heat come from?

2. What carries the continental plates around over the Earth's surface?

3. (a) Explain the origin of tides. (b) If the Moon were farther away from the Earth than it actually is, how would tides be affected?

4. Draw a diagram showing the positions of the Earth, Moon, and Sun at a time when there is the least difference between high and low tides.

†5. Calculate your weight if you were standing on the Moon.

6. Look at a globe and make a list or sketches of which pieces of the various continents probably lined up with each other before the continents drifted apart.

7. To what locations, relative to the Earth-Sun line, does the Earth's terminator correspond?

8. What does cratering tell you about the age of the surface of the Moon, compared to that of the Earth's surface?

9. Why are we more likely to learn about the early history of the Earth by studying the rocks from the Moon than those on the Earth?

10. Why may the near side and far side of the Moon look different?

11. Discuss one of the proposed theories for the origin of the Moon. List points both pro and con.

12. How can we get lunar material to study on Earth?

13. Assume that on a given day, Mercury sets after the Sun. Draw a diagram, or a few diagrams, to show that the height of Mercury above the horizon depends on the angle that the Sun's path in the sky makes with the horizon as the Sun sets. Discuss how this depends on the latitude or longitude of the observer.

14. If Mercury did always keep the same side toward the Sun, would that mean that the night side would always face the same stars? Draw a diagram to illustrate your answer.

15. Explain why a day/night cycle on Mercury is 176 Earth day/night cycles long.

16. What did radar tell us about Mercury? How did it do so?

17. If ice has an albedo of 70–80 per cent, and volcanic rocks typically have albedos of 5–20 per cent, what can you say about the surface of Mercury based on its measured albedo?

18. If you increased the albedo of Mercury, would its surface temperature increase or decrease? Explain.

19. How would you distinguish an old crater from a new one?

20. What evidence is there for erosion on Mercury? Does this mean there must have been water on the surface?

21. List three major discoveries of Mariner 10. List three improved results found by MESSENGER.

22. Make a table displaying the major similarities and differences between the Earth and Venus.

23. Why does Venus have more carbon dioxide in its atmosphere than does the Earth?

24. Why do we think that there have been significant external effects on the rotation of Venus?

25. Suppose a planet had an atmosphere that was opaque in the visible but transparent in the infrared. Describe how the effect of this type of atmosphere on the planet's temperature differs from the greenhouse effect.

26. Why do radar observations of Venus provide more data about the surface structure than a flyby with close-up optical cameras?

27. Why do we say that Venus is the Earth's "sister planet"?

28. Describe the most current radar observations of Venus.

29. What are some signs of volcanism on Venus?

30. List three of the features of Mars that made scientists think that it was a good place to search for life.

†31. Compare the tallest volcanoes on Earth and Mars relative to the diameters of the planets. Give the ratios.

32. What evidence exists that there is, or has been, water on Mars?

33. Describe the composition of Mars's polar caps.

34. Consult an atlas and compare the sizes of the Grand Canyon in Arizona and the Rift Valley in Africa. How do they compare in size with the giant canyon on Mars?

35. Compare the temperature ranges on Venus, Earth, and Mars.

36. List the evidence from Mars landers for and against the existence of life on Mars.

37. Plan a set of experiments or observations that you, as a martian scientist, would have an uncrewed spacecraft carry out on Earth to find out if life existed here. What data would your spacecraft radio back if it landed in a cornfield? In the Sahara? In the Antarctic? In New York's Times Square?

38. **True or false?** All four terrestrial planets show evidence of water flowing on their surfaces, either now or in the distant past.

39. **True or false?** From Earth, neither Mercury nor Venus can be seen high in the sky at midnight.

40. **True or false?** Earth's Moon has a composition similar but not identical to that of the Earth's crust, consistent with the hypothesis that the Moon formed after a large object collided with Earth.

41. **True or false?** At any given coastal location on Earth, high tide occurs only once per day, when the Moon is overhead.

42. **True or false?** Venus's surface is exceptionally hot primarily because its atmosphere traps x-rays and ultraviolet radiation from the Sun.

43. **True or false?** The terrestrial planets have iron cores and rocky outer parts; the iron sank when they were young and molten.

44. **True or false?** Of the four terrestrial planets, Earth is the only one that now has obvious plate tectonics.

45. **Multiple choice:** Astronomers have observed that the Moon is more heavily cratered than the Earth. This is primarily because **(a)** the Moon is more distant than the Earth, causing incoming rocks to hit it first; **(b)** the Moon is much more deficient in iron relative to the Earth, causing craters to form more easily; **(c)** the Moon has been around longer than the Earth and has been exposed to more periods of meteor bombardment; **(d)** the Moon experiences almost no erosion compared to the Earth, causing craters to remain much longer; or **(e)** the Moon experiences frequent volcanic eruptions that cause craters.

46. **Multiple choice:** Which one of the following statements about the greenhouse effect, Venus, and Earth is *false*? **(a)** An extreme, possibly runaway, greenhouse effect occurred on Venus, making its planetary surface the hottest in the Solar System. **(b)** Venus's atmosphere is much thicker than that of Earth – but some of Earth's gases are trapped in rocks and oceans. **(c)** If we dump much more carbon dioxide into Earth's atmosphere, Earth might become significantly hotter because of the greenhouse effect. **(d)** The greenhouse effect occurs when an atmosphere is transparent to optical (visible) light but opaque to infrared light. **(e)** No greenhouse effect currently occurs on Earth, and this is a good thing for humans.

†47. **Multiple choice:** If the Earth's radius suddenly shrank by a factor of 2 but the Earth's mass remained unchanged, how much would you weigh while standing on the new (smaller) surface of the Earth? **(a)** 16 times as much. **(b)** 4 times as much. **(c)** Twice as much. **(d)** Half as much. **(e)** Your weight would remain unchanged.

48. **Multiple choice:** Which one of the following statements about the Earth-Moon system is *true*? **(a)** The same half of the Moon's surface is perpetually dark (craters on that side never see sunlight), leading us to call it the "dark side of the Moon." **(b)** At a given location on Earth, there are two high tides each day – one caused by the gravitational pull of the Sun, and the other by the gravitational pull of the Moon. **(c)** High tide occurs on the side of the Earth nearest to the Moon, while low tide occurs on the opposite side of the Earth. **(d)** By observing the Moon long enough from Earth, night after night, we are able to draw a map of its entire surface. **(e)** The orbital period of the Moon around the Earth and the rotation period of the Moon around its axis are equal.

49. **Multiple choice:** Why does Mars appear reddish-orange from Earth? **(a)** Its surface temperature is lower than that of the Earth, which appears blue, and according to Wien's law Mars is therefore redder. **(b)** Its thick atmosphere consists of reddish-orange clouds. **(c)** Its rocks have suffered "rusting" and contain reddish-orange iron oxides. **(d)** It is moving away from us rapidly and hence is Doppler redshifted. **(e)** There is a large population of martians wandering around, and they tend to have a reddish-orange skin color.

50. **Multiple choice:** The surface of Venus is best observed using **(a)** ultraviolet satellites; **(b)** radar; **(c)** large optical telescopes; **(d)** large radio telescopes; or **(e)** infrared telescopes.

51. **Multiple choice:** Which one of the following statements about the terrestrial planets is *false*? **(a)** Earth's atmosphere plays a central role in keeping our planet from freezing cold temperatures. **(b)** Venus is the hottest planet in our Solar System, mainly due to a greatly elevated greenhouse effect. **(c)** Long ago, Mars probably had a thicker atmosphere than it does now, with sufficiently high temperatures and pressures to allow liquid water to exist on its surface. **(d)** Earth's atmosphere allows the Sun's infrared radiation in but doesn't allow much visible light to escape, resulting in the greenhouse effect. **(e)** Despite being the closest planet to the Sun, some parts of Mercury have a surface temperature far below the freezing point of water.

52. **Fill in the blank:** Earth's _____ occur in regions where Earth's magnetic field lines intersect its atmosphere, as a result of collisions between charged particles and air molecules.

53. **Fill in the blank:** The lunar _____ must have formed more recently than the cratered highlands, covering up older craters.

54. **Fill in the blank:** _____ is often seen as the brightest evening or morning "star" in the sky, in part because it is shrouded in highly reflective clouds.

55. **Fill in the blank:** A planet is said to be in _____ when it appears to move across the face of the Sun.

56. **Fill in the blank:** The process of forming layers within a planet, because of differences in density between materials, is called _____ .

†This question requires a numerical solution.

TOPICS FOR DISCUSSION

1. It is sometimes said that the U.S. mission to the Moon was entirely motivated by the Soviet Union's launch of the Sputnik satellite in 1957. Do you think the scientific benefits of lunar landings would have been sufficient reason to take the risks and spend the funds? Do you think the plans of China and India to land people on the Moon in about a decade should change NASA's plans?

2. How much does your understanding of the runaway greenhouse effect on Venus affect your evaluation of the global warming of Earth? How do you evaluate the risk of waiting for further consequences against the risk of acting more decisively early, limiting the rise in global temperature?

3. How likely do you think it is that humans will eventually "terraform" Mars so that its climate becomes suitable for humans?

This natural-color close-up view of Saturn, its rings (seen nearly edge-on), and its largest moon was taken in 2012 from the Cassini spacecraft. Titan (5150 km = 3,200 miles across), larg[...] than the planet Mercury, is covered with organic smog. We are looking toward the northern, sunlit side of the rings from just above the ring plane. Shadows of the rings produce the dark ba[...] on the planet.

NASA/JPL/Space Science Institute

7

The Jovian Planets: Windswept Giants

ORIGINS

We find that the properties of giant planets differ from those of Earth because they formed much farther out from the center of the solar nebula. Comparison of features on these planets and their moons leads to a greater understanding of the origin of Earth's properties.

Jupiter, Saturn, Uranus, and Neptune are **giant planets**; they are also called the **jovian planets**. They are much bigger, more massive, and less dense than the inner, terrestrial planets; see *A Closer Look 7.1: Comparative Data for the Major Worlds*. Their internal structure is entirely different from that of the four inner planets. All the giant planets have atmospheres that are almost entirely hydrogen and helium. The small percentage of heavier elements is greater in Uranus and Neptune than in Jupiter and Saturn.

In this chapter, we also discuss a set of moons of these giant planets, some of which range in diameter between $1/2$ and $1/4$ the size of the Earth, as large as Mercury or the dwarf planet Pluto. Close-up space observations have shown that these moons are themselves interesting objects for study.

Jupiter is the largest planet in our Solar System (■ Fig. 7–1). Four of its very numerous moons are bright and were discovered by Galileo; they show fascinating surface structure.

Saturn has long been famous for its beautiful rings. We now know, however, that each of the other giant planets also has rings. When seen close-up, as on the opposite page, the astonishing detail in the rings is very beautiful.

Uranus and Neptune were known for a long time to us as mere points in the sky. Spacecraft views have transformed them into objects with more character.

AIMS

1. Understand the giant planets of our Solar System, their moons, and their rings (Sections 7.1 to 7.4).

2. Describe similarities and differences among Earth, the giant planets, and their moons (Sections 7.1 to 7.4).

3. See how spacecraft have vastly improved our understanding of the giant planets, their moons, and their rings (Sections 7.1 to 7.4).

4. Understand how the location at which the giant planets formed made them very different from the terrestrial planets (Section 7.5).

■ FIGURE 7–1 Crescent Jupiter, in a farewell view as the Cassini spacecraft left the Jupiter system in 2001.

The age of first exploration of the giant planets, with spacecraft that simply flew by the planets, is over. We now are in the stage of space missions to orbit the planets, with the Juno spacecraft en route to orbit Jupiter, and a Saturn orbiter that started collecting data in 2004. These missions study the planets, their rings, their moons, and their magnetic fields in a much more detailed manner than before.

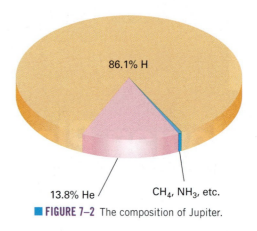

■ FIGURE 7–2 The composition of Jupiter.

7.1 JUPITER

Jupiter, by far the largest and most massive planet, dominates the Sun's planetary system. It alone contains two-thirds of the mass in the Solar System outside of the Sun, 318 times as much mass as the Earth (but only 0.001 times the Sun's mass). Jupiter has at least 64 moons of its own and is like a miniature "planetary system" (that is, several planet-like objects orbiting a central object) in itself. It is often seen as a bright object in our night sky, and observations with even a small telescope reveal bands of clouds across its surface and show four of its moons, the Galilean satellites. (See *Star Party 7.1: Observing the Giant Planets.*)

Jupiter is more than 11 times greater in diameter than the Earth (see *Figure It Out 7.1: The Size of Jupiter*). From its mass and volume, we calculate its density to be 1.3 g/cm³, not much greater than the 1 g/cm³ density of water. This low density tells us that any core of heavy elements (such as iron) makes up only a small fraction of Jupiter's mass. Jupiter, rather, is mainly composed of the light elements hydrogen and helium. Jupiter's chemical composition is closer to that of the Sun and stars than it is to that of the Earth (■ Fig. 7–2), so its origin can be traced directly back to the solar nebula with much less modification than the terrestrial planets underwent.

Jupiter has no crust. At deeper and deeper levels, its gas just gets denser and denser, turning mushy and eventually liquefying about

A CLOSER LOOK 7.1   Comparative Data for the Major Worlds

Planet	Semimajor Axis	Orbital Period	Equatorial Radius ÷ Earth's	Mass ÷ Earth's
Jupiter	5.2 au	12 years	11.2	318
Saturn	9.5 au	29 years	9.4	95
Uranus	19.2 au	84 years	4.0	15
Neptune	30.1 au	164 years	3.9	17

More detailed information appears in Appendices 3 and 4.

Observing the Giant Planets

Use a telescope, perhaps the one on your campus, to view the giant planets. Jupiter and Saturn are among the brightest objects in the sky, and are easy to locate, but Uranus and especially Neptune are more challenging. (To get current positions of planets, consult *Sky & Telescope* or *Astronomy* magazines, J.M.P.'s *Field Guide to the Stars and Planets*, planetarium software, or various smart-phone apps.)

You should be able to see a few of the main bands of Jupiter, and maybe even the Great Red Spot, in addition to the bright Galilean satellites. Sometimes one or more of these moons is behind Jupiter or casts a shadow on its disk. Their relative positions change from night to night; the innermost moon, Io, takes only 1.8 days to complete an orbit.

Saturn's bands are less distinct than those of Jupiter, but the rings look magnificent. You may be able to see the dark gap in the rings, known as Cassini's division, as well as the shadow of the rings on the disk of the planet. The rings appear edge-on to our line of sight twice during Saturn's 29.5-year orbital period, making them difficult to detect at those times. They will be edge-on next in 2016. Only one moon, Titan, is readily visible through a small telescope; it looks like a faint star.

Uranus and Neptune are too small and too far away to reveal detail, but you may be able to tell that they have resolved disks, unlike the much more distant, point-like stars. Try to decide whether Uranus and Neptune look greenish to your eyes.

20,000 km (15 per cent of the way) down. Jupiter's core, inaccessible to direct study, is calculated to be made of heavy elements and to be larger and perhaps 10 times more massive than Earth.

Jupiter's "surface" (actually, the top of the clouds that we see) rotates in about 10 hours, though different latitudes rotate at slightly different speeds. Regions with different speeds correspond to different bands; Jupiter has a half-dozen jet streams while Earth has only one in each hemisphere. Jupiter's clouds are in constant turmoil; the shapes and distribution of bands can change within days. The bright bands are called "zones" and the dark bands are called "belts," but the strongest winds appear on the boundaries between them. The zones seem to be covered by a uniformly high cloud deck. The belts have both towering convective clouds and lightning, as well as clear spaces that allow glimpses of the deeper atmosphere.

The most prominent feature of Jupiter's surface is a large reddish oval known as the **Great Red Spot**. It is two to three times larger in diameter than the Earth. Other, smaller spots are also present.

Jupiter emits radio waves, which indicates that it has a strong magnetic field and strong "radiation belts." Actually, these are belts of magnetic fields filled with trapped energetic particles – large-scale versions of the Van Allen belts of Earth (see the discussion in Chapter 6).

7.1a Spacecraft to Jupiter

Our understanding of Jupiter was revolutionized in the 1970s, when first Pioneer 10 (1973) and Pioneer 11 (1974) and then Voyager 1 and Voyager 2 (both in 1979) flew past it. The Galileo spacecraft arrived at Jupiter in 1995, when it dropped a probe into Jupiter's atmosphere and went into orbit in the Jupiter system. The spacecraft plunged into Jupiter's atmosphere on September 21, 2003, ending a tremendously successful mission. (It was sent on that course in large part to avoid the possibility that it could eventually hit and contaminate the Galilean satellite Europa, on which some scientists speculate

The Size of Jupiter

It is easy and instructive to estimate the physical diameter of Jupiter from its angular size. Jupiter's disk appears to be about 50 arc seconds across when it is closest to the Earth. (You can actually measure this yourself, by looking through a telescope with an eyepiece whose field of view you have already calibrated – perhaps by looking at the Moon, which is 30 arc minutes in diameter.) Since Jupiter is roughly 5 au from the Sun and the Earth is 1 au from the Sun, Jupiter must be about 4 au from Earth when at its closest.

Now we need to convert an angular diameter into a physical one, measured in kilometers. If the angular size (φ, the Greek letter "phi") of an object is measured in radians, and if the object appears small

in the sky, then its physical size (s) is given by its distance (d) multiplied by its angular size. This relation, very useful in astronomy and known as the small-angle formula, is expressed as

$$s = d\varphi.$$

One radian is equal to 206,265 arc seconds, because in a full circle 2π radians correspond to $360°$. Converting Jupiter's angular diameter of 50 arc seconds into radians, we have $\varphi = 50/206,265 = 0.00024$ radians. Since 1 au $= 1.5 \times 10^8$ km, the distance of Jupiter is 6.0×10^8 km. According to the small-angle formula, then, Jupiter's physical diameter must be $s = d\varphi = (6.0 \times 10^8$ km$) \times (0.00024) = 1.4 \times 10^5$ km. This size is about 11 times larger than Earth's diameter of 12,800 km.

(a)

NASA/JPL/University of Arizona

(b)

NASA, ESA, and E. Karkoschka (University of Arizona)

■ **FIGURE 7–3** (a) Jupiter from the Cassini spacecraft in 2000. Jupiter's bands and Great Red Spot are visible; Jupiter's moon Europa casts its shadow. (b) Jupiter's moon Ganymede is visible just before it went behind Jupiter in 2007, in this view with the Hubble Space Telescope. Ganymede is larger than the planet Mercury.

that life may exist, as we will discuss in Section 7.1g(ii).) The Cassini spacecraft flew by Jupiter, en route to Saturn, in 2000–2001 (■ Fig. 7–3). In 2011, NASA launched the Juno spacecraft on its five-year journey to Jupiter, which it will then orbit in an effort to further probe under the planet's thick cloud cover. Its orbit will be elliptical, from pole to pole and sometimes skimming the cloud tops. It carries three LEGO figurines, each about 4 cm high: Galileo (carrying a telescope) and the Roman mythological figures Jupiter and Juno, husband and wife.

7.1b The Great Red Spot

The Great Red Spot is a gaseous "island" 2–3 times larger across than the Earth (■ Fig. 7–4). It is the vortex of a violent, long-lasting storm, similar to large storms on Earth, and drifts about slowly with respect to the clouds as the planet rotates. From the sense of its rotation (counterclockwise rather than clockwise in its hemisphere), measured from time-lapse photographs, we can tell that it is a pressure high rather than a low. We also see how it interacts with surrounding clouds and smaller spots. The Great Red Spot has been visible for at least 150 years, and maybe even 300 years. Sometimes it is relatively prominent and colorful, and at other times the color may even disappear for a few years.

Table 7–1 Comparative planetology: rings
Jupiter: One relatively narrow ring, discovered by Voyager.
Saturn: Glorious ring system with many rings and divisions.
Uranus: Eleven narrow rings, discovered during a stellar occultation.
Neptune: Several rings discovered during a stellar occultation; one ring has several clumps.

Why has the Great Red Spot lasted this long? Heat, energy flowing into the storm from below it, partly maintains its energy supply. The storm also contains more mass than hurricanes on Earth, which makes it more stable. Furthermore, unlike Earth, Jupiter has no continents or other structure to break up the storm. Also, we do not know how much energy the Spot gains from the circulation of Jupiter's upper atmosphere and eddies (rotating regions, swirls) in it. Until we can sample lower levels of Jupiter's atmosphere, we will not be able to decide definitively.

Studying the eddies in Jupiter's atmosphere helps us interpret features on Earth. For example, one hypothesis to explain Jupiter's spots holds that they are similar to circulating rings that break off from the Gulf Stream in the Atlantic Ocean.

7.1c Jupiter's Atmosphere

Heat emanating from Jupiter's interior churns the atmosphere. (In the Earth's atmosphere, on the other hand, most of the energy comes from the outside – from the Sun.) Pockets of gas rise and fall, through the process of convection, as described for the Sun in Chapter 10. The bright bands ("zones") and dark bands ("belts") on Jupiter represent different cloud layers (■ Fig. 7–5).

Wind velocities show that each hemisphere of Jupiter has a half-dozen currents blowing eastward or westward. The Earth, in contrast, has only westerly (west-to-east) polar and subtropical jets and a seasonal easterly tropical jet over Africa.

On December 7, 1995, a probe dropped from the Galileo spacecraft transmitted data for 57 minutes as it fell through Jupiter's atmosphere. It gave us accurate measurements of Jupiter's composition; the heights of the cloud layers; and the variations of temperature, density, and pressure. It went through about 600 km of Jupiter's atmosphere, only about 1 per cent of Jupiter's radius. The probe found that

(a)

NASA/JPL-Caltech

(b)

NASA/JPL-Caltech

■ **FIGURE 7–4** (a) The Great Red Spot, in an image from the Galileo spacecraft. It rotates in the anticyclonic (counterclockwise) sense. The Great Red Spot is now about 26,000 km by 134,000 km and has been gradually shrinking for 50 years. It goes 20 to 40 km deep. (b) This Galileo image reveals clouds of ammonia ice, shown in light blue to the upper left of the Great Red Spot, which appears in false color. The ammonia-ice clouds are produced by powerful updrafts of ammonia-laden air caused by turbulence in Jupiter's atmosphere that results from the Great Red Spot. Red-orange shows high-level clouds, yellow shows mid-level clouds, and green shows low-level clouds.

(a)

Jay Pasachoff, Glenn Schneider, and colleagues/NASA/STScI

(b)

NASA/JPL/University of Arizona

■ **FIGURE 7–5** (a) Jupiter in August 2012 from the Hubble Space Telescope, in a color composite made from a pair of ultraviolet and infrared images as part of the research of one of us (J.M.P.) to study the transit of Venus that was to be visible from Jupiter and which we attempted to view in reflection. A band of clouds was very active. (b) Bright and dark bands on Jupiter, as seen from the Cassini spacecraft as it flew by. Computer processing has unrolled the surface.

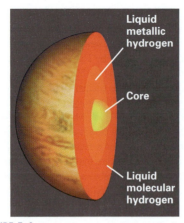

■ FIGURE 7–6 The current model of Jupiter's interior.

Jupiter's winds were stronger than expected and increased with depth, which shows that the energy that drives them comes from below.

Extensive lightning storms, including giant-sized lightning strikes called "superbolts," were discovered by the Voyagers. The Galileo spacecraft photographed giant thunderclouds on Jupiter, which indicates that some regions are relatively wet and others relatively dry. The probe found less water vapor than expected, probably because it fell through a dry region.

7.1d Jupiter's Interior

Most of Jupiter's interior is in liquid form; thus, Jupiter is more appropriately called a "gas/liquid giant" than simply a "gas giant." (The same is true of Saturn.) Jupiter's central temperature may be between 13,000 and 35,000 K. The central pressure is 100 million times the pressure of the Earth's atmosphere measured at our sea level

due to Jupiter's great mass pressing in. (The Earth's central pressure is 4 million times its atmosphere's pressure, and Earth's central temperatures are several thousand degrees.) Because of this high pressure, Jupiter's interior is probably composed of ultra-compressed hydrogen surrounding a rocky core consisting of perhaps 10 Earth masses of iron and silicates (■ Fig. 7–6).

Jupiter radiates 1.6 times as much energy as it receives from the Sun. It must have an internal source – perhaps the energy remaining from its collapse from a primordial gas cloud 20 million km across or from the accretion of matter long ago. Jupiter is undoubtedly still contracting inside and this process also liberates energy. It lacks the mass necessary by a factor of about 75, however, to have heated up enough to become a star, generating energy by nuclear processes (see Chapter 12). It is therefore not "almost a star," contrary to some popular accounts.

7.1e Jupiter's Magnetic Field

The space missions showed that Jupiter's tremendous magnetic field is even more intense than many scientists had expected (■ Fig. 7–7a). At the height of Jupiter's clouds, the magnetic field strength is 10 times that of the Earth, which itself has a rather strong field.

The inner field is shaped like a doughnut, containing several shells of charged particles, like giant versions of the Earth's Van Allen belts (■ Fig. 7–7b). The outer region of Jupiter's magnetic field interacts with the particles flowing outward from the Sun. When this solar wind is strong, Jupiter's outer magnetic field (shaped like a pancake) is pushed in. When the high-energy particles interact with Jupiter's magnetic field, radio emission results. Jupiter's magnetic field leads Jupiter to have giant auroras (■ Fig. 7–7c).

■ FIGURE 7–7 (a) Cassini could detect neutral atoms expelled from the magnetosphere. A black dot shows Jupiter's size. Lines of Jupiter's magnetic field and a cross section of the doughnut-shaped ring of material from Jupiter's moon Io are sketched. (b) Cassini's radio instrument, operating at a wavelength of 2.2 cm, detected high-energy electrons in the radiation belts. A picture of Jupiter to scale is superimposed. (c) The northern and southern auroral ovals on Jupiter, imaged in the ultraviolet by the Hubble Space Telescope. They are about 500 km above the level in Jupiter's atmosphere where the pressure is the same as Earth's surface pressure.

Mark Showalter, NASA's Ames Research Center

■ **FIGURE 7–8** Jupiter's ring (overexposed), with Jupiter's moon Europa behind it. The Galileo spacecraft was looking at light scattered forward through the ring, and at the side of Europa away from the Sun, illuminated by reflected light off Jupiter itself (the analogue of "earthshine" on the Moon; see Chapter 4).

7.1f Jupiter's Ring

Though Jupiter wasn't expected to have a ring, Voyager 1 was programmed to look for one just in case; Saturn's rings, of course, were well known, and Uranus's rings had been discovered only a few years earlier during ground-based observations. A single Voyager 1 photograph indeed showed a wispy ring of material around Jupiter at about 1.8 times Jupiter's radius, inside the orbit of its innermost moon.

As a result, Voyager 2 was targeted to take a series of photographs of the ring. From the far side looking back, the ring appeared unexpectedly bright, probably because small particles in the ring scattered the light toward the spacecraft. Within the main ring, fainter material appears to extend down to Jupiter's cloud tops (■ Fig. 7–8). The ring particles were knocked off Jupiter's inner moons by micrometeorites. Whatever their origin, the individual particles probably remain in the ring only temporarily.

7.1g Jupiter's Amazing Satellites

Four of the innermost satellites were discovered by Galileo in 1610 when he first looked at Jupiter through his small telescope. These four

moons (Io, Europa, Ganymede, and Callisto) are called the **Galilean satellites** (■ Fig. 7–9). One of these moons, Ganymede, 5276 km in diameter, is the largest satellite in the Solar System and is larger than the planet Mercury.

The Galilean satellites have played a very important role in the history of astronomy. The fact that these particular satellites were noticed to be going around another planet, like a solar system in miniature, supported Copernicus's Sun-centered model of the Solar System. Not everything revolved around the Earth! It was fitting to name the Galileo spacecraft after the discoverer of Jupiter's moons.

Jupiter also has dozens of other satellites, some known or discovered from Earth and others discovered by the Voyagers. None of these other satellites is even 10 per cent the diameter of the smallest Galilean satellite.

Through first Voyager-spacecraft and then Galileo-spacecraft close-ups (■ Fig. 7–10), the satellites of Jupiter have become known to us as worlds with personalities of their own. The four Galilean satellites, in particular, were formerly known only as dots of light. Not only the Galilean satellites, which range between 0.9 and 1.5 times the size of our own Moon, but also the smaller ones that have been imaged in detail, turn out to have interesting surfaces and histories.

7.1g(i) Pizza-like Io

Io, the innermost Galilean satellite, provided the biggest surprises. Scientists knew that Io gave off particles as it went around Jupiter, and other scientists had predicted that Io's interior would be heated by its flexing (to be discussed below). Voyager 1 discovered that these particles resulted from active volcanoes on the satellite, a nice confirmation of the earlier ideas. Eight volcanoes were seen actually erupting, more than erupt on the Earth at any one time. When Voyager 2 went by a few months later, most of the same volcanoes were still erupting.

Though the Galileo spacecraft could not go close to Io during most of its mission, for fear of getting ruined because of Jupiter's strong radiation field in that region, it could obtain high-quality images of Io and its volcanoes (■ Fig. 7–11a). Finally, it went within a few hundred kilometers of Io's surface, and found that 100 volcanoes were erupting simultaneously. Io's surface (■ Fig. 7–11b and c)

NASA/JPL/Malin Space Science Systems

■ **FIGURE 7–9** Three of the four Galilean satellites of Jupiter alongside Jupiter. Callisto (extreme bottom left), Ganymede (left center), and Europa (right) show in this image taken from a camera on Mars Global Surveyor orbiting Mars.

■ **FIGURE 7–10** The four Galilean satellites of Jupiter, alongside Jupiter's Great Red Spot, in a montage made from Galileo images. From top to bottom, matching the order from close to Jupiter and outward, they are Io, Europa, Ganymede, and Callisto.

has been transformed by the volcanoes, and is by far the youngest surface we have observed in the Solar System.

Why does Io have so many active volcanoes? Gravitational forces from Ganymede and Europa distort Io's orbit slightly, which changes the tidal force on it from Jupiter in a varying fashion. This changing tidal force flexes Io, creating heat from friction that heats the interior and leads to the volcanism.

The surface of Io is covered with sulfur and sulfur compounds, including frozen sulfur dioxide, and the thin atmosphere is full of sulfur dioxide. It certainly wouldn't be a pleasant place to visit! Io's surface, orange in color and covered with strange formations because of the sulfur, led Brad Smith, the head of the Voyager imaging team, to remark that "It's better looking than a lot of pizzas I've seen."

Galileo imaging shows many mountains too tall to be supported by sulfur, so stronger types of rock must be involved, with a crust at least 30 km thick above the molten regions. Also, Galileo's infrared observations show that some of the volcanoes are too hot to be sulfur volcanism. The surface changed substantially even as the Galileo spacecraft watched it. By the time Galileo made its sixth, last, and closest pass of Io in 2001, it had raised the total of identified volcanoes to 120. Additional flybys of Io by Cassini, en route to Saturn, and New Horizons, en route to Pluto, have raised that number to 150.

7.1g(ii) Europa, a Possible Abode for Life

Europa, Jupiter's Galilean satellite with the highest albedo (reflectivity), has a very smooth surface and is covered with narrow, dark stripes. The lack of surface relief, mapped by the Galileo spacecraft to be no more than a couple of hundred meters high, suggests that the surface we see is ice.

The markings may be fracture systems in the ice, like fractures in the large fields of sea ice near the Earth's north pole, as apparently

■ **FIGURE 7–11** (a) Io, from the Galileo spacecraft, when Io was in Jupiter's shadow. This false-color image uses red to show the brightest features, with the intensity range descending through yellow and green to blue. The diffuse glow on the left limb extends 800 km, 10 times greater in height than the volcano Prometheus below it, which emitted the particles causing the glow. The yellow and red surface spots show erupting magma. Close flybys have revealed over 100 erupting volcanoes. Many of the volcanic plumes active now are different from ones viewed by the Voyagers in 1979. (b) A close-up view of Io, from Galileo's last pass. This 2001 optical image shows Tupan Caldera. (c) An infrared image of Tupan Caldera, showing the temperature in false color; white is hottest, followed by red and yellow. The hottest regions, probably hot lavas, are darkest in the visible-light image. The cool region in the center of the crater may be an island. Tupan was named for the Brazilian god of thunder.

■ **FIGURE 7–12** Close-up views of icy regions on Europa, almost certainly the crust of a global ocean. In 2013, magnesium sulfate salt was detected on a small part of Europa's surface, using an infrared spectrograph and adaptive optics on the Keck II telescope to give the best-ever spectral resolution of a small area. The presence of magnesium on the surface indicates an interplay of the surface with the ocean, allowing us to understand the ocean better even without sampling it directly.

verified in Galileo close-ups (■ Fig. 7–12). Some longer ridges can be traced far across Europa's surface. Few craters are visible, suggesting that the ice was soft enough below the crust to close in the craters. Either internal radioactivity or, more likely, a gravitational tidal heating like that inside Io provides the heat to soften the ice.

Because Europa possibly has a liquid-water ocean and extra heating, many scientists consider it a worthy location to check for signs of life. We can only hope that the ice crust, which may be about 10–50 km thick, is thin enough in some locations for us to be able to penetrate it to reach the ocean that may lie below – and see if life has ever existed there.

7.1g(iii) Giant Ganymede

The largest satellite in the Solar System, Ganymede, shows many craters (■ Fig. 7–13) alongside weird, grooved terrain (■ Fig. 7–14). Ganymede is bigger than Mercury but less dense; it contains large amounts of water ice surrounding a rocky core. But an icy surface is as hard as steel in the cold conditions that far from the Sun, so

■ **FIGURE 7–13** Ganymede, Jupiter's largest satellite. Many impact craters, some with systems of bright rays, are visible. The large, dark region has been named Galileo Regio. Low-albedo features on satellites are named for astronomers.

■ **FIGURE 7–14** A computer-generated view, with exaggerated perspective, of ridges on Ganymede. Ganymede's icy surface has been fractured and broken into many parallel ridges and troughs. Such bright grooved terrain covers over half of Ganymede's surface.

it retains the craters from perhaps 4 billion years ago. The grooved terrain is younger.

Ganymede shows many lateral displacements, where grooves have slid sideways, like those that occur in some places on Earth (for example, the San Andreas fault in California). It is the only place besides the Earth where such faults have been found. Thus, further studies of Ganymede may help our understanding of terrestrial earthquakes. The Galileo spacecraft found a stronger magnetic field around Ganymede than expected, so perhaps Ganymede is more active inside than previously supposed.

7.1g(iv) Pockmarked Callisto

Callisto, the outermost of Jupiter's Galilean satellites, has so many craters (■ Fig. 7–15) that its surface must also be the oldest. Callisto, like Europa and Ganymede, is covered with ice. A huge bull's-eye formation, Valhalla, contains about 10 concentric rings, no doubt resulting from an enormous impact. Perhaps ripples spreading from the impact froze into the ice to make Valhalla.

Callisto had been thought to be old and uninteresting, but observations from the Galileo spacecraft have revised the latter idea, by showing changes: There are fewer small craters than expected, so the small craters that must have once been there were probably covered by dust that meteorite impacts eroded from larger craters, or disintegrated by themselves through electrostatic charges.

The Galileo spacecraft's measurements of Callisto's gravity from place to place show that its mass is concentrated more toward its center than had been thought. This concentration indicates that heavier materials inside have sunk, and perhaps even indicates that there is an ocean below Callisto's surface. Callisto's interactions with Jupiter's magnetic field have been interpreted to back up the idea of an internal ocean.

7.1g(v) Other Satellites

Galileo's last pass near a Jupiter satellite occurred in 2002 at Amalthea, a small, inner, potato-shaped satellite. (See *A Closer Look 7.2: Jupiter and Its Satellites in Mythology*.) The cameras weren't used; Jupiter's magnetosphere especially close to the planet was primarily studied.

NASA/JPL/Caltech

■ **FIGURE 7–15** The side of Callisto that leads in its orbit around Jupiter, imaged from the Galileo spacecraft. (Callisto and the other Galilean satellites are locked in synchronous orbit, so that their same sides always face Jupiter, similar to the case of Earth's Moon.) The impact region Asgard (top center) is surrounded by concentric rings that are up to 1700 km in diameter. The bright icy materials excavated by the younger craters contrast with the darker and redder coatings on older surfaces. Galileo images show, surprisingly, that Callisto has very few craters smaller than 100 m in diameter.

By tracking the spacecraft, Amalthea's mass was measured from its gravitational attraction. The mass coupled with the observed volume gives the density, which turned out to be unexpectedly low, close to that of water ice. Jet Propulsion Laboratory scientists deduced that Amalthea seems to be a loosely packed pile of rubble. Amalthea is thus probably mostly rock with perhaps a little ice, rather than a mix

A CLOSER LOOK 7.2 — Jupiter and Its Satellites in Mythology

All of Jupiter's moons except Amalthea are named after lovers of Zeus in Greek mythology, Zeus being the Greek equivalent of Jupiter. Amalthea, a goat-nymph, was Zeus's nurse, and out of gratitude he made her into the constellation Capricorn, a goat. Zeus changed Io into a heifer to hide her from his jealous wife, Hera; in honor of Io, Earth's crescent moon has horns, as do bulls.

Ganymede was a Trojan youth carried off by an eagle to be Jupiter's cup bearer (the constellation Aquarius). Callisto was changed into a bear as punishment for her affair with Zeus. She was then slain by mistake, and rescued by Zeus by being transformed into the Big Bear in the sky, the constellation Ursa Major. Jealous Hera persuaded the sea god to forbid Callisto to ever bathe in the sea, which is why Ursa Major never sinks below the horizon as seen from the latitude of Greece.

Europa was carried off to Crete on the back of Zeus, who took the form of a white bull. She became King Minos's mother. Pasiphae (also the name of one of Jupiter's moons) was the wife of Minos and the mother of the Minotaur.

NASA/JPL/Space Science Institute

■ **FIGURE 7–16** A high-resolution view of Saturn, imaged with NASA's Cassini spacecraft. The original digital image can be blown up to wall size without showing noticeable fuzziness.

Table 7–2 Comparative planetology: spots
Great Red Spot on Jupiter: Lasts hundreds of years.
Great Dark Spot on Neptune: There when Voyager flew by (1989) and gone a few years later as shown by Hubble images. But nobody knows how long it had been there.

■ **FIGURE 7–17** Since Saturn's density is lower than that of water, it would float, like a famous soap in commercials dating back over 100 years, if we could find a big enough bathtub. But it would leave a ring!

of rock and iron, which would be denser. Amalthea, and presumably other irregular satellites, seems to have been broken apart, with the pieces subsequently drawn roughly together.

Many much smaller satellites of Jupiter are being discovered from the ground, given the existence of mosaics of sensitive CCDs (electronic detectors; see Chapter 3) that cover larger regions of the sky than previously possible and of computer processing methods to analyze the data. In recent years, dozens of small moons, some only 2 km across, have been discovered around Jupiter, bringing Jupiter's total of moons up to at least 67.

These "irregular" satellites are, no doubt, captured objects that were once in orbit around the Sun. This origin is different from that of the large, regular satellites (like the Galilean moons), and from that of the set of smaller satellites in close orbits around Jupiter that apparently are remnants of collisions. Both of these latter types of moons are thought to have formed from a disk of gas and dust around the planet. The other jovian planets also have these three types of satellites.

Studies of Jupiter's moons tell us about the formation of the Jupiter system, and help us better understand the early stages of the entire Solar System. NASA's Juno spacecraft, named after the wife of Roman god Jupiter, is expected to help us understand much more about the origins and evolution of our Solar System's largest planet.

ESA's Jupiter Icy Moon Explorer (JUICE) is scheduled for blast-off from Earth in 2022 to study Ganymede, Callisto, and Europa. It is to arrive in 2030. All these moons are thought to have internal oceans.

7.2 SATURN

Saturn, like Jupiter, Uranus, and Neptune, is a giant planet. Its diameter, without its rings, is 9 times that of Earth; its mass is 95 Earth masses. It is a truly beautiful object in a telescope of any size. The glory of its system of rings makes it stand out even in small telescopes (see *Star Party 7.1: Observing the Giant Planets*). The view from the Cassini spacecraft, now in orbit around it, is breathtaking (■ Fig. 7–16).

The giant planets have low densities. Saturn's is only 0.7 g/cm³, 70 per cent the density of water (■ Fig. 7–17). The bulk of Saturn is hydrogen molecules and helium, reflecting Saturn's formation directly from the solar nebula. Saturn is thought to have a core of heavy elements, including rocky material, making up about the inner 20 per cent of its diameter.

Voyagers 1 and 2 flew by Saturn in 1980 and 1981, respectively. Cassini, a joint NASA/European Space Agency mission, arrived at Saturn in 2004. It is orbiting the Saturn system, going up close in turn to various of Saturn's dozens of moons. We will be discussing Cassini observations throughout the following sections.

7.2a Saturn's Rings

The rings extend far out in Saturn's equatorial plane, and are inclined to the planet's orbit. Over a 30-year period, we sometimes see them from above their northern side, sometimes from below their southern side, and at angles in between (■ Fig. 7–18). When seen edge-on, they are almost invisible.

The rings of Saturn consist of material that was torn apart by Saturn's gravity, or material that failed to collect into a moon at the

(a)

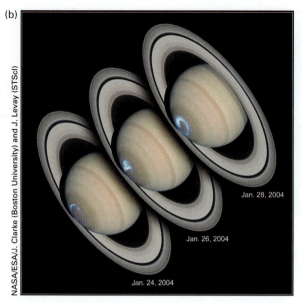

(b)

■ **FIGURE 7–18** (a) Saturn, from the Hubble Space Telescope, in a series of views over a five-year period. (b) These auroras on Saturn, seen in Hubble's visible-light images, show significant changes over the course of a few days.

time when the planet and its moons were forming. However, the rings may have formed fairly recently, within the past few hundred million years.

Every massive object has a sphere, called its **Roche limit**, inside of which blobs of matter do not hold together by their mutual gravity. The forces that tend to tear the blobs apart from each other are tidal forces. They arise, like the Earth's tides, because some parts of an object are closer to the planet than others and are thus subject to higher gravity. The difference between the gravity force farther in and the gravity force farther out is the tidal force.

The radius of the Roche limit is usually $2^{1}/_{2}$ to 3 times the radius of the larger body, closer to the latter for the relative densities of Saturn and its moons. The Sun also has a Roche limit, but all the planets lie outside it. The natural moons of the various planets lie outside their respective Roche limits. Saturn's rings lie inside Saturn's Roche limit, so it is not surprising that the material in the rings is spread out rather than collected into a single orbiting satellite.

Artificial satellites that we send up to orbit the Earth are constructed of sufficiently rigid materials that they do not break up even though they are within the Earth's Roche limit; they are held together by forces much stronger than gravity.

Saturn has several concentric major rings visible from Earth. The brightest ring is separated from a fainter broad outer ring by an apparent gap called **Cassini's division**. (The 17th-century astronomer Jean-Dominique Cassini (1625–1712), who moved from Italy to France in 1671, discovered several of Saturn's moons as well as the division in the rings, the latter in 1675. It was very appropriate to name not only the ring division but also NASA's spacecraft to Saturn after him.) Another ring is inside the brightest ring. We know that the rings are not solid objects, because the rotation speed of the outer rings is slower than that of rings closer to Saturn.

Radar waves bounced off the rings show that the particles in the rings are at least a few centimeters, and possibly a meter, across. Infrared studies show that at least their outer parts consist of ice.

The images from the Voyagers revolutionized our view of Saturn, its rings, and its moons. The Cassini mission is providing even more detailed views, as it orbits for years instead of merely flying by (■ Fig. 7–19). Only from spacecraft can we see the rings from a vantage point different from the one we have on Earth. Backlit views showed that Cassini's division, visible as a dark (and thus apparently empty) band from Earth, appeared bright, so it must contain some particles.

The rings are thin, for when they pass in front of stars, the starlight easily shines through. Studies of the changes in the radio signals from the Voyagers when they went behind the rings showed that the rings are only about 20 m thick. Relative to the diameter of the rings, this is equivalent to a CD (compact disc) that is 30 km across though still its normal thickness!

The closer the spacecraft got to the rings, the more individual rings became apparent. Each of the known rings was actually divided into many thinner rings. The number of these rings (sometimes called "ringlets") is in the hundreds of thousands. The images from the Cassini spacecraft (■ Fig. 7–20) surpassed even Voyager's views of ringlets.

The outer major ring turns out to be kept in place by a tiny satellite orbiting just outside it (■ Fig. 7–21). At least some of the rings are kept narrow by "shepherding" satellites that gravitationally affect the ring material, a concept that we can apply to rings of other planets.

A post-Voyager theory said that many of the narrowest gaps may be swept clean by a variety of small moons. These objects would be embedded in the rings in addition to the icy snowballs that make up most of the ring material. The tiny moon Pan has been observed

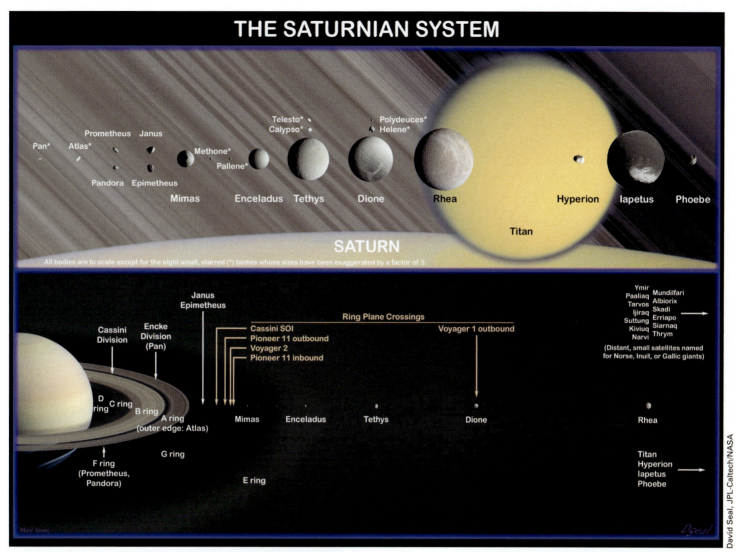

THE SATURNIAN SYSTEM

SATURN

All bodies are to scale except for the eight small, starred (*) bodies whose sizes have been exaggerated by a factor of 5.

Pan* Atlas* Prometheus Janus Methone* Telesto* Polydeuces*
Pandora Epimetheus Pallene* Calypso* Helene*

Mimas Enceladus Tethys Dione Rhea Titan Hyperion Iapetus Phoebe

Cassini Division Encke Division (Pan) Janus Epimetheus Ring Plane Crossings

Cassini SOI Voyager 1 outbound
Pioneer 11 outbound
Voyager 2
Pioneer 11 inbound

Ymir Mundilfari
Paaliaq Albiorix
Tarvos Skadi
Ijiraq Erriapo
Suttung Siarnaq
Kiviuq Thrym
Narvi

(Distant, small satellites named for Norse, Inuit, or Gallic giants)

D ring C ring B ring A ring (outer edge: Atlas)

Mimas Enceladus Tethys Dione Rhea

F ring (Prometheus, Pandora) G ring E ring

Titan
Hyperion
Iapetus
Phoebe

May 2005

David Seal, JPL–Caltech/NASA

■ **FIGURE 7–19** Saturn and many of its moons, photographed from the Cassini spacecraft (except for a few Voyager or generic images, whose labels are marked with *). In the lower half, the positions of the rings and moons are shown to scale.

clearing out the Encke Gap in just this way. Theorists suppose that smaller moons probably clear out several other gaps in Saturn's rings, and Cassini is finding some of these moons.

7.2b Saturn's Atmosphere

Like Jupiter, Saturn rotates quickly on its axis; a complete period is only 10 hours, in spite of Saturn's diameter being over 9 times greater than Earth's. The rapid rotation causes Saturn to be larger across the equator than from pole to pole. This equatorial bulging makes Saturn look slightly "flattened." Jupiter also looks flattened, or oblate, for this reason.

The structure in Saturn's clouds is of much lower contrast than that in Jupiter's clouds. It is not a surprise that the chemical reactions can be different; after all, Saturn is colder than Jupiter. Saturn has extremely high winds, up to 1800 km/hr, 4 times faster than the winds on Jupiter. Cassini is tracking the winds with higher precision than was previously possible. On Saturn, the variations in wind speed

do not seem to correlate with the positions of bright and dark bands, unlike the case with Jupiter (■ Fig. 7–22). As on Jupiter, but unlike the case for Earth, the winds seem to be driven by rotating eddies, which in turn get most of their energy from the planet's interior. Such differences provide a better understanding of storm systems in Earth's atmosphere.

7.2c Saturn's Interior and Magnetic Field

Saturn radiates about twice as much energy as it absorbs from the Sun, a greater factor than for Jupiter. One interpretation is that only $2/3$ of Saturn's internal energy remains from its formation and from its continuing contraction under gravity. The rest would be generated by the gravitational energy released by helium sinking through the liquid hydrogen in Saturn's interior. The helium that sinks has condensed because Saturn, unlike Jupiter, is cold enough.

NASA/JPL/Space Science Institute

■ **FIGURE 7–20** The several rings we see from Earth break up into thousands upon thousands of narrow ringlets when viewed up close with the Cassini mission's cameras.

NASA/JPL/Space Science Institute

■ **FIGURE 7–21** Cassini imaged these density waves in Saturn's rings, caused by the gravity of Saturn's satellites.

(a)

redrawn from NASA/JPL-Caltech

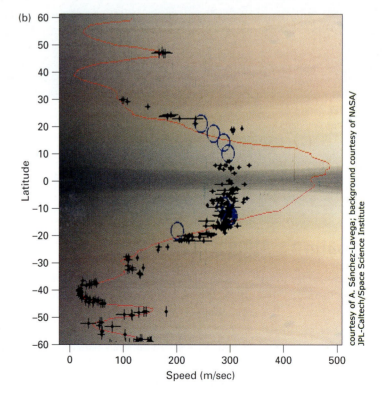

(b)

courtesy of A. Sánchez-Lavega; background courtesy of NASA/JPL-Caltech/Space Science Institute

■ **FIGURE 7–22** (a) The winds of Jupiter, a graph of their speed on a background that shows Jupiter's surface and how it varies with latitude. (b) The winds of Saturn. The solid line shows the Voyager data and the points show newer data from the Hubble Space Telescope. Note how the winds appear much less strong at the equator in the newer data. The situation is under study to see if the change is global or if a wind-shear effect is showing.

FIGURE 7–23 Saturn's auroras, showing the effect of the planet's magnetic field. This ultraviolet image was taken with the Hubble Space Telescope.

Saturn gives off radio signals, as does Jupiter, a pre-Voyager indication to earthbound astronomers that Saturn also has a magnetic field. The Voyagers found that the magnetic field at Saturn's equator is only $2/3$ of the field present at the Earth's equator. Remember, though, that Saturn is much larger than the Earth and so its equator is much farther from its center.

Saturn's magnetic field contains belts of charged particles (analogous to Van Allen belts), which are larger than Earth's but smaller than Jupiter's. (Saturn's surface magnetic field is 20 times weaker than Jupiter's.) These particles interact with the atmosphere near the poles and produce auroras (■ Fig. 7–23).

7.2d Saturn's Moon Titan

At 40 per cent the diameter of the Earth, Titan (see *A Closer Look 7.3: Saturn's Satellites in Mythology*) is an intriguing body for a number of reasons. Titan has an atmosphere that was detected from Earth pre-Voyager. Studies of how the radio signals faded when Voyager 1 went behind Titan showed that Titan's atmosphere is denser than Earth's. The surface pressure on Titan is $1\frac{1}{2}$ times that on Earth.

Titan's atmosphere is opaque, apparently because of the action of sunlight on chemicals in it, forming a sort of "smog" and giving it its reddish tint. Smog on Earth forms in a similar way. The Voyagers showed (■ Fig. 7–24a) several layers of haze. They detected nitrogen, which makes up the bulk of Titan's atmosphere, as it does Earth's. Methane is a minor constituent, perhaps 1 per cent. A greenhouse effect is present, making some scientists wonder whether Titan's surface may have been warmed enough for life to have evolved there. There is also an "anti-greenhouse effect" caused by the smog.

The temperature near the surface, deduced from measurements made with Voyager's infrared radiometer, is only about –180°C (93

K), somewhat warmed (12°C) by a combination of the greenhouse effect and the anti-greenhouse effect. This temperature is near that of methane's "triple point," at which it can be in any of the three physical states – solid, liquid, or gas. So methane may play the role on Titan that water does on Earth.

Scientists wanted to know whether Titan is covered with lakes or oceans of methane mixed with ethane, and whether other parts are covered with methane ice or snow. Below, we will see how a recent spacecraft gave some answers. Also, using filters in the near-infrared, first the Hubble Space Telescope and then ground-based telescopes have been able to penetrate Titan's haze to reveal some structure on its surface and methane clouds (■ Fig. 7–24b,c), though no unambiguous lakes were found.

Some of the organic molecules formed in Titan's atmosphere from lightning storms and other processes may rain down on its surface. Thus the surface, largely hidden from our view, may be covered with an organic crust about a kilometer thick, perhaps partly dissolved in liquid methane. These chemicals are similar to those from which we think life evolved on the primitive Earth. But it is probably too cold on Titan for life to begin.

Titan is so intriguing, and potentially so important, that the lander of the Cassini mission was sent to plunge through its atmosphere. So while the lander of Jupiter's Galileo mission went into the planet itself, this lander, known as Huygens, penetrated the clouds around Saturn's largest moon when it arrived on January 14, 2005. (The lander is named after Christiaan Huygens, who first realized that Saturn was surrounded by a ring and who also, in 1655, discovered Titan.) At higher levels, the Huygens probe was buffeted by strong winds that ranged up to 400 km/hour, though the winds calmed near the surface. Radio telescopes on Earth measured the winds by following the Doppler shifts in the probe's signals, since an onboard Doppler experiment failed.

As the Huygens lander drifted downward on its parachute for $2\frac{1}{2}$ hours, it measured the chemical composition of Titan's atmosphere. It then imaged the surface below it when it got sufficiently far below the haze. Dunes were seen, and the shape of a shoreline was visible (■ Fig. 7–24d,e). Huygens did not see liquid substances, though the Cassini orbiter's radar eventually detected hydrocarbon lakes (methane and/or ethane), especially in the far north, and its infrared imager even saw a glint of reflection of the type that occurs off a shiny body. The dark material is thought to be tar-like and has probably settled out of the atmosphere. Some "islands" in it may be raised material that diverted the flow.

The branching of systems that were visible clearly shows that some liquid flowed in the past. Though few clouds were seen, perhaps methane or ethane rainstorms occurred intensely though rarely. A metal rod attached to the bottom of the Huygens lander

A CLOSER LOOK 7.3 — Saturn's Satellites in Mythology

Saturn's moons are named after the Titans. In Greek mythology, the Titans were the children and grandchildren of the early supreme god Uranus and of Gaea, the goddess of the Earth. The Olympic gods like Zeus (Roman: Jupiter) followed the Titans.

■ **FIGURE 7–24** (a) Titan was disappointingly featureless even to Voyager 1's cameras because of its thick, smoggy atmosphere. Its northern polar region was relatively dark. The southern hemisphere was lighter than the northern. New imaging with adaptive optics on the Keck II and Gemini North telescopes showed (b) a bright continent on Titan and (c) methane clouds near its south pole. (d) Dunes on Titan (left) imaged with Cassini's radar compared with similar dunes in Saudi Arabia on Earth (right). (e) A shoreline and a branching riverbed were imaged by the Huygens probe's camera as it drifted downward through Titan's atmosphere on January 14, 2005. (f) Eroded blocks of ice on Titan's surface, imaged during the 90 minutes that Huygens survived on Titan's surface.

A CLOSER LOOK 7.4

Saturn's Rings and Moons from Cassini

The Cassini mission is orbiting Saturn, passing close to various of its dozens of moons.

FIGURE ACL 7.4–3 Mimas (about 200 km in radius, though not quite round) has an impact crater named Herschel, after the discoverer of Uranus. The crater, which makes Mimas resemble *Star Wars'* Death Star, is 130 km across.

FIGURE ACL 7.4–1 Giant Titan, the largest moon (2575 km in radius), seems hazy from afar; elsewhere, we show the views from Cassini's Huygens lander. (a) A true-color image from August 21, 2005. (b) Infrared penetrates the haze, as seen in this August 22, 2005, image. (c) Radar on the Cassini orbiter penetrated the smog to make this image, about 175 km across, with half-kilometer resolution, better than the typical resolution of the optical cameras.

FIGURE ACL 7.4–4 Rhea (765 km in radius) has craters as large as 300 km across. This 2013 close flyby only 1,000 km above the surface of this battered, icy moon will be Cassini's last close encounter.

FIGURE ACL 7.4–5 Iapetus (720 km in radius), strangely, has one very dark side and one bright side. The dark side faces forward as Iapetus orbits. The low albedo may result from sunlight reacting with methane that wells up, forming hydrocarbons. A newer model is that the low albedo is mostly from dark dust from Saturn's moon Phoebe.

FIGURE ACL 7.4–2 (a) Enceladus (about 250 km in radius, though not quite round) shows smooth regions, indicating that its surface had been melted, a sign of internal heating. It also has regions covered with impact craters. Linear grooves may be geologic faults. Its south polar region is venting water vapor and icy particles, giving it a tenuous atmosphere. (b) Tiny Enceladus in silhouette as it gives off particles to make Saturn's E-ring.

FIGURE ACL 7.4–6 Saturn's rings take on unusual structure when affected by the orbiting moons. Here we see the wake of tiny Prometheus (70 km × 50 km × 40 km in radius).

penetrated 15 cm into the surface, but it apparently had to break through a hard crust to do so. The crust might have formed from a recent inundation. A device on the lander measured a puff of methane, so perhaps there had been methane frost on the surface before Huygens landed there.

A final triumph came when the probe survived for 90 minutes on the surface, sending back pictures of ice blocks (■ Fig. 7–24f). These pieces of ice are rounded, apparently also revealing the past presence of flowing liquid.

Later analysis of infrared images seems to show a 30-km-wide structure that may be a volcanic dome. Such an ice volcano could be caused by the energy generated by tidal stresses within Titan caused by its elliptical orbit. It would release methane to the atmosphere. So the methane long measured in Titan's atmosphere might not be from a methane-rich hydrocarbon ocean after all.

7.2e Saturn's Other Satellites

So many of Saturn's other moons proved to have interesting surfaces when seen close up from Cassini that we show a variety of images in *A Closer Look 7.4: Saturn's Rings and Moons from Cassini.* Perhaps the biggest surprise is the activity of Enceladus, just 500 km across, only about one-tenth the diameter of Titan (and therefore, since volume is proportional to diameter cubed, one-thousandth Titan's volume). Enceladus gives off plumes of water vapor and ices that include organic compounds (■ Fig. 7–25), which Cassini has even been able to fly through and sample directly. Enceladus is now at the top of many scientists' lists of where we are most likely to first find signs of life outside Earth (as we discuss further in Chapter 20).

The Cassini spacecraft continues to fly by a variety of Saturn's moons, making images as well as other measurements. Also, as was the case for Jupiter, dozens of small, irregular moons continue to be found from space images and from Earth-based telescopes.

7.3 URANUS

The two other giant planets beyond Saturn – Uranus (pronounced "U´ran-us"; see *A Closer Look 7.5: Uranus and Neptune in Mythology*) and Neptune – are each about 4 times the diameter of (and about 15 times more massive than) the Earth. They reflect most of the sunlight that hits them, which indicates that they are covered with clouds.

Like Jupiter and Saturn, Uranus and Neptune don't have solid surfaces. Their atmospheres are also mostly hydrogen and helium, but they have a higher proportion of heavier elements. Below the outer atmosphere, it is thought that the composition changes to ices of water, methane, and ammonia; Uranus and Neptune are thus sometimes called "ice giants" instead of "gas giants" or "gas/liquid giants." At the planets' centers, a rocky core contains mostly silicon and iron, probably surrounded by ices. From their average densities, we have deduced that the cores of Uranus and Neptune make up substantial parts of those planets, differing from the relatively more minor cores of Jupiter and Saturn.

Uranus was the first planet to be discovered that had not been known to the ancients. The English astronomer and musician William Herschel reported the discovery in 1781. Actually, Uranus had been plotted as a star on several sky maps during the hundred years prior to Herschel's discovery, but had not been singled out as anything other than an ordinary star.

Uranus revolves around the Sun in 84 years at an average distance of more than 19 au. Uranus appears so tiny that it is not much bigger than the resolution we are allowed by Earth's atmosphere. Uranus is apparently surrounded by thick clouds of methane ice crystals (■ Fig. 7–26), with a clear atmosphere of molecular hydrogen above them. The trace of methane gas mixed in with the hydrogen makes Uranus look greenish blue.

Uranus is so far from the Sun that its outer layers are very cold. Studies of its infrared radiation give a temperature of –215°C (58 K).

(a)
NASA/JPL/Space Science Institute

(b)
NASA/JPL/Space Science Institute

(c)
NASA/JPL/Space Science Institute

■ **FIGURE 7–25** Enceladus is a surprising favorite in the search for life – surprising because of its relatively small size. But tidal heating provides the jets of water vapor and icy particles that include organic compounds, seen by Cassini [(a) and (b)]. (c) The jets come from "tiger stripes" in the south polar region, in which we find Baghdad Sulcis with its fractures.

(a)   (b)

Imke de Pater (UC Berkeley) and
Heidi Hammel (AURA)/W. M. Keck
Observatory

Imke de Pater (UC Berkeley) and
Heidi Hammel (AURA)/W. M. Keck
Observatory

NASA/JPL/Caltech

■ **FIGURE 7–26** (a) Uranus, as the Voyager 2 spacecraft approached. The picture at top shows Uranus as the human eye would see it. At bottom, false colors bring out slight differences in contrast. The "donut" shapes are out-of-focus dust spots on the camera. (b) Uranus from the Keck Observatory, with adaptive optics (AO) off and on, showing the obvious advantage of AO.

There is no evidence for an internal heat source, unlike the case for Jupiter, Saturn, and Neptune.

The other planets rotate such that their axes of rotation are very roughly parallel to their axes of revolution around the Sun. Uranus is different (as is Pluto), for its axis of rotation is roughly perpendicular to the other planetary axes, lying only 8° from the plane of its orbit (■ Fig. 7–27). Sometimes one of Uranus's poles faces the Earth, 21 years later its equator crosses our field of view, and then another 21 years later the other pole faces the Earth. Polar regions remain alternately in sunlight and in darkness for decades.

The strange seasonal effects that result on Uranus became obvious in 1999, when a series of Hubble Space Telescope views taken over the preceding years revealed the activity in the clouds of Uranus's once-every-84-years springtime. When we understand just how the seasonal changes in heating affect the clouds, we will be closer to understanding our own Earth's climate.

Voyager 2 reached Uranus in 1986. It revealed most of our current understanding of Uranus, its rings, and its moons.

(■ Fig. 7–28 shows more recent views.) Its moon Miranda, for example, though relatively small, has a surface that is extremely varied and interesting (■ Fig. 7–29).

7.3a Uranus's Atmosphere

Even though Voyager 2 came very close to Uranus's surface, as close as 107,000 km (a quarter of the distance from the Earth to the Moon), it saw very little detail on it. Thus Uranus's surface is very bland. Apparently, chemical reactions are more limited than on Jupiter and Saturn because it is colder. Uranus's clouds form relatively deep in the atmosphere.

A dark polar cap was seen on Uranus, perhaps a result of a high-level photochemical haze added to the effect of sunlight scattered by hydrogen molecules and helium atoms. At lower levels, the abundance of methane gas (CH_4) increases. It is this gas that absorbs the orange and red wavelengths from the sunlight that hits Uranus. Thus most of the light that is reflected back at us is blue-green.

Tracking some of the ten clouds that were detected by Voyager 2 revealed the rotation periods of their levels in Uranus's atmosphere. The larger cloud, at 35° south latitude, rotated in 16.3 hours. The smaller, fainter cloud, at 27° south latitude, rotated in 16.9 hours. Observations through color filters give evidence that these clouds are higher than their surroundings by 1.3 km and 2.3 km, respectively. Very good wind profiles can now be obtained from Earth using adaptive optics.

It was a surprise to find that both of Uranus's poles, even the one out of sunlight, are about the same temperature. The equator is nearly as warm. Comparing such a strange atmosphere with our own will help us understand Earth's weather and climate better.

7.3b Uranus's Rings

In 1977, astronomers on Earth watched as Uranus occulted (passed in front of) a star. Predictions showed that the occultation would be visible only from the Indian Ocean southwest of Australia. The scientists who went to study the occultation from an instrumented airplane turned on their equipment early, to be sure they caught the event. Surprisingly, about half an hour before the predicted time of occultation, they detected a few slight dips in the star's brightness (■ Fig. 7–30). They recorded similar dips, in the reverse order, about half an hour after the occultation.

The dips indicated that Uranus is surrounded by several rings, some of which have since been photographed by Voyager 2 and with the Hubble Space Telescope. Each time a ring went between us and

---

A CLOSER LOOK 7.5   **Uranus and Neptune in Mythology**

In Greek mythology, Uranus was the personification of Heaven and ruler of the world, as well as the son and husband of Gaea, the

Earth. Neptune, in Roman mythology, was the god of the sea, and the planet Neptune's trident symbol reflects that origin.

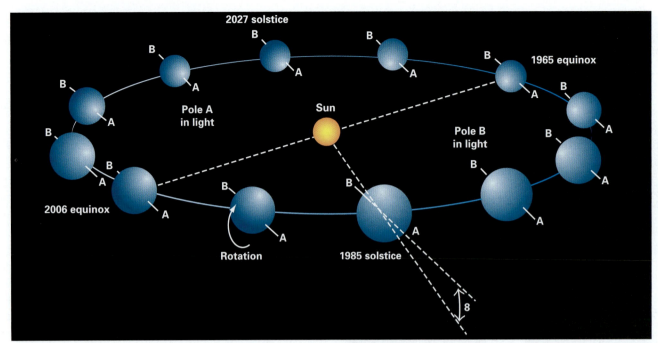

FIGURE 7–27 Uranus's axis of rotation lies roughly in the plane of its orbit. Notice how the planet's poles come within 8° of pointing toward the Sun, while ¼ of an orbit before or afterward, the Sun is almost over Uranus's equator.

FIGURE 7–28 (a) Uranus and its rings, imaged in the near-infrared with the "Antu" Unit Telescope of the European Southern Observatory's Very Large Telescope in Chile. Using a filter at the wavelength of the methane absorption from Uranus's disk diminishes the intensity of the disk. Since we see the rings by reflected sunlight, which does not have a diminished intensity at the methane wavelength, the rings appear relatively bright. We also see several of Uranus's moons. (b) A false-color infrared composite of Uranus, taken on July 9, 2004, with the adaptive optics system at the Keck Observatory. The rings and moons show better at 2.2 μm, because the methane absorption of the atmosphere makes the planet's disk appear dark. The atmospheric clouds show better at 1.6 μm, while the rings are barely visible.

FIGURE 7–29 Uranus's moon Miranda, apparently broken up by impacts and solidified in new configurations, providing the obviously jumbled surface.

FIGURE 7–30 A graph of the intensity of starlight during an occultation of a star by one of Uranus's rings. Each color image, recorded through a different filter, is shown in a different color ink.

NASA/JPL-Caltech

■ **FIGURE 7–31** When the Voyager spacecraft looked through the rings back toward the Sun, the backlit view (left) revealed new dust lanes between the known rings, which also show in this view. The streaks are stars. The forward-lighted view (right) is provided for comparison. One of the new rings discovered from Voyager is marked with an arrow. The originally discovered rings were given Greek letters; newer discoveries were numbered.

the distant star, the ring blocked some of the starlight, making a dip. Eleven rings are now known. They are quite dark, reflecting only about 2 per cent of the sunlight that hits them.

The rings have radii 1.7 to 2.1 times the radius of the planet. They are very narrow from side to side; some are only a few km wide. How can narrow rings exist, when we know that colliding particles tend to spread out? The discovery of Uranus's narrow rings led to the suggestion that a small unseen satellite (a "shepherd moon") near each ring keeps the particles together. As we saw, this model turned out to be applicable to at least some of the narrow ringlets of Saturn later discovered by the Voyagers.

Voyager provided detailed ring images (■ Fig. 7–31). Interpreting the small color differences is important for understanding the composition of the ring material. Quite significant was the single long-exposure, backlit view taken by Voyager. Study of these data has shown that less of the dust in Uranus's rings is very small particles compared with the dust in the rings of Saturn and Jupiter.

The particles in the main ring are large, from about 10 cm up to meters. Dust sheets are interspersed, not to mention an inner, dusty ring, the zeta ring.

We are now realizing that ring systems change more over time than had been thought, meaning that the ring system we see now is relatively young.

7.3c Uranus's Interior and Magnetic Field

Voyager 2 detected Uranus's magnetic field. It is intrinsically about 50 times stronger than Earth's. Surprisingly, it is tipped 60° with respect to

Uranus's axis of rotation. Even more surprising, it is centered on a point offset from Uranus's center by $^1/_3$ its radius. Our own Earth's magnetic field is nothing like that! Since Uranus's field is so tilted, it winds up like a corkscrew as Uranus rotates. Uranus's magnetosphere contains belts of protons and electrons, similar to Earth's Van Allen belts.

Voyager also detected radio bursts from Uranus every 17.24 hours. These bursts apparently come from locations carried in Uranus's interior by the magnetic field as the planet rotates. Thus Uranus's interior rotates slightly more slowly than its atmosphere. Since Uranus's and Neptune's interiors are similar, we will defer additional comments about the former for a few paragraphs.

7.4 NEPTUNE

Neptune is even farther from the Sun than Uranus, 30 au compared to about 19 au. Neptune takes 164 years to orbit the Sun. Its discovery was a triumph of the modern era of Newtonian astronomy. Mathematicians analyzed the amount that Uranus (then the outermost known planet) deviated from the orbit it would follow if gravity from only the Sun and the other known planets were acting on it. The small deviations could have been caused by gravitational interaction with another, as yet unknown, planet.

The first to work on the problem successfully was John C. Adams in England. In 1845, soon after he graduated from Cambridge University (■ Fig. 7–32), he predicted positions for the new planet. But neither of the two main astronomers in England made and analyzed observations to test this prediction quickly enough.

A year later, the French astronomer Urbain Leverrier independently worked out the position of the undetected planet. The French astronomers didn't test his prediction right away either. Leverrier sent his predictions to an acquaintance at the observatory in Berlin, where a star atlas

Master and Fellows of St. Johns College of Cambridge

■ **FIGURE 7–32** From John Couch Adams's diary, kept while he was in college. "1841, July 3. Formed a design in the beginning of this week, of investigating, as soon as possible after taking my degree, the irregularities in the motion of Uranus which are yet unaccounted for."

had recently been completed. The Berlin observer, Johann Galle, discovered Neptune within hours by comparing the sky against the new atlas.

Neptune has made a little more than one full orbit since it was located in 1846. But it now seems that Galileo inadvertently observed Neptune in 1613 and recorded its position, which more than doubles the period of time over which it has been studied. Galileo's observing records from January 1613 (when calculations indicate that Neptune had passed near Jupiter) show stars that were very close to Jupiter (■ Fig. 7–33), yet one of them does not appear in modern catalogues. Galileo himself once noted that one of the "stars" actually seemed to have moved from night to night, as a planet would. The object that Galileo saw was very close to but not quite exactly where our calculations of Neptune's orbit show that Neptune would have been at that time. Presumably, Galileo saw Neptune. We have used the positions he measured to improve our knowledge of Neptune's orbit!

7.4a Neptune's Atmosphere

Neptune, like Uranus, appears greenish blue in a telescope because of its atmospheric methane (■ Fig. 7–34). Some faint markings could be detected on Neptune even before adaptive optics systems became available on Earth (■ Fig. 7–35). It was thus known before Voyager that Neptune's surface was more interesting than Uranus's. Still, given its position in the cold outer Solar System, nobody was prepared for the amount of activity that Voyager discovered.

As Voyager approached Neptune, active weather systems became apparent (■ Fig. 7–36). An Earth-sized region that was soon called the **Great Dark Spot** (■ Fig. 7–37) became apparent. Though colorless, the Great Dark Spot seemed analogous to Jupiter's Great Red Spot in several ways. For example, it was about the same size relative to its planet, and it was in the same general position in its planet's southern

Charles Kowal, Space Telescope Science Institute

■ **FIGURE 7–33** Galileo's notebook from late December 1612, showing a * marking an apparently fixed star to the side of Jupiter and its moons. The "star" was evidently Neptune. The episode shows the importance of keeping good lab notebooks, a lesson we should all take to heart.

hemisphere (■ Fig. 7–38). Putting together a series of observations into a movie, scientists discovered that it rotated counterclockwise, as does the Great Red Spot. Thus it was anti-cyclonic, which made it a high-pressure region. Clouds of ice crystals, similar to Earth's cirrus but made of methane, form at the edge of the Great Dark Spot as the

■ **FIGURE 7–34** Each of the four graphs shows a spectrum divided by the Sun's spectrum. The spectra dip at wavelengths where there is a lot of absorption. The dark, broad troughs are from methane absorption; these wavelengths (the right side of the graphs) are in the infrared. Notice how strong the methane absorption is for Uranus and Neptune, and how it leaves mainly blue and green radiation for us to see.

Tobias Owen, after R. Canhy

FIGURE 7–35 Neptune, imaged from Earth in the near-infrared light (8900 Å) that is strongly absorbed by methane gas in Neptune's atmosphere, making most of the planet appear dark. The bright regions are clouds of methane ice crystals above the methane gas, reflecting sunlight. The northern hemisphere appeared brighter than the southern, a reversal of the earlier situation. Neptune's disk was 2.3 arc seconds across; the angular resolution observable was less than 1 arc second.

Heidi Hammel, Space Science Institute, Boulder

NASA/JPL/Caltech

FIGURE 7–36 Neptune, from Voyager 2, with its Great Dark Spot, a giant storm in Neptune's atmosphere.

NASA/JPL/Caltech

FIGURE 7–37 The Great Dark Spot, as seen by Voyager in 1989. It was about the size of the Earth, but has since disappeared.

NASA/JPL/Caltech

FIGURE 7–38 Cloud systems in Neptune's southern hemisphere. Neptune's south pole is at the center of this polar projection.

high pressure forces methane-rich gas upward. But this Great Dark Spot had disappeared when the Hubble Space Telescope photographed Neptune a few years later, so it was much less long lived (at least since the time we first saw it) than Jupiter's Great Red Spot.

Several other cloud systems were also seen by Voyager 2. Clouds can, in addition, be viewed with the Hubble Space Telescope and now even with ground-based telescopes using adaptive-optics systems (■ Fig. 7–39).

7.4b Neptune's Interior and Magnetic Field

Voyager 2 measured Neptune's average temperature: 59 K, that is, 59°C above absolute zero (–214°C). This temperature, though low, is higher than would be expected on the basis of solar radiation alone. Neptune gives off about 2.7 times as much energy as it absorbs from the Sun. Thus there is an internal source of heating, unlike the case of Uranus, which otherwise seems like a similar planet.

The average densities of Uranus and Neptune are higher than those of Jupiter and Saturn. Why? The discrepancy may reflect slight differences

in the origins of those planets. Their densities show that Uranus and Neptune have a higher percentage of heavy elements than Jupiter and Saturn. Perhaps the rocky cores they built up from the solar nebula were smaller, giving them less gravity and thus attracting less hydrogen and helium. Or, perhaps the positions of Uranus and Neptune farther out in the solar nebula put them in a region where either the solar nebula was less dense, or their slower orbital motion moved them through less gas.

Voyager detected radio bursts from Neptune every 16.11 hours. Thus Neptune's interior must rotate at this rate. On Neptune and Uranus, as on the Earth, equatorial winds blow more slowly than the interior rotates. By contrast, equatorial winds on Venus, Jupiter, Saturn, and the Sun blow more rapidly than the interior rotates. We now have quite a variety of planetary atmospheres to help us understand the basic causes of circulation.

Voyager discovered and measured Neptune's magnetic field. The field, as for Uranus, turned out to be both greatly tipped and offset from Neptune's center (■ Fig. 7–40). Thus the tentative explanation for Uranus that its field was tilted by a collision is not plausible; such a rare event would not be expected to happen twice, fitting with the fact that only one such planet is so tilted. Astronomers favor the explanation that the magnetic field is formed in an electrically conducting shell outside the planets' cores. The fields of Earth and Jupiter, in contrast,

■ **FIGURE 7–39** Neptune, (a) observed with the 5-m Palomar Hale telescope and with the adaptive optics (AO) system turned off (left) and on (right). (b) Neptune, observed with adaptive optics on the Keck II telescope. A storm is prominent at lower right. (c) Neptune from the Hubble Space Telescope's relatively new Wide Field Camera 3, a series of images taken every 4 hours on June 25–26, 2011, to mark Neptune's first birthday – that is, when it completed 1 Neptune year (one 164-Earth-year orbit around the Sun) since its discovery. The images show that the clouds, which are composed of crystals of methane ice, have shifted toward the northern hemisphere, where it is winter. In the false-color images, Neptune appears bluish because methane absorbs reddish light, and the clouds appear pinkish because they are reflecting in the near-infrared.

are thought to be formed deep within the core. The tilted and offset magnetic fields are some of the biggest surprises found by Voyager 2.

7.4c Neptune's Rings

Before Voyager's arrival, astronomers wanted to know whether Neptune has rings, like the other giant planets. There was no obvious reason why it shouldn't. Some observations from Earth detected dips during occultations of stars, similar to those produced by Uranus's rings, while others didn't. It was thought that perhaps Neptune had incomplete rings, "ring arcs."

As Voyager came close to Neptune, it radioed back images that showed conclusively that Neptune has narrow rings. (See *A Closer Look 7.6: Naming the Rings of Neptune.*) Further, it showed the rings going all the way around Neptune. The material in the densest of Neptune's rings is very clumpy. The clumps had led to the incorrect idea of "ring arcs." The clumpy parts of the ring had blocked starlight, while the other parts and the other rings were too thin to do so (■ Figs. 7–41 and 7–42). The rings can now be studied from the Hubble Space Telescope and with ground-based telescopes having adaptive optics (see discussion in Chapter 3).

The fact that Neptune's rings are so much brighter when seen backlit tells us about the sizes of particles in them. The most detectable parts of the rings have at least a hundred times more dust-sized grains than most of the rings of Uranus and Saturn. Since dust particles settle out of the rings, new sources must continually be active. Probably moonlets collide and are destroyed. Though much less dusty, Saturn's outer ring and Uranus's rings are similarly narrow.

A big surprise came in 2005: The clumps in Neptune's rings are fading and one of the clumps, *Liberté*, is now much weaker. At this rate, it would disappear within 100 years. Neptune's clumpy rings are a much more transitory system than had been realized.

7.4d Neptune's Moon Triton

Neptune's largest moon, Triton, is a little larger than our Moon and has a retrograde (backward) orbit. It is named after a sea god who was a son of Poseidon. It is massive enough to have a melted interior. Its density is 2.07 g/cm³, so it is probably about 70 per cent rock and 30 per cent water ice. It is denser than any jovian-planet satellite except Io and Europa.

Even before Voyager 2 visited, it was known that Triton has an atmosphere. Since Triton was Voyager 2's last objective among the planets and their moons, the spacecraft could be sent very close to Triton without fear of its destruction. The scientists waited eagerly to learn if Triton's atmosphere would be transparent enough to see the surface. It was. The atmosphere is mostly nitrogen gas, like Earth's.

Triton's surface is incredibly varied. Much of the region Voyager 2 imaged was near Triton's south polar cap (■ Fig. 7–43). The ice appeared slightly reddish. The color probably shows the presence of organic material formed by the action of solar ultraviolet light and particles from Neptune's magnetosphere hitting methane in Triton's atmosphere and surface. Nearer to Triton's equator, nitrogen frost was seen.

Many craters and cliffs were seen. They could not survive if they were made of only methane ice, so water ice (which is stronger) must be the major component. Since Neptune's gravity captures

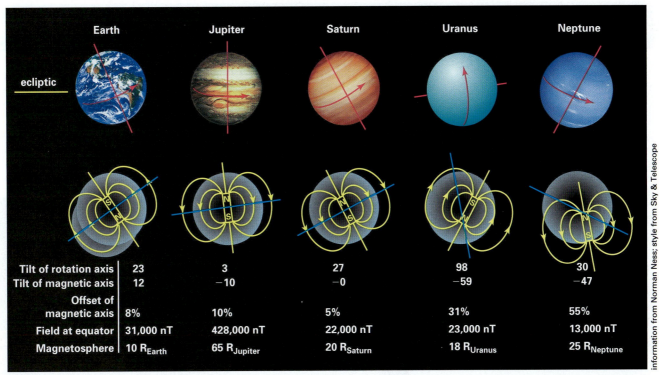

FIGURE 7–40 The magnetic fields of the planets. Mercury, whose magnetic field is only 1 per cent that of Earth's, is not shown. The fields at the planets' equators are given in nanoteslas (nT). 100,000 nT = 1 gauss (1 G), $1/100$ the strength of a toy magnet.

FIGURE 7–41 Three of Neptune's rings, showing the clumpy structure in one of them that led ground-based scientists to conclude from occultation data that Neptune had ring arcs. The Leverrier and Adams rings are 53,000 and 63,000 km, respectively, from Neptune's center. These photographs were taken as Voyager 2 left Neptune, 1.1 million km later, and so are backlit. The main clumps are 6° to 8° long.

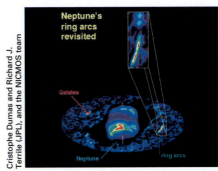

FIGURE 7–42 Neptune, its moon Galatea, and some of its apparent ring arcs (clumps in the outer ring), imaged with the Hubble Space Telescope.

many comets in that part of the Solar System, most of the craters are thought to result from collisions with comets.

Triton's surface showed about 50 dark streaks parallel to each other as is readily visible near the bottom of Figure 7–43. They are apparently dark material vented from below. The material is spread out by winds. A leading model is that the Sun heats darkened methane ice on Triton's surface. This heating vaporizes the underlying nitrogen ice, which escapes through vents in the surface. A couple of these "ice volcanoes" were erupting when Voyager 2 went by. Since the streaks are on top of seasonal ice, they are all probably less than 100 years old.

Much of Triton is so puckered that it is called the "cantaloupe terrain." It contains depressions 30 km in diameter, crisscrossed by ridges (■ Fig. 7–44).

Triton has obviously been very geologically active. Since it is in a retrograde (backward) orbit around Neptune, it was probably born elsewhere in the Solar System and later captured by Neptune. Tidal forces from Neptune would have kept Triton molten until its orbit became circular. While it was molten, the heavier rocky material would have settled to form a 2000-km-diameter core.

One of the authors (J.M.P.) was fortunate to participate in a series of Earth-based expeditions to observe occultations of stars by Triton. Only rarely does Triton go in front of a star bright enough to study, but James Elliot, who discovered Uranus's rings during an occultation, organized a few such expeditions. In 1997, several collaborating expeditions of astronomers went first to Australia and later to Hawaii to try to get in Triton's shadow cast onto the Earth.

A CLOSER LOOK 7.6

Naming the Rings of Neptune

The rings of Neptune were named by a committee of the International Astronomical Union. The major rings are named after the two theoretical predictors and the actual observational discoverer of Neptune. The three clumps of the outer ring, at the suggestion of the French member of the committee, were named Liberté, Egalité, and Fraternité after the slogan of the French Revolution (1789). A fourth clump, named later, was called Courage (pronounced in the French way), in part because that is another worthwhile value that should be used to defend the other values.

■ FIGURE 7–43 Triton's southern region, including its south polar ice cap. Most of the cap may be nitrogen ice. The pinkish color may come from the action of ultraviolet light and magnetospheric radiation upon methane in the atmosphere and surface. Most of the dark streaks represent material spread downwind from eruptions. One erupting plume is the long, thin, dark streak marked with arrows. Though the streaks are dark relative to other features on Triton, they are still ten times more reflective than the surface of our own Moon. The most detailed information on the photograph was taken in black and white through a clear filter; color information was added from lower-resolution images.

NASA/JPL/Caltech, reprocessed by USGS

It turns out that on the second occasion our group got the best observations with the Hubble Space Telescope, which happened to pass the right place at the right time. When Triton blocked out the star, which was about 6 times brighter, the brightness of the merged image dropped. From the way in which it dropped, we could analyze what Triton's atmosphere was like. The temperature turned out to be a few degrees warmer than it was when Voyager flew by, so our joint publication, in the journal *Nature,* was called "Global Warming on Triton."

7.5 THE FORMATION OF THE GIANT PLANETS

Why are the jovian planets so different from the terrestrial planets? Although Section 9.1b discusses models of planet formation in more detail, it is relevant to consider this question here.

The inner planets formed close to the Sun, where temperatures are so high that volatile substances (those that tend to vaporize at

NASA/JPL/Caltech

■ FIGURE 7–44 A computer-generated perspective view of one of Triton's caldera-like depressions.

relatively low temperatures) were not able to condense, remaining in the gaseous state. Thus, only dust grains consisting of the much less abundant, heavier, less volatile elements and compounds (silicates, metals) were able to form, and these gradually coalesced into small, rocky planets. By this time, the gases had already vanished from the inner parts of the Solar System, having been heated and blown away by sunlight and charged particles from the Sun.

The jovian planets, on the other hand, were able to grow to much larger sizes because they are beyond the so-called **frost line** (sometimes called the "snow line") in the Solar System, a distance of about 3 au from the Sun. Clumps of abundant ices – primarily water (H_2O),

ammonia (NH_3), and methane (CH_4) – easily condensed at the low local temperatures, as did silicate grains. The grains and ices collided and grew larger, eventually coming together and forming a solid Earth-like planet. Once a critical size was reached, the gravitational pull of the growing planet gathered a large quantity of hydrogen and helium from its surroundings. Jupiter, in particular, quickly became huge by this process.

As already mentioned, Uranus and Neptune have a smaller proportion of hydrogen and helium than do Jupiter and Saturn, and their interior structure is more dominated by compressed ices. They must not have attracted as much hydrogen and helium, probably forming after Jupiter and Saturn used up most of the nearby material.

CONCEPT REVIEW

» Jupiter (318 times more massive than Earth), Saturn (95 times more massive), Uranus (15 times more massive), and Neptune (17 times more massive) are **giant planets**, also known as **jovian planets** (introductory section). They have extensive gaseous atmospheres, liquid (Jupiter, Saturn) or icy (Uranus, Neptune) interiors, and rocky cores. Each of them has many moons and at least one ring. All were visited by at least one Voyager spacecraft.

» Jupiter's diameter is 11 times that of Earth (Sec. 7.1). We learned a tremendous amount about the planet and its moons from the two Voyager (both in 1979) and Galileo (1995–2003) spacecraft (Sec. 7.1a). Jupiter's **Great Red Spot** (Sec. 7.1b), a giant circulating storm, is a few times larger than Earth. Jupiter's atmosphere shows bright and dark bands, which are different cloud layers (Sec. 7.1c). A probe dropped from the Galileo spacecraft measured the atmosphere's composition, density, and other properties. Jupiter has an internal source of energy, perhaps the energy remaining from its contraction (Sec. 7.1d). Its magnetic field is very strong (Sec. 7.1e), and it has a thin ring (Sec. 7.1f).

» Jupiter's **Galilean satellites** (Sec. 7.1g) are comparable in size to our Moon. They were studied in detail from the Voyagers and the Galileo spacecraft. Io has over 100 volcanoes, many of which are erupting at any given time (Sec. 7.1g(i)). Europa is covered with smooth ice, perhaps over an ocean (Sec. 7.1g(ii)). Ganymede (Sec. 7.1g(iii)) and Callisto (Sec. 7.1g(iv)) are heavily cratered. Jupiter has dozens of additional moons (Sec. 7.1g(v)), making it look like a miniature planetary system of its own.

» Saturn has a very low density, lower even than water (Sec. 7.2). Saturn's rings were produced by tidal forces that prevented the formation of a moon, and are within Saturn's **Roche limit** (Sec. 7.2a). The rings are so thin that stars can be seen through them. **Cassini's division** is the major break between inner and outer rings. Saturn's atmosphere shows fewer features than Jupiter's, probably because it is colder (Sec. 7.2b). Saturn has an internal source of heat, perhaps both energy from its collapse and energy from sinking helium (Sec.

7.2c). Saturn's moon Titan has a thick, smoggy atmosphere (Sec. 7.2d). The Cassini spacecraft, in 2004, arrived at the Saturn system, which it is still studying in detail. It dropped the Huygens probe down to Titan's surface. On the way down, it photographed channels that were cut by running liquid, undoubtedly methane or ethane. Saturn has many additional satellites, some of which were imaged by the Cassini spacecraft (Sec. 7.2e).

» Uranus was the first planet to be discovered that had not been known to the ancients (Sec. 7.3). It consists largely of hydrogen and helium, although its rocky and icy core makes up a more substantial part of the planet than in the case of Jupiter or Saturn. Surrounding the core is an extensive layer of various compressed ices, primarily water, methane, and ammonia; thus, Uranus can be thought of as an "ice giant," while Jupiter and Saturn are often called "gas giants" or "gas/liquid giants." There is no internal energy source. Uranus rotates on its side, so it is heated by the Sun in a strange way over an 84-year period of revolution.

» Its atmosphere appears very bland (Sec. 7.3a); chemical reactions are more limited than on Jupiter and Saturn because it is colder. Uranus's extensive clouds form relatively deep in the atmosphere. Methane is a major constituent of Uranus's atmosphere, including the clouds, and it accounts for the planet's greenish blue color. Uranus has several thin rings, discovered during an occultation of a star observed from Earth (Sec. 7.3b). The magnetic field of Uranus is greatly tipped and offset from the planet's center (Sec. 7.3c).

» Neptune was discovered in 1846, after mathematicians predicted its existence based on discrepancies between the observed and expected positions of Uranus in the sky (Sec. 7.4). Apparently, Galileo saw it in 1613, but did not recognize it as a new planet. Like Uranus, Neptune is an ice giant and appears greenish blue in a telescope because of its atmospheric methane (Sec. 7.4a). Methane-crystal clouds also cover Neptune. When Voyager 2 flew by in 1989, Neptune's surface showed a **Great Dark Spot,** a high-pressure region analogous to Jupiter's Great Red Spot. It disappeared between the Voyager visit and a later Hubble Space Telescope view. Neptune has a substantial core, as does Uranus; unlike Uranus, however, Neptune has an internal energy source (Sec. 7.4b). Neptune's magnetic field is strange, resembling that of Uranus. These magnetic fields are thus probably formed in electrically conducting shells of gas. Neptune's rings are clumpy (Sec. 7.4c). The clumps are dynamic, fading faster than had been expected.

» Neptune's moon Triton is very large (Sec. 7.4d). Voyager 2 took high-resolution images of Triton that showed a varied surface with plumes from ice volcanoes. Triton has been very geologically active. Occultation studies show that it is gradually warming.

» The jovian planets are much larger than the terrestrial planets because they formed beyond the **frost line** (or snow line) in the Solar System, accumulating nearby grains and clumps of ices that had condensed out of the gas (Sec. 7.5). After reaching a substantial size, they were able to gravitationally attract very large amounts of hydrogen and helium, unlike the terrestrial planets.

QUESTIONS

1. Why does Jupiter appear brighter than Mars despite Jupiter's greater distance from the Earth?

2. Even though Jupiter's atmosphere is very active, the Great Red Spot has persisted for a long time. How is this possible?

3. What advantages over the 5-m Palomar telescope on Earth did Voyagers 1 and 2 have for making images of the outer planets?

4. Other than photography, what are two types of observations made from the Voyagers?

5. How does the interior of Jupiter differ from the interior of the Earth?

†6. Given that Jupiter's diameter is 11.2 times that of Earth, about how many Earths could fit inside of Jupiter? (Neglect the empty spaces that exist below closely packed spheres.)

7. Why did we expect Io to show considerable volcanism?

8. Contrast the volcanoes of Io with those of Earth.

9. Compare the surfaces of Callisto, Io, and the Earth's Moon. Explain what this comparison tells us about the ages of features on their surfaces.

10. What are the similarities between Jupiter and Saturn?

11. What is the Roche limit, and how does it apply to Saturn's rings?

12. How did we know, even before Voyager, that Saturn's rings are very thin?

13. What did the Voyagers reveal about Cassini's division?

14. How are narrow rings thought to be kept from spreading out?

15. What is strange about the direction of rotation of Uranus?

16. Which of the giant planets are known to have internal heat sources?

†17. How many orbits has Neptune traversed since it was discovered? Since it was first seen by Galileo?

18. Compare Neptune's Great Dark Spot with Jupiter's Great Red Spot and with the size of the Earth.

19. Compare the rings of Jupiter, Saturn, Uranus, and Neptune.

20. Describe the major developments in our understanding of Jupiter, from ground-based observations to the Voyagers to the Galileo and Cassini spacecraft.

21. Describe the major developments in our understanding of Saturn, from ground-based observations to the Voyagers and Cassini.

22. Describe the major developments in our understanding of Uranus, from ground-based observations to the Voyagers.

23. Discuss Miranda's appearance and explain why it is so surprising.

24. Describe the major developments in our understanding of Neptune, from ground-based observations to the Voyagers.

25. Compare Triton in size and surface with other major moons, like Ganymede and Titan.

26. **True or false?** Through a telescope on Earth, we can sometimes see Jupiter and Saturn in their crescent phases, just as the Moon is sometimes a crescent.

27. **True or false?** Saturn's rings are at such a large distance from the planet that it was difficult or impossible for the sparse material to coalesce into a moon.

28. **True or false?** The rings of Uranus are kept narrow by the presence of adjacent small moons that gravitationally confine the particles to the rings.

29. **True or false?** Uranus has extreme seasons because its rotation axis is nearly in its orbital plane.

30. **True or false?** Tidal forces from Jupiter heat its large moons Io and Europa, leading to extensive volcanism on Io and an icy slush beneath Europa's surface.

31. **True or false?** The existence of Neptune was correctly predicted after astronomers noted that Uranus's observed and expected positions in the sky differed.

32. **Multiple choice:** The low average densities of Jupiter and Saturn compared with Earth suggest that **(a)** they are hollow; **(b)** their gravitational attraction has squeezed material out of their cores; **(c)** they consist mostly of water; **(d)** they contain large quantities of light elements, such as hydrogen and helium; or **(e)** volcanic eruptions have ejected all the iron that was originally in their cores.

33. **Multiple choice:** Jovian planets have rings because **(a)** their thick gaseous atmospheres would disintegrate any small rock that enters them; **(b)** there is too much material to have fit into the ball of each planet; **(c)** tidal forces prevent the material in rings from forming into moons; **(d)** Jovian planets rotate very rapidly, and some material near the equator of these planets was flung outward, forming the rings; or **(e)** tidal forces cause volcanic eruptions on some moons, and part of this material subsequently escaped the gravity of the moons, forming the rings.

34. **Multiple choice:** Which one of the following statements about the jovian planets is *false*? **(a)** They generally have the thickest atmospheres of all planets in the Solar System. **(b)** They are the most massive of all the planets in the Solar System since they are mainly composed of heavy elements like iron. **(c)** Despite their size, they rotate about their axis very rapidly, in fewer than 24 hours. **(d)** They have many moons, probably due to their large gravitational fields. **(e)** They are the largest of all planets in the Solar System, with diameters up to nearly $1/10$ the diameter of the Sun.

35. **Multiple choice:** The rings of Uranus were discovered by accident, during observations of its **(a)** phases, **(b)** occultation of a star, **(c)** rotation period, **(d)** transit across the Sun's disk, or **(e)** moons.

36. **Multiple choice:** Which one of the following statements about the giant planets in our Solar System is *true*? **(a)** Neptune and Uranus appear greenish blue because they are covered by a liquid water ocean. **(b)** Of the four giant planets, only Saturn and Uranus have rings. **(c)** Jupiter's volume is roughly 10 times Earth's volume. **(d)** The giant planets consist mostly of hydrogen and helium, similar to the Sun. **(e)** Saturn's rings rotate as a solid body, like a bicycle wheel and its spokes; all particles have the same orbital period.

37. **Multiple choice:** Jupiter and Saturn radiate more energy than they receive from the Sun. The most likely source of this excess energy is **(a)** gravitational contraction or settling, **(b)** hydrogen fusion, **(c)** chemical reactions, **(d)** radioactive decay, or **(e)** the magnetic fields of these planets.

38. **Fill in the blank:** The distance from a planet within which it is not possible to form a moon through gravitational attraction between particles is called _____.

39. **Fill in the blank:** The _____ spacecraft dropped a probe into Titan's atmosphere in 2005.

40. **Fill in the blank:** Perhaps the most likely moon of the Solar System on which life could have developed is _____, given its warm interior and partially melted water ice.

41. **Fill in the blank:** The _____ of Uranus and Neptune are strange, with tilted axes very offset from the planet centers.

42. **Fill in the blank:** Small moons that keep the material in narrow rings from spreading apart are called _____.

†This question requires a numerical solution.

TOPICS FOR DISCUSSION

1. Why do you think the chemical compositions of Jupiter and Saturn are so close to that of the Sun, but those of the terrestrial planets are not?

2. Given that we have already learned so much about the outer planets with uncrewed spacecraft, should we next send humans to them, or do you think using uncrewed spacecraft and robots (like the Spirit, Opportunity, and Curiosity rovers on Mars) makes more sense? Why?

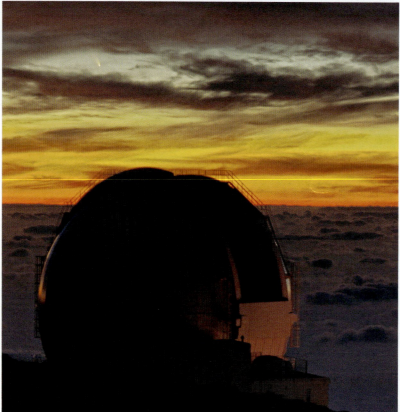

Top: Comet Lovejoy, whose tail extended high in the sky in this view over Santiago, Chile, in 2011. The Moon is visible at left.

Y. Beletski/ESO

Bottom: Comet PanSTARRS was visible in both northern and southern hemispheres in 2013. Here we see it past the dome of the William Herschel Telescope on La Palma, Canary Islands, along with a crescent moon. Comet ISON is predicted to be very bright late in 2013. The Moon looks squashed because of refraction from Earth's atmosphere so close to the horizon.

Babak Tafreshi/TWAN

8

Pluto, Comets, and Space Debris

ORIGINS

We see how objects in the outer part of the Solar System, including comets, are left over from the origin of our Solar System and so can be studied to learn about conditions in our Solar System's early years. We find that asteroids, meteoroids, and comets may lead to our demise, just as they produced at least one episode of major extinction in the past when they collided with Earth.

AIMS

1. Explore the properties of distant Pluto, recently promoted from being a diminutive "planet" to being the first "dwarf planet." Learn about the other dwarf planets (Section 8.1).

2. Learn about the space debris that makes up the outer part of our Solar System, including Kuiper-belt objects and comets (Sections 8.2 and 8.3).

3. Discuss some other minor constituents of our Solar System, such as asteroids and meteoroids (Sections 8.4 and 8.5).

4. See how collisions between space debris and the Earth can lead to profound changes in life on Earth (Section 8.5).

We have learned about the Solar System's giant planets, which range in size from about 4 to 11 times the diameter of the Earth. We have seen that our Solar System has a set of terrestrial planets, which range in size from the Earth down to 40 per cent the diameter of the Earth. This size range includes the four inner planets as well as seven planetary satellites.

The remaining object that had long been called a "planet," Pluto, is only 20 per cent the diameter of Earth but is still over 2300 km across, so there is much room on it for interesting surface features. Recently, additional objects like it, all but perhaps one smaller, have been found in the outer reaches of the Solar System. We shall see how we determined Pluto's odd properties, and what the other, similar objects are.

Besides the planets and their moons, many other objects are in the family of the Sun. The most spectacular, as seen from Earth, are comets (■ Fig. 8–1). Bright comets have been noted throughout history, instilling great awe of the heavens. Comets have long been seen as omens, usually bad ones. As Shakespeare wrote in *Julius Caesar*, "When beggars die, there are no comets seen; The heavens themselves blaze forth the death of princes."

Asteroids, which are minor planets, along with chunks of rock known as meteoroids are other residents of our Solar System. We shall see how they and the comets are storehouses of information about the Solar System's origin. People are worried

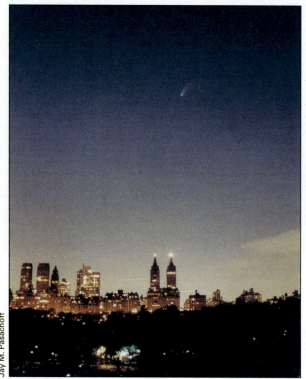

Jay M. Pasachoff

■ **FIGURE 8–1** Comet Hale-Bopp was so bright it was even visible over Central Park in the middle of New York City.

about collisions with such bodies, so astronomers are now monitoring the skies better. Plans are being made to divert an oncoming object if we were to find one.

8.1 PLUTO

Pluto, known as the outermost planet from 1930 until 2006, is a deviant. Compared with the eight planets now officially accepted, its elliptical orbit is the most out-of-circular and is inclined by the greatest angle with respect to the Earth's orbital plane (the "ecliptic plane," defined in Chapter 4), near which the other planets revolve.

Pluto's elliptical orbit is so eccentric that part lies inside the orbit of Neptune. Pluto was closest to the Sun (perihelion) in 1989 and moved farther away from the Sun than Neptune in 1999. So Pluto is still relatively near its perihelion out of its 248-year period, and it appears about as bright as it ever does to viewers on Earth. It hadn't been as bright for over 200 years. It is barely visible through a medium-sized telescope under dark-sky conditions.

The discovery of Pluto was the result of a long search for an additional planet that, together with Neptune, was believed to be slightly distorting the orbit of Uranus. Finally, in 1930, Clyde Tombaugh, hired at age 23 to search for a new planet because of his experience as an amateur astronomer, found the dot of light that is Pluto (■ Fig. 8–2). It took him a year of diligent study of the photographic plates he obtained at the Lowell Observatory in Arizona. From its slow motion with respect to the stars over the course of many nights, he identified Pluto as a new planet.

8.1a Pluto's Mass and Size

Even such basics as the mass and diameter of Pluto are very difficult to determine. It had been hard to deduce the mass of Pluto because to do so was, at first, thought to require measuring Pluto's effect on Uranus, a far more massive body. (The orbit of Neptune, known for less than a hundred years at the time Pluto was discovered, was too poorly determined to be of much use.) Moreover, Pluto has made less than one revolution around the Sun since its discovery, thus providing little of its path for detailed study. As recently as 1968, it was mistakenly concluded that Pluto had 91 per cent the mass of the Earth, instead of the correct value of 0.2 per cent.

The situation changed drastically in 1978 with the surprise discovery (■ Fig. 8–3) that Pluto has a satellite. The moon was named Charon, after the boatman who rowed passengers across the

Lowell Observatory photographs

Lowell Observatory photographs

■ **FIGURE 8–2** Small sections of the photographic plates on which Clyde Tombaugh discovered Pluto. On February 18, 1930, Tombaugh noticed that one dot among many had moved between January 23, 1930 (left), and January 29, 1930 (right). He didn't have the arrows on the plates, of course, to aid his search!

U.S. Naval Observatory/James W. Christy, U.S. Navy photograph

■ **FIGURE 8–3** The image of Pluto and Charon on which Charon was discovered.

■ FIGURE 8–4 An artist's conception showing Pluto and Charon as they might appear if we were on the surface of one of that system's smaller moons. Another of the small moons appears as the dot near the left edge.

NASA, ESA and G. Bacon (STScI)

River Styx to the realm of Pluto, god of the underworld in Greek mythology. (Its name is informally pronounced "Shar'on," similarly to the name of the discoverer's wife, Charlene, by astronomers working in the field.) The presence of a satellite allows us to deduce the mass of the planet by applying Newton's form of Kepler's third law (Chapter 5). Charon is 12 per cent of Pluto's mass, and Pluto is only $^1/_{500}$ the mass of the Earth, ten times less than had been suspected just before the discovery of Charon.

Pluto's rotation axis is nearly in the ecliptic, like that of Uranus. This is also the axis about which Charon orbits Pluto every 6.4 days. Consequently, there are two five-year intervals during Pluto's 248-year orbit when the two objects pass in front of (that is, occult) each other every 3.2 days, as seen from Earth. Such mutual occultations were the case from 1985 through 1990. When we measured their apparent brightness, we received light from both Pluto and Charon together (their separation is so small that they appeared as a single point in the sky). Their blocking each other led to dips in the total brightness we received.

From the duration of fading, we deduced how large they are. Pluto is 2300 km in diameter, smaller than expected, and Charon is 1200 km in diameter. Charon is thus half the size of Pluto. Further, it is separated from Pluto by only about 8 Pluto diameters, compared with the 30 Earth diameters that separate the Earth and the Moon. So Pluto/Charon is even more nearly a "double-planet" system than the Earth/Moon is.

The rate at which the light from Pluto/Charon faded also gave us information that revealed the reflectivities (albedos) of their surfaces, since part of the surface of the blocked object remained visible most of the time. The surfaces of both vary in brightness (■ Fig. 8–4). Pluto seems to have a dark band near its equator, some markings on that band, and bright polar caps.

In 1990, the Hubble Space Telescope took an image that showed Pluto and Charon as distinct and separated objects for the first time, and they can now be viewed individually by telescopes on Mauna Kea in Hawaii (■ Fig. 8–5) and elsewhere where the "seeing" is exceptional. The latest Hubble views show that Pluto has a dozen

(a)

Keith S. Noll, STScI, and NASA

(b)

Dr. R. Albrecht, ESA/ESO Space Telescope European Coordinating Facility, NASA

■ FIGURE 8–5 (a) This series of ground-based views of Pluto and Charon, showing them orbiting around each other, was taken in 1999 at the highest possible resolution at near-infrared wavelengths, with adaptive optics (mirror-shape distortion in real time, to compensate for atmospheric turbulence; see discussion in Chapter 3) using the new Gemini North telescope on Mauna Kea in Hawaii. (b) Pluto and Charon, clearly distinguished with the Hubble Space Telescope.

Pluto • Hubble Space Telescope ACS/HRC

1400 mi
2300 km   0".1

NASA, ESA, and Z. Levay (STScI); observations by
Marc Buie (SwRI)

■ **FIGURE 8–6** A series of Hubble Space Telescope images, the most detailed ever made of Pluto's surface. The spatial resolution of the Hubble image of Pluto is as though you were standing in Los Angeles and looking at the spots on a soccer ball in San Francisco. Dark areas may be dirty water-ice and brighter areas are frosts of various molecules. Methane ice is shown as orange.

areas of bright and dark, the finest detail ever seen on Pluto, whose diameter is smaller than that of the United States (■ Fig. 8–6). But we don't know whether the bright areas are bright because they are high clouds near mountains or low haze and frost. We merely know that there are extreme contrasts on Pluto's surface.

If we were standing on Pluto, the Sun would appear over a thousand times fainter than it does to us on Earth. Consequently, Pluto is very cold; infrared measurements show that its temperature is less than 60 K. From Pluto, we would need a telescope to see the solar disk, which would be about the same size that Jupiter appears from Earth.

8.1b Pluto's Atmosphere

Pluto occulted – passed in front of and hid – a star on one night in 1988. Astronomers observed this occultation to learn about Pluto's atmosphere. If Pluto had no atmosphere, the starlight would wink out abruptly; any atmosphere would make the starlight diminish more gradually. The observations showed that the starlight diminished gradually and unevenly. Thus Pluto's atmosphere has layers in it. Another such occultation wasn't observed until 2002, when (again) Pluto was seen to make the star wink out for a minute or so which did it on two separate occasions (July and August).

From the 1988 occultation, astronomers were also able to conclude that the bulk of Pluto's atmosphere is nitrogen. A trace of methane must also be present, since the methane ice on Pluto's surface, detected from its spectrum, must be evaporating.

Still, Pluto's atmospheric pressure is very low, only 1/100,000 of Earth's. The data from the first occultation seemed to show a change at a certain height in Pluto's atmosphere, leading to the deduction that either the atmosphere had a temperature inversion or that the lower atmosphere contained a lot of dust. The lone high-quality scan obtained in July 2002 showed no such change at a certain height in the rate at which the star's light was dimming as it passed through Pluto's atmosphere.

Then, in August 2002, a group of scientists, of which one of the authors (J.M.P.) was a member, succeeded in observing an occultation of a star by Pluto on ten different telescopes, several of them

on Mauna Kea (■ Fig. 8–7). J.M.P.'s team from Williams College obtained a thousand data points in a 5-minute interval of the occultation, part of a 20-minute data run. Further work in 2005 on a similar occultation of a star but this time by Pluto's moon, Charon, gave the MIT-Williams College consortium success on all but one of the five telescopes in South America they used. We have since observed occultations of stars by Pluto and by Charon several times, including a double occultation in 2011 that had only a few minutes' difference between the two objects' occultations, allowing their mutual orbit to be determined better.

Our Pluto results showed an expansion of its atmosphere, which would result from a global warming since 1988. Perhaps some contribution to that warming comes from the changing orientation of Pluto's darker spots with respect to incoming solar radiation. We also saw some bright spikes in the light curve, which could be signs of waves or turbulence in Pluto's atmosphere. Moreover, observations from several telescopes showed that Pluto's atmosphere is not quite round, undoubtedly resulting from strong winds.

Our Charon results pinned down its size, and therefore density, better than ever before, but even the high-time-resolution observations did not show an atmosphere.

As Pluto goes farther from the Sun, as it is now doing, it was feared that its atmosphere would freeze out and snow onto the surface, making it undetectable. But three groups of astronomers have used occultations to monitor Pluto's atmosphere every year or so, and as of this writing it hasn't started cooling yet. Probably the cooling from an increasing distance from the Sun is still being offset by sublimating ice into its atmosphere as its Sun-facing surface changes.

In 2011, not only was there a double occultation, but also the new NASA-German **S**tratospheric **O**bservatory **f**or **I**nfrared **A**stronomy (SOFIA) was able to use one of its first research flights to observe the occultation of a star by Pluto from very near its centerline, based on adjustments in the predictions only hours before the event. The maneuver meant that the scientists could capture the starlight focused around Pluto by Pluto's atmosphere, allowing the atmosphere to be studied at larger depths than ever before.

NASA's New Horizons mission is a small satellite that is en route to Pluto, for an anticipated arrival on July 14, 2015. Its investigators

■ **FIGURE 8–7** (a) A series showing the images as Pluto occulted (hid; note that "occulted" is not "occluded") a star. Earth's atmospheric conditions were so good that Charon was easily visible alongside Pluto (though the images are not distinct from each other). (b) A light curve from the Williams College observations made with the 2.2-m University of Hawaii telescope on Mauna Kea. The five-minute drop in total intensity, as Pluto occulted the background star, is obvious. (c) A set of false-color images from an occultation of a star by Charon: (*left*) all objects visible; (*center*) Charon hiding the star, dropping the total intensity; (*right*) all objects visible again. The occultation took less than a minute.

used the Hubble Space Telescope to find three additional, small (under 100 km) moons of Pluto. The detailed views of Pluto and its moons that we will get from New Horizons, and the other measurements from 10 different experiments aboard, will truly transform our understanding of Pluto, with spectacularly detailed images of its surface and that of Charon being the most obvious.

8.1c What Is Pluto?

From Pluto's mass and radius, we calculate its density. It turns out to be about 2 g/cm³, twice the density of water and less than half the density of Earth. Since ices have even lower densities than Pluto, Pluto must be made of a mixture of ices and rock. Its composition is more similar to that of the satellites of the giant planets, especially Neptune's large moon Triton, than to that of Earth or the other inner planets.

Ironically, now that we know Pluto's mass, we calculate that it is far too small to cause the deviations in Uranus's orbit that originally led to Pluto's discovery. The discrepancy probably wasn't real: The wrong mass had been assumed for Neptune when predicting the orbit of Uranus. The discovery of Pluto was the reward of Clyde Tombaugh's hard work in conducting a thorough search in a zone of

the sky near the ecliptic, but he was also lucky that Pluto happened to be in that zone at the time.

Pluto, with its moon and its atmosphere, has some similarities to the more familiar planets. But it is so small, so low in mass, and in such an inclined and eccentric orbit with respect to the eight inner planets that many people have long thought that it should only be called a type of asteroid, a "Kuiper-belt object," or a "Trans-Neptunian Object." As we will see in the next section, another such object that seemed to be even bigger than Pluto turned up in 2005, and several slightly smaller objects have also been found.

The discovery of an object arguably bigger than Pluto, or at least comparable in size, brought up the question: Should both be called planets, leaving the possibility that we may soon know of even more, and that the third-graders of the world would have to memorize 10 or even 50 or 100 planets? (Of course, it would be better for third graders to learn the science of planetary atmospheres, or to distinguish how things are categorized, but realistically, they will wind up being asked to memorize the names of the planets, however many there are.) Or should Pluto be demoted to asteroid or the mere status of a Trans-Neptunian Object? The situation led the International Astronomical Union, at its 2006 triennial General Assembly, to no longer officially count Pluto as a genuine planet, and to set up a new category, **dwarf planets**, as we shall now see.

8.2 KUIPER-BELT OBJECTS AND DWARF PLANETS

Beyond the orbit of Neptune, a population of icy objects with diameters of a few tens or hundreds of kilometers is increasingly being found. The planetary astronomer Gerard Kuiper (pronounced kai'per or koy'per) suggested a few decades ago that these objects may have formed there long ago – but, ironically, he thought that they were subsequently ejected by gravitational interactions with the planets, leaving few if any remaining at this time. In any case, these objects are now known as the **Kuiper-belt objects**, or, less often, **Trans-Neptunian Objects**.

The main part of the Kuiper belt (also sometimes called the Edgeworth-Kuiper belt, to recognize Kenneth Edgeworth's earlier hypothesis of its existence) is roughly 10 au thick and extends from 30 au (the orbit of Neptune) to about 50 au (■ Fig. 8–8a,b), although many objects extend beyond these limits and populate the "scattered disk" discussed later in this chapter. More than 1000 Kuiper-belt objects have been found so far, and tens of thousands larger than 100 km across are thought to exist. The objects are left over from the formation of the Solar System.

They are generally very dark, with albedos of only about 4 per cent. Pluto, by contrast, has an albedo of about 60 per cent. Still, Pluto is one of the largest of the Kuiper-belt objects, so much larger that it is covered with frost. Triton may have initially been a similar object, subsequently captured by Neptune.

A Kuiper-belt object larger than Pluto's moon Charon was found in 2001, about half of Pluto's diameter, and was named Varuna. One that may be even somewhat larger was found in 2002, though the uncertainty limits of these two Kuiper-belt objects overlap. The newer one, named Quaoar (pronounced "kwa-whar") after the Indian tribe that inhabited today's Los Angeles, was even imaged with the Hubble Space Telescope, so we have a firmer grasp of its diameter, 1300 km, slightly over half that of Pluto. The size, in turn, gives us the albedo (12 per cent), which is larger than had been assumed for Kuiper-belt objects (KBOs). Our group has recently been observing occultations of stars by these largest KBOs to pin down their sizes and to tell if they have atmospheres.

David Jewitt, then of the University of Hawaii and now at UCLA, and Jane Luu, now at MIT's Lincoln Lab, were the discoverers of most of the first Kuiper-belt objects, starting in 1992.

Michael Brown of Caltech and his colleagues stunned the world in 2005 with their discovery of an outer-Solar-System object that, with a reasonable estimate for its albedo, seemed to be even larger than Pluto (■ Fig. 8–8c). Initially named 2003 UB$_{313}$, it was first sighted in 2003 but not confirmed until 2005. The object is now 97 au out from the Sun, more than twice as far out as Pluto. It takes over 500 years to orbit the Sun. Its orbit is tilted an incredible 44°, taking it so high out of the ecliptic that no previous planet hunter found it. It was almost certainly thrown into that highly inclined orbit after a close gravitational encounter with Neptune.

Faced with an object apparently bigger than Pluto, and a steadily growing number of objects at least half Pluto's size, the astronomers at the International Astronomical Union's General Assembly in Prague in 2006 discussed the problem all week, and a resolution came to the floor for a vote at the closing session. The motion was actually amended from the floor as discussion went on, perhaps a sign that making a decision was premature. But a decision was made: a new category was set up, that of "dwarf planet." The word "planet" had never been formally defined by astronomers – you knew one when you saw one – but a definition was made: A planet orbits the Sun, is massive enough to be round, and is also massive enough to "clear its neighborhood," which basically means that a planet's gravity makes it dominant in its own orbit. (Note that Jupiter and other giant planets have gravitationally trapped some asteroids at specific locations near them, but this does not disqualify them from being planets.) Dwarf planets, on the other hand, orbit the Sun, are massive enough to be round, but are not massive enough to clear their neighborhoods. For example, Pluto orbits in the Kuiper belt, so it has not cleared its orbital region of other objects.

The new object was given a name, Eris, after the Greek goddess of discord and strife, given all the fighting over its status. A moon had been discovered around it, and it was named Dysnomia, after the goddess of lawlessness, Eris's daughter. (This name was a bit of an inside joke, since Mike Brown and his group had used Xena, the Warrior Princess, a cable TV show's hero, and sidekick Gabrielle,

---

A CLOSER LOOK 8.1 — Dwarf Planets

As of 2012, there are 5 dwarf planets known. The ones far out in the Solar System, also called plutoids, are Pluto (after a Roman god), Eris (after a Greek god), Haumea (after a Hawaiian god), and Makemake (after an Easter Island god). The asteroid (1) Ceres, the first asteroid to be discovered, is massive enough to be round yet doesn't clear its neighborhood gravitationally, so it is also a dwarf planet. (2) Vesta is known not to be round, so it doesn't qualify

as a dwarf planet. NASA's Dawn spacecraft is visiting both Ceres and Vesta.

Now that Pluto is considered a dwarf planet, a different mnemonic is needed to remember the names of the planets in order. For example, if you previously used the initial letters of "**M**y **V**ery **E**nergetic **M**other **J**ust **S**erved **U**s **N**ine **P**izzas," perhaps a suitable alternative is "**M**y **V**ery **E**nergetic **M**other **J**ust **S**erved **U**s **N**achos."

■ **FIGURE 8–8** (a) Locations of all known Kuiper-belt objects with positions accurately shown as of March 2013, viewed from above the plane of the Solar System. The light-blue circles, from inner to outer, are the orbits of Jupiter, Saturn, Uranus, and Neptune. Directions in which there appears to be a deficit of objects have not been searched as carefully (for example, because of the confusion caused by the large number of stars in the plane of the Milky Way Galaxy). The "classical" Kuiper-belt objects (those in the main belt) are shown as red circles. "Plutinos" in a "2:3 resonance" with Neptune (that is, orbiting the Sun twice for every three Neptune orbital periods) are white circles, and Pluto itself is the white symbol circled in a crosshair, at bottom center. Unusual high-eccentricity objects are shown as cyan triangles, centaur objects as orange triangles, and scattered-disk objects as magenta circles. Objects observed in multiple years and only in a single year are respectively denoted by filled and unfilled symbols. Numbered periodic comets and other comets are respectively shown as filled and unfilled light-blue squares; note the very large number of them that are in the so-called Jupiter family. (b) The spatial distribution of Kuiper-belt objects with diameters greater than about 130 km, viewed from above the plane of the Solar System (top) and from the side (bottom). Data similar to those shown in part (a) were used to create these plots, but various biases (such as confusion from stars in the plane of the Milky Way Galaxy) were removed in order to obtain the true distribution. Circles are the orbits of Jupiter (blue), Saturn (green), Uranus (purple), and Neptune (red). Objects that extend beyond the dense part of the Kuiper belt in both directions constitute part of the "scattered disk." (c) A comparison of the sizes of our Moon with Pluto and Eris, which are about the same size, and with the other dwarf planets in the outer solar system, which are known as plutoids.

as nicknames for the objects until they got formal names, and the actress Lucy Lawless had played Xena.) An attempt to add the word "classical" ahead of "planet" failed in the voting, eliminating the proffered interpretation that there would be two types of planets: classical and dwarf. With that vote, it turned out that "dwarf planets" weren't planets, even though "dwarf stars" are stars (as we shall see in Chapter 12) and "dwarf galaxies" are galaxies (as we shall see in Chapter 16). In other words, "dwarf" is not an adjective for "planet"; rather, "dwarf planet" is a compound noun.

Pluto and Eris weren't the only original dwarf planets. Long-known Ceres, the first asteroid to be discovered, was in the asteroid belt rather than the Kuiper belt, but it was round so it qualified. Two other Kuiper-belt objects were found to be round; they were named Makemake, after an Easter Island god, and Haumea, after a Hawaiian goddess.

The decision to set up a category of "dwarf planets" received worldwide attention and scorn. It was widely said that Pluto was demoted, though in some sense it was actually promoted to being an important object in a class of similar objects instead of a minor one among the planets. Also, one day an announcement was made without fanfare that the dwarf planets in the outer Solar System would be called "plutoids," giving Pluto back some of its heroic status.

The matter remained controversial for years, with meetings and petitions on the status of Pluto. But by the time of the next International Astronomical Union General Assembly, in Rio de Janeiro in 2009, the fight was essentially over. People had calmed down enough that nobody, then or at the following General Assembly, in Beijing in 2012, tried to reverse the decision of setting up a category of "dwarf planets."

In late 2010, a large international team headed by Bruno Sicardy of the Paris Observatory succeeded in capturing an occultation of a star by Eris. Eris turns out to be indeed spherical, but only 1163±6 km in radius. Ironically, it is now Eris's size that is better known than Pluto's, since occultation studies are prevented by Pluto's opaque lower atmosphere from determining the size of Pluto's surface, estimated to be between 1150 and 1200 km in radius. So Eris is on the low end of the range for Pluto's size. If it hadn't been thought to be bigger than Pluto, would all that fuss have been made at the IAU General Assembly in 2006? Possibly not, but we think the creation of a new category for Pluto would have occurred within a few years in any case, given the growing number of large known Kuiper-belt objects.

Moreover, Eris has a moon whose orbital period and distance give an accurate mass for Eris, and it is definitely more massive than Pluto. So, as Mike Brown put it, people can fight over which is more important: mass or size.

There is no Nobel Prize for planetary astronomy (nor for astronomy in general), but two major prizes were given in 2012 for work on Kuiper-belt objects. Jewitt and Luu received the Shaw Prize "for their discovery and characterization of trans-Neptunian bodies, an archaeological treasure dating back to the formation of the solar system and the long-sought source of short period comets." Jewitt, Luu, and Brown received the Kavli Prize, "for discovering and

characterizing the Kuiper Belt and its largest members, work that led to a major advance in the understanding of the history of our planetary system."

Pluto and Eris appear to be almost twins, but with Pluto at 40 au and Eris at 96 au. So Eris's atmosphere has frozen and snowed to the surface, much as Pluto's might in dozens of years as it moves farther from the Sun. And in two hundred years or so, Eris will be closer to the Sun and have its surface frost sublimate into the atmosphere. We just have to accept the idea of all those Kuiper-belt objects, some large enough to be dwarf planets.

A few objects may once have been Kuiper-belt objects but now come somewhat closer to the Sun, crossing the orbits of the outer planets. About 100 of these "centaur" objects a few hundred kilometers across may exist; in the next section, we will read about the first one, Chiron. Since they are larger and come closer to the Earth and Sun than most Kuiper-belt objects, we can study them better. On at least one, a coma (typical of comets, as we will soon see) was seen, so these centaurs are intermediate between comets and asteroids.

NASA's New Horizons mission is to go to two Kuiper-belt objects, yet to be selected, after it visits Pluto. Our MIT-Williams consortium so far has picked up one occultation of a star by a Kuiper-belt object, surprisingly finding it so small that its albedo is over 90 per cent, making it the shiniest known object in the Solar System. We hope that astronomers will find more Kuiper-belt objects that we can characterize through occultations.

8.3 COMETS

Nearly every decade, a bright **comet** appears in our sky. From a small, bright area called the **head**, a **tail** may extend gracefully over one-sixth (30°) or more of the sky. The tail of a comet is always directed roughly away from the Sun, even when the comet is moving outward through the Solar System.

Although the tail may give an impression of motion because it extends out only to one side, the comet does not move noticeably with respect to the stars as we casually watch during the course of a night. With binoculars or a telescope, however, an observer can accurately note the position of the comet's head and after a few hours can detect that the comet is moving at a slightly different rate than the stars. Still, both comets and stars rise and set more or less together (■ Fig. 8–9). Within days, weeks, or (even less often) months, a bright comet will have become too faint to be seen with the naked eye, although it can often be followed for additional months with binoculars and then for more months with telescopes.

Most comets are much fainter than the one we have just described. About two dozen new comets are discovered each year, and most become known only to astronomers. If you should ever discover a comet and are among the first three people to report it to the International Astronomical Union Central Bureau for Astronomical Telegrams at the Smithsonian Astrophysical Observatory in Cambridge, Massachusetts, it will be named after you. Keep their

■ **FIGURE 8–9** A false-color view of a sungrazing comet, the same Comet Lovejoy (C/2011 W3) whose later image graces the photograph opening this chapter, about to go behind an "occulting disk" that hides the inner two solar radii of space around the Sun (whose actual silhouette is drawn in). This false-color red image is pasted onto a false-color blue image from a coronagraph that shows a greater region of sky around the Sun but also has a larger occulting disk. Observers looking at images from the Solar and Heliospheric Observatory have detected over 2000 of these sungrazers, most of which either whip around the Sun very close to it or are vaporized before they actually hit the Sun. Coronal mass ejections, as seen at upper left, occur so often that they misleadingly look as though they were caused by the incoming comet, though they aren't. Among many other reasons, the comet is too insubstantial to cause such a large effect as a coronal mass ejection. Little white specks are background stars.

Akira Fujii

■ **FIGURE 8–10** The coma and inner tail of Halley's Comet in 1986.

website in mind: www.cfa.harvard.edu/iau/cbat.html. Check the procedures for reporting your discovery at www.cbat.eps.harvard.edu/CometDiscovery.html.

About two thousand comets that go very close to the Sun or even hit it, destroying themselves, have been discovered by (and named after) the **So**lar and **H**eliospheric **O**bservatory (SOHO) spacecraft, since it can uniquely monitor a region of space too close to the Sun to be seen from Earth given our bright daytime skies. We saw images of such comets in Figures 8–9 and 8–13.

Chiron, the first astronomical object called a centaur, was named after the wisest of the centaurs in Greek mythology; the mythical Chiron was the tutor of the heroes Achilles and Hercules. With its 51-year period, Chiron's semimajor axis is between those of Saturn and Uranus. Astronomers think it escaped from the Kuiper belt. When it was discovered it was classified as asteroid number 2060, but when it was seen to be active by showing a coma, it was reclassified as periodic comet 95 (that is, 95P). The research group of one of us (J.M.P.) obtained an occultation of a star by Chiron in 2011, with signs that some dips in brightness could be from jets of gas and dust emitted by the object.

8.3a The Composition of Comets

At the center of a comet's head is its **nucleus**, which is composed of chunks of matter. The most widely accepted theory of the composition of comets, advanced in 1950 by Fred L. Whipple of the Harvard and Smithsonian Observatories, is that the nucleus is like a "dirty snowball." It may be made of ices of such molecules as water (H_2O), carbon dioxide (CO_2), ammonia (NH_3), and methane (CH_4), with dust mixed in.

The nucleus itself is so small that we cannot observe it directly from Earth. Radar observations have verified in several cases that it is a few kilometers across. The rest of the head is the **coma** (pronounced coh'ma), which may grow to be as large as 100,000 km across (■ Fig. 8–10). The coma shines partly because its gas and dust are reflecting sunlight toward us and partly because gases liberated from the nucleus get enough energy from sunlight to radiate.

The tail can extend 1 au (150,000,000 km), so comets can be the largest objects in the Solar System. But the amount of matter in the tail is very small – the tail is a much better vacuum than we can make in laboratories on Earth.

Many comets actually have two tails (■ Fig. 8–11). The **dust tail** is caused by dust particles released from the ices of the nucleus when they are vaporized. The dust particles are left behind in the comet's orbit, blown slightly away from the Sun by the pressure of sunlight hitting the particles. As a result of the comet's orbital motion, the dust tail usually curves smoothly behind the comet. The **gas tail** (also known as the "ion tail," since the gas is ionized) consists of gas blown outward from the comet, at high speed, by the "solar wind" of particles emitted by the Sun (see our discussion in Chapter 10). It follows the interplanetary magnetic field. As puffs of gas are blown out and as the solar wind varies, the gas tail takes on a structured appearance. Each puff of matter can be seen.

■ **FIGURE 8–11** Comet Hale-Bopp, technically C/1995 O1, was the brightest comet in decades at its peak in March and April 1997. Its nucleus, 40 km across, gave off large quantities of gas and dust that made its long tails. The broad, yellowish dust tail curved behind the comet's head; we see it by the sunlight that it reflects toward us. The much fainter, blue tail is made of ions (an ion is an atom that has lost an electron) given off by the comet's nucleus and is blown in the direction opposite to the Sun by the solar wind. We detect this "gas tail" by its own emission, though it was much less visible to the eye than it was to the camera. Comet Hale-Bopp was so bright that scientists discovered a third type of tail: a faint one consisting of neutral sodium, not visible in this image.

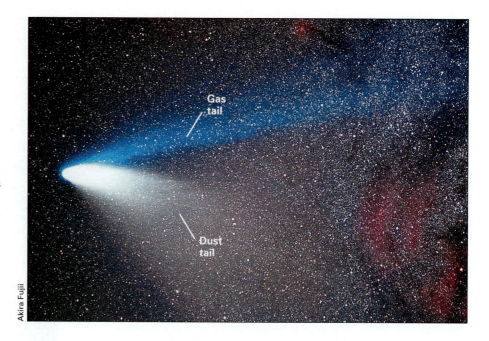

A comet – head and tail together – contains less than a billionth of the mass of the Earth. It has jokingly been said that comets are as close as something can come to being nothing.

8.3b The Origin and Evolution of Comets

It is now generally accepted that trillions of tail-less comets surround the Solar System in a sphere perhaps 50,000 au (that is, 50,000 times the distance from the Sun to the Earth, or almost 1 light-year) in radius. This sphere, far outside Pluto's orbit, is the **Oort comet cloud** (named after the Dutch scientist Jan Oort). The total mass of matter in the cloud may be only 1 to 10 times the mass of the Earth. In current models, most of the Oort cloud's mass is in the inner 1000 to 10,000 au.

Occasionally one of these incipient comets leaves the comet cloud, because the gravity of either a nearby star or the disk of the Milky Way tugged it out of place. In any case, the comet generally gets directly ejected from the Solar System, but in some cases the comet can approach the Sun. The comet's orbit may be altered, sometimes into an elliptical orbit, if it passes near a giant planet, most frequently Jupiter (see the blue squares in Figure 8–8a). Because the comet cloud is spherical, comets are not limited to the plane of the ecliptic, which explains why one major class of comets ("nearly isotropic") comes in almost randomly from all directions.

Another group of comets has orbits that are much more concentrated around the plane of the Solar System (Earth's orbital plane). Although it was originally thought that these "ecliptic comets" come from the Kuiper belt, the vast majority probably originate in what is now called the **scattered disk**, which goes out to greater distances (roughly 100 au, or somewhat more) and is thicker than the Kuiper belt (see the bottom part of Fig. 8–8b again). The overall classification scheme for comets is complicated and has changed with time (■ Fig. 8–12), but the basic subdivision into nearly isotropic and ecliptic comets seems robust, and there are a number of subdivisions in each class.

■ **FIGURE 8–12** (a) The widely used classification of comets. (b) A newer classification of comets, though it has not taken over from the former scheme. The symbol *a* stands for the semimajor axis of the comet's orbit. Encke's comet has a small orbit and has many returns. Chiron, a so-called centaur, is in the outer parts of the Solar System.

Naval Research Laboratory, LASCO on the Solar and Heliospheric Observatory; SOHO is a project of international cooperation between ESA and NASA

2011/10/01 21:24

2011/10/01 18:00

■ FIGURE 8–13 The outer solar corona extends outward near the solar equator in this view from the **So**lar and **H**eliospheric **O**bservatory (SOHO) spacecraft's outer coronagraph. A comet plunging into the Sun is also seen. A disk that blocks the bright everyday Sun and the rod that holds the disk in place appear in silhouette; the size of the ordinary solar disk is drawn on. In the background, we see stars and the Milky Way. Over 2000 comets have been discovered on SOHO images. You, too, can help in the search for SOHO comets, scanning images online. For information on how, see http://sungrazer. nrl.navy.mil and http://www.cometary.net/searching_for_soho_comets.htm.

But the sometimes apparent link between a sungrazer coming in and a coronal mass ejection going out isn't real. There are so many coronal mass ejections within the couple of hours the comet spends behind the telescope's occulting disk that the two types of events randomly appear close in time.

Notice that comets on highly eccentric orbits spend most of their time far away from the Sun, an excellent example of Kepler's second law (Chapter 5). As a comet gets closer to the Sun than those distant regions, the solar radiation begins to vaporize the ice in the nucleus. The tail forms, and grows longer as more of the nucleus is vaporized. Even though the tail can be millions of kilometers long, it is still so tenuous that only about $1/500$ of the mass of the nucleus may be lost each time it visits the solar neighborhood. Thus a comet may last for many passages around the Sun. But some comets hit the Sun and are destroyed (■ Fig. 8–13).

We shall see in the following section that meteoroids can be left in the orbit of a comet that has disintegrated or is in the process of doing so. Some of the asteroids, particularly those that cross the Earth's orbit, may be dead comet nuclei. In recent years, a handful of asteroids – notably Chiron in the outer Solar System – have shown comas or tails, making them comets; conversely, a few comets have died out and seem like asteroids. So we may have misidentified some of each in the past.

How did incipient comets get to where they are, during and shortly after the formation of the Solar System? As we will discuss in Chapter 9, there were many small particles that clumped together in the early eras. Some of these clumps interacted gravitationally with Jupiter, and those that were flung into highly eccentric orbits wound up forming the Oort comet cloud. Other clumps were already beyond the orbits of giant planets, where fewer interactions took place; they formed the

Kuiper belt. The outward migration of Neptune early in the history of the Solar System pushed the Kuiper belt outward and flung many of these objects to even greater distances, forming the scattered disk.

Because new comets come from the places in the Solar System that are farthest from the Sun and thus coldest, they probably contain matter that is unchanged since the formation of the Solar System. So the study of comets is important for understanding the birth of the Solar System. Moreover, some astronomers have concluded that early in Earth's history, the oceans formed when an onslaught of water-bearing comets collided with Earth, although this view is still controversial.

8.3c Halley's Comet

In 1705, the English astronomer Edmond Halley (Halley is pronounced to rhyme with "Sally," and not with "say´lee") (■ Fig. 8–14) applied a new method developed by his friend Isaac Newton to determine the orbits of comets from observations of their positions in the sky. He reported that the orbits of the bright comets that had appeared in 1531, 1607, and 1682 were about the same. Moreover, the intervals between appearances were approximately equal, so Halley suggested that we were observing a single comet orbiting the Sun, and he correctly attributed the slightly different periods to the effects of gravitational interactions with planets.

Halley predicted that this bright comet would again return in 1758. Its reappearance on Christmas night of that year, 16 years after Halley's death, confirmed Halley's hypothesis (and Newton's method). The comet has thereafter been known as Halley's Comet. Since it was the first known "periodic comet" (i.e., the first comet found to repeatedly visit the inner parts of the Solar System), it is officially called 1P, number 1 in the list of periodic (P) comets.

It seems probable that the bright comets reported every 74 to 79 years since 240 BCE were earlier appearances of Halley's Comet. The fact that it has been observed dozens of times endorses the calculations that show that less than 1 per cent of a cometary nucleus's mass is lost at each passage near the Sun.

Halley's Comet came especially close to the Earth during its 1910 return (■ Fig. 8–15), and the Earth actually passed through its tail. Many people had been frightened that the tail would somehow

from Halley's *Miscellanea Curiosa* (1726), collection of Jay M. Pasachoff

■ FIGURE 8–14 Edmond Halley.

■ **FIGURE 8–15** Halley's Comet during its 1910 closest approach to the Sun. The image was subsequently colorized at the National Optical Astronomy Observatories.

NOAO/AURA/NSF

damage the Earth or its atmosphere, but the tail had no noticeable effect. Even then, most scientists knew that the gas and dust in the tail were too tenuous to harm our environment.

The most recent close approach of Halley's Comet was in 1986. It was not as spectacular from the ground in 1986 as it was in 1910, for this time the Earth and comet were on opposite sides of the Sun when the comet was brightest. Since we knew long in advance that the comet would be available for viewing, special observations were planned for optical, infrared, and radio telescopes. For example, spectroscopy showed many previously undetected ions in the coma and tail.

When Halley's Comet passed through the plane of the Earth's orbit, it was met by an armada of spacecraft. The best was the European Space Agency's spacecraft Giotto (named after the 14th-century Italian artist who included Halley's Comet in a painting), which went right up close to Halley. Giotto's several instruments also studied Halley's gas, dust, and magnetic field from as close as 600 km from the nucleus.

The most astounding observations were undoubtedly the photographs showing the nucleus itself (■ Fig. 8–16), which turns out to be potato-shaped (■ Fig. 8–17). It is about 16 km in its longest dimension, half the size of Manhattan Island. The "dirty snowball" theory of comets was confirmed in general, but the snowball is darker than expected. It is as black as velvet, with an albedo of only about 3 per cent. Moreover, the evaporating gas and dust is localized into jets that are stronger than expected. They come out of fissures in the dark crust. We now realize that comets may shut off not when they have lost all their material but rather when the fissures in their crusts close.

Giotto carried 10 instruments in addition to its camera. Among them were mass spectrometers to measure the types of particles present, detectors for dust, equipment to listen for radio signals that revealed the densities of gas and dust in the coma, detectors for ions, and a magnetometer to measure the magnetic field.

About 30 per cent of Halley's dust particles are made only of hydrogen, carbon, nitrogen, and oxygen (■ Fig. 8–18). This simple

■ **FIGURE 8–16** The arrow points to the nucleus of Halley's Comet from the Giotto spacecraft. The dark, potato-shaped nucleus is visible in silhouette. It is about 16 km by 8 km in size; the frame is 30 km across. Two bright jets of gas and dust reflecting sunlight are visible to the left of Halley's nucleus. The small, bright region in the center of the nucleus is probably a raised region of its surface that is struck by sunlight.

© Max-Planck-Institut für Aeronomie, Lindau/Hartz, FRG; photographed with the Halley Multicolour Camera aboard the European Space Agency's Giotto spacecraft, courtesy of H. U. Keller, additional processing by Harold Reitsema, Ball Aerospace

The next appearance of Halley's Comet, in 2061, again won't be spectacular. Not until the one after that, in 2134, will the comet show a long tail to earthbound observers. Fortunately, though Halley's Comet is predictably interesting, a more spectacular comet appears every 10 years or so. When you read in the newspaper that a bright comet is here, don't wait to see it another time. Some bright comets are at their best for only a few days or a week.

8.3d Comet Shoemaker-Levy 9

A very unusual comet gave thrills to people around the world. In 1993, Eugene Shoemaker, Carolyn Shoemaker, and David Levy discovered their ninth comet in a search with a wide-field telescope at the Palomar Observatory. (The authors of this book would prefer to give each Shoemaker individual credit for the discovery, calling the comet Shoemaker-Shoemaker-Levy 9, though the comet is formally known as Shoemaker-Levy 9.)

This comet looked weird – it seemed squashed. Higher-resolution images taken with other telescopes, including the Hubble Space

■ **FIGURE 8–17** This potato looks remarkably like the nucleus of Halley's Comet, except it is 100,000 times shorter.

composition resembles that of the oldest type of meteorite. It thus indicates that these particles may be from the earliest years of the Solar System.

Many valuable observations were also obtained from the Earth. For example, radio telescopes were used to study molecules. Water vapor is the most prevalent gas, but carbon monoxide and carbon dioxide were also detected. The comet was bright enough that many telescopes obtained spectra (■ Fig. 8–19).

■ **FIGURE 8–18** The mass spectrometer aboard the Vega spacecraft showed the number of nuclear particles (the atomic mass number) of each element in Halley's dust particles. (a) A CHON dust particle, composed only of carbon, hydrogen, oxygen, and nitrogen. (b) Other dust particles contain other elements as well.

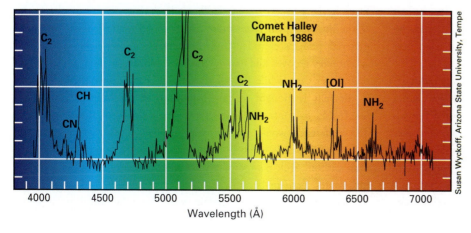

■ **FIGURE 8–19** The visible spectrum of Halley's Comet on March 14, 1986, when it was at its brightest. We see several spectral lines and bands from molecules. Their relative strengths are approximately the same as they are in other comets, except for NH_2, which was slightly stronger in Halley.

FIGURE 8–20 Comet Shoemaker-Levy 9 viewed with the Hubble Space Telescope. Over twenty fragments, each with its own tail, appear in this image taken six months before their collisions with Jupiter.

Telescope (■ Fig. 8–20), showed that the comet had broken into bits, forming a chain that resembled beads on a string. Even stranger, the comet was in orbit not around the Sun but around Jupiter, and would hit Jupiter a year later. Apparently, several decades earlier the comet was captured in a highly eccentric orbit around Jupiter, and in 1992, during its previous close approach, it was torn apart into more than 20 pieces by Jupiter's tidal forces.

Telescopes all around the world and in space were trained on Jupiter when the first bit of comet hit. The site was slightly around the back side of Jupiter, but rotated to where we could see it from Earth after about 15 minutes. Even before then, scientists were enthralled by a plume rising above Jupiter's edge. When they could view Jupiter's surface, they saw a dark spot and ring (■ Fig. 8–21a,b). Infrared telescopes detected a tremendous amount of radiation from the heated gas (■ Fig. 8–21c). Over a period of almost a week, one bit of the comet after another hit Jupiter, leaving a series of Earth-sized rings and spots as Jupiter rotated. The largest dark spots could be seen for a few months even with small backyard telescopes. (On one of the April 2005 solar eclipse cruises, David Levy sometimes wore a T-shirt that said "My comet crashed.")

The dark material showed us the hydrocarbons and other constituents of the comet. Spectra showed sulfur and other elements, presumably dredged up from lower levels of Jupiter's atmosphere than we normally see.

The biggest comet chunk released the equivalent of 6 million megatons of TNT – 100,000 times more than the largest hydrogen bomb. Had any of the fragments hit Earth, they would have made a crater as large as Rhode Island, with dust thrown up to much greater distances. Had the entire comet (whose nucleus was 10 km across) hit Earth at one time, much of life could have been destroyed. So Comet Shoemaker-Levy 9 made us even more wary about what may be coming at us from space.

Indeed, 15 years later, in 2009, an amateur astronomer discovered a dark spot on Jupiter similar to those produced by Comet Shoemaker-Levy 9, and the Hubble Space Telescope imaged it with its new Wide Field Camera 3 (Fig. 8–21d). The object, a comet nucleus or asteroid, was calculated to be the size of several football fields, and the explosion would have been thousands of times more powerful than the Tunguska event over Siberia in 1908, presumably from an explosion of a meteor or comet in the air, that leveled huge regions of forest.

8.3e Recently Observed Comets

In 1995, Alan Hale and Thomas Bopp independently found a faint comet, which was soon discovered to be quite far out in the Solar System. Its orbit was to bring it into the inner Solar System, and it was already bright enough that it was likely to be spectacular when it came close to Earth in 1997. It lived up to its advance billing (■ Fig. 8–22a).

Telescopes of all kinds were trained on Comet Hale-Bopp, and hundreds of millions of people were thrilled to step outside at night and see a comet just by looking up. Modern powerful radio telescopes were able to detect many kinds of molecules that had not previously been recorded in a comet.

Occasionally, other bright comets with long tails, such as C/2009 R1, known as Comet McNaught (■ Fig. 8–22b), and C/2011 W3, called Comet Lovejoy (■ Fig. 8–22c,d), turn up and are fun to watch. Both McNaught and Lovejoy were best seen from the southern hemisphere. As of this writing, Comet ISON is predicted to be spectacularly bright in late 2013.

8.3f Spacecraft to Comets

NASA's Deep Space 1 mission flew close to Comet 19P/Borrelly in 2001. It obtained more detailed images of the bowling-pin-shaped nucleus (■ Fig. 8–23) than even Giotto's views of Halley's nucleus. This comet's surface, and therefore probably the surfaces of comet nuclei in general, was rougher and more dramatic than expected. Deep Space 1 found smooth, rolling plains that seem to be the source of the dust jets, which are more concentrated than Halley's. Darkened material, perhaps extruded from underneath, covers some regions and accentuates grooves and faults. Borrelly's albedo in these places is less than 1 per cent, while Borrelly's overall albedo is only 4 per cent.

Borrelly is thought to have originated in the Kuiper belt or scattered disk, in contrast to Halley's Comet's origin in the Oort cloud. This difference would explain why Halley's Comet gives off many carbon compounds while Borrelly gives off more water and ammonia than carbon. Still, compared with Halley, Borrelly gives off relatively little water, perhaps because so much of its surface is inactive.

Scientists have yet to explain why the solar wind is deflected around Borrelly's nucleus in an asymmetric fashion. The center of

(a)

(b)

Heidi Hammel (AURA)/NASA

(c)

Heidi Hammel (AURA)/NASA

W. M. Keck Observatory/Inke de
Pater, James Graham, Garrett
Jernigan (University of
California, Berkeley)

(d)

NASA, ESA, and H. Hammel (AURA), and the Jupiter Impact Team

■ FIGURE 8–21 (a) Jupiter, with scars from the July 1994 crash of Comet Shoemaker-Levy 9. The image was taken with the Hubble Space Telescope. (b) The shape of the spot and ring left on Jupiter's surface showed that the fragment of Comet Shoemaker-Levy 9 that caused it entered at an angle. The image was taken with the Hubble Space Telescope. (c) This infrared image, taken with one of the Keck telescopes, shows bright plumes of hot gas from the collisions of comet fragments with Jupiter. (d) Another object crashed into Jupiter 15 years later, in 2009. The scar shows in this image from the Hubble Space Telescope's new camera.

the plasma in Borrelly's coma is some 2000 km off to the side, as strange as if a supersonic jet's shock wave were displaced far to the airplane's side.

NASA's Deep Impact spacecraft crashed a 370-kg projectile into Comet Tempel 1 in 2005. The remainder of the spacecraft studied the impact, which should have formed a football-field-sized crater some 7 stories deep. Astronomers were at telescopes all around the Earth, and were using telescopes in space like Hubble, to record the impact (■ Fig. 8–24); see *A Closer Look 8.2: Deep Impact*.

NASA's Stardust mission, launched in 1999, went to Comet Wild 2 (pronounced Vilt-too), a periodic comet with a six-year orbit. When the spacecraft got there in 2004, it not only photographed the comet but also gathered some of its dust, less than a milligram. It carried an

extremely lightweight material called aerogel (■ Fig. 8–25), and flew through the comet with the aerogel exposed so that the comet dust could stick in it. Stardust's orbit brought it back to Earth in 2006, when it parachuted the aerogel down to the Utah desert. Since its parachute didn't open, and the payload crashed to Earth, scientists were worried about the condition of the comet dust, but eventually they were able to carry out the desired analysis.

The scientists expected to find primitive material, from the era of the formation of the Solar System, preserved in ice. They even expected to find dust from other stars that had been incorporated in the Solar System during its formation, which was why the mission was named Stardust. But the scientists were surprised to find, instead, that most of the rocky material had formed inside

(a)

Jay M. Pasachoff

(b)

S. Deiries/ESO

■ **FIGURE 8–22** (a) Comet Hale-Bopp (C/1995 O1) in the sky, showing both a whitish dust tail and a bluish gas tail. (b) Comet McNaught (C/2009 R1) passing the Andromeda Galaxy. (c) Comet Lovejoy (C/2011 W3), a "Christmas comet" because of the dates when it was at its brightest. McNaught and Lovejoy were at their best only when seen from sites in the southern hemisphere, while Hale-Bopp was nicely visible from the United States. (d) Comet Lovejoy was imaged approaching and departing from the Sun in the SOHO coronagraph, but also, surprisingly, through filters on the **S**olar **D**ynamics **O**bservatory (SDO) that normally viewed only hot, million-kelvin coronal gas. In this SDO image, we see the comet leaving the solar vicinity; the fact that its tail is pushed out along magnetic lines of force shows such magnetic fields interacting with a comet for the first time.

(c)

ESO/Gabriel Brammer

(d)

Lockheed Martin Solar and Astrophysics Laboratory and Smithsonian Astrophysical Observatory, and NASA

(a) (b)

NASA/JPL/Caltech

NASA/JPL/Caltech

■ **FIGURE 8–23** Comet Borrelly (19P), 8 km long, viewed by NASA's Deep Space 1 probe. (a) Seen from a distance of only 3,500 km, a variety of terrains is visible, including mountains and faults. Sunlight is coming from the bottom of the image; the comet's narrower end is tipped away from us. We see features as small as 50 m across. (b) Dust jets are seen coming from a region at least 3 km wide on the nucleus, which is overexposed in this image. The brightest jet is at lower left; material seen on the dark (night) side at upper right may have been swept around the nucleus from the day side.

our Solar System at very high temperatures. So even though they confirmed that comets had formed billions of years ago with lots of ice, some of which formed at extremely cold temperatures at the edge of the Solar System, other particles had formed in high temperatures, presumably much closer to the Sun. Somehow, those particles had been transported outward to the edge of the Solar System, where they were incorporated into the incipient comets. The scientists did find some of the expected pre-solar "stardust" grains, as they could tell from the unusual ratios of isotopes in them, but only a few. They even found some glycine, an amino acid and therefore a building block of life, verifying the idea that comets delivered at least some of the amino acids to early Earth along with water. They found that a fairly steady 4.5 billion years of space weathering has deposited some iron grains on the surface of Comet Wild 2, making it appear slightly reddish.

The main part of the Stardust spacecraft remained in space, and was sent to fly by first asteroid Annefrank and then, in 2011, Comet Tempel 1, into which Deep Impact had crashed. The mission was called Stardust-NExT (**N**ew **Ex**ploration of **T**empel). The spacecraft successfully imaged the region where Deep Impact had hit, showing changes in the terrain over time even in other regions of the comet's surface. The region hit by Deep Impact didn't show much change, presumably because the dust that had been excavated settled back.

A major European Space Agency spacecraft, Rosetta, was launched in 2004 to orbit with a comet for some years and to land a probe on the comet's nucleus in 2014. It is heading for Comet 67P/Churyumov-Gerasimenko. It has used three gravity assists from Earth and one from Mars to reach the comet, along the way passing asteroids (2867) Steins in 2008 and (21) Lutetia in 2010, both in the asteroid belt. (Asteroids are discussed in Section 8.5.) Rosetta will drop a lander, Philae, onto the comet's nucleus. Just as the Rosetta Stone, now in the British Museum, enabled Egyptian hieroglyphics to be deciphered by having the same text in three scripts (hieroglyphics, Demotic, and Greek), scientists hope that the Rosetta spacecraft will prove to be the key to deciphering comets. (Philae was an island in the Nile on which an obelisk was found that helped to decipher the hieroglyphics of the Rosetta Stone.)

Rosetta is to orbit the comet at an altitude of only a few kilometers, mapping its surface and making other measurements, for 18 months, including the comet's closest approach to the Sun and therefore, it is hoped, its increasing activity. The lander is to work for some weeks, taking photographs and drilling into the surface.

Jay M. Pasachoff

■ **FIGURE 8–24** NASA's Deep Impact probe crashed into a comet in July 2005.

NASA/JPL/UMD

■ **FIGURE 8–25** Aerogel, a frothy, extremely light substance. Even a large amount weighs very little, so a rocket can carry more of it into space than it could of a denser material.

A CLOSER LOOK 8.2

Deep Impact

On July 4, 2005, NASA's Deep Impact spacecraft went right up to Comet Tempel 1, also known as 9P, the ninth periodic comet on astronomers' official list. The comet is 14 km by 4 km in size. The day before, the spacecraft had ejected a 370-kg copper ball, and as cameras on the flyby probe and on the impactor watched, the ball smashed into the comet's surface. Debris flew, forming a crater.

The Hubble Space Telescope, the Chandra X-ray Observatory, and many other telescopes on the ground and in space photographed the event. When the dust cleared, the fact that a crater was visible disproved the idea that the impactor might merely disappear into a pile of rubble. The mission resulted in scientists gaining a much improved understanding of the internal composition of this comet.

As the dust from Deep Space's impact settled, Mike A'Hearn of the University of Maryland and colleagues watched the edge of the ejecta broadening day by day as debris fell back on the comet's surface. On the assumption that we are seeing "ballistic trajectories" – that is, paths following only Newton's laws of gravity and motion – the scientists were able to measure the comet's gravity and thus its mass. It turned out that its density is so low that 75 per cent of its volume must be empty. The outer tens of meters seem to be dirty snowflakes, an elaboration of Whipple's old, standard dirty-snowball model. The comet's nucleus resembles a handful of fluffy snow that you haven't compressed to make a snowball.

Another surprise is that more organic material – material containing carbon-hydrogen molecules – was released than expected, compared with water vapor, carbon monoxide, and carbon dioxide. This result is support for the idea that our Earth's atmosphere and the atmospheres of other planets around our Sun and other stars were seeded with organics, leading to the development of life.

"I've gone from working 60 hours a week to 70 hours a week, but the fun part has gone up from 1 per cent to 50 per cent," said Mike A'Hearn.

■ **Figure ACL 8.2–1** The surface of Comet 9P/Tempel 1 before impact, taken from the camera on the impactor.

■ **Figure ACL 8.2–2** A close-up view from the impactor only 20 seconds before the collision, showing small detail on the comet's surface, including the impact site.

■ **Figure ACL 8.2–3** The comet imaged from the fly-by probe 67 seconds later.

■ **Figure ACL 8.2–4** A Chandra X-ray Observatory view from 4 days before the collision, showing that the x-rays come primarily from the interaction between charged oxygen ions in the solar wind and the comet's neutral gases.

■ **Figure ACL 8.2–5** A Hubble Space Telescope series of views. The first is before the impact, showing light reflected off the nucleus. The others, up to an hour and a half after the impact, show the plume that formed, reflecting sunlight toward Earth.

Red oval: Site of Deep Impact collision
Yellow arrows: Selected features in both pre- and post-impact images
Dotted yellow circle: Region centered on impact site

■ **Figure ACL 8.2–6** When NASA's Stardust-NExT mission flew by Comet Tempel 1 in 2011, it was able to image both the part of the comet where Deep Impact had hit and also previously unseen parts. The white arrow shows the impact probe's trajectory and the red oval shows the area of impact.

8.4 METEOROIDS

There are many small (like grains of sand) or medium-sized (up to about 50 m across) chunks of matter, known as **meteoroids**, orbiting in the Solar System. (Larger objects are generally called asteroids; see Section 8.5.) When one hits the Earth's atmosphere, friction (more correctly, the compression of air in front of it) heats it up – usually at a height of about 100 km – until all or most of it is vaporized. Such events result in streaks of light in the sky, which we call **meteors** (popularly, and incorrectly, known as shooting stars or falling stars).

On any clear night a naked-eye observer with a dark sky may see a few **sporadic meteors** each hour – that is, meteors that are not part of a shower (see below). Just try going out to a field in the country and watching the sky for an hour, preferably on a dark, moonless night. Meteors are best viewed after midnight, when the Earth is rotated so that we are plowing into the particles that produce the meteors. Most meteors result from meteoroids roughly the size of sand grains, but the brighter ones are produced by small pebbles or larger rocks.

When a fragment of a meteoroid survives its passage through the Earth's atmosphere, the remnant that we find on Earth is called a **meteorite**. Counting even tiny meteorites, whose masses are typically a milligram, some 10,000 tons of this interplanetary matter land on Earth's surface each year.

8.4a Types and Sizes of Meteorites

Space is full of meteoroids of all sizes, with the smallest being most abundant. Most of the small particles, less than 1 mm across, may come from comets. The large particles, more than 1 cm across, may generally come from collisions of asteroids in the asteroid belt (see Section 8.5).

Tiny meteorites less than a millimeter across, **micrometeorites**, are the major cause of erosion (what little there is) on the Moon. Micrometeorites also hit the Earth's upper atmosphere all the time, and remnants can be collected for analysis from balloons or airplanes or from deep-sea sediments. They are often sufficiently slowed down by Earth's atmosphere to avoid being vaporized before they reach the ground.

Some of the meteorites that are found have a very high iron content (about 90 per cent); the rest is nickel. These iron meteorites are thus very dense – that is, they weigh quite a lot for their volume (■ Fig. 8–26a).

Most meteorites that hit the Earth are stony in nature. Because they resemble ordinary rocks (■ Fig. 8–26b) and disintegrate with weathering, they are not easily discovered unless their fall is observed. That difference explains why most meteorites discovered at random are made of iron. But when a fall is observed, most meteorites recovered are made of stone. Some meteorites are rich in carbon, and some of these even have complex molecules like amino acids.

A large terrestrial crater that is obviously meteoritic in origin is the Barringer Meteor Crater in Arizona (■ Fig. 8–27). It resulted from what was perhaps the most recent large meteoroid to hit the Earth, for it was formed only about 50,000 years ago.

Every few years a meteorite is discovered on Earth immediately after its fall. The chance of a meteorite landing on someone's house or car is very small, but it has happened (■ Fig. 8–28)! Often the positions in the sky of extremely bright meteors are tracked in the hope of finding fresh meteorite falls. The newly discovered meteorites are rushed to laboratories in order to find out how long they have been in space by studying their radioactive elements. For the dramatic Russian meteor explosion of February 15, 2013, for which broken glass from its shock wave injured over 1,000 people, many bits of meteorite have been found (see *A Closer Look 8.3*).

Over 10,000 meteorites have been found in the Antarctic, where they have been well preserved as they accumulated over the years.

Fred Bruenjes

■ **FIGURE 8–26** A compound exposure covering six hours of the Perseid meteor shower, which, weather permitting, is easy to see in the sky every August 12 and a day or two to either side of that date. There are 51 meteors visible in this image. The radiant in Perseus is a perspective effect; the meteors actually travel through space in parallel tracks.

(a)

(b)

■ **FIGURE 8–27** (a) A 15-ton iron meteorite, the Willamette meteorite, at the Rose Center for Earth and Space at the American Museum of Natural History in New York City. Note how dense it has to be to have this high a mass, given its size. (b) The largest mass of the Cape York meteorite, a 35-ton piece known as "Ahnighito," or "the Tent." This stony meteorite was discovered by Inuits in Greenland in the early 1800s and was brought to New York by William Peary and Matthew Henson in 1897. It is the largest on display in a museum (second in size only to a meteorite still in the ground in Namibia) and is also in the American Museum of Natural History in New York City.

Although the Antarctic ice sheets flow, the ice becomes stagnant in some places and disappears, revealing meteorites that had been trapped for over 10,000 years.

Some odd Antarctic meteorites are now known to have come from the Moon or even from Mars. Recall that in Chapter 6 we even discussed controversial evidence for ancient primitive life-forms on Mars, found in one such meteorite. Few scientists accept it. As the late Carl Sagan said, "Extraordinary claims require extraordinary evidence," and the evidence from this meteorite does not seem compelling.

Meteorites that have been examined were formed up to 4.6 billion years ago, the beginning of the Solar System. The relative abundances of the elements in meteorites thus tell us about the solar nebula from which the Solar System formed. In fact, up to the time of the Moon landings, meteorites and cosmic rays (charged particles from outer space) were the only extraterrestrial material we could get our hands on.

A CLOSER LOOK 8.3   February 15, 2013 – An Exploding Meteor; A Nearby Asteroid

Front-page headlines and extensive coverage on the web and on television showed the world's surprise on February 15, 2013, when a meteor exploded over central Russia and then left a long trail in the sky. It had been over 100 years since a similar event had been recorded, but this one was over a populated area and about 1,500 people were injured.

The object's arrival at Earth was completely unexpected. The first anybody knew about it was when there was a bright flash, and people rushed to their windows to look out. Then, a few minutes later, a shock wave arrived and blew out many thousands of windows in and around Chelyabinsk in the Ural Mountains, which divide European Russia from Siberia. Most of the injuries were from shards of glass from the windows. Fortunately, nobody died of the injuries, though a few were serious.

Analysis shows that the meteoroid was perhaps 60 feet across and had a mass of around 12,000 tons; it was traveling toward us at about 10 miles per second. Experts from the Russian Academy of Sciences reported that the meteoroid exploded when it hit our lower atmosphere, mostly disintegrating about 15 miles up. Apparently, it was made of iron and stone, and about 5 per cent of its mass

reached the Earth's surface. Days afterward, people were picking up bits of meteorite, often by digging down where they saw holes in the snow. One piece made a hole yards across in the ice on a lake. Many of the recovered pieces were charred pea-size, while others were larger.

The event was the most powerful of its type to be recorded since the Tunguska event in 1908. This time, many car dashboard cameras were operating, something that is common in Russia, so there are a large number of videos of the flash and of the meteor trail. Also, the explosion led to infrasound waves, very low frequency pressure waves, which were recorded worldwide by sets of detectors set up to detect nuclear-bomb tests. For a selection of videos, see http://www.nytimes.com/interactive/2013/02/15/world/europe/16meteorite_videogrid.html?ref=europe.

Coincidentally, the event came in the morning of a day in which the passage of an asteroid was expected very close to Earth. (The two objects were on very different trajectories and completely unrelated.) Asteroid 2012 DA14, about half a football field across (similar in size to the objects that caused the Tunguska explosion and Meteor Crater in Arizona), had been discovered about a year

■ **FIGURE ACL 8.3–1** A composite of newspaper headlines the day after a meteoroid streaked through the skies of central Russia and caused a powerful shock wave that broke thousands of windows. (inset) A 52-gram bit of Chelyabinsk meteorite.

■ **FIGURE ACL 8.3–2** The asteroid 2012 DA14, which came so close to Earth on February 15, 2013, that it was even within the orbits of our communications satellites, is centered here in a composite of seven 10-second images with intermediate gaps. It had been discovered only a year previously. The telescope of the Clay Center Observatory in Brookline, MA, tracked the asteroid, allowing the stars to trail. The asteroid was visible even in small telescopes or binoculars.

earlier; it was to come only 17,200 miles above the Earth's surface, within the orbits of geosynchronous satellites, closer than any other asteroid had come for decades. The orbits of 99 per cent of asteroids that size haven't yet been mapped, though if we knew of any that were to hit Earth decades in advance, we might have time to deflect them.

The twin events have led to increased efforts to map our region of the Solar System with telescopes on the ground and in space to discover smaller and smaller asteroids and meteoroids that could hit us. One effort is from the B612 Foundation (b612foundation.org), a private organization named after the asteroid that the author Antoine de Saint Exupéry's *The Little Prince* inhabited in a 1943 book well known in its French original and in its English translation, not least for its wonderful watercolors showing the asteroid, the Prince, his rose, and a fox. The B612 Foundation is building a 20-inch space telescope for launch in around 2018 to extensively map asteroids in our part of the Solar System, especially to find any that could threaten us.

Near-Earth Asteroid experts estimate that an event like the "Chelyabinsk" meteor (named after the most affected city in Russia) occurs about once per century, while an event like the explosion of the somewhat larger meteorite that affected Tunguska, Siberia, in 1908 occurs about once per thousand years. Current surveys generally don't detect asteroids less than about football-field-size across.

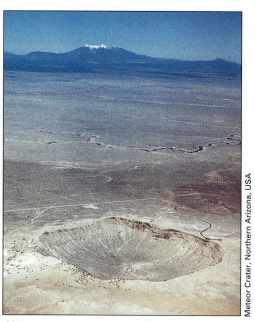

■ **FIGURE 8–28** The Barringer "meteor crater" (actually a meteorite crater) in Arizona. (The family pronounces its name with a hard "g.") It is 1.2 km in diameter. Dozens of other terrestrial craters are now known, many from aerial or space photographs. The largest may be a depression over 400 km across under the Antarctic ice pack, comparable with lunar craters. Another very large crater, in Hudson Bay, Canada, is filled with water. Most are either disguised in such ways or have eroded away.

Meteor Crater, Northern Arizona, USA

8.4b Meteor Showers

Meteors sometimes occur in **showers**, when meteors are seen at a rate significantly above average, perhaps one every few minutes. Meteor showers are named after the constellation in which the **radiant**, the point from which the meteors appear to come, is located. The most widely observed – the Perseids, whose radiant is in Perseus – takes place each summer around August 12 and the nights on either side of that date (■ Fig. 8–29). The best winter show is the Geminids, which takes place around December 14 and whose radiant is in Gemini.

Meteor showers generally result from the Earth's passing through the orbits of defunct or disintegrating comets and hitting the meteoroids left behind. Two known meteor showers, however, appear to be produced by debris from asteroids.

Though the Perseids and Geminids can be counted on each year, the Leonid meteor shower (whose radiant is in Leo) peaks every 33 years, when the Earth crosses the main clump of debris from Comet Tempel-Tuttle. On November 17/18, 1998, one fireball (a meteor brighter than Venus) was visible each minute for a while (■ Fig. 8–30), and on November 17/18, 1999 through 2001, thousands of meteors were seen in the peak hour. We will now have to wait until about 2031 for the next Leonid peak.

The visibility of meteors in a shower depends in large part on how bright the Moon is; you want as dark a sky as possible. See *Star Party 8.1: Observing a Meteor Shower*, and *A Closer Look 8.3: Meteor Showers* for more details. Meteors are best seen with the naked eye; using a telescope or binoculars merely restricts your field of view.

A CLOSER LOOK 8.4   Meteor Showers

Shower	Date of Maximum	Extreme Limits	Number Per Hour at Maximum	Radiant Constellation	Parent Object
Quadrantids	Jan. 3–4	Dec. 29–Jan. 7	60–100	Bootes	Asteroid 2003 EH1
Lyrids	Apr. 21–22	Apr. 16–26	10	Lyra	Comet 1861 G1 (Thatcher)
Eta Aquarids	May 5–6	Apr. 21–May 12	10–30	Aquarius	1P/Halley
Delta Aquarids	July 29–30	July 14–Aug. 18	15–20	Aquarius	Unknown
	Aug. 13–14	4 July 16–Sep. 10	10	Pisces	(double shower)
Perseids	Aug. 12	July 23–Aug. 20	70–80	Perseus	107P/Swift-Tuttle
Orionids	Oct. 20–22	Oct.16–27	20–30	Orion	1P/Halley
Taurids	Nov. 8	Nov. 5–12	5–8	Taurus	2P/Encke
Leonids	Nov. 17	Nov. 14–20	10	Leo	55P/Tempel-Tuttle
Geminids	Dec. 13–14	Dec. 7–17	60–80	Gemini	Asteroid 3200 Phaethon
Ursids	Dec. 22–23	Dec. 17–25	10	Ursa	Minor 8P/Tuttle

(a)

(b)

John Bortle

Meteorite Magazine/Grant Christie

■ **FIGURE 8–29** (a) A Peekskill, New York, woman heard a noise on October 9, 1992, and went outside to find extensive damage to her car. Under the car, she found a just-fallen meteorite. (b) This 1.3-kg meteorite crashed through a roof in New Zealand in 2004 and landed in someone's living room. A chip was taken out of the meteorite as it plunged through the roof of the house and interior ceilings.

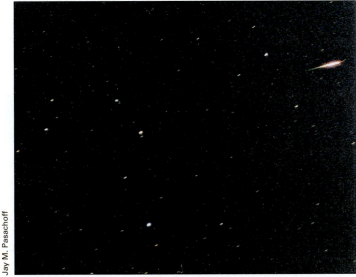

Jay M. Pasachoff

■ **FIGURE 8–30** A "fireball" – an extremely bright meteor – streaking away from the backward-question-mark shape of the constellation Leo, photographed during the 1999 Leonid meteor shower. Thousands of fainter meteors were seen in the peak hour.

8.5 ASTEROIDS

The eight known genuine planets, together with objects in the Kuiper belt, the scattered disk, and the Oort cloud, were not the only bodies to result from the gas and dust cloud that collapsed to form the Solar System 4.6 billion years ago. Thousands of **minor planets**, called **asteroids**, also resulted. We detect them by their small motions in the sky relative to the stars (■ Fig. 8–31). Most of the asteroids have elliptical orbits between the orbits of Mars and Jupiter, in a zone called the **asteroid belt**. It is thought that Jupiter's gravitational tugs perturbed the orbits of asteroids, leading to collisions among them that were too violent to form a planet.

Asteroids are assigned a number in order of discovery and then a name: (1) Ceres, (16) Psyche, and (433) Eros, for example. Often the number is omitted when discussing well-known asteroids. Though the concept of the asteroid belt may seem to imply a lot of asteroids close together, asteroids rarely come within a million kilometers of each other. Occasionally, collisions do occur, producing the small chips that make meteoroids.

8.5a General Properties of Asteroids

Only 6 asteroids have been measured to be larger than 300 km in diameter. Hundreds are over 100 km across (■ Fig. 8–32), roughly the size of some of the moons of the planets, but most are small, less than 10 km in diameter. (Objects smaller than about 50 m across are usually called meteoroids, as mentioned in Section 8.4.) Perhaps 100,000 asteroids could be detected with Earth-based telescopes; automated searches are now discovering asteroids at a prodigious rate. Yet all the asteroids together contain less mass than the Moon.

Thierry Pauwels, Koninklijke Sterrenwacht van Belgie (Observatoire Royal de Belgique—Royal Observatory of Belgium)

■ **FIGURE 8–31** Asteroid (5100) Pasachoff moves significantly against the background of stars in this series of exposures spanning six minutes. The asteroid appears elongated only because of its motion during the individual exposures.

Meteor showers can be fun to watch. You might see a "shooting star" nearly every minute or two. You need to wait until the radiant is on or above the horizon; the higher it is in the sky, the more meteors you will generally see. It is also very helpful if the Moon is down or at most a thin crescent; a bright, moonlit sky tends to drown out the fainter meteors. A list of annual meteor showers is given in *A Closer Look 8.3: Meteor Showers*.

The Perseids around August 12 is the most commonly observed meteor shower. (For most others, bundle up against the cold.) Bring a reclining chair, blanket, or sleeping bag if possible, and find as dark a site as you can with a wide-open sky. Lie back, relax, and gaze up at the sky for at least half an hour. Any part of the sky is fine, but you will probably see the most meteors in the general direction of the radiant (for example, Perseus, in the case of the Perseids). Try to see how many meteors you can count. Note any especially bright ones. Sometimes they leave a vapor "trail" (or "train") for a few seconds. Occasionally, colors are seen in the streak of light.

On Earth 65 million years ago, long before there were humans, dinosaurs and about ²/₃ of the other species died in a catastrophe known as the Cretaceous/Tertiary (K/T) extinction. (K is the traditional symbol for the Cretaceous; in many recent publications, the T for Tertiary has been superseded by Pg for Paleogene.) This "mass extinction" marked the end of the Cretaceous period, the third part of the Mesozoic era. In the movie *Jurassic Park* (the Jurassic period – 208 to 146 million years ago – was the second part of the Mesozoic era), dinosaurs were cloned from their DNA, something we can't do – yet. No human has ever been contemporaneous with a dinosaur.

Evidence has been accumulating that this extinction was sudden and was caused by a Near-Earth Object hitting the Earth. Among the signs is the fact that the rare element iridium, which is more common in asteroids and comets than in Earth's crust, is widely distributed around the Earth in a thin layer laid down 65 million years ago, presumably after it was thrown up in the impact.

The impact would have raised so much dust into the atmosphere that sunlight could have been shut out for months; plants and animals would not have been able to survive. Moreover, there is evidence for widespread fires 65 million years ago: Charcoal is found in the corresponding strata. The impact would have raised the Earth's temperature enough to start such fires.

A giant buried crater in Mexico that was probably the site of the impact (see the figure) has been discovered. It is one of the largest craters on Earth, about 200 km across. Its age has been measured to be 65 million years – precisely coinciding with the extinction event. The idea of an asteroid or comet impact therefore seems compelling.

Research is continuing in this exciting area. Revised dating of another giant crater has led to the still controversial idea that perhaps several asteroids or comets hit the Earth within a few hundred thousand years, and that they provided the "one-two punch" that knocked out the various species over a relatively short but not as abrupt time interval. An even more devastating extinction event took place 250 million years ago, terminating the Permian period and beginning the Triassic, but we do not yet know whether it was caused by the collision of a Near-Earth Object.

Virgil Sharpton, Lunar and Planetary Institute

FIGURE ACL 8.5 The impact structure known as Chicxulub, at the north end of Mexico's Yucatán Peninsula, was formed by an asteroid or comet 12 to 15 km in diameter. In this false-color map, which shows structure under a top layer of a kilometer of sediment, the blue areas are low-density rocks broken up by the impact. The green mound at the center is denser and probably represents a rebound at the point of impact.

(a)

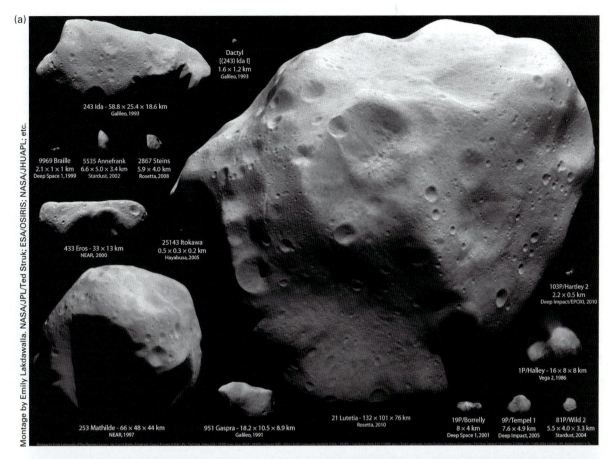

Montage by Emily Lakdawalla. NASA/JPL/Ted Struk; ESA/OSIRIS; NASA/JHUAPL; etc.

Dactyl
[(243) Ida I]
1.6 × 1.2 km
Galileo, 1993

243 Ida - 58.8 × 25.4 × 18.6 km
Galileo, 1993

9969 Braille
2.1 × 1 × 1 km
Deep Space 1, 1999

5535 Annefrank
6.6 × 5.0 × 3.4 km
Stardust, 2002

2867 Steins
5.9 × 4.0 km
Rosetta, 2008

433 Eros - 33 × 13 km
NEAR, 2000

25143 Itokawa
0.5 × 0.3 × 0.2 km
Hayabusa, 2005

103P/Hartley 2
2.2 × 0.5 km
Deep Impact/EPOXI, 2010

1P/Halley - 16 × 8 × 8 km
Vega 2, 1986

253 Mathilde - 66 × 48 × 44 km
NEAR, 1997

951 Gaspra - 18.2 × 10.5 × 8.9 km
Galileo, 1991

21 Lutetia - 132 × 101 × 76 km
Rosetta, 2010

19P/Borrelly
8 × 4 km
Deep Space 1, 2001

9P/Tempel 1
7.6 × 4.9 km
Deep Impact, 2005

81P/Wild 2
5.5 × 4.0 × 3.3 km
Stardust, 2004

(b)

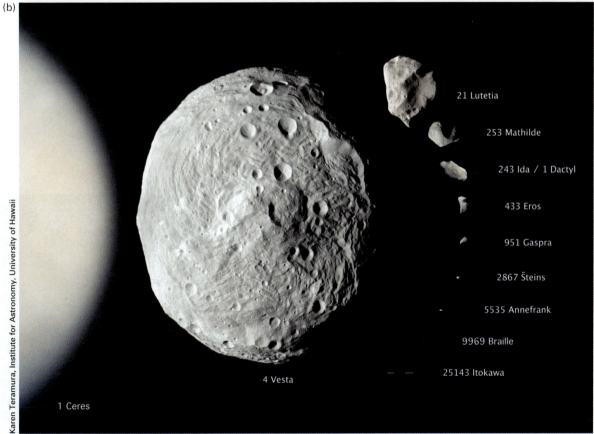

Karen Teramura, Institute for Astronomy, University of Hawaii

21 Lutetia

253 Mathilde

243 Ida / 1 Dactyl

433 Eros

951 Gaspra

2867 Šteins

5535 Annefrank

9969 Braille

25143 Itokawa

4 Vesta

1 Ceres

■ **FIGURE 8–32** (a) The asteroid belt's asteroids that have been visited close up by spacecraft. (b) All the asteroids that have been imaged close up, most recently Vesta from NASA's Dawn mission. Also included is a Hubble Space Telescope image, also to scale, of Ceres, Dawn's current destination.

Spacecraft en route to Jupiter and beyond travelled through the asteroid belt for many months and showed that the amount of dust among the asteroids is not much greater than the amount of interplanetary dust in the vicinity of the Earth. So the asteroid belt is not a significant hazard for space travel to the outer parts of the Solar System.

Asteroids are made of different materials, and represent the chemical compositions of different regions of space. The asteroids at the inner edge of the asteroid belt are mostly stony in nature, while the ones at the outer edge are darker (because they contain more carbon). Most of the small asteroids that pass near the Earth belong to the stony group. Three of the largest asteroids belong to the high-carbon group. A third group is mostly composed of iron and nickel. The differences may be telling us about conditions in the early Solar System as it was forming and how the conditions varied with distance from the young Sun. Many of the asteroids must have broken off from larger, partly "differentiated" bodies in which dense material sank to the center (as in the case of the terrestrial planets; see our discussion in Chapter 6).

8.5b Asteroids Viewed Close Up

The path of the Galileo spacecraft to Jupiter sent it near the asteroid (951) Gaspra in 1991 (■ Fig. 8–33a). It detected a magnetic field from Gaspra, which means that the asteroid is probably made of metal and is magnetized. Galileo passed the asteroid (243) Ida in 1993, and discovered that the asteroid has an even smaller satellite (■ Fig. 8–33b), which was then named Dactyl. Other double asteroids have since been discovered, and astronomers newly recognize the frequency of such pairs. For example, ground-based astronomers found a 13-km satellite orbiting 200-km-diameter (45) Eugenia every five days.

The European Space Agency's Rosetta spacecraft, en route to orbit a comet, passed asteroids (2867) Steins in 2008 and (21) Lutetia in 2010 as it went through the asteroid belt (Fig. 8–34). Lutetia's density, calculated from the mass it was found to have by its gravitational deflection of the spacecraft and its observed size, 121 km long, turned out to be sufficiently high that scientists concluded that it is one of the original planetesimals left over from the early stage of the Solar System (see Chapter 9 for a discussion of planetesimals).

Our studies of (4) Vesta took a jump in quality with the arrival of NASA's Dawn spacecraft at Vesta in 2011. Dawn went into orbit around Vesta, with the height of the orbit gradually decreasing so that the detail in the images continually improved (Fig. 8–35). Unlike the chemical-exhaust rockets that took people to the Moon and spacecraft to Mars and beyond, Dawn has an ion-drive engine. Its original thrust isn't as high as that of a chemical exhaust but it can run for a long time, gradually building up high speed. Also, though it was turned off when Dawn reached Vesta, it was turned on again in mid-2012 to take Dawn to asteroid (1) Ceres. This would make Dawn the first spacecraft to go into orbit around two separate bodies in sequence.

(a)

U.S. Geological Survey

(b)

NASA/JPL/Caltech

■ **FIGURE 8–33** (a) (951) Gaspra, in this high-resolution image taken by the Galileo spacecraft, looks like a fragment of a larger body. The illuminated part seen here is 16 km by 12 km in size. Galileo discovered that Gaspra rotates every 7 hours. We see a natural color image of Gaspra at left. At right, the color-enhanced image brings out variations in the composition of Gaspra's surface. (b) Ida and its moon (or satellite) Dactyl, photographed from the Galileo spacecraft at a range of 10,500 km. Filters passing light at 4100 Å (violet), 7560 Å (near-infrared), and 9860 Å (infrared) are used, enhancing the color. Subtle color variations show differences in the physical state and composition of the soil. The bluish areas around some of the craters suggest a difference in the abundance or composition of iron-bearing minerals there. Dactyl differs from all areas of Ida in both infrared and violet filters. Asteroid pairs are turning out to be fairly common.

ESA 2010 MPS for OSIRIS Team MPS/ UPD/LAM/IAA/RSSD/INTA/UPM/DASP/IDA

■ **FIGURE 8–34** (21) Lutetia, 121 km long, imaged from the European Space Agency's Rosetta spacecraft in 2010.

(a)

NASA/JPL-Caltech/UCLA/MPS/DLR/IDA

(b)

NASA/JPL-Caltech/UCLA/MPS/DLR/IDA

10 km

(c)

Lucille Le Corre/ David P. O'Brien/ NASA/ JPL-Caltech/ UCLA/ MPS/ DLR/ IDA

■ **FIGURE 8–35** (4) Vesta, from NASA's Dawn spacecraft. (a) An overall view of Vesta shows not only many craters but also a set of grooves parallel to the equator. (b) A crater on Vesta, taken from only 700 km above it, has a high resolution of 65 m per pixel. (c) Perspective view of the crater Oppia and its red/orange ejecta using Dawn color ratio images wrapped on a high-resolution shape model. Oppia, located in the center of the image, is approximately 34 km in diameter. The two bright craters visible at the bottom of the image are the 11.2-km crater Canuleia (left) and the 7-km crater Justina (right). To the north of Oppia, at the top of the image, is an area called Feralia Planitia with the crater Lepida (diameter of 44 km).

The spacecraft showed a huge mountain on Vesta's southern hemisphere, one of the largest mountains in the Solar System. A set of grooves looks as though it was formed as part of a giant impact. Some areas in the southern hemisphere may be as young as 1 or 2 billion years old, much younger than the surface of the northern hemisphere. Vesta is a big rocky body but it looks as though a chip was once knocked off it.

At almost 1,000 km across, icy (1) Ceres is even bigger than Vesta, massive enough that its gravity pulled it into a round shape, which makes it also a dwarf planet. Analysis of Hubble Space Telescope images of Ceres, matched with occultation results to give its absolute dimensions, have shown that its shape indicates that its interior is differentiated. Its bulk density is less than that of our Earth's crust, leading scientists to conclude that it may contain water ice under the surface, surrounding a rocky core. Spectra of the surface support this idea by finding water-bearing minerals. We eagerly await the results of NASA's Dawn mission, when it orbits Ceres in 2016.

A Japanese space mission, Hayabusa (which means "Peregrine Falcon"), formerly MUSES-C, was launched in 2003 in order to rendezvous with a near-Earth Apollo-type asteroid, (25143) Itokawa, about 600 m by 300 m by 260 m in size, in 2005 (Fig. 8–36). It returned about 1,500 micron-size grains of the asteroid to Earth in 2010, landing by parachute in the Australian outback. Detailed analysis of the dust grains showed that they were chondrites, like meteorites, though in a special subgroup with low ratios of iron. Since measurements showed that the dust particles had been annealed at high temperatures, they must have been in Itokawa's interior, and Itokawa must have been 40 times bigger for the temperature to get so high. The conclusion was that Itokawa as we now see it is the reassembled interior parts of a larger asteroid.

The Japanese space agency is now planning Hayabusa-2 for launch in 2014 toward a carbonaceous-type asteroid, the kind that is of interest because it is especially likely to carry organic molecules. Also to bring back samples from a carbonaceous asteroid, NASA is planning the OSIRIS-REx (**O**rigins, **S**pectral **I**nterpretation, **R**esource **I**dentification, **S**ecurity–**R**egolith **Ex**plorer) mission for launch in 2016, with the samples to be returned to Earth in 2023. The European Space Agency is considering MarcoPolo-R, an asteroid-sample-return mission for still later.

8.5c Near-Earth Objects

Some asteroids wander far from the main asteroid belt; their orbits approach or cross that of Earth. (In fact, one asteroid, discovered in 2003, is known to have an orbit that is entirely within Earth's.) We have observed only a small fraction of such **Near-Earth Objects**, bodies (asteroids and comets) that come within 1.3 au of Earth.

The **N**ear **E**arth **A**steroid **R**endezvous (NEAR) mission passed and photographed the main-belt asteroid (253) Mathilde in 1997. The existence of big craters that would have torn a solid rock apart, and the asteroid's low density, led scientists to conclude that Mathilde is a giant "rubble pile," rocks held together by mutual gravity.

Japan Aerospace Exploration Agency (JAXA)

■ **FIGURE 8–36** (25143) Itokawa, the asteroid visited by Hayabusa, which brought back some of its grains for examination on Earth. It looks like a "rubble pile," reassembled from a larger parent body that broke up.

(a)

(b)

■ **FIGURE 8–37** (a) The rotation of asteroid (433) Eros, imaged from the Near Earth Asteroid Rendezvous mission (renamed NEAR Shoemaker, after Eugene Shoemaker, who helped train Apollo astronauts, discovered many craters, showed that the Barringer crater in Arizona is from a meteorite impact, and co-discovered many comets including Shoemaker-Levy 9, which hit Jupiter). Eros's surface looks a uniform butterscotch color in visible light, but shows more variation in the infrared. (b) The final two photographs of Eros taken from NEAR Shoemaker as it descended on February 12, 2001. The last photograph, taken from 120 m above the surface, shows a field of view only 6 m across and resolves features as small as 1 cm across! At the bottom of the final image, the vertical streaks show where the transmission of the photo was cut when NEAR Shoemaker impacted Eros.

NEAR went into orbit around (433) Eros on Valentine's Day, 2000 (■ Fig. 8–37a), when it was renamed NEAR Shoemaker after the planetary geologist Eugene Shoemaker. Eros was the first near-Earth asteroid that had been discovered. It is 33 km by 13 km by 13 km in size. NEAR Shoemaker photographed craters, grooves, layers, house-sized boulders, and a 20-km-long surface ridge.

The existence of the craters and ridge, which indicates that Eros must be a solid body, disagrees with the previous suggestions of some scientists that most asteroids are mere rubble piles as Mathilde seems to be. The impact that formed the largest crater, 8 km across and now named Shoemaker, is thought to have formed most of the large boulders found across Eros's surface.

Eros's density, 2.4 g/cm³, is comparable to that of the Earth's crust, about the same as Ida's, and twice Mathilde's. From orbit, NEAR Shoemaker's infrared, x-ray, and gamma-ray spectrometers measured how the minerals vary from place to place on Eros's surface. The last of these even survived the spacecraft's landing on Eros (■ Fig. 8–37b), and radioed back information about the composition of surface rocks.

Scientists analyzing the data have found abundances of elements similar to those of the Sun (apart from the Sun's hydrogen and helium), and also of a type of primitive meteorite known as chondrites that are the most common type found on Earth. They have concluded that Eros is made of primitive material, unchanged for 4.5 billion years, so we are studying the early eras of the Solar System with it. NEAR Shoemaker's observations show that Eros was probably broken off billions of years ago from a larger asteroid as a uniformly dense fragment. This solidity contrasts with Mathilde's rubble-pile nature.

Besides providing much detailed information, the close-up studies of these objects are allowing us to verify whether the lines of reasoning we use with ground-based asteroid observations give correct results.

Near-Earth asteroids may well be the source of most meteorites, which could be debris of collisions that occurred when these asteroids visit the asteroid belt. Eventually, most Earth-crossing asteroids will probably collide with the Earth. It is estimated that about 1000 of them are greater than 1 km in diameter, and over 90 per cent of these have been found; only a few are known to be larger than 10 km across. Statistics show that there is a 0.2 per cent chance of a collision of this tremendous magnitude per millennium, a pretty high rate on a cosmic scale. Such collisions would have drastic consequences for life on Earth.

Smaller objects are much more common, but a smaller percentage of them have been discovered. There is a roughly 1 per cent chance that an asteroid greater than 300 m in diameter will hit the Earth in the next century. Such a collision could kill thousands or millions of people, depending on where it lands.

The question of how much we should worry about Near-Earth Objects hitting us is being increasingly discussed, including at a meeting sponsored by the United Nations. Even Hollywood movies have been devoted to the topic, though at present we can't send out astronauts to deflect or break up the objects the way the movies showed.

Several projects are under way to find as many Near-Earth Objects as possible. Current plans are to map 90 per cent of those larger than about 140 m in the next decade, and the pace of discovery is accelerating. Several projects use CCD detectors, repetitive scanning, and computers to locate asteroids and are discovering thousands each year, some of which are Near-Earth Objects.

In 2011, the asteroid 2005 YU55, 400 m across, flew within the Moon's orbit, close enough that ground-based radars could make images of it with resolutions down to about 2 m. It is the largest object to come this near to Earth since 1976, and should remain the largest until another known asteroid comes that close in 2028.

CONCEPT REVIEW

» Pluto, which emerged in 1999 from inside Neptune's orbit, is so small and so far away that we know little about it (Sec. 8.1). The discovery of its moon, Charon, allowed us to calculate that Pluto contains only $^1/_{500}$ the mass of the Earth (Sec. 8.1a). Mutual occultations of Pluto and Charon revealed sizes and surface structures of each. Pluto has a very thin nitrogen atmosphere, at least when it is closest to the Sun in its orbit (Sec. 8.1b).

» Because Eris was found, at first thought to be bigger than Pluto, Pluto lost its designation as a genuine planet (Sec. 8.1c). It is now a **dwarf planet**, along with Eris, Haumea, and Makemake. These dwarf planets are the largest of the **Kuiper-belt objects**, which are also called **Trans-Neptunian Objects** (Sec. 8.2). In 2011, it was found that Eris and Pluto are virtual twins, and it is not clear which is larger, without considering Pluto's atmosphere. More than 1000 Kuiper-belt objects are now known, and they appear to be left over from the formation of the Solar System.

» In a **comet**, a long **tail** that points away from the Sun may extend from a bright **head** (Sec. 8.3). The head consists of the **nucleus**, which is like a dirty snowball, and the gases of the **coma** (Sec. 8.3a). The **dust tail** contains dust particles released from the dirty ices of the nucleus. The **gas tail**, also known as an ion tail, is blown out behind the comet by the solar wind. Comets we see come from a huge spherical **Oort comet cloud** far beyond Pluto's orbit or from a more flattened belt of icy objects called the **scattered disk**, somewhat beyond Neptune's orbit (Sec. 8.3b).

» Spacecraft imaged the nucleus of Halley's Comet up close in 1986 and showed it to be about 16 km long and 8 km across (Sec. 8.3c). The impact of the tidally disrupted Comet Shoemaker-Levy 9 with Jupiter in 1994 produced temporary Earth-sized scars in Jupiter's atmosphere (Sec. 8.3d). Comet Hale-Bopp in 1997 was not only spectacular for the general public but also allowed many scientific studies (Sec. 8.3e). The Deep Impact mission crashed into a comet in 2005 (Sec. 8.3f), providing information on its composition.

» **Meteoroids** are chunks of rock in space up to about 50 m across, but most are small, like grains of sand (Sec. 8.4). When one hits the Earth's atmosphere, we see the streak of light from its vaporization as a **meteor**, often incorrectly called a **shooting star** or a falling star. A fragment that survives and reaches Earth's surface is a **meteorite**. Bits of space dust are **micrometeorites** (Sec. 8.4a). Most meteorites that are found are made of an iron–nickel alloy, but when meteorites are seen to fall, they are most often stony. The meteorites bring us primordial material to study and occasionally pieces of the Moon or Mars.

» Many meteors are seen in **showers**, which occur when the Earth crosses the path of a defunct or disintegrating comet (Sec. 8.4b); these meteors appear to come from a single radiant in the sky. **Sporadic meteors** are not associated with showers and appear at the rate of a few each hour.

» **Asteroids** are **minor planets** (Sec. 8.5). Most asteroids are in the **asteroid belt** between the orbits of Mars and Jupiter. They were prevented from forming a planet by Jupiter's gravitational tugs. Asteroids range up to about 1,000 km across (Sec. 8.5a), but the vast majority are much smaller, down to about 50 m. Most of them are stony, while some contain carbon and others consist largely of iron and nickel. The Galileo, Hyabusa, Rosetta, Dawn, and Near Earth Asteroid Rendezvous (renamed NEAR Shoemaker) missions have studied asteroids up close, and NEAR Shoemaker landed on the asteroid Eros (Sec. 8.5b). It found a solid object of primitive composition, in contrast with a rubble pile like the asteroid Mathilde.

» It is plausible that an impact of a large asteroid or comet threw so much dust into the Earth's atmosphere that it led to the extinction of most living species, including dinosaurs, 65 million years ago (Sec. 8.5b and *A Closer Look 8.4*). We worry about a **Near-Earth Object** (an asteroid or a comet) hitting the Earth, which could lead to the demise of civilization, or of the more likely collision with the Earth of a smaller object, which could cause substantial damage, killing thousands or millions of people.

QUESTIONS

1. What fraction of its orbit has Pluto traversed since it was discovered?

2. What evidence suggested quite some time ago (before the discovery of the Kuiper belt) that Pluto is not a "normal" planet?

3. Describe what mutual occultations of Pluto and Charon are and what they have told us.

4. Compare Pluto to the giant planets.

5. Compare Pluto to the terrestrial planets.

†6. If Pluto's surface temperature is 60 K, use Wien's law (Chapter 2) to calculate the wavelength at which it gives off most of its thermal (blackbody) radiation. In what part of the electromagnetic spectrum is this?

7. Which of Kepler's laws has enabled the discovery of Charon to help us find the mass of Pluto?

8. Describe the observations of Pluto's atmosphere when Pluto occulted a star.

9. What are Kuiper-belt objects, and how is Pluto thought to be one of them?

10. What is the main reason Pluto is now considered a dwarf planet rather than a genuine planet?

11. What creates the tail of a comet? Is something chemically burning?

12. Which part of a comet has the most mass?

13. From where are the comets that we see thought to come?

†14. How far is the Oort comet cloud from the Sun, relative to the distance from the Sun to Pluto?

15. Describe some of the Giotto spacecraft's discoveries about Halley's Comet.

†16. Use Kepler's third law (Chapter 5) and the known period of Halley's Comet (about 76 years) to deduce its semimajor axis. Which planet has a comparable semimajor axis?

17. Discuss the significance of the collision of Comet Shoemaker-Levy 9 with Jupiter, and subsequent comet collisions.

18. What is the relation of meteoroids and asteroids?

19. What, physically, is a shooting star?

20. Why don't most meteoroids reach the Earth's surface?

21. Explain the process by which a meteor shower occurs.

22. Why might some meteor showers last only a day while others can last several weeks?

23. Why are meteorites important in our study of the Solar System?

24. Compare the sizes and surfaces of asteroids with the moons of the planets.

25. How might the study of the apparent brightness of a star during an occultation by an asteroid tell us whether the asteroid has an atmosphere?

26. Why is the study of certain types of asteroids important for understanding our long-term survival on Earth?

27. What have the Galileo, NEAR Shoemaker, Deep Space 1, Hyabusa, Rosetta, Deep Impact, and Dawn spacecraft found out about comets and asteroids?

28. **True or false?** Comets are made of mostly rocky material from the asteroid belt.

29. **True or false?** The retrograde motion of Pluto can be attributed to the relatively high eccentricity of its orbit around the Sun, as compared to normal planets.

30. **True or false?** Periodic comets generally grow brighter each time they appear, since they gradually accumulate more material from the outer parts of the Solar System.

31. **True or false?** A "shooting star" or "falling star" is not a star at all, but rather a rock from space that enters Earth's atmosphere at a high speed and burns up due to the sudden compression of air (friction).

32. **True or false?** The Giotto spacecraft observations of Comet Halley showed the evaporating gas and dust to come out of fissures in the comet's dark crust, creating jets.

33. **Multiple choice:** Most known comets do not reappear in our sky very often because **(a)** comets are not gravitationally bound to our Solar System; **(b)** comets burn quite rapidly, using up all their fuel

within a few decades; **(c)** the orbits of comets are highly eccentric and their semimajor axes are large; **(d)** most comets only pass close to the Sun two or three times, and then they disintegrate; or **(e)** it is impossible to tell when most comets will return close to the Sun.

34. **Multiple choice:** Pluto **(a)** is about as large as the planet Mercury; **(b)** has no moons; **(c)** has a density intermediate between that of the terrestrial planets and the giant planets; **(d)** has a nearly circular orbit around the Sun, due to its interactions with other objects in its vicinity; or **(e)** is massive enough to substantially affect the orbit of Uranus, which led to its discovery.

35. **Multiple choice:** Which one of the following statements about the main asteroid belt is *true*? **(a)** The main asteroid belt is probably debris that would have formed a planet had Jupiter not prevented it. **(b)** The mass of all the asteroids together is roughly that of the Earth. **(c)** The iron/nickel asteroids are the most common kind. **(d)** There is no evidence that the material from which asteroids formed was ever part of one or more larger bodies that could have experienced differentiation. **(e)** All asteroids appear to be quite solid objects, rather than collections of smaller rocks held together by gravity.

†36. **Multiple choice:** Suppose you discover a comet having an orbital period of 1000 years around our Sun. Roughly what is its semimajor axis? **(a)** 10 au. **(b)** 32 au. **(c)** 10^2 au. **(d)** $10^{4.5}$ au. **(e)** 10^6 au.

37. **Multiple choice:** If an asteroid with a diameter of 30 km were to collide with the Earth, it would probably **(a)** burn up in the atmosphere before reaching the ground; **(b)** only crush a car, if one happened to be at the point of impact; **(c)** devastate an area no larger than 30 km in diameter; **(d)** create a crater more than 300 km in diameter, but have little or no effect on the rest of the Earth and its inhabitants; or **(e)** bring an end to human civilization, or at least destroy much of the human race.

38. **Fill in the blank:** Pluto is a large member of the _____, and thus is no longer considered to be a genuine planet.

39. **Fill in the blank:** A meteoroid that has landed on the surface of a planet or a moon is called a _____.

40. **Fill in the blank:** When Earth passes through the orbit of an old, disintegrating comet, a _____ can occur.

41. **Fill in the blank:** A thin layer of unusually high levels of the element _____ in the strata of the Earth's crust deposited 65 million years ago provides evidence for the impact theory of the Cretaceous/Tertiary extinction.

42. **Fill in the blank:** It is now generally thought that billions, or even trillions, of tail-less comets surround the Solar System in a spherical shell nearly one light-year in radius called the _____.

†This question requires a numerical solution.

TOPICS FOR DISCUSSION

1. Now that many objects, some of them quite large, have been discovered in the Kuiper belt, do you think Pluto should still be called a planet? Does discovering that Eris may be slightly smaller than Pluto change your thinking?

2. Do you worry about asteroid or comet collisions with Earth? What could be done to save the Earth if an asteroid were discovered sufficiently far in advance of the collision?

3. Some scientists have suggested that the dinosaurs were about to become extinct anyway, without an asteroid or comet collision, because of changes in Earth's climate and other reasons. If so, does this detract from the impact theory, in view of evidence that $^2/_3$ of all species perished quite suddenly and a large crater having the right age has been found?

1ˢᵀ
SUN

2ᴺᴰ
JUPITER

3ᴿᴰ
SATURN

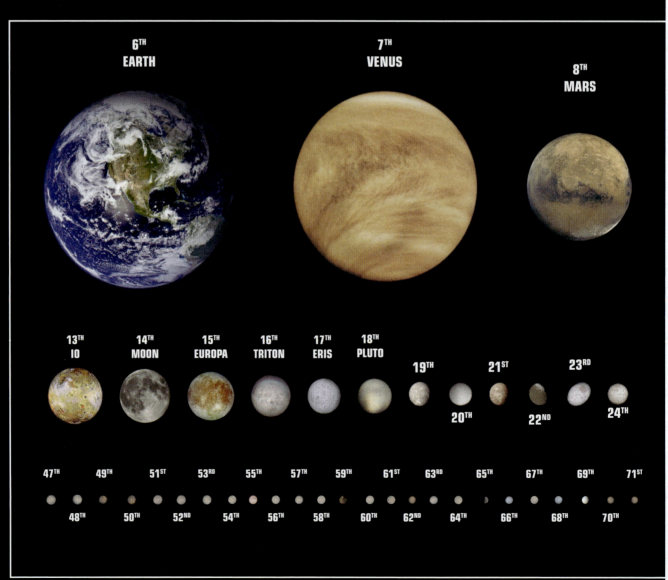

6ᵀᴴ
EARTH

7ᵀᴴ
VENUS

8ᵀᴴ
MARS

13ᵀᴴ
IO

14ᵀᴴ
MOON

15ᵀᴴ
EUROPA

16ᵀᴴ
TRITON

17ᵀᴴ
ERIS

18ᵀᴴ
PLUTO

19ᵀᴴ

21ˢᵀ

23ᴿᴰ

20ᵀᴴ

22ᴺᴰ

24ᵀᴴ

47ᵀᴴ   **49ᵀᴴ**   **51ˢᵀ**   **53ᴿᴰ**   **55ᵀᴴ**   **57ᵀᴴ**   **59ᵀᴴ**   **61ˢᵀ**   **63ᴿᴰ**   **65ᵀᴴ**   **67ᵀᴴ**   **69ᵀᴴ**   **71ˢᵀ**

48ᵀᴴ   **50ᵀᴴ**   **52ᴺᴰ**   **54ᵀᴴ**   **56ᵀᴴ**   **58ᵀᴴ**   **60ᵀᴴ**   **62ᴺᴰ**   **64ᵀᴴ**   **66ᵀᴴ**   **68ᵀᴴ**   **70ᵀᴴ**

THE 100 LARGEST OBJECTS IN THE SOLAR SYSTEM

as known in january 2012

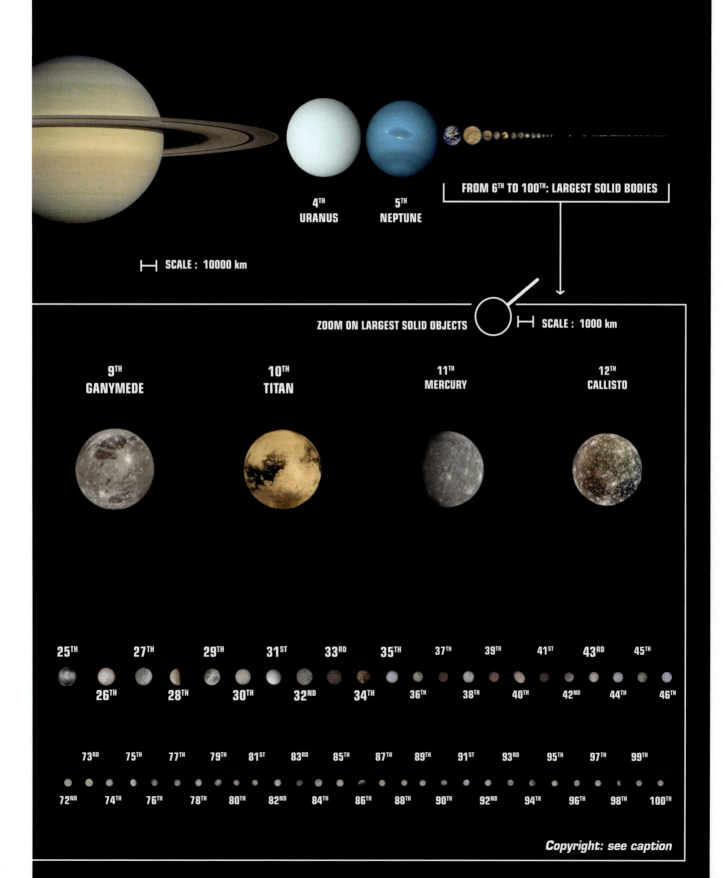

4TH URANUS

5TH NEPTUNE

FROM 6TH TO 100TH: LARGEST SOLID BODIES

⊢⊣ SCALE : 10000 km

ZOOM ON LARGEST SOLID OBJECTS ⊢⊣ SCALE : 1000 km

9TH GANYMEDE

10TH TITAN

11TH MERCURY

12TH CALLISTO

25TH 27TH 29TH 31ST 33RD 35TH 37TH 39TH 41ST 43RD 45TH

26TH 28TH 30TH 32ND 34TH 36TH 38TH 40TH 42ND 44TH 46TH

73RD 75TH 77TH 79TH 81ST 83RD 85TH 87TH 89TH 91ST 93RD 95TH 97TH 99TH

72ND 74TH 76TH 78TH 80TH 82ND 84TH 86TH 88TH 90TH 92ND 94TH 96TH 98TH 100TH

Rank	Body	Type	Equivalent diameter D (km)	Accuracy on D (*)	Image credit (**)	Rank	Body	Type	Equivalent diameter D (km)	Accuracy on D (*)	Image credit (**)
1	Sun	Star	1392000	good	[i]	51	2005 RN43	TNO	610	poor	[h]
2	Jupiter	Giant planet	139822	good	[d]	52	2010 KZ39	TNO	610	poor	[h]
3	Saturn	Giant planet	116464	good	[d]	53	2004 XA192	TNO	580	poor	[h]
4	Uranus	Giant planet	50724	good	[f]	54	2010 RE64	TNO	560	poor	[h]
5	Neptune	Giant planet	49244	good	[f]	55	2001 UR163	TNO	560	poor	[h]
6	Earth	Terrestrial planet	12742	good	[j]	56	2010 RF43	TNO	560	poor	[h]
7	Venus	Terrestrial planet	12104	good	[b]	57	2004 PF115	TNO	560	poor	[h]
8	Mars	Terrestrial planet	6780	good	[c]	58	Salacia	TNO	560	poor	[h]
9	Ganymede	Moon of Jupiter	5268	good	[d]	59	Pallas	Asteroid	545	fair	[h]
10	Titan	Moon of Saturn	5152	good	[e]	60	2003 UZ413	TNO	540	poor	[h]
11	Mercury	Terrestrial planet	4880	good	[a]	61	Huya	TNO	540	poor	[h]
12	Callisto	Moon of Jupiter	4820	good	[d]	62	2002 KX14	TNO	530	poor	[h]
13	Io	Moon of Jupiter	3643	good	[d]	63	2010 FX86	TNO	530	poor	[h]
14	Moon	Moon of the Earth	3474	good	[j]	64	2008 ST291	TNO	530	poor	[h]
15	Europa	Moon of Jupiter	3122	good	[d]	65	Vesta	Asteroid	522	good	[g]
16	Triton	Moon of Neptune	2707	good	[f]	66	2005 RM43	TNO	520	poor	[h]
17	Eris	Dwarf planet	2326	good	[h]	67	2004 NT33	TNO	520	poor	[h]
18	Pluto	Dwarf planet	2306	fair	[h]	68	2004 XR190	TNO	520	poor	[h]
19	Titania	Moon of Uranus	1578	good	[f]	69	Enceladus	Moon of Saturn	504	good	[e]
20	Rhea	Moon of Saturn	1529	good	[e]	70	2004 TY364	TNO	500	poor	[h]
21	Oberon	Moon of Uranus	1523	good	[f]	71	2004 UX10	TNO	500	poor	[h]
22	Iapetus	Moon of Saturn	1472	good	[e]	72	2010 VK201	TNO	500	poor	[h]
23	Haumea	Dwarf planet	1450	poor	[h]	73	2008 AP129	TNO	500	poor	[h]
24	Makemake	Dwarf planet	1420	poor	[h]	74	2004 PR107	TNO	480	poor	[h]
25	Charon	Moon of Pluto	1212	good	[h]	75	Miranda	Moon of Uranus	471	good	[f]
26	Sedna	TNO	1180	poor	[h]	76	1999 DE9	TNO	460	poor	[h]
27	Umbriel	Moon of Uranus	1169	good	[f]	77	2003 QX113	TNO	460	poor	[h]
28	Ariel	Moon of Uranus	1158	good	[f]	78	2007 JH43	TNO	460	poor	[h]
29	Dione	Moon of Saturn	1123	good	[e]	79	Vanth	Moon of Orcus	460	poor	[h]
30	2007 OR10	TNO	1080	poor	[h]	80	Chaos	TNO	450	poor	[h]
31	Tethys	Moon of Saturn	1066	good	[e]	81	2003 FY128	TNO	450	poor	[h]
32	Quaoar	TNO	1050	fair	[h]	82	2000 YW134	TNO	440	poor	[h]
33	2002 TC302	TNO	1040	poor	[h]	83	2004 PG115	TNO	440	poor	[h]
34	Ceres	Dwarf planet	945	good	[h]	84	2007 XV50	TNO	440	poor	[h]
35	Orcus	TNO	850	poor	[h]	85	2005 QU182	TNO	430	poor	[h]
36	2003 MW12	TNO	730	poor	[h]	86	Hygiea	Asteroid	430	fair	[h]
37	2002 AW197	TNO	730	poor	[h]	87	2010 VZ98	TNO	420	poor	[h]
38	2005 UQ513	TNO	730	poor	[h]	88	2008 OG19	TNO	420	poor	[h]
39	2003 VS2	TNO	725	poor	[h]	89	2002 KW14	TNO	420	poor	[h]
40	Varuna	TNO	700	poor	[h]	90	2000 CN105	TNO	420	poor	[h]
41	2002 XW93	TNO	690	poor	[h]	91	2002 XV93	TNO	420	poor	[h]
42	2002 UX25	TNO	680	poor	[h]	92	2001 KA77	TNO	420	poor	[h]
43	2004 GV9	TNO	680	poor	[h]	93	2011 FW62	TNO	420	poor	[h]
44	2003 AZ84	TNO	680	poor	[h]	94	2001 QF298	TNO	410	poor	[h]
45	2007 JJ43	TNO	660	poor	[h]	95	2010 EL139	TNO	410	poor	[h]
46	2002 MS4	TNO	660	poor	[h]	96	2011 GM27	TNO	410	poor	[h]
47	2010 EK139	TNO	640	poor	[h]	97	2002 WC19	TNO	410	poor	[h]
48	2006 QH181	TNO	640	poor	[h]	98	Proteus	Moon of Neptune	404	fair	[f]
49	2007 UK126	TNO	620	poor	[h]	99	1996 GQ21	TNO	400	poor	[h]
50	Ixion	TNO	620	poor	[h]	100	2008 QY40	TNO	400	poor	[h]

Global image and table: courtesy of Thales Alenia Space, Joel Poncy, Thomas Briot

(*) Accuracy on diameter knowledge:

good = better than +/−3%

fair = typically between 3% and 10%

poor = worse than +/−10%

(**) Image credits per object:

[a] NASA – Messenger mission

[b] NASA – Pioneer mission

[c] NASA – Viking Orbiter

[d] NASA – Galileo mission

[e] NASA – Cassini mission

[f] NASA – Voyager mission

[g] NASA – Dawn mission

[h] Graphic art by Thales Alenia Space, Thomas Briot

[i] NASA – SDO mission

[j] Image Istockphoto

A CLOSER LOOK 8.6

Images from Curiosity on Mars

NASA/JPL–Caltech/Malin Space Science Systems

■ **FIGURE ACL 8.6–1** ACL 8.6–1 NASA's Mars rover Curiosity carries a Mastcam, a camera held high on a vertical mast, to photograph the Martian surface in Gale Crater. The image at left is the raw, unprocessed, color, as it is received from Mars. The central image has been adjusted to show approximately what we would see as the colors if we were ourselves on Mars. The right image uses "white balancing" to show what the surface would look like if it was on Earth with everyday lighting here, instead of in light filtered by the pinkish martian sky.

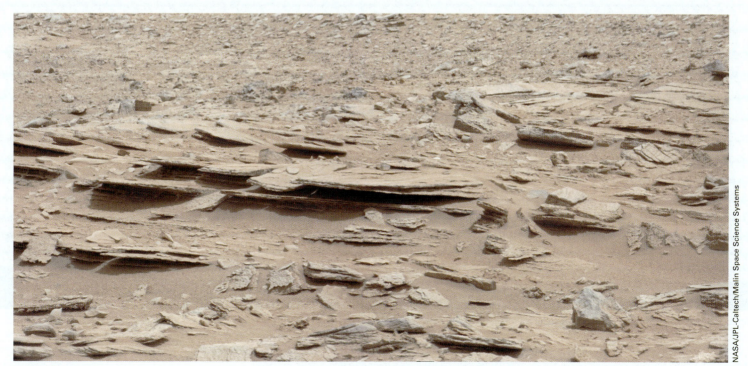

NASA/JPL–Caltech/Malin Space Science Systems

■ **FIGURE ACL 8.6–2** Outcroppings inside Gale Crater on Mars, imaged by Mastcam on Curiosity. The image has been white-balanced, so we see the crossbedding as it would appear to us if it were on Earth. The outcrop is informally named Shaler, and is near a location, Gleneig, where three types of terrain meet.

A sample of Kepler planetary candidate parent stars, out of an even larger total number, each with its transiting companion or companions in silhouette. The sizes of the stars and transiting companions are properly scaled. The colors of the stars show very roughly how the eye would see the star outside of the Earth's atmosphere (although white stars, like the Sun, are indicated with a yellowish tinge). Stars have been properly limb darkened. The largest star is 6.1 times larger than the Sun, and the smallest stars are estimated to be only 0.3 times the radius of the Sun. The Sun is shown below the top row on the right by itself with the Earth and Jupiter in transit.

Jason Rowe

9

Our Solar System and Others

ORIGINS

Discoveries of planetary systems around other stars are providing insights into the process by which planets form. The exoplanets are also possible abodes for life outside our own Solar System.

AIMS

1. Discuss the formation of the Solar System (Section 9.1).

2. Learn the methods by which we have discovered over 1000 planets orbiting other stars, and consider the properties of these planets (Section 9.2).

3. See how the discovery of planets and protoplanetary disks around other stars affects our thinking about the formation of our own Solar System (Section 9.3).

Astronomers have long speculated about the origin of our Solar System. They have noted regularities in the way the planets orbit the Sun and in the spacing of the planetary orbits. But until recently, astronomers have been limited to studying one planetary system: our own.

In 1600, Giordano Bruno was burned, naked, at the stake, in part for wondering about the "plurality of worlds," to use the term of that day. Finally, after hundreds of years of wondering whether they existed, astronomers discovered another planetary system in 1992, though it was around a very strange type of star. And in the last two decades, planetary systems galore have been discovered around stars like the Sun. While twenty years ago we knew of only the handful of planets orbiting our Sun, as of this writing (2012) we know of about 1000 planets around other stars, as well as more than 2000 additional good candidates (see the Chapter Opener), the latter mostly from NASA's Kepler spacecraft (■ Fig. 9–1).

In this chapter, we will first discuss our own Solar System and its formation. Then we will describe how we find other planetary systems and how they may have been formed. This discussion, in turn, should give us insights into how well we understand the origin of our own Solar System.

■ **FIGURE 9–1** The view into our Galaxy with NASA's Kepler spacecraft from its vantage point in Earth orbit; it measures the brightness of about 150,000 stars over and over again, searching for transiting exoplanets.

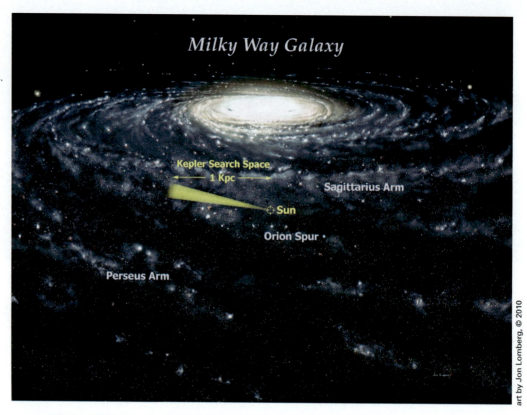

9.1 THE FORMATION OF THE SOLAR SYSTEM

Many scientists studying the Earth and the planets are particularly interested in an ultimate question: How did the Earth and the rest of the Solar System form? We can accurately date the formation by studying the oldest objects we can find in the Solar System and allowing a little more time. For example, astronauts found rocks on the Moon older than 4.4 billion years, and some meteorites originating from the asteroid belt have been dated to be 4.56 billion years old. We conclude that the Solar System formed about 4.6 billion years ago.

9.1a Collapse of a Cloud

Our best current idea is that 4.6 billion years ago, a huge cloud of gas and dust in space collapsed, pulled together by the force of gravity. What triggered the collapse isn't known. It might have been gravity pulling together a random cloud, or it might have been a shock wave from a nearby supernova (■ Fig. 9–2). As the gas pressure increased, eventually the rate of collapse decreased to a slower contraction.

You have undoubtedly noticed that ice skaters spin faster when they pull their arms in. Similarly, the cloud that ultimately formed the Solar System began to spin faster as it collapsed and contracted. The original gas and dust may have had some spin, and this spin is magnified in speed by the collapse, because the total amount of "angular momentum" doesn't change.

The contracting cloud spins faster because the quantity known as angular momentum stays steady in a rotating system, unless the system is acted on by an outside force (assuming the force is not directed at the center of the object). The amount of angular momentum depends on the speed of the spin and on how close each bit of spinning mass is to the axis of spin; see our discussion at the end of Chapter 5.

■ **FIGURE 9–2** Larger and smaller shock waves formed by an F/A-18 Hornet jet as it broke the sound barrier are visible. The clouds of condensation formed as the shocked air cooled.

Alan Boss, Dept. of Terrestial Magnetism, Carnegie Institute of Washington

■ FIGURE 9–3 A false-color representation of density clumps in a model of the early solar nebula with a radius of 30 au. A cross section through the midplane of the nebula is shown. Red corresponds to the highest density, and black to the lowest density. Five clumps have formed in only 338 years. The model is by Alan Boss.

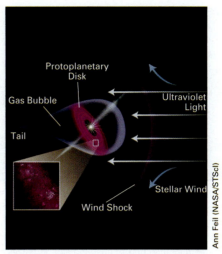

Ann Feil (NASA/STScI)

■ FIGURE 9–4 Artist's conception of the effect of strong ultraviolet (UV) light from nearby stars on a young star and its planet-forming disk. The UV light heats the gas so much (to 10,000 K) that it effectively evaporates from the disk and forms a comet-like tail downstream.

Objects that are spinning around tend to feel an apparent outward force known as the centrifugal force (physicists explain this fictitious force with Newton's laws of motion), and in our case this force eventually became strong enough to counteract the effect of gravity pulling inward. Thus the Solar System stopped contracting in one plane. Perpendicular to this plane, there was no spin to stop the contraction, so the solar nebula ended up as a disk. The central region became hot and dense, eventually becoming hot enough for the gas to undergo nuclear reactions (see Chapter 12), thus forming our Sun.

In the disk of gas and dust, we calculate that the dust began to clump (■ Fig. 9–3). Smaller clumps joined together to make larger ones, and eventually **planetesimals**, bodies that range from about 1 kilometer up to a few hundred kilometers across, were formed. Gravity subsequently pulled many planetesimals together to make **protoplanets** (pre-planets) orbiting a **protosun** (pre-Sun).

The protoplanets then contracted and cooled to make the planets we have today, and the protosun contracted to form the Sun. Some of the planetesimals may still be orbiting the Sun; that is why we are so interested in studying small bodies of the Solar System like comets, meteoroids, and asteroids. Most of the unused material was either swallowed by the Sun or evaporated away by the young Sun's ultraviolet radiation, or by the ultraviolet radiation from nearby stars (■ Fig. 9–4).

Note that, early on, the Sun would have been spinning very fast. But much of the excess angular momentum was transported upward by a "bipolar outflow"; see the discussion in Chapter 12.

9.1b Models of Planet Formation

In the currently most popular model for the formation of planets, the terrestrial planets are believed to result from the accumulation of planetesimals (■ Fig. 9–5a). The rocky material, mainly silicates, condensed at the temperatures at these planets' distances from the Sun and was not blown outward by particles or radiation from the forming Sun. These terrestrial planets formed 50 to 100 million years

after the giant planets were made. By that late stage, the hydrogen and helium gas had long since vanished, lasting only about 10 million years, leaving a debris disk of rocky material that subsequently formed planetesimals and finally the terrestrial planets, devoid of substantial atmospheres. Jupiter and the other giant planets, on the other hand, assembled huge atmospheres of hydrogen and helium, matching the initial composition of the Sun.

For each of the outer planets, silicates (rocks) and ice grains (largely of water, ammonia, and methane) in the early Solar System collided and grew at the same time, forming planetesimals. The ices had condensed out of the surrounding gas because of the cold temperatures beyond the "frost line" in the solar nebula; recall the discussion in Chapter 7. A solid core (resembling a terrestrial planet, but having perhaps 10 Earth masses) formed from the runaway growth of planetesimals (■ Fig. 9–5b). After becoming sufficiently large, the planet gravitationally attracted large amounts of nearby hydrogen and helium gas in the nebula, growing dramatically, especially in the cases of Saturn and (more spectacularly) Jupiter.

Uranus and Neptune probably formed in the general vicinity of Jupiter and Saturn, with Neptune perhaps being closer to the Sun than Uranus (but this idea is still controversial), and then migrated outward because of gravitational interactions with the remaining planetesimals before they had a chance to gather very much hydrogen and helium. This would explain why the atmospheres of Jupiter and Saturn are very much more massive than those of Uranus and Neptune. Eventually, a complex series of gravitational interactions induced by Jupiter and Saturn caused Neptune to cross beyond the orbit of Uranus (if it indeed was initially closer to the Sun), and then it gravitationally catapulted some of the nearby objects outward to form the present-day Kuiper belt and the extended scattered disk. A majority of the interactions sent the icy bodies inward, however, where they were captured by Jupiter or (in most cases) ejected into the Oort cloud, thereby explaining why there is so little mass in the Kuiper belt and the scattered disk.

Updated by Alan Boss from Joseph A. Burns, et al., An Integrated Strategy for the Planetary Sciences: 1995–2010; reprinted with permission from the National Academy of Sciences, courtesy of the National Academies Press, Washington, D.C.

■ **FIGURE 9–5** Proposed model for the formation of our Solar System. (a) The four inner, terrestrial planets. The first panel shows the initial disk of dust and gas, where dust grains stick together to become 10-km-diameter planetesimals. In the second panel, planetesimals grow over about 10,000 years to form 2000-km-diameter planetary embryos. In the third panel, gas disappears over the next million to 10 million years. In the fourth panel, the gravity of the planetary embryos changes their orbits and brings them together to form the current planets over about 100 million years. Other objects are ejected. (b) The four outer, giant planets. The first panel shows the initial disk of dust and gas. In the second panel, the runaway growth of planetesimals forms very large (10-Earth-mass) cores in a few million to 10 million years. In the third panel, Jupiter and Saturn capture gas from the solar nebula before it dissipates. Uranus and Neptune started out as cores that formed between Jupiter and Saturn but migrated outward to their present orbits without gathering much gas. In the fourth panel, we see some comets as well as bodies like Pluto and Triton forming as Kuiper-belt objects that weren't incorporated in the giant planets.

Any successful model of Solar-System formation must explain not only the range of planets we see but also the relative emptiness of the asteroid belt. Many people mistakenly think that the asteroids are the remains of a large planet that was shattered by a collision or tidally disrupted, but the asteroid belt actually contains only about 4 per cent of the mass of Earth's Moon. It is now thought that planetesimals formed everywhere throughout the solar nebula, including the asteroid belt, yet only in some regions were they able to coalesce into planets. Specifically, Jupiter "bullied" the asteroid belt, gravitationally kicking out material that orbited an integer number of times in the time that Jupiter itself made one orbit (or some other integer number of orbits). As Jupiter migrated to various locations in the Solar System, it was able to eject a significant fraction of the original material in the asteroid belt. Saturn had the same effect, though to a lesser extent. Additional depletion occurred because of gravitational interactions among some large planetesimals and small planets that had formed in the asteroid belt. Moreover, collisions among asteroids produced dust, which was subsequently expelled from the Solar System by the Sun's radiation pressure. Together, all of these processes left the asteroid belt with only a few per cent of its initial mass.

However, many (but not all) experts now think that an additional 90–95 per cent of the remaining mass was lost roughly 500–600 million years later, when Jupiter and Saturn achieved a "2:1 resonance," meaning that Jupiter orbits the Sun twice in the time it takes Saturn to orbit the Sun once. The gravitational effect of such a resonance was to eject asteroids, most of them inward toward the Sun, and some of the objects would have collided with the terrestrial planets and the Moon, producing an era of "late heavy bombardment." Indeed, we saw in Chapter 6 that the Moon's cratered highlands are thought to have formed about 4 billion years ago, consistent with this idea.

Another prevalent (but still unproven) idea is that much of the water on Earth came from asteroids that were gravitationally dislodged from the main belt. This process may explain how we got our oceans, something that in the past was more often attributed to comets.

It is interesting that nothing about our current model of planetary-system formation implies that the Solar System is unique. As we will see next, we are increasingly finding systems of planets around other stars. To our surprise, their properties generally aren't like those of our Solar System. Apparently, planetary systems can form with a wide variety of properties, something we didn't know when we had only our own Solar System to study.

9.2 EXTRA-SOLAR PLANETS (EXOPLANETS)

People have been looking for planets around other stars for decades. Many times in the last century, astronomers reported the discovery of a planet orbiting another star, but for a long time each of these reports had proven false. Finally, in the 1990s, the discovery of **extra-solar planets** (planets outside, "extra-," the Solar System) seemed valid. They have also become known as **exoplanets**.

Since exoplanets shine only by reflecting a small amount of light from their "parent stars" (i.e., the stars that they orbit; also called "host stars"), they are very faint and extremely difficult to see in the glare of their parent stars with current technology. So the search for planets has not concentrated on visible sightings of these planets (although a few have been made; see Section 9.2e). Rather, in most cases it has depended on watching for motions in the star that are caused by an exoplanet orbiting it, or by noticing that the apparent brightness of the star dims a bit when an exoplanet passes directly along our line of sight to it.

We now describe the main methods used to find exoplanets, and the status of the discoveries at the time of writing (2013). However, progress in the search for exoplanets is so rapid that you should keep up by looking at the websites www.exoplanet.eu and www.exoplanets.

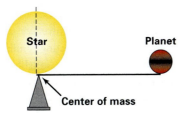

Table 9–1 Comparative planetology: Comparative planetary systems

Our Solar System: Massive planets in almost circular orbits beyond 5 au, with orbital periods of many years.

Other Planetary Systems: Massive planets often in very eccentric orbits much smaller than 1 au and with orbital periods of months, or in circular orbits with periods of days.

■ **FIGURE 9–6** The center of mass for a two-body system is defined to be the point at which the two objects balance each other, as on a seesaw. In the case of a star-planet system, the center of mass is very close to the massive star, and far from the planet. Both objects orbit their common center of mass.

org. Various iPad/iPhone apps, such as "Kepler" and "exoplanets," can also be consulted. Our discussion is roughly chronological; the first two methods below have given only a handful of planets compared with the thousands by the later methods.

9.2a Astrometric Method

The earlierst reports of exoplanets, now rejected, were based on tracking the motion of the nearest stars across the sky with respect to other stars. The precise measurement of stellar positions and motions is called **astrometry**, so this method is known as the **astrometric method**.

If a star wobbled from side to side, it would reveal that a planet was wobbling invisibly the other way, so that the star/planet system was moving together in a straight line. (Technically, the "center of mass" of the system has to move in a straight line, unless its motion is distorted by some outside force. The center of mass is illustrated in ■ Fig. 9–6, and is described more fully in Chapter 11, when we discuss binary stars. Both the star and the planet orbit their common center of mass, though the star is much closer to the center of mass than the planet is. Thus, the star moves very slightly, in a kind of "reflex motion" caused by the orbiting planet.)

Astrometric measurements have been made over the last hundred years or so, and a few of the nearby stars whose motions in the sky were followed seemed to show such wobbling. But the effects always turned out to be artifacts of the measuring process.

The parent star of one of the planets discovered from its radial-velocity variations (periodic Doppler shifts; see below) has also been observed, with the Hubble Space Telescope, to move to-and-fro-and-to in the sky. These astrometric measurements of the star GJ876 were coupled with the radial-velocity measurements to deduce the angle at which we are looking at the system. Thus, we determined the planet's mass pretty accurately: 1.9 to 2.4 times Jupiter's mass. Further work discovered two other planets around this star.

9.2b Timing of Radio Pulsars

The first exoplanet was discovered in 1992 around a pulsar, a weird kind of collapsed star (see our discussion in Chapter 13) that gives off extremely regular pulses of radio waves with a period that is a fraction of a second. The pulses came more frequently for a time

and then less frequently in a regular pattern. So the planet around this pulsar must be first causing the pulsar to move in our direction, making the pulses come more rapidly, and then causing it to move in the opposite direction, making the pulses come less often. (As described above, the planet and star are actually orbiting their common center of mass.)

Even when the existence of two more planets orbiting that pulsar was established, all with masses comparable to those of the terrestrial planets in our Solar System, the pulsar system seemed too unusual to think much about, except by specialists. It didn't help that another pulsar planet, discovered slightly earlier, turned out to be a mistaken report.

The existing system consists of two "super-Earth"-mass planets with an inner one that has twice the mass of the Moon. But these planets must have formed later than the catastrophic explosion that destroyed most of the star after its inner core collapsed; the planets could not have survived the stellar explosion. Thus, these are not the kind of planets that, like Earth, are born at the same time as the star.

9.2c Periodic Doppler Shifts: The Doppler-Wobble Method

In the 1990s, techniques were developed using the Doppler effect. Recall that in Chapter 2 we described how the Doppler effect is a shift in the wavelengths of light that have been emitted, caused by motion of the source or the receiver along the line of sight (that is, by the "radial velocity").

In the case of a star–planet system, the planet has a large orbit around the center of mass, moving rapidly. But the parent star, being much more massive than the planet, is much closer to the center of mass (Fig. 9–6) and therefore moves much more slowly in a smaller orbit. With sufficiently good spectrographs and numerous observations, this slight "wobble" can be detected as a periodically changing Doppler shift in the star's spectrum (■ Fig. 9–7); the radial velocity of the star varies in a periodic way. But actual detections of exoplanets using this **radial-velocity method**, also known as the **Doppler-wobble method**, eluded astronomers for many years.

The breakthrough came because new computer methods were developed that measure the changing Doppler shifts very precisely. On a computer, the star's spectrum can be simulated including changes in the wavelengths of light, just as happens in the Doppler effect. These simulated spectra can be compared to the observed spectrum of that star, until an excellent match is found. This method allows very small Doppler shifts to be detected, and the speeds of

Doppler shift due to stellar wobble

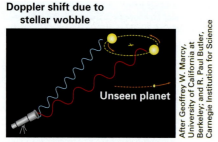

Unseen planet

After Geoffrey W. Marcy, University of California at Berkeley; and R. Paul Butler, Carnegie Institution for Science

■ **FIGURE 9–7** The technique used to detect planets around Sun-like stars by searching for Doppler shifts in the stars' spectra. The unseen planet and the star orbit their common center of mass (marked with a white cross), with the star having the much smaller (but nonzero) orbital distance because of its much larger mass. As the star comes toward us, the light it emits appears blueshifted. Half an orbit later, when it is going away from us, its light appears redshifted. The cycle then repeats itself. By observing a star over many days, weeks, months, or years, periodic Doppler shifts like these are revealed, signaling the presence of the unseen planet through its gravitational tug; see Figure 9–11 for an example.

stars toward or away from us can be measured to a precision of 1 m/s, a leisurely walking speed. (This very high precision was achieved with the Keck-I telescope and the HARPS (**H**igh **A**ccuracy **R**adial velocity **P**lanet **S**earcher) spectrometer on ESO's 3.6-m telescope; at most other sites, the precision has typically been 3 m/s or worse.)

The first surprising report came in 1995 from Swiss astronomer Michel Mayor and his student Didier Queloz. They found a planet around a nearby star, 51 Pegasi (a star in the constellation Pegasus bright enough that it was numbered centuries ago). One strange thing about the planet is that it seemed to be a giant planet, at least half as massive as Jupiter, but with an orbit far inside what would be Mercury's orbit in our Solar System. Known as a **hot Jupiter**, 51 Peg b (as it is now called, since the star itself is "A," though the "A" is usually omitted) has an orbital period of only 4.2 days, much shorter than any planet in our Solar System. (Mercury orbits our Sun in 88 days.)

Two American astronomers, Geoffrey Marcy and R. Paul Butler, then at San Francisco State University and the University of California, Berkeley, had been collecting similar data on dozens of other stars. But they had assumed, reasonably, that a planet like Jupiter around another star would take years to orbit, so they were collecting years of data while perfecting their analysis techniques to measure exceedingly small speeds. They hadn't run their spectra through the computer programs they were writing to measure the Doppler shifts. When they heard of the Mayor and Queloz results, they quickly examined their existing data and also observed 51 Peg. They soon verified the exoplanet, 51 Peg b, and discovered planets around several other stars.

A majority of the new objects turned out to be giant planets either in highly eccentric elliptical orbits or in circular orbits very close to the parent stars (■ Fig. 9–8). Most of these planets are orbiting stars within 50 light-years from us, not extremely far away but not the very closest few dozen stars either. Stars this close are bright enough for us to carry out the extremely sensitive spectroscopic measurements.

One limitation of the method is that we generally don't know the angle of the plane in which the planets are orbiting their parent stars. The Doppler-shift method works only for the part of the star's motion that is toward or away from us (that is, the "radial motion"), and not for the part that is from side to side. So the planets we discover can be more massive than our measurements suggest; we are able to find only a *minimum* value for their masses.

In the first few years of exoplanet discovery, this problem left the nagging question of whether the objects were really planets or merely low-mass companion stars. They might even be objects called "brown dwarfs," which have between about 10 and 75 times Jupiter's mass, not quite enough to make it to "star status" (see our discussion in Chapter 12, and in Section 9.2c below). Nevertheless, most

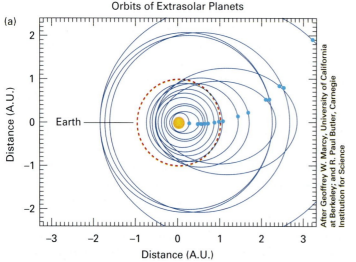

(a) Orbits of Extrasolar Planets

Distance (A.U.)

Earth

Distance (A.U.)

After Geoffrey W. Marcy, University of California at Berkeley; and R. Paul Butler, Carnegie Institution for Science

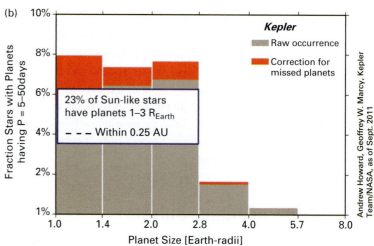

(b) *Kepler*

Fraction Stars with Planets having P = 5–50days

23% of Sun-like stars have planets 1–3 R_Earth

– – – Within 0.25 AU

Raw occurrence

Correction for missed planets

Planet Size [Earth-radii]

Andrew Howard, Geoffrey W. Marcy, Kepler Team/NASA, as of Sept. 2011

■ **FIGURE 9–8** (a) The orbits of some of the exoplanets overlaid on the same scale. We see "hot Jupiters" close in, eccentric orbits farther out, and circular orbits exceeding the size of Earth's. (b) The distribution of sizes of the known exoplanet candidates discovered by the Kepler mission as of 2013; the change in the number of discoveries over the preceding year is much greater for the smaller sizes.

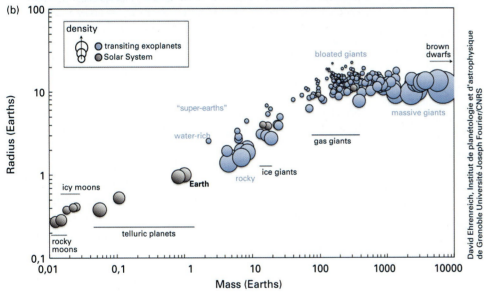

■ **FIGURE 9–9** (a) The radii of all Kepler's exoplanet candidates known by April 2013 compared with the Earth's radius are shown vs. their orbital periods in Earth days. A surprisingly large number of planets are only a few times larger than Earth, with others even smaller than Earth. Color represents the succession of discovery during the years since launch: first blue, then red, and then yellow, showing the increasing discovery of smaller exoplanets. Kepler's mission was extended through 2016 to find more planets in the habitable zone, but acquiring new data was apparently cut short in 2013 by the failure of the second of four "reaction wheels," of which three are needed to keep the spacecraft accurately pointed. An image of Jupiter from NASA's Cassini mission en route to Saturn, of Neptune from Voyager 2, and an image of Earth from NASA's Suomi NPP (National Polar-orbiting Partnership) mission, launched in 2011 to study Earth's ozone layer, appear at their corresponding masses at right. (b) A figurative description of the types of objects in planetary systems, with radius graphed as a function of mass, and density denoted by the size of the symbol (low-density objects have smaller circles than dense ones). Many known exoplanets are included. "Telluric" planets resemble the terrestrial planets in our Solar System. "Super-Earths" are objects having a mass less than about 10 Earth masses, and typical radii less than twice Earth's radius (but sometimes more, in bloated, low-density cases). "Ice giants" are somewhat more massive, and "gas giants" have even greater masses and generally low densities. But very massive giants are dense, and at still higher masses (about 13 times that of Jupiter, or roughly 4000 Earth masses) they transition to brown dwarfs.

astronomers believed these objects are planets, because there is a large gap in mass between them and low-mass stars. Very few intermediate-mass objects (10–75 Jupiter masses) had been found, yet they should have been easily detected if they existed. This gap suggested that the new objects are much more numerous than brown dwarfs.

The discovery, in 1999, of a system of three planets around the star upsilon Andromedae clinched the case that at least most of the objects are planets. Many other multiple-exoplanet systems were found thereafter. It seems most unlikely that a system would have formed with four closely spaced stars (or brown dwarfs) in it, while a system with one star and three planets is reasonable. One of upsilon Andromedae's planets even has an orbit that corresponds to Venus's in our Solar System, one not as elliptical as the orbits of the planets around other stars. This planet is in a zone that may be neither too hot for life nor too cold for it. Though such a massive planet would be gaseous, and so not have a surface for life to live on, it could have a moon with a solid surface, just as the giant planets in our Solar System have such moons.

The exciting announcements from the Doppler-wobble method continue. We now know of many systems that each contain a few (sometimes up to 5 or 6) planets. Our measurements and analysis techniques are approaching the sensitivity to find "minor" bodies like our own Earth. For example, a planet whose mass is as low as only 7 to 8 times Earth's, a "super-Earth" (a planet having fewer than 10 times Earth's mass), was found by a group of scientists that included Marcy and Butler. This exoplanet orbits a star, known as Gliese 876 or GJ876, that is only 15 light-years from Earth. Astonishingly, the planet orbits its parent star with a period of only two days, meaning that it is only 1/50 of an au out, just 10 times the star's radius. It is so close to its star's surface that its temperature is probably 200°C to 400°C, so it isn't a candidate for bearing life as we know it. Being so hot, yet having relatively low mass and thus moderately weak gravity, it could not have retained a lot of gas. It was therefore apparently the first rocky, terrestrial-type planet ever found orbiting another star. Many additional super-

■ **FIGURE 9–10** A stellar spectrum, so spread out (by the telescope's spectrograph) that it appears in many strips. The detailed measurements that led to the detection of relatively low-mass "super-Earths" were made possible by recent improvements to the spectrograph on the Keck-I telescope, and with the HARPS spectrometer on ESO's 3.6-m telescope.

■ **FIGURE 9–11** The actual velocity measurements over time of the star Gliese 876, whose rocky exoplanet has only about 7 to 8 times the Earth's mass. The measurements also reveal the masses of the two outer planets to be 2 and 0.5 times Jupiter's mass, respectively, where Jupiter's mass is about 318 times Earth's mass. The points are fit by the wobbling line; the newly discovered, low-mass exoplanet causes the small, frequent wiggles that form a band going through the plot (e.g., from the lower left to the upper right, days 1000 to 2500). At the bottom, near velocity –400 km/s, a straight line shows the tiny deviations of the points from the curve; the fit is nearly perfect.

Earths having even smaller masses have subsequently been found with the Doppler-wobble method, and especially with the transit method described below (■ Fig. 9–9).

Finding such low-mass planets with the radial-velocity method required improving the Keck telescope's system for detecting small Doppler shifts (■ Fig. 9–10). The detailed radial-velocity observations of Gliese 876 (■ Fig. 9–11), together with astrometric measurements, allowed a computer model to account for the angle at which we are viewing the system, which turns out to be inclined to our line of sight by 40°. That measurement allowed the mass of the orbiting planets themselves to be determined, and not merely lower limits as before. The measurements revealed that the slight discrepancy in the orbits of the two already-measured planets could be resolved by the presence of the third body, the newer exoplanet.

9.2d Transiting Planets: The Blink Method

Since 1999, astronomers have not had to resort only to periodically changing Doppler shifts to detect exoplanets. One planet was discovered to have its orbit aligned so that the planet went in front of the star each time around – that is, it underwent a transit, causing a slight temporary drop in the star's apparent brightness. Since that first discovery using the **transit method** or **blink method**, so many additional examples of transiting planets have been found that the field of exoplanets has become perhaps the most exciting and vital in all of astronomy.

The dip in the star's brightness of a few per cent can be measured not only by professional but also by amateur astronomers (■ Fig. 9–12a,b). Such transits have analogues in our Solar System: the transits of Venus that happen in 8-year pairs alternately with 105.5 or 121.5 years between pairs (since 2012, not again until 2117 and 2125), and the transits of Mercury that happen 15 or so times

per century (next in 2016). Venus blocks 0.1 per cent of the Sun's light when it transits (Fig. 9–12c), which we can measure; Mercury takes up $1/30$ of that, which is currently imperceptible.

Since a transiting planet's orbital plane is along our line of sight, its mass can be accurately determined; the inclination of the orbital plane is known to be about 90°. For the first such planet to be discovered, it turned out to be 63 per cent of Jupiter's mass. This result confirmed our conclusion that at least some (and probably most) of the "exoplanets" really are planets rather than more massive brown dwarfs (see Sec. 9.5).

The transit method has also revealed, through spectroscopy, sodium and other gases in the atmosphere of a few exoplanets. The atmosphere produced some absorption lines in the starlight passing through it, and these were detected in very-high-quality spectra. In the future, it is possible that astronomers will detect evidence for life on other planets by analyzing the composition of their atmosphere using the transit method.

The French started taking data from the CoRoT spacecraft (**Co**nvection **Ro**tation et **T**ransits Planétaires; English: **Co**nvection **Ro**tation and Planetary **T**ransits) in 2007. It soon discovered fascinating planets, so different from the ones we are used to in our Solar System. For example, it found many "hot Jupiters" like those found by the Doppler-wobble method – planets that are so big that they are nearly as big as (or bigger than) Jupiter, yet so close to their parent stars that their surfaces must be very hot from the stars' radiation.

At the other extreme, CoRoT 7b has a radius only 1.6 times that of Earth and only 4.8 times Earth's mass, giving it a rocky density higher than Earth's. But it is so close to its parent star, only 2.8 stellar radii out, that it is too hot for life to have formed on the hot side

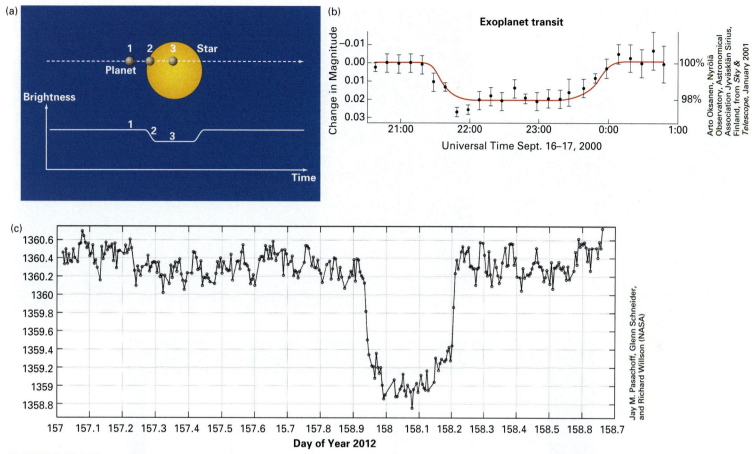

■ **FIGURE 9–12** (a) Schematic showing the dips that would result from an exoplanet. (b) Observations made by amateur astronomers detecting the 2-hour-long passage of an exoplanet across its parent planet by a 2% dip in the measured brightness. (c) The light curve of the transit of Venus on June 5/6, 2012, measured with NASA's ACRIMsat spacecraft in Earth's orbit, showing the drop in the total solar irradiance in units of watts per square meter at a distance of 1 au from the Sun.

(it is in synchronous rotation with its star) and too cold on its cool side. Could the temperate zone in the narrow strip at the terminator have allowed the origin of life?

Cheops (Characterizing Exoplanet Satellite) is Europe's next exoplanet mission, a "small" mission for launch in 2017. Cheops will monitor known exoplanets found with radial velocities in order to search for their transits. Cheops will thus yield planets transiting *bright* stars.

NASA had received a proposal for a small spacecraft to monitor the brightnesses of about 150,000 stars all the time to look for transits, but at first the reviewers had been skeptical that it would work. Still, in 2009, NASA's **Kepler spacecraft** (Fig. 9–13a) was launched, obviously named after the astronomer who had figured out the laws of planetary orbits. Kepler's 0.95-m telescope stares at one region of the sky, in the constellations Cygnus, Lyra, and Draco, about the size of your fist held at the end of your arm against the sky (Fig. 9–13b). The data are analyzed on Earth with the aid of computers, and "planet candidates" are noted. They are at first called only "candidates," or Kepler Objects of Interest (KOIs), because scientists want to rule out other possible causes of changes in brightness, such as the equivalent of sunspots rotating or accidental background objects. Moreover, the data are imperfect in part because the stars themselves have inherent

surface turmoil, which led to an extension of the mission lifetime, up to its 2013 reaction-wheel failure.

In 2010, the Kepler team announced the discovery of several exoplanets, based on truly exquisite transit data and confirmed with radial-velocity variations. Discovery of thousands of exoplanet candidates was announced each year, and the total has grown to more than 2,700 candidates (see the image opening this chapter, and also Figure 9–9a again). It is thought that at least 90 per cent of these candidates are genuine exoplanets (Fig. 9–13c). Many are in multi-planet systems (Fig. 9–13d). Remarkably, more than 200 of them are similar in size to Earth; it is estimated that more than 50 per cent of all stars have candidate Earth-sized exoplanets within 1 au, greater than was previously thought. About 700 of the known candidates are super-Earths having radii roughly twice that of Earth, nearly 1200 are comparable to the size of Neptune, 200 are the size of Jupiter, and more than 50 exceed Jupiter's size. About 20 per cent are in planetary systems containing more than a single planet.

Progressively smaller exoplanets are being found in the Kepler spacecraft's data (Figures 9–9a and 9–13). In 2013, the discovery of several small exoplanets was announced, one (Kepler-37b) smaller than Mercury, with the size of the parent star determined to a remarkable 3% accuracy by asteroseismology. Large planets still dominate the

(a)

Kepler team/NASA, courtesy of Geoffrey W. Marcy (UC Berkeley)

(b)

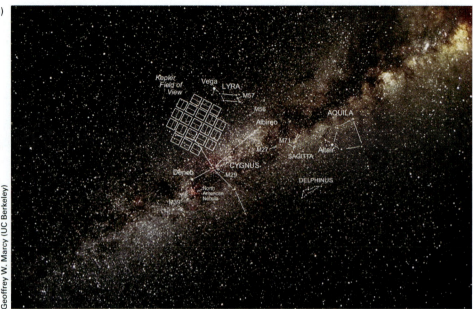

Carter Roberts/Kepler team/NASA, courtesy of Geoffrey W. Marcy (UC Berkeley)

(c)

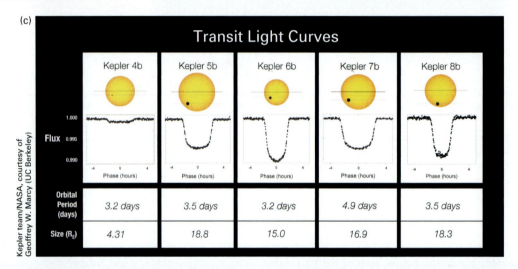

Transit Light Curves

	Kepler 4b	Kepler 5b	Kepler 6b	Kepler 7b	Kepler 8b
Orbital Period (days)	3.2 days	3.5 days	3.2 days	4.9 days	3.5 days
Size (R_E)	4.31	18.8	15.0	16.9	18.3

Kepler team/NASA, courtesy of Geoffrey W. Marcy (UC Berkeley)

(d)

Kepler's Transiting Planet Systems

NASA Ames/Jason Steffen, Fermilab Center for Particle Astrophysics

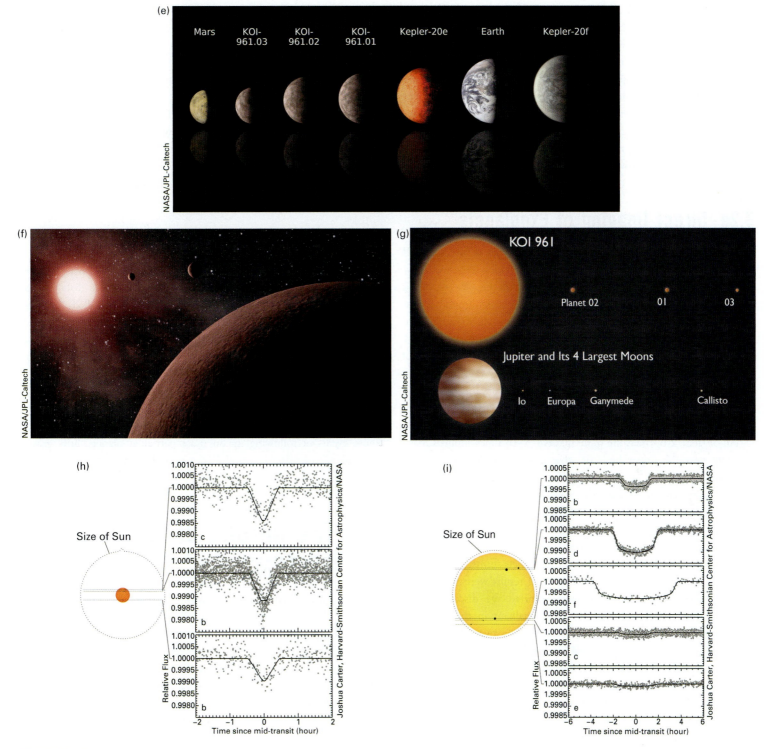

■ **FIGURE 9–13** (a) The Kepler satellite, before its 2009 launch. (b) The region of the sky that Kepler monitors, showing the grid of CCDs that cover many square degrees. The brightness of more than 150,000 stars in this region is monitored continuously. (c) Light curves (brightness vs. time) of several stars measured by the Kepler satellite; a dip in the brightness is apparent as the planet passes (transits) across the star, with one detected planet per star in these systems. These data also provide information about how far from the star's center the planet transits. The very short orbital periods indicate that these exoplanets are close to their parent stars. (d) Kepler's transiting planetary systems each of which contains multiple planets, with the 8 planets of our Solar System shown in a column in blue at left, in order of distance from our Sun. (e) A comparison of the sizes of three exoplanets, one as small as Mars, around star KOI-961 (KOI stands for Kepler Object of Interest), discovered by John Johnson of Caltech and colleagues by carefully examining Kepler data. KOI-961 is a 16th-magnitude red dwarf, spectral type M4 V (see Chapter 11 for definitions of these terms), in the constellation Cygnus, and is about 130 light-years from Earth. The exoplanets' periods are less than 2 days, meaning that they are separated from their parent star by less than $1/100$ au. Also shown are two Earth/Venus-sized exoplanets, which get Kepler numbers (Kepler-20e and Kepler-20f) instead of KOI numbers because they were discovered by the main Kepler team. Exoplanets in Kepler-37 are even smaller, with Kepler-37b only slightly larger than our Moon and 37c also smaller than Earth. (f) An artist's conception of the KOI-961 system. (g) The KOI-961 system of a star with its exoplanets compared with our Solar System's planet Jupiter with some of its moons. (h) Light curves for the KOI-961 system. (i) Exoplanet light curves for the Kepler-20 system.

total inventory of known exoplanets, but the detection techniques for terrestrial-like exoplanets have rapidly been improving.

The range of discoveries that can be made with the transits shown in Kepler data is astounding. We can find out for some of the planets whether they have atmospheres, and what the constituent gases may be. The orientation of the axis of the parent star's rotation can be found in some cases. Even planets that don't transit their stars have been discovered by their gravitational effects on the transiting planets that Kepler observed, by sometimes delaying the start and the end of a transit compared with what was expected.

9.2e Direct Imaging of Exoplanets

In late 2008, astronomers released infrared images of the star HR 8799, about 130 light-years away in the constellation Pegasus, showing the presence of three planets orbiting it (■ Fig. 9–14a). This was the first time a multi-planet system of exoplanets around a relatively normal star had been directly imaged. The data were obtained at infrared wavelengths with the adaptive optics systems of the Keck and Gemini telescopes on Mauna Kea, but new careful analysis of Hubble Space Telescope images obtained in 1998 also showed an exoplanet and confirmed that it is orbiting the star. In 2009–2010, additional observations of HR 8799 revealed a fourth planet, even closer to the parent star than the other three.

All of these exoplanets are quite massive, 5–10 times Jupiter's mass, probably near the upper limit of what can be considered a planet. Moreover, they orbit at 2–3 times the distance of Jupiter, Saturn, Uranus, and Neptune in our own Solar System, so they have orbital periods of about 45 to 460 years. Only about 30 million years old, they still glow brightly at infrared wavelengths, releasing their residual energy built up through gravitational contraction. The star is still surrounded by the dusty disk of debris from which its planets formed.

The same day that the original HR 8799 results were announced, an independent team of astronomers reported the discovery of a possible planet orbiting the bright star Fomalhaut in the constellation Pisces Australis, on direct images obtained in two different years with the Hubble Space Telescope (■ Fig. 9–14b), and images at more epochs have been obtained since. The debris disk (a ring of dust) is visible as well; such disks are frequently being found around young stars, as we discuss in the next section.

The putative planet appears only in red images obtained with Hubble, but not in images obtained at infrared wavelengths with large ground-based telescopes where all evolutionary models predict a young cooling planet should be brightest. Therefore, it has been suggested that the Hubble image is not of radiation from a planet itself but of multitudes of tiny dust particles, either in orbit around an unseen planet or produced as the aftermath of a collision of two large Kuiper-belt-like objects. The jury is still out.

Work continues on the development of telescopes and instruments designed to find and study exoplanets through direct imaging. For example, the Gemini Planet Imager will be a very high-resolution instrument to be used with the Gemini 8-m telescope on Mauna Kea. Installation of interferometric equipment at the four 8-m units of the Very Large Telescope in Chile has been completed, and it may lead to the direct detection of some young, giant planets that are luminous at infrared wavelengths, as in the case of HR 8799. The European Space Agency's Gaia mission is to measure positions for a billion stars, and while doing so it may discover 10,000 planets! It should be launched in 2013. NASA's Transiting Exoplanet Survey Satellite (TESS), to be launched in 2017, will carry out an all-sky survey to discover transiting exoplanets around the nearest and brightest stars in our sky. While Kepler's exoplanets are too faint to follow up with ground-based telescopes, TESS's should be accessible to the Giant Magellan Telescope and other huge new telescopes.

9.2f Gravitational Microlensing

Thousands of stars are being monitored to see if their gravity temporarily focuses and brightens the light from other, more distant stars behind them equally at all wavelengths, a process called "gravitational lensing" that we will discuss more fully in Chapters 16 and 17. Occasionally, not only does a background star brighten because of lensing by a foreground star for a few months to a year, but also, during that time, an exoplanet orbiting the foreground star briefly (e.g., for a day or two) causes an increase in brightness. Worldwide groups are searching for such "microlensing" events, and many have already been found. Unfortunately, we then have no way of repeating those events or following up on those particular planets. Still, gravitational microlensing is a useful way to find exoplanets.

9.3 THE NATURE OF EXOPLANET SYSTEMS

We have discovered enough exoplanets to be able to study their statistics. About 1 per cent of nearby solar-type stars have jovian planets in circular orbits that take between 3 and 5 Earth days. These "hot Jupiters" are so close to their parent stars (within $1/10$ au) that temperatures are very high. Another 7 per cent of these nearby stars have jovian planets whose orbits are very eccentric. As we observe for longer and longer periods of time, we have better chances of discovering planets with lower masses or with larger orbits.

The discovery of so many planetary systems instead of just our own will obviously change the models for how planetary systems are formed. Since giant planets couldn't form in the torrid conditions close in to the parent star, theorists work from the idea that the exoplanets formed far out, as Jupiter did, and migrated inward, bringing them closer to their parent stars. Maybe their orbits shrink as the planets encounter debris in the dusty disk from which they formed; when they fling planetesimals outward, for example, the planets themselves move inward. The idea that the planetary orbits start larger but migrate in was quantified at an observatory in Nice (pronounced "neece"), France, so it is a main point of what is known as the Nice model.

FIGURE 9–14 (a) An infrared image of HR 8799 obtained with the Keck and Gemini telescopes using the technique of adaptive optics. Four exoplanets are visible, and their orbital motion is shown with arrows. (b) An optical Hubble Space Telescope image of Fomalhaut (Pisces A), only 25 light-years from us, revealing a dust ring and a small speck that appears to be an exoplanet, Fomalhaut b. The jury is still out, though, as to whether the object is an exoplanet surrounded by a large dusty ring or cloud, a giant exocomet, or a collisionally disrupted planetesimal. The enlarged box shows Fomalhaut b's motion at four epochs over eight years. The orbit of Fomalhaut b is highly eccentric, suggesting that it was gravitationally scattered by another, yet-to-be-detected planet lurking in the system. In the near future, Fomalhaut b will appear to cross through the belt. It is 10 times farther out than Saturn is in our Solar System, with an orbital period over 2,000 years, and comes about as close to its star as Pluto does to ours, with its elliptical orbit taking it up to 8 times farther out. Structure in the ring probably results from the gravity of exoplanets, seen and unseen. Fomalhaut itself and a region around it are hidden by an occulting disk, which is surrounded from some starlight "noise." A background star appears at bottom center. (c) The dust ring (red and yellow false color) around Fomalhaut, about 140 au in radius and 16 au across, in the first science image from ALMA, the Atacama Large Millimeter/submillimeter Array. The false-color blue image is the earlier Hubble Space Telescope image. Since we can see that both inner and outer edges of the disk are sharp, the astronomers' conclusions are that the disk's dust particles, about 1 mm in size, are kept in place by shepherding satellites (the inner one corresponds with Hubble's Fomalhaut b), as with Saturn's and Uranus's rings. The particles would have resulted from collisions of asteroids or comets. The planets, calculations show, would be larger than Mars but no larger than a few times larger than Earth; larger planets would have destroyed the ring. (d) A Herschel Space Observatory far-infrared image of Fomalhaut and its dust ring. Both the narrowness and asymmetry that the belt shows are thought to result from the possible exoplanet. The discrepancy between the size and temperature of the dust particles suggested by the Hubble observations and the size and temperature suggested by the Herschel observations has led the Herschel scientists to the idea that the dust belt is continuously replenished by collisions among comets orbiting the star.

Perhaps the highly eccentric orbits didn't start out that way, but were produced by gravitational interactions that completely ejected some planets from the system. The gravitational interactions between two planets can lead to the ejection of one planet, leaving the other in a very eccentric orbit. Another idea is that interactions between a planet and the protoplanetary disk can cause high eccentricities. Once a planet has part of its orbit close to its parent star, tidal forces can circularize the orbit. Other planets can spiral all the way into the star and be destroyed.

The most accepted explanation for the hot Jupiters, which orbit so close to their stars, is that they were formed farther out and migrated in. But the discovery of a hot Jupiter in a triple star system complicated matters, since the two farther-out stars would have disrupted any protoplanetary disk. Perhaps the triple system had a very different type of protoplanetary disk than the one with which we are familiar from our own system, or the planet was captured.

Since Tatooine was Luke Skywalker's planet in the *Star Wars* movies, the discovery of an exoplanet orbiting a double-star system led to its being nicknamed Tatooine.

9.4 GOLDILOCKS PLANETS

Most people don't seem to be satisfied with such exotic things as exoplanets. Instead, they want astronomers to find exoplanets on which there are signs of life. We aren't yet up to finding life there, but we can at least find planets that are in the "habitable zone," a region around a star that is neither too hot nor too cold for water to be liquid, since water is thought to be a vital ingredient for life as we know it.

Since the region around the star is not too cold and not too hot, it is "just right" for life to exist on a planet, just as Goldilocks found her porridge not too hot and not too cold, but just right. These planets in the habitable zone are thus known as **Goldilocks planets**.

There are controversies as to which kind of star would be best. Much of the work has concentrated on "sunlike stars" – stars about the size, mass, and surface temperature of our Sun. But there are many more "red dwarf stars" (spectral types K and M; see Chapter 11) than there are "yellow dwarf stars" (spectral type G) like our Sun. Since the red dwarfs are less powerful than our Sun, their habitable zone would be closer in, and exoplanets within them would have shorter orbital periods that are easier for the Kepler spacecraft to detect using the blink method. A few exoplanets in or near the habitable zone have already been found, and some are even small enough to have rocky surfaces.

One thing is guaranteed: over the next years, we will be finding more and more fascinating exoplanets, with many different reasons for getting excited about them and about the chance of their harboring life.

9.5 BROWN DWARFS

As mentioned in Section 9.2c, some of the objects found with the Doppler shift technique might actually be too massive to be true "planets" (more than 13 Jupiter masses). If so, they are **brown dwarfs**, which are in some ways "failed stars."

Each of the brown dwarfs has less than 75 Jupiter masses (or 7.5 per cent the mass of the Sun), which is not enough for them to become normal stars, shining through sustained nuclear fusion of ordinary hydrogen, as we shall discuss in Chapter 12. Their central temperatures and pressures are just not high enough for that. (However, they do fuse a heavy form of hydrogen known as "deuterium," so they are not complete failures as stars. We discuss the origin of deuterium, all of which was formed in the first few minutes after the big bang, in Chapter 19.) Brown dwarfs can be thought of as the previously "missing links" between normal stars and planets.

Though no detection of a brown dwarf was accepted until 1995, hundreds have now been observed. Many of these are in the Orion Nebula, while others are alone in space. The current best model for brown dwarfs is that they are formed similarly to the way that normal stars are: in contracting clouds of gas and dust. A disk forms, perhaps even with planets in it, though the material in it contracts onto the not-quite-star.

This idea is backed up by observations with the European Southern Observatory's Very Large Telescope, which has detected an excess of near-infrared radiation from many brown dwarfs. The scientists involved interpret their observations as showing that the radiation is from dusty disks. Moreover, they conclude that since both regular stars and brown dwarfs have such disks, they must also form in similar ways.

Most exciting, a planetary-mass object next to a 25-Jupiter-mass brown dwarf (presumably orbiting it) was imaged (■ Fig. 9–15), the first planetary-mass object to be imaged around any self-luminous body other than our Sun. The almost-star with its possible-planet (since it may have formed by fragmentation rather than by building up over a core, so it may be very different from planets in our Solar System) is 200 light-years away from us; observations from the ground and from space have shown that the two objects are moving through space together, confirming that they are in one system together. The almost-planet is 55 au away from its parent almost-star.

9.6 PLANETARY SYSTEMS IN FORMATION

We are increasingly finding signs that planetary systems are forming around other stars. One of the first signs was the discovery of an apparent disk of material around a southern star in the constellation Pictor (■ Fig. 9–16). The best observations of it were made with the Hubble Space Telescope, and may show signs of orbiting planets. An even nearer planetary disk has been found around the star AU Microscopii, enabling observations with higher resolution (■ Fig. 9–17).

Other images of regions in space known as "stellar nurseries" show objects that appear to be protoplanetary disks (■ Fig. 9–18). Observations have shown that these objects contain about as much mass as a planetary system, clinching the idea that they are locations where planets are forming. Locations where there are planets in formation glow in the infrared, as in the case of HR 4796A

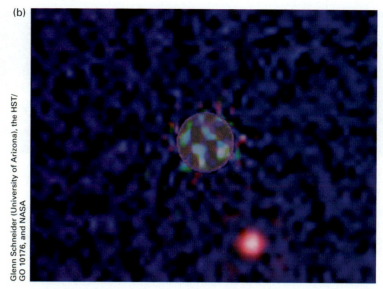

■ FIGURE 9–15 The first images ever of an exoplanet. This giant object orbits a brown dwarf at a distance of 55 au. (a) An image taken with the Very Large Telescope in Chile. (b) A Hubble Space Telescope view, with the central circle showing where the star's relatively bright image is hidden.

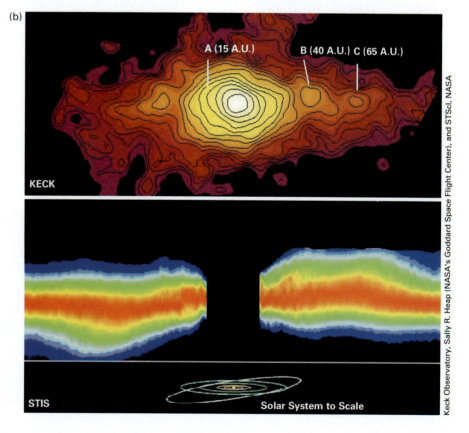

■ FIGURE 9–16 (a) This picture of what may be a planetary system in formation was taken with a ground-based telescope. A circular "occulting disk" blocked out the star β (beta) Pictoris; shielding its brightness and observing in the infrared revealed the material that surrounds it. (b) The region around the southern star β Pictoris. (top) A view of the dust from the Keck-II telescope at the medium infrared wavelength of 18 micrometers. The inner contours are misaligned with respect to the outer disk; this warp is labelled *A*. Labels *B* and *C* may show the position of tilted dust rings. (middle) A Hubble view with its Space Telescope Imaging Spectrometer. We apparently see a side view of a disk of dust, thought to be a planetary system in formation. (bottom) Our Solar System to the same scale as the β Pictoris images.

The beta Pic system is known to have one planet (at ~ 8 au) that has been imaged in thermal emission at near-infrared wavelengths, but is too faint to see in visible light.

Glenn Schneider (University of Arizona) and the HST/GO 12228 Team

■ **FIGURE 9–17** The star AU Microscopii is closer to Earth than β Pictoris, so even more detail can be observed in the disk that has been discovered around it. This Hubble image shows the disk's dust particles reflecting the starlight. The image is 100 au wide; the black mask hides the inner 5 au. Features as small as 0.5 au are resolved in the image, revealing clumps of dust in the inner disk. These clumps are presumably formed by the gravitational influence of planets, though we cannot see the planets themselves. The clumps are 25 to 40 au from the central star, corresponding for our own Sun to the orbits of Neptune and Pluto. AU Microscopii itself, only 33 light-years away from us, is a dim red star, with half the Sun's mass and $1/10$ its energy output. It is 12 million years old, much younger than the Sun's 4.6 billion years, and is apparently in its phase of planetary formation.

AU Mic is the same age as beta Pic; they are both members of the eponymous "beta Pic moving group." Both presumably initially formed from the same interstellar-medium cloud. AU Mic, though, is the least massive (and latest spectral type) star for which we have imaged a scattered-light debris disk.

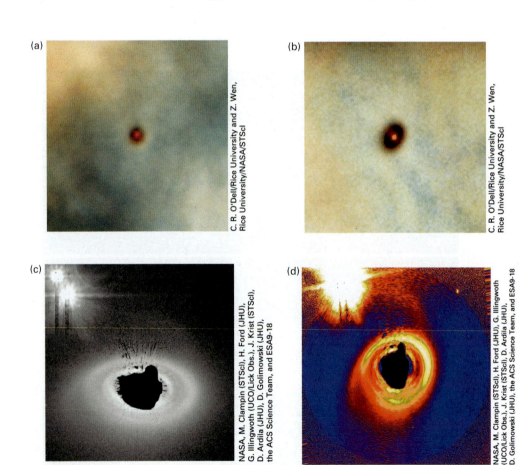

(a)

(b)

C. R. O'Dell/Rice University and Z. Wen, Rice University/NASA/STScI

(c)

NASA, M. Clampin (STScI), H. Ford (JHU), G. Illingworth (UCO/Lick Obs.), J. Krist (STScI), D. Ardila (JHU), D. Golimowski (JHU), the ACS Science Team, and ESA9-18

(d)

NASA, M. Clampin (STScI), H. Ford (JHU), G. Illingworth (UCO/Lick Obs.), J. Krist (STScI), D. Ardila (JHU), D. Golimowski (JHU), the ACS Science Team, and ESA9-18

■ **FIGURE 9–18** (top row, a and b) The Hubble Space Telescope has revealed protoplanetary disks around several stars, such as these in the Orion Nebula in the constellation Orion. (bottom row, c and d) Detail in the circumstellar dust disk around a young star, seen in black and white and in false color. The star, only 5 million years old, is 320 light-years from us and is in the constellation Libra. It has long been known to be a candidate for having a disk because of its excess infrared emission. The Hubble Space Telescope's Advanced Camera for Surveys, using its coronagraph that has a dark spot to block out the bright central region, shows that the disk is a tightly wound spiral. One of the spiral arms seems to be associated with a nearby double star (upper left edge of frame), so interaction with the double star may be causing the structure. It had previously been known that the inner 3 billion km in radius around the star, about 20 au (roughly the orbit of Uranus, for our Solar System), is relatively clear of dust. Perhaps that region was swept clear by a planet or planets.

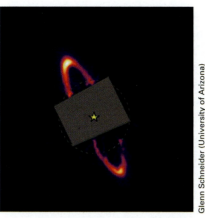

Glenn Schneider (University of Arizona) and NASA

■ FIGURE 9–19 An almost edge-on dust ring around the star HR 4796A. It is thought that planets are forming, or will soon be forming, in this disk. But the star is 4 times farther away than Fomalhaut (Fig. 9–14b), so any planet would be too faint to detect with current equipment.

(■ Fig. 9–19), because of the wavelength of the peak of the blackbody curves for the temperature of warm dust (see our discussion of blackbodies in Chapter 2); thus, the Spitzer Space Telescope, the Herschel Space Observatory, and, eventually (we hope), the James Webb Space Telescope, should give many insights into planetary formation.

As noted earlier, the Hubble Space Telescope has imaged such a ring of dust around the nearby star Fomalhaut (see Fig. 9–14b again),

even possibly detecting an exoplanet and showing its motion over time. From what we think we know about the dynamics of disks, we might infer the possible existence of a planet orbiting interior to the ring shown for HR 4796A in Figure 9–19, though it remains unseen in this more distant, difficult target.

A Wesleyan University team has found a young Sun-like star that winks on and off, apparently being eclipsed by dust grains, rocks, or asteroids that are orbiting it in a disk. The star is near the Cone Nebula, a prolific nursery of young stars. On a changing schedule, it fades drastically over 2 or 3 days to only a few per cent of its maximum brightness, stays dim for another two or three weeks, and then brightens for another two or three days out of every 48.3 days. No single object could provide such a long eclipse, and the scientists have concluded that we are seeing the 48.3-day orbital period of a pair of stars, each about half a solar mass. The actual eclipse is from a circumbinary disk that is precessing (visible as a changing orientation of its axis) on a timescale of about 1000 years. The system, only 3 million years old, is changing from month to month. When the system was first discovered, one star periodically peeked above the edge of the disk, corresponding to coming out of eclipse. By 2010, the disk's precession led to both stars being blocked all the time; the previously unseen star reemerged in 2012. We see the stars only by their reflections off the disk. The system may be a precursor to the several systems known in which an exoplanet circles a double star, with Kepler 16b as an example, as in Tatooine from the *Star Wars* movies.

CONCEPT REVIEW

» The conservation of angular momentum explains why planetary systems in formation contract into a disk (Sec. 9.1a). We think that small clumps of dust joined to make **planetesimals**, and planetesimals combined to make **protoplanets** orbiting the **protosun**. In one of the main models for the formation of the Solar System's outer planets (Sec. 9.1b), silicates (rocks) and clumps of various ices that had formed beyond the "frost line" collided and grew. These planetesimals subsequently came together to form a solid core resembling a terrestrial planet, but more massive, and its gravity then attracted large amounts of hydrogen and helium. Terrestrial planets, on the other hand, formed much later from a debris disk that consisted of rocks but not ice grains, since temperatures were too high inside the frost line. Also, by that time, the powerful young Sun had already blown away the gas in the disk.

» The search for planets around stars other than the Sun, known as **extra-solar planets** or **exoplanets**, has been going on for many decades (Sec. 9.2). With current technology, they are extremely difficult to see in the glare of their "parent star" (or "host star," the star they orbit), though a few have been directly imaged. Instead,

we watch for motions in the star that are caused by something orbiting it, or we look for planetary transits.

» The **astrometric method** depends on **astrometry**, the precise measurement of stellar positions and motions; the presence of an unseen planet is deduced from the star's observed wobble in the sky (Sec. 9.2a). However, so far no exoplanets have been discovered this way. Some planets were finally found in 1992 by timing the radio pulses from a "pulsar," a weird kind of collapsed star to be discussed in Chapter 13, but these are very unusual planets that have little to do with normal planets orbiting Sun-like stars (Sec. 9.2b).

» Most exoplanets discovered with ground-based telescopes since 1995 were found with the **radial-velocity method**, also known as the **Doppler-wobble method** (Sec. 9.2c), in which periodic changes in the Doppler shift (radial velocity) of a star (the subtle "reflex motion" caused by the orbiting exoplanet) are measured with high-quality spectrographs. But one limitation of this method is that we generally don't know the angle of the plane in which the planets are orbiting their parent stars; thus, only the *minimum* mass of each planet can be derived. This problem left the nagging

» question of whether the objects were really planets or merely low-mass companion stars.

» The discovery of a three-planet system helped dispel these concerns, as did the large observed gap in mass between the purported exoplanets and the lowest-mass stars. Now many such planetary systems are known. Moreover, we have found many "super-Earths" in which the planet appears to be rocky (rather than a gas giant) and has a mass fewer than 10 Earth masses (down to just a few Earth masses).

» Another very successful technique for finding exoplanet candidates (called the **transit method** or **blink method**) is by looking for the very slight drop in a star's light as an exoplanet **transits** across the face of its parent star (Sec. 9.2d). These candidates must then be verified with the Doppler-wobble technique, and most of them end up being genuine exoplanets. For these systems, we know that the inclination of the orbital plane is close to 90 degrees, so the mass of the exoplanet can be accurately determined. With the launch of the CoRoT spacecraft and especially the Kepler mission, plus some observations from ground-based telescopes, thousands of exoplanet candidates have now been observed. In a few cases, the transit method has revealed, through spectroscopy, sodium and other gases in the exoplanet's atmosphere.

» With sophisticated equipment, we are now beginning to obtain direct images of exoplanets, especially young systems viewed at infrared wavelengths (Sec. 9.2e). When the images of a given system are separated by a few years, the orbital motion of exoplanets can even be detected. Another method for finding exoplanets involves gravitational microlensing (Sec. 9.2f), though it is extremely difficult to conduct additional follow-up studies of them.

» The first exoplanets discovered were giant planets, though less-massive, terrestrial planets (with rocky surfaces) are being increasingly found in those planetary systems. Exoplanets are often in quite eccentric (elongated) orbits, perhaps as a result of previous gravitational encounters with other planets or the protoplanetary disk (Sec. 9.3). Some of the planets are in circular orbits very close to their parent stars, which probably means (according to the Nice model) that they were formed elsewhere in the systems and later drifted inward.

» Astronomers are now intensively looking for **Goldilocks planets** – that is, exoplanets in the habitable zone, in which water can exist in the liquid state, being neither too hot nor too cold (Sec. 9.4). Exoplanets in or near the habitable zone have already been found, and a few are even small enough to have rocky surfaces.

» Some of the objects found with the Doppler-shift technique might actually be **brown dwarfs**, which are sometimes called "failed stars" because they don't fuse ordinary hydrogen into helium (Sec. 9.5). They can be thought of as links between planets and normal stars. Planetary companions to some brown dwarfs have been imaged at infrared wavelengths.

» We are increasingly finding signs that planetary systems are forming around other stars (Sec. 9.6). A probable disk around the star β (beta) Pictoris and apparent protoplanetary disks observed with the Hubble Space Telescope seem to be direct observations of planetary systems in formation.

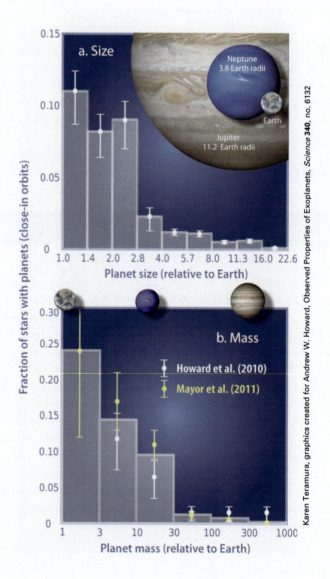

Karen Teramura, graphics created for Andrew W. Howard, Observed Properties of Exoplanets, *Science* **340**, no. 6132

QUESTIONS

1. What is the difference between a protoplanet and a planet?

2. What role do planetesimals play in the origin of the planets?

3. Discuss similarities and differences in the way terrestrial planets and giant planets formed in our Solar System.

4. How does the pulsar-planet system differ from the exoplanet systems discovered subsequently?

5. Explain the method by which most planets have been discovered around other stars.

6. Does the "Doppler-wobble method" for deducing the existence of a planet orbiting a star depend on the star's distance from Earth? (Assume the star's apparent brightness is independent of distance.)

†7. It turns out that the ratio of the speed of a planet to the speed of its parent star (as they orbit their common center of mass) is equal to the inverse of the ratios of their masses. If Earth orbits the Sun with a speed of 30 km/s, what is the Sun's corresponding orbital speed (i.e., its "reflex motion," induced by the Earth's motion)?

8. Why may it be that at first only giant planets and not terrestrial planets were discovered around other stars?

9. Why can we measure only the minimum of the mass of an exoplanet using the Doppler-wobble method?

10. Many of the giant exoplanets are quite close to their parent stars. If they initially formed farther out, how could they have ended up at their observed positions?

11. What was the special importance of the discovery of a system of several (rather than just one) planets orbiting a star?

12. What are two reasons that the discovery of a planet transiting a Sun-like star is desirable?

†13. If a transiting planet has 10 per cent the diameter of its parent star, the observed brightness of the star will dim by what percentage during the transit?

14. Give two reasons why the discoveries of "super-Earth" exoplanets instead of much more massive planets are important.

15. How do brown dwarfs differ from planets and from stars?

16. Why is β (beta) Pictoris an interesting object?

†17. If the temperature of a dust ring around a star is 30 K, at what wavelength does its blackbody spectrum peak? In what spectral region is this wavelength?

18. Discuss how the Kuiper belt in our Solar System may be related to some of the observed phenomena found around other stars.

19. Why do infrared observations help us learn about planetary systems in formation?

20. Describe the basic idea of gravitational lensing and its application to exoplanets.

21. **True or false?** Many of the known exoplanets are peculiar compared with planets in our own Solar System; these exoplanets have very eccentric orbits, or they orbit very close to the star.

22. **True or false?** Exoplanets can sometimes be detected by monitoring the brightness of a star, and seeing it decrease as the planet blocks part of the star's light.

23. **True or false?** When viewing an exoplanet orbit edge-on (i.e., your line of sight is along the orbital plane), the star's measured radial-velocity curve can be used to determine the star's actual orbital speed.

24. **True or false?** The small number of low-mass, Earth-like exoplanets found thus far around normal stars is surprising; with current technology, they should have dominated our discoveries, if they are indeed plentiful.

25. **True or false?** Many exoplanets have already been found with the astrometric method, in which a star's measured position in the sky moves back and forth very slightly.

26. **True or false?** The periodic Doppler-shift method is biased toward finding planets whose orbital plane is along the line of sight, rather than closer to the plane of the sky.

27. **Multiple choice:** Using ground-based telescopes, most massive exoplanets orbiting other stars have been found by **(a)** looking at a massive planet's Doppler-shifted light as it orbits the star; **(b)** taking spectra of the star to look for contamination from the light of a massive planet; **(c)** looking for a periodic Doppler shift in the star's spectrum as a massive planet causes the star to wobble slightly; **(d)** pointing the telescope to a spot about 5 to 10 au away from the star, knowing that massive planets can only orbit near that distance; or **(e)** measuring the Doppler shift of the star, and seeing whether it keeps on increasing with time, as would be expected if a massive planet were pulling on it.

28. **Multiple choice:** The giant (jovian) planets are large compared with the terrestrial planets because the giant planets **(a)** have higher densities; **(b)** formed by the same process as the Sun, and can be considered miniature stars; **(c)** have many moons; **(d)** obtained and retained more gas and ice because of their large distance from the Sun; or **(e)** did not suffer as many major, destructive collisions as the terrestrial planets (like Earth, when the Moon formed).

†29. **Multiple choice:** Suppose exoplanets Vladimir and Alexandra both orbit the same star. The orbital period of exoplanet Vladimir is 8 times that of exoplanet Alexandra. Compared with exoplanet Alexandra, exoplanet Vladimir is _____ from its star. **(a)** $1/4$ as far; **(b)** $1/2$ as far; **(c)** 2 times as far; **(d)** 4 times as far; or **(e)** 8 times as far.

30. **Multiple choice:** Jupiter's chemical composition is most similar to that of **(a)** Earth; **(b)** the Sun; **(c)** Pluto; **(d)** Venus; or **(e)** meteorites.

31. **Multiple choice:** Since 1995, nearly a thousand exoplanets around normal main-sequence stars have been detected with ground-based optical telescopes. Which one of the following statements about these planets is *true*? **(a)** All of them are less massive than Jupiter. **(b)** In some cases, the planet's mass is as small as Mercury's mass. **(c)** The measured masses of the planets are generally only upper limits; that is, their true masses might be smaller, but not larger. **(d)** In some cases, they are so close to their parent stars that they complete a full orbit in only a few days. **(e)** They all have circular or nearly circular orbits.

32. **Multiple choice:** At the present time, most of the known exoplanets or exoplanet candidates have been found by using **(a)** astrometric measurements; **(b)** the Doppler-wobble method; **(c)** direct imaging; **(d)** gravitational microlensing; or **(e)** the transit method.

33. **Fill in the blank:** Objects having low masses compared with normal stars, but higher masses than planets, are known as _____, and are sometimes called "failed stars."

34. **Fill in the blank:** Theories of the formation of the Solar System suggest that planets formed through the accumulation of _____, much smaller bodies.

35. **Fill in the blank:** A planet and its parent star orbit their common _____, though it is far closer to the star than to the planet.

36. **Fill in the blank:** In 2004 and 2012, astronomers measured a slight dimming of the Sun when _____ transited across it.

37. **Fill in the blank:** A spinning, collapsing cloud of gas and dust forms a disk because of the conservation of _____.

38. **Fill in the blank:** The _____ satellite has found a huge number of transiting exoplanet candidates.

†This question requires a numerical solution.

TOPICS FOR DISCUSSION

1. Consider the properties of the various types of exoplanets found thus far. Do you think life as we know it would be common or rare in these different cases? Might life be present on some of the moons of these exoplanets?

2. In your opinion, how important is it to spend Federal funds on the development of spacecraft that could detect Earth-like planets around other stars?

The mirror of NASA's Kepler spacecraft, in a test facility before its 2009 launch. The spacecraft worked fabulously well until 2013, when the second malfunctioning reaction wheel (out of the four it carried) stopped the fine guidance that was necessary for its monitoring star brightnesses to the required accuracy. Data reduction will continue for years.

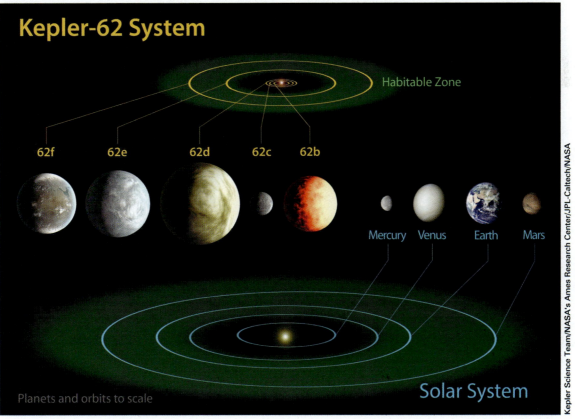

Of the five confirmed exoplanets around the star Kepler-62, two seem to be in the habitable zone, as least as it is defined with our knowledge of life on Earth (and thus specifying that the temperature must be such that water can exist in solid, liquid, and gaseous stages). These planets are shown to scale alongside the inner planets of our solar system, and their orbits are also shown to the same scale as the orbits of the inner planets of our solar system. The star Kepler-62 is $2/3$ the size of the Sun, cooler, older, and only $1/5$ as bright. Exoplanet Kepler-62f and Kepler-62e orbit in the star's habitable zone, with periods of 267 and 122 days, respectively. The three inner planets orbit in 5, 12, and 18 days, respectively, so they are too hot for life as we know it.

The Sun viewed in the ultraviolet wavelength of eleven-times ionized iron, a high-resolution image from the Atmospheric Imaging Assembly on NASA's Solar Dynamics Observatory space.
Most of its filters show us only the hottest gas, at millions of degrees. We clearly see how the coronal gas is shaped into loops, which are kept in that form by magnetic fields.

Lockhead Martin Solar and Astrophysics Lab, Smithsonian Astrophysical Observatory, and NASA

10

Our Star: The Sun

ORIGINS

The Sun, the source of light and heat on Earth, is critical to the existence of humans. Its properties are typical of stars.

N ot all stars are far away; one is very close at hand. By studying the Sun, we not only learn about the properties of a particular star but also can study processes that undoubtedly take place in more distant stars as well (see the Chapter Opener). We will first discuss the quiet Sun, the solar phenomena that typically appear every day. Afterward, we will discuss the active Sun, solar phenomena that appear nonuniformly on the Sun and vary over time.

We study the Sun from Earth's surface not only from telescopes permanently set up on mountaintops and elsewhere but also from temporary observation sites set up to observe total solar eclipses (■ Fig. 10–1). Moreover, several important spacecraft send back varied and high-resolution images of the Sun from outside our atmosphere, allowing us to study aspects of the Sun and parts of its spectrum that were previously unavailable to us.

In this chapter, we discuss mainly the outer layers of the Sun, which are visible to us. In Chapter 12, we will consider the deep-down source of energy of the Sun and of the other stars – nuclear fusion in the core, where temperatures and pressures are extremely high.

AIMS

1. Learn about the structure of our Sun, the only star we can study in detail (Sections 10.1 and 10.2).

2. Describe sunspots, flares, prominences, and other solar phenomena that vary with the 11-year solar-activity cycle (Section 10.2).

3. See how Einstein's general theory of relativity, first verified at a total solar eclipse, is a theory of gravity (Section 10.3).

Jay M. Pasachoff, Muzhou Lu, and Craig Malamut/ Williams College Solar Eclipse Expedition

■ **FIGURE 10–1** The diamond-ring effect, the pinkish solar chromosphere, and the inner pearly white solar corona, photographed by the team of one of the authors (J.M.P.) at the 2010 total solar eclipse on Easter Island.

10.1 WHAT IS THE SUN'S BASIC STRUCTURE?

We think of the Sun as the bright ball of gas that appears to travel across our sky every day. We are seeing only one layer of the Sun, part of its atmosphere. The layer that we see is called the photosphere (■ Fig. 10–2), which simply means the sphere from which the light comes (from the Greek *photos*, "light"); it is typically called the Sun's "surface," though any object placed there would fall through it, of course, if it didn't burn.

The photosphere is about 110 Earth diameters across, so over a million Earths (110 cubed) could fit inside the Sun! We find the size of the Sun by knowing its distance and measuring its angular diameter. (We can now find the distance by bouncing radio waves off the Sun's outer atmosphere directly, but historically, the scale of the Solar System was first found reasonably accurately by measuring transits of Venus – when Venus is silhouetted against the Sun. Such events take place very rarely; we discussed the transits of 2004 and 2012 in Chapters 6 and 9, and now we have to wait for 2117 for the next one.)

As is typical of many stars, about 92 per cent of the atoms and nuclei in the outer parts are hydrogen, just under 8 per cent are helium, and a mixture of all the other elements makes up the remaining approximately two-tenths of one per cent (see *A Closer Look 10.1: The Most Common Elements in the Sun's Photosphere*). The overall composition of the Sun's interior is not very different. We basically find out about the chemical composition of the Sun and stars by studying their spectra. Recently, the hard-to-find abundance of helium has been determined most accurately with the technique of solar seismology that we discuss below (Section 10.1a).

The Sun is sometimes considered a typical star, in the sense that stars much hotter and much cooler, and stars intrinsically much brighter and much fainter, exist. (Actually, there are many more stars that are fainter and cooler than there are stars like our Sun or hotter.) Radiation from the photosphere peaks (is brightest) in the middle of the visible spectrum; after all, our eyes evolved over time to be sensitive to that region of the spectrum because the greatest amount of the solar radiation is emitted there. If we lived on a planet orbiting an object that emitted mostly x-rays, we, like Superman, might have x-ray vision. (Of course, unlike Superman, we would be only passively receiving x-rays, not sending them out.)

Note that the Sun is white, not yellow, though it is often thought of as being yellow. It only appears yellow, or even orange or red, when it is close to the horizon: The blue and green light is selectively absorbed and scattered by Earth's atmosphere. (The scattered blue light produces the color of the daytime sky, as we discussed in Chapter 4.) Also, on many photographs, the filter used to cut down the solar brightness to a safe level, a factor of about a hundred thousand, favors the yellow/orange.

Beneath the photosphere is the solar **interior** (■ Fig. 10–3). All the solar energy is generated there at the solar **core**, which is about 10 per cent of the solar diameter at this stage of the Sun's life. The temperature there is about 15,000,000 K and the density is sufficiently high to allow nuclear fusion to take place, as we will discuss in Chapter 12.

The photosphere is the lowest level of the **solar atmosphere**. Though the Sun is gaseous throughout, with no solid parts, we still use the term "atmosphere" for the upper part of the solar material because it is relatively transparent.

■ **FIGURE 10–2** The solar photosphere, the everyday surface of the Sun, in an image taken in January 2012, showing several sunspots during the rising stage in the sunspot cycle. The artificial grid lines allow the positions of sunspots to be measured. The photosphere is displayed in a yellow color that most people associate with the Sun, the totality of whose light is actually white. (right) The magnetogram corresponds to the visible-light image at left.

Atmospheric Imaging Assembly (AIA) on Solar Dynamics Observatory/Stanford Lockheed Institute for Space Research, NASA

HMI on Solar Dynamics Observatory/Stanford Lockheed Institute for Space Research, NASA

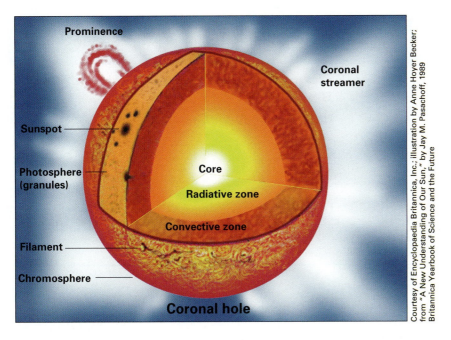

Prominence

Coronal streamer

Sunspot

Photosphere (granules)

Core

Radiative zone

Convective zone

Filament

Chromosphere

Coronal hole

Courtesy of Encyclopaedia Britannica, Inc.; illustration by Anne Hoyer Becker; from "A New Understanding of Our Sun," by Jay M. Pasachoff, 1989 Britannica Yearbook of Science and the Future

■ **FIGURE 10–3** The parts of the solar atmosphere and interior. The solar surface is depicted as it appears through a hydrogen filter; thus, here we actually see the chromosphere, except in the narrow sliver labeled "photosphere (granules)." The photosphere is the visible surface of the Sun. The interior has a core, where the energy is generated, surrounded by a zone in which the dominant method of energy transfer is radiation and a zone in which convection (a "boiling" effect) dominates. On the photosphere we find sunspots, and just above it loops of gas that are dark filaments when seen in projection against the disk and that are bright prominences when seen off the Sun's edge. The chromosphere, a thin layer above the photosphere, shows best in hydrogen light and is composed of small spikes known as spicules. The white corona above it extends far into space. Its shape, including its huge streamers, is determined by the Sun's magnetic field.

Just above the photosphere is a jagged, spiky layer about 10,000 km thick, only about 1.5 per cent of the solar radius. This layer glows colorfully pinkish when seen during an eclipse (■ Fig. 10–4a), when the photosphere is hidden, and is thus called the **chromosphere** (from the Greek *chromos*, "color").

Above the chromosphere, a ghostly white halo called the corona (from the Latin, "crown") extends tens of millions of kilometers into space (Fig. 10–4b). The corona is continually expanding into interplanetary space and in this form is called the solar wind. The Earth is bathed in the solar wind, a stream of particles having many different speeds.

The "slow solar wind" is from a general expansion of the corona. Particles in it usually take about 20 days to reach Earth. The "fast solar wind" particles travel at twice that speed or greater. They come from coronal holes. Occasionally, solar flares can expel particles moving so rapidly that they reach Earth in only 20 or 25 minutes,

much faster than the usual few hours or days from a flare, giving even less time for astronauts in space to protect themselves.

10.1a The Photosphere

The Sun is a normal star with a surface (photospheric) temperature of about 5800 K, neither the hottest nor the coolest star. (One way we know the temperature is that the Sun's visible spectrum closely resembles the blackbody curve, discussed in Chapter 2, for that temperature.) The Sun is the only star sufficiently nearby to allow us to study its surface in detail. We can resolve (that is, distinguish) parts of the surface about 700 km across (70 km with adaptive optics from the best sites), roughly the distance from Boston to Washington, D.C.

When we study the solar surface in **white light** – all the visible radiation taken together – with typical good resolution (about

(a)

(b)

Jay M. Pasachoff, Muzhou Lu, and Craig Malamut/ Williams College Solar Eclipse Expedition

Jay M. Pasachoff, Muzhou Lu, and Craig Malamut/ Williams College Solar Eclipse Expedition

■ **FIGURE 10–4** (a) The chromosphere of the Sun and, the same color, solar prominences, photographed during a total solar eclipse, adjacent to the diamond ring and surrounding the Moon's silhouette. Note how pink the chromosphere and prominences look, since they are largely H-alpha (Hα) but with an admixture of H-beta (Hβ) and other spectral emission lines. (b) The inner solar corona surrounding the dark silhouette of the Moon at the same eclipse.

A CLOSER LOOK 10.1   The Most Common Elements in the Sun's Photosphere

For each 1,000,000	atoms of	hydrogen, there are	Symbol H	Atomic Number 1	Atomic Weight 1
85,000	atoms of	helium	He	2	4
850	atoms of	oxygen	O	8	16
400	atoms of	carbon	C	6	12
120	atoms of	neon	Ne	10	20
100	atoms of	nitrogen	N	7	14
47	atoms of	iron	Fe	26	56

Note: This table gives the relative numbers of atoms. To calculate the relative masses, multiply each relative number by the atomic weight of that element before summing the column. One interesting result is that 25% of the mass of the Sun is helium.

1 arc second), we see a salt-and-pepper texture called **granulation** (■ Fig. 10–5). The effect, caused by **convection**, is similar to that seen in boiling liquids on Earth: Hot pockets (cells) of gas are more buoyant than surrounding regions, so they rise and deposit their energy at the surface. This causes them to become denser, and so they subsequently sink. But the granules, or convective cells of gas, are about the same size as the limit of our resolution, so they are difficult to study. No spacecraft has yet exceeded the best resolution obtained from the ground.

On close examination, the photosphere as a whole oscillates – vibrating up and down slightly – as can be studied using the Doppler effect. The first period of vibration discovered was 5 minutes long, and for many years astronomers thought that 5 minutes was a basic duration for oscillation on the Sun. However, astronomers have since realized that the Sun simultaneously vibrates with many different periods, and that studying these periods tells us about the solar interior. Indeed, we are able to test the standard model of the solar interior (that is, the way temperatures, densities, and chemical compositions vary with distance from the center) through studies of the vibrations. The method works similarly to the way terrestrial geologists investigate the Earth's interior by measuring seismic waves on the Earth's surface; the studies of the Sun are thus called, by analogy, "solar seismology" or **helioseismology** (after Helios, the Greek Sun god).

To find the longest periods, astronomers have observed the Sun from the Earth's south pole, where the Sun stays above the horizon for months on end. Now, even better, they use the Global Oscillation Network Group (GONG), a program centered at the National Solar Observatory that has erected a network of telescopes around the world to study solar oscillations. In addition, a European Space Agency/NASA spacecraft called SOHO (the **So**lar and **H**eliospheric **O**bservatory) is stationed in space at a location from which it continuously views the Sun, and has also assembled long runs of observations lasting many months and covering over a decade. The helioseismology telescope on the **S**olar **D**ynamics **O**bservatory (SDO) has higher resolution for its series of images.

■ **FIGURE 10–5** (a) A visible-light view of a tiny part of the solar surface, showing the salt-and-pepper granulation at the highest detail available from observatories on Earth. (b) A magnetic-field map of the same region.

(a)

(b)

National Solar Observatory/AURA

National Solar Observatory/AURA

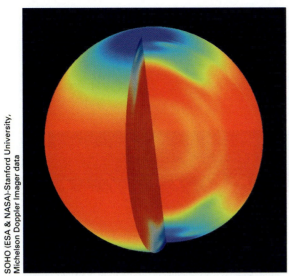

SOHO (ESA & NASA)-Stanford University, Michelson Doppler Imager data

■ **FIGURE 10–6** The rotation periods of different layers and regions of the Sun, based on helioseismology results, shown in false color. Measurements reveal that the surface rotation periods, which vary from 26 days at the equator to 36 days at the poles, persist inward through the solar convection zone. Further toward the center, the data show that the Sun rotates as a solid body with a 27-day period.

■ **FIGURE 10–7** (a) Fraunhofer's original spectrum, shown on a German postage stamp on the occasion of the 200th anniversary of Fraunhofer's birth. We see a continuous spectrum, from red on the left to violet on the right, crossed from top to bottom by the dark lines that Fraunhofer discovered, now known as "Fraunhofer lines" or "absorption lines." (b) A 2012 German postage stamp honoring the 225th anniversary of Fraunhofer's birth.

Studies of solar vibrations thus far have told us about the temperature and density at various levels in the solar interior, and about how fast the interior rotates (■ Fig. 10–6). We have used these studies to test the standard model of the Sun's interior to unprecedented accuracy. Since other stars probably behave like the Sun, we are learning about the interiors of stars in general. Helioseismology can even be used to image, though not with high resolution, what the *back* side of the Sun looks like – a good trick, since we are essentially "seeing" through over a million kilometers of opaque gas!

The spectrum of the solar photosphere, like that of almost all stars, is a continuous spectrum crossed by absorption lines (■ Fig. 10–7; see also the discussion in Chapter 2). Hundreds of thousands of these absorption lines, which are also called Fraunhofer lines after Joseph von Fraunhofer, who first systematically studied the strongest ones, have been photographed and catalogued. They represent sets of spectral lines from most of the chemical elements. Iron has many lines in the spectrum. The hydrogen lines are strong but few in number. From the spectral lines, we can figure out the relative abundances (the percentages) of the elements. Not only the Sun but also all other ordinary stars have only absorption lines in their spectra.

10.1b The Chromosphere

When we look at the Sun through a filter that passes only the red light from hydrogen gas, the chromosphere is opaque. Thus, through a hydrogen-light filter (■ Fig. 10–8a), our view is of the chromosphere. The view has brighter and darker areas, with the brighter areas in the same regions as "sunspots" (see Section 10.2a), but the chromosphere looks different from the photosphere below it. The brighter areas are called "plages," from the French word for "beaches," since they are relatively bright.

An important theme valid throughout much of astronomy is that when you look at astronomical objects in detail, you see that their properties may vary quite a lot from the average you see with the image blurred out. For example, under high resolution, we see that the chromosphere is not a spherical shell around the Sun but rather is composed of small "spicules." These jets of gas rise and fall, and have been compared in appearance to blades of grass or burning prairies (Fig. 10–8b,c).

Spicules are more or less cylinders about 700 km across and 7000 km tall. They seem to have lifetimes of about 5 to 15 minutes, and there may be approximately half a million of them on the surface of the Sun at any given moment. They are about the same size as the granules, and are matter ejected into the chromosphere, presumably from the boiling effect that also makes the granules.

Chromospheric matter appears to be at a temperature of 7000 to 15,000 K, somewhat higher than the temperature of the photosphere. The extra energy comes in part from the mechanical motions of convective granules, and in part from magnetic fields, although the relative proportion of these contributions is not yet known. Ultraviolet spectra of stars recorded by spacecraft have shown unmistakable signs of chromospheres in Sun-like stars. Thus, by studying the solar chromosphere, we are also learning what the chromospheres of other stars are like.

(a)

(b)

(c)

Kanzelhoehe Observatory/University of Graz

Jay M. Pasachoff, David L. Butts, Joseph W. Gangestad, Owen W. Westbrook, and Jennifer Yee, with NASA support

From Angelo Secchi, *Die Sonne*, courtesy of Jay M. Pasachoff

■ **FIGURE 10–8** (a) The Sun, photographed in hydrogen radiation, shows bright areas called plages and dark filaments around the many sunspot regions. The image matches the sunspot diagram of Figure 10–2 and the magnetic-field map of Figure 10–5b. (b) A high-resolution image of the Sun, showing spicules, taken by one of the authors (J.M.P.) and his students with the Swedish 1-m Solar Telescope in the Canary Islands. (c) An original, 19th-century drawing showing how the spicules look like burning prairies.

Studying the solar chromosphere once led to a major discovery: that of helium. A yellow spectral line was seen in the chromosphere at a nineteenth-century total solar eclipse, and it didn't quite match the known yellow lines from sodium. The gas was called "helium," after the Greek Sun god, Helios, since it was known only on the Sun. It took decades before helium was isolated on Earth by chemists. The Sun's photosphere isn't hot enough for helium absorption lines to show in its visible-light spectrum.

10.1c The Corona

During total solar eclipses, when first the photosphere and then the chromosphere are completely hidden from view, a faint white halo around the Sun becomes visible. This corona is the outermost part of the solar atmosphere, and technically extends throughout the Solar System. At the lowest levels, the corona's temperature is about 2,000,000 K. The heating mechanism of the corona is magnetic, though the details are hotly debated.

The **E**xtreme-**U**ltraviolet **I**maging **T**elescope (EIT) on the SOHO spacecraft has imaged the corona since 1995, every few minutes for a long time, and since 2010 NASA's Atmospheric Imaging Assembly on SDO has imaged the corona at even more wavelengths every 12 seconds! By looking through filters that pass only light given off by gas at a very high temperature, the spacecraft can make images of the corona even in the center of the Sun's disk (■ Fig. 10–9). On the occasional dates of solar eclipses, these images can be lined up with the eclipse images, to allow astronomers to trace many of the coronal streamers seen during an eclipse back to their roots on the Sun's surface (■ Fig. 10–10).

Even though the temperature of the corona is so high, the actual amount of energy in the solar corona is not large. The temperature quoted is actually a measure of how fast individual particles (electrons, in particular) are moving. There aren't very many coronal particles, even though each particle has a high speed. The corona has less than one-billionth the density of the Earth's atmosphere, and would be considered a very good vacuum in a laboratory on Earth. In fact, assuming the photosphere were somehow blocked, you would freeze to death (aside from being in a near vacuum) if you were immersed in the corona; the hot particles would hit you too rarely to replenish the energy your body radiates into space!

The corona serves as a unique and valuable celestial laboratory in which we may study gaseous "plasmas" at extremely low densities (nearly a vacuum). Plasmas are electrically neutral gases consisting of positively and negatively charged particles and can be shaped by magnetic fields. We are trying to learn how to use magnetic fields on Earth to control plasmas, retaining charged particles within a small volume, in order to provide energy through nuclear fusion.

Photographs of the corona show that it is very irregular in form. Beautiful long **streamers** extend away from the Sun in the equatorial regions. The shape of the corona varies continuously and is thus different at each successive eclipse. The structure of the corona is maintained by the magnetic field of the Sun.

NASA's SDO spacecraft makes extremely high-resolution observations of the solar corona by observing in the ultraviolet. It shows clearly how the corona is made up of loops of gas, which are held in their shapes by the Sun's magnetic field (■ Fig. 10–11). It is part of NASA's "Living with a Star" program. A joint Japanese-American-British spacecraft known (until its launch in 2006) as Solar-B, and now as Hinode (meaning "sunrise"), has high-resolution imagers in the x-ray, extreme-ultraviolet, and optical parts of the spectrum.

The corona is normally too faint to be seen except during an eclipse of the Sun because it is fainter than the everyday blue sky. But at certain locations on mountain peaks on the surface of the Earth, the sky is especially clear and dust-free, and the innermost part of the corona can be seen with special telescopes (■ Fig. 10–12). The limited extent of the corona seen in this way shows how valuable the observations at eclipses and from space are.

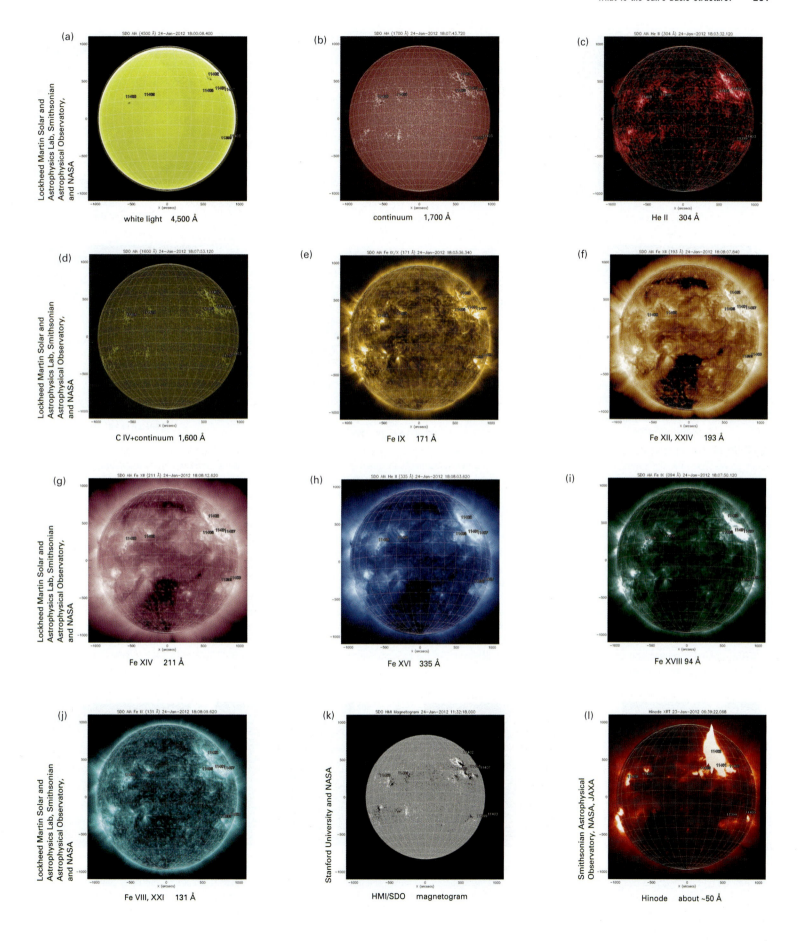

(a) white light   4,500 Å

(b) continuum   1,700 Å

(c) He II   304 Å

(d) C IV+continuum   1,600 Å

(e) Fe IX   171 Å

(f) Fe XII, XXIV   193 Å

(g) Fe XIV   211 Å

(h) Fe XVI   335 Å

(i) Fe XVIII 94 Å

(j) Fe VIII, XXI   131 Å

(k) HMI/SDO   magnetogram

(l) Hinode   about ~50 Å

Lockheed Martin Solar and Astrophysics Lab, Smithsonian Astrophysical Observatory, and NASA

Stanford University and NASA

Smithsonian Astrophysical Observatory, NASA, JAXA

■ **FIGURE 10–9** A set of images taken almost simultaneously on or near a date in January 2012 through the several filters of the Atmospheric Imaging Assembly (AIA) on NASA's Solar Dynamics Observatory in space, and from other instruments in space and on the ground at about the same day. They are listed in order of ascending temperature of the solar layer. (a) The solar photosphere in white light, about 5,800 K. (b) 1,700 Å, the temperature minimum just above the photosphere, roughly 5,000 K. (c) 304 Å, He II, ionized helium gas at a temperature of about 60,000 K, with a small contribution from ionized silicon. This wavelength shows the chromosphere. (d) 1,600 Å, the upper photosphere and the transition region to the corona, about 100,000 K. (e) 171 Å, Fe IX, quiet corona, roughly 600,000 K. (f) 193 Å, Fe XII, ionized iron at a temperature of about 1,200,000 K, with some contribution from Fe XXIV, typical of 20 million K, that shows only during a flare. (g) 211 Å, Fe XIV, ionized iron at the active-region temperature of about 2,000,000 K (and matching the ion of the visible spectrum's coronal green line). (h) 335 Å, Fe XVI, ionized iron at the active-region temperature of about 2,500,000 K. (i) 94 Å, Fe XVIII, an ion produced in hotter conditions typical of solar flares, 6 million K; (j) 131 Å, Fe VIII and Fe XXI, sampling even hotter regions in solar flares, 10 and 16 million K, respectively. All but the first wavelengths show the corona. Images like this are taken by SDO every 12 seconds each day of the year with the exception of brief periods when the Sun is hidden from the satellite by the Earth. This image ends the sequence from the AIA on SDO. (k) A magnetogram from the Helioseismic and Magnetic Imager (HMI) also on SDO. (l) An x-ray image from the XRT on Hinode. (m) A ground-based Hα image. (n) 174 Å, a Sun Watcher (Belgian) image, overlapping in coverage with SDO's similar band. (o) A computed image of the Sun's far side from Earth, based on observations from the ground-based Global Oscillation Network Group (GONG). (p) GONG's phototospheric, white-light view. (q) STEREO-Ahead's image, from halfway around Earth's orbit, in 195 Å radiation, at about a million K. (r) STEREO-Behind's coronagraph image, from halfway around Earth's orbit in the opposite sense. As shown in part (n): ESA's SWAP/PROBA2 coronal image, similar to AIA's 10-9e but with a wider field of view when uncropped.

■ **FIGURE 10–10** The Sun at the time of the total solar eclipse of July 11, 2010. (a) A composite image made from photographs of the solar corona taken during the eclipse on Easter Island by one of the authors (J.M.P.) and his students surrounds the lunar disk. It shows the equatorial streamers and polar plumes in the corona over a wide range of intensity. (b) A coronagraph on the SOHO spacecraft takes images of the corona every few minutes, but it has to block out the whole photosphere plus the inner corona to do so. Here is an image from eclipse day 2010, matching the image in the first panel. (c) Another coronagraph on SOHO hides more of the inner and middle coronae but shows the outer corona, again here from eclipse day 2010. (d) One of the dozen possible filters used to image the Sun from the Solar Dynamics Observatory at about the time of the eclipse in 2010, showing gas at a temperature of over a million kelvins.

(a)

Alan Title, Lockheed Martin Advanced Technology Center, and Leon Golub, Harvard-Smithsonian Center for Astrophysics

(b)

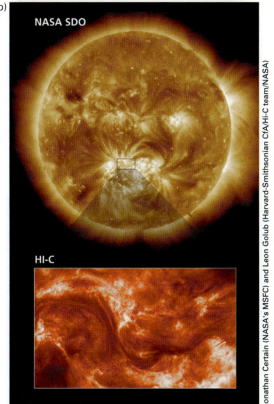

NASA SDO

HI-C

Jonathan Certain (NASA's MSFC) and Leon Golub (Harvard-Smithsonian CfA/Hi-C team/NASA)

■ **FIGURE 10–11** (a) Loops of hot coronal gas imaged in the ultraviolet from the Solar Dynamics Observatory spacecraft. See also the photo opening this chapter. (b) Even higher-resolution ultraviolet coronal observations, with pixels $1/5$ the size of those on SDO, were taken of a small bit of the Sun during the few minutes that a NASA rocket launched in New Mexico was above the Earth's atmosphere in 2012, the so-called "Hi-C observation."

(a)

Sacramento Peak Observatory, National Solar Observatory/ AURA/NSF

(b)

National Center for Atmospheric Research (NCAR)/NSF

■ **FIGURE 10–12** From a few mountain sites, the innermost corona can be photographed without need for an eclipse on many days of the year. (a) The corona shows up best in its green emission lines from thirteen-times ionized iron, iron that has lost thirteen electrons. (b) Another technique, used at Mauna Loa, Hawaii, detects the "polarization" of the corona (like the effect seen with polarizing sunglasses) and so distinguishes it from the sky, seen here on February 9, 2012.

Several spacecraft, both crewed and robotic, have used devices that made a sort of artificial eclipse to photograph the corona hour by hour in visible light. One of the instruments aboard SOHO makes such observations very well at present. SOHO studies the corona to much greater distances from the solar surface than can be studied with coronagraphs on Earth. (SOHO's coronagraphs cannot see the innermost part, which we can still study best at eclipses seen from Earth's surface.)

Among the major conclusions of this research is that the corona is much more dynamic than we had thought. For example, many blobs of matter were seen to be ejected from the corona into interplanetary space, one per day or so. These "coronal mass ejections" (see Sec. 10.2d) sometimes even impact the Earth, causing surges in power lines and zapping – and even occasionally destroying the capabilities of – satellites that bring you television or telephone calls. SOHO and other spacecraft far above the Earth in the direction of the Sun give us early warning when solar particles pass them. NASA has a major program called the Sun–Earth Connection, studying what we increasingly call "space weather" and unifying the study of the Sun, the Earth, and the space between them.

The visible region of the coronal spectrum, when observed at eclipses, shows continuous radiation, absorption lines, and emission lines (■ Fig. 10–13). The presence of emission lines, which we see standing out brightly against the dark sky, is in contrast to the

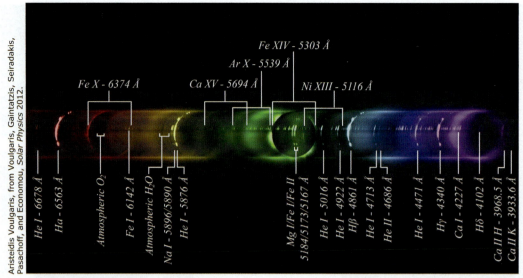

Aristeidis Voulgaris, from Voulgaris, Gaintatzis, Seiradakis, Pasachoff, and Economou, *Solar Physics* 2012.

Fe XIV - 5303 Å

Ar X - 5539 Å

Fe X - 6374 Å   *Ca XV - 5694 Å*   *Ni XIII - 5116 Å*

He I - 6678 Å
Hα - 6563 Å
Atmospheric O₂
Fe I - 6142 Å
Atmospheric H₂O
Na I - 5896/5890 Å
He I - 5876 Å
Mg I/Fe I/Fe II 5184/5173/5167 Å
He I - 5016 Å
He I - 4922 Å
Hβ - 4861 Å
He I - 4713 Å
He II - 4686 Å
He I - 4471 Å
Hγ - 4340 Å
Ca I - 4227 Å
Hδ - 4102 Å
Ca II H - 3968.5 Å
Ca II K - 3933.6 Å

■ **FIGURE 10–13** A spectrum of the prominences, chromosphere, and inner corona at the July 2010 total solar eclipse. The last traces of photospheric spectrum show as the band of color. The bright points are the emission lines from the chromosphere and prominences. The yellow helium line at 5876 Å was the emission line from which helium was first identified over a hundred years ago. The element was named helium because at that time it was found only in the Sun (*helios* in Greek). Faintly visible in the green and in the red are complete circles that are emission lines in the spectrum of the inner corona. The continuum color with absorption lines, which was visible in Figure 10–7, has been hidden by the Moon during totality.

photosphere's continuum and absorption lines (but never emission lines), which appear in the everyday spectrum of the Sun.

The coronal emission lines do not correspond in wavelength to any normal spectral lines known in laboratories on Earth or in other stars, and for decades in the late 19th and early 20th centuries, their identification was one of the major problems in solar astronomy. The lines were even given the name of an element: "coronium." (After all, the element helium was first discovered on the Sun, so why not another one?) In the late 1930s, it was discovered that the emission lines arose in atoms that had lost about a dozen electrons each. This was the major indication that the corona was very hot (and that coronium doesn't exist as a separate element). The corona must be very hot indeed, millions of degrees, to have enough energy to strip that many electrons off atoms. This very hot corona also reflects the photospheric spectrum to us, but the Doppler shift produced by the very rapidly moving plasma particles broadens the absorption lines so they are no longer visible, thus leaving only a continuum for the component reflected by the hot gas.

While the coronal emission lines tell us about the coronal gas, the visible coronal absorption lines are mere reflections of the absorption lines in the spectrum of the Sun's photosphere. To provide these absorption lines (not significantly Doppler-broadened, unlike the case discussed above), the photospheric spectrum is reflected toward us by slowly moving *dust* in interplanetary space far closer to the Earth than to the Sun.

The gas in the corona is so hot that it emits mainly x-rays, photons of high energy. The photosphere, on the other hand, is too cool to emit x-rays. As a result, when photographs of the Sun are taken in the x-ray region of the spectrum (from satellites, since x-rays cannot pass through the Earth's atmosphere), they show the corona and its structure rather than the photosphere (■ Fig. 10–14a).

(a)

Leon Golub/Smithsonian Astrophysical Observatory

(b)

Yohkoh image courtesy of Loren Acton, Montana State University

■ **FIGURE 10–14** (a) The corona in relatively low-energy ("soft") x-rays observed on November 25, 2011, with the XRT from the Hinode spacecraft, an American x-ray telescope on a Japanese spacecraft. (b) Another low-energy x-ray view of the corona, this time as observed from the Yohkoh spacecraft, during a partial eclipse. Structure in the corona is visible, including several coronal holes.

■ **FIGURE 10–15** An ultraviolet image shows small coronal loops. The image was made with the Atmospheric Imaging Assembly on the Solar Dynamics Observatory spacecraft. The expeditions of one of the authors (J.M.P.) have observed such loops from the ground during eclipses, with lower spatial resolution but higher temporal resolution.

at its locations as it orbits. Both missions are especially aimed to find out why the corona is so much hotter than the photosphere, as well as what causes and propels the solar wind.

10.1d The Scientific Value of Eclipses

We discussed solar eclipses and eclipse expeditions in Chapter 4. In these days of orbiting satellites, why is it worth making an expedition to observe a total solar eclipse for scientific purposes?

There is much to be said for the benefits of eclipse observing. Eclipse observations are a relatively inexpensive way, compared to space research, of observing the outer layers of the Sun. Artificial eclipses made by spacecraft hide not only the photosphere but also the inner corona. And for some kinds of observations, space techniques have not yet matched ground-based eclipse capabilities. We were on the ground for the total solar eclipses visible from Greece in 2006, Siberia in 2008, China in 2009, Easter Island in 2010, and Australia in 2012, coordinating with non-eclipse and spacecraft observations.

The x-ray images also reveal very dark areas at the Sun's poles and extending downward across the center of the solar disk. These dark locations are **coronal holes**, regions of the corona that are particularly cool and quiet (■ Fig. 10–14b). The density of gas in those areas is lower than the density in adjacent areas. There is usually a coronal hole at one or both of the solar poles. Less often, we find additional coronal holes at lower solar latitudes. The regions of the coronal holes seem very different from other parts of the Sun. The fast streams in the solar wind flow to Earth mainly out of the coronal holes, so it is important to study the coronal holes to understand our environment in space.

The most detailed x-ray images support the more recent ultraviolet high-resolution images in showing that most, if not all, the radiation appears in the form of loops of gas joining separate points on the solar surface (■ Fig. 10–15). We must understand the physics of coronal loops in order to understand how the corona is heated. It is not sufficient to think – as was done for decades, to simplify calculations – in terms of a uniform corona, since the corona is obviously not uniform.

NASA is planning to launch the Solar Probe Plus mission in 2018, to fly within 8.5 solar radii ($1/25$ au) of the solar photosphere; its closest passes near the Sun will occur in 2024 after seven gravity maneuvers with Venus. It should be able to make observations of exceedingly high spatial resolution and, at each perihelion of its very elliptical orbit, also sample the solar wind closer to the Sun than has ever been done. The European Space Agency is planning Solar Orbiter for a 2017 launch, to orbit inside Mercury's orbit, at a distance of 60 solar radii, or about 0.3 au. It will carry a variety of imagers, as well as devices to sample the solar wind and to measure the magnetic field

10.2 SUNSPOTS AND OTHER SOLAR ACTIVITY

The Sun's photosphere and its immediate surroundings exhibit a variety of phenomena, including sunspots, flares, coronal mass ejections, filaments, plages, and prominences, as we describe in this section. Their frequency appears to vary with an 11-year cycle.

10.2a What Are Those Blemishes on the Sun?

If you examine the Sun through a properly filtered telescope, you may notice some **sunspots** (■ Fig. 10–16), which appear relatively dark when seen in white light. Sunspots were discovered in 1610, independently by Galileo and by others, shortly after the telescope was first put to astronomical use; an occasional sunspot had been seen with the naked eye previously, and Kepler even saw one in 1607 with a pinhole projection.

Sunspots look dark because they are giving off less radiation per unit area than the photosphere that surrounds them. Thus they are relatively cool (about 2000 K cooler than the photosphere), since cooler gas radiates less than hotter gas (recall our discussion in Chapter 2). Actually, if we could somehow remove a sunspot from the solar surface and put it off in space, it would appear bright against the dark sky; a large one would give off as much light as the full moon seen from Earth.

STAR PARTY 10.1

Observing Sunspots

Sunspots seem dark in contrast with the rest of the Sun, but would be as bright as the Moon if they could be in the dark sky by themselves. A couple of types of sunspot telescopes were invented in the last decade that are available for only about $350. They can be taken outdoors and set up in a few seconds so that anyone can see the day's sunspots on a screen a few inches across.

Any telescope can be used to project an image of the Sun, including the sunspots, on a piece of paper or on a screen. (*Don't* look up through the telescope at the Sun, however, as emphasized in Chapter 4.) Furthermore, inexpensive but safe solar filters are available that can be put over the front end of a telescope, cutting down the Sun's intensity by a factor of about 100,000, to allow you to look through the telescope and filter at the Sun. (Do not use a filter only at the eyepiece end, as it can heat up and crack.) It is interesting to see how the sunspots change from day to day and as the Sun rotates.

FIGURE SP 10.1 A small telescope for viewing sunspots, with the beam folded by reflection off several mirrors to make the system compact and easy to set up.

(a)

(b)

(c)

■ **FIGURE 10–16** (a) A sunspot group, in which each of the major spots is about the size of our Earth. (b) A sunspot, also viewed with the Swedish 1-m Solar Telescope on La Palma, Canary Islands, Spain. Astronomers from the Institute for Solar Physics of the Royal Swedish Academy of Sciences used adaptive optics and subsequent computing to improve this and the previous image over the raw observations. (c) A sunspot, showing the dark umbra surrounded by the lighter penumbra. This sunspot, including umbra and penumbra, is slightly larger than the Earth. The sunspot is surrounded by the solar granulation in this image using adaptive optics, taken with the 1-m vacuum New Solar Telescope of the Big Bear Solar Observatory in California, operated by the New Jersey Institute of Technology.

A sunspot includes an apparently dark central region, called the **umbra** (from the Latin for "shadow"; plural: umbrae). The umbra is surrounded by a **penumbra** (plural: penumbrae), which is not as dark. Some large sunspots are quite long-lived, lasting for over a month. If you follow them day after day, you can see how the Sun rotates at their latitudes.

To explain the origin of sunspots, we must understand magnetic fields. When iron filings are put near a simple bar magnet, the filings show a pattern (■ Fig. 10–17a). The magnet is said to have a north pole and a south pole, and the magnetic field linking them is characterized by what we call **magnetic-field lines** (after all, the iron filings are spread out in what look like curved lines). The Earth (as well as some other planets) has a magnetic field having many characteristics in common with that of a bar magnet. The structure seen in the solar corona, including the streamers, shows matter being held by the Sun's magnetic field.

(a)

Jay M. Pasachoff

(b)

NSO/AURA/NSF

(c)

PROBA2/SWAP image, Processed by D. B. Seaton

(d)

eclipse: Jay M.Pasachoff and Hana Druckmülerová; SDO/AIA: Alan Title, Lockheed Martin Advanced Technology Center, and Leon Golub, Harvard-Smithsonian Center for Astrophysics

■ **FIGURE 10–17** (a) Lines of force from a bar magnet are outlined by iron filings. One end of the magnet is called a north pole and the other is called a south pole. Similar poles ("like poles," where "like" means "similar") – a pair of norths or a pair of souths – repel each other, and unlike (1 north and 1 south) attract each other. Lines of force go between opposite poles. (b) The magnetic field of the Sun on the day of the July 2010 total solar eclipse. Black and white, respectively, show opposite polarities. (c) The solar corona from the Sun Watcher with Active Pixel (SWAP) instrument aboard ESA's PROBA2 mission, through a filter that passes only the radiation from highly ionized iron, corresponding to a temperature over a million kelvins, at the time of the July 2010 total solar eclipse. (d) A composite including a Solar Dynamics Observatory high-temperature image pasted into the middle of a composite total-eclipse image from the Williams College team.

The strength of the solar magnetic field is revealed in spectra. George Ellery Hale showed, in 1908, that the sunspots are regions of very high magnetic-field strength on the Sun, thousands of times more powerful than the Earth's magnetic field (■ Fig. 10–17b). Sunspots usually occur in pairs, and often these pairs are part of larger groups. In each pair, one sunspot will be typical of a north magnetic pole and the other will be typical of a south magnetic pole.

The strongest magnetic fields in the Sun occur in sunspots. The magnetic fields in sunspots are able to restrain charged matter; they keep hot gas from being carried upward to the surface. As a result, sunspots are relatively cool and appear dark compared with the surrounding photosphere.

10.2b The Solar-Activity Cycle

In about 1850, it was realized that the number of sunspots varies with roughly an 11-year cycle (■ Fig. 10–18), the **sunspot cycle**, which is one aspect of the more general **solar-activity cycle**. Every 11-year cycle, the north magnetic pole and south magnetic pole on the Sun reverse; what had been a north magnetic pole is then a south magnetic pole and vice versa. So it is 22 years before the Sun returns to its original state, making the real period (including magnetism) of the solar-activity cycle 22 years.

We passed through the maximum of the sunspot cycle – the time when there is the greatest number of sunspots – in 2001–2002 (Cycle 23). The number of sunspots then decreased, but it reached an unusually prolonged low period, not starting to increase (Cycle 24) until after the 2009 solar eclipse. It is predicted to reach its maximum in late 2013, though a relatively low maximum is generally expected.

Careful studies of the solar-activity cycle are now increasing our understanding of how the Sun affects the Earth, part of the Sun-Earth Connection. Although, for many years, scientists were skeptical of the idea that solar activity could have a direct effect on the Earth's weather, today's scientists currently seem to be accepting more and more the possibility of such a relationship.

An extreme test of the interaction may be provided by the interesting probability that there were essentially no sunspots on the Sun from 1645 to 1715! This period, called the **Maunder minimum**, was largely forgotten until the late 20th century. An important conclusion is that the solar-activity cycle may be considerably less regular than we had thought.

Much of the evidence for the Maunder minimum is indirect, and has been challenged, as has the specific link of the Maunder minimum with colder climate during that period. It would be reasonable for several mechanisms to affect the Earth's climate on this timescale, rather than only one.

Precise measurements made from spacecraft have shown that the total amount of energy flowing out of the Sun varies slightly, by up to plus or minus 0.2 per cent. On a short timescale, the dips in energy seem to correspond to the existence of large sunspots. Astronomers are now trying to figure out what happens to the blocked energy. On a longer timescale, the effect goes the other way. Spacecraft observations have shown that as sunspot minimum is reached, the Sun becomes overall slightly fainter. As the last maximum was approached, the Sun brightened back up, suggesting the correlation.

As the sunspot cycle progresses, the sunspots appear at latitudes closer and closer to the solar equator. As they reach about 20° of latitude, precursors of the next sunspot cycle have appeared at high solar latitudes (Fig. 10–18c). But, in 2011, most of those precursors

(a)

(b)
DAILY SUNSPOT AREA AVERAGED OVER INDIVIDUAL SOLAR ROTATIONS

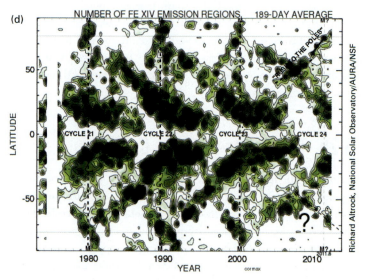

■ **FIGURE 10–18** (a) The 11-year sunspot cycle is but one manifestation of the solar-activity cycle. Black shows yearly means; red shows monthly means. (b) If you plot the latitudes of sunspots over time, the graph is called the "butterfly diagram" for obvious reasons. You can see that sunspots approach lower latitudes as the sunspot cycle evolves. The beginning of Cycle 24 is visible around 2010. (c) As sunspots from Cycle 23 (1995–2008) neared solar latitude 20° in about 1999, helioseismology revealed jet streams under the solar surface at about 60° solar latitude; these jet streams eventually became the current Cycle 24. Similar jet streams should have appeared around 2010, as Cycle 24 reached solar latitude 20°, to form the future Cycle 25. But no such high-latitude jet streams have been detected, indicating that the sunspot cycle may be temporarily shutting down. (d) The latitudes of regions of emission in the coronal green line of thirteen-times-ionized iron, typical of about 2 million K, showing through early 2013, continuing the diagram, the tentative, delayed beginning of a rush to the poles in at least the Sun's northern hemisphere that could mean the onset of a new sunspot cycle. In the top graphs, yearly means are given in black for dates prior to 1750; monthly means are given in red for dates 1750 onward.

Big Bear Observatory/New Jersey Institute of Technology

Big Bear Observatory/New Jersey Institute of Technology

■ **FIGURE 10–19** The brightening of regions of the Sun, seen in the Hα line, marking the position of a solar flare. To make a flare, magnetic-field lines change the way their north and south magnetic poles are connected. This "reconnection" releases high-energy particles, which follow the magnetic-field lines back down to lower levels of the solar atmosphere. This pair of images is separated by nearly a day, showing that the continual flaring at upper left follows persistent magnetic-field lines.

of Cycle 25 aren't there! Joined by two additional lines of evidence, the data indicate that the next sunspot cycle may be very weak or even that there may be no next sunspot cycle but instead perhaps another Maunder-type minimum that could last decades. So we can all watch, especially around and after 2022, to see if the Sun surprises us yet again one way or the other.

10.2c Fireworks on the Sun, and Space Weather

Violent activity sometimes occurs in the regions around sunspots. Tremendous eruptions called **solar flares** (■ Fig. 10–19) can eject particles and emit radiation from all parts of the spectrum into space. These solar storms begin abruptly, over a few seconds, and then can last up to four hours. Temperatures in the flare can reach 10 to 20 million kelvins, even hotter than the quiet corona.

Until a decade or so ago, scientists thought that most of the particles affecting the Earth came from solar flares. But spacecraft revealed that the corona was ejecting puffs of mass with a frequency of once every day or so. These **coronal mass ejections** cause many of the magnetic storms on Earth. At first, it was thought that flares cause coronal mass ejections, but that is not definitely the case. The link between coronal mass ejections and solar flares is being debated; they often but not always occur together.

Ultraviolet radiation and x-rays that are given off by flares reach Earth at the speed of light in a little over 8 minutes and can disrupt radio communications, because they ionize the upper part of Earth's atmosphere. Flare particles that are ejected reach the Earth in a few hours or days and can cause the auroras (see the discussion in Chapter 6) and even surges on power lines that lead to blackouts of electricity. Particles from coronal mass ejections usually take a day or so to reach Earth, but, in some cases, these particles can reach Earth in less than an hour. So astronauts of the future are now thought to have less time to seek safety in their interplanetary spacecraft than had been assumed.

The study of space weather – the effect of the Sun on interplanetary space and the Earth – has increasing importance, as robotic and crewed spacecraft spend more time outside the protection of the Earth's atmosphere. Solar flares can even affect the accuracy of GPS (Global Positioning System) devices that control not only our cars' directions but also airplanes' routings. The website spaceweather.com can keep you up-to-date on a daily basis.

Because of the solar-terrestrial relationships, high priority is placed on understanding solar activity and being able to predict it. Better predictions are now being made because NASA's **S**olar **Te**rrestrial **Re**lations **O**bservatory (STEREO) has two spacecraft that can see parts of the Sun that are not facing Earth. Following Kepler's laws, STEREO-Ahead (STEREO-A) orbits the Sun slightly inside the Earth's orbit, and therefore moves slightly faster than Earth, while STEREO-Behind (STEREO-B) orbits the Sun slightly outside the Earth's orbit, and therefore moves slightly slower than Earth. Thus they respectively drift ahead of or behind the Earth as our planet orbits the Sun; in fact, at the time of writing (2012) they were viewing the Sun's far side (■ Fig. 10–20). By seeing, for example, that many coronal mass ejections are shooting out of the far side of the Sun, we might expect strong ones to be heading toward Earth about two weeks later, when that side is roughly facing us.

No specific theory for explaining the details of the eruption of solar flares and coronal mass ejections is generally accepted, though it is agreed that the tangled solar magnetic fields in sunspot regions are the only places where enough energy to power these phenomena could be stored. A good analogy is a bunch of rubber bands stretched tight and tangled up, but the solar phenomena are on a far greater energy scale. There are multiple possible triggers, likely not the same for every eruption.

10.2d Filaments and Prominences

Studies of the solar atmosphere through filters that pass only hydrogen radiation also reveal other types of solar activity. Relatively dark

(a)

■ **FIGURE 10–20** Two views of the Sun obtained from NASA's two **S**olar **Te**rrestrial **R**elations **O**bservatory (STEREO) spacecraft on July 11, 2010, the day of a total solar eclipse. (a) STEREO-A (ahead of Earth) and (b) STEREO-B (behind Earth). By obtaining images from different perspectives, we get a more complete picture of the three-dimensional structure of the solar corona and coronal mass ejections.

filaments are seen threading their way for up to 100,000 km across the Sun, often in the vicinity of sunspots. They mark the locations of zero vertical field strength (that is, of horizontal magnetic field) separating opposite polarities. There are also bright areas, called "plages," often associated with sunspots.

When filaments happen to be on the Sun's visible edge, they project into space, often in beautiful shapes; these are called **prominences** (■ Fig. 10–21). Prominences can be seen with the eye during solar eclipses and glow pinkish at that time because of their emission in hydrogen and a few other spectral lines (■ Fig. 10–22). They can be observed from the ground even without an eclipse, if a filter that passes only light emitted by hydrogen gas is used.

Prominences appear to be composed of matter in a condition of temperature and density similar to matter in the quiet chromosphere, somewhat hotter than the photosphere. Sometimes prominences can hover above the Sun, supported by magnetic fields, for weeks or months. Other prominences change rapidly.

10.3 THE SUN AND THE THEORY OF RELATIVITY

The intuitive notion we have of gravity corresponds to the theory of gravity advanced by Isaac Newton in 1687. We now know, however, that Newton's theory and our intuitive ideas are not sufficient to explain the Universe in detail. Theories advanced by Albert Einstein in the first decades of the 20th century provide us with a more accurate understanding (see *Lives in Science 10.1: Albert Einstein*).

Einstein was a young clerk in the patent office in Switzerland when, in 1905, he published five scientific papers that revolutionized three major subfields of physics. One of the papers dealt with the effect of light on metals, two dealt with molecules and the irregular motions of small particles in liquids, and the final two dealt with motion.

■ **FIGURE 10–21** A giant solar prominence sticks off the Sun. This is a view in the fundamental ultraviolet line of ionized helium.

■ **FIGURE 10–22** Prominences and the chromosphere appear pinkish in this image from a solar eclipse. Their color comes mainly from the red Hα emission line that they radiate, but with some other emission lines mixed in. Note that the spectral lines are in emission only because we see them off the edge of the Sun silhouetted against the dark sky; spectral lines in the bright photosphere, on the other hand, are absorption lines. Prominences (as well as the corona) change with time over minutes or hours.

LIVES IN SCIENCE 10.1   Albert Einstein

Albert Einstein was born in Ulm, Germany, in 1879. He graduated from a university in Switzerland. Unable to get an academic position, he went to work in the Patent Office as an examiner. Out of this lemon, he made lemonade: He later stated that this job with relatively few working hours left him all the rest of his time to do physics unconstrained.

In 1905, Einstein published papers on five major ideas of physics – a record never matched in the history of the subject. First, he explained the mechanism by which light falling on a metal surface causes electrons to be given off. This work on the "photoelectric effect" turned out to be basic to our current ideas of light and how it interacts with atoms. (Later, it was the official reason for his being awarded the Nobel Prize in Physics.) It showed that particles of light exist, the particles we now call photons (see our discussion in Chapter 2).

Next, for his doctoral thesis, he found a way to estimate the dimensions of molecules by statistically studying their motion in liquids. In his third, somewhat related paper, he explained the jiggling motion (known as "Brownian motion") of tiny particles that a careful observer can see moving around inside liquids. Einstein said that the particles are hit by atoms over and over again, and are displaced from their original positions by distances proportional to the square root of time.

In his fourth paper, he advanced his special theory of relativity, which described motion and gave a central role to the speed of light. He showed that time, space, and motion are all relative to the observer; absolute measurements of these quantities cannot be made. This work culminated in a fifth paper, in which Einstein presented $E = mc^2$, probably the most famous equation in all of physics. (We will discuss it in Chapter 12.)

Einstein's work made him well known in scientific circles, and he was offered university professorships. Over the next decade, he worked incessantly on incorporating gravity into his theories, and by 1914 had a prediction of the angle by which light passing near the Sun would be bent. He was then a professor in Berlin. German scientists went to Russia to try to observe this effect during a total solar eclipse, but were interned during the war and could not make their observations. The fact that they did not observe the eclipse in 1914 turned out to be a blessing, for Einstein had not yet found an adequate form of the theory, and his prediction was too low by a factor of 2. By 1916, Einstein had revised his theory, and his prediction was the value that we now know to be correct. When the British scientist Arthur Eddington verified the prediction at the 1919 eclipse, Einstein triumphed.

The coming to power of the Nazis in the 1930s forced Einstein to renounce his German citizenship. He was persecuted as a Jew and his work was attacked as "Jewish physics," as though one's religion had some bearing on scientific truth. He accepted an offer to come to America to be the first professor at the Institute for Advanced Study, which was being set up in Princeton, New Jersey, not far from Princeton University.

During his years in Princeton, Einstein continued to work on scientific problems. However, his ideas were far from the mainstream and, partly as a result, his work on unifying the fundamental forces of nature did not succeed. Of course, Einstein had long worked far from the mainstream, but this time he was avoiding the basic physics of atoms – quantum mechanics – which he felt was an incomplete representation of nature.

Einstein was a celebrity in spite of his modesty; he was so much photographed that he once gave his profession as "artist's model." He devoted himself to pacifist and Zionist causes and was very influential.

Einstein was a strong backer of the State of Israel and was even once asked to be its president. His understanding of the Nazi peril forced him to put aside his pacifism in that context. On one important occasion, scientists wishing to warn President Roosevelt that atomic energy could lead to a Nazi bomb enlisted Einstein to help them reach the president's ear.

Einstein died in 1955. A project is now well under way to publish all his papers in a uniform set of books. The volume containing his childhood documents has given insights into the formation of his thought. For his achievements and their influence on modern science and technology, Einstein was named "Person of the Century" in the December 31, 1999, issue of TIME magazine.

Courtesy of the Huntington Library, San Marino, CA, and the Albert Einstein Archives, courtesy of the Hebrew University and Library

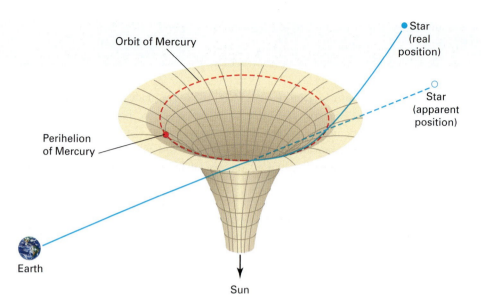

■ **FIGURE 10–23** According to Einstein's general theory of relativity, the presence of a massive body warps the space nearby, here illustrated with the curved grid. The solid line shows the actual path of light. We on Earth, though, trace back the light in the direction from which it came. At any given instant, we have no direct knowledge that the light's path has curved, so we assume that the light has always been travelling straight. Also shown is the orbit of Mercury and its perihelion, or point of closest approach to the Sun. The location of the perihelion slowly shifts with time as a result of several effects, including one that was unexplained without Einstein's general theory of relativity.

Einstein's work on motion became known as his "special theory of relativity." It is "special" in the sense that it does not include the effect of gravity or accelerations and is therefore not "general." "Special" thus means "limited" in this case. Einstein's special theory of relativity assumes that the speed of light – about 300,000 km/s – is an important constant that cannot be exceeded by real objects. (More precisely, no particle or information can go *through* space faster than the speed of light.) This theory must be used to explain the motion of objects moving very fast – near the speed of light – and has been thoroughly tested and verified.

In 1916, Einstein came up with his **general theory of relativity**, which explains gravity and also deals with accelerations. In this theory, objects simply move freely in a curved (warped) space, making it look like they are pulled by gravity. In fact, Einstein showed that time is a dimension, almost but not quite like the three spatial dimensions, and it can also be warped. Thus Einstein worked in four-dimensional space-time.

Picture a ball rolling on a golf green. If the green is warped, the ball will seem to curve. If the surface could be flattened out without distortions, though, or if we could view it from a perspective in which the surface were flat, we would see that the ball is rolling in a straight line (defined to be the shortest path between two points). This analogy shows the effect of a two-dimensional space (a surface) curved into an extra spatial dimension. Einstein's general theory of relativity treats mathematically what happens as a consequence of the curvature (warping) of space-time.

In Einstein's mathematical theory, the presence of mass or energy curves space-time, just as your bed's surface ceases to be flat when you put your weight on it. Light travelling on Einstein's curved space tends to fall into the dents, just as a ball rolling on your bed would fall into its dents. Einstein, in particular, predicted that light travelling near the Sun would be slightly bent because the Sun's gravity must warp space (■ Fig. 10–23).

The Sun, as the nearest star to the Earth, has been very important for testing some of the predictions of Einstein's general theory

of relativity. The theory could be checked by three observational tests that depended on the presence of a large mass like the Sun for experimental verification.

First, Einstein's theory showed that the closest point to the Sun (the "perihelion") of Mercury's elliptical orbit would move slightly around the sky over centuries – that is, it would gradually "precess" (have its orientation rotate). Precession is produced mainly by gravitational interactions with the other planets, but even after all of them had been taken into account, a small amount of the observed effect remained unexplained (only 43 arc seconds per century, out of the observed total of 570 arc seconds per century produced by all relevant phenomena). Calculations with Einstein's theory accounted precisely for the amount of this residual precession of the perihelion of Mercury. Though it was a plus for Einstein to have explained it, the true test of a scientific theory is really whether it predicts *new* things rather than simply explaining old ones.

The second test arose from a major new prediction of Einstein's theory: Light from a star would act as though it were bent toward the Sun by a very small amount (■ Fig. 10–24), though twice the

■ **FIGURE 10–24** The prediction, in Einstein's own handwriting, of the deflection of starlight by the Sun. In this early version of his theory, Einstein predicted half the value he later calculated. The number he gives is slightly further wrong, because of his arithmetic error. (There is a story that one of his fellow chamber-music players once said, when Einstein missed a beat, "Einstein, can't you count?") We also see his signature with the name of his university.

amount that could be predicted by pre-Einsteinian theory. We on Earth, looking back past the Sun, would see the star from which the light was emitted as though it were shifted slightly away from the Sun. Only a star whose radiation grazed the observed edge ("limb") of the Sun would seem to undergo the full deflection; the effect diminishes as one considers stars farther away from the solar limb.

To see the effect, one has to look near the Sun at a time when the stars are visible, and this could be done only during a total solar eclipse. The effect in the amount predicted by Einstein was verified at the total solar eclipse of 1919. Scientists hailed this confirmation of Einstein's theory, and from the moment of its official announcement, Einstein was recognized by scientists and the general public alike as the world's greatest scientist (■ Fig. 10–25).

Similar observations have been made at subsequent eclipses, though they are very difficult to conduct. Fortunately, the effect is constant throughout the electromagnetic spectrum, and the test can now be performed more accurately by observing how the Sun bends radiation from radio sources, especially quasars (which are discussed in Chapter 17). The results agree with Einstein's theory to within 1 per cent, enough to make the competing theories very unlikely. As we shall see later on, the effect is now very well tested farther out in the Universe.

The third traditional test of general relativity was to verify the prediction of Einstein's theory that gravity would cause the spectrum of light going outward from a massive object to be redshifted, an effect known as the "gravitational redshift." This effect is exceedingly slight for the Earth and very small for the Sun, but has been detected in both cases. It has best been verified for extremely dense stars known as white dwarfs (see Chapter 13), in which mass is very tightly packed together.

As a general rule, scientists try to find theories that not only explain the data at hand but also make predictions that can be tested. This is an important part of the scientific method. Because the bending of radiation by a certain amount was a prediction of the general theory of relativity that had not been measured before the prediction, its verification provided more convincing evidence of the theory's validity than the theory's ability to explain the previously known rotation (precession) of Mercury's orbit. Now, decades later, the general theory of relativity is a standard part of the arsenal of tools of a theoretical astronomer. We will meet it in several places later in this book.

In particular, in Chapter 13 we will introduce the concept of gravitational waves, one of Einstein's own predictions based on the general theory of relativity. Astronomers have detected the effects of gravitational waves by studying several unusual pulsar systems. Major facilities have recently been constructed in several places around the globe to try to detect gravitational waves directly. In Chapter 14, we will see how the general theory of relativity is integral for understanding black holes.

In Chapter 16, we will formally meet gravitational lensing (see also the brief discussion of microlensing in Chapter 9), a phenomenon in which the mass of a nearer object, such as a cluster of

ECLIPSE SHOWED GRAVITY VARIATION

Diversion of Light Rays Accepted as Affecting Newton's Principles.

HAILED AS EPOCHMAKING

British Scientist Calls the Discovery One of the Greatest of Human Achievements.

Copyright, 1919, by The New York Times Company.
Special Cable to The New York Times.

LONDON, Nov. 8.—What Sir Joseph Thomson, President of the Royal Society, declared was "one of the greatest —perhaps the greatest—of achievements in the history of human thought" was discussed at a joint meeting of the Royal Society and the Royal Astronomical Society in London yesterday, when the results of the British observations of the total solar eclipse of May 29 were made known.

There was a large attendance of astronomers and physicists, and it was generally accepted that the observations were decisive in verifying the prediction of Dr. Einstein, Professor of Physics in the University of Prague, that rays of light from stars, passing close to the sun on their way to the earth, would suffer twice the deflection for which the principles enunciated by Sir Isaac Newton accounted. But there was a difference of opinion as to whether science had to face merely a new and unexplained fact or to reckon with a theory that would completely revolutionize the accepted fundamentals of physics.

The discussion was opened by the Astronomer Royal, Sir Frank Dyson, who described the work of the expeditions sent respectively to Sobral, in Northern Brazil, and the Island of Principe, off the west coast of Africa. At each of these places, if the weather were propitious on the day of the eclipse, it would be possible to take during totality a set of photographs of the obscured sun and a number of bright stars which happened to be in its immediate vicinity.

The desired object was to ascertain whether the light from these stars as it passed by the sun came as directly toward the earth as if the sun were not there, or if there was a deflection due to its presence. And if the deflection did occur the stars would appear on the

Copyright © 1919 The New York Times Co.

■ **FIGURE 10–25** The first report of the eclipse results, from *The New York Times* in 1919, showing how the eclipse captured the public's attention.

galaxies, affects light from farther objects, sometimes smearing it into beautiful arcs. In Chapter 17, we will learn about the giant black holes thought to be at the centers of galaxies and quasars. And in Chapters 18 and 19, we will discuss cosmology, which is based on Einstein's general theory of relativity applied to the Universe on the largest observable scales.

CONCEPT REVIEW

» The everyday layer ("surface") of the quiet Sun we see is the **photosphere**; phenomena in it that appear nonuniformly and change with time are signs of the **active Sun**. Beneath it is the solar **interior**, with the energy generated at the solar **core**, whose temperature is 15 million kelvins (Sec. 10.1). The **solar atmosphere** above the photosphere contains the pinkish **chromosphere** and the white **corona**. The corona expands into space as the **solar wind**, a stream of particles that bathes the Earth.

» The Sun's photosphere, observed in all the visible light together (which is known as **white light**), is covered with tiny **granulation** (Sec. 10.1a). Each granule is a pocket of rising hot gas or falling cool gas driven by the process of **convection**, which is similar to water boiling. Oscillations (vibrations) of the surface reveal conditions in the solar interior; the study of the Sun in this manner is called **helioseismology** by analogy with Earth seismology. The spectrum of the photosphere shows a multitude of Fraunhofer (absorption) lines; as with other normal stars, it does not produce emission lines.

» The Sun's chromosphere is a very thin layer, somewhat hotter than the photosphere below it (Sec. 10.1b). It consists of many spicules, which are jets of gas that rise and fall, and it shows an emission-line spectrum rather than absorption lines, since the gas is silhouetted against the dark sky rather than against the bright photosphere.

» The corona, best seen at total solar eclipses, contains gas with a temperature of 2 million kelvins (Sec. 10.1c); thus, it is actually a plasma, consisting of positively and negatively charged particles, and it produces emission lines. The magnetic field shapes the coronal gas into **streamers**. Regions where the corona is less dense and cooler than average are **coronal holes**. The slow solar wind is a general coronal expansion, and the fast solar wind comes from coronal holes. Although instruments on spacecraft can be used to block the photosphere, some chromospheric and coronal phenomena are still best observed from the ground during solar eclipses (Sec. 10.1d).

» **Sunspots** are regions in the photosphere with strong, tangled magnetic fields and are cooler than the surrounding photosphere; the magnetic fields inhibit the rise of hot, ionized gases from beneath the sunspot (Sec. 10.2a). Each sunspot has a dark **umbra** surrounded by a lighter **penumbra**. Sunspots usually appear in pairs having a north pole, a south pole, and **magnetic-field lines** linking them; these pairs are often in larger groups.

» The **solar-activity cycle**, including the **sunspot cycle**, lasts about 11 years (or about 22 years, if magnetic polarity is included); the average number of sunspots periodically rises and falls (Sec. 10.2b). The **Maunder minimum** was a 17th- and 18th-century period when sunspots were essentially absent. The next solar maximum is expected in late 2013 and will probably be weaker than average, but there is some evidence that the following sunspot cycle might not occur at all.

» **Solar flares** and **coronal mass ejections** are eruptions of tremendous amounts of energy (Sec. 10.2c). Electromagnetic radiation and particles from flares can disrupt radio communications or produce electrical surges, affecting satellites and leading to blackouts of electricity. **Prominences** are pinkish protrusions off the edge of the Sun and can be seen during total solar eclipses; when viewed in silhouette on the face of the Sun, they appear as relatively dark **filaments** (Sec. 10.2d).

» In 1905, Albert Einstein published five very important papers, including two on the "special theory of relativity," which assumes that the speed of light is an important constant that cannot be exceeded by real objects moving through space (Sec. 10.3). In 1916, Einstein published his new theory of gravity, the **general theory of relativity**; mass and energy are assumed to warp (curve) space and time, and objects move freely within curved space-time. Some basic tests of this theory involved the large mass of the Sun. Specifically, it was verified that the Sun's mass affects the orbit of Mercury; the perihelion (point of closest approach to the Sun) rotates, or precesses, slowly with time, showing the changing orientation of Mercury's elliptical orbit. Observational verification that the Sun bends starlight passing near it, as predicted by Einstein, instantly made him world famous.

QUESTIONS

1. Sketch the Sun, labeling the interior, the photosphere, the chromosphere, the corona, sunspots, and prominences.

2. Draw a graph showing the Sun's approximate temperatures, starting with the core and going upward through the corona.

3. Define and contrast a prominence and a filament.

4. Why are we on Earth particularly interested in coronal holes?

5. List three phenomena that vary with the solar-activity cycle.

6. Discuss what can be learned from studies of the vibration (oscillation) of the Sun's atmosphere.

7. Why can't we observe the corona every day from the Earth's surface?

8. How do we know that the corona is hot?

9. If the corona is so much hotter than the photosphere, why isn't it much brighter than the photosphere, per unit area?

10. Describe relative advantages of ground-based eclipse studies and of satellite studies of the corona.

11. What is the process of convection? Give an example in everyday life.

12. Describe the sunspot cycle.

13. Why do we say that the true solar-activity cycle actually has a period of 22 years, rather than 11 years?

†14. Large groups of sunspots can be relatively long-lived (a few months), and they remain essentially fixed at the same physical location on the photosphere throughout their lives. Suppose a sunspot group appears to move from the center to the edge (limb) of the Sun's disk in 7 days. What is the approximate rotation period of the Sun? (A rough estimate will suffice; don't worry about not being able to see the sunspot if it is exactly at the edge of the Sun.)

†15. (a) From the table in *A Closer Look 10.1: The Most Common Elements in the Sun's Photosphere*, calculate the percentage (by number) of helium atoms in the Sun and the percentage of iron atoms. (b) Calculate the percentage of mass taken up by helium atoms and by iron atoms.

16. What are solar flares and coronal mass ejections?

17. What feature in the corona shows the magnetic field there?

18. What are two ways that the Sun's corona can be studied from the Solar Dynamics Observatory?

19. What is the difference between the special theory of relativity and the general theory of relativity?

20. Why was the Sun useful for checking the general theory of relativity?

21. Explain in your own words how Einstein's general theory of relativity accounts for the Sun's gravity.

22. **True or false?** The hottest region of the entire Sun is the core, even though the corona emits profusely at x-ray wavelengths.

23. **True or false?** During times of sunspot maximum, an unusually large number of prominences, solar flares, and coronal mass ejections also occur on the Sun.

24. **True or false?** It is safe and easy to view sunspots on a bright sunny day with the naked eye, without filters or other equipment.

25. **True or false?** According to Einstein's general theory of relativity, mass causes the surrounding space-time to curve.

26. **True or false?** If a dark sunspot could be viewed alone, without the glare of the surrounding photosphere, it would still appear quite dark because it emits very little light.

27. **True or false?** The solar activity cycle has a period of about 22 years, including the magnetic pole reversal.

28. **True or false?** The Sun's corona can typically be seen by blocking our view of the Sun's photosphere, for example with a coin.

29. **True or false?** The temperature of the Sun decreases from its core to the photosphere, and out through the corona.

30. **Multiple choice:** Sunspots appear dark because (a) they are patches of the photosphere that occasionally burn up, creating soot; (b) the changing magnetic polarity of the Sun causes gas in the sunspot to cool down substantially; (c) they are regions in which strong magnetic fields make it difficult for fresh supplies of hot, ionized gas to reach the photosphere; (d) they are much hotter than the surrounding area, so their emission peaks at ultraviolet wavelengths, which our eyes cannot see; or (e) they are holes in the photosphere through which the cooler interior of the Sun is visible.

31. **Multiple choice:** Which one of the following is *not* influenced by the Sun's magnetic field? (a) Solar prominences. (b) Sunspots.

(c) Solar flares. (d) Prominences and filaments. (e) The Sun's blackbody spectrum.

32. **Multiple choice:** As the solar atmosphere expands outward from the Sun, into interplanetary space, it becomes the (a) chromosphere; (b) corona; (c) spicules; (d) prominences; or (e) solar wind.

33. **Multiple choice:** Suppose the temperature of a sunspot is 4,000 K and that of the surrounding photosphere is about 6,000 K. What is the ratio (sunspot to photosphere) of the wavelengths at which the roughly blackbody spectra are brightest? (Hint: See Chapter 2.) (a) $2/3$. (b) $3/2$. (c) $(2/3)^2$. (d) 3. (e) 2,000.

†34. **Multiple choice:** Suppose the temperature of a sunspot is 4,000 K and that of the surrounding photosphere is about 6,000 K. Per unit area, about how much energy per second does the sunspot emit, compared with the photosphere? (Hint: see Chapter 2.) (a) $2/3$. (b) $3/2$. (c) 2000. (d) $(2/3)^2$. (e) $(2/3)^4$.

†35. **Multiple choice:** If you observe a sunspot whose diameter is about $1/20$ that of the Sun, it is roughly _____ times the diameter of the Earth. (a) $1/20$. (b) $1/5$. (c) 1. (d) 5. (e) 20.

36. **Multiple choice:** The Sun doesn't have a perfect blackbody spectrum because (a) its spectrum consists of emission lines of the elements in the Sun, so there are some gaps; (b) its core is much hotter than its surface, so the shape of the curve is distorted; (c) the coronal mass ejections interfere with the photons, adding too much noise to the spectrum; (d) the Sun is too hot to radiate at all wavelengths; or (e) the relatively cool outer layers absorb photons emitted in the hot inner parts of the Sun, producing absorption lines.

37. **Fill in the blank:** During a total solar eclipse, stars near the Sun's edge are observed to be a bit _____ the Sun than they would have been had the Sun not been present, in the amount predicted by Einstein's general theory of relativity.

38. **Fill in the blank:** In the few days after a powerful solar flare, we might expect to see _____ at night in Earth's atmosphere.

39. **Fill in the blank:** By far the most abundant element in the Sun is _____.

40. **Fill in the blank:** The Sun's color, when it is high in the sky and there isn't much pollution, is _____.

41. **Fill in the blank:** The gradual _____ of Mercury's orbit provided one test of Einstein's general theory of relativity.

42. **Fill in the blank:** The _____ satellite allows us to obtain three-dimensional views of coronal mass ejections and other phenomena, even on the far side of the Sun.

†This question requires a numerical solution.

TOPICS FOR DISCUSSION

1. Try to put yourself in the position of the ancients. What would you think if someone suggested that the stars are simply very distant Suns?

2. Do you think it is worthwhile to spend federal funding on the solar-terrestrial connection? Consider both short-term effects, such as electricity blackouts caused by particles from solar flares and distortions in GPS results, and long-term effects, such as those potentially affecting global climate change produced by events similar to the Maunder minimum.

A CLOSER LOOK 10.2

Solar Eclipses of 2013

During 2013, we on Earth had one total solar eclipse and one annular eclipse (when an annulus, a ring, of everyday sunlight – the solar photosphere – remained visible, largely because the Moon was relatively far from Earth). Both were visible only from Australia and the Pacific Ocean.

■ **FIGURE ACL 10.2a** A composite image of the November 14 (local time), 2012, total solar eclipse, showing the coronal streamers, which are held in place by the solar magnetic field. As is typical of the maximum of the solar-activity cycle, the streamers appear all around the lunar silhouette, and not only at equatorial latitudes. Pinkish prominences and chromosphere also show.

Jay M. Pasachoff, composited by Muzhou Lu '13, Williams College

■ **FIGURE ACL 10.2b** On May 10, 2013, from the Australian Outback north of Tennant Creek, the Sun rose in clouds, but the clouds dispersed during the hour-long partial phases and the sky was clear during the over 4 minutes of annularity, when a ring (annulus) of solar photosphere remained visible around the Moon's silhouette.

Stars are forming in this region of our Galaxy; we see the stellar nursery NGC 3324, in which radiation from hot stars has carved out a cavity. These new stars' radiation has also ca various gas atoms to glow in their respective colors. The image was taken in visible light with a European Southern Observatory telescope in Chile, using filters that emphasize light hydrogen (red) and ionized oxygen (green).

ESO

11

Stars: Distant Suns

ORIGINS

The stars are simply distant versions of our Sun. By studying their properties, such as surface temperature and intrinsic brightness, we can achieve a deeper understanding of the Sun and how its light and other forms of energy originate.

The properties of all stars, including the Sun, depend mostly on their mass, which in special cases can be determined from studies of binary stars. Star clusters can be used to determine the ages of stars, and to study stellar evolution.

The thousands of stars in the sky that we see with our eyes, and the millions more that telescopes reveal, are glowing balls of gas. Their bright surfaces send us the light that we see. Though we learn a lot about a star from studying its surface, we can never see through to a star's interior, where the important action goes on.

In this chapter, we will discuss the surfaces of stars and what they tell us. First we explain how we tell the surface temperatures of stars and what we observe to study them. We also explain how stars move, and how we determine their distances. Then we will learn about stars that come with friends: other stars or groups of stars. Only when we finish these useful studies will we go on to discuss the stellar interiors, in Chapter 12.

AIMS

1. Learn how the colors and spectral types of stars tell us their surface temperatures (Section 11.1).

2. Discuss the most fundamental way of measuring the distances of stars, and see how their intrinsic brightnesses (luminosities) are determined (Sections 11.2 and 11.3).

3. See how stars can be classified into several different groups when their surface temperatures and luminosities are plotted on a graph (Section 11.4).

4. Explore the motions of stars along and perpendicular to our line of sight (Section 11.5).

5. Investigate binary stars, and understand how they are important for measuring stellar masses (Section 11.6).

6. Learn about stars whose brightness changes with time, some of which are used to measure distances (Section 11.7).

7. Discuss star clusters and how we determine their ages (Section 11.8).

■ **FIGURE 11–1** When a narrow beam of light is dispersed (spread out) by a prism, we see a spectrum. Incoming light is usually passed through an open slit to provide the narrow beam.

11.1 COLORS, TEMPERATURES, AND SPECTRA OF STARS

11.1a Taking a Star's Temperature

Let us continue with the discussion of spectra that we began in Chapter 2. When you heat an iron poker in a fire, it begins to glow and then becomes red hot. If we could make it hotter still, it would become white hot, and eventually bluish-white. To understand the temperature of the poker or of a hot gas, we measure its spectrum (■ Fig. 11–1). A dense, opaque gas or a solid gives off a continuous spectrum – that is, light changing smoothly in intensity (brightness) from one color to the next.

Recall from Chapter 2 that a "blackbody" is a perfect emitter: Its spectrum depends only on its temperature, not on chemical composition or other factors. We can approximate the spectrum of the visible radiation from the outer layer of a star as a **blackbody curve**. The blackbody curve is also called the **Planck curve**, in honor of the physicist Max Planck. Its derivation about 100 years ago was a triumph in the early development of quantum physics.

A different blackbody curve corresponds to each temperature (■ Fig. 11–2). Note that as the temperature increases, the gas gives off more energy at *every* wavelength. Indeed, per unit of surface area, a hot blackbody emits much more energy per second than a cold one.

Moreover, the wavelength at which the spectrum reaches its maximum value is farther and farther toward the blue as the temperature increases. The wavelength of this peak energy is shown with a dashed

■ **FIGURE 11–2** The brightness (intensity) of radiation at different wavelengths for different temperatures. "Blackbodies" – ideally radiating matter – give off radiation that follows these curves. Stars follow these curves fairly closely, but not exactly.

■ **FIGURE 11–3** The series of spectral lines of hydrogen that appears in the visible part of the spectrum. The strongest line in this series, Hα (H-alpha), is in the red. This "Balmer series" is labeled with the first letters of the Greek alphabet. Note how the series converges toward the ultraviolet end of the spectrum; that location is known as the "Balmer limit."

line in Figure 11–2. At temperatures of 4,000 K, 5,000 K, 6,000 K, and 7,000 K, the peak of the blackbody curve is at wavelengths of 7,200 Å (red), 5,800 Å (yellow), 4,800 Å (green), and 4,100 Å (violet), respectively. Thus, the hottest stars look bluish and the coolest ones are reddish. Those of intermediate temperature (like the Sun) appear white, despite having spectra that peak around yellow or green wavelengths, because of the physiological response of our eyes.

Astronomers rely heavily on the quantitative expression of the two radiation laws noted in the preceding paragraphs to determine various aspects of stars (see Chapter 2, *Figure It Out 2.2: Blackbody Radiation and Wien's Law*; *Figure It Out 2.3: Blackbody Radiation and the Stefan-Boltzmann Law*). For example, by simply measuring the wavelength of the peak brightness of a star's spectrum, astronomers can take the star's temperature (though not with great accuracy, because the exact wavelength of the peak can be difficult to measure and the star isn't a perfect blackbody).

11.1b How Do We Classify Stars?

All the spectral lines of normal stars are absorption lines, also called dark lines. The absorption lines cause a star's spectrum to deviate from that of a blackbody, but not by much. Studying the absorption lines has been especially fruitful in understanding the stars and their composition.

We can duplicate on Earth many of the contributors to the spectra of the stars. Since hydrogen has only one electron, hydrogen's spectrum (Chapter 2) is particularly simple (■ Fig. 11–3). Atoms with more electrons have more possible energy levels, resulting in more choices for jumps between energy levels. Consequently, elements other than hydrogen have more complicated sets of spectral lines.

When we look at a variety of stars, we see many different sets of spectral lines, usually from many elements mixed together in the star's outer layers. Hydrogen is usually prominent, though often iron, magnesium, sodium, or calcium lines are also present. The different elements can be distinguished by looking for their distinct patterns of spectral lines.

Early in the 20th century, Annie Jump Cannon at the Harvard Observatory classified hundreds of thousands of stars by their visible-light spectra (■ Fig. 11–4). She first classified them by the strength of their hydrogen lines, defining stars with the strongest lines as "spectral type A," stars of slightly weaker hydrogen lines as "spectral type B," and so on. (Many of the letters wound up not being used.)

It was soon realized that hydrogen lines were strongest at some particular temperature, and were weaker at hotter temperatures (because in many atoms the electrons escaped completely from the atoms) or at cooler temperatures (because in many atoms the electrons were only in the lowest possible energy levels, which do not allow the visible-light hydrogen lines to form).

■ **FIGURE 11–4** Part of the "computing staff" of the Harvard College Observatory in 1917, when the word "computer" meant a person and not a machine. Annie Jump Cannon, fifth from the right, classified over 500,000 spectra in the decades following 1896. In this chapter we also discuss the work of Henrietta Leavitt, fifth from the left, who studied variable stars and discovered the period-luminosity relation that applies to Cepheid variable stars. Our basic ideas of cosmology are based on that discovery, which, we emphasize, applies only to one of the hundreds of types of variable stars: the Cepheid variables.

Rearranging the spectral types in order of temperature – from hottest to coolest – gave O B A F G K M. Recently, even cooler objects – stars and brown dwarfs – have been found, and are designated L-type and (still more recently) T-type and Y-type. (However, all of the Y-type objects, most of the T-type objects, and even some of the L-type objects are actually brown dwarfs, and hence, in a sense, aren't fully stars; see the discussion in Chapter 12.) Generations of astronomy students have memorized the order using the mnemonic "**O**h, **b**e **a f**ine **g**irl [or **g**uy], **k**iss **m**e." The authors invite you to think of your own mnemonic for O B A F G K M, including the newer spectral types L T Y mentioned above and that we discuss later in this chapter, and we will be glad to receive your entries.

Looking at a set of stellar spectra in order (■ Fig. 11–5a) shows how the hydrogen lines are strongest at spectral type A, which corresponds to a surface temperature of roughly 10,000 K.

A pair of spectral lines from calcium becomes strong in spectral type G (a type that includes the Sun), at about 6,000 K (■ Fig. 11–5b). In very cool stars, those of spectral type M, the temperatures are so low (only about 3,000 K) that molecules can survive, and we see complicated sets of spectral lines from them. At the other extreme, the hottest stars, of spectral type O, can reach 50,000 K. Some stars of type O have shells of hot gas around them and give off emission lines, though stars generally show only absorption lines in their spectra.

Astronomers subdivide each spectral type into ten subtypes ranging from hottest (0) to coolest (9); thus, for example, we have A0 through A9, and then F0 through F9. One of the first things most astronomers do when studying a star is to determine its spectral type and thus its surface temperature. The Sun is a type G2 star, corresponding to a surface temperature of 5,800 K.

We have known only since the 1920s that stars are made mainly of hydrogen and helium. When Cecilia Payne (later Cecilia Payne-Gaposchkin) at Harvard first suggested the idea that stars consist largely of hydrogen, it seemed impossible. But this was what her analyses of stellar spectra implied, and she was right. It was years before other astronomers accepted this conclusion.

11.1c The Coolest Stars

From the Planck curves of stars, we know that the cooler a star is, the farther toward the longer-wavelength end of the spectrum its peak radiation is found. Stars cooler than type M peak in the infrared, which until recent years was not very accessible to observers. But advances in electronics have made it possible to observe the infrared much better; infrared surveys like 2MASS (at 2 micrometers, but also commemorating its origin at the University of Massachusetts) and high-altitude, dry sites like Mauna Kea have provided wider windows of transparency to allow infrared radiation down to Earth.

The first additional spectral type was given the letter L, since it was adjacent to M, the previously coolest stars. The surfaces of these stars are only 1,300 to 2,000 K. Recall that a 2,000 K star has the peak of its Planck curve at 3 times the peak in the yellow/green of a 6,000 K star like the Sun. Some of these objects have interior nuclear fusion, but some are a bit too cool inside for fusion to have started,

so they aren't even technically stars. Their atmospheric spectra have many spectral lines and bands, including some from metal hydrides.

Even cooler are brown dwarfs that have been given the name of spectral type T. Their surface temperatures range from 1,300 K down to only 700 K. They show methane molecules in their spectra.

Astronomers were running out of letters of the alphabet that wouldn't be confused with other astronomical topics when they assigned spectral class Y to even cooler star-like objects. Still, a handful of such extremely cool brown dwarfs (Fig. 11–6) have been detected with WISE, NASA's **W**ide-**F**ield **I**nfrared **S**urvey **E**xplorer (run out of UCLA), which was in full operation for about 9 months until its coolant ran out in late 2010.

11.2 HOW DISTANT ARE THE STARS?

When you spot your friends on campus, you might use your binocular vision to judge how far away they are. Your brain interprets the slightly different images from your two eyes to give such nearby objects some three-dimensionality and to assess their distance. Also, we often unconsciously judge how far away an object is by assessing its apparent size compared with the sizes of other objects.

But the stars are so far away that they appear as points, so we cannot judge their size. And our eyes are much too close together to give us binocular vision at that great distance. Only for the nearest stars can we reliably measure their distances fairly directly even with our best methods.

The distances of nearby stars can be determined by **triangulation** (also known as "trigonometric parallax") – a sort of binocular vision obtained by taking advantage of our location on a moving platform (the Earth). The basic idea is that the position of a nearby object shifts relative to distant objects when viewed from different lines of sight (■ Fig. 11–7).

For example, if you put a finger in front of your eyes, and close one eye, the finger will appear to be projected against a particular background object. If you close this eye and open the other one, the position of the finger will shift toward a different background object. The amount of shift is smaller if the finger is farther away.

Here's how we apply this method to nearby stars (see *Figure It Out 11.1: Stellar Triangulation* for details). At six-month intervals, the Earth carries us entirely across its orbit, halfway around (■ Fig. 11–8a). Since the average radius of the Earth's orbit is 1 astronomical unit (au), we move by 2 au. This distance is enough to give us a somewhat different perspective on the nearest stars. These stars appear to shift very slightly against the background of more distant stars. Clearly, the more distant the star, the smaller is its angular shift (■ Fig. 11–8b). The nearest star – known as Proxima Centauri – appears to shift by only the apparent diameter of a dime seen from a distance of 2.4 km! It turns out to be about 4.2 light-years away. (It is in the southern constellation of Centaurus, and is not visible from most of the United States.)

(a)

(b)

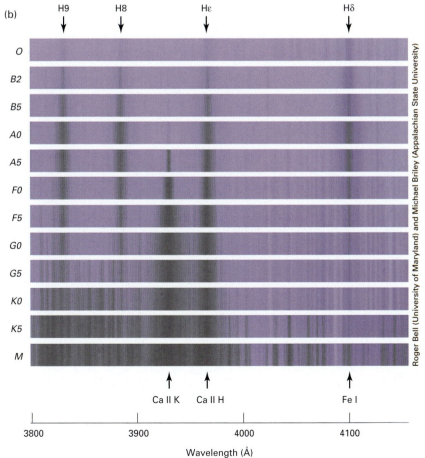

■ **FIGURE 11–5** (a) A computer simulation of the spectra of stars from a wide range of spectral types. The hottest stars are at the top and the coolest at the bottom. Notice how the hydrogen lines decline in darkness for types either above or below A-type stars. (An astronomer would often simply say "A stars.") The original lettering was in decreasing order of the strength of the hydrogen lines, starting with "A"; only subsequently was the temperature ordering discovered. (b) The violet and blue parts of the previous image, displayed to show more clearly the change in visibility of the H and K lines of ionized calcium (labeled Ca II, spoken "calcium 2") versus the Hε (H-epsilon) line of hydrogen as spectral types go from cooler to hotter. (Roger Bell, University of Maryland, and Michael Briley, Appalachian State University.)

FIGURE IT OUT 11.1

Stellar Triangulation

The technique of measuring the distance of a nearby star by triangulation is also known as "trigonometric parallax." We photograph the star twice, about 6 months apart (that is, when the Earth has moved a considerable distance in its orbit around the Sun). We then measure the angular shift of the star relative to galaxies or very distant stars, whose positions are stationary or nearly so.

The "parallax" (p) of a star is defined to be half of the angular shift produced over a 6-month baseline (2 au, the diameter of Earth's orbit – see Figure 11–7).

The distance of a star whose parallax is 1" (1 second of arc) is called 1 parsec, abbreviated 1 pc. (This term comes from a parallax of one second of arc. See Chapter 3, *Figure It Out 3.3: Angular Resolution of a Telescope*, for a review of angular measure.) In the more familiar unit of light-years (lt-yr), 1 pc is about 3.26 lt-yr.

It turns out that with this definition of a parsec, the relationship between distance and parallax is very simple. A star's distance d (in parsecs) is simply the inverse (reciprocal) of its parallax p (in seconds of arc):

$$d = 1/p.$$

For example, a star whose measured parallax is 0.5" has a distance of $1/0.5 = 2$ pc, or about 6.5 lt-yr. The nearest star, Proxima Centauri, has a parallax of 0.77" (its full angular shift over a 6-month period is twice this value, 1.54"), and hence a distance of $1/0.77 = 1.3$ pc (or 4.2 lt-yr).

(a)

WISE Team/UCLA/NASA

(b)

M. Liu, IfA/University of Hawaii

■ **FIGURE 11–6** (a) An infrared image of part of the constellation Lyra from NASA's Wide-Field Infrared Survey Explorer, with an object circled whose surface temperature is less than about 300 K. Within 30 or so light-years from us, it is one of the first stars of spectral class Y to be noted. (b) A pair of spectral class Y brown dwarfs, at temperatures of only about 800 K, so cool that the objects have methane molecules in their atmospheres, imaged with the aid of adaptive optics using a laser guide star at the Keck Observatory. Each of the brown dwarfs has only about 3 per cent of the Sun's mass, or about 30 times Jupiter's mass. The pair is in Libra, about 45 light-years from us; their separation is about 2 au, twice that of the Earth from the Sun, meaning their mutual orbit takes about 10 years.

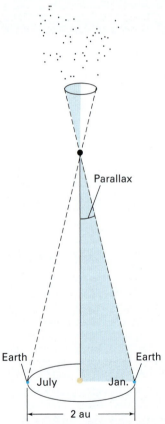

Parallax

Earth   Earth

July   Jan.

2 au

■ **FIGURE 11–7** The nearer stars seem to be slightly displaced with respect to the farther stars when viewed from different locations in the Earth's orbit.

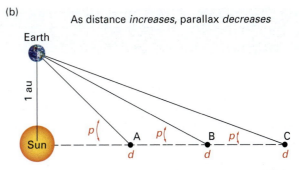

■ **FIGURE 11–8** (a) A schematic illustrating parallax. Relative to very distant stars (here, a set of 5 distant stars is shown), a nearby star appears to shift in position when viewed from Earth in January versus July. The total angular shift over a 6-month interval (when Earth's position has changed by 2 au) is defined to be twice the parallax, 2p. Equivalently, the parallax p of a star is the angle subtended (i.e., covered) by 1 au as seen from the star. (b) Stars with larger distances from Earth have smaller values for the parallax. This is the same effect as when you look at your finger with one eye, and then the other; the angular shift decreases as the distance between your finger and your eyes increases.

By calculating the length of the long side of a giant triangle – by "triangulating" – we can measure distances in this way out to a few hundred light-years. But there are only a few thousand stars that are so close to us, and at the farthest distances the results are not very accurate.

The European Space Agency lofted a satellite, Hipparcos, in 1989 to measure the positions of stars. Measuring the positions and motions of stars is known as astrometry. Based on its data, in 1997 the scientists involved released the Hipparcos catalogue, containing distances for 118,000 stars, relatively accurate to about 300 light-years away. A secondary list, the Tycho-2 catalogue, contains less-accurate distances for a million stars but provides accurate "proper motions" (motion across the sky, which we will discuss in Section 11.5) for a total of two and a half million stars.

These catalogues greatly improve our fundamental knowledge of the distances of nearby stars. Moreover, our understanding of distant stars and even galaxies is often based on the distances to these closer stars (see the discussion in Chapter 16), so the improvement is important for much of astronomy.

Our Milky Way Galaxy is perhaps 100,000 light-years across, so even the 600 light-years (diameter) spanned by the Hipparcos catalogue takes up less than 1 per cent of the diameter of our Galaxy. A successor European Space Agency spacecraft (Gaia, named after the Greek goddess of the Earth) is being readied for launch in 2013 to make a 3-D map of about 1 per cent of stars in our Galaxy, with an accuracy of 0.001 per cent close to us and 20 per cent even at the distance of the center of our Galaxy. During its expected 5-year lifetime, it will provide measurements of the positions of a billion (!) stars in our Galaxy with unprecedented accuracy, thereby leading to much improved distances. In each of its approximately 70 measurements of each star, it will record not only the star's position but also its brightness and color. Over time, comparison of the positions will also reveal proper motion – that is, the angular motion of stars across the sky. Gaia's twin cameras each contain 1.5 gigapixels – that is, 1,500 megapixels – over 100 times more pixels than ordinary digital cameras in common use for amateur photography.

11.3 HOW POWERFUL ARE THE STARS?

Automobile headlights appear faint when they are far away but can almost blind us when they are up close. Similarly, stars that are intrinsically the same brightness appear to be of different brightnesses to us, depending on their distances. A star's intrinsic brightness is its **luminosity**, or "power." If we use units that show the amount of energy a star gives off in a given time, the Sun's luminosity, for example, is 4×10^{26} joules/second, where a joule (symbol J, a capital letter since it is named after a person) is a unit of energy. (Sometimes you will see energy given in ergs, a smaller unit of energy. One joule is 10 million ergs.)

We have previously described (in Chapter 4) how we give stars' apparent brightnesses in "apparent magnitude." To tell how inherently bright stars are, astronomers often use a modification of the magnitude scale known as **absolute magnitude**; see *A Closer Look 11.1: Using Absolute Magnitudes*.

To do so, they have set a specific distance at which to compare stars. Basically, if something is intrinsically fainter, it has a higher absolute magnitude. We label our graphs in this book in both absolute magnitude and in luminosity. Some people find one easier to use; others prefer the other.

You can determine how far away a star is by comparing how bright it looks with its intrinsic brightness (its luminosity, whether measured in joules/second or as absolute magnitude). The method works because we understand how a star's energy spreads out with distance (■ Fig. 11–9).

The energy received from a star changes with the square of the distance: the energy decreases as the distance increases. Since one value goes up as the other goes down, it is an inverse relationship. We call it the **inverse-square law**. Using this law, there is a simple way to express the luminosity of a star in terms of its apparent brightness and distance (see *Figure It Out 11.2: The Inverse-Square Law*).

A CLOSER LOOK 11.1 — Using Absolute Magnitudes

To tell how intrinsically bright stars are, astronomers have set a specific distance at which to compare stars. The distance, which is a round number (10) in the special units astronomers use for triangulation (parsecs; see *Figure It Out 11.1: Stellar Triangulation*), comes out to be about 32.6 light-years. (Ten parsecs corresponds to a parallax of 0.1", the parallax being what is actually measured at a telescope.)

We calculate how bright a star would appear on the magnitude scale if it were 32.6 light-years away from us. In this way, we are comparing stars as though they were all at that distance.

The apparent brightness that a star would have at this standard distance of 32.6 light-years is its "absolute magnitude." For a star that is actually at the standard distance, its absolute magnitude and apparent magnitude are the same. For a star that is farther from us than that standard distance, we would have to move it closer to us to get it to that standard distance. This would make its magnitude at the standard distance (its absolute magnitude) brighter than the apparent magnitude we saw for it before it moved. (A brighter star has a lower numerical value, 4th magnitude instead of 6th magnitude, for example.) Then, by comparing its absolute magnitude and its apparent magnitude (which can easily be measured at a telescope), astronomers calculate how far away the star must be.

The method links absolute magnitude, apparent magnitude, and distance. Thus, it is particularly valuable, since astronomers have several ways of finding a star's absolute magnitude.

FIGURE IT OUT 11.2 — The Inverse-Square Law

The magnitude scale is intuitive but from another era; still, astronomers often use it, and it appears in amateur astronomy magazines. It has the advantage of using small numbers that are relatively easy to remember.

A different approach is to use the apparent brightness itself, b, instead of apparent magnitude. Its unit of measure is energy received per unit area, per unit time. For example, b can be measured in joules/m^2/s, where a joule is a small unit of energy (formally, one kg-m^2/s^2). Similarly, instead of absolute magnitude, we can use luminosity L to denote a star's intrinsic brightness or power (the amount of energy it emits per unit time, joules/s). A star's apparent brightness depends on its distance, because the light spreads out uniformly in all directions over a sphere whose radius is the distance. The larger the distance, the greater the surface area of the sphere, and so the smaller the amount of light passing through any small section of the sphere (see Figure 11–9). Since the surface area of a sphere is just 4π times the square of the radius, and in this case the radius is the star's distance d, the surface area is $4\pi d^2$. So, we find that the star's apparent brightness b is related to its luminosity L and distance d as follows:

$$b = L/(4\pi d^2).$$

(Note that we can verify that the equation will work by making sure that the units turn out correctly: b is in joules/m^2/s, L is in joules/s, and d is in meters. Multiplying out the units and cancelling equal terms at top and bottom shows that the equation is plausible.) In other words, each unit area of the sphere (for example, 1 m^2) intercepts only a fraction, $1/(4\pi d^2)$, of the total L.

This equation is just the well-known inverse-square law of light. For example, if a star's distance is doubled, it appears four times fainter (i.e., one-fourth as bright), because a given amount of light has spread over a surface area four times larger. As an analogy, recall that if you are spraying water out of a hose by putting your thumb over part of the opening, you can either get many people slightly wet (by being relatively far from them) or one person very wet (by being close).

We now have the tools needed to determine the luminosity of a nearby star. First, measure its apparent brightness, b (for example, with a CCD, the type of electronic detector now in universal use by astronomers); this procedure is called "photometry." Also measure its parallax, p (units: seconds of arc), and determine its distance in parsecs from the equation $d = 1/p$ (recall *Figure It Out 11.1: Stellar Triangulation*). Then convert parsecs to meters. (The conversion factor appears in Appendix 2.) Finally, use the inverse-square law to compute $L = 4\pi d^2 b$.

It is often convenient to express the luminosity in terms of the Sun's luminosity, $L_{Sun} = 3.83 \times 10^{26}$ joules/s. For example, a star whose luminosity is 7.66×10^{26} joules/s is said to have $L = 2L_{\odot}$. (In astronomy, the symbol $\odot$ denotes the Sun. Thus, the Sun's luminosity is often written as an L with this symbol as a subscript, $L_{\odot}$.) The Sun's luminosity is equivalent to an absolute magnitude of 4.8 – that is, the Sun would just barely be visible to the naked eye if it were at a distance of about 32.6 light-years (10 pc).

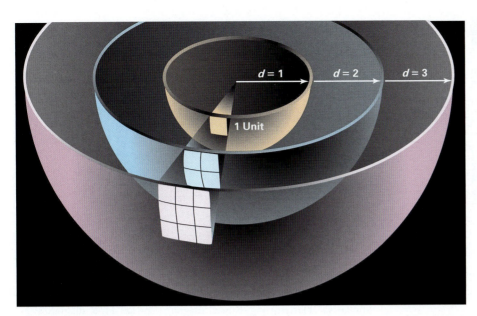

FIGURE 11–9 The inverse-square law of apparent brightness. Radiation passing through a sphere twice as far away as another sphere from a central point (radius, or distance, $d = 2$ units instead of $d = 1$ unit) has spread out so that it covers $2^2 = 4$ times the area. The central point thus appears $1/4$ as bright to an observer located on the second sphere. Similarly, radiation passing through a sphere of radius (distance) $d = 3$ units has spread out so that it covers $3^2 = 9$ times the area, so the central point appears $1/9$ as bright to an observer located on the third sphere. In general, if it were n times farther away, it would cover n^2 times the area, and appear $1/n^2$ as bright.

11.4 TEMPERATURE-LUMINOSITY DIAGRAMS

If you plot a graph that has the surface temperature of stars on the horizontal axis (typically the x-axis) and the luminosity (intrinsic brightness) of the stars on the vertical axis (typically the y-axis), the result is called a **temperature-luminosity** diagram. The luminosity (vertical axis) can also be expressed in absolute magnitudes, so the term "temperature-magnitude diagram" is synonymous. Sometimes the horizontal axis is labeled with spectral type, since temperature is often measured from spectral type (Section 11.1). Also, the colors of stars are correlated with their surface temperatures, so color is occasionally used on the horizontal axis.

If we plot such a diagram for the nearest stars (■ Fig. 11–10), we find that most of them fall on a narrow band that extends downward

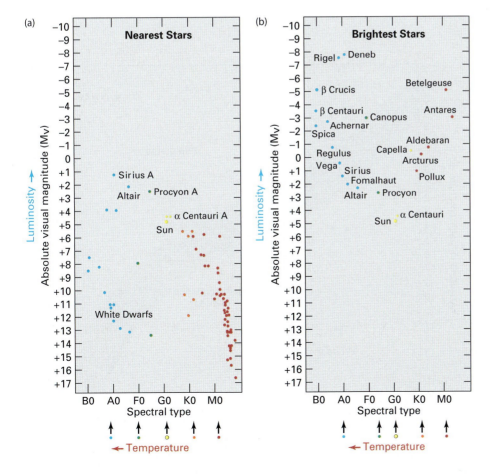

FIGURE 11–10 Temperature-luminosity (or temperature-magnitude) diagrams (a) for the nearest stars in the sky and (b) for the brightest stars in the sky. The brightness scale is given in absolute magnitude – the apparent magnitude a star would have at a standard distance of about 32.6 light-years. The upward arrow indicates increasing luminosity (intrinsic brightness). Because the effect of distance has been removed, the intrinsic properties of the stars can be compared directly on such a diagram.

Note that there are no stars near the top of the nearest-stars diagram (a). Thus none of the nearest stars is intrinsically very luminous. Note further that there are no stars near the bottom of the brightest-stars diagram (b). Thus none of the stars that appear brightest to us is intrinsically very faint.

American Institute of Physics, Emilio Segrè Visual Archives, Margaret Russell Edmondson Collection

■ **FIGURE 11–11** Henry Norris Russell and his family, circa 1917.

Dorrit Hoffleit—Yale University Observatory; courtesy of American Institute of Physics; Emilio Segrè Visual Archives

■ **FIGURE 11–12** Ejnar Hertzsprung in the 1930s.

(fainter) from the Sun. If we plot such a diagram for the brightest stars we see in the sky, we find that the stars are more scattered, but that all are intrinsically brighter than the Sun.

The idea of such plots came to two astronomers in the early years of the 20th century. Henry Norris Russell (■ Fig. 11–11), at Princeton University in the United States, plotted the absolute magnitudes (equivalent to luminosities) as his measure of brightness. Ejnar Hertzsprung (■ Fig. 11–12), in Denmark, plotted apparent magnitudes but did them for a group of stars that were all at the same distance. He could do this by considering a cluster of stars (Section 11.8), since all the stars in the cluster are essentially the same distance away from us. The two methods were very similar, in that

the magnitudes found for different stars could be directly compared. Temperature-luminosity (temperature-magnitude) diagrams are often known as "Hertzsprung–Russell diagrams" or as "H–R diagrams."

When we plot quite a lot of stars on a graph, or put together both nearby and farther stars (■ Fig. 11–13), we can see clearly that most stars fall in a narrow band across the temperature-luminosity diagram. This band is called the **main sequence**. Normal stars in the longest-lasting phase of their lifetime are on the main sequence.

The position of a star on the main sequence turns out to be determined by its *mass*: Massive stars are hotter and more powerful than low-mass stars. For reasons we will discuss in Chapter 12, the position of a star does not change much while it is on the main

■ **FIGURE 11–13** The temperature-luminosity (temperature-magnitude) diagram with both the nearest and the brightest stars included. The spectral-type axis (bottom) is equivalent to the surface-temperature axis (top). The absolute-magnitude axis (left) is equivalent to the luminosity axis (right), which shows the visual luminosity compared with that of the Sun.

The stars at the upper left of the main sequence (the gray diagonal band) are more luminous than those at the lower right. They are both hotter and larger. We mostly determine the sizes of stars indirectly: We get the luminosity of each bit of surface from their temperatures and then figure out how much surface they must have to appear as luminous as they are.

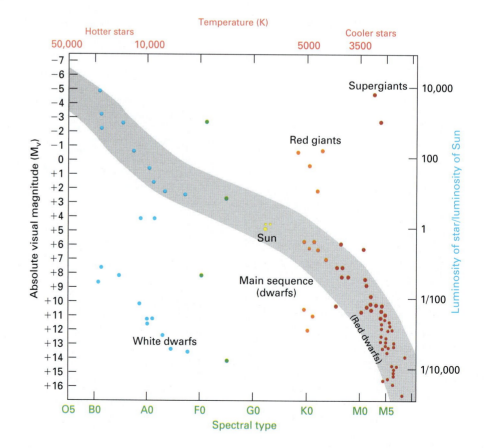

FIGURE IT OUT 11.3

A Star's Luminosity

We can easily relate the luminosity (intrinsic brightness, or power), radius, and surface temperature of a star. Recall from Chapter 2 (*Figure It Out 2.3: Blackbody Radiation and the Stefan-Boltzmann Law*) the Stefan-Boltzmann law, $E = \sigma T^4$, where E is the energy emitted per unit area (for example, cm^2) per second, T is the surface temperature, and σ (the Greek letter "sigma") is a constant. To get the luminosity (joules/s) of a star, we must multiply E by the surface area, $4\pi R^2$ for a sphere of radius R, or

$$L = 4\pi R^2 \sigma T^4.$$

Thus, if we know the luminosity and surface temperature of a star, we can derive its radius R. (The luminosity may have been derived with the inverse-square law by using the apparent brightness and distance,

as in *Figure It Out 11.2: The Inverse-Square Law*. The surface temperature can be found from Wien's law; see Chapter 2, *Figure It Out 2.2: Blackbody Radiation and Wien's Law.*)

Suppose instead we have two stars, labeled by subscripts 1 and 2, with $L_1 = 4\pi R_1^2 \sigma T_1^4$ and $L_2 = 4\pi R_2^2 \sigma T_2^4$. If we divide one equation by the other, constants such as 4π and σ cancel, and we get

$$(L_1/L_2) = (R_1/R_2)^2 (T_1/T_2)^4.$$

So, for example, if Star 1 has twice the radius of Star 2, but only half the surface temperature of Star 2, then the luminosity of Star 1 will be $^1/_4$ that of Star 2: $(L_1/L_2) = (2)^2(^1/_2)^4 = (4)(^1/_{16}) = {}^1/_4$. This example illustrates the advantage of using ratios when solving certain kinds of problems: Nowhere do we have to actually use the values of 4π and σ.

A CLOSER LOOK 11.2

Proxima Centauri: The Nearest Star Beyond the Sun

Scientists using a pair of the Very Large Telescope array's four 8.2-m telescopes made interferometric measurements of Proxima Centauri, the nearest star to Earth other than the Sun. They measured its diameter to be 1.02 ± 0.08 thousandths of a second of arc, or about the size of an astronaut on the surface of the Moon as seen from the Earth (or the head of a pin on the surface of the Earth as seen from

the International Space Station). Proxima is only one-seventh the size and one-seventh the mass of the Sun. It is on the extreme low-mass end of M-type stars; if it had only half its mass, it would be a brown dwarf (see Chapter 12) instead of a normal star. Proxima is 4.24 light-years from us. Alpha Centauri B, a Sun-like star 4.37 light-years from us, has an Earth-size planet.

sequence: The Sun stays at more or less the same position on the main sequence for about 10 billion years. Stars on the main sequence are called **dwarfs**, so the Sun is a dwarf. (Note that although the Sun is white, and also a dwarf, it is not a "white dwarf" – that specific term is reserved for a different type of star, as we will see below.)

A few stars are more luminous (intrinsically brighter) than main-sequence stars of the same surface temperature. Since the same surface area of gas at the same temperature gives off the same amount of energy per second, these stars must be bigger than the main-sequence stars (see *Figure It Out 11.3: A Star's Luminosity*). They are thus called **giants** or even, when they are much larger, **supergiants**. The reddish star Betelgeuse in the shoulder of the constellation Orion is a supergiant. Scientists are increasingly able to use interferometers in the infrared and even in the visible to measure the sizes and shapes of very large or nearby stars directly (■ Figs. 11–14 and 11–15); see *A Closer Look 11.2: Proxima Centauri: The Nearest Star Beyond the Sun*.

A few stars are intrinsically fainter than dwarfs (main-sequence stars) of the same color. These less luminous stars are called **white**

dwarfs. Do not confuse white dwarfs with normal dwarfs (main-sequence stars); white dwarfs are smaller than normal dwarfs. The Sun is 1.4 million kilometers across, while the white dwarf Sirius B (a companion of the bright star Sirius, the first entry in the appendix list of the brightest stars in the sky) is only about 10 thousand kilometers across, roughly the size of the Earth. We shall see in Chapter 13 that giants, supergiants, and white dwarfs are later stages of life for stars.

If the spectral type (that is, surface temperature) of a given star is measured, and if a detailed analysis of the star's spectrum reveals that it is a main-sequence star (this can be determined by examining subtle aspects of the spectrum), then we know its luminosity. Comparison with its observed apparent brightness then yields its distance (see *Figure It Out 11.2: The Inverse-Square Law*). This method for finding distances is known as "spectroscopic parallax." Although it works best for main-sequence stars, it can also be used with white dwarfs, giants, and supergiants, as long as the star can be confidently placed in one of these categories through detailed spectral analysis.

(a)

CHARA INTERFEROMETER: LARGEST INFRARED TELESCOPE IN THE WORLD

Size of Hubble Space Telescope compared to CHARA

A million times farther than the Sun (15 lightyears)

250 m

Mt. Wilson, CA

6 telescopes total
1 m diameter mirror in each
Effective Mirror diameter: 250 m

Altair
2 times as wide as the Sun
(Artist's conception)

100 miles away

ANALOGY

A

2 mm

Newspaper

John Monnier/Ming Zhao (University of Michigan)

(b)

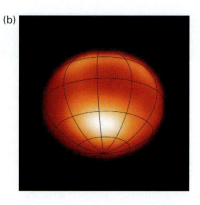

■ **FIGURE 11–14** (a) We can now measure the diameters of the nearest and brightest stars directly, using a technique known as interferometry. For example, measurements made with an interferometer on Palomar Mountain show that Altair, in the constellation Aquila, the 12th brightest star in the sky, is substantially out of round ("aspherical"). The Sun, by comparison, is so round that nobody has ever been reliably able to measure any oblateness, in spite of trying hard. Altair is 1.8 times the Sun's mass, yet it spins with a period of only 10 hours, about 60 times faster than our Sun. The spin makes Altair's equatorial diameter 14 per cent larger than its polar diameter. (b) Alpha Cephei, also known as Alderamin, is oblate and the hot polar regions appear as a bright spot shown in this CHARA image.

Interferometer measurements surprisingly showed that Vega, one of the brightest stars in the sky, is so bloated that we must be looking down nearly (within 5 per cent) at the pole of a rapidly rotating star. The measurement is particularly significant because Vega has been used by astronomers for decades as a standard star with which other stars are compared. The new result leads to the conclusion that Vega is half a magnitude brighter than expected from its spectral type. It is 37 times more luminous than our Sun, a third lower than expected based on previous measurements and on its distance (from its parallax).

11.5 HOW DO STARS MOVE?

11.5a Proper Motions of Stars

The stars are so far away that they hardly move across the sky relative to each other. Don't confuse this very slow *relative* motion of stars with the *nightly* motion of stars across the sky, which is merely a reflection of Earth's rotation about its axis (as discussed in Chapter 4), or with the *seasonal* changes of constellations, which is a reflection of Earth's orbit around the Sun (as discussed in Chapter 1).

Only for the nearest stars can we easily detect any such relative motion among the stars, which is called **proper motion** (■ Fig. 11–16). However, for the most precise work astronomers have to take the effect of the proper motions (over decades) into account. The Hipparcos satellite has improved our measurements of many stars' proper motions, and the future Gaia mission will provide enormous amounts of additional data.

Note that the proper motion is an angular movement of stars across the sky, relative to each other; the units are typically seconds of arc per century. To get the physical velocity (speed, in km/s) of each star in the plane of the sky, perpendicular to our line of sight, we need to also know its distance. After all, a given angular speed could be the result of a nearby star whose physical speed is low, or of a more distant star whose physical speed is higher. For a given physical speed, the angular speed (proper motion) decreases with increasing distance. A good example of this relationship is an airplane taking off from the airport: when it is close, it appears to zoom by, whereas when it is far away, it just crawls across the sky, even though its physical speed is actually greater than at takeoff.

11.5b Radial Velocities of Stars

Astronomers can actually measure motions toward and away from us (that is, along our line of sight) much better than they can measure motions from side to side, in the plane of the sky (perpendicular to our line of sight). Recall (Chapters 2 and 9) that a motion toward or away from us on a line joining us and a star (a radius) is called a **radial velocity**. A radial velocity shows up as a Doppler shift, a change of the spectrum so that anything that originally appeared at a given wavelength now appears at another wavelength. We briefly discussed the Doppler shift in Chapter 2, but here we will give a more thorough explanation.

Doppler shifts in sound are more familiar to us than Doppler shifts in light. You can easily hear the pitch of a car's engine drop as the car passes us. We are hearing the wavelength of its sound waves

FIGURE IT OUT 11.3 — A Star's Luminosity

We can easily relate the luminosity (intrinsic brightness, or power), radius, and surface temperature of a star. Recall from Chapter 2 (*Figure It Out 2.3: Blackbody Radiation and the Stefan-Boltzmann Law*) the Stefan-Boltzmann law, $E = \sigma T^4$, where E is the energy emitted per unit area (for example, cm^2) per second, T is the surface temperature, and σ (the Greek letter "sigma") is a constant. To get the luminosity (joules/s) of a star, we must multiply E by the surface area, $4\pi R^2$ for a sphere of radius R, or

$$L = 4\pi R^2 \sigma T^4.$$

Thus, if we know the luminosity and surface temperature of a star, we can derive its radius R. (The luminosity may have been derived with the inverse-square law by using the apparent brightness and distance,

as in *Figure It Out 11.2: The Inverse-Square Law*. The surface temperature can be found from Wien's law; see Chapter 2, *Figure It Out 2.2: Blackbody Radiation and Wien's Law.*)

Suppose instead we have two stars, labeled by subscripts 1 and 2, with $L_1 = 4\pi R_1^2 \sigma T_1^4$ and $L_2 = 4\pi R_2^2 \sigma T_2^4$. If we divide one equation by the other, constants such as 4π and σ cancel, and we get

$$(L_1/L_2) = (R_1/R_2)^2 (T_1/T_2)^4.$$

So, for example, if Star 1 has twice the radius of Star 2, but only half the surface temperature of Star 2, then the luminosity of Star 1 will be $1/4$ that of Star 2: $(L_1/L_2) = (2)^2(1/2)^4 = (4)(1/16) = 1/4$. This example illustrates the advantage of using ratios when solving certain kinds of problems: Nowhere do we have to actually use the values of 4π and σ.

A CLOSER LOOK 11.2 — Proxima Centauri: The Nearest Star Beyond the Sun

Scientists using a pair of the Very Large Telescope array's four 8.2-m telescopes made interferometric measurements of Proxima Centauri, the nearest star to Earth other than the Sun. They measured its diameter to be 1.02 ± 0.08 thousandths of a second of arc, or about the size of an astronaut on the surface of the Moon as seen from the Earth (or the head of a pin on the surface of the Earth as seen from

the International Space Station). Proxima is only one-seventh the size and one-seventh the mass of the Sun. It is on the extreme low-mass end of M-type stars; if it had only half its mass, it would be a brown dwarf (see Chapter 12) instead of a normal star. Proxima is 4.24 light-years from us. Alpha Centauri B, a Sun-like star 4.37 light-years from us, has an Earth-size planet.

sequence: The Sun stays at more or less the same position on the main sequence for about 10 billion years. Stars on the main sequence are called **dwarfs**, so the Sun is a dwarf. (Note that although the Sun is white, and also a dwarf, it is not a "white dwarf" – that specific term is reserved for a different type of star, as we will see below.)

A few stars are more luminous (intrinsically brighter) than main-sequence stars of the same surface temperature. Since the same surface area of gas at the same temperature gives off the same amount of energy per second, these stars must be bigger than the main-sequence stars (see *Figure It Out 11.3: A Star's Luminosity*). They are thus called **giants** or even, when they are much larger, **supergiants**. The reddish star Betelgeuse in the shoulder of the constellation Orion is a supergiant. Scientists are increasingly able to use interferometers in the infrared and even in the visible to measure the sizes and shapes of very large or nearby stars directly (■ Figs. 11–14 and 11–15); see *A Closer Look 11.2: Proxima Centauri: The Nearest Star Beyond the Sun.*

A few stars are intrinsically fainter than dwarfs (main-sequence stars) of the same color. These less luminous stars are called **white**

dwarfs. Do not confuse white dwarfs with normal dwarfs (main-sequence stars); white dwarfs are smaller than normal dwarfs. The Sun is 1.4 million kilometers across, while the white dwarf Sirius B (a companion of the bright star Sirius, the first entry in the appendix list of the brightest stars in the sky) is only about 10 thousand kilometers across, roughly the size of the Earth. We shall see in Chapter 13 that giants, supergiants, and white dwarfs are later stages of life for stars.

If the spectral type (that is, surface temperature) of a given star is measured, and if a detailed analysis of the star's spectrum reveals that it is a main-sequence star (this can be determined by examining subtle aspects of the spectrum), then we know its luminosity. Comparison with its observed apparent brightness then yields its distance (see *Figure It Out 11.2: The Inverse-Square Law*). This method for finding distances is known as "spectroscopic parallax." Although it works best for main-sequence stars, it can also be used with white dwarfs, giants, and supergiants, as long as the star can be confidently placed in one of these categories through detailed spectral analysis.

(a)

(b)

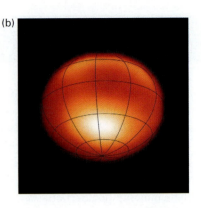

John Monnier/Ming Zhao (University of Michigan)

■ **FIGURE 11–14** (a) We can now measure the diameters of the nearest and brightest stars directly, using a technique known as interferometry. For example, measurements made with an interferometer on Palomar Mountain show that Altair, in the constellation Aquila, the 12th brightest star in the sky, is substantially out of round ("aspherical"). The Sun, by comparison, is so round that nobody has ever been reliably able to measure any oblateness, in spite of trying hard. Altair is 1.8 times the Sun's mass, yet it spins with a period of only 10 hours, about 60 times faster than our Sun. The spin makes Altair's equatorial diameter 14 per cent larger than its polar diameter. (b) Alpha Cephei, also known as Alderamin, is oblate and the hot polar regions appear as a bright spot shown in this CHARA image.

Interferometer measurements surprisingly showed that Vega, one of the brightest stars in the sky, is so bloated that we must be looking down nearly (within 5 per cent) at the pole of a rapidly rotating star. The measurement is particularly significant because Vega has been used by astronomers for decades as a standard star with which other stars are compared. The new result leads to the conclusion that Vega is half a magnitude brighter than expected from its spectral type. It is 37 times more luminous than our Sun, a third lower than expected based on previous measurements and on its distance (from its parallax).

11.5 HOW DO STARS MOVE?

11.5a Proper Motions of Stars

The stars are so far away that they hardly move across the sky relative to each other. Don't confuse this very slow *relative* motion of stars with the *nightly* motion of stars across the sky, which is merely a reflection of Earth's rotation about its axis (as discussed in Chapter 4), or with the *seasonal* changes of constellations, which is a reflection of Earth's orbit around the Sun (as discussed in Chapter 1).

Only for the nearest stars can we easily detect any such relative motion among the stars, which is called **proper motion** (■ Fig. 11–16). However, for the most precise work astronomers have to take the effect of the proper motions (over decades) into account. The Hipparcos satellite has improved our measurements of many stars' proper motions, and the future Gaia mission will provide enormous amounts of additional data.

Note that the proper motion is an angular movement of stars across the sky, relative to each other; the units are typically seconds of arc per century. To get the physical velocity (speed, in km/s) of each star in the plane of the sky, perpendicular to our line of sight, we need to also know its distance. After all, a given angular speed could be the result of a nearby star whose physical speed is low, or of a more distant star whose physical speed is higher. For a given physical speed, the angular speed (proper motion) decreases with increasing distance. A good example of this relationship is an airplane taking off from the airport: when it is close, it appears to zoom by, whereas when it is far away, it just crawls across the sky, even though its physical speed is actually greater than at takeoff.

11.5b Radial Velocities of Stars

Astronomers can actually measure motions toward and away from us (that is, along our line of sight) much better than they can measure motions from side to side, in the plane of the sky (perpendicular to our line of sight). Recall (Chapters 2 and 9) that a motion toward or away from us on a line joining us and a star (a radius) is called a **radial velocity**. A radial velocity shows up as a Doppler shift, a change of the spectrum so that anything that originally appeared at a given wavelength now appears at another wavelength. We briefly discussed the Doppler shift in Chapter 2, but here we will give a more thorough explanation.

Doppler shifts in sound are more familiar to us than Doppler shifts in light. You can easily hear the pitch of a car's engine drop as the car passes us. We are hearing the wavelength of its sound waves

FIGURE IT OUT 11.4 — Doppler Shifts

The wavelength emitted by a source of light that is *not* moving relative to the observer (it is "at rest") is known as its "rest wavelength" or "laboratory wavelength," λ_0 (pronounced "lambda nought"). If the source is moving relative to the observer, the observed wavelength λ will differ from this. (Don't make the mistake of thinking that λ_0 is the observed wavelength; the "0" does not stand for "observed," but rather for the rest frame.) For the Doppler shift of a moving source, we can write

$$\frac{\text{Change in wavelength}}{\text{Rest wavelength}} = \frac{\text{Speed of emitter}}{\text{Speed of light}}, \text{ or } \frac{\Delta\lambda}{\lambda_0} = \frac{v}{c},$$

where v is the speed of the source toward or away from us, c is the speed of light (about 3×10^5 km/s), and $\Delta\lambda = \lambda - \lambda_0$ is the change in wavelength. (The uppercase Greek letter "delta," Δ, usually stands for "the change in" or "the difference in.") This equation is valid only if v is much less than c; a more complicated expression from the special theory of relativity should be used at high speeds (say, $v > 0.2c$), if accuracy is desired. The greater the speed, the greater the error when using the simple equation $\frac{\Delta\lambda}{\lambda_0} = \frac{v}{c}$.

Note that when λ is larger than λ_0 (that is, $\lambda > \lambda_0$), the light is redshifted, so the source and observer are receding from each other. Conversely, when $\lambda < \lambda_0$, the light is blueshifted, so the source and observer are approaching each other.

The procedure to use with a star, for example, is to obtain its spectrum, recognize a familiar pattern of lines (such as the Balmer series, which marks the presence of hydrogen), measure the observed wavelength of an absorption or emission line, and compare it with the known wavelength at rest to get $\Delta\lambda$.

For example, if the Hα (H-alpha) absorption line is observed at $\lambda = 6,565$ Å, and its rest wavelength is known to be $\lambda_0 = 6,563$ Å, we find that

$$\Delta\lambda = \lambda - \lambda_0 = 2 \text{ Å, so } \Delta\lambda/\lambda_0 = 2 \text{ Å}/6,563 \text{ Å} = 3 \times 10^{-4}.$$

But we know that this result must be equal to v/c, so

$$v = (3 \times 10^{-4})c = (3 \times 10^{-4})(3 \times 10^5 \text{ km/s}) = 90 \text{ km/s}.$$

Thus, we and the star are moving away from each other at about 90 km/s. (Note that the relative radial motion determines the Doppler shift; it doesn't matter whether the emitter or receiver is actually moving.)

Capricorni (actually only the third brightest star in the constellation Capricornus, contrary to the usual convention of α referring to the brightest star) consists of two unrelated stars separated by about 6.6 arc minutes, and they are easily resolved (that is, seen individually) with the naked eye.

However, over 50 per cent of the apparently single "stars" you see in the sky are actually multiple systems bound by gravity and orbiting each other. (More precisely, the stars orbit their mutual, or common, center of mass; see *Figure It Out 11.5: Binary Stars*.) Astronomers take advantage of such **binary stars** (or simply "binaries") to find out how much mass stars contain.

Visual Binaries

If you look up at the Big Dipper, you might be able to see, even with your naked eye, that the middle star in the handle (Mizar) has a fainter companion (Alcor). Native Americans called these stars a horse and rider. Though they were long thought to be only an optical double, recent studies indicate that they are probably bound by gravity and orbit each other, forming a genuine **visual binary**.

When you look at Mizar through a telescope, even after separating it from Alcor you can see that it is split in two. In fact, Mizar was the first double star to have been discovered telescopically (in

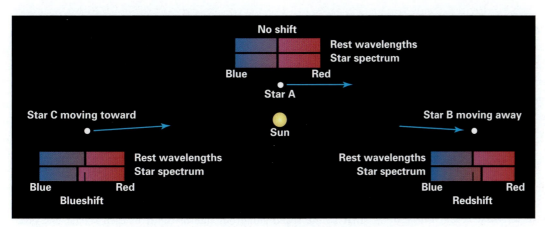

■ **FIGURE 11–18** The Doppler effect in stellar spectra. In each pair of spectra, the position of a spectral line in the laboratory on Earth is shown at top, and the position observed in the spectrum of a star is shown below it. Lines from approaching stars appear blueshifted and lines from receding stars appear redshifted. Lines from stars that are not moving toward or away from us, even if they are moving transversely, are not shifted. A short, dashed, vertical line marks the unshifted position on the spectrum showing shifts.

FIGURE IT OUT 11.5 **Binary Stars**

The two stars in a binary system generally have elliptical orbits. For simplicity, however, consider the orbits to be circular (see the figure). One can define the center of mass of the system with the equation

$$m_1 r_1 = m_2 r_2,$$

where r_1 and r_2 are the distances of Stars 1 and 2 (respectively) from the center of mass. Note that the center of mass must be along a line joining the two stars.

This behavior is analogous to two weights on opposite ends of a long plank that balances on a point, like a child's seesaw. The larger weight must be closer to the balance point than the smaller weight.

If the two stars have equal mass, they are the same distance from the center of mass. Their orbits have equal size, the stars move with equal speed, and their orbital periods are equal.

If $m_1 > m_2$, then Star 1 is closer to the center of mass than Star 2 is. The orbit of Star 1 is smaller than that of Star 2, and Star 1 moves around the center of mass at a lower speed than Star 2. Their orbital periods remain equal.

If m_1 is much, much larger than m_2 (that is, $m_1 \gg m_2$), then Star 1 is very close to the center of mass. The orbit of Star 1 is tiny, and Star 1 moves very slowly around the center of mass. Their orbital periods remain equal. A good example is the Sun–Earth system, ignoring the other planets. The Sun is nearly (but not quite) stationary. (This slight movement of a star makes it possible to detect planets indirectly, as discussed in Chapters 2 and 9.) We often say that the "Earth orbits the Sun," but a formally more correct statement is that the Sun and Earth each orbit their common center of mass.

Most of these basic principles remain valid even when we consider elliptical orbits. The main difference is that the orbital speed of each star varies with position in the orbit; recall Kepler's second law (equal areas in equal times; Chapter 5).

■ FIO 11–5

1650). These two stars orbit each other, and are individually visible through a telescope; thus they also constitute a visual binary.

We see visual binaries best when the stars are relatively far away from each other in their mutual orbits (■ Fig. 11–19). Because of Kepler's third law, these relatively large orbits correspond to long orbital periods – even hundreds of years or longer. We can detect visual binaries better when the star system is relatively near to us compared with other star systems.

Spectroscopic Binaries

The two components of Mizar are known as Mizar A and Mizar B. If we look at the spectrum of either one (■ Fig. 11–20), we can see

that, over a period of days, the spectral lines seem to split and come together again. The spectrum shows that two stars are actually present in Mizar A, and another two are present in Mizar B. (Alcor, too, is a binary star; thus the Mizar-Alcor system actually consists of 6 stars bound together by gravity!)

Mizar A and Mizar B are each spectroscopic binaries, since they are distinguished by their spectra. Even when only the spectrum of one star is detectable (■ Fig. 11–21), the other star being too faint, we can still tell that it is part of a spectroscopic binary if the spectral lines shift back and forth in wavelength over time. These shifts arise from the stars' motions; they are Doppler shifts. The radial velocity (that is, the speed along our line of sight) of each star at any given time can be determined from the Doppler formula.

We give examples of the use of this method in Chapter 9, when we discuss how we find the masses of exoplanets orbiting stars, and in Chapter 14, when we discuss how we conclude that the masses of objects in certain binary systems show that they are black holes. Both are 21st-century uses of a 19th-century technique. Unfortunately, since we don't usually know the angle at which we are seeing these orbiting systems, we get only lower limits to the stars' masses rather than actual values for the masses. Since faster-moving stars have greater Doppler shifts, we tend to detect spectroscopic binaries when the stars in them are close to each other. Following Kepler's third law, such large velocities correspond to small orbits.

Eclipsing Binaries

Yet another type of binary star is detectable when the light from a star periodically dims because one of the components passes in front of (eclipses) the other (■ Fig. 11–22). From Earth, we graph the "light curves" of these **eclipsing binaries** by plotting their brightnesses over time.

Eclipsing binaries are detected by us only because they happen to be aligned so that one star passes between the Earth and the other star. Thus all binaries would be eclipsing if we could see them at the proper angle. (As discussed in Chapter 9, many planets around distant stars, and thousands of additional candidate exoplanets, have

■ **FIGURE 11–19** The double star Albireo, also known as β (the Greek letter "beta") Cygni, contains a B-type star and a K-type star. They make a particularly beautiful pair even for amateur observers using small telescopes because of their brightness and contrasting colors (blue and gold, which unfortunately do not appear clearly in most photographs, including this one; the effect is much more striking when viewed visually through a telescope). Albireo is high overhead on summer evenings.

■ **FIGURE 11–20** Two spectra of this star taken 2 days apart show that it is a spectroscopic binary. The lines of both stars are superimposed in the upper stellar absorption spectrum (red arrow). They are separated in the lower spectrum (blue arrow) by an amount that shows that the stars were then moving 140 km/s with respect to each other.

■ **FIGURE 11–21** Two spectra of Castor B, taken at different times. The star's spectrum we see shows a change in the amount of Doppler shift caused by its changing speed toward us. Thus the star must be moving around another star. We conclude that it is a spectroscopic binary, even though only one of the components can be seen. Castor and Pollux are Gemini, the Twins in Greek mythology and a northern constellation.

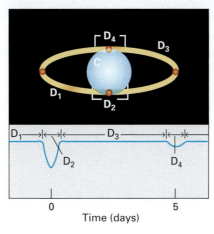

■ **FIGURE 11–22** The shape of the light curve of an eclipsing binary star depends on the sizes of the components, the surface temperature, and the angle from which we view them. At lower left, we see the light curve that would result for Star B orbiting Star A, as pictured above it. When Star B is in the positions shown at top with subscripts, the regions of the light curve marked with the same subscripts result. The eclipse at B_4 is total. At right, we see the appearance of the orbit and the light curve for Star D orbiting Star C, with the orbit inclined at a greater angle than at left. The eclipse at D_4 is partial.

■ **FIGURE 11–23** The appearance of the spectrum of a binary star and whether the lines shift in wavelength over time depend on the angle from which we view it. From far above or below the plane of the orbit, we might see a visual binary, as at top and at lower right. From close to the plane of the orbit, even if the stars are so close to each other in angle in the sky (closer than shown here at left) that they can't be seen as separate stars, we might see only a spectroscopic binary, as at left. Similarly, for stars exactly in the plane of the orbit but too close in angle to see as separate objects, we could still see the system as an eclipsing binary, as at lower left.

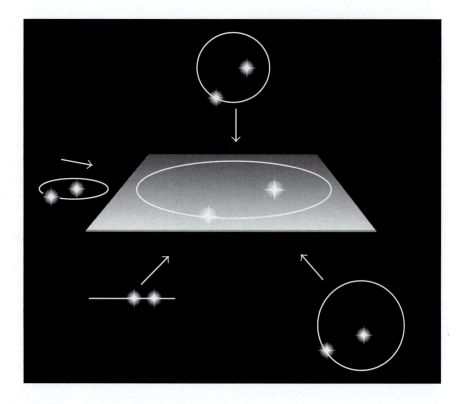

been detected in a similar fashion, by a dimming of the total light presumably when the planet passes in front of, or transits, its parent star.) Since one star can eclipse another only when we are in the plane of its orbit, we know that we are seeing the orbit edge-on and can calculate the masses of the component stars. We can accurately measure stellar masses only for eclipsing binaries.

The same binary system might be seen as different types depending on the angle at which we happen to view the system (■ Fig. 11–23).

Astrometric Binaries

One more type of double star can be detected when we observe a wobble in the path of a star as its proper motion, its motion with respect to other stars in our Galaxy, takes it very slowly across the sky (■ Fig. 11–24). The laws of physics hold that a moving mass must travel in a straight line unless affected by a force. The wobble away from a straight line tells us that the visible star must be orbiting an invisible object, pulling the visible star from side to side. Since measurement of the positions and motions of stars is known as astrometry, these stars are called **astrometric binaries**.

A similar method was used to try to detect planets around distant stars, though it has been unsuccessful. The Hipparcos spacecraft, for example, was designed to measure such small motions of stars, and Gaia will, from 2013 onward, give even better precision. However, the new planets that have been located around other stars (Chapter 9) were found at first by using the Doppler effect (and, later on, by

■ **FIGURE 11–24** (a) Sirius A and B, an astrometric binary. From studying the motion of Sirius A (usually called simply Sirius), astronomers deduced the presence of Sirius B before it was seen directly. (b) The relative positions of Sirius A and B at two extremes of their orbits, when they are closest together (top) and farthest apart (bottom).

A Sense of Mass: Weighing Stars

The amount of matter in a body is its mass. The standard kilogram resides at the International Bureau of Weights and Measures near Paris and is the ultimate standard of mass. To determine the mass of an object on Earth, we compare it, in some ultimate sense, with the standard kilogram or, more commonly, with secondary standard kilograms held by the various countries. (Work is under way to define the kilogram in terms of some standard that doesn't depend on an actual object, and a meeting was held on the subject in 2011, but a new definition is still in the future.)

Kepler's third law, as modified by Newton (Chapter 5), gives us a way of determining the mass of an astronomical body from studying an object orbiting it. But this determination requires knowledge of the universal constant of gravitation, usually called G, that appears in Newton's law of gravity. This constant is measured by laboratory experiments, which limits the resulting accuracy. The first determination of the constant nearly two hundred years ago appeared in a paper entitled "On Weighing the Earth."

Note that weight is actually different from mass. An object's weight on Earth is the gravitational force of the Earth's mass on the object's mass. The object would have a different weight on another planet, but the same mass.

The Earth's mass seems so large to us that we think we can jump up and down without moving the Earth. Actually, when we jump up, the Earth moves away from us, according to Newton's third law of motion ("For every action, there is an equal and opposite reaction"). The Earth's mass is about 10^{23} times the mass of a person, though, so it doesn't move much!

The Earth and the rest of the Solar System may be important to us, but they are only minor companions to the Sun. In "Captain Stormfield's Visit to Heaven," by Mark Twain, the Captain races with a comet and gets off course. He comes into heaven by a wrong gate, and finds that nobody there has heard of "the world." ("The world! H'm! There's billions of them!" says a gatekeeper.) Finally, someone goes up in a balloon to try to detect "the world" on a huge map. The balloonist rises into clouds and after a day or two of searching comes back to report that he has found it: an unimportant planet named "the Wart."

Indeed, the Earth is inconsequential in terms of mass next to Jupiter, which is 318 times more massive. And the Sun, an ordinary star, is about 300,000 times more massive than the Earth. Some stars are up to 100 times more massive than the Sun, but others can be as little as $1/_{12}$ as massive. We shall see that a star's mass is the key factor that determines how the star will evolve and how it will end its life.

Hundreds of billions of stars bound together by gravity make up a galaxy, the fundamental unit of the Universe as a whole. Our own Milky Way Galaxy contains a few hundred billion stars, and about a trillion (a million million) times as much mass as the Sun; we will see in Chapter 18 that a large fraction of this mass is in the form of "dark matter" rather than stars.

Since there are billions of galaxies in the Universe, if not an infinite number, the total mass itself in the Universe is not a meaningful number. More often, we discuss the *density* of a region of the Universe – the region's mass divided by its volume. The Universe is so spread out, with so much almost empty space between the galaxies, that the Universe's average density is very low. In Chapter 18 we will see that determining the value of this average density is important for predicting the future of the Universe, whether it will eventually collapse or expand forever. (However, we now know that the expansion is *accelerating*, driven by mysterious "dark energy," so the density of gravitationally attractive matter is not the only important variable.)

their transits of their parent stars) rather than by astrometry. The planets are not massive enough to cause side-to-side shifts that are measurable at the distances to their parent stars.

11.6b How Do We Weigh Stars?

A primary use of binary stars is the measurement of stellar masses (informally, "weighing" stars, though physicists actually distinguish between "mass" and "weight"). We can sometimes determine enough about the stars' orbits around each other that we can calculate how much mass the stars must have to stay in those orbits (see *A Closer Look 11.3: A Sense of Mass: Weighing Stars*).

The most accurate masses are derived from spectroscopic binaries in which the absorption lines of both stars are visible, and which are additionally eclipsing binaries. For these cases, we can also describe the radii of the two stars reliably. Such binary systems are relatively rare, so we know accurately only a few hundred stellar masses.

11.6c The Mass-Luminosity Relation

It turns out that the mass of a star is its single most important characteristic: Essentially everything else about its properties and evolution depends on mass. In *Figure It Out 11.6: The Mass-Luminosity Relation*, you can learn how the mass is connected to how powerfully the star shines. In particular, massive main-sequence stars are far more luminous and have much shorter lives than low-mass main-sequence stars. Though massive stars start with more fuel, they consume it at a huge rate.

When we discuss the evolution of stars, we will see that matter often flows from one member of a binary system to another. This

FIGURE IT OUT 11.6 The Mass-Luminosity Relation

For main-sequence stars (that is, not giants, supergiants, or white dwarfs), we find that the luminosity L (intrinsic brightness) is roughly proportional to the fourth power of mass M, or

$$L \propto M^4,$$

the "mass-luminosity relation." This proportionality ($\propto$) is only approximate; the exponent of M actually varies to some extent along the main sequence. However, it is sufficient for our purposes. Thus, for example, a main-sequence star with twice the Sun's mass is a factor of $2^4 = 16$ times more luminous (powerful) than the Sun.

The mass-luminosity relation for main-sequence stars implies that massive stars have much shorter lives than low-mass stars. Massive stars have more fuel, but they consume it at a disproportionately rapid rate. For example, a main-sequence star with twice the Sun's mass has twice as much fuel, so if all else were equal it would live twice as long. But it actually uses its fuel 16 times more rapidly! So, it can sustain itself for only $^2/_{16}$ (which is equal to $^1/_8$) as long as the Sun.

interchange of matter can drastically alter a star's evolution. And the flowing matter can heat up by friction to the very high temperatures at which x-rays are emitted. Our satellite observatories above the Earth's atmosphere detect many such sources of celestial x-rays; the Chandra X-ray Observatory and the XMM-Newton spacecraft, both launched in 1999, Suzaku, launched in 2005, and NuSTAR, launched in 2012, are the most powerful and sensitive x-ray telescopes ever available.

11.7 STARS THAT DON'T SHINE STEADILY

Some stars vary in brightness over hours, days, or months. Eclipsing binaries are one example of such **variable stars**, but other types of variable stars are actually individually changing in brightness. Many professional and amateur astronomers follow the "light curves" of such stars, the graphs of how their brightness changes over time.

A very common type of variable star changes slowly in brightness with a period of months or up to a couple of years. The first of these **long-period variables** to be discovered was the star Mira in the constellation Cetus, the Whale (■ Fig. 11–25). At its maximum brightness, it is 3rd magnitude, quite noticeable to the naked eye. At its minimum brightness, it is far below naked-eye visibility (■ Fig. 11–26). Such stars are giants whose outer layers actually change in size and temperature.

A particularly important type of variable star for astronomers is the **Cepheid variable**, of which the star δ (the Greek letter "delta") Cephei is a prime example (■ Fig. 11–27). A strict relation exists linking the period over which the star varies with the star's average luminosity (average intrinsic brightness). This "period-luminosity relation" (■ Fig. 11–28) is the key to why Cepheid-variable stars are so phenomenally important, possibly the most important type of star known from the point of view of cosmology. (To mark this fact, we are using the term "Leavitt's law," to parallel what it led to: "Hubble's law," which we will discuss in Chapter 16.)

Jay M. Pasachoff

■ **FIGURE 11–25** The constellation Cetus, the Whale (or Sea Monster). It is the home of the long-period variable star Mira, also known as o (the Greek letter "omicron") Ceti.

After all, we can observe a star's light curve easily, with just a telescope, so we know how long its period is. From the period of a newly observed Cepheid, we know its intrinsic brightness (its luminosity). By comparing its average luminosity with how bright it appears on average, we can tell how far away the star is. We can do so because a given star would appear fainter if it were farther away, according to the inverse-square law of light (see *Figure It Out 11.2: The Inverse-Square Law*). In essence, we are using the Cepheids as **standard candles**, a modern-day use of an old term for a source of known brightness. (Technically, they should instead be called "standardizable candles" or "calibratable candles"; their intrinsic brightness is not standard but rather depends on their period, and it can thus be standardized or calibrated.)

How do we pick out a Cepheid variable star? If we follow the light curves of a lot of stars, only a few of them will match the two characteristics of Cepheids. To be a Cepheid, (1) the shape of the light curve must have a specific form, with a sharp rise and a slower fall, and (2) the period of the variation must be days, weeks, or a few months long. If only one of those two criteria is met, then the star isn't a Cepheid, and we therefore can't use it for the period-luminosity relation, which applies only to Cepheids.

Cepheids are supergiant stars, powerful enough that we can detect them in some of the nearby galaxies. Finding the distances to these

(a)

Year

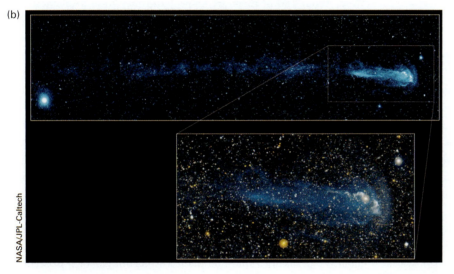

(b)

NASA/JPL–Caltech

American Association of Variable Star Observers

■ FIGURE 11–26 (a) The light curve for Mira, the first known example of a long-period variable. The horizontal axis shows time and the vertical axis shows apparent magnitude (related to apparent brightness). Notice that both the light curve's shape and the maximum brightness vary. The Sun may become a Mira-like star in about 5 billion years. The data were gathered and plotted by the American Association of Variable Star Observers (AAVSO). They often use Julian Days (JD), a consecutive count of days, to keep track of time (American Association of Variable Star Observers). (b) Ultraviolet views of Mira with GALEX (Galaxy Evolution Explorer), a 2003 spacecraft lent to Caltech by NASA in 2012, unexpectedly show that Mira, a red giant star, has a huge turbulent tail, 13 light-years in length, released over the past 30,000 years or so. The tail is shedding carbon, oxygen, and other elements that will eventually be incorporated in stars that are not yet born and perhaps even in life that could eventually arise in them. Perhaps the ultraviolet light's fluorescence makes it visible, while the 400 years of visible-light imaging didn't reveal the tail. Mira itself is a small dot in the middle of the apparent head of the comet-like object, at right, shown better in the enlargement at bottom. The bright spot at extreme left is a foreground star.

Cepheids gives us the distances to the galaxies they are in. This chain of reasoning, linking the period of variation of Cepheid variables with their average luminosities, is perhaps the most accurate method we have of finding the distances to nearby galaxies (see our discussions below and in Chapter 16, in which we describe how a main purpose for the launch of the Hubble Space Telescope was to use its especially clear images to detect Cepheid variables in a few dozen galaxies, making Cepheids visible farther away from us than previously), and so is at the basis of most of our measurements of the size of the Universe.

The period-luminosity relation was discovered in the early years of the 20th century by Henrietta Leavitt in the course of her study of stars in the Magellanic Clouds (■ Fig. 11–29). She knew only that all the stars in each of the Magellanic Clouds were about the same distance from us, so she could compare their apparent brightnesses without worrying about distance effects. The Magellanic Clouds turn out to be among the nearest galaxies to us.

The Hubble Space Telescope has been used to detect Cepheids in galaxies farther away than those in which Cepheids could be detected with ground-based telescopes (■ Fig. 11–30), since these Cepheids were blurred with adjacent stars in ground-based observations. The high resolution of Hubble has enabled these Cepheids to be picked out and to have their brightnesses followed over time. The results, as we shall see in the cosmic-distance-scale discussion (Chapter 18), are at the basis of our knowledge of the size and age of the Universe.

But we need a good way, now lacking, to measure the distances to a few Cepheids reliably in a manner independent of the period-luminosity relation. We expect that the European Space Agency's Gaia spacecraft will give us trigonometric distances to some Cepheids in our Galaxy, to improve our basis for the period-luminosity relation (Leavitt's law).

Cepheid variables are rather massive stars, more massive than the Sun, at an advanced and unstable stage of their lives. They are supergiant stars, not on the main sequence. Cepheid variables are changing in size, which leads to their variations in brightness. They contract a bit, which makes them hotter, and because of a peculiarity in the structure of the star, less energy escapes. Thus the pressure rises, and they expand. They overshoot, expanding too far, and the expansion cools them. The energy easily escapes, the pressure decreases, and they respond by contracting again. A cycle is thus set up.

Stars of a related type, **RR Lyrae variables**, have shorter periods – only several hours long – than regular Cepheids. RR Lyrae variables are often (but not exclusively) found in clusters of stars, groupings of stars we will discuss next. Since all RR Lyrae stars have the same average luminosity (average intrinsic brightness), observing an RR Lyrae star and comparing its average apparent brightness with its average luminosity enables us to tell how far away the cluster is.

How do we pick out RR Lyrae stars? Again, as for Cepheids, we follow the light curves of many stars. Only the few that have a sharp rise and a slower fall (like Cepheids), though periods less than a day (unlike Cepheids), are RR Lyrae stars.

11.8 CLUSTERS OF STARS

Clusters, or physical groups of stars, in the Milky Way Galaxy come in two basic varieties, as summarized in *A Closer Look 11.4: Star Clusters in Our Galaxy*. Clusters in other galaxies have similar properties, though there may be a few differences in detail (e.g., some globular clusters in other galaxies may be young, unlike the case in our Galaxy).

11.8a Open and Globular Star Clusters

The face of Taurus, the Bull, is outlined by a "V" of stars, which are close together out in space. On Taurus's back rides another group of stars, the Pleiades (pronounced Plee'a-deez). Both are examples of **open clusters**, groupings of dozens or a few hundred stars. We can often see 6 of the Pleiades stars with the naked eye (which is why Subaru automobiles – "Subaru" is the Japanese name for what we call the Pleiades – have badges with 6 stars showing), but binoculars reveal dozens more and telescopes show the rest.

Open clusters (■ Figs. 11–31 and 11–32) are irregular in shape. About 1,000 open clusters are known in our Galaxy. They are found near the band of light called the Milky Way (see Chapter 15), which means that they are in the plane of our Galaxy. Specifically, they are generally in or near our Galaxy's spiral arms. (They are often known as

"galactic clusters," not to be confused with "clusters of galaxies," which we shall study in Chapter 16.) They are only loosely bound by gravity, gradually dissipating ("evaporating") away as individual stars escape.

In some places in the sky, we see clusters of stars with spherical symmetry. Each such **globular cluster** looks like a faint, fuzzy ball when seen through a small telescope. Larger telescopes reveal many individual stars. Globular clusters (■ Fig. 11–33) contain tens of thousands or hundreds of thousands of stars, many more than typical open clusters. They are more strongly bound than open clusters, although their stars do occasionally get ejected through gravitational encounters with other stars.

Globular clusters usually appear far above and below the Milky Way. About 180 of them are known, and they form a spherical halo around the center of our Galaxy. We find their distances from Earth by measuring the apparent brightnesses of RR Lyrae stars in them.

Star clusters are important because each is a grouping of stars that were all formed at the same time and out of the same material. A further advantage is that all those stars are at essentially the same distance from us. Thus, by comparing members of a star cluster, we can often find out what factors make stars evolve differently or at different rates. Below, we will see how using a star cluster enables us to tell the age of its members.

When we study the spectra of stars in open clusters to determine what elements are in them, we find that all the stars (including the Sun; see *A Closer Look 10.1: The Most Common Elements in the Sun's Photosphere*) are about 92 per cent hydrogen (that is, 92 per cent of the number of atoms are hydrogen; by mass, this corresponds to about 73 per cent). Almost all the rest is helium (about 8 per cent by number, or 25 per cent by mass, since each helium atom has about four times the mass of a hydrogen atom). Generally, elements heavier

■ **FIGURE 11–27** (a) The light curve for δ (the Greek letter "delta") Cephei, the first example of a Cepheid variable. (b) Light curves for several variable stars. Note that though many types of variable stars have regular periods, while others vary irregularly over time, Cepheid variables are distinctive by the shape of their light curves, with a rapid rise and a slower fall. So merely by examining the shapes and periods of light curves of variable stars, astronomers can pick out the Cepheids. We see, left to right, CY Aquarii (popular with amateur astronomers since its short, 88-minute period makes it easy to follow), a type of variable whose prototype is δ Scuti; RR Lyrae, with a period of 0.44 days; SU Cygni, a Cepheid with a period of 3.85 days; Mira, with a period of 213.5 days; AV Delphini, an eclipsing binary (like Algol, a more widely observed eclipsing binary and another favorite of amateur astronomers); V2252 Ophiuchi, a very irregular star (like R Coronae Borealis, the prototype of the class); and SS Cygni, a flickering "cataclysmic variable."

(a)

Observations obtained for Jay M. Pasachoff with the Automatic Photoelectric Telescope

← 5.37 days →
Time

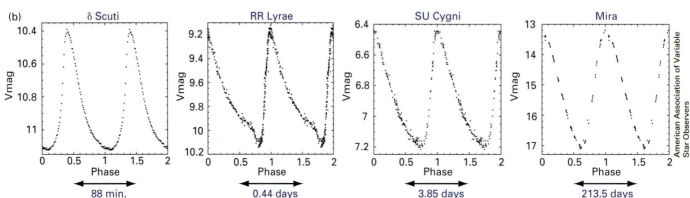

(b)

American Association of Variable Star Observers

(a)

(b)

Period in days

© G. F. Benedict 2012

■ **FIGURE 11–28** The period-luminosity relation for Cepheid variables. The stars' average intrinsic brightnesses (average luminosities) are shown on the vertical axis. The upward arrow shows increasing brightness. Note that the Cepheids were found because of the shape of their light curves and their range of periods (greater than 1 day); only after they were found in this way could Henrietta Leavitt plot them on a graph of period versus brightness to discover the period-luminosity relation, or "Leavitt's law." (a) Visual (yellow) magnitudes. (b) Infrared magnitudes, which give smaller deviations from the slanted line that represents Leavitt's law.

(a)

(b)

■ **FIGURE 11–29** (a) From the southern hemisphere, the two Magellanic Clouds are sometimes high in the sky. Since all stars in a given Magellanic Cloud are essentially the same distance from us, their intrinsic brightnesses (luminosities) are related to each other in the same way as their relative apparent brightnesses when we observe them from Earth. (b) Here we see a 2011 image of the Large Magellanic Cloud from the Spitzer Space Telescope, a 3-false-color overlay (3.6, 8, and 24 μm), with dots showing the hundreds of stars (about 5 per cent of the stars individually seen) that have been found to have escaped from the Small Magellanic Cloud and landed in its Large cousin. They were located because they are rotating in a skew direction compared with most of the stars in the Large Magellanic Cloud; they later turned out also to have different chemical compositions.

(a)

Cepheid Variable Star in Galaxy M100

Wendy L. Freedman, Observatories of the Carnegie Institution of Washington, and NASA

(b)

Cepheid Variable Star V1 in M31
HST WFC3/UVIS

F475X
F600LP

5"

N
E

Dec. 17, 2010   Dec. 21, 2010

Dec. 30, 2010   Jan. 26, 2011

NASA, ESA, the Hubble Heritage Team (STScI/AURA), and Z. Levay (STScI)

■ **FIGURE 11–30** (a) Brightness variations over time (images are shown for three different times) of a Cepheid variable in a distant galaxy, imaged with the Hubble Space Telescope. (b) Because of the historic interest in Edwin Hubble's 1923 discovery of a Cepheid in the Andromeda Galaxy, astronomers looked back at that galaxy in 2011 with the Hubble Space Telescope to image that historic star, labeled "VAR!" on the photographic plate by Hubble, at its brightest and faintest. (c) Using a light curve for the VAR! star measured by members of the American Association of Variable Star Observers (red dots), the Hubble Space Telescope's images were taken at the times shown with large cartoon stars. See Figure 16–6 for a view of the original plate.

(c)

Light curve of Cepheid variable star V1 in galaxy M31

Brightness

Period
31.4 days

Jul. '10   Aug. '10   Sep. '10   Oct. '10   Nov. '10   Dec. '10   Jan. '11

NASA, ESA, the Hubble Heritage Team (STScI/AURA), Z. Levay (STScI), and the American Association of Variable Star Observers

A CLOSER LOOK 11.4   Star Clusters in Our Galaxy

Open Clusters	Globular Clusters
No regular shape	Shaped like a ball, stars more closely packed toward center
Many are young	All are old
Temperature-luminosity diagrams have long main sequences	Temperature-luminosity diagrams have short main sequences
Where stars leave the main sequence tells the cluster's age	All clusters have nearly the same temperature-luminosity diagram and thus nearly the same age
Stars have similar composition to the Sun	Stars have much lower abundances of heavy elements than the Sun
Hundreds of stars per cluster	10,000–1,000,000 stars per cluster
Found in Galactic plane	Found in Galactic halo

■ **FIGURE 11–31** An open cluster of stars in the southern constellation Puppis.

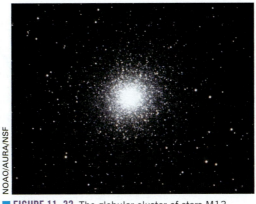

■ **FIGURE 11–32** The globular cluster of stars M13, which is the most prominent globular cluster in the northern night sky.

■ **FIGURE 11–33** The globular cluster 47 Tucanae in the southern constellation Tucana.

than helium constitute under 0.2 per cent of the number of atoms (or 2 per cent by mass). But the abundances of these "heavy elements" in the stars in globular clusters are roughly ten times lower.

We now think that the globular clusters formed about 11 to 13 billion years ago, early in the life of our Galaxy, from the original gas in the Galaxy. This gas had a very low abundance of elements heavier than hydrogen and helium. As time passed, stars in our Galaxy "cooked" lighter elements into heavy elements through fusion in their interiors, died, and spewed off these heavy elements into space (see our discussion in Chapters 12 and 13). The stars in open clusters formed later on from chemically enriched gas. Thus the open clusters have higher abundances of these heavier elements.

X-ray observations have detected bursts of x-rays coming from a few regions of sky, including a half-dozen globular clusters, and the Chandra X-ray Observatory, XMM-Newton, Suzaku (after a mythical Vermilion bird), and NuSTAR are studying more. We think that among the many stars in those clusters, some have lived their lives, have died, and are attracting material from nearby companions to flow over to them. We are seeing x-rays from this material as it "drips" onto the dead star's surface and heats up. So globular clusters still have action going on inside them, in spite of their great age.

11.8b How Old Are Star Clusters?

Large, young star clusters contain many stars with a wide range of masses spanning the entire main sequence, from type O to type L. We can determine the ages of such star clusters by looking at their temperature-luminosity (temperature-magnitude) diagrams. For the youngest open clusters, almost all the stars are on the main sequence. But the stars at the upper-left part of the main sequence – which are relatively hot, luminous, and massive – live on the main sequence for much shorter times than the cooler stars, which are fainter and less massive. The points representing their temperatures and magnitudes begin to lie off the main sequence (■ Fig. 11–34).

Thus the length of the main sequence on the temperature-luminosity diagram of a cluster gets progressively shorter: It is like the

wick of a candle burning down. Since we can calculate the main-sequence lifetimes of stars of different masses, we can tell the age of an open cluster by which stars do not appear on the main sequence. This calculation is similar to estimating how long a candle has been burning if the candle's original length and burn rate are known.

For example, suppose a populous cluster does not have any O and B main-sequence stars, but does have A-, F-, G-, K-, M-, and L-type main-sequence stars. We deduce that the cluster is about 100 million years old, because that's how long it takes B-type stars to leave the main sequence. The cluster cannot be a billion years old, because A-type stars would have left the main sequence by that stage. As time passes, the end of the main sequence will move to F stars, then G, and so on – just as the wick of a candle gradually burns down. Even in a cluster whose main sequence appears relatively short, consisting only of low-mass stars, we know that massive main-sequence stars were originally present: We can detect their descendants as, for example, red giants.

The ages of open clusters range from "only" a few million years up to billions of years. The two clusters in the closely spaced pair of clusters known as h and χ (the Greek letter "chi") Persei are among the youngest known (■ Fig. 11–35).

Though different open clusters have different-looking temperature-luminosity diagrams because of their range of ages, almost all globular clusters have very similar temperature-luminosity diagrams (■ Fig. 11–36). Thus almost all the globular clusters in our Galaxy are of roughly the same age, about 12 billion years based on the absence of main-sequence stars more massive than the Sun. Some of the clusters that seem younger than others are in a nearby dwarf galaxy that seems to be disintegrating. Several globular clusters have two turn-offs leading to two giant branches, though closely spaced ones on the H-R diagram and therefore in age; they must have formed in two stages.

The ages of globular clusters set a minimum value for the age of the Universe: They must be younger than the Universe itself! The distances set from measurements made with the Hipparcos spacecraft have led to a revision in the calculated ages of some of the globular clusters, and the resulting calculation of their ages was a controversial

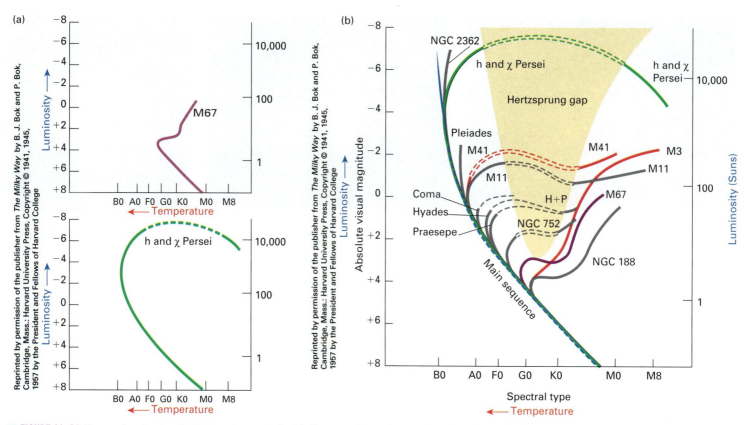

■ FIGURE 11–34 Temperature-luminosity (temperature-magnitude) diagrams of several open clusters. (a) Individual temperature-luminosity curves showing the temperature and luminosity points on the graph for all the stars in an old open cluster, M67 (top left), and for a pair of young open clusters, h and χ (the Greek letter "chi") Persei (bottom left). (b) The temperature-luminosity points of all the stars in one cluster make a curve, and several curves are superimposed here. The two individual curves at left are included. Note how the lower parts of the main sequence coincide. Few stars appear in the region marked as the "Hertzsprung gap," because stars with these combinations of luminosity and temperature are apparently unstable, so they do not stay in this region of the diagram for long.

If all stars recently reached the main sequence, they would all lie on the curve that extends highest and to the left at the top. The older the stars are, the more likely they are to have evolved to the right. Thus the clusters with more stars matching the original main sequence, like h and χ Persei, are younger. The curve for M3, an old globular cluster, is also shown.

■ FIGURE 11–35 (a) The double cluster in Perseus, h and χ Persei, a pair of open clusters that is readily visible in a small telescope. They are separated in the sky by approximately the Moon's diameter. Perseus is a northern constellation that is most prominent in the winter sky. In Greek mythology, Perseus slew the Gorgon Medusa and saved Andromeda from a sea monster. (b) A false-color Chandra X-ray Observatory image of h Persei, part of the research of one of the authors (J.M.P.). Colors represent brightness in various x-ray energy ranges, corresponding to different x-ray wavelengths. Many of the x-rays come from the hot outer layers (coronas) of cool stars like the Sun.

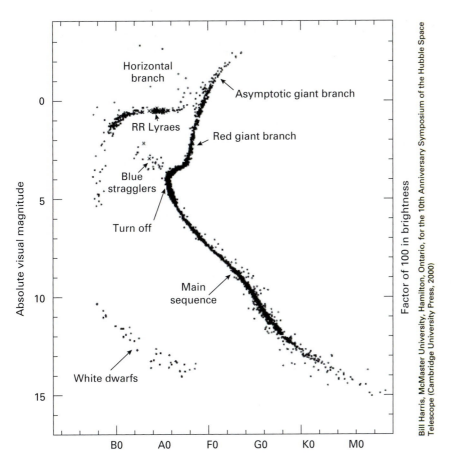

Bill Harris, McMaster University, Hamilton, Ontario, for the 10th Anniversary Symposium of the Hubble Space Telescope (Cambridge University Press, 2000)

■ **FIGURE 11–36** A temperature-luminosity diagram for a globular cluster in which the horizontal axis shows spectral type (hotter to the left, cooler to the right). All globular clusters in the Milky Way Galaxy have very similar temperature-luminosity diagrams, indicating that they are all the same age, and because they are old, many of the stars have evolved off the main sequence. This modern diagram for a globular cluster was put together with real data from 6 clusters, including M3, M55, M68, NGC 6397, and NGC 2419. Current CCD technology allows much fainter stars to be measured than previously, extending the main sequence downward and also (at left) showing the white dwarfs. The measurements are also more accurate than previously, showing that the main sequence and the branches of the diagram are each intrinsically narrow.

We clearly see the turnoff point, with the subgiant branch just above it to the right and the red-giant branch high to the right. Once the stars travel up the red-giant branch, they reverse and proceed to the left over the horizontal branch. Note the "RR Lyrae stars" in the horizontal branch, where the RR Lyrae short-period variable stars lie. The white dwarfs are at lower left. Note a few "blue stragglers" just above and to the left of the turnoff point from the main sequence; these are stars whose properties were altered by interactions with companions.

part of the study of the age of the Universe. For a time around the mid-1990s, the ages determined for globular clusters seemed too old for the age of the Universe deduced otherwise – after all, the Universe can't be younger than something in it. That discrepancy, however, has now been cleared up with further measurements.

The ages of globular clusters are also considered by measuring the brightnesses of RR Lyrae stars in them. There have long been known to be two types of globular clusters distinguished by the average periods of their RR Lyrae stars – 13.2 hours and 15.6 hours. An interpretation gaining favor is that one of the groups includes globular clusters that were formed in another galaxy and that were captured by our own Galaxy. In this model, these captured clusters are about a billion years younger than the globular clusters that were formed in our Galaxy, but no clear observational differences in age have been measured. We hope for improvements in spacecraft, such as Gaia, that will allow the distances to globular clusters to be measured directly by their trigonometric parallaxes; this will lead to more accurate derived ages.

In *A Closer Look 11.5: How We Measure Basic Stellar Parameters*, we summarize many of the concepts introduced in this chapter.

A CLOSER LOOK 11.5 · How We Measure Basic Stellar Parameters

Here is a summary of some vital types of data for stars more distant than the Sun and how we measure them.

Mass – Usually from studies of the orbits of binary stars.

Size – Usually from the star's apparent brightness, given its surface temperature and distance. (In some cases, the size is determined directly with the technique of interferometry.)

Luminosity – From the star's apparent brightness, given its distance.

Rotation rate – From the Doppler shift, which broadens the star's spectral lines, as one side of the star approaches us relative to the other. (In some cases, the rotation rate is determined from brightness variations as a hot spot or a dark spot rotates to the side of the star facing us, but this technique was not discussed in the text.)

Surface temperature – From the star's spectral lines or from its blackbody curve, which can be found accurately enough by measuring the brightness through different filters such as blue, yellow, and red.

Chemical composition – From detailed analysis of the star's spectral lines.

CONCEPT REVIEW

» Heating dense, opaque matter causes it to grow much brighter and to have the maximum in its continuous spectrum shift to shorter wavelengths (Sec. 11.1a). The continuous radiation approximately follows a **blackbody curve**, also called a **Planck curve**. Classifying stars by their spectra led to O B A F G K M L T Y, with spectral-type A stars having the strongest hydrogen lines (Sec. 11.1b). The spectral sequence is one of progressively decreasing surface temperature, with O-type stars being the hottest. Some L-type objects, and most of the even cooler T-type and Y-type objects, are actually brown dwarfs.

» We find the distance to the nearest stars by **triangulation**, or "trigonometric parallax" (Sec. 11.2), a kind of binocular vision obtained by taking advantage of our location on a moving platform (the Earth). The more distant a star, the smaller is its parallax. Measuring the positions and motions of stars is known as **astrometry**. The Hipparcos spacecraft has greatly improved the accuracy of the astrometry for a hundred thousand stars, and with a lesser improvement for about a million more, but only within about 1 per cent of the diameter of our Galaxy.

» A star's intrinsic brightness, or power, is its **luminosity** (Sec. 11.3). **Absolute magnitudes** are the magnitudes that stars would appear to have if they were at a standard distance equivalent to about 32.6 light-years from us. The apparent brightness of a star decreases with increasing distance from it, following the **inverse-square law**.

» Plotting temperature (often as spectral type) on the horizontal axis and luminosity (intrinsic brightness) on the vertical axis gives a **temperature-luminosity diagram** (Sec. 11.4), also known as a temperature-magnitude diagram or a Hertzsprung-Russell diagram. Most stars appear in a band known as the **main sequence**, which goes from hot, bluish, luminous stars to cool, reddish, dim stars. Stars on the main sequence are known as **dwarfs**. Some brighter stars exist and are **giants** or even **supergiants**. Stars less luminous than the main sequence for a given temperature are **white dwarfs**.

» Stars have small motions across the sky, known as **proper motions** (Sec. 11.5a). Proper motion is an angular movement of stars across the sky, relative to each other; to get the physical velocity (in km/s) of each star in the plane of the sky, perpendicular to our line of sight, we need to also know its distance. Studies of their Doppler shifts show their **radial velocities** (motions toward or away from us along the line of sight) as **blueshifts** or **redshifts** (Sec. 11.5b). The larger the blueshift or redshift, the proportionally larger is the object's speed toward or away from you.

» **Optical doubles** appear close together by chance (Sec. 11.6a). Most stars are in multiple-star systems, **binary stars**, bound by gravity. **Visual binaries** are revolving around each other and can be detected as double from Earth (Sec. 11.6a). **Spectroscopic binaries** show their double status from spectra (Sec. 11.6a); one or two sets of spectral lines show periodically changing Doppler shifts. **Eclipsing binaries** pass in front of each other, as shown in their plots of brightness over time, or "light curves" (Sec. 11.6a). Since astrometry is the study of the positions and motions of stars, **astrometric binaries** can be detected by their wobbling from side to side in their paths across the sky (Sec. 11.6a).

» The primary use of binary stars is the determination of stellar masses (Sec. 11.6b). In the case of main-sequence stars, it is found that massive stars are much more luminous (intrinsically bright) than low-mass stars (Sec. 11.6c). Thus, massive stars use up their fuel faster, and have shorter lives, than low-mass stars.

» Among **variable stars**, stars that change considerably in brightness over time, **long-period variables**, like Mira, are common (Sec. 11.7). **Cepheid variables** are rare but valuable; the period of a Cepheid variable's variation is linked to its average intrinsic brightness (luminosity), so studying its variation shows its luminosity. Measuring luminosity allows Cepheid variable stars to be used as **standard candles**. Comparing intrinsic brightness with apparent brightness gives the distances of stars, and they are valuable tools for determining distances to other galaxies. **RR Lyrae variables** also can be used to find distances, particularly to globular clusters in our Milky Way Galaxy.

» Clusters, or physical groups of stars, come in two basic varieties. **Open clusters** contain up to a few thousand stars irregularly spread in a small region of sky, generally in the plane of our Galaxy (Sec. 11.8a). They are loosely bound by gravity, but gradually dissipate away as individual stars escape. **Globular clusters**, consisting of 10,000 to a million stars more strongly bound by gravity than in open clusters, have spherical shapes and form a **halo** around the center of our Galaxy. They contain low abundances of heavy elements, so they are very old.

» The temperature-luminosity diagrams for open clusters show the ages of the clusters by the length of the cluster's main sequence; massive, short-lived stars leave the main sequence faster than the low-mass, long-lived stars (Sec. 11.8b). The temperature-luminosity diagrams for globular clusters in our Galaxy are all very similar, with short main sequences, indicating that the globular clusters are all about the same age – very old – providing a minimum value for the age of our Universe.

QUESTIONS

1. The Sun's spectrum reaches its maximum brightness at a wavelength of 5,000 Å. Would the spectrum of a star whose surface temperature is higher than that of the Sun peak at a longer or a shorter wavelength?

†2. The Sun's surface temperature is about 5,800 K and its spectrum peaks at 5,000 Å. An O-type star's surface temperature may be 40,000 K. (a) According to Wien's law, at what wavelength does its spectrum peak? (b) In what part of the spectrum might that peak be? (c) Can the peak be observed with the Keck telescopes? Explain.

3. One blackbody peaks at 2,000 Å. Another, of the same surface area, peaks at 10,000 Å. (a) Which emits more radiation per second at 2,000 Å? (b) Which emits more radiation per second at 10,000 Å?

†4. If a star has a surface temperature four times higher than that of the Sun, but the same surface area, by what factor is the star more luminous than the Sun?

5. Star Albert appears to have the same brightness through red and blue filters. Star Bohr appears brighter in the red than in the blue. Star Curie appears brighter in the blue than in the red. Rank these stars in order of increasing surface temperature.

6. (a) What is the difference between continuous radiation and an absorption line? (b) What is the difference between continuous radiation and an emission line? (c) Graph a spectrum that shows both continuous radiation and absorption lines. (d) Can you draw absorption lines without continuous radiation? (e) Can you draw emission lines without continuous radiation? Explain.

7. (a) Make up your own mnemonic for spectral types, through M. (b) Make up your own mnemonic for spectral types, through Y.

†8. (a) What is the distance of a star whose parallax is 0.2 seconds of arc? (b) What is the parallax of a star whose distance is 100 pc?

9. If a star that is 100 light-years from us appears to be 10th magnitude, would its absolute magnitude be a larger or a smaller number? Explain.

†10. If the Sun were 10 times farther from us than it is, how many times less light would we get from it?

11. Sketch a temperature-luminosity diagram, and distinguish among giants, dwarfs, and white dwarfs. Be sure to label the axes.

12. Compare the surface temperatures of white dwarfs with those of dwarfs, and explain their relative luminosities.

†13. If a red giant has half the Sun's surface temperature but 100 times its radius, what is the giant's luminosity relative to that of the Sun?

14. Two stars the same distance from us are the same temperature, but one is a giant while the other is a dwarf. Which appears brighter?

15. Does measuring Doppler shifts depend on how far away an object is, assuming its motion is entirely radial? Explain.

16. Consider two stars of the same spectral type and subtype, but not necessarily both on the main sequence. Describe whether this information is sufficient to tell you how the stars differ in (a) surface temperature, (b) size, and (c) distance.

17. Suppose you find an object that has a Doppler shift corresponding to 20 per cent of the speed of light. Is the object likely to be a star in our Galaxy? Why or why not?

18. Use the explanation of Doppler shifts to show what happens to the sound from a fire truck as the truck passes you.

19. (a) Sketch a star's spectrum that contains two spectral lines. (b) Then sketch the spectrum of the same star if the star is moving toward us. (c) Finally, sketch the spectrum if the star is moving away from us.

†20. Suppose a star is moving toward you with a speed of 200 km/s. (a) Will its spectrum appear blueshifted or redshifted? (b) At what wavelength will you observe the Hα (H-alpha) line (rest wavelength 6,563 Å) in the star's spectrum?

21. Describe the appearance, orientation, and motion (as seen from our point of view) of a double star that is simultaneously a visual, an eclipsing, and a spectroscopic binary.

22. If the light in a binary star system is greatly dominated by one of the stars (the other star is very faint), how can you tell from its spectrum that the star is a binary?

23. Assume that an eclipsing binary contains two identical stars, but visually to us looks like a single star. (a) Sketch the apparent brightness of light received as a function of time. (b) Sketch to the same scale another light curve to show the result if both stars were much larger (though still appearing to us as a single star) while the orbit stayed the same.

†24. Suppose you find a binary system in which one star is 6 times as massive as the other star. Which one is closer to the center of mass, and by what factor?

25. Sketch the path in the sky of the visible (with a solid curve) and invisible (with a dotted curve) components of an astrometric binary.

26. (a) Explain how astrometric methods were used to search for planets around other stars. (b) How were such planets finally found?

27. Explain why massive main-sequence stars have shorter lives than low-mass main-sequence stars.

†28. (a) If one main-sequence star is 3 times as massive as another one, what is the ratio of their luminosities? (b) What is the ratio of their main-sequence lifetimes? Be sure to state which one is more luminous and which one lives longer.

29. (a) Explain briefly how observations of a Cepheid variable in a distant galaxy can be used to find the distance to the galaxy. (b) Why can't we use triangulation instead?

†30. If you find a Cepheid variable star in a star cluster, and the Cepheid appears 100 times fainter (on average) than another Cepheid with the same period but whose distance is known to be 400 light-years, what is the distance of the star cluster?

31. What ability of the Hubble Space Telescope is enabling astronomers to study Cepheid variables farther out in space than ever before?

32. (a) What are the Magellanic Clouds? (b) How did we measure the distances to them?

33. What are two distinctions between open and globular star clusters?

34. Why is it more useful to study the temperature-luminosity diagram of a cluster of stars instead of one for stars in a field chosen at random?

35. Suppose you find two clusters, one whose main sequence doesn't have type O, B, and A stars, and the other whose main sequence doesn't have any type O, B, A, and F stars. Which one is older? Why?

36. Which are generally older: open clusters or globular clusters? Explain.

37. **True or false?** Thoroughly studying certain types of binary stars is the main way in which astronomers directly measure stellar masses.

38. **True or false?** A star of spectral type K must be less luminous than a star of spectral type F.

39. **True or false?** It would be easier to accurately measure the parallax of a given star from Jupiter than from Earth (though it would take longer).

40. **True or false?** A star cluster whose main sequence contains only type F, G, K, M, and L stars is probably older than a star cluster whose main sequence contains only type K, M, and L stars.

41. **True or false?** "Open" star clusters in the Milky Way Galaxy generally have hundreds of thousands of densely packed stars.

42. **True or false?** We usually determine the chemical composition of stars by examining absorption lines produced in their relatively cool outer layers.

43. **True or false?** A star having a small proper motion must be moving in the plane of the sky at a physically low speed.

44. **True or false?** Massive stars live significantly longer lives than low-mass stars, since they have much more fuel available to them.

45. **Multiple choice:** Albireo, the "Cal Star" (that is, the official star of the University of California, Berkeley, because of its colors), is a physical binary system of a bright, yellow ("gold") star and a fainter, blue star. Which one of the following statements must be *true*? (a) The gold star is significantly larger than the blue star. (b) The blue star is significantly farther away from Earth than the gold star. (c) The blue star is significantly more massive than the gold star. (d) The blue star is significantly younger than the gold star. (e) The gold star orbits the center of the blue star.

†46. **Multiple choice:** Treating the Sun as a "blackbody," suppose its surface temperature were doubled from 5,800 K to 11,600 K. It would have the same luminosity if, at the same time, its radius (a) doubles; (b) remains the same; (c) decreases to one-half its former size; (d) decreases to one-quarter its former size; or (e) decreases to one-sixteenth its former size.

†47. **Multiple choice:** If Stars Reka and Ivan have the same apparent brightness, but Star Reka is 5 times farther away from us than Star Ivan, then Star Reka's luminosity (power) is _____ times the luminosity of Star Ivan. (a) 25. (b) 5. (c) 1. (d) $^1/_5$. (e) $^1/_{25}$.

48. **Multiple choice:** A binary star that varies in apparent brightness as one member of the binary passes in front of the other is (a) an astrometric binary; (b) a visual binary; (c) an eclipsing binary; (d) a spectroscopic binary; or (e) impossible, unless at least one of the two stars is itself intrinsically variable.

†49. **Multiple choice:** Suppose Star Ben is 10 light-years from Earth and its spectrum peaks at 4,000 Å, while Star Monica is 20 light-years from Earth and its spectrum peaks at 1,000 Å. Which one of the following statements *must* be *true*? (a) Star Monica is more luminous than Star Ben. (b) The apparent brightnesses of the two stars (measured over the entire spectrum) are equal because their differences in luminosity are compensated by their differences in distance. (c) Star Monica emits more light per second and per

unit surface area at 4,000 Å than does Star Ben. **(d)** The parallax angle of Star Ben is 4 times that of Star Monica. **(e)** Star Ben has a smaller radius than Star Monica.

50. Multiple choice: One can determine the radius of a nearby star (that is not necessarily on the main sequence) knowing only its **(a)** chemical composition, distance, and surface temperature; **(b)** radial velocity and luminosity; **(c)** apparent brightness and distance; **(d)** apparent brightness, parallax, and surface temperature; or **(e)** apparent brightness, parallax, and luminosity.

†51. Multiple choice: If the amount of nuclear fuel available during the main-sequence stage of a star's life is proportional to the star's mass (M), and the luminosity of a main-sequence star is proportional to the fourth power of the star's mass (M^4), then to what is a star's main-sequence lifetime proportional? **(a)** $1/M^3$. **(b)** M^3. **(c)** M. **(d)** $1/M^5$. **(e)** Insufficient information is given.

52. Multiple choice: Which one of the following is a *true* statement for a globular cluster and a *false* statement for an open cluster? **(a)** All stars in the cluster formed at about the same time. **(b)** There are usually about a few hundred stars loosely held together by gravity. **(c)** All stars in the cluster had approximately the same initial chemical composition. **(d)** They are usually found away from the disk and arms of a spiral galaxy. **(e)** A temperature-luminosity diagram can reveal the cluster's age.

†53. Multiple choice: Suppose stars Obi-Wan and Luke are two main-sequence stars in a binary system, with circular orbits around their common center of mass. Star Obi-Wan has a luminosity 16 times that of Star Luke. The ratio of Star Obi-Wan's orbital radius to Star Luke's orbital radius is **(a)** 4; **(b)** 2; **(c)** $^1/_2$; **(d)** 16; or

(e) $^1/_{16}$ has a luminosity 16 times that of Star Luke. The ratio of Star Obi-Wan's orbital radius to Star Luke's is (a) 4; (b) 2; (c) $^1/_{12}$; (d) 16; or (e) $^1/_{16}$.

†54. Fill in the blank: Suppose you measure an absorption line from Star Hal to be at $\lambda = 5,005$ Å. If you know from laboratory experiments that the same line from a stationary gas is at $\lambda_0 = 5,000$ Å, the radial velocity of Star Hal must be _____ km/s, in the direction _____ us.

†55. Fill in the blank: A star whose parallax is 0.04 second of arc is _____ as distant as a star whose parallax is 0.08 second of arc.

56. Fill in the blank: _____ are stars that vary in brightness in a regular way, with the period of variation of the most luminous cases being the longest.

57. Fill in the blank: Most stars within a few hundred light-years from Earth are in the broad category of _____, though this is not generally true of the brightest visible stars.

†58. Fill in the blank: Star Sam is 8 times farther away than Star Lily but 16 times as luminous. The apparent brightness of Star Sam compared with Star Lily (Sam/Lily) is _____.

59. Fill in the blank: The luminosities of massive main-sequence stars are much _____ than those of low-mass main-sequence stars.

60. Fill in the blank: In a binary star system, the two stars actually orbit their _____ rather than each other.

†This question requires a numerical solution.

TOPICS FOR DISCUSSION

1. Do you find it surprising that we can learn so many fundamental characteristics of stars, despite their enormous distances? Explain.

2. Why do you think massive main-sequence stars use up their fuel much faster than low-mass main-sequence stars, despite having more fuel?

3. Why might planetary systems (containing at least several planets) be more likely or less likely to form around binary stars than around single stars? What observational evidence is there?

The region of eta (η) Carinae in our Milky Way Galaxy, in this near-infrared false-color image made with the Very Large Telescope in Chile. The original is a giant mosaic, which shows not only the brightest stars in the region but also hundreds of thousands of faint ones, including brown dwarfs. The star eta Carinae itself, surrounded by a nebula that is energized by the star's ultraviolet radiation, is in a corner of the image. Dark globules impenetrable even by infrared appear across the image. The image was taken in the filter bands known as J, H, and K (1.25, 1.65, and 2.2 μm, respectively).

12

How the Stars Shine: Cosmic Furnaces

ORIGINS

The formation of stars like the Sun is critical to our existence. With its large supply of fuel, the Sun exists for a long time as it produces energy through nuclear fusion, making possible the development of advanced life on Earth.

Even though individual stars shine for a relatively long time, they are not eternal. Stars are born out of the gas and dust that exist within a galaxy (see the Chapter Opener); they then shine brightly on their own for a long time. Eventually, they die. Though we can directly observe only the outer layers of stars, we can deduce that the temperatures at their centers must be millions of kelvins. We can even figure out what it is deep down inside that makes the stars shine.

To determine the probable life history of a typical star, we observe stars having many different ages and assume that they evolve in a similar manner. However, we must take into account the different masses of stars; some aspects of their evolution depend critically on mass.

We start this chapter by discussing the birth of stars. We see how new capabilities of observing in the infrared in addition to the visible are helping us understand star formation (Chapter Opener and ■ Fig. 12–1). We then consider the processes that go on inside a star during its life on the main sequence. Finally, we begin the story of the evolution of stars when they finish the main-sequence stage of their lives. Chapters 13 and 14 will continue the story of what is called "stellar evolution," all the way to the deaths of stars.

Near the end of this chapter we will see that the most important experiment to test whether we understand how stars shine is the search for elusive particles, called neutrinos, from the Sun. For several decades a search for

AIMS

1. Understand how stars form out of gigantic clouds of gas and dust (Section 12.1).

2. Study the way in which stars generate their energy through nuclear fusion deep in their cores (Sections 12.2 to 12.5).

3. Discuss brown dwarfs, which do not fuse ordinary hydrogen to helium (Section 12.6).

4. Learn how neutrino detection experiments have moved from a test of our most fundamental understanding of how stars shine to a test of the fundamental physics of small particles (Section 12.7).

5. Briefly introduce the endpoints of stellar evolution (Section 12.8).

■ **FIGURE 12–1** For comparison, a visible-light image of approximately the same region of space as in the infrared image opening the chapter.

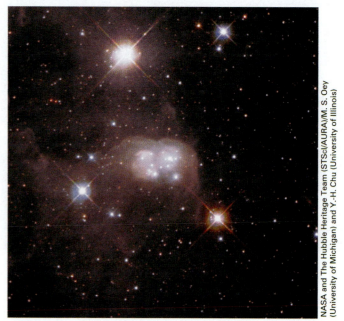

■ **FIGURE 12–2** A Hubble Space Telescope image of a reflection nebula of gas and dust around a cluster of young, hot stars. The image shows a region about 60 light-years across in the Large Magellanic Cloud.

them had been made, but only about a third to half of those expected had been found. Recent experiments have provided better ways of detecting neutrinos than we previously had. It turns out that the neutrinos were there all along, though transformed and thus hidden! The results indicate that we did not understand neutrinos as well as we had thought. These astronomical results therefore have added important knowledge about fundamental physics in addition to our understanding of the stars.

12.1 STAR BIRTH

The birth of a star begins with a nebula – a large region of gas and dust (■ Figs. 12–2 and 12–3, as well as the earlier Chapter Opener and Fig. 12–1). The dust (tiny solid particles) may have escaped from the outer atmospheres of giant stars. The regions of gas and dust (often called clouds, or "giant molecular clouds") from which stars are forming are best observed in the infrared and radio regions of the

spectrum, because most other forms of radiation (such as optical and ultraviolet) cannot penetrate them. Here we largely discuss the infrared observations, including the capabilities of the European Space Agency's Herschel Infrared Observatory and NASA's new instrumented airplane called the Stratospheric Observatory for Infrared Astronomy (SOFIA). We leave the radio observations to Chapter 15 on the Milky Way Galaxy.

Stars forming at the present time incorporate this previously cycled gas and dust, which gives them their relatively high abundances of elements heavier than helium in the periodic table (still totaling less than a per cent). In contrast, the oldest stars we see formed long ago when only primordial hydrogen and helium were present, and therefore have lower abundances of the heavier elements, as we discussed near the end of the previous chapter.

12.1a Collapse of a Cloud

Consider a region that reaches a higher density than its surroundings, perhaps from a random fluctuation in density or – in a leading theory of why galaxies have spiral arms – because a wave of compression passes by. Still another possibility is that a nearby star explodes (a "supernova"; see Chapter 13), sending out a shock wave that compresses the gas and dust. In any case, once the cloud gains a higher-than-average density, the gas and dust continue to collapse because of the pull of gravity. Energy is released, and the material accelerates inward. Magnetic fields may resist the infalling gas, slowing the infall, though the role of magnetic fields is not well understood in detail.

Eventually, the cloud fragments and many dense cores, each with a mass comparable to that of a star, form and grow like tiny seeds

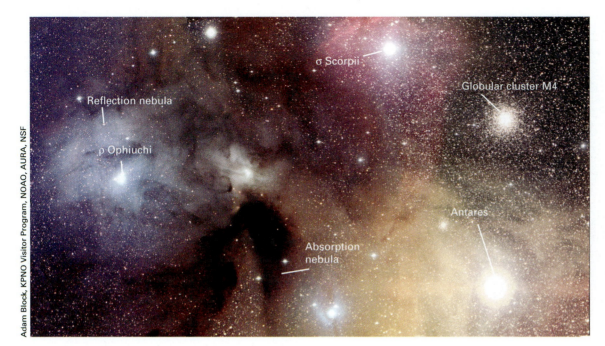

■ **FIGURE 12–3** The clouds of gas and dust around the star ρ (the Greek letter "rho") Ophiuchi (in the blue reflection nebula). The bright star Antares is surrounded by yellow and red nebulosity. The complex is about 500 light-years away. M4, the nearest globular cluster to us, is also seen.

within the vast cloud. These **protostars** (from the prefix of Greek origin meaning "primitive"), which will collectively form a star cluster, continue to collapse, almost unopposed by internal pressure.

But when a protostar becomes sufficiently dense, frequent collisions occur among its particles; hence, part of the gravitational energy released during subsequent collapse goes into heating the gas, increasing its internal pressure. (In general, compression heats a gas; for example, a bicycle tire feels warm after air is vigorously pumped into it.) The rising internal pressure, which is highest in the protostar's center and decreases outward, slows down the collapse until it becomes very gradual and more accurately described as "contraction" rather than "collapse." The object is now called a **pre-main-sequence star** (■ Fig. 12–4).

By this time, the object has contracted by a huge fraction, a factor of 10 million, from about 10 trillion km across to about a million km

across – that is, something initially larger than the whole Solar System collapses until most of its mass is in the form of a single star.

During the contraction phase, a disk tends to form because the original nebula was rotating slightly. We discussed this process when considering the "nebular hypothesis" for the formation of our Solar System (Chapter 9). Dusty disks have been found around young stars and pre-main-sequence stars in nebulae such as the Orion Nebula (■ Fig. 12–5) and the Eagle Nebula (■ Fig. 12–6). These are sometimes called protoplanetary disks ("proplyds"), and they support the theoretical expectation that planetary systems are common.

As energy is rapidly radiated from the surface of the pre-main-sequence star, its internal pressure becomes insufficient to hold the star up, and it gradually contracts. This release of gravitational energy heats the interior, thereby again increasing the internal temperature and pressure. It is also the source of the large amount of radiated

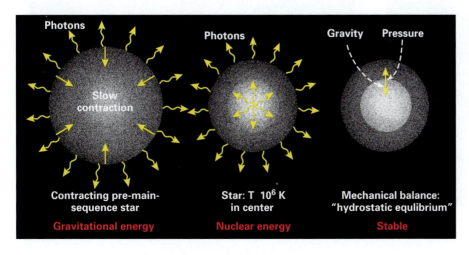

■ **FIGURE 12–4** Stars form out of collapsing clouds of gas and dust. The contraction of the pre-main-sequence star (left) supplies energy that eventually allows nuclear fusion (center) to begin, at which time we say that the star is born. (The central temperature is at least a few million K, but typically more like 10^7 K for low-mass stars.) The outward pressure of the hot gas matches the inward force of gravity (right), so the star is in "hydrostatic equilibrium."

C. R. O'Dell (Rice University/NASA)

■ **FIGURE 12–5** The Hubble Space Telescope has revealed protoplanetary disks around several stars, as we saw in Section 9.1. Here we view such "proplyds" in a portion of the Orion Nebula only 0.1 light-year across.

energy. Gravitational energy was released in this way in the early Solar System. As the temperature in the interior rises, the outward force resulting from the outwardly decreasing pressure increases, and eventually it balances the inward force of gravity, a condition known as "hydrostatic equilibrium." As we shall discuss later, this mechanical balance is the key to understanding stable stars; see Figure 12–4.

Theoretical analysis shows that the dust surrounding the stellar embryo we call a pre-main-sequence star should absorb much of the radiation that the luminous object emits. The radiation from the

pre-main-sequence star should heat the dust to temperatures that cause it to radiate primarily in the infrared. Infrared astronomers have found many objects that are especially bright in the infrared but that have no known optical counterparts. These objects seem to be located in regions where the presence of a lot of dust, gas, and young stars indicates that star formation might still be going on.

Imaging in the visible (with the Hubble Space Telescope) and in the infrared (not only with previous infrared space telescopes but now especially with the Spitzer Space Telescope) has shown how young stars are born inside giant pillars of gas and dust inside certain nebulae. The Eagle Nebula is the most famous example because of the beautiful Hubble image showing exquisite detail, with false colors assigned to different filters (see Figure 12–8 again, and ■ Fig. 12–7). As hot stars form, their intense radiation evaporates the gas and dust around them, freeing them from the cocoons of gas and dust in which they were born. We see this "evaporation" taking place at the tops of the Eagle Nebula's "pillars." The stars are destroying their birthplaces as they become independent and more visible from afar.

12.1b The Birth Cries of Stars

To their surprise, astronomers have discovered that rotating pre-main-sequence stars send matter out in oppositely directed beams from their poles, while they had expected to find only evidence of infall. This "bipolar ejection" (■ Fig. 12–8) may imply that a disk of matter orbits such pre-main-sequence stars, blocking an outward flow of gas in the equatorial direction and later coalescing into planets. Thus the flow of gas is channeled toward the poles.

Sometimes clumps of gas appear, but only recently have they been identified with ejections from stars in the process of collapsing. The clumps seem like spinning bullets, though what makes them

(a)

Jeff Hester and Paul Scowen (Arizona St. University), and NASA/STScI/ESA

(b)

NASA/ESA, and The Hubble Heritage Team (STScI/AURA)

■ **FIGURE 12–6** (a) The orientation of the Hubble Space Telescope visible-light image of the Eagle Nebula on a near-infrared image taken with a ground-based telescope. (b) A Hubble detail of one of the pillars of gas and dust.

Far-infrared: ESA/Herschel/PACS/SPIRE/Hill, Motte, HOBYS Key Programme
Consortium; ESA/XMM-Newton/EPIC/XMM-Newton-SOC/Boulanger; optical: MPG/ESO;

Far Infrared

Visible

X-rays

Far Infrared & X-rays

Visible & Infrared & X-rays

Visible

Near Infrared

Mid-Infrared

■ **FIGURE 12–7** A Herschel Space Observatory infrared view in false color of the Eagle Nebula region.

spin is uncertain (perhaps connected with the magnetic field). The ejection of these spinning clumps helps slow the star's rate of spin, since they carry away angular momentum from what had been a rapidly spinning pre-main-sequence star. At the same time, some gas with low angular momentum is falling in toward the star. Hidden here is perhaps the main unsolved problem in star formation at the

moment: How do individual stars figure out what their final mass will be?

The bipolar ejection appears as "Herbig-Haro objects," clouds of interstellar gas heated by shock waves from jets of high-speed gas. (They are named after George Herbig, formerly of the Lick Observatory and now of the University of Hawaii, and Guillermo

near-infrared: VLT/ISAAC/
McCaughrean & Andersen/AIP/ESO

■ **FIGURE 12–8** Hubble Space Telescope visible-light observations of Herbig-Haro objects #1 and #2 (HH-1 and HH-2).

■ **FIGURE 12–9** An outflow of gas in the Herbig-Haro objects HH-46/47, imaged in the infrared with the Spitzer Space Telescope and displayed in false color. At these infrared wavelengths, we look through a dark cloud (a "Bok globule") to see molecular gas flowing outward from a bright newborn star. This low-mass star is ejecting supersonic jets of gas in opposite directions. It is 1140 light-years away from Earth in the direction of the southern constellation Vela.

Spitzer's 3.6 μm image, from its Infrared Array Camera (IRAC), is shown as blue, 4.5 μm and 5.8 μm as green, and 8 μm, which reveals organic molecules made of carbon and hydrogen, as red. The organic molecules are glowing in a shape that matches that of the dark cloud.

Infrared spectra of the water molecule from the Herschel Space Observatory's Photodetector Array Camera and Spectrometer (PACS) also appear. They show the high velocities deep inside.

Haro of the Mexican National Observatory.) The jets are being ejected from a pre-main-sequence star, a star in the process of being born. Since the pre-main-sequence star is hidden in visible light by a dusty cocoon of gas, infrared observations of Herbig-Haro objects most clearly reveal what is going on (■ Fig. 12–9).

The jets of gas were formed as the pre-main-sequence star contracted under the force of its own gravity. Because a thick disk of cool gas and dust surrounds the pre-main-sequence star, the gas squirts outward along the pre-main-sequence star's axis of rotation at speeds of perhaps 1 million km/hour.

HH-1 and HH-2 (Fig. 12–8) are more irregular in shape than many other Herbig-Haro objects, perhaps because the bow shock wave we are seeing (a shock wave like those formed by the bow of a boat plowing through the water) has broken up. These objects are about 1,500 light-years from us, in a star-forming region of the

constellation Orion. The smallest features resolved are about the size of our Solar System, and the whole image is only about 1 light-year across.

Several classes of stars that vary erratically in brightness have been found. One of these classes, called T Tauri, contains pre-main-sequence stars as massive as or less massive than the Sun. Presumably, these stars are so young that they have not quite settled down to a steady and reliable existence on the main sequence. (T Tauri stars always have the word "stars" in their name, though technically they haven't reached the main sequence, so they are not yet fully formed stars.)

In astronomical teaching, we have the question of whether to first consider the formation of stars in the star section of the book, as here, or in the section about the gas and dust between the stars from which the stars form. We choose to do some of each, and will continue our discussion of stars in formation in that latter location, Chapter 15 on the Milky Way Galaxy.

12.2 WHERE STARS GET THEIR ENERGY

If the Sun got all of its energy from gravitational contraction, it could have shined for only about 50 million years, not very long on an astronomical timescale. Yet we know of meteorites that are nearly 4.6 billion years old, so the Sun and the Solar System have been around at least that long. Moreover, fossil records of plants and animals, which presumably used the Sun's light and heat, date back hundreds of millions of years, and single-celled organisms that depended on the Sun's output have existed billions of years. Some other source of energy must hold up the Sun and other stars against their own gravitational pull.

Actually, a pre-main-sequence star will heat up until its central portions become hot enough (typically 10 million K) for **nuclear fusion** to take place, at which time it reaches the main sequence of the temperature-luminosity (temperature-magnitude, or Hertzsprung-Russell; see Chapter 11) diagram. Using this process, which we will soon discuss in detail, the star can generate enough energy to support it during its entire lifetime on the main sequence. A star's luminosity and temperature change little while it is on the main sequence; nuclear reactions provide the stability. (Stars gradually become slightly more luminous as they age.)

The energy makes the particles in the star move around rapidly. Such rapid, random motions in a gas are the definition of high temperature. The **thermal pressure**, the force from these moving particles pushing on each area of gas, is also high. The varying pressure, which decreases outward from the center, produces a force that pushes outward on any given pocket of gas. This outward force balances gravity's inward pull on the pocket ("hydrostatic equilibrium," which we illustrated in Figure 12–4).

The basic fusion process in main-sequence stars fuses four hydrogen nuclei into one helium nucleus. In the process, tremendous amounts of energy are released. (Hydrogen bombs on Earth fuse hydrogen nuclei into helium, but start with a different form of hydrogen and thus use different fusion sequences. The fusion sequences that occur in stars are far too slow for bombs.)

A hydrogen nucleus is but a single **proton**. A helium nucleus is more complex; it consists of two protons and two **neutrons** (■ Fig. 12–10). The mass of the helium nucleus that is the final product of the fusion process is slightly less than the sum of the masses of the four hydrogen nuclei (protons) that went into it. (Two of the protons turn into neutrons during the fusion process, but the masses of protons and neutrons can be considered equal for our purposes.) A small amount of the mass, m, "disappears" in the process: 0.007 (0.7 per cent) of the mass of the four protons.

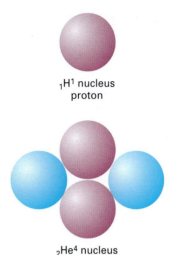

$_1H^1$ nucleus
proton

$_2He^4$ nucleus
alpha particle

■ **FIGURE 12–10** The nucleus of hydrogen's most common form (isotope) is a single proton, while the nucleus of helium's most common isotope consists of two protons and two neutrons. Protons and neutrons are made of smaller particles called quarks (Chapter 19).

The mass difference does not really disappear, but rather is converted into energy, E, according to Albert Einstein's famous formula

$$E = mc^2,$$

where c is the speed of light; see *Figure It Out 12.1: Energy Generation in the Sun*. Even though m is only a small fraction of the original mass, the amount of energy released is prodigious; in the formula, c is a very large number. This energy is known as the "binding energy" of the nucleus, here specifically that of helium.

The loss of only 0.7 per cent of the central part of the Sun, for example, is enough to allow the Sun to radiate as much as it does at its present rate for a period of about ten billion (10^{10}) years; see *Figure It Out 12.1: Energy Generation in the Sun*. This fact, not realized until 1920 and worked out in more detail in the 1930s, solved the longstanding problem of where the Sun and the other stars get their energy.

All the main-sequence stars are approximately 90 per cent hydrogen (that is, 90 per cent of the atoms are hydrogen), so there is a lot of raw material to fuel the nuclear "fires." We speak colloquially of "nuclear burning," although, of course, the processes are quite different from the chemical processes that are involved in the "burning" of logs or of autumn leaves. In order to be able to discuss these processes, we must first review the general structure of nuclei and atoms.

12.3 ATOMS AND NUCLEI

As we mentioned in Chapter 2, an atom consists of a small nucleus surrounded by electrons. Most of the mass of the atom is in the nucleus, which takes up only a very small volume in the center of

FIGURE IT OUT 12.1 — Energy Generation in the Sun

We can easily calculate the amount of hydrogen needed each second for the nuclear fusion in the Sun to account for its observed luminosity (power). The energy produced by the Sun comes from the following net reaction:

$$4\ _1H^1 \rightarrow\ _2He^4 + \text{energy}.$$

The energy is given by Einstein's famous formula, $E = mc^2$, where m is the mass difference between the final helium nucleus and the original four protons – that is, the "binding energy" of the helium nucleus. (In the nuclear fusion process, two protons are converted into two neutrons, but here we will ignore the mass difference between neutrons and protons, which is very small.)

The mass of a proton is 1.6726×10^{-27} kg; thus, the mass of the original four protons is 6.6904×10^{-27} kg. The mass of a helium nucleus is 6.6448×10^{-27} kg, which is less by 0.0456×10^{-27} kg (roughly 0.7 per cent of the original mass). Multiplying this value, m, by c^2, we have

$$\begin{aligned} E = mc^2 &= (0.0456 \times 10^{-27}\ \text{kg})(3.00 \times 10^8\ \text{m/s})^2 \\ &= 4.10 \times 10^{-12}\ \text{kg}\cdot\text{m}^2/\text{s}^2 \\ &= 4.10 \times 10^{-12}\ \text{J}, \end{aligned}$$

since 1 J (1 joule) $= 1$ kg·m²/s². So, 4.10×10^{-12} J of energy is liberated every time 4 protons fuse to form a helium nucleus (call this "one reaction").

The luminosity of the Sun is 3.83×10^{26} J/s. To account for this, we need

$$(3.83 \times 10^{26}\ \text{J/s})/(4.10 \times 10^{-12}\ \text{J/reaction})$$
$$= 9.33 \times 10^{37}\ \text{reactions/s}.$$

But each reaction used 4 protons, whose mass is about 6.69×10^{-27} kg. Thus, the total mass of protons used per second is

$$(9.33 \times 10^{37}\ \text{reactions/s})(6.69 \times 10^{-27}\ \text{kg/reaction})$$
$$= 6.24 \times 10^{11}\ \text{kg/s}.$$

Let's put this into more familiar units. 1 kg = 2.2 pounds, so about 1.37×10^{12} pounds of protons fuse each second. But 1 ton = 2,000 lbs, so this is equivalent to 6.86×10^8 tons per second. The Sun fuses nearly *700 million tons* of fuel each second!

Roughly how long will the Sun continue to fuse hydrogen in its core? (This will be its total main-sequence lifetime.) The Sun's mass is 2.0×10^{30} kg, of which about 70 per cent is hydrogen (protons), or 1.4×10^{30} kg. Only about the central 14 per cent will be used for fusion, so the amount of fuel is

$$(0.14)(1.4 \times 10^{30}\ \text{kg}) \approx 2.0 \times 10^{29}\ \text{kg}.$$

Above, we found that the Sun uses about 6.2×10^{11} kg/s, and at this rate it could shine on the main sequence for

$$(2.0 \times 10^{29}\ \text{kg})/(6.2 \times 10^{11}\ \text{kg/s}) \approx 3.2 \times 10^{17}\ \text{s}.$$

But there are about 3.2×10^7 s/year, so the main-sequence lifetime of the Sun should be about $(3.2 \times 10^{17}\ \text{s})/(3.2 \times 10^7\ \text{s/year}) \approx 10^{10}$ years, about 10 billion years! The Sun is currently about halfway through its main-sequence life, so it is a middle-aged star.

The above calculation assumed that the current rate of energy generation stays constant throughout the Sun's lifetime. Actually, the Sun is gradually becoming more luminous, so the fusion rate will be somewhat faster in the future. However, the fusion rate used to be slower (the Sun is currently about 40 per cent brighter than when it was born), so the current rate is a good approximation of the overall average.

---

the atom. The effective size of the atom, the chemical interactions of atoms to form molecules, and the nature of spectra are all determined by the electrons.

12.3a Subatomic Particles

The nuclear particles with which we need to be most familiar are the proton and neutron. Both of these particles have nearly the same mass, 1836 times greater than the mass of an electron, though still tiny. The neutron has no electric charge and the proton has one unit of positive electric charge. The electrons, which surround the nucleus, have one unit each of negative electric charge. When an atom loses an electron, it has a net positive charge of 1 unit for each electron lost. The atom is now a form of **ion** (■ Fig. 12–11).

Since the number of protons in the nucleus determines the charge of the nucleus, it also dictates the quota of electrons that the neutral state of the atom must have. To be neutral, after all, there must be

equal numbers of positive and negative charges. Each **element** (sometimes called "chemical element") is defined by the specific number of protons in its nucleus. The element with only one proton is hydrogen, that with two protons is helium, that with three protons is lithium, and so on.

12.3b Isotopes

Though a given element always has the same number of protons in a nucleus, it can have several different numbers of neutrons. (The number of neutrons is usually somewhere between 1 and 2 times the number of protons. The most common form of hydrogen, just a single proton, is the main exception to this rule.) The possible forms of the same element having different numbers of neutrons are called **isotopes**.

For example, the nucleus of ordinary hydrogen contains one proton and no neutrons. An isotope of hydrogen (■ Fig. 12–12)

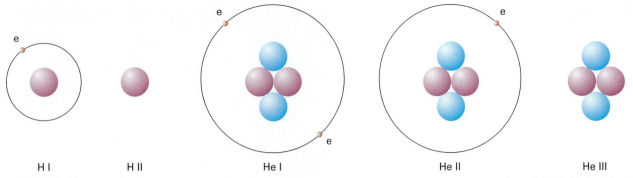

■ **FIGURE 12–11** Hydrogen and helium ions. The sizes of the nuclei are greatly exaggerated with respect to the sizes of the "orbits" of the electrons.

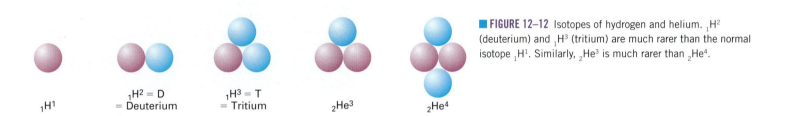

■ **FIGURE 12–12** Isotopes of hydrogen and helium. $_1H^2$ (deuterium) and $_1H^3$ (tritium) are much rarer than the normal isotope $_1H^1$. Similarly, $_2He^3$ is much rarer than $_2He^4$.

called deuterium (and sometimes "heavy hydrogen") has one proton and one neutron. Another isotope of hydrogen, called tritium, has one proton and two neutrons.

Most isotopes do not have specific names, and we keep track of the numbers of protons and neutrons with a system of superscripts and subscripts. The subscript before the symbol denoting the element is the number of protons (called the atomic number), and a superscript after the symbol is the total number of protons and neutrons together (called the mass number, or atomic mass).

For example, $_1H^2$ is deuterium, since deuterium has one proton, which gives the subscript, and an atomic mass of 2, which gives the superscript. (Note that $_1^2H$ is also correct notation.) Deuterium has atomic number equal to 1 and mass number equal to 2. Similarly, $_{92}U^{238}$ is an isotope of uranium with 92 protons (atomic number = 92) and mass number of 238, which is divided into 92 protons and 238–92 = 146 neutrons.

Each element has only certain isotopes. For example, most of the naturally occurring helium is in the form $_2He^4$, with a much lesser amount as $_2He^3$.

12.3c Radioactivity and Neutrinos

Sometimes an isotope is not stable, in that after a time it will spontaneously change into another isotope or element; we say that such an isotope is **radioactive**. The most massive elements, those past uranium, are all radioactive, and have average lifetimes that are very short.

It had been theoretically predicted that around element 114, elements should begin being somewhat more stable again. The handful of atoms of elements 114 and 116 discovered in 1998 and 1999

are more stable than those of slightly lower mass numbers – lasting even about 5 seconds instead of a small fraction of a second.

In 2011, element 114 was named florovium (symbol: Fv), after the director of the Russian research institute in Dubna where it and many other elements were discovered, and element 116 was named livermorium (Lv), after the Lawrence Livermore National Laboratory in Livermore, California, also where many of these newer elements were first made. For example, the two teams collaborated in making livermorium by crashing calcium, with 20 protons, into curium, with 96 protons. The livermorium quickly decayed into the element florovium, which they also made directly by crashing calcium into plutonium with its 94 protons. (A claim that element 118 was also discovered has been withdrawn.)

During certain types of radioactive decay, as well as when a free proton and electron combine to form a neutron, a particle called a **neutrino** is given off. A neutrino is a neutral particle having almost no mass (its name comes from the Italian for "little neutral one").

Neutrinos have a very useful property for the purpose of astronomy: They rarely interact at all with matter. Thus, when a neutrino is formed deep inside a star, it can usually escape to the outside without interacting with any of the matter in the star. A photon of electromagnetic radiation, on the other hand, can travel only about 0.5 mm (on average) in a stellar interior before it is absorbed, and it takes about a hundred thousand years for a photon to zig and zag its way to the surface.

The elusiveness of the neutrino not only makes it a valuable messenger – indeed, the only possible direct messenger – carrying news of the conditions inside the Sun at the present time, but also makes it very difficult for us to detect on Earth. A careful experiment carried out over several decades found only about $^1/_3$ the expected number of neutrinos, as we shall soon see.

12.4 STARS SHINING BRIGHTLY

Let us now use our knowledge of atomic nuclei to explain how stars shine. For a pre-main-sequence star, the energy from the gravitational contraction goes into giving the individual particles greater speeds; that is, the gas temperature rises. When atoms collide at high temperature, electrons get knocked away from their nuclei, and the atoms become fully ionized. The electrons and nuclei can move freely and separately in this "plasma."

For nuclear fusion to begin, atomic nuclei must get close enough to each other so that the force that holds nuclei together, the "strong nuclear force" (to be discussed in Chapter 19), can play its part. But all nuclei have positive charges, because they are composed of protons (which bear positive charges) and neutrons (which are neutral). The positive charges on any two nuclei cause an electrical repulsion between them, which tends to prevent fusion from taking place.

However, at the high temperatures (millions of kelvins) typical of a stellar interior, some nuclei occasionally have enough energy to overcome this electrical repulsion. They come sufficiently close to each other that they almost collide, and the strong nuclear force takes over. Fusion on the main sequence proceeds in one of two ways, as will be discussed below.

Once nuclear fusion begins, enough energy is generated to maintain the pressure and prevent further contraction. Differences in pressure (pressure decreases with increasing distance from the center) provide a force that pushes outward strongly enough to balance gravity's inward pull.

In the center of a star, the fusion process is self-regulating. The star finds a balance between thermal pressure pushing out and gravity pushing in. It thus achieves stability on the main sequence (at a nearly constant temperature and luminosity). When we learn how to control fusion in power-generating stations on Earth, which currently seems decades off (and has long seemed so), our energy crisis will be over, since deuterium, the potential "fuel," is so abundant in Earth's oceans. A huge international project to show that it is feasible to produce fusion energy on Earth is based at ITER (International Thermonuclear Experimental Reactor) in France, with a related facility in Japan. The United States and the European Union are important parts of the consortium. The National Ignition Facility in Livermore, California, is another major center for fusion studies on Earth.

The greater a star's mass, the hotter its core becomes before it generates enough pressure to counteract gravity. The hotter core gives off more energy, so the star becomes brighter (■ Fig. 12–13), explaining why main-sequence stars of large mass have high luminosity. In fact, it turns out that more massive stars use their nuclear fuel at a very much higher rate than less massive stars. Even though the more massive stars have more fuel to burn, they go through it relatively quickly and live shorter lives than low-mass stars, as we discussed in Chapter 11. The next two chapters examine the ultimate fates of stars, with the fates differing depending on the masses of the stars.

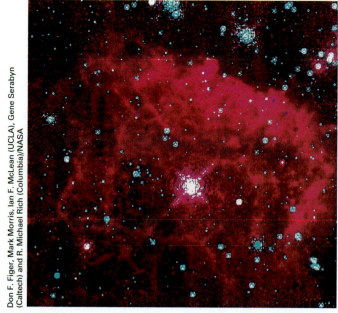

Don F. Figer, Mark Morris, Ian F. McLean (UCLA), Gene Serabyn (Caltech) and R. Michael Rich (Columbia)/NASA

■ **FIGURE 12–13** The Pistol Nebula, gas glowing because of radiation from a massive, highly luminous, main-sequence star. The star, perhaps the most massive known, shines 10 million times more powerfully than our Sun and would fill the Earth's orbit. Hidden behind the dust that lies between us and the center of our Galaxy, the star and nebula are revealed here by the infrared camera aboard the Hubble Space Telescope.

12.5 WHY STARS SHINE

Several chains of nuclear reactions have been proposed to account for the fusion of four hydrogen nuclei into a single helium nucleus. Hans Bethe of Cornell University suggested some of these procedures during the 1930s. The different chain reactions prevail at different temperatures, so chains that are dominant in very hot stars may be different from the ones in cooler stars.

When the temperature of the center of a main-sequence star is less than about 20 million kelvins, the **proton-proton chain** (■ Fig. 12–14) dominates. This sequence uses six hydrogen nuclei (protons), and winds up with one helium nucleus plus two protons. The net transformation is four hydrogen nuclei into one helium nucleus. (Though two of the protons turn into neutrons, here this isn't the main point.)

But the original six protons contained more mass than do the final single helium nucleus plus two protons. The small fraction of mass that disappears is converted into an amount of energy that we can calculate with the formula $E = mc^2$; see *Figure It Out 12.1: Energy Generation in the Sun.* According to Einstein's special theory of relativity, mass and energy are equivalent and interchangeable, linked by this equation.

For stellar interiors significantly hotter than that of the Sun, the **carbon-nitrogen-oxygen (CNO)** cycle dominates. This cycle begins with the fusion of a hydrogen nucleus (proton) with a carbon nucleus. After many steps, and the insertion of four protons, we are left with one helium nucleus plus a carbon nucleus. Thus, the carbon allows the chain of reactions to proceed and participates in it, but as

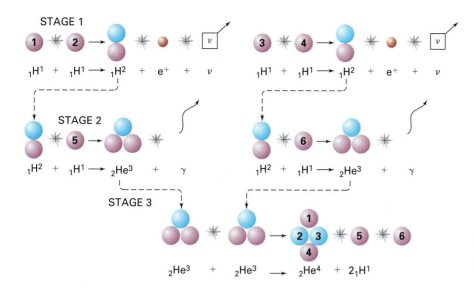

STAGE 1

$_1H^1$ + $_1H^1$ ⟶ $_1H^2$ + e^+ + ν $_1H^1$ + $_1H^1$ ⟶ $_1H^2$ + e^+ + ν

STAGE 2

$_1H^2$ + $_1H^1$ ⟶ $_2He^3$ + γ $_1H^2$ + $_1H^1$ ⟶ $_2He^3$ + γ

STAGE 3

$_2He^3$ + $_2He^3$ ⟶ $_2He^4$ + 2_1H^1

■ FIGURE 12–14 The proton-proton chain; e^+ stands for a positron (the equivalent of an electron, but with a positive charge; an "antielectron"), ν (the Greek letter "nu") is a neutrino, and γ (the Greek letter "gamma") is electromagnetic radiation at a very short wavelength. (Gamma rays are similar to x-rays but correspond to photons of even higher energies.)

In the first stage, two nuclei of ordinary hydrogen (protons) fuse to become a deuterium (heavy hydrogen) nucleus, a positron, and a neutrino. The neutrino immediately escapes from the star, but the positron soon collides with an electron. They annihilate each other, forming gamma rays.

Next (stage 2), the deuterium nucleus fuses with yet another proton to become an isotope of helium with two protons and one neutron. More gamma rays are released.

Finally (stage 3), in the most common version or "branch" of the proton-proton chain (a second branch occurs 15 per cent of the time, and a third branch is exceedingly rare), two of these helium isotopes fuse to make one nucleus of ordinary helium plus two protons. The protons are numbered to help you keep track of them.

much carbon remains at the end as there was at the beginning of a complete cycle (it is neither produced nor destroyed), and the carbon nucleus can start the cycle again.

As in the proton-proton chain, four hydrogen nuclei have been converted into one helium nucleus during the CNO cycle, 0.7 per cent of the mass has been transformed, and an equivalent amount of energy has been released according to $E = mc^2$. Main-sequence stars more massive than about 1.1 times the Sun are dominated by the CNO cycle.

Later in their lives, when they are no longer on the main sequence, stars can have even higher interior temperatures, above 10^8 K. They then fuse helium nuclei to make carbon nuclei. The nucleus of a helium atom is called an "alpha particle" for historical reasons. Since three helium nuclei ($_2He^4$) go into making a single carbon nucleus ($_6C^{12}$), the procedure is known as the **triple-alpha process**. After this process produces a carbon nucleus, there is a good chance that it will fuse with yet another alpha particle, forming a nucleus of $_8O^{16}$; thus, typically a mixture of C and O results from the triple-alpha process.

A series of other processes can build still heavier elements inside very massive stars. These processes, and other element-building methods, are called **nucleosynthesis** (new'clee-oh-sin'tha-sis). The theory of nucleosynthesis in stars can account for the abundances (proportions) we observe of the elements heavier than helium. Currently, we think that the synthesis of isotopes of the lightest elements (hydrogen, helium, and lithium) took place in the first few minutes after the origin of the Universe (Chapter 19), though some of the observed helium was produced later by stars.

12.6 BROWN DWARFS

When a pre-main-sequence star has at least 7.5 per cent of the Sun's mass (that is, it has about 75 Jupiter masses), nuclear reactions begin

and continue, and it becomes a normal star. But if the mass is less than 7.5 per cent of the Sun's mass, the central temperature does not become high enough for nuclear reactions using ordinary hydrogen (protons) to be sustained. (Masses of this size do, however, fuse deuterium into helium, but this phase of nuclear fusion doesn't last long because there is so little deuterium in the Universe relative to ordinary hydrogen.)

These objects shine dimly, shrinking and dimming as they age. They came to be called **brown dwarfs**, mainly because "brown" is a mixture of many colors and people didn't agree how such supposedly "failed stars" would look, and also because they emit very little light (■ Fig. 12–15). When old, they have all shrunk to roughly the same radius, about that of the planet Jupiter. We have met them already in Section 9.2c.

For decades, there was a debate as to whether brown dwarfs exist, but finally some were found in 1995. We now know of about 1,000, because of the great advances in astronomical imaging and in spectroscopy, not only in the visible but also in the infrared. The coolest ones, of spectral types T and Y, show methane and water in their spectra, like giant planets but unlike normal stars.

It is difficult to tell the difference between a brown dwarf and a small, cool, ordinary star, unless the brown dwarf is exceptionally cool. One way is to see whether an object has lithium in its spectrum. Lithium is a very fragile element, and undergoes fusion in ordinary stars, which converts it to other things. So if you detect lithium in the spectrum of a dim star, it is probably a brown dwarf (which isn't sustaining nuclear fusion using protons) rather than a cool, ordinary dwarf star of spectral class M or L, which are the coolest stars on the main sequence (and thus have begun to sustain their nuclear fusion). A complication is that very young M and L stars might not be old enough to have burned all their lithium, leading to potential confusion with brown dwarfs.

How do we tell the difference between brown dwarfs and giant planets in cases where they are orbiting a more normal star? Some astronomers would like to distinguish between them by the way

Palomar Observatory, Caltech

T. Nakajima and S. Kulkami (Caltech), S. Durance and F. Golimowski (Johns Hopkins University), and NASA

■ **FIGURE 12–15** The first brown dwarf to be unambiguously discovered, Gliese 229B (a number in a catalogue by a Swiss professor who studied nearby stars). A ground-based image (top) barely shows the brown dwarf orbiting the larger star, because the atmosphere blurs the larger star's image, but the Hubble Space Telescope shows the brown dwarf clearly (bottom). (The straight streak at 7 o'clock orientation is an artifact in the imaging.)

that they form: While planets form in disks of dust and gas as the central star is born, brown dwarfs form like the central star, out of the collapse of a cloud of gas and dust. But we can't see the history of an object when we look at it, so it is hard to translate the distinction into something observable.

All of the proposed tests are difficult to make. So, currently, for lack of definitive methods, the distinction is usually made on the basis of mass: Any orbiting object with a mass less than 13 times Jupiter's is called a planet, while the range 13 to 75 Jupiter masses corresponds to brown dwarfs. There are also many isolated brown dwarfs within the mass range 13–75 Jupiters, not orbiting stars. Objects less massive than 13 Jupiter masses not orbiting stars are sometimes called "free-floating planets" since they are not planets in the conventional sense of the word.

The rationale for using 13 Jupiter masses as the dividing line between planets and brown dwarfs is that above this mass, fusion of deuterium occurs for a short time, whereas below this mass, no fusion ever occurs. Thus, although brown dwarfs are not normal stars, they do fuse nuclei for a short time, and hence aren't completely "failed stars" as many people call them.

Brown dwarfs are being increasingly studied, especially in the infrared. Hubble Space Telescope images show that one of the nearby ones is double, with the components separated by 5 AU. By watching it over a few years, we should be able to measure its orbit and derive

accurate masses of the components. Also recall the Keck image (Figure 11–6b) of a brown-dwarf pair 45 light-years away whose component masses are about 30 Jupiters and whose orbital period is 10 years.

12.7 THE SOLAR-NEUTRINO EXPERIMENT

Astronomers can apply the equations that govern matter and energy in a star, and make a model of the star's interior in a computer. Though the resulting model can look quite nice, nonetheless it would be good to confirm it observationally.

However, the interiors of stars lie under opaque layers of gas. Thus we cannot directly observe electromagnetic radiation from stellar interiors. Only neutrinos escape directly from stellar cores. Neutrinos interact so weakly with matter that they are hardly affected by the presence of the rest of the Sun's mass. Once formed, they zip right out into space, at (or almost at) the speed of light. Thus they reach us on Earth about 8.3 minutes after their birth.

Neutrinos should be produced in large quantities by the proton-proton chain in the Sun, as a consequence of protons turning into neutrons, positrons, and neutrinos; see Figure 12–14. (A positron is an "antielectron," an example of antimatter. Whenever a particle and its antiparticle meet, they annihilate each other.)

12.7a Initial Measurements

For over three decades, astrochemist Raymond Davis carried out an experiment to search for neutrinos from the solar core, set up in consultation with the theorist John Bahcall, whose calculations long drove the theory. Davis set up a tank containing 400,000 liters of a chlorine-containing chemical (■ Fig. 12–16). To shield the experiment from cosmic rays, it was set up deep underground in the Homestake Gold Mine in Lead (pronounced "leed"), South Dakota.

One isotope of chlorine can, on rare occasions, interact with one of the passing neutrinos from the Sun. It turns into a radioactive form of argon, which Davis (then at the Brookhaven National Lab and later at the University of Pennsylvania) and his colleagues could detect. He needed such a large tank because the interactions are so rare for a given chlorine atom. In fact, he detected fewer than 1 argon atom formed per day, despite the huge size of the tank.

Over the years, Davis and his colleagues detected only about $1/3$ the number of interactions predicted by theorists. Where is the problem? Was it that astronomers don't understand the temperature and density inside the Sun well enough to make proper predictions? Or was it that the physicists don't completely understand what happens to neutrinos after they are released?

The answer turned out to be that neutrinos actually change after they are released. According to a theoretical model, the neutrinos of the specific type produced inside the Sun change, before they

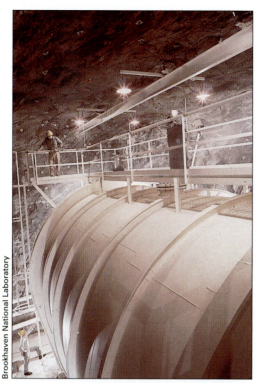

■ **FIGURE 12–16** The original neutrino telescope, deep underground in the Homestake Gold Mine in South Dakota to shield it from other types of particles from space. The telescope is mainly a tank containing 400,000 liters of perchloroethylene (dry-cleaning solution).

reach Earth, into all three types (called "flavors") of neutrinos that are known. But Davis's experiment is sensitive only to the specific flavor ("electron neutrinos") released by the Sun. Thus only a fraction of the original prediction is expected, and that is what we detect.

The Homestake Gold Mine no longer mines gold, but it has been converted to the Sanford Underground Research Facility for similar experiments shielded from cosmic-ray backgrounds.

12.7b Further Solar-Neutrino Experiments

The chlorine experiment, though it ran the longest by far, was not the only way to detect solar neutrinos. An experiment in Japan (■ Fig. 12–17), led by Masatoshi Koshiba, first set up to study protons and whether they decay, has used a huge tank of purified water to verify that the number of neutrinos is less than expected. Since the water detection method picks up some neutrinos from types other than the electron neutrinos, it gives a different ratio of detection than the chlorine experiment did. For their pioneering work in the detection of astrophysical neutrinos, Davis and Koshiba were together awarded half the 2002 Nobel Prize in Physics.

(a)

(b)

■ **FIGURE 12–17** (a) The SuperKamiokande (that is, super-Kamioka [a mine] nucleon decay experiment [NDE]) neutrino detector in Japan. We see the empty stainless steel vessel, 40 m across, and some of the 13,000 photomultiplier tubes, each 50 cm across, surrounding it. Neutrinos interacting within molecules of the 50,000 tons of highly purified water (note the boat, photographed as the detector was partly filled) lead to the emission of blue flashes of light that are being detected. The Kamiokande and SuperKamiokande experiments detected neutrinos from the Sun and provided a test not only of how stars shine but also of whether our basic understanding of fundamental physics is correct. In one of the most dramatic and expensive accidents ever in science, about half of these tubes "popped" and were destroyed in 2001, when the tank was being refilled after being emptied for the first time. Apparently, one defective tube broke, and sent out a shock wave that destroyed thousands of other tubes. (b) It took until 2006 to restore capabilities, making the current version into SuperKamiokande II.

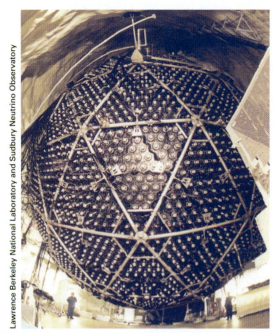

Lawrence Berkeley National Laboratory and Sudbury Neutrino Observatory

■ **FIGURE 12–18** The outside of the light sensors surrounding the tank of deuterated water ("heavy water") that was the heart of the Sudbury Neutrino Observatory.

Other sets of experiments in Italy (in the Gran Sasso auto tunnel) and in Russia that use gallium, an element that is much more sensitive to neutrinos than chlorine, also show that up to half of the expected neutrinos are missing. The chlorine was sensitive only to neutrinos of very high energy, which come out of only a very small fraction of the nuclear reactions in the Sun and not out of the two main branches of the proton-proton chain. Gallium is sensitive to a much wider range of interactions, including the most basic ones, so it picks up a larger fraction of the predicted neutrino rate from the Sun than the chlorine experiment.

A Canadian–U.S.–U.K. experiment deep underground in a mine in Sudbury, Ontario, Canada, collected data from 1999 to 2006 (■ Fig. 12–18). It had even more sensitive detection capability than earlier experiments. It used a large quantity of "heavy water," water whose molecules contain deuterium instead of the more normal hydrogen isotope. This experiment, like those in Japan, looked for light given off when neutrinos hit the water.

The Sudbury Neutrino Observatory (SNO) observations apparently resolved the remaining mysteries about the missing solar neutrinos. They aren't missing after all, since SNO has been able to detect the correct rate in one of its configurations, which is sensitive to all flavors of neutrinos. In another mode, sensitive only to the electron neutrinos that the Sun gives off, it confirms the deficit found by other detectors. Aided by adding salt to its heavy water to enhance its sensitivity to all three types of neutrinos, its scientists found a value of 0.30 ± 0.03 for the ratio of electron neutrinos to all neutrino types. So SNO confirmed that most of the neutrinos change in flavor ("oscillate") en route from the

Sun to the Earth. The solar-neutrino problem was indeed in the neutrino physics rather than in our understanding of the temperature of the Sun's core. Their measurements of the total neutrino flux also confirmed that the detailed models of the Sun by Bahcall and others are very accurate.

Still other experiments use a chemical liquid, usually just known as the "scintillator," since it scintillates, giving off light, when it interacts with a neutrino. The "Borexino" experiment reported in 2011 the detection of 46 neutrino interactions per day per 100 tons in their 278-ton scintillator. They verify the idea of neutrino oscillations by detecting neutrinos at a lower energy than many of the previous experiments, thus being sensitive to a large fraction of the emitted neutrinos.

Neutrinos were first thought of theoretically as particles that have no rest mass – that is, particles that would have no mass if they weren't moving ("at rest"). The various recent experiments described above have shown that neutrinos probably have a tiny rest mass after all: Theoretically, neutrinos can oscillate from one type to another only if they have some rest mass, so the detection of oscillations implies a non-zero rest mass.

The SNO experiment in Canada is now being modified to make SNO+, also with a liquid scintillator. Besides being sensitive to most of the reactions in the proton-proton chain, it could also detect solar CNO-cycle neutrinos, aided by its lower background of cosmic rays because of its greater depth underground.

The various neutrino experiments have caused a revolution in fundamental ideas of physics. In a similar case of astronomical discoveries affecting a broader range of basic science, there are also important repercussions for physics from measurements of distant supernovae (exploding stars) indicating that the expansion of the Universe is now accelerating (Section 18.5). The relation between physics and astronomy is very close (indeed, most of what we call "astronomy" these days is actually "astrophysics"), though the point of view that physicists and astronomers have in tackling problems may be different.

12.7c Beyond Solar Neutrinos

Techniques similar to those pioneered to find solar neutrinos have been and are being used to search for other types of strange particles. For example, in Princeton University's Darkside-50 experiment in the Gran Sasso tunnel, Italy, 50 kg of liquid argon scintillator is used to search for particles suspected of making up the so-called dark matter (which we discuss in Chapter 16 on galaxies). Many other experiments to detect dark matter exist elsewhere, such as in the Homestake, Sudbury, and Kamioka underground laboratories, but so far, only limits have been set on the interactions of dark-matter particles with ordinary matter.

A cubic kilometer of clear Antarctic ice at the South Pole has been made into the IceCube neutrino detector (■ Fig. 12–19). With its huge volume of detector, it is looking for neutrinos from violent events in the distant universe, such as supernovae and

NSF/B. Gudbjartsson/IceCube Collaboration

■ **FIGURE 12–19** The IceCube neutrino observatory at the South Pole contains 86 strings of over 5,000 photomultipliers lowered into a cubic kilometer of ice, as far as 2.5 km down. In spite of the hardships involved in working at the South Pole, searching for neutrinos and other particles there is desirable because the strings of photomultipliers are more stable in ice than they were in the water versions aimed at the same astronomical detection problem. (And no fish swim by.) Further, sea water has a radioactive calcium component whose decays raise the background level; the radioactivity has decayed away in the ancient Antarctic ice.

gamma-ray bursts. Some of these neutrinos are far more energetic than those produced by accelerators on Earth, so the basic study of the physics of neutrinos is also aided by IceCube's sensitivity. Moreover, the scientists involved hope they may find signs of the dark matter.

12.8 THE END STATES OF STARS

The next two chapters discuss the various end states of stellar evolution. The mass of an isolated, single star determines its fate and we provide a figure here as a type of "coming attractions" (■ Fig. 12–20). This brief section and diagram thus set the stage for what is coming next, which is so interesting that it occupies two separate chapters.

In Chapter 13, we discuss low-mass stars, like the Sun, which wind up as white dwarfs. We also examine the more massive stars, which use up their nuclear fuel much more quickly. These high-mass stars are eventually blown to smithereens in supernova explosions, becoming neutron stars or even black holes! Chapter 14 is devoted entirely to exotic black holes and their properties.

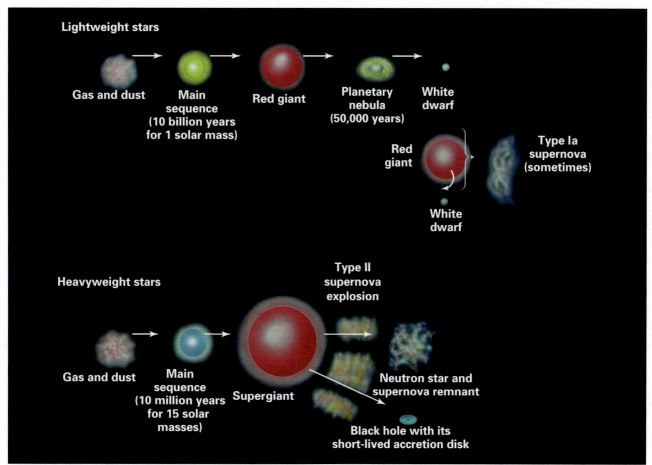

■ **FIGURE 12–20** A summary of the stages of stellar evolution for stars of different masses, using the boxing terms "lightweight" and "heavyweight" to mean low mass and high mass, respectively. The various types of objects shown will be defined and discussed in the next two chapters.

CONCEPT REVIEW

» Stars are formed from dense regions in nebulae, "clouds" of gas and dust that are best observed at infrared and radio wavelengths (Sec. 12.1). The collapsing cloud becomes a cluster of **protostars** and subsequently luminous **pre-main-sequence stars**, which are powered by gradual gravitational contraction (Sec. 12.1a). The dust heats up and becomes visible in the infrared. Some planetary systems in formation may have been discovered around young stars (Sec. 12.1b).

» Stars get their energy from **nuclear fusion** (Sec. 12.2), keeping the interior gas hot. The outward force from the **thermal pressure** balances the inward force of gravity, a condition known as "hydrostatic equilibrium." The basic fusion process in the Sun and other main-sequence stars is a merger (fusion) of four hydrogen nuclei (**protons**) into one helium nucleus (two protons and two **neutrons**; two of the protons turned into neutrons during the process), with the difference in mass between the original four protons and the helium nucleus (the "binding energy" of the helium nucleus) transformed to energy according to Einstein's famous equation, $E = mc^2$.

» Atoms that have lost electrons are **ions** (Sec. 12.3a). Each **element** is defined by the number of protons in its nucleus. Varieties of the same element with different numbers of neutrons are **isotopes** (Sec. 12.3b). Isotopes that decay spontaneously are **radioactive** (Sec. 12.3c). **Neutrinos** are extremely light, neutral particles given off in some radioactive decays as well as during nuclear fusion in stars. They are so elusive that they escape from a star immediately.

» In the hot cores of stars, collisions have stripped electrons away from the atomic nuclei, leaving the atoms fully ionized (Sec. 12.4). Nuclear fusion requires very high temperatures (millions of kelvins); only at high speeds can some of the protons overcome their electrical repulsion and get sufficiently close to each other

for the "strong nuclear force" (which holds protons and neutrons together in nuclei) to be effective. The greater a star's mass, the hotter its core becomes before it generates enough pressure to counteract gravity, so the star is more luminous.

» The **proton-proton chain** is the basic fusion process in the Sun, while the **carbon-nitrogen-oxygen cycle** dominates in hotter, more massive stars (Sec. 12.5). At still higher temperatures, when stars have left the main sequence, the **triple-alpha process** is dominant, where an alpha particle is a helium nucleus. Element-building processes are what is meant by nucleosynthesis.

» **Brown dwarfs** are sometimes called "failed stars," with insufficient mass to begin nuclear fusion of protons, ordinary hydrogen (Sec. 12.6). However, they do fuse deuterium (a rare, heavy form of hydrogen) to helium for a short time, so in a sense they do act like stars. Brown dwarfs have a mass range of 13 to 75 Jupiter masses (which is 1.3 to 7.5 per cent of the Sun's mass). Being very cool and dim, brown dwarfs were not discovered until 1995, but new infrared surveys of the sky have revealed large numbers of them.

» Theorists predict a certain number of neutrinos that should result from fusion in the Sun. Careful experiments had generally found only about one-third to one-half that number (Sec. 12.7a). To explain this deficit, physicists suggested that most of the neutrinos change from one type to several different types on their journey from the Sun to the Earth. Recently, measurements at the Sudbury Neutrino Observatory showed that the predicted number of neutrinos is indeed being produced by the Sun, and that they change type as they travel (Sec. 12.7b).

» Stars have several different endpoints to their evolution after the main sequence (Sec. 12.8). These will be discussed in the next two chapters.

QUESTIONS

1. Since individual stars can live for millions or billions of years, how can observations taken at the current time tell us about stellar evolution?

2. **(a)** Why are regions of star formation difficult to study at optical and ultraviolet wavelengths? **(b)** Which wavelengths are better for this purpose, and why?

3. What is the source of energy in a pre-main-sequence star?

4. At what point does a pre-main-sequence star become a normal star?

5. Arrange the following in order of development: pre-main-sequence star, nebula, protostar, Sun.

6. What are T Tauri stars and Herbig-Haro objects?

†7. Give the number of protons, the number of neutrons, and the number of electrons in ordinary hydrogen ($_1H^1$), lithium ($_3Li^7$), and iron ($_{26}Fe^{56}$) atoms.

8. **(a)** If you remove one neutron from helium, the remainder is what element? **(b)** If you remove one proton from helium, what element is left?

9. Why is four-times-ionized helium never observed?

10. Explain why nuclear fusion takes place only in the center of stars rather than on their surface.

11. What is the major fusion process that takes place in the Sun?

12. What is meant by the term "nucleosynthesis"?

13. What forces are in mechanical balance for a star to be on the main sequence?

14. **(a)** Qualitatively, how do the main-sequence lifetimes of massive stars compare with those of low-mass stars? **(b)** Explain why this is so.

†15. Use the speed of light and the distance between the Earth and the Sun to verify that it takes neutrinos 8.3 minutes to reach us from the Sun.

16. What does it mean for the temperature of a gas to be higher?

†17. If 0.7 per cent of the mass of four protons gets emitted as energy when they fuse to form a helium nucleus, how many joules of energy are produced? Show your work. (Note that a joule is a $kg·m^2/s^2$.)

18. Why do neutrinos give us different information about the Sun than light does?

19. What is a brown dwarf?

20. Why are the results of the solar neutrino experiment so important?

21. Summarize the current status of our understanding of solar neutrinos.

22. **True or false?** Nuclear reactions occur throughout almost the entire Sun, interior to the photosphere, because of the high temperature of the gas everywhere within the Sun.

23. **True or false?** Neutrinos were long thought to have zero rest mass, but it turns out that they actually have a small but non-zero rest mass.

†24. **True or false?** If a helium nucleus has 0.7 per cent less mass than four free protons (each of mass m_p), then the equation $E = (0.007)(4m_p)c^2$ represents the amount of energy generated by the Sun each time a new helium nucleus forms through nuclear fusion. (Assume neutrons and protons have equal mass.)

25. **True or false?** As a main-sequence star ages, its surface becomes much hotter and it moves to a very different location on the main sequence in a temperature-luminosity diagram.

26. **True or false?** Brown dwarfs are formed by the same initial process that forms main-sequence stars, but they are less massive.

27. **True or false?** Stars generate their energy through nuclear fission, the breaking apart of unstable heavy elements into lighter ones.

28. **True or false?** Unlike a planet, a brown dwarf does produce energy through nuclear fusion, at least for a short time.

29. **True or false?** The main way in which the puzzling early results of the solar neutrino experiment were resolved was by adopting a modified, more correct value for the Sun's central temperature.

30. **Multiple choice:** As a protostar is formed from a cloud of gas and becomes a pre-main-sequence star, which one of the following does *not* occur? **(a)** The cloud contracts due to gravity. **(b)** The density of the gas increases. **(c)** The pressure of the gas increases. **(d)** The temperature of the gas increases. **(e)** The protons begin to fuse.

31. **Multiple choice:** In a stable star, the gravitational forces are balanced by **(a)** gravitational forces acting in the opposite direction; **(b)** thermal pressure; **(c)** the strong nuclear force; **(d)** chemical reactions; or **(e)** electromagnetic forces.

†32. **Multiple choice:** The number of protons, neutrons, and electrons in a neutral atom of $_{26}Fe^{56}$ is, respectively, **(a)** 26, 30, 26; **(b)** 30, 26, 30; **(c)** 26, 30, 56; **(d)** 30, 26, 26; or **(e)** 26, 56, 26.

33. **Multiple choice:** The amount of energy released in a nuclear fusion reaction is directly proportional to the **(a)** mass of the initial components; **(b)** mass of the final products; **(c)** mass difference between the initial components and final products; **(d)** mass of the proton; or **(e)** mass of the photon.

34. **Multiple choice:** Which one of the following statements about the birth of stars is *false*? **(a)** During gravitational contraction, the temperature in the core of a pre-main-sequence star increases. **(b)** Gravitational contraction of a pre-main-sequence star ceases once nuclear reactions begin. **(c)** Stars form from giant clouds of gas and dust whose collapse and fragmentation are due to gravity. **(d)** A pre-main-sequence star does not release any energy (i.e., shine) until nuclear reactions begin. **(e)** Contracting balls of gas with mass below about 75 Jupiter masses actually end up as brown dwarfs, not main-sequence stars.

35. **Multiple choice:** Which one of the following statements regarding stars is *false*? **(a)** Stars form from large clouds of gas and dust that fragment as they collapse. **(b)** Even though they are gravitationally contracting, pre-main-sequence stars do not emit electromagnetic radiation; they have not yet started fusing hydrogen to helium. **(c)** Brown dwarfs, sometimes called "failed stars," undergo fusion for only a short time compared with M-type main-sequence stars.

(d) Once the core of a sufficiently massive contracting star gets hot enough, hydrogen starts fusing to helium and the star settles on the main sequence. **(e)** The electromagnetic radiation (gamma rays) emitted by the nuclear reactions in the core of a main-sequence star does not immediately escape the star because the ionized gases are quite opaque.

36. **Multiple choice:** In main-sequence stars, the pressure to counteract gravity comes primarily from **(a)** energy released through the fusion of all elements lighter than iron, since fusion of elements heavier than iron doesn't release energy; **(b)** energy released through the chemical burning (combustion) of hydrogen, since hydrogen is so flammable; **(c)** energy released through the fusion of hydrogen into helium, since hydrogen is readily available and fuses at the central temperatures of stars; **(d)** energy released through gravitational contraction of the star, since it must contract before fusion of heavier elements can occur; or **(e)** energy released through collisions between matter and antimatter, since this is a very efficient way to produce energy.

37. **Fill in the blank:** Brown dwarfs emit electromagnetic radiation primarily at _____ wavelengths.

38. **Fill in the blank:** Forms (varieties) of the same element having different numbers of neutrons are called _____.

39. **Fill in the blank:** Pre-main-sequence stars generate energy predominantly through the process of _____.

40. **Fill in the blank:** Theorists predicted that the Sun should be producing a large number of _____, but for a long time their measured rate on Earth was substantially less than expected.

†41. **Fill in the blank:** The number of neutrons in a nucleus of $_7N^{15}$ is _____.

42. **Fill in the blank:** The Sun produces most of its energy through a set of reactions known as the _____.

†This question requires a numerical solution.

TOPICS FOR DISCUSSION

1. Can you convince yourself that radio waves are not much affected by gas and small particles (dust)? Consider whether you can hear a radio on a foggy or smoggy day.

2. Do you think astronomers were overly bold in predicting that the solar neutrino problem would be resolved by changes in our theories of fundamental particles, rather than by abandoning the standard model of solar energy production? Defend your response.

3. A solitary star that is still forming must get rid of its angular momentum through the ejection of spinning jets (and, to some extent, by storing it in planets that form in the disk). Is there as much of a problem for a cloud from which a binary star forms?

SN 2011fe

Supernova 2011fe, the nearest Type Ia supernova (exploding white dwarf star) in the past several decades, in the beautiful Pinwheel Galaxy, M101, about 21 million light-years from Earth. The rapidly brightening star was first noticed less than a day after it exploded, more than two weeks before reaching peak brightness. For a few weeks, it could be seen through a good pair of binoculars.

B.J. Fulton for creating the image with data collected at the Byrne Observatory at Sedgwick, a LCOGT facility (optical data), and at the Palomar Observatory 48-inch telescope (H-alpha channel)

13

The Death of Stars: Recycling

ORIGINS

We discover how the heavy elements in the Earth, so necessary for life itself, came into existence through nuclear reactions in stars. We discuss the ultimate fate of our Sun, billions of years into the future. And, we learn about the origins of interesting classes of stars, including red giants, white dwarfs, and neutron stars, as well as the origins of exploding stars.

AIMS

1. Follow the evolution of low-mass stars like the Sun after they live their main-sequence core hydrogen-burning lives (Section 13.1).

2. Learn about red giants, planetary nebulae, white dwarfs, and novae (Section 13.1).

3. Understand supernovae, the catastrophic explosions of certain types of stars, and how they eject heavy elements into the cosmos (Section 13.2).

4. Describe the origin, properties, and utility of neutron stars, often visible as pulsars (Section 13.3).

The more massive a star is, the shorter its stay on the main sequence. The most massive stars may be there for only a few million years. A star like the Sun, on the other hand, is not especially massive and will live on the main sequence for about ten billion years. Since it has taken over four billion years for humans to evolve, it is a good thing that some stars can be stable for such long times. A few types of stars explode violently at the ends of their lives (see the Chapter Opener), but our Sun will expire in a less sensational manner.

In this chapter, we will first discuss what will happen when the Sun dies: It will follow the same path as other single lightweight stars, stars born with up to about 10 (but possibly as low as 8) times the mass of the Sun. They will go through planetary nebula (■ Figs. 13–1, 13–2) and white-dwarf stages. Then we will discuss the death of more massive stars, greater than about 8 or 10 times the Sun's mass, which we can call heavyweight stars. (We use the boxing terms "lightweight" and "heavyweight" colloquially, though we are referring to mass rather than weight.) The most massive stars go through spectacular stages. Some wind up in such a strange final state – a black hole – that we devote the entire next chapter to it.

■ **FIGURE 13–1** (a) The Cat's Eye Nebula, a planetary nebula, the late stage of evolution of a lightweight star like the Sun. It is imaged here in the Hubble Heritage Project. (b) The Ring Nebula (M57), a planetary nebula, in an image from the Hubble Heritage Project. As in all planetary nebulae, radiation from the hot central star provides the energy for the nebula to shine.

(a)

(b)

13.1 THE DEATH OF THE SUN

Though the details can vary, most solitary (that is, single) stars containing less than about 8–10 times the Sun's mass will have the same fate. (Later we will see that stars in tightly bound binary systems can end their lives in a different manner.) Since the Sun is a typical star in this mass range, we focus our attention on stars of one solar mass when describing the various evolutionary stages of stars. We quote only approximate timescales for each stage; more massive stars will evolve faster, and less massive stars will evolve more slowly.

13.1a Red Giants

As fusion exhausts the hydrogen in the center of a star (after about 10 billion years on the main sequence, for the Sun), its core's internal pressure cannot be maintained through energy from nuclear reactions because temperatures are not yet sufficiently high to fuse helium into heavier elements. Gravity pulls the helium core and its surrounding hydrogen shell inward, and half of the gravitational energy released by this process goes into heating the gas (the other half is radiated away). Hydrogen begins to "burn" more vigorously in the now hotter

(a)

Infrared

Visible light

(b)

■ **FIGURE 13–2** Ground-based images of the Helix Nebula (NGC 7293), a very nearby planetary nebula, 700 light-years from Earth. The main ring is about 2 light-years across, about half the distance from our Sun to the nearest star. (a) The left image is in the near-infrared, taken from the European Southern Observatory's 2009 wide-field VISTA (Visible and Infrared Survey Telescope for Astronomy) 4.1-m telescope on Cerro Paranal in Chile. The red shows molecular gas, which clearly clumps into filaments. At right, we see a true-color image, in which the red light has contributions from both hydrogen and ionized nitrogen, the blue-green is from doubly ionized oxygen, and the interior blue is ionized helium. The image was taken with a 2.2-m telescope at the European Southern Observatory on La Silla in Chile. (b) A Hubble Space Telescope close-up of part of the Helix Nebula shows comet-shaped features formed by the outward flow of stellar wind from the central star. This wind pushes on knots that are shielded by dust and gas from being broken up by short-wavelength radiation. The heads of the knots are roughly the size of our Solar System (out to the orbit of Neptune), and the tails are perhaps 1,000 au long, ten times longer. Observations from the Hubble Space Telescope and ground-based observatories have shown that the Helix Nebula isn't a helix after all. It consists of a 6,000-year-old inner disk and a 6,600-year-old outer ring that are almost at right angles to each other. One way that such a complicated shape could form is in a binary system.

(a) (b)

■ **FIGURE 13–3** (a) A Hubble Space Telescope view of the planetary nebula NGC 6826 shows the basic parts of elliptical planetary nebulae. We see the central cavity and shell caused by the fast wind and, around it, the material given off earlier. Colors show degrees of ionization and thus the energy of photons: singly ionized (red), doubly ionized (green), and triply or more ionized (purple). (b) A Hubble Space Telescope view of the planetary nebula NGC 7662, another elliptical planetary nebulae.

shell around the helium core. (The process is once again nuclear fusion, not the chemical burning we have on Earth.)

The newly released energy from the shell causes the outer layers of the star to swell by a factor of 10 or more. They become very large, so large that when the Sun reaches this stage, its diameter will be roughly 30 per cent the size of the Earth's orbit, about 60 times its current diameter. The solar surface will be relatively cool for a star, only about 4,000 K, so it will appear reddish. Such a star is called a **red giant**. Red giants appear at the upper right of temperature-luminosity (Hertzsprung-Russell) diagrams. The Sun will be in this stage, or on the way to it after the main sequence, for about a billion years, only 10 per cent of its lifetime on the main sequence.

Red giants are so luminous that we can see them at quite a distance, and a few are among the brightest stars in the sky. Arcturus in Boötes and Aldebaran in Taurus are both red giants.

The contracting core eventually becomes so hot that helium will start fusing into carbon (via the triple-alpha process; see our discussion in Chapter 12) and oxygen nuclei, but this stage will last only a brief time (for a star), perhaps 100 million years. During this time, the star becomes slightly hotter (at its surface) and fainter. We end up with a star whose core is carbon and oxygen, and is surrounded by shells of helium and hydrogen that are undergoing nuclear fusion.

Not being hot enough to fuse carbon nuclei, the core once again contracts, releasing gravitational energy that heats it up, causing the surrounding shells of helium and hydrogen to fuse even faster than before. This input of energy makes the outer layers expand again, and the star becomes an even larger and cooler (roughly 3,500 K) red giant. (Note that stars less massive than 0.45 times the mass of the Sun don't ever produce a carbon-oxygen core; their fusion process creates only helium, which remains in the core.)

13.1b Planetary Nebulae

As the star grows larger during the second red-giant phase, its helium-burning shell is unstable, and undergoes periodic pulses every 10,000

to 100,000 years. These invisible pulses produce large bursts of energy that serve to expand the star's envelope in a "slow wind" followed by a "fast wind," a "superwind" accelerated by the now-exposed stellar core. The loosely bound outer layers can continue to drift outward until they leave the star. We see glowing gas energized by a central hot star, a "planetary nebula" (■ Fig. 13–3).

The best-known planetary nebula is the Ring Nebula in the constellation Lyra (Fig. 13–1b). It is visible in even a medium-sized telescope as a tiny apparent smoke ring in the sky. Only photographs reveal the vivid colors. The Helix Nebula (Fig. 13–2) is so close to us that it covers about half the apparent diameter of the full moon in the sky, though it is much fainter.

We know of about a thousand such glowing shells of gas in our Milky Way Galaxy. Each shell contains roughly 20 per cent of the Sun's mass. They are exceptionally beautiful. In the small telescopes of a hundred years ago, though, they appeared as faint greenish objects, similar to the planet Uranus. These objects were thus named **planetary nebulae**, though in reality they are unrelated to planets.

The remaining part of the star in the center of the planetary nebula is the star's exposed hot core, which reaches temperatures of 100,000 K and so appears bluish. It is known as the "central star of a planetary nebula." It is on its way to becoming a white dwarf (see Section 13.1c). Ultraviolet radiation from this hot star partly ionizes gas in the planetary nebula, causing it to glow: electrons cascade down various energy levels after recombining with the positive ions. Also, collisions kick the bound electrons into higher energy levels, and photons are emitted as the electrons in the ions jump back down to lower energy levels.

We now know that planetary nebulae generally look greenish because the gas in them emits mainly a few strong spectral emission lines that include greenish ones, specifically lines of doubly ionized oxygen; see, for example, Figure 13–3 again. Uranus seems green for an entirely different reason (principally because the methane molecule absorbs the red). But the name "planetary nebulae" remains. (Note: Planetary nebulae are almost never called just "nebulae"; always use the adjective "planetary.")

■ **FIGURE 13–4** A Hubble Space Telescope view of a planetary nebula, the Egg Nebula, being born. (a) In this visible-light view, the central star is hidden by a dense lane of dust. We are seeing light from the central star scattered toward us by dust farther away. Some of the light escapes through relatively clear places, so we see beams coming out of the polar regions. The circular arcs are presumably the shells of gas and dust that were irregularly ejected from the central star. (b) The false-color view from Hubble's **N**ear **I**nfrared **C**amera and **M**ulti-**O**bject **S**pectrometer (NICMOS) shows starlight reflected by dust particles (blue) and radiation from hot molecular hydrogen (red). The collision between material ejected rapidly along a preferred axis and the slower, outflowing shells causes the molecular hydrogen to glow.

(a)

Raghvendra Sahai and John Trauger (JPL), the WFPC2 Science Team, and NASA

(b)

Rodger Thompson, Marcia Rieke, Glenn Schneider, Dean Hines (University of Arizona); Raghvendra Sahai (JPL); NICMOS Instrument Definition Team; and NASA

Astronomers are very interested in "mass loss" from stars, and planetary nebulae certainly provide many of the prettiest examples, even if hotter and more massive stars give off mass at a more prodigious rate.

The Hubble Space Telescope has viewed planetary nebulae with a resolution about 10 times better than most images from the ground, and has revealed new glories in them. Its infrared camera provided views of different aspects of some of them (■ Fig. 13–4). To our surprise, planetary nebulae turn out to be less round than previously thought. The stars aren't losing their mass symmetrically in all directions. In some cases, the star shedding mass is only one member of a binary system, and the presence of the secondary star affects the direction of mass loss.

The planetary-nebula stage in the life of a Sun-like star lasts only about 50,000 years; after that time, the nebula is so spread out that it fades too much to be seen at our great distance. But there can be episodic mass ejections, leading to the multiple shells that are sometimes seen.

13.1c White Dwarfs

Through a series of winds and planetary-nebula ejections, all stars that are initially up to 8 (or some perhaps even 10) times the Sun's mass manage to lose most of their mass. The remaining stellar core is less than 1.4 times the Sun's mass. (The Sun itself will have only 0.6 of its current mass at that time, in about 7 billion years.)

When this contracting core reaches about the size of the Earth, 100 times smaller in diameter than the star had been on the main sequence, a new type of pressure succeeds in counterbalancing gravity so that the contraction stops. This "electron degeneracy pressure" is the consequence of processes that can be understood only with quantum physics. It comes from the resistance of electrons to being packed too closely together; they become "degenerate" (indistinguishable from each other in certain respects), and end up differing mainly

in their energy levels. This resistance results in a type of star called a **white dwarf** (■ Fig. 13–5).

The Sun is 1.4 million km (nearly a million miles) across. When its remaining mass (about 0.6 solar masses) is compressed into a volume 100 times smaller across, which is a million times smaller in volume, the density of matter goes up to an incredible value, a million times higher than before. On Earth, a single teaspoonful of white-dwarf material would weigh about 5 tons! Such a high density may have been momentarily achieved in a terrestrial laboratory experiment.

A white dwarf's mass cannot exceed 1.4 times the Sun's mass; it would become unstable and, in principle, gravitationally collapse. This theoretical maximum was worked out by an Indian university

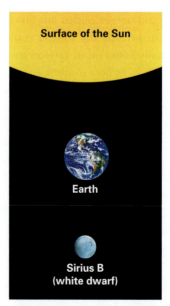

■ **FIGURE 13–5** The sizes of white dwarfs are not very different from that of the Earth. A white dwarf contains about 300,000 times more mass than the Earth, however. The Sun's curvature is exaggerated; the Sun's diameter is actually about 109 Earth diameters.

(a)

Jimmy Westlake

(b)

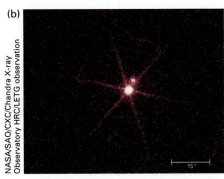

NASA/SAO/CXC/Chandra X-ray
Observatory HRC/LETG observation

15"

student, S. Chandrasekhar (usually pronounced "chan-dra-sek′har" in the United States), en route to England in 1930. It is called the "Chandrasekhar limit." In a long career in the United States, Chandrasekhar became one of the most distinguished astronomers in history, and shared the 1983 Nobel Prize in Physics with William A. Fowler for this early research. A major NASA spacecraft, the Chandra X-ray Observatory, is named after him.

Because they are so small, white dwarfs are very faint and therefore hard to detect. Only a small number of single ones are known. We find most of them as members of binary systems. Even the brightest star, Sirius (the Dog Star), has a white-dwarf companion, which is named Sirius B and sometimes called "The Pup" (■ Fig. 13–6).

A very odd and interesting system of white dwarfs was discovered with the Chandra X-ray Observatory. From the periodicity of the x-ray observations (■ Fig. 13–7a), the system is thought to contain two white dwarfs orbiting each other so closely that the orbit has a period of only five minutes (■ Fig. 13–7b). That makes the stars' separation only 80,000 km, roughly a fifth of the distance between Earth and the Moon! The white dwarfs are thought to be

roughly 1,600 light-years from us, in the direction of the constellation Cancer. We will see in Section 13.3f that there are testable consequences from such an extreme, violent wrenching around of these objects.

White-dwarf stars have all the energy they will ever have. Over billions of years, they will gradually radiate the energy stored in the motions of primarily their hot positive ions (carbon and oxygen nuclei, which are not degenerate). We can call them "retired stars," since they are spending their life savings of stored energy; they are not undergoing nuclear fusion as normal stars do. White dwarfs eventually become so cool and dim that they can no longer be seen, at least not easily. Some astronomers refer to such objects as "black dwarfs," but there is no clear boundary between a white dwarf and a black dwarf. The Sun will wind up in this state.

From the above discussion, we see that the main-sequence stage is the longest of a star's active lifetime. All subsequent stages are much shorter, except for the white-dwarf stage. But white dwarfs do not create nuclear energy; they simply use their life savings of stored energy. So, they are not considered an "active" phase of a star's life.

(a)   LIGHT CURVE

NASA/CXC/GSFC/T. Strohmayer

(b)

ILLUSTRATION

NASA/GSFC/D. Berry

■ **FIGURE 13–7** (a) The x-ray light curve (x-ray brightness vs. time) of a binary white-dwarf system, measured with the Chandra X-ray Observatory. (b) An artist's impression of the two orbiting white dwarfs, whose separation is much less than that of the Earth-Moon system. Gravitational waves are emitted, making the two white dwarfs gradually spiral in toward each other. Eventually, they will merge.

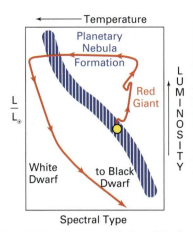

■ FIGURE 13–8 The post-main-sequence evolution of the Sun, roughly sketched on a temperature-luminosity diagram (Hertzsprung-Russell diagram); see text for details. From left to right, the spectral sequence is O, B, A, F, G, K, M; it corresponds to decreasing surface temperature. The broad hatched area represents the main sequence; massive stars are at the upper left, while low-mass stars are at the bottom right. The evolutionary tracks corresponding to different masses are not shown, but in general low-mass stars follow curves similar to that of the Sun, while stars more massive than 8 to 10 Suns become red supergiants before exploding as supernovae.

13.1d Summary of the Sun's Evolution

The entire "post-main-sequence" evolution of the Sun, a representative solitary low-mass star, can be tracked in a temperature-luminosity diagram (■ Fig. 13–8), or Hertzsprung-Russell diagram (recall our discussion in Chapter 11). We will say informally that the Sun "moves" through the diagram, but of course we really mean that the combination of luminosity and surface temperature changes with time, and that this changing set of values is reflected by a "trajectory" in the diagram.

Referring to Figure 13–8, the basic features are as follows (we will ignore small wiggles that represent details in the evolution). Initially the Sun will become a red giant, growing much more luminous (moving up in the diagram) and a bit cooler (moving slightly to the right), as its helium core contracts while surrounded by a hydrogen-burning shell. This stage will take about a billion years.

When the core becomes sufficiently hot to ignite helium, producing carbon (through the triple-alpha process) and oxygen nuclei, the Sun will briefly (for only about 100 million years) become a bit less luminous and hotter, thus accounting for the slight wiggle to the left and down visible in the temperature-luminosity diagram. It will subsequently become an even larger red giant, growing even more luminous (moving up in the diagram) and slightly cooler (moving a bit to the right) as the carbon-oxygen core contracts while surrounded by helium-burning and hydrogen-burning shells.

The Sun will then lose its outer envelope of gases through winds and gentle ejections, producing planetary nebulae and revealing a hotter surface (since those layers used to be in the hot interior). The luminosity, however, will stay roughly the same, as the burning shells of hydrogen and helium generate the energy. Thus, the Sun will quickly (over the course of 10^5 to 10^6 years) move to the left in the temperature-luminosity diagram.

As the supply of hydrogen and helium dwindles, the Sun will contract, but at the same time its temperature will decrease; thus, it will move steeply down toward the lower right of the diagram.

When the contracting Sun reaches roughly the size of the Earth, pressure from electron degeneracy will dominate completely, thwarting further contraction. However, the Sun will continue to cool, so its luminosity will continue to drop, though not as rapidly as when it was contracting as well. Thus, the Sun will move less steeply toward the lower right of the diagram. This stage will last tens of billions of years, eventually producing a very dim white dwarf (a "black dwarf," according to some astronomers).

13.1e Binary Stars and Novae

Single stars evolve in a simple manner. In particular, their main-sequence lifetime depends primarily on their mass. Most stars, however, are in binary systems, and the stars can exchange matter.

Surrounding each star is a region known as the **Roche lobe** (pronounced "Rohsh"), in which its gravity dominates over that of the other star (■ Fig. 13–9a). The Roche lobes of the two stars join at a point between them, forming a "figure-8" shape. (Édouard Roche was a 19th-century French mathematician.)

Consider two main-sequence stars. As the more massive star evolves to the red-giant phase, it fills its Roche lobe, and gas can flow from this "donor" star toward the lower-mass companion (■ Fig. 13–9b). The recipient star can gain considerable mass, and it subsequently evolves faster than it would have as a single star. Note that the flowing matter forms an **accretion disk** around the recipient star because of the rotation of the system (■ Fig. 13–10).

If one star is already a white dwarf and the companion (donor) fills its Roche lobe (for example, on its way to the red-giant phase), a **nova** can result (■ Fig. 13–11). For millennia, apparently new stars (**novae**, pronounced "no´vee" or "no´vay," the plural of nova) have occasionally become visible in the sky. Actually, however, a nova is an old star that brightens by a factor of a hundred to a million (corresponding to 5 to 15 magnitudes) in a few days or weeks. It then fades over the course of weeks, months, or years. The ejected gas may eventually become visible as an expanding shell.

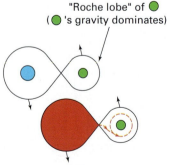

■ FIGURE 13–9 (a) The Roche lobe of each star in a physical binary system is the region where the gravity of that star dominates. As the two stars orbit each other, the "figure-8" pattern also rotates. (b) When the initially more massive star evolves to fill its Roche lobe, its outer gases can be transferred to the other, lower-mass star, thereby increasing its mass and making it evolve faster.

■ **FIGURE 13–10** A binary system in which a normal star fills its Roche lobe and transfers material to a white-dwarf companion. The material forms an "accretion disk" around the white dwarf, and then slowly falls onto the white dwarf. These systems are called "cataclysmic variables" because they undergo two types of outbursts. The accretion disk can periodically brighten as the material suddenly spirals inward, producing a "dwarf nova" outburst. Dwarf-nova outbursts typically repeat on timescales of weeks to months. On much longer timescales, usually centuries to millennia, the hydrogen gas accumulated on the surface of the white dwarf can eventually reach a sufficiently high temperature to undergo nuclear fusion, triggering a thermonuclear runaway on the surface – a spectacular nova explosion.

The brightening of a nova can occur in at least two ways. First, as gas, largely hydrogen, from the donor accumulates in the accretion disk, the disk becomes unstable and suddenly dumps material onto the white dwarf. The gravitational energy is converted to emitted light. Alternatively, matter can accumulate on the white dwarf's surface and suddenly undergo nuclear fusion, when it gets hot and dense enough. This mechanism can brighten the system far more than the release of gravitational energy. Matter is ejected at hundreds or thousands of kilometers per second. The fusion produces a few of the elements heavier than helium (such as carbon and oxygen, and very rarely up to silicon and sulfur) in the periodic table of the elements. This process involves only about $1/10,000$ of a solar mass, so it can happen many times. We do indeed see some novae repeat their outbursts after an interval of years or decades.

An especially peculiar, and still poorly understood, eruption was V838 Monocerotis. This object brightened by a large amount, but probably for a different physical reason than normal novae. It was surrounded by many shells of dust that were "lit up" by the nova outburst (■ Fig. 13–12). These "light echoes" evolved with time, as shells at different distances from the nova were successively illuminated. A consensus is emerging that its outbursts were from a violent merger of the two components of a binary star.

An interesting question in stellar evolution has long been the "blue stragglers," stars whose temperature and luminosity on the H–R diagram left them behind (further left, that is) when other stars of similar mass have turned off the main sequence to move up toward the red-giant stage. We now realize that for the most part they are stars in binary systems that have merged; in some cases, partners have become white dwarfs, having given the primary stars a new lease on main-sequence life by transferring additional mass to them.

13.2 SUPERNOVAE: STELLAR FIREWORKS!

Though most stars with about the mass of the Sun gradually puff off faint planetary nebulae, ending their lives as white dwarfs, some of them join more massive stars in going off with a spectacular bang – a **supernova** (plural: supernovae). Let us consider these celestial fireworks, and show how some explosions come from low-mass stars and others from high-mass stars.

What about intermediate-mass stars, those whose initial mass is in the range of 8 to 10 solar masses? They can fuse carbon to magnesium and neon; they subsequently become oxygen-neon-magnesium white dwarfs. However, some of them might explode as core-collapse supernovae like high-mass stars before losing their hydrogen envelope; the theory and observations are still uncertain.

Fritz Zwicky and Walter Baade coined the word "supernova" in 1931, although several other astronomers had previously pointed out the concept of an exceptionally luminous class of nova. A few years later, Zwicky and Baade correctly, at least in part, suggested that supernovae represent the collapse of ordinary stars into neutron stars, releasing tremendous amounts of energy and producing cosmic rays.

Supernovae are named in order of discovery in a given calendar year. Thus, the first supernova discovered in 2012 was SN 2012A, the

(a)

(b)

■ **FIGURE 13–11** Nova Cygni 1975. (a) During its outburst, in which it brightened by over 50,000 times and easily reached naked-eye visibility, and (b) after it faded (marked with an arrow). The star that became the nova was very faint, and was not noticeable except in deep images.

■ **FIGURE 13–12** V838 Monocerotis, a peculiar nova surrounded by shells of gas that were "lit up" by the nova outburst. As shells at progressively larger distances from the nova were illuminated, we saw the "light echoes" evolve with time. (a) The images were taken with the Hubble Space Telescope's Advanced Camera for Surveys, using blue, visible, and infrared filters. The exposure times are equivalent, as shown by the background stars, so the fading of V838 Mon itself is real. (b) A later observation of V838 Mon from November 2005, with the images scaled to show the nebulosity better, which also brightens the star images. Only visible and infrared filters were used. (c) An observation from September 2006. (d) Observations from January 2008, though the Advanced Camera for Surveys was broken, so the less-sensitive WFC2 was used. V838 Mon has become so faint that color images were not obtained from this time onward. (e) Observations from September 2009, using the repaired Advanced Camera for Surveys. We are now seeing primarily light backscattered off dust *behind* the star. (f) Observations from five Hubble orbits in February 2011. The echo is now very noticeably off center, suggesting that the dust has an elongated morphology extending from the star into the background on the NE side; note also the deep indentation on the SSE side.

second was SN 2012B, and so on through SN 2012Z. Subsequent supernovae are named SN 2012aa, SN 2012ab, and so on through SN 2012az, followed by SN 2012ba, SN 2012bb, etc.

13.2a Core-Collapse Supernovae

Stars that are more than 8–10 times as massive as the Sun whip through their main-sequence lifetimes at a rapid pace. These prodigal stars use up their store of hydrogen very quickly. A star containing 15 times as much mass as the Sun may take only 10 million years from the time it reaches the main sequence until it fully uses up the hydrogen in its core. This timescale is 1,000 times faster than that of the Sun.

For these massive stars, the outer layers expand as the helium core contracts. The star has become so large and cool (about 3,000 K) that we call it a **red supergiant**. Betelgeuse, the star that marks one shoulder of Orion, is the best-known example (■ Fig. 13–13).

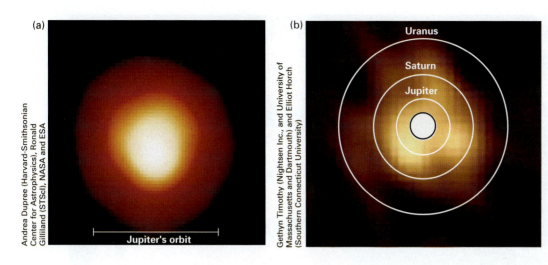

(a)

Jupiter's orbit

Andrea Dupree (Harvard-Smithsonian Center for Astrophysics), Ronald Gilliland (STScI), NASA and ESA

(b)

Uranus

Saturn

Jupiter

Gethyn Timothy (Nightsen Inc., and University of Massachusetts and Dartmouth) and Elliot Horch (Southern Connecticut University)

■ **FIGURE 13–13** (a) The red supergiant star Betelgeuse, α Orionis, is revealed by an ultraviolet image with the Hubble Space Telescope to have an atmosphere the size of Jupiter's orbit. A huge, hot, bright spot is visible on the surface; it is 2,000 K hotter than the surrounding gas. (b) When viewed farther in the ultraviolet with Hubble's imaging spectrograph, Betelgeuse is even larger, by a factor of 1.5. (The images are not reproduced at the same scale. The scale of this image is shown by an overlay of the orbits of planets in our Solar System.) Betelgeuse's atmosphere shows more structure at this far-ultraviolet wavelength. The central white disk shows the size of Betelgeuse's image in visible light.

Eventually, the core temperature reaches 100 million degrees, and the triple-alpha process begins to transform helium into carbon. Some of the carbon nuclei then fuse with a helium nucleus (alpha particle) to form oxygen. The carbon-oxygen core of a supergiant contracts, heats up, and begins fusing into still heavier elements. The ashes of one set of nuclear reactions become the fuel for the next set. Each stage of fusion gives off energy.

Finally, even iron builds up. Layers of elements of progressively lower mass surround the iron core, somewhat resembling the shells of an onion, and the outermost layer consists mostly of hydrogen. But when iron fuses into heavier elements, it takes up energy instead of giving it off. No new energy is released to make enough pressure to hold up the star against the force of gravity pulling in; thus, the iron doesn't fuse.

Instead, the mass of the iron core increases as nuclear fusion of lighter elements takes place, and its temperature increases. Eventually the temperature becomes so high that the iron begins to break down (disintegrate) into smaller units like helium nuclei. This breakdown soaks up energy and reduces the pressure. The core can no longer counterbalance gravity, and it collapses. The core's density becomes so high that electrons are squeezed into the nuclei. They react with the protons there to produce neutrons and neutrinos. Additional neutrinos are emitted spontaneously at the exceedingly high temperature (10 to 100 billion kelvins) of the collapsing core. All of these neutrinos escape within a few seconds, carrying large amounts of energy.

The collapsing core of neutrons overshoots its equilibrium size and rebounds outward, like someone jumping on a trampoline. The rebounding core collides with the inward-falling surrounding layers and propels them outward (■ Fig. 13–14). (For a stunning analogy demonstrating the mechanical "rebound" mechanism in a core-collapse supernova, hold a tennis ball on top of a basketball several feet above the ground, and drop both of them simultaneously. The tennis ball will rebound to a great height.) In addition, the surrounding layers get an important extra push from the plentiful neutrinos, only a very tiny fraction of which actually interact with the gas.

The star explodes as a supernova, achieving within one day a stupendous optical luminosity rivaling the brightness of more than a hundred million normal Sun-like stars. It will continue to shine for several years, gradually fading away. So much energy is available that very heavy elements, including those heavier than iron, form in the ejected layers. The core remains as a compact sphere of neutrons called a neutron star (to be discussed in more detail in Section 13.3a). There is even some evidence that occasionally the neutron star further collapses to form a black hole (Chapter 14).

Such supernovae, known as **Type II** (they show hydrogen lines in their spectra, unlike Type I supernovae), mark the violent death of heavyweight stars that have retained at least part of their outer layer of hydrogen. Since the fundamental physical mechanism is the collapse of the iron core, they are also one type of **core-collapse supernova**.

Two physically related kinds of supernovae are known as Type Ib and Type Ic. Both lack hydrogen lines (hence the "Type I" classification), but those of Type Ib show helium lines whereas those of Type Ic do not. We think they arise from massive stars undergoing iron core collapse (and the subsequent rebound). Prior to this, however, they got rid of their outer atmosphere of gases (■ Fig. 13–15), either through winds and small eruptions or by transferring material to a companion star. We shall see in Chapter 14 that some Type Ic supernovae are associated with stupendous bursts of gamma rays.

13.2b White-Dwarf Supernovae (Type Ia)

Another type of supernova, **Type Ia**, comes from carbon-oxygen white dwarfs in closely spaced (tightly bound) binary systems (Fig. 13–14). (We mentioned Types Ib and Ic, variations of explosions of heavyweight stars, in the previous section.) Gas can be transferred from the companion to the white dwarf, via an accretion disk, similarly to what was discussed for novae (Section 13.1e) and illustrated in Figure 13–10. However, for some reason the white dwarf avoids surface explosions (novae), which tend to eject all of the material

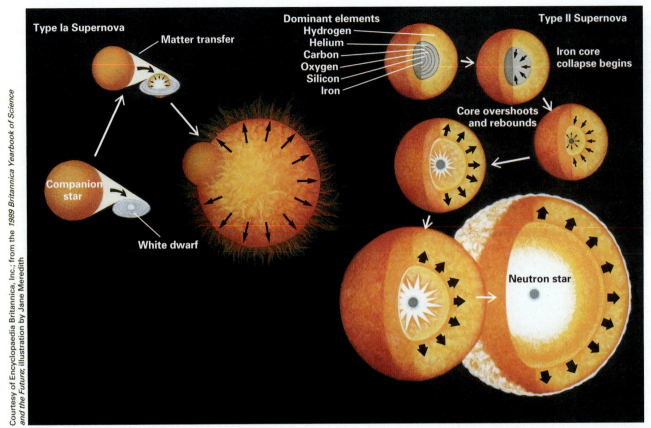

Courtesy of Encyclopaedia Britannica, Inc.; from the 1989 Britannica Yearbook of Science and the Future; illustration by Jane Meredith

■ **FIGURE 13–14** Type Ia supernovae (left), which show no hydrogen in their spectra, come from the incineration of a white dwarf that is gaining matter from a neighboring star and approaches the Chandrasekhar limit, 1.4 times the Sun's mass. Type II supernovae (right), whose spectra exhibit lines of hydrogen, are the explosions of massive stars, usually from the supergiant phase. When iron forms at the center of the onion-like layers of heavy elements, the iron core of the star collapses. In this model of the collapse, the core overshoots its final density and rebounds. Neutrinos also push outward. The shock wave from the rebound and the neutrinos blast off the star's outer layers.

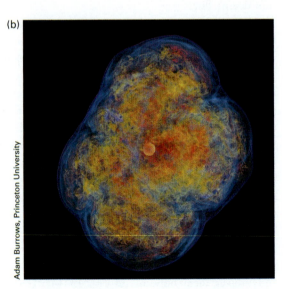

Adam Burrows, Princeton University

■ **FIGURE 13–15** (a) Massive stars usually have an extensive hydrogen envelope (left), but they can sometimes lose their outer envelopes of hydrogen (middle) and even helium (right), through winds or transfer to a binary companion. Yet their iron cores implode in the manner shown in Fig. 13–14 for Type II supernovae; a neutron star (or black hole) is formed, and the outer layers are ejected. Thus, these are "core-collapse" supernovae, as are Type II supernovae, but their spectra lack hydrogen. "Type Ib" supernovae show helium in their spectra, while "Type Ic" supernovae lack helium. They come from stars like those shown at middle and right, respectively. (b) The exploding ejecta in a three-dimensional model of a core-collapse supernova explosion, radius around 1,500 km. The remaining object in the interior is the proto-neutron star.

that had settled on the surface; in so doing, the white dwarf's mass can grow.

If the gas-donating companion adds too much matter to the white dwarf, causing the white dwarf to approach the Chandrasekhar limit of 1.4 solar masses, it becomes very hot inside and unstable, resulting in a runaway chain of nuclear-fusion reactions that literally explodes the star. (We now know that the white dwarf actually undergoes slow fusion of carbon for several centuries before blowing up. A thermonuclear runaway starts to occur when the energy generation rate through fusion exceeds the energy loss rate through the emission of neutrinos – and almost coincidentally, this occurs just shy of the Chandrasekhar limit.)

During the runaway, heavy elements are synthesized from carbon and oxygen, thereby liberating a huge amount of energy. The entire white dwarf explodes, leaving no compact remnant (unlike the case in a core-collapse supernova, which produces a neutron star or a black hole). After exploding, the supernova takes about 2.5 weeks to brighten, but then fades to obscurity over the course of a year. Because they come from white dwarfs, Type Ia supernovae are also sometimes called **white-dwarf supernovae** or "thermonuclear supernovae" (meaning that nuclear reactions occur in a very hot gas of atomic nuclei).

Such Type Ia supernovae become very luminous, typically as powerful as 4 billion Suns, the energy coming from the decay of radioactive heavy elements produced by the incineration. Moreover, since they all occur at nearly the same mass (a bit below the Chandrasekhar limit), Type Ia supernovae all have nearly the same absolute brightness at their peak. (There is actually a relatively small range of peak luminosities, but we have ways of correcting for this effect, as in the case of the period-luminosity relation for Cepheid variable stars.) Therefore, they can be used to find distance by comparing their absolute brightness with their apparent brightness. Because they are so luminous, they are detectable far into space and can be used to measure distances of galaxies billions of light-years away, and hence billions of years back in time. This distant visibility allows astronomers to determine the expansion history of the Universe, its age, and possibly even its fate, as we will see in Chapters 16 and 18.

One of the most surprising discoveries in quite a long time came from studies of the distances of Type Ia supernovae: the farthest ones are farther than we had thought. Thus the Universe appears to be expanding at an accelerating rate, propelled by a mysterious form of "dark energy." Einstein postulated "cosmic antigravity" of this type in 1917 in order to balance gravity to make what seemed obvious – a static Universe; ironically, he later retracted the idea, believing it to be nonsense (Chapter 18). He may well have been too hasty in his retraction.

Theoretical models can now account fairly well for the spectrum and the amount of light from a Type Ia supernova. This basic physical picture for the explosion seems fairly secure, but the exact nature of the companion is unclear. One problem is that if the donor were a normal star donating hydrogen to the white dwarf, why do we generally not see any hydrogen in the spectrum? Although the hydrogen

on the white dwarf's surface could have fused into helium prior to the explosion, the hydrogen blown away from the atmosphere of the donor companion star might be visible, especially if the donor is a red giant. Perhaps there is just too little hydrogen to be seen; only a small amount of hydrogen is stripped off the companion.

But another possibility for the "progenitors" of Type Ia supernovae (that is, for the kinds of stars that explode) is that the companion to the white dwarf is itself another white dwarf. If the two stars are sufficiently close together, they will gradually spiral toward each other as they emit gravitational radiation (see Section 13.3f). Eventually, the less massive white dwarf will be ripped apart by tidal forces, forming a disk of material around the more massive white dwarf. As material in the disk falls onto the white dwarf, its mass grows, and the white dwarf explodes when the mass gets close to the Chandrasekhar limit.

Some evidence for the first hypothesis (slow transfer of gas from a relatively normal donor star to a white dwarf) was recently found by one of the authors (A.F.) and his collaborators in the case of several well-studied Type Ia supernovae. By analyzing the optical spectra in detail, they found evidence that gas was recently expelled from the binary star system, which is not possible in the case of a white-dwarf pair. In one object, PTF 11kx, a large amount of hydrogen was found in a disk surrounding the Type Ia supernova; it probably came from the atmosphere of the hydrogen-rich donor star.

On the other hand, there is also strong evidence for the binary-white-dwarf model for the progenitors of Type Ia supernovae. In a few cases, careful searches for a surviving donor star near the explosion site of a nearby Type Ia supernova have not revealed any compelling candidates, suggesting that nothing remains from the original binary system, as expected if two white dwarfs merged together before exploding. Moreover, the x-rays expected from white dwarfs accreting gas from a normal donor star have not been detected in the expected amounts, again implying that the companion is another white dwarf.

Thus, it is quite possible that there are two ways to make Type Ia supernovae – that is, two different kinds of progenitors. In some Type Ia supernovae, the donor is a relatively normal star having a hydrogen atmosphere, but in others the companion appears to be another white dwarf.

13.2c Observing Supernovae

Only in the 1920s was it realized that some of the "novae" – apparently new stars – that had been seen in other galaxies (■ Fig. 13–16) were really much more luminous than ordinary novae seen in our own Milky Way Galaxy. These supernovae are very different kinds of objects. Whereas novae are small eruptions involving only a tiny fraction of a star's mass, supernovae involve entire stars. A supernova may appear about as bright as the entire galaxy it is in.

Unfortunately, we have seen very few supernovae in our own Galaxy, and none since the invention of the telescope. The most

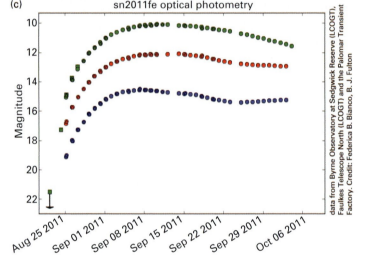

■ **FIGURE 13–16** (a) Supernova 2004dj is visible near the edge of this Hubble Space Telescope image of the galaxy NGC 2403 (right), only 11 million light-years away. A Type II supernova (that is, showing hydrogen in its spectrum), it is thought to have been produced by a star having an initial mass of 15 Suns. It was discovered near its peak at magnitude 11, about 100 times fainter than the faintest star you can see with your unaided eye. At left, a ground-based image, taken months before the eruption, from the Wisconsin-Indiana-Yale-National Optical Astronomy Observatory telescope. (b) The supernova PTF 11kly before and shortly after the explosion. (PTF stands for the **P**alomar **T**ransient **F**actory, which uses the 1.2-m Oschin Schmidt telescope at Palomar Observatory. The numbers indicate the year 2011, and the letters show that this was one object among a very large number of objects that suddenly brightened and were discovered by PTF that year.) Later assigned the more official name SN 2011fe, it is in the nearby galaxy M101, only 21 million light-years away. (c) Light curves (brightness vs. time) of SN 2011fe during the first 1.5 months after its discovery, obtained through three different filters: green (top), red (middle, offset downward for clarity), and near-infrared (bottom, also offset downward, shown in false-color blue); the specific filters were the so-called g′ band, r′ band, and i′ band of the Sloan Digital Sky Survey (SDSS), respectively. The 2-m Faulkes Telescope North in Hawaii was used, as was the Byrne Observatory in the Sedgwick Reserve, California, as part of the Las Cumbres Observatory Global Telescope Network, with some points from the 1.2-m Palomar Oschin Schmidt telescope.

recent ones in our Galaxy definitely noticed were observed by Kepler in 1604 and Tycho in 1572. A relatively nearby supernova might appear as bright as the full moon, and be visible night and day. Tycho's was visible to the naked eye for 15 months. Since studies in other large galaxies show that supernovae erupt every 30 to 50 years on average, we appear to be due, although a few supernovae have probably occurred in distant, obscured parts of our Galaxy. Maybe the light from a nearby supernova will reach us tonight. Meanwhile, scientists must remain content with studying supernovae in other galaxies (see *A Closer Look 13.1: Searching for Supernovae*).

A relatively nearby supernova, PTF 11kly or SN 2011fe (Fig. 13–16b), was discovered in the galaxy M101 (21 million light-years away) by the Palomar Transient Factory in August 2011, just about 11 hours after it exploded. At its peak, about 2.5 weeks after it was first detected, it reached a bit brighter than 10th magnitude at visual

wavelengths (Fig. 13–16c), about 40 times fainter than the faintest stars typically visible with the unaided eye, but easily seen through a good pair of binoculars or a small telescope. Since it is a Type Ia supernova, arising from the thermonuclear incineration of a white dwarf, studying it in detail should help us better understand how such stars explode and refine their utility in determining distances across the Universe. For example, spectra of the supernova showed carbon and oxygen, making it plausible that the star that exploded was indeed a carbon-oxygen white dwarf.

The research team of one of the authors (A.F.) analyzed Hubble Space Telescope images of the site of SN 2011fe that had been obtained for other reasons prior to the explosion. At first they were disappointed not to detect anything at that location, but they soon realized that a red giant would have been visible had it been present. So the companion star that fed mass onto the white dwarf must have

(a)

David F. Malin (Anglo-Australian Observatory) and Jay M. Pasachoff, from Palomar Observatory plates made available by the California Institute of Technology, courtesy of Robert Brucato

(b)

N. A. Sharp, REU program/NOAO/AURA/NSF

■ **FIGURE 13–17** (a) The Crab Nebula, a prominent supernova remnant, the result of the great supernova of the year 1054; it is about 6 light-years across. See also Figure 13–31. (b) The supernova optical remnant in Cygnus known as the Veil Nebula. The supernova that formed it exploded over 20,000 years ago, and the remnant covers more than 100 light-years.

been a subgiant, a main-sequence star, or possibly even another white dwarf. These observations thus eliminated one popular model for Type Ia supernova progenitors, that of a red-giant donor star feeding a white dwarf, at least for this particular object.

13.2d Supernova Remnants

Photography of the sky has revealed some two dozen regions of gas in our Galaxy that are **supernova remnants**, the gas spread out by the explosion of a supernova. The most studied supernova remnant is the Crab Nebula in the constellation Taurus (■ Fig. 13–17a). The explosion was noticed widely in China, Japan, and Korea in the year 1054; there is still debate as to why Europeans did not see it. An even older explosion produced the Veil Nebula in Cygnus (■ Fig. 13–17b).

If we compare photographs of the Crab Nebula taken decades apart, we can measure the speed at which its filaments of gas are expanding. Tracing them back shows that they were together at one point, at about the time the bright "guest star" was seen in the sky by the observers in Asia, confirming the identification. The rapid speed of expansion – thousands of kilometers per second – also confirms that the Crab Nebula comes from an explosive event. Moreover, spectra of the expanding gases reveal the presence of unusually high abundances (proportions) of heavy elements, as expected if such elements are indeed produced by the explosion.

The Chandra X-ray Observatory is giving us high-resolution x-ray images of supernova remnants (■ Fig. 13–18). The x-rays reveal exceptionally hot gas produced by the collision of the supernova with gas surrounding it.

13.2e Supernovae and Us

The heavy elements that are formed and thrown out by both core-collapse supernovae and white-dwarf supernovae are necessary for life

as we know it. Directly or indirectly, supernovae are the only known source of most heavy elements, especially those past iron (Fe) on the periodic table of the elements. They are spread through space and are incorporated in stars and planets that form later on.

Specifically, the Sun and our Solar System were made from the debris of many previous generations of stars. So we humans, who depend on heavy elements for our existence, are here because of supernovae and this process of recycling material.

Think about it: The carbon in your cells, the oxygen that you breathe, the calcium in your bones, and the iron in your blood are just a few examples of the elements produced long ago by stars and their explosions. (Other examples are the silver, gold, and platinum in jewelry – but these are not vital for the existence of life!) Thus, as the late Carl Sagan was fond of saying, "We are made of star stuff [or stardust]."

13.2f Supernova 1987A!

An astronomer's delight, a supernova quite bright but at a safe distance, appeared in 1987. On February 24 of that year, Ian Shelton, then of the University of Toronto, was photographing the Large Magellanic Cloud, a small galaxy 170,000 light-years away (on the far side of our Galaxy), with a telescope in Chile. Fortunately, he chose to develop his photograph that night.

When he looked at the photograph still in the darkroom, he saw a bright star where no such star belonged (■ Fig. 13–19). He went outside, looked up, and again saw the star in the Large Magellanic Cloud, this time with his naked eye. He had discovered the nearest supernova to Earth seen since Kepler saw one in 1604. By the next night, the news was all over the world, and all the telescopes that could see Supernova 1987A (the first supernova found in 1987) were focused on it. Some of these telescopes, as well as the Hubble Space Telescope, continue to observe the supernova on a regular basis to this day.

A CLOSER LOOK 13.1 Searching for Supernovae

A group led by one of the authors (A.F.) is currently running one of the world's most successful searches for nearby supernovae (up to a few hundred million light-years away), with a 0.76-m robotic telescope (KAIT, the Katzman Automatic Imaging Telescope; see the lower-left figure) at Lick Observatory in California. KAIT has been used to discover over 1,000 supernovae. An example of a KAIT discovery is shown in the lower-right figure: SN 1998dh in the galaxy NGC 7541.

The telescope automatically obtains CCD images of more than 1,000 galaxies per night (on average), cycling through about 8,000 galaxies in one week. It then repeats the list of observations. Over the course of a whole year, about 14,000 galaxies are monitored; the available galaxies change with the seasons, along with the changing constellations (recall our discussion in Chapter 1, *Star Party 1.1: Using the Sky Maps*). The computer software processes the images, and then it compares new images of galaxies with old images of the same galaxies, performing a digital subtraction.

Most of the time, nothing new is seen in the most recent images, but in about 50 images each night there is a new spot. These are supernova candidates – but they could also be poorly subtracted stars, pixels affected by cosmic rays, or asteroids that happened to be in the field of view. Thus, undergraduate students scan the supernova candidates (each student is responsible for one particular night of the week), determining which are the best supernova candidates and requesting that KAIT obtain confirmation images the next night. When they are correct, they are officially credited with the supernova discovery.

With the assurance that some new supernovae will be discovered each month, time on larger telescopes can be booked to follow these supernovae in some detail. In particular, a spectrum is obtained of each supernova to determine its type. Sometimes a whole series of spectra is obtained, over many months, to study in detail the expanding gases. One goal is to determine which chemical elements are produced by different kinds of explosions. Another is to understand the supernovae sufficiently well to be able to use them for cosmological distance determinations, as we will discuss in Chapters 16 and 18.

FIGURE ACL 13.1–1 The Katzman Automatic Imaging Telescope (KAIT), a 0.76-m robotic telescope operated by one of the authors (A.F.) and his team at Lick Observatory, a 2-hour drive from San Francisco, California. KAIT automatically searches for supernovae in about 14,000 galaxies, obtaining CCD images of more than 1,000 galaxies on a typical night.

(a)

(b)

FIGURE ACL 13.1–2 A supernova found by KAIT in a spiral galaxy, about 130 million light-years away from Earth. (a) An image of the galaxy NGC 7541 before the supernova went off. (b) Another image, clearly showing the supernova (SN 1998dh). Though only a single star, it is comparable in brightness to the entire galaxy consisting of billions of stars. It appears as bright as foreground stars (at left) in our own Milky Way Galaxy, which are only about 1,000 light-years away from us.

Hubble's high resolution shows clear views of an inner ring of material produced prior to the supernova, slowly expanding around the supernova (■ Fig. 13–20). The supernova debris is in the process of meeting up with the ring, and the collision has caused the supernova debris and the ring to brighten substantially over the last few years. In particular, many individual "hot spots" have brightened in the ring of material (■ Fig. 13–21). Two outer rings are also visible.

One exciting thing about such a close supernova is that we even know with certainty which star erupted! Pre-explosion photographs showed that a blue supergiant star had been where the supernova now is (■ Fig. 13–22a). It had been thought that Type II supernovae always erupt from red supergiants, not blue ones. Most do come from red supergiants (as shown by observations of a few other relatively nearby supernovae), but perhaps this particular star was a blue supergiant because the Large Magellanic Cloud is deficient in heavy elements relative to large galaxies like the Milky Way, affecting the star's structure. Another idea is that the star had merged with a binary companion prior to exploding, making it turn into a blue supergiant.

(a)

NASA, ESA, CXO, and P. Ruiz-Lapuente (University of Barcelona)

(b)

NASA/CXC/NCSU/S.Reynolds et al.

(c)

NASA/CSC/SAO

■ **FIGURE 13–18** (a) Tycho's supernova remnant, expanding since 1572, gives off gamma rays (magenta), as imaged from NASA's Fermi Gamma-ray Space Telescope. Also shown is the remnant in x-rays (yellow, green, and blue), as imaged from NASA's Chandra X-ray Observatory, superimposed on an infrared image (red) and an optical image showing stars along the line of sight. (The apparent offset between the magenta gamma-ray emission and the other images of the remnant is probably not real; the positional accuracy of the gamma-ray measurements is relatively low.) The Fermi observations support previous suggestions that supernova remnants can produce cosmic rays: Collisions between high-energy protons from the supernova remnant and protons in the surrounding interstellar matter produce particles called pions, which then decay to form the detected gamma rays. (b) An image of Kepler's supernova remnant, expanding since 1604, made with data from the Chandra X-ray Observatory. Chandra's observations of the abundance ratio of oxygen and iron showed that it was a Type Ia supernova. (c) A Chandra view of the Cassiopeia A supernova remnant, expanding since about 1680.

(a)

© 1987 Anglo-Australian Observatory, photographs by David Malin

(b)

© 1987 Anglo-Australian Observatory, photographs by David Malin

■ **FIGURE 13–19** The region of the Tarantula Nebula (upper left) in the Large Magellanic Cloud. We see the region (a) before and (b) after February 23, 1987, the explosion date. Supernova 1987A shows clearly at lower right in (b). It was a Type II supernova, having hydrogen in its spectrum.

In the leading current model, the progenitor star for the supernova formed about 10 million years ago. A million years ago, a slow stellar wind blew off its outer layers and formed a large cloud of cool gas, exposing a hot stellar surface. Before the star exploded, a high-speed stellar wind from the hot surface carved out a cavity in the cool gas, though leaving some fingers of cooler gas extending inward. We now see the edge of this cavity as the ring imaged by the Hubble Space Telescope, which lit up from the strong ultraviolet light given off as the supernova exploded. The shock wave from the supernova reached the ring in 2005, and the x-ray strength increased greatly. The Chandra X-ray Observatory image (■ Fig. 13–22b) shows hot gas, millions of kelvins, matching the optical bright spots. The optical and x-ray spots result from a collision of the shock waves with the fingers of cool gas. The new instruments installed on the Hubble Space Telescope in 2009 have been used to continue to follow the evolution of the supernova (■ Fig. 13–22c).

Supernova 1987A initially did not brighten as much as expected, and the fact that it was from a blue supergiant explains why. Blue supergiants are smaller than red ones, so they are not able to radiate as much light in the early stages of the explosion, and more of the explosion energy is used up in expanding the star. Theoretical models coupled with the observations indicate that the star had once

FIGURE 13–20 (a) A Hubble view of the inner and outer rings of ejected gas around Supernova 1987A. This gas was lost from the star prior to its explosion. (b) Light curves through blue (B) and red (R) filters of the rings and the supernova debris. After the first $1/3$ of a year, the light curve of the supernova debris followed the radioactive decay of cobalt-56, with a mean life of 111 days. But after about 1 year, the optical brightness declined more quickly because of the formation of dust in the expanding debris that absorbed the optical light. After about 3 years, the decay rate slowed down because it was then the longer-lived isotope cobalt-57 that was providing most of the heating. The current light curve of the debris is almost flat because the dominant heating is from the slow decay of titanium-44; some slight brightening may be due to contamination from the rings. The brightness of the rings is now increasing significantly because of energy emitted during the ongoing collision between the supernova debris and the ring material.

FIGURE 13–21 A series of images of the inner ring of gas around Supernova 1987A obtained with the Hubble Space Telescope. The ring consists of gas thrown off by the star prior to the explosion itself. Debris expanding from the explosion is reaching the ring, making spots on it brighten. (a) From 1995. (b) From 1998. (c) From 2003. (d) From 2005. (e) From 2011. (f) Contours show the highest-resolution image of SN1987A in radio wavelengths, made with the Australia Telescope Compact Array, as an overlay on a Hubble view (blue/white color scale) from the same time period.

(a)

© 1989 Anglo-Australiian Observatory, photograph by David Malin

(b)

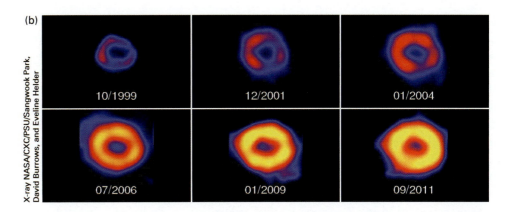

10/1999   12/2001   01/2004

07/2006   01/2009   09/2011

X-ray NASA/CXC/PSU/Sangwook Park, David Burrows, and Eveline Helder

(c)

Richard McCray, University of Colorado at Boulder

■ **FIGURE 13–22** (a) A composite photo in which a pre-explosion photograph of the blue supergiant star that exploded is superimposed as a negative (thus appearing black) on an image of Supernova 1987A. Several stars appear as black dots in the negative. (b) Chandra's images of x-rays from the inner ring around Supernova 1987A brightening over several years. Chandra's x-ray spectra show that the hot spots result from the shock wave hitting dense fingers of gas protruding inward from the ring. The false-color Chandra images are from October 1999, December 2001, January 2004, July 2006, January 2009, and September 2011. (c) A recent spectrum of SN 1987A, in the region of the hydrogen Lyman-alpha line, using the Cosmic Origins Spectrograph installed on the Hubble Space Telescope during the 2009 Servicing Mission 4. We see narrow (v about 300 km/s) emission lines from N V (4-times-ionized nitrogen) in the circumstellar ring, broad (v about 10,000 to 20,000 km/s) emission lines in hydrogen's Lyman-alpha and N V produced by a shock wave resulting from the expanding material's interaction with the surrounding interstellar medium, and continuum emission. The narrow lines marked in green are due to absorption by the respective ions in interstellar gas between us and the supernova. Likewise, interstellar hydrogen blocks the central part of the broad Lyman-alpha emission line. The bright, narrow emission line at the center of this Lyman-alpha absorption line is produced by hydrogen in the Earth's geocorona, while geocoronal oxygen produces the emission line at the far right.

contained 20 times the mass of the Sun, with 6 times the mass of the Sun in the form of a helium core. Through winds and gentle ejections, it had already lost 4 solar masses by the time it exploded.

The rate at which the supernova faded matched the rate of decay of radioactive cobalt into stable iron. Its brightness corresponded to the 0.07 solar mass of cobalt that theory predicted would be formed (initially as radioactive nickel, which decayed into radioactive cobalt). Moreover, gamma-ray telescopes observing Supernova 1987A detected photons having the specific energies emitted only by a radioactive, short-lived isotope of cobalt. The radioactive cobalt must have been formed by the supernova; it could not have been present in the 10-million-year-old star prior to the explosion.

When the core of a massive star collapses, creating a neutron star and a supernova, theorists tell us that many neutrinos are produced. In fact, about 99 per cent of the energy is carried by neutrinos. The solar neutrino experiment using chlorine in liquid form (Chapter 12) was not sensitive enough to the energy range of neutrinos emitted by Supernova 1987A. Fortunately, at least two other experiments that had been set up for other purposes were operating during this event. Both experiments contained large volumes of extremely pure water surrounded by sensitive phototubes to measure any light given off as a result of interactions in the water.

One of the detectors, in a salt mine in Ohio, reported that 8 neutrinos had arrived and interacted within a 6-second period on February 23, 1987 (■ Fig. 13–23). Normally, an interaction of some kind with a neutrino was seen only about once every five days. Another detector, in a zinc mine in Japan, detected a burst of 11 neutrinos. (The detectors were in mines to shield them from other types of particles.) These few neutrinos marked the emergence of a new observational field of astronomy: extra-solar neutrino astronomy.

■ **FIGURE 13–23** A graph showing the detection of neutrinos from two detectors. A short "burst" of neutrinos, substantially above the background "noise" (some of which is produced by other particle interactions), is visible.

The 2002 Nobel Prize in Physics was awarded in part for the discovery of neutrinos from Supernova 1987A.

The fact that the neutrinos arrived three hours before we saw any signs of the optical burst matches our theoretical ideas about how a star collapses and rebounds to make a supernova. (The neutrinos escape from the stellar core at essentially the speed of light, whereas the surface of the exploding star doesn't brighten until a few hours later, when the shock wave from the rebounding material reaches the star's surface and heats it.) Also, the amount of energy carried by neutrinos, taking account of the tiny fraction that we detect, matches that of theoretical predictions. Our basic ideas of how Type II supernovae occur were validated: A neutron star had indeed been created, at least temporarily. We are now awaiting the emergence of definite signs of a pulsar, the type of star we shall discuss in the next section.

The observations have also given us important basic knowledge about neutrinos themselves. If neutrinos have mass, they will travel at different speeds depending on how much energy they were given. The fact that the neutrinos arrived so closely spaced in time placed a sensitive limit on how much mass they could have. So though the current observations of neutrinos show that they have some mass, since otherwise they wouldn't solve the solar neutrino problem for us by oscillating between different types of neutrinos, we also know that they don't have *much* mass. Knowing the mass of neutrinos is important for understanding how much matter there is in the Universe, as we shall see in our discussion of galaxies and cosmology (Chapters 16 and 18).

Note that the observed closeness of the travel speed of the SN 1987A neutrinos to the speed of light is a major reason why so many scientists were skeptical about the announcement in 2011 that some neutrinos may have travelled faster than light in going from the CERN accelerator in Geneva, Switzerland, to a detector at Gran Sasso in Italy. Such a finding, if true, would either violate Einstein's special theory of relativity or open the doors to exotic new physics such as neutrinos travelling in dimensions different from those that we experience. But it now appears that errors in the clock readings were the source of the apparently superluminal speed; there is no convincing evidence for the claimed result.

13.2g Cosmic Rays

So far, our study of the Universe in this book has relied on information that we get by observing electromagnetic radiation – including not only visible light but also gamma rays, x-rays, ultraviolet, infrared, and radio waves. Moon rocks and meteorites have given important insights as well. But we also receive a few high-energy particles from space.

These **cosmic rays** (misnamed historically; we now know that they aren't rays at all) are nuclei of atoms moving at tremendous speeds. Some of the weaker cosmic rays come from the Sun, while other cosmic rays come from farther away. Cosmic rays provide about $1/5$ of the radiation environment of Earth's surface and of the people on

it. (Almost all the radiation we are exposed to comes from cosmic rays, from naturally occurring radioactive elements in the Earth or in our bodies, and from medical x-rays.)

For a long time, scientists have debated the origin of the non-solar "primary cosmic rays," the ones that actually hit the Earth's atmosphere as opposed to cosmic rays that hit the Earth's surface. (Our atmosphere filters out most of the primary cosmic rays. When they hit the Earth's atmosphere, the collisions with air molecules generate "secondary cosmic rays.")

Because cosmic rays are charged particles – we receive mostly protons and also some nuclei of atoms heavier than hydrogen (primarily helium), and many fewer electrons – our Galaxy's magnetic field bends them. Thus we cannot trace back the paths of cosmic rays we detect to find their origin. It seems that most middle-energy cosmic rays were accelerated to their high speeds in supernova explosions.

Primary cosmic rays can be captured with high-altitude balloons or satellites. Many images taken with CCDs, whether those on the Hubble Space Telescope, on the Solar Dynamics Observatory, or on Earth, show streaks from cosmic rays (■ Fig. 13–24). When studying your CCD data, you have to eliminate the pixels that show these "cosmic-ray hits."

Stacks of suitable plastics (the observations were formerly made with thick photographic emulsions) show the damaging effects of cosmic rays passing through them. Scientists are now worried about cosmic rays damaging computer chips vital for navigation in airplanes as well as in spacecraft, and, for safety, engineers are providing chips that can work even when slightly damaged in this way.

When primary cosmic rays hit the Earth's atmosphere, they cause flashes of light that can be detected with telescopes on Earth. IceCube is a project that can observe secondary cosmic rays by studying light they generate as they plow through a cubic kilometer of clear natural ice near the South Pole, in addition to its work on neutrinos. A similar project, using clear Mediterranean water, is also progressing.

A huge cosmic-ray-detector project, the Pierre Auger Cosmic Ray Observatory (named after the discoverer of cosmic-ray air showers), covering more than 3,000 square kilometers of ground, has been built on the Pampas in Argentina. The detectors are 1,600 water tanks, each containing 3,000 gallons (12,000 liters), separated from each of its neighbors by 1.5 km. It is dark inside the tanks except when a cosmic ray interacts with one of the water molecules, and then the resulting burst of light is detected by a photomultiplier inside the tank. The Auger Observatory has a second detection method for cosmic rays, observing the sky overhead for ultraviolet fluorescence from interacting cosmic rays. Scientists using the Auger Observatory have determined the spectrum of ultra-high-energy cosmic rays over a wide range of energy; with unprecedented precision. Already it is known that there are fewer of the highest-energy cosmic rays detectable than had been extrapolated from lower energies.

Cosmic rays may even affect the Earth's weather, by encouraging clouds to form in the Earth's atmosphere. The flux of cosmic rays that reach Earth varies over the solar-activity 11-year cycle, with fewer cosmic rays reaching Earth at solar maximum, when more magnetic field threads through the heliosphere, diverting the charged particles.

(a)

(b)

Chien Peng (University of Arizona) and Alex Filippenko (UC Berkeley)

Chien Peng (University of Arizona) and Alex Filippenko (UC Berkeley)

■ **FIGURE 13–24** The central part of the galaxy M77, made by combining Hubble Space Telescope images obtained through red, green, and blue filters. In (a), many cosmic-ray hits can be seen as streaks and specks. (See arrows; the shape of each one depends on the angle at which the cosmic ray hits the CCD.) In (b), the cosmic-ray hits are gone. They were removed by comparing pairs of images obtained through each filter: Pixels that were much brighter in one image than the other were assumed to have been hit by cosmic rays, and were replaced by the unaffected pixels.

The Auger Observatory has been following the variation of cosmic rays over the solar-activity cycle with unprecedented accuracy.

When neutron stars were first discussed theoretically in the 1930s, the chances of observing one seemed hopeless. But we currently can detect signs of them in several independent and surprising ways, as we now discuss.

13.3 PULSARS: STELLAR BEACONS

We have examined the fate of the outer layers of a massive star that explodes as a core-collapse supernova. But what about the core itself? Let us now discuss cores that wind up as super-dense stars. In the next chapter, we will see what happens when the core is too massive to ever stop collapsing.

13.3a Neutron Stars

The collapsed core in a core-collapse supernova is a very compact object, generally a neutron star consisting mainly of neutrons, as already mentioned in Section 13.2a. They are like a single, giant atomic nucleus, without the protons (■ Fig. 13–25). Neutron stars have measured masses of about 1.4 Suns, but some might exist with up to 2 or 3 solar masses; astronomers don't understand the limit for neutron stars as accurately as they know that 1.4 solar masses is the limit for white dwarfs.

Neutron stars are only about 20 or 30 km across (Fig. 13–25); a teaspoonful would weigh a billion tons. At such high densities, the neutrons resist being further compressed; they become "degenerate," as we can explain using laws of quantum physics, acting similarly to the way electrons behave in a white dwarf (see our discussion in Section 13.1c). A pressure ("neutron-degeneracy pressure") is created, which counterbalances the inward force of gravity.

When an object contracts, its magnetic field is compressed. As the magnetic-field lines come together, the field gets stronger. A neutron star is so much smaller than the Sun that its magnetic field should be about a trillion times stronger.

13.3b The Discovery of Pulsars

Recall that the light from stars twinkles in the sky because the stars are point-like objects, and the Earth's atmospheric turbulence bends the light rays in various directions with time. Similarly, point-like radio sources (radio sources that are so small or so far away that they have no apparent length or breadth) fluctuate in brightness on timescales of a second because of variations in the density of electrons in interplanetary space. In 1967, a special radio telescope was built to study this radio twinkling; previously, radio astronomers

Background: NASA's Johnson Space Center

■ **FIGURE 13–25** A neutron star is about the size of a city, even though it may contain a solar mass or more of material. Here we see the ghost of a neutron star superimposed on a photograph of New York City. A neutron star might have a solid, crystalline crust about a hundred meters thick. Above these outer layers, its atmosphere probably takes up only another few centimeters. Since the crust is crystalline, there may be irregular structures like mountains, which would only poke up a few centimeters through the atmosphere.

■ **FIGURE 13–26** Jocelyn Bell Burnell, the discoverer of pulsars. She did so with a radio telescope – actually a field of aerials – at Cambridge, England. (a) This photo shows her in 2006 in her role as Vice President of the International Astronomical Union presenting the resolutions dealing with the reclassification of Pluto. She had brought a stuffed Disney Pluto as well as a box of cereal to represent objects like the asteroid Ceres. (b) A portrait.

(a)

(b)

■ **FIGURE 13–27** (a) The up-and-down variations on this chart led Jocelyn Bell Burnell to suspect that something interesting was going on. (b) The chart record showing the discovery of the individual pulses from CP 1919. Here downward blips actually represent increases in brightness.

(a)

(b)

had mostly ignored and blurred out the effect to study the objects themselves.

In 1967, Jocelyn Bell (now Jocelyn Bell Burnell) was a graduate student working with Professor Antony Hewish's special radio telescope (■ Fig. 13–26). As the sky swept over the telescope, which pointed in a fixed direction relative to Earth's surface, she noticed that the signal occasionally wavered a lot in the middle of the night, when radio twinkling was usually low.

Her observations eventually showed that the position of the source of the signals remained fixed with respect to the stars rather than constant in terrestrial time (for example, always occurring at exactly midnight). This timing implied that the phenomenon was celestial rather than terrestrial or solar.

Bell and Hewish found that the signal, when spread out, was a set of regularly spaced pulses, with one pulse every 1.3373011 seconds (■ Fig. 13–27). The source was briefly called LGM, for "**L**ittle **G**reen **M**en," because such a signal might come from an intelligent extraterrestrial civilization!

But soon Bell located three other sources, pulsing with regular periods of 0.253065, 1.187911, and 1.2737635 seconds. Though they could be LGM2, LGM3, and LGM4, it seemed unlikely that extraterrestrials would have put out four such beacons at widely spaced locations in our Galaxy. The objects were named **pulsars** – to indicate that they gave out pulses of radio waves – and announced to an astonished world. It was immediately apparent that they were an important discovery, but what were they?

13.3c What Are Pulsars?

Other observatories set to work searching for pulsars, and dozens were soon found. They were all characterized by very regular periods, with the pulse itself taking up only a small fraction of a period. When the positions of many of the roughly 2,000 known pulsars are plotted on a map of the whole sky (■ Fig. 13–28), we easily see that they are concentrated along the relatively flat plane of our Galaxy. Thus they must be in our Galaxy; if they were located outside our Galaxy we would see them distributed uniformly around the sky or even partly obscured near our Galaxy's plane where the Milky Way might block something behind it.

The question of what a pulsar is can be divided into two parts. First, we want to know why the pulses are so regular – that is, what the "clock" is. Second, we want to know where the energy comes from.

We can get pulses from a star in two ways: if the star oscillates in size ("vibrates") or if it rotates. (The only other possibility – collapsed stars orbiting each other – would result in progressively decreasing pulse periods as the objects release energy and spiral inward, but the

R. N. Manchester, CSIRO Astronomy
and Space Science, Australia

■ **FIGURE 13–28** The distribution of nearly 2,000 pulsars (all those known and published as of 2012) on a map that shows the entire sky, with the plane of the Milky Way along a horizontal line at center. From the concentration of pulsars along the plane of our Galaxy, we can conclude that the pulsars are members of our Galaxy; otherwise, we would have expected to see as many near the poles of the map. The uneven concentration of pulsars in Galactic longitude on this map merely represents the fact that this section of the sky has been especially carefully searched. Twenty one of the pulsars are in the Large Magellanic Cloud, and so are "extragalactic."

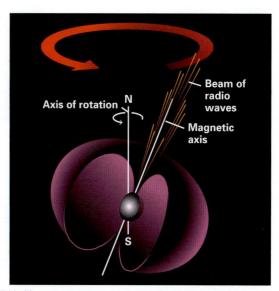

■ **FIGURE 13–29** In the lighthouse model for pulsars, which is now commonly accepted, a beam of radiation flashes by us once every pulsar period. (A beam pointing in the opposite direction gives an additional pulse, if the magnetic axis is perpendicular to the rotation axis.) Similarly, a lighthouse beam appears to flash by a ship at sea.

periods were not observed to decrease.) The theory worked out for ordinary variable stars had shown that the speed with which a star oscillates depends on its average density. Ordinary stars would oscillate much too slowly to be pulsars, and even white dwarfs would oscillate somewhat too slowly. Moreover, neutron stars would oscillate too rapidly to be pulsars. So oscillations were excluded.

That left only rotation as a possibility. But it can be calculated that a white dwarf is too large to rotate fast enough to cause pulsations as rapid as those that occur in a pulsar; it would be torn apart. So, the only remaining possibility is the rotation of a neutron star. We have solved the problem, or at least come to a plausible hypothesis, by the process of elimination. Over the years, more and more evidence has accumulated to back the **lighthouse model** for pulsars (■ Fig. 13–29), and there is general agreement on it. Just as a lighthouse seems to flash light every time its beam points toward you, a pulsar is a rotating neutron star.

How is the energy generated? There is much less agreement about that, and the matter remains unsettled. Remember that the magnetic field of a neutron star is extremely high. This high magnetic field can lead to a powerful beam of radio waves. If the magnetic axis is tilted with respect to the axis of rotation (which is also true for the Earth, whose magnetic north pole is quite far from true north), the beam from stars oriented in certain ways will flash by us at regular intervals. (We wouldn't see other neutron stars if their beams were oriented in other directions.) But the longest-known period of a pulsar is over 8 seconds, longer than can be easily accounted for by any of the current models of how the energy for the pulses is generated; thus, the details of what is going on are still not understood.

In 1974, Hewish received the Nobel Prize in Physics, largely for his discovery of pulsars. Given the crucial role played by Jocelyn Bell, it is unfortunate that she was not honored in this way as well. At that time, it was not the custom of the Nobel Prize committees to honor work done by a graduate student while also honoring the

advisor, but the custom has changed largely as a result of the omission in this case.

13.3d The Crab, Pulsars, and Supernovae

Several months after the first pulsars had been discovered, strong bursts of radio energy were found to be coming from the direction of the Crab Nebula. Observers detected that the Crab pulsed 30 times per second, almost ten times more rapidly than the fastest other pulsar then known. This very rapid pulsation definitely excluded white dwarfs from the list of possible explanations.

The discovery of a pulsar in the Crab Nebula made the theory that pulsars were neutron stars look more plausible, since neutron stars should exist in supernova remnants like this one. And the case was clinched when it was discovered that the clock in the Crab pulsar was not precise – it was slowing down slightly. The energy given off as the pulsar slowed down was precisely the amount of energy needed to keep the Crab Nebula shining. The source of the Crab Nebula's energy had been discovered!

Astronomers soon found, to their surprise, that an optically visible star in the center of the Crab Nebula could be seen apparently to turn on and off 30 times per second. Actually the star only appears "on" when its beamed light is pointing toward us as it sweeps around. Long photographic exposures had always hidden this fact, though the star had been thought to be the remaining core because of its spectrum, which oddly doesn't show any emission or absorption lines. Later, similar observations of the star's blinking on and off in x-rays were also found (■ Fig. 13–30). Even more recently, the

■ **FIGURE 13–30** The central region of the Crab Nebula. Images by the Hubble Space Telescope. Wisps of gas take on an apparent whirlpool shape.

NASA/CXC/ASU/J. Hester et al.

■ **FIGURE 13–31** The central region of the Crab Nebula. Images by the Chandra X-ray Observatory, showing an outer ring 2/3 light-year across and an inner ring.

NASA/CXC/ASU/J. Hester et al.

high-resolution observations by the Hubble Space Telescope and Chandra X-ray Observatory of the Crab Nebula revealed interesting structure near its core (■ Fig. 13–31).

Although every pulsar is thought to be a neutron star, not every neutron star will be visible as a pulsar. The spinning beam of radiation (in other words, the lighthouse beam) might not intersect our line of sight, and the neutron star might not even have a beam of radiation if it is rotating too slowly or has too weak a magnetic field.

Until 2011, it wasn't possible to tell which kind of supernova had led to which pulsars, though most neutron stars were thought to come from a couple of possible sources. But then a report on x-ray pulsars showed two different classes, at least for the x-ray pulsars found in the Small Magellanic Cloud, though there is some evidence that our Milky Way should be similar. The ones from the collapse stemming from electron capture in a lower-mass oxygen-neon-magnesium core, losing pressure support, seem to have shorter spin periods compared with those from iron-core-collapse supernovae.

13.3e Slowing Pulsars and Fast Pulsars

The Crab, when discovered, was the most rapidly pulsing known pulsar, and it is slowing down by the greatest amount. But most other pulsars have also been found to be slowing gradually. The theory had been that the younger the pulsar, the faster it was spinning and the faster it was slowing down. After all, the Crab came from a supernova explosion only about 1,000 years ago.

So, scientists were surprised to find a pulsar spinning 20 times faster – 642 times per second. Even a neutron star rotating at that speed would be on the verge of being torn apart. And this pulsar is hardly slowing down at all; it may be useful as a long-term time standard to test even the atomic clocks that are now the best available to scientists. The object, which is in a binary system, is thought to be old – over a hundred million years old – because of its gradual slowdown rate. Astronomers conclude that its rotation rate has been sped up in an interaction with its companion.

This pulsar's period is 1.56 milliseconds (0.00156 second), so it became known as "the millisecond pulsar." Dozens more millisecond pulsars have since been discovered. Each pulses rapidly enough to sound like a note in the middle of a piano keyboard (■ Fig. 13–32) if its rotation frequency were converted to an audio signal.

Many of the millisecond pulsars we have detected are in globular star clusters. So many stars are packed together in globular clusters that the former companion star might have been stripped off in most of the cases. But recently, observations with the Chandra X-ray Observatory revealed a closely spaced binary consisting of a millisecond pulsar and a normal star. In this system, known as 47 Tuc W (it is in the constellation Tucana, and after 47 Tuc A, B, and so on, it was the Wth to be discovered), the neutron star makes a complete rotation every 0.00235 second (2.35 milliseconds). It has almost certainly been spun up by accretion of material from its companion star.

13.3f Binary Pulsars and Gravitational Waves

Joseph Taylor and Russell Hulse set out to discover lots of pulsars. One of the millisecond pulsars they found, in the constellation Aquila, seemed less regular in its pulsing than expected. It turned out that another neutron star in close orbit with the pulsar, which pulsed every 59 milliseconds, was pulling it to and fro every 8 hours, shortening and lengthening the interval between the pulses that reached us on Earth. The system (technically, B1913+16 from its right ascension and declination in the sky) is called "the binary pulsar," though one of its neutron stars is not a pulsar. It has an elliptical orbit that can be traced out by studying small differences in the time of arrival of the pulses. The pulses come a little later when the pulsar is moving away from us in its orbit, and a little earlier when the pulsar is moving toward us.

Einstein's general theory of relativity explained the slight change over decades in the orientation of Mercury's orbit around

(a)

Walter Brisken (NRAO/VLA)

(b)

R. N. Manchester, CSIRO Astronomy and Space Science, Australia

* Millisecond pulsars
+ Magellanic Cloud pulsars
· Other pulsars

■ **FIGURE 13–32** (a) Many of the millisecond pulsars have periods fast enough to hear as musical notes when we listen to a signal at the frequency of their pulse rate. The display, of the first set of millisecond pulsars discovered, is in order of celestial longitude. (b) All the known pulsars as of 2012, except the ones in globular clusters. Of these, 240 of the pulsars are millisecond pulsars; 141 pulsars, mostly millisecond pulsars, are in globular clusters.

the Sun. The gravity in the binary-pulsar system is much stronger, and the effect is much more pronounced. Calculations show that the orientation of the pulsar's orbit should change by 4° per year (■ Fig. 13–33), which is verified precisely by measurements. From these and other relativistic measurements, the two stars' masses are known to incredible accuracy: 1.4398 ± 0.0002 and 1.3886 ± 0.0002 Suns, respectively, in work done by Joel Weisberg of Carleton College and his collaborators. This is near the Chandrasekhar limit, matching theoretical expectations. (The core of a massive star collapses to form a neutron star when its mass grows close to the Chandrasekhar limit.)

Another prediction of Einstein's general theory of relativity is that gravitational waves, wiggles in the curvature of space-time caused by fluctuations of the positions of masses, should travel through space. The process would be similar to the way that electromagnetic radiation, caused by fluctuations of the positions of electric charges, travels through space.

Gravitational waves have never been detected directly. But the motion of the binary pulsar in its orbit is speeding up by precisely the rate that would be expected if the system were giving off gravitational waves (■ Fig. 13–34). (The two stars' orbital period is decreasing, so according to Kepler's third law they are getting closer and closer together at a rate of 10 mm/day.) Therefore, scientists consider that the existence of gravitational waves has been verified in this way. The Nobel Prize in Physics was awarded in 1993 to Taylor and Hulse for their discovery and analysis of the binary pulsar.

An even better system for testing general relativity, called the "double pulsar" (technically, J0737-3039), was discovered in 2003. One of the pulsars spins every 23 milliseconds, while the other spins every 2.8 seconds, much more slowly. While we can detect pulses from only one of the Taylor-Hulse "binary pulsar" pair of neutron stars, for the double pulsar in the southern constellation Puppis, pulses were detected from both neutron stars until one disappeared in 2008. Because both pulsars could be detected, more relativistic

tests have been possible with the double pulsar than with the binary pulsar. In particular, the system's edge-on orbit permits interesting tests of the curvature of space-time when one star is almost directly behind the other, so that the radio signals have to pass very near to the closer one, where space-time is very curved, to reach us. These stars have masses of 1.3381 ± 0.0007 Suns and 1.2489 ± 0.0007 Suns, very accurately but not yet quite as accurately known as the masses of the earlier-discovered, binary pulsar.

The two pulsars in the double pulsar system orbit each other every 2.4 hours (■ Fig. 13–35) compared with the 8-hour orbital period of the Taylor-Hulse neutron stars, making them much closer together. Calculations based on measurements of the period change show that the double pulsar's orbit is shrinking – as gravitational waves carry off energy – by 7 millimeters per day. (The rates by which the orbits of the binary-pulsar and double-pulsar systems shrink are both incredibly

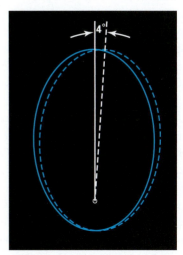

■ **FIGURE 13–33** In the binary pulsar's orbit, the major axis of the ellipse rotates by 4° per year. This measurement matches, and thus endorses, the prediction of Einstein's general theory of relativity.

■ **FIGURE 13–34** The rate at which the Taylor-Hulse pair of pulsars is orbiting (black points with error bars) slows in a way that precisely matches the predictions of Einstein's general theory of relativity (solid curve). The cumulative shift of periastron time (the time of closest approach of the two pulsars) in seconds is shown over time.

Other kinds of binaries besides pulsars are now also showing orbital decay as a consequence of a general-relativity effect. Just five binary systems with periods under 15 minutes are known. The only one that doesn't undergo mass transfer contains 2 white dwarfs orbiting every 13 minutes, each eclipsing the other. Einstein's theory predicts that the period should diminish by 0.25 ms per year, causing the eclipses to occur sooner and sooner. A University of Texas-led team confirmed such an effect in 2012.

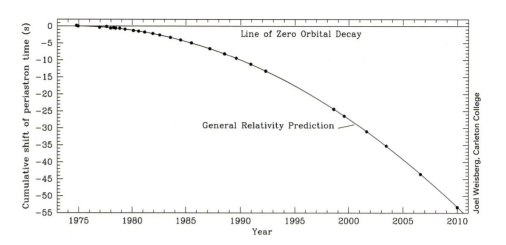

Line of Zero Orbital Decay

General Relativity Prediction

Joel Weisberg, Carleton College

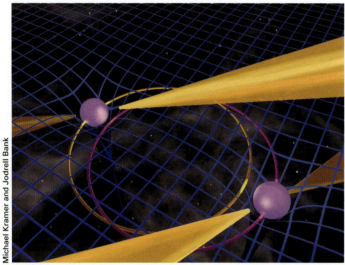

■ **FIGURE 13–35** An artist's conception of the double pulsar in Puppis, whose orbit is shrinking by 7 millimeters per day.

Michael Kramer and Jodrell Bank

Jay M. Pasachoff

■ **FIGURE 13–36** One of Joseph Weber's original bars with which over 40 years ago he searched for vibrations, waiting to see if it would start changing in length slightly, by 10^{-16} m at about 1.6 kHz, due to passing gravitational waves. Though his results were negative, he is credited with inspiring the field. This bar is on display at LIGO Hanford in Washington state.

tiny to be able to measure.) The measurements are in excellent agreement with the predictions of Einstein's theory.

The pulsar-orbit method of detecting gravitational waves was indirect. Only the speeding up of the orbit caused by the waves was measured, rather than the waves themselves. Efforts to sense gravitational waves directly by whether they set large metal bars vibrating (■ Fig 13–36) have failed, because the experiments lacked the sensitivity that would be necessary.

A major pair of observational facilities has been built to try to detect gravitational waves directly in another manner. This **L**aser **I**nterferometer **G**ravitational-wave **O**bservatory (LIGO) uses complicated lasers, optics, and electronics to see whether a several-kilometer-long length is distorted as a gravitational wave goes by. Nearly identical LIGOs have been set up in Hanford, Washington, and in Livingston, Louisiana (■ Fig. 13–37), to make sure that the system isn't tricked by local effects such as vibrations from nearby logging or traffic. The Livingston site has perpendicular evacuated tubes 4 km long, while the Hanford site has 2.5-km and 4-km tubes within

the same vacuum. Any signal detected must affect all three interferometers to be believed. A similar site, VIRGO, has been built in Italy by a French and Italian group of scientists with 3-km evacuated tubes. Thus far, no clear cases of gravitational waves have been detected.

LIGO is being upgraded to improve its sensitivity. Calculations show that by extending its range by a factor of 10, and therefore the volume within which it is sensitive by a factor of $10^3 = 1,000$, which is expected by 2014, scientists may detect a gravitational-wave "event" every week or two. We wish them luck, but we will believe it when we see it.

Scientists had long been planning for LISA (**L**aser **I**nterferometer **S**pace **A**ntenna), a 3-dimensional interferometer to be launched into space. (The 3 dimensions are laid out by 3 perpendicular arms each 5 million kilometers long (!), coming from pairs of 3 individual spacecraft.) LISA was to be in an Earth-trailing orbit around the Sun and should have been able to search even more sensitively for gravitational waves, detecting all kinds of mergers of compact objects like pulsars or black holes. But, in 2011, NASA dropped out for financial reasons, and the revamped, less capable, European-only project is known as the **N**ew **G**ravitational-wave **O**bservatory (NGO), though this newer concept lost out in a 2012 round of funding to the Jupiter

(a)

(b)

(c)

(d)

■ **FIGURE 13–37** (a) The 4-km-long perpendicular arms of the **L**aser **I**nterferometer **G**ravitational-wave **O**bservatory (LIGO) at Livingston, Louisiana, and at (b) Hanford, Washington. We see the outer part of long concrete tubes in which light travels 4 km and back through a vacuum. (c) The mounting of the reflecting mirrors demonstrated at LIGO Livingston; the mirrors will be hung with 14 levels of isolation from outside vibrations. (d) The entrance to the LIGO site in Hanford, Washington.

Icy Moons Explorer (JUICE). A proof-of-concept LISA Pathfinder may still happen, but this remains to be seen.

The prime system we know of for the space mission to detect is the pair of rapidly orbiting white dwarfs referred to in Section 13.1c. Calculations based on the observed decrease of the x-ray period show that gravitational waves should be given off at such a high rate that the orbit shrinks by about 1 centimeter per hour! The waves' frequency would be prime for NGO's sensitivity, though not a good frequency for even the improved LIGO.

13.3g A Pulsar with a Planet

Earlier in this book, in Chapters 2 and 9, we described the discovery of planets around other stars. But these extra-solar planets were not the first to be discovered. In 1991, the first ones were discovered by observing a pulsar.

The detections were from observations of a pulsar that pulses very rapidly – 162 times each second. The arrival time of the pulsar's radio pulses varied slightly (■ Fig. 13–38), indicating that something is orbiting the pulsar and pulling it slightly back and forth. Alex Wolszczan, now at Pennsylvania State University, has concluded that the variations in the pulse-arrival time are caused by three planets in orbit around the pulsar. These planets are 0.19, 0.36, and 0.47 au from the pulsar, within about the same distance that Mercury is

from the Sun. They revolve in 25.3-, 66.5-, and 98.2-day periods, respectively. The system is 2,000 light-years from us, too faint for us to detect optically.

The presence of the planets was conclusively verified when they interacted gravitationally as they passed by each other. The two most massive planets are calculated from the observations to be somewhat larger than Earth, each containing about 4 times its mass. The innermost planet is much less massive than Earth.

Astronomers think that neutron stars are formed in supernova explosions, so any original planets almost certainly didn't survive the explosion. Most likely, the planets formed after the supernova explosion, from a disk of material in orbit around the neutron star remnant. These pulsar planets are not the ones on which we expect life will have arisen!

13.3h X-ray Binaries

Neutron stars are now routinely studied in a way other than their existence as radio pulsars. Many neutron stars in binary systems interact with their companions. As gas from the companion forms an accretion disk much like those discussed in Section 13.1e (Fig. 13–10), or is funneled toward the neutron star's poles by a strong magnetic field, the gas heats up (because of friction and compression) and gives off x-rays.

After a Cornell University visualization, imaging by Chris Hildreth and Wayne Lytle; courtesy of Alex Wolszczan, Pennsylvania State University

■ **FIGURE 13–38** The pulses from the pulsar, shown in blue schematically on a chart, sometimes arrive slightly before or after regular pulses would (red dots). The average pulse period is 6.2 milliseconds. The difference is graphed at the bottom. It reveals that there must be planets orbiting the pulsar.

X-ray telescopes in orbit (such as the Chandra X-ray Observatory) detect the x-rays, sometimes in pulses if there is a rotating beam of emission. But in many of these binary systems, unlike the case for normal pulsars, the pulse rate usually speeds up. Gas from the accretion disk spins up the neutron star as it lands on the surface.

Although white dwarfs and neutron stars are indeed peculiar endpoints of stellar evolution, the last remaining possibility – black holes – are so interesting and bizarre that we devote the entire next chapter to them. Many of them are found in x-ray binary systems similar to those harboring neutron stars, but apparently the stellar remnant was too massive to be a neutron star. Meanwhile, however, it would be useful to go back to Chapter 12 and review Figure 12–20, in which we gave a preview of stellar evolution; with the knowledge you have gained in this chapter, you should now understand that diagram better.

CONCEPT REVIEW

» Stars have different fates depending on the mass with which they were born. **Lightweight stars** are those with mass up to about 10 (but perhaps only 8) times the Sun's mass, while **heavyweight stars** have larger masses (introductory section). When the Sun and other low-mass stars exhaust their central hydrogen, they will swell and become luminous **red giants** as their cores contract and the outer layers expand (Sec. 13.1a). The outer layers subsequently drift off as **planetary nebulae**, which glow because their gases are ionized by radiation from the hot dying star (Sec. 13.1b). The remaining core continues to contract until electrons won't be compressed further (they become "degenerate"), and the core becomes a **white dwarf** (Sec. 13.1c). The entire post-main-sequence evolution of the Sun can be conveniently traced on a temperature-luminosity (Hertzsprung-Russell) diagram (Sec. 13.1d). The evolution of other solitary (single) stars is similar, but differs in detail.

» If a star is in a binary system, its evolution can be sped up by the transfer of matter from a companion star filling its **Roche lobe**, the region in which the companion's gravity dominates (Sec. 13.1e). The flowing gas forms an **accretion disk** around the recipient star because of the rotation of the system. Matter falling onto a white dwarf from a companion star can release gravitational energy suddenly, or even undergo rapid nuclear fusion on the white dwarf's surface, flaring up to form a **nova** (plural **novae**).

» Heavyweight stars, having more than 8 to 10 times the Sun's mass, become **red supergiants** (Sec. 13.2a). Heavy elements build up in layers inside these stars. When the innermost core builds up enough iron, it collapses and the surrounding layers rebound (explode); the result is a **Type II supernova** (plural supernovae), a type of **core-collapse supernova** that has hydrogen in its spectrum. The remaining compact object is a very dense neutron star, consisting almost entirely of neutrons. If the evolved massive star had previously lost its outer envelope of gases, it would explode in a similar manner, but hydrogen lines would be absent in its spectrum; these are known as Type Ib (helium present) and Type Ic (no helium) supernovae.

» Hydrogen-deficient **Type Ia supernovae**, in contrast, occur when a carbon-oxygen white dwarf in a binary star system receives too much mass from its companion to remain in that state and undergoes a nuclear runaway that completely destroys the white dwarf (Sec. 13.2b). This happens a little bit below the Chandrasekhar limit, the theoretical maximum mass of a white dwarf (about 1.4 solar masses). In some of these **white-dwarf supernovae**, the companion appears to be a relatively normal star, but in others it may be a second white dwarf that merges with the first one. In both scenarios for white-dwarf supernovae, the explosion does not produce a compact remnant (neutron star or black hole), unlike the case in core-collapse supernovae.

» The last bright supernovae to have been seen in our Galaxy were in 1572 and 1604 (Sec. 13.2c). Many supernovae are discovered each year in other galaxies, however.

» Supernovae in our Galaxy have left detectable **supernova remnants**, the expanding debris of the explosions. The gases ejected by supernovae are enriched in heavy elements produced prior to, and during, the explosions (Sec. 13.2d). We owe our existence to supernovae, being made of their debris.

» The Type II Supernova 1987A, in a satellite galaxy known as the Large Magellanic Cloud, was the brightest supernova since 1604 and provided many valuable insights (Sec. 13.2e). The star that exploded was a massive, evolved star, as predicted, although its exact nature (blue, instead of red, supergiant) was a surprise. Neutrinos from the supernova, as well as the detection of radioactive heavy elements, show that we had the basic ideas of the theory of Type II supernovae correct. **Cosmic rays** are particles accelerated to high energy, in many cases from supernova explosions (Sec. 13.2f).

» The cores of massive stars, after the supernova explosions, consist of degenerate neutrons that cannot be compressed further; they are called neutron stars (Sec. 13.3a). Some give off beams of radio radiation (and sometimes other electromagnetic waves as well) as they rotate like a lighthouse, and we detect pulses in their brightness. We now know of thousands of these **pulsars** (Sec. 13.3b). They are explained by the **lighthouse model**, in which two oppositely directed beams along the magnetic axis are seen only when they rotate into our line of sight (Sec. 13.3c). The discovery of a very rapid pulsar in the Crab Nebula, a young supernova remnant, provided strong support for the hypothesis that pulsars are rotating, magnetized neutron stars (Sec. 13.3d).

» Most pulsars slow down as they radiate energy (Sec. 13.3e). Some pulsars, however, spin very rapidly, up to a few hundred times per second. They were probably spun up by accretion of gas from a companion star.

» Binary-star systems consisting of a pair of neutron stars, at least one of which is a pulsar, are proving very useful for testing the general theory of relativity (Sec. 13.3f). Their orbital periods are getting shorter at a rate that agrees with the idea that energy in the form of **gravitational waves** is given off. Observational facilities have been built to directly detect gravitational waves, so far without success.

» Strange planets have been discovered around one pulsar, but they must have formed after the supernova explosion (Sec. 13.3g). Neutron stars in binary systems can be studied in a way other than their existence as radio pulsars: They give off x-rays emitted by hot gas in an accretion disk fed by the companion star (Sec. 13.3h).

QUESTIONS

1. What happens to the core and envelope of a star at the end of its main-sequence stage?

2. Why doesn't the helium core of a red giant immediately start fusing to heavier elements?

3. Why does a red giant appear reddish?

4. Sketch a temperature-luminosity diagram, label the axes, and point out the location of red giants.

5. If you compare a photograph of a nearby planetary nebula taken 100 years ago with one taken now, how would you expect them to differ?

6. Why is the surface of a star hotter after the star sheds a planetary nebula?

7. What forces balance to make a white dwarf?

8. What is the difference between the Sun and a 1-solar-mass white dwarf?

9. When the Sun becomes a white dwarf, about how much mass will it contain? Where will the rest of the mass have gone?

10. Which has a higher surface temperature, the Sun or a white dwarf having the same spectral type?

11. What is the relation of novae and white dwarfs?

12. Is a nova really a new star? Explain.

13. If we see a massive main-sequence star (a heavyweight star), what can we conclude about its age, relative to the ages of most stars? Why?

14. Distinguish between what is going on in novae and supernovae.

15. In what way do we distinguish observationally between Type Ia and Type II supernovae?

16. What are the physical differences between Type Ia and Type II supernovae, in terms of the kinds of stars that explode and their explosion mechanisms?

17. What is special about the iron core that forms at the end of the life of a massive star? Why does it collapse and then produce a supernova?

†18. In a supernova explosion of a 15-solar-mass star, about how much material is ejected (blown away)?

19. Explain how a very massive star might explode as a Type I supernova (that is, with no hydrogen in its spectrum), yet still be a core-collapse supernova like the Type II variety.

20. From where did the heavy elements in your body come?

21. Why do we think that the Crab Nebula is a supernova remnant?

22. Would you expect the appearance of the Crab Nebula to change in the next 1,000 years? How?

23. What are two reasons why Supernova 1987A was significant?

24. From what event in Supernova 1987A did the neutrino burst come?

25. What is the difference between cosmic rays and x-rays?

26. What keeps a neutron star from collapsing?

†27. Compare the Sun, a white dwarf, and a neutron star in size. Include a sketch.

28. In what part of the spectrum do most or all pulsars give off the energy that we study? (A few pulsars also emit significant energy at other wavelengths.)

29. How do we know that pulsars are in our Galaxy?

30. Why do we think that the lighthouse model explains pulsars?

31. How did studies of the Crab Nebula pin down the explanation of pulsars?

32. How has the Taylor-Hulse binary pulsar been especially useful?

33. Why is the binary pulsar in Puppis even more useful than the Taylor-Hulse pulsar?

34. Compare the discovery of gravitational waves with the method of detecting them now being worked on.

35. **True or false?** The decreasing orbital period of a neutron star binary system provides astronomers with strong evidence for the validity of Einstein's general theory of relativity.

36. **True or false?** A white dwarf shines because of ongoing nuclear fusion of heavy elements in its core.

37. **True or false?** A red giant expands because its helium core contracts, accelerating the fusion of hydrogen in a surrounding layer.

38. **True or false?** Neutron stars are commonly seen in star clusters, where they are among the brightest red stars.

39. **True or false?** The Sun will initially shrink in size, becoming a white dwarf, before moving on to the red-giant stage of its evolution.

40. **True or false?** When a star having a mass of more than about 10 solar masses explodes, most of the energy is carried away in the form of energetic neutrinos, not visible light.

41. **True or false?** A white dwarf in a binary star system can undergo nuclear reactions on its surface, producing a nova but not destroying the entire star.

42. **True or false?** A supernova whose spectrum shows no sign of hydrogen may have been produced by collapse of the core of a very massive star.

43. **Multiple choice:** When a main-sequence star runs out of hydrogen fuel in its core, **(a)** the core expands and thus heats up; **(b)** the core expands and thus cools down; **(c)** the core contracts and thus heats up; **(d)** the core contracts and thus cools down; or **(e)** the core remains about the same size, but heats up as fusion of helium to carbon begins immediately after the hydrogen fuel is gone.

44. **Multiple choice:** Which one of the following statements about pulsars is *true*? **(a)** All neutron stars are pulsars, if viewed from the right direction. **(b)** Pulsars could very likely be signals from intelligent aliens. **(c)** Once a pulsar forms, it typically lasts for billions of years. **(d)** If we observe a Type II supernova and do not see a pulsar afterward at the same location, we know a black hole must be present there. **(e)** Pulsars cannot be rotating white dwarfs because their rotation rate is so high that a white dwarf would be ripped apart.

†45. **Multiple choice:** Suppose white dwarf Yoda is the size of the Earth and has a surface temperature of 3,000 K. If the Earth's surface temperature is 300 K, how many times more luminous than the Earth is white dwarf Yoda? (a) 10^2. (b) 10^3. (c) 10^4. (d) 10^8. (e) $3,000 \times 300$, which is 9×10^5.

46. **Multiple choice:** White dwarf stars (a) are the end states only of stars whose initial mass is much greater than that of the Sun; (b) in some cases consist largely of carbon and oxygen; (c) in some cases consist largely of uranium and other very heavy elements; (d) shine only while nuclear reactions continue within them; or (e) support themselves against the pull of gravity in the same way as normal stars like the Sun, using the pressure exerted by hot gases within them.

47. **Multiple choice:** Which one of the following statements about Supernova 1987A is *true*? (a) Most of the energy from the explosion was released in the form of tiny, uncharged particles called neutrinos. (b) A newly formed white dwarf surrounded by a planetary nebula was left over after the explosion. (c) The star that exploded was a red giant with about the same mass as our Sun. (d) Heavy elements were not detected in the ejected material because they were ripped apart by the explosion. (e) The star exploded because of a runaway chain of nuclear reactions in the star's core.

48. **Multiple choice:** Which one of the following does *not* provide a test of Einstein's general theory of relativity? (a) The slow rotation (precession) of Mercury's orbit around the Sun. (b) The gradually decreasing orbital period of a binary neutron star. (c) The redshifting of light coming from a white dwarf's surface. (d) The bending of starlight in the vicinity of the Sun. (e) The presence of a pulsar in a supernova remnant.

49. **Multiple choice:** Typical novae occur when (a) a red-giant star ejects a planetary nebula; (b) two neutron stars merge, forming a more massive neutron star; (c) an extremely massive star collapses, and also ejects its outer atmosphere; (d) matter accreted from a companion star unstably ignites on the surface of a white dwarf;

or (e) a neutron star's magnetic field becomes strong enough to produce two oppositely directed jets of rapidly moving particles.

50. **Multiple choice:** Which one of the following statements about supernovae is *false*? (a) At optical wavelengths, a supernova can appear about as bright as the entire galaxy of stars in which it's located. (b) A supernova produces an expanding volume of gas that is rich in heavy elements. (c) Nobody has seen a bright supernova in our own Milky Way Galaxy during the past century. (d) A neutron star is produced when a white dwarf exceeds a certain mass limit and becomes a supernova, experiencing a runaway chain of nuclear fusion reactions. (e) One kind of supernova occurs when the iron core of a massive star collapses.

51. **Fill in the blank:** Mass exchange from one star to the other can occur if one star extends into the _____ of its companion.

52. **Fill in the blank:** A white dwarf is supported against gravitational collapse by _____, a weird kind of quantum-mechanical pressure.

53. **Fill in the blank:** A red giant forms a _____ when its outer atmosphere becomes unstable, producing a gentle ejection.

54. **Fill in the blank:** One kind of _____ occurs when a clump of material is suddenly dumped onto the surface of a white dwarf in a binary system.

55. **Fill in the blank:** A rapidly rotating neutron star having a strong magnetic field can sometimes be observed as a _____.

56. **Fill in the blank:** High-energy charged particles produced by supernova remnants are called _____.

57. **Fill in the blank:** Two neutron stars in a binary system gradually spiral toward each other because they emit energy in the form of _____.

†This question requires a numerical solution.

TOPICS FOR DISCUSSION

1. How can we be so confident in our theory of the Sun's future evolution?

2. Does the fact that you are made of stardust ("star stuff") give you a sense of unity with the cosmos? Discuss.

3. If only about 10 neutrinos from SN 1987A were detected by each underground tank containing several thousand tons of water, and if a typical human consists of 100 pounds of water, what are the odds that a given human body directly detected a neutrino from SN 1987A?

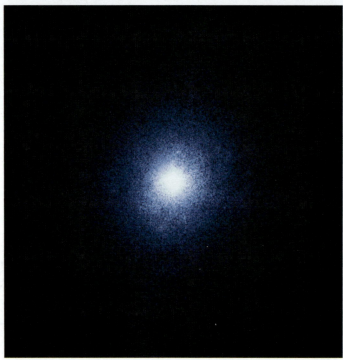

Top: Cygnus X-1, a stellar-mass black hole determined in 2009 with a variety of x-ray and optical observations to contain 15 times the mass of the Sun. In this artist's conception, gas from a companion blue star is falling toward the black hole, forming an accretion disk of orbiting particles with some matter being ejected as jets.

The image at bottom shows a 16-hour Chandra X-ray Observatory exposure of the x-rays. Observations with not only Chandra but also two other x-ray satellites – the Rossi X-ray Timing Explorer and Suzaku, a Japanese x-ray satellite – enabled astronomers to measure how fast Cygnus X-1 is spinning. At its limit, called the event horizon, it is spinning at over 800 times a second, close to the maximum that it could support.

14

Black Holes: The End of Space and Time

ORIGINS

Gravity is the dominant force shaping the structure and evolution of the Universe. Black holes provide a laboratory for testing the most extreme predictions of Einstein's general theory of relativity, our best existing description of gravity.

The peculiar forces of electron and neutron degeneracy pressure (an effect of quantum physics) support dying lightweight stars and some heavyweight stars against gravity. The strangest case of all occurs at the death of the most massive stars, which contained much more than 10 and up to about 100 solar masses when they were on the main sequence. After these stars undergo supernova explosions, some may retain cores of over 2 or 3 solar masses. Nothing in the Universe is strong enough to hold up the remaining mass against the force of gravity, so it collapses.

The result is a **black hole**, in which the matter disappears from contact with the rest of the Universe (Chapter Opener and ■ Fig. 14–1). Later, we shall discuss much more massive black holes than those that result from the collapse of a star. Most astronomers now agree that black holes are widespread and very important in our Universe.

AIMS

1. Understand the properties of black holes, regions in which gravity is so incredibly strong that nothing, not even light, can escape (Sections 14.1 to 14.6).

2. Discuss the observational evidence for stellar-mass black holes in binary systems, and for supermassive black holes in the centers of many galaxies (Sections 14.7 and 14.8).

3. Learn about the new category of intermediate-mass black holes (Section 14.9).

4. Explore enigmatic gamma-ray bursts, the probable birth cries of black holes (Section 14.10).

5. Introduce the concept of mini black holes that evaporate (Section 14.11).

■ **FIGURE 14–1** Directions to a small, deep cove on the shore of the Bay of Fundy. It got its name (probably a century or two ago) because it appears very dark as seen from the sea.

■ **FIGURE 14–2** Mass warps (bends) space and time in its vicinity. Here we show just two of the three warped spatial dimensions, analogous to the effect of a billiard ball on a stretched sheet of rubber. Note that the surface warps around a third, perpendicular dimension, not part of the surface itself.

14.1 THE FORMATION OF A STELLAR-MASS BLACK HOLE

Astronomers had long assumed that the most massive stars would somehow lose enough mass to wind up as white dwarfs. When the discovery of pulsars (neutron stars) ended this prejudice, it seemed more reasonable that black holes could exist. If more than 2 or 3 times the mass of the Sun remains after the supernova explosion, the star collapses through the neutron-star stage. We know of no force that can stop the collapse. (In some cases, the supernova explosion itself may fail, so the remaining mass can be much larger than 2 or 3 solar masses.)

We may then ask what happens to an evolved 5-, 10-, or 50-solar-mass star as it collapses, if it retains more than 2 or 3 solar masses. As far as we know, it must keep collapsing, becoming denser and denser. Einstein's general theory of relativity suggests that the presence of mass or energy warps (bends) space and even time in its vicinity (■ Fig. 14–2); recall our discussion of the Sun's gravity in Section 10.3. The greater the density of the material, the more severe is the warp.

As we explore in more detail below, a strong gravitational field appears to bend light; thus, radiation leaving the star other than perpendicularly to the surface is bent more and more as the star contracts and the strength of gravity at the star's surface increases. In addition, according to general relativity, the light we see from a distance will become progressively more redshifted as the star contracts. Eventually, when the mass has been compressed to a certain size, light from the star can no longer escape into space. The star has withdrawn from our observable Universe, in that we can no longer receive radiation from it. We say that the star has become a black hole, specifically a **stellar-mass black hole**.

Why do we call it a black hole? We think of a black surface as a surface that reflects none of the light that hits it. Similarly, any radiation that hits a black hole continues into its interior and is not reflected or transmitted out. In this sense, the object is perfectly black.

What is the actual minimum mass of a black hole? In 2011, a team from Northwestern University revealed the results of their survey of 20 known stellar-mass black holes (whose mass we can measure because they are found in binary systems, as we will discuss in Section 14.7). From the distribution of masses they calculated, they concluded that the masses of 99 per cent of them would be greater than 4.5 solar masses. (What happens to remnants of stars with more than the neutron-star upper limit of 2 or 3 solar masses and this apparent black-hole lower limit of 4.5 solar masses is not known.)

14.2 THE PHOTON SPHERE

Consider what happens to radiation emitted by the surface of a star as it contracts. Although what we will discuss applies to radiation of all wavelengths, let us simply visualize standing on the surface of the collapsing star while holding a flashlight.

On the surface of a supergiant star, if we shine the beam at any angle, it seems to go straight out into space. As the star collapses, two effects begin to occur. (We will ignore the star's outer layers, which are unimportant here. We also ignore the star's rotation.) Although we on the surface of the star cannot notice the effects ourselves, a friend on a planet revolving around the star could detect them and radio information back to us about them. For one thing, our friend could see that our flashlight beam is redshifted.

Second, our flashlight beam would be bent by the gravitational field of the star (■ Fig. 14–3). If we shine the beam straight up, it would continue to go straight up. But the further we shine it away from the vertical, the more it would be bent from the vertical. When the star reaches a certain size, a horizontal beam of light would not escape (■ Fig. 14–4).

From this time on, only if the flashlight is pointed within a certain angle of the vertical does the light continue outward. This angle forms a cone, with its apex at the flashlight, and is called the **exit cone** (■ Fig. 14–5). As the star grows smaller yet, we find that the flashlight has to be pointed more directly upward in order for its light to escape. The exit cone grows narrower as the star shrinks.

FIGURE 14–3 As a star contracts, a light beam emitted other than straight outward, along a radius, will be bent. It will also become gravitationally redshifted – that is, the wavelength of the light will increase as the light recedes from the star.

FIGURE 14–4 Light can be bent so that it circles a star, but in this case there is no gravitational redshifting because the light is at a constant distance from the star's center. If the light is bent so much that it approaches the star, it would become blueshifted.

FIGURE 14–5 When a star (shown by the inner sphere) has contracted enough, only light emitted within the "exit cone" escapes. Light emitted on the edge of the exit cone goes into the photon sphere. The further the star contracts within the photon sphere, the narrower the exit cone becomes.

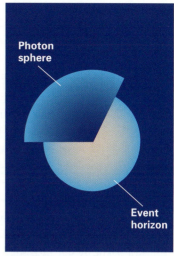

FIGURE 14–6 When a star becomes smaller than its Schwarzschild radius, we can no longer observe it. It has passed its event horizon.

When we shine our flashlight in a direction outside the exit cone, the light is bent sufficiently that it falls back to the surface of the star. When we shine our flashlight exactly along the side of the exit cone, the light goes into orbit around the star, neither escaping nor falling onto the surface.

The sphere around the star in which light can orbit is called the **photon sphere**. Its radius is calculated theoretically to be 4.5 km for each solar mass present. It is thus 45 km in radius for a star of 10 solar masses, for example.

As the star continues to contract, theory shows that the exit cone becomes narrower and narrower. Light emitted within the exit cone still escapes. The photon sphere remains at the same height even though the matter inside it has contracted further, since the total amount of matter within has not changed.

14.3 THE EVENT HORIZON

We might think that the exit cone would simply continue to become narrower as the star shrinks. But Einstein's general theory of relativity predicts that the cone vanishes when the star contracts beyond a certain size. Even light travelling straight up can no longer escape into space, as was worked out in 1916 by Karl Schwarzschild while solving Einstein's equations of general relativity.

The radius of the star at this time is called the **Schwarzschild radius**, and the imaginary surface at that radius is called the **event horizon** (Fig. 14–6). (A horizon on Earth, similarly, is the limit to which we can see.) Its radius is exactly $1/3$ times that of the photon sphere, or 3 km for each solar mass. Formally, the equation for the

Schwarzschild radius is $R = 2GM/c^2$, where M is the star's mass, G is Newton's constant of gravitation, and c is the speed of light.

14.3a A Newtonian Argument

We can visualize the event horizon in another way, by considering a classical picture based on the Newtonian theory of gravitation. The picture is essentially that conceived in 1783 by John Michell and in 1796 by Pierre Laplace.

A projectile launched from rest must be given a certain minimum speed, called the **escape velocity**, to escape from the other body, to which it is gravitationally attracted. For example, we would have to launch rockets at 11 km/s (40,000 km/hour) or faster in order for them to escape from the Earth, if they got all their velocity at launch. (From a more massive body of the same size as Earth, the escape velocity would be higher.)

Note that the escaping object still feels the gravitational pull of the other body; one cannot "cut off" or block gravity. However, the object's speed is sufficiently high that it will never turn around and fall back down.

Now imagine that this body contracts; we are drawn closer to the center of the mass. As this happens, the escape velocity rises. If the body contracts to half of its former radius, for instance, the gravitational force on its surface increases by a factor of four, since gravity follows an inverse-square law; the corresponding escape velocity therefore also increases (although not by a factor of four).

When all the mass of the body is within its Schwarzschild radius, the escape velocity becomes equal to the speed of light. Thus, even light cannot escape. If we begin to apply the special theory of relativity, which deals with motion at very high speeds, we might then reason that since nothing can go faster than the speed of light, nothing can escape.

14.3b Black Holes in General Relativity

Now let us return to the picture according to the general theory of relativity, which explains gravity and the effects caused by large masses. A black hole curves space (actually, space-time) to such a large degree that light cannot escape; it remains trapped within the event horizon.

The size of the Schwarzschild radius of the event horizon is directly proportional to the amount of mass that is collapsing: $R = 2GM/c^2$. A star of 3 solar masses, for example, would have a Schwarzschild radius of 9 km. A star of 6 solar masses would have a Schwarzschild radius of 18 km.

One can calculate the Schwarzschild radii for less massive stars as well, although the less massive stars would be held up in the white dwarf or neutron-star stages and not collapse to their Schwarzschild radii. The Sun's Schwarzschild radius is 3 km. The Schwarzschild radius for the Earth is only 9 mm; that is, the Earth would have to be compressed to a sphere only 9 mm in radius in order to form an event horizon and be a black hole.

Anyone or anything on the surface of a star as it passes its event horizon would not be able to survive. An observer would be stretched out and torn apart by the tremendous difference in gravity between his head and feet (■ Fig. 14–7). This resulting force is called a **tidal force**, since this kind of difference in gravity also causes tides on Earth (recall our discussion in Chapter 6). The tidal force on an observer would be smaller near a more massive black hole than near a less massive black hole: Just outside the event horizon of a massive black hole, the observer's feet would be only a little closer to the center (in comparison with the Schwarzschild radius) than the head. The stretching process near a stellar-mass black hole is sometimes informally called "spaghettification"!

If the tidal force could be ignored, the observer on the surface of the star would not notice anything particularly wrong locally as the star passed its event horizon, except that the observer's flashlight signal would never get out. (On the other hand, the view of space outside the event horizon would be highly distorted.)

Once the star passes inside its event horizon, it continues to contract, according to the general theory of relativity. Nothing can ever stop its contraction. In fact, the classical mathematical theory predicts that it will contract to zero radius – it will become a **singularity**. Quantum effects, however, probably prevent it from reaching exactly zero radius.

Even though the mass that causes the black hole has contracted further, the event horizon doesn't change. It remains at the same radius forever, as long as the amount of mass inside is constant.

Note that if the Sun were turned into a black hole (by unknown forces; the Sun should actually become a white dwarf at the end of its life), Earth's orbit would not be altered: The masses of the Sun and Earth would remain constant, as would the distance between them,

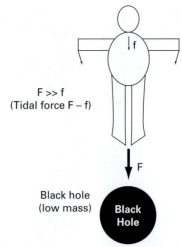

■ FIGURE 14–7 If you are near a black hole, the force of gravity on your feet (long, bold arrow marked "F") exceeds that on your head (short arrow marked "f"), so you get "tidally" stretched, analogous to the Moon causing the Earth's tides (see our discussion in Chapter 6).

so the gravitational force would be unchanged. Indeed, the gravitational field would remain the same everywhere outside the current radius of the Sun. Only at smaller distances than the current radius would the force be stronger.

14.4 TIME DILATION

According to the general theory of relativity, if you were far from a black hole and watching a space shuttle carrying some of your friends fall into it, your friends' clocks would appear to run progressively slower as they approached the event horizon. (Eventually they would get torn apart by tidal forces, as discussed above, but here we will ignore this complication.) From your perspective, time would be slowing down (or "dilated") for them; indeed, it would take an infinite amount of time (as measured by your clock) for them to reach the horizon. From your friends' perspective, in contrast, no such **time dilation** occurs; it takes a finite amount of time for them to reach and cross the event horizon, and shortly thereafter they hit the singularity.

On the other hand, if your friends were to approach the event horizon and subsequently escape from the vicinity of the black hole, they would have aged less than you did (for example, only 3 months instead of 30 years). This is a method for jumping into the future while aging very little!

However, this strategy doesn't increase longevity – your friends' lives would not be extended. Locally, they would not read more books or see more movies than you would in the same short time interval (3 months in our case). In contrast, while viewing you from the vicinity of the black hole, they would see you read many books and watch many movies, signs that you are aging much more than they are (30 years in our case).

Because of time dilation, even a collapsing star never seems to reach the event horizon, if you are an outside observer. However, the star very quickly fades from sight: It emits light infrequently because of time dilation, the light becomes progressively more redshifted, and only the photons heading radially outward can eventually escape (that is, the exit cone is very narrow). Thus, everything that has ever fallen into the black hole seems to be in a thin, dark "membrane" slightly outside the event horizon.

14.5 PROPERTIES OF BLACK HOLES

Once matter is inside a black hole and reaches the singularity, it loses its identity in the sense that from outside a black hole, all we can tell is the mass of the black hole, the rate at which it is spinning (more precisely, its angular momentum), and what total electric charge it has. These three quantities are sufficient to completely describe the black hole, assuming it has reached an equilibrium configuration (that is, matter is not still falling asymmetrically toward the singularity). And since any charged black hole would quickly attract enough opposite charge to become neutral, only mass and spin matter in practice.

Thus, in some respects, black holes are simple objects to describe physically, because we only have to know three numbers to characterize each one. The theorem that summarizes the simplicity of black holes is often colloquially stated by astronomers active in the field as "a black hole has no hair."

14.5a Rotating Black Holes

The theoretical calculations about black holes we have discussed in previous sections are based on the assumption that black holes do not rotate. But this assumption is only a convenience; we think, in fact, that the rotation of a black hole is one of its important properties. It took decades before Einstein's equations were solved for a black hole that is rotating. (The realization that the solution applied to a rotating black hole came after the solution itself was found.)

In this more general case, black holes have some additional interesting properties. For example, there are two event horizons: the usual outer horizon, and an inner one as well. Moreover, a special boundary – the **stationary limit** – appears beyond the outer event horizon, with somewhat different properties from the original event horizon. Within the stationary limit, no particles can remain at rest even though they are outside the outer event horizon.

The equator of the stationary limit of a rotating black hole has the same diameter as the event horizon of a nonrotating black hole of the same mass. But a rotating black hole's stationary limit is squashed. The outer event horizon touches the stationary limit at the poles. Since the outer event horizon remains a sphere, it is smaller than the event horizon of a nonrotating black hole (■ Fig. 14–8).

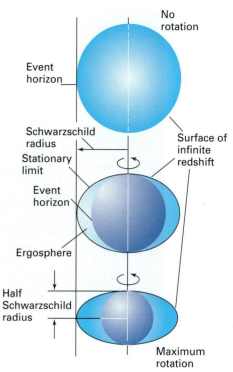

■ **FIGURE 14–8** Side views of black holes of the same mass and different rotation speeds: low, medium, and high. The region between the stationary limit and the event horizon of a rotating black hole is called the ergosphere (from the Greek word ergon, meaning "work") because, in principle, work can be extracted from it.

The space between the stationary limit and the outer event horizon is the **ergosphere**. This is the region in which particles cannot be at rest. In principle, we can get energy and matter out of the ergosphere. For example, if one sends an object into the ergosphere along an appropriate trajectory, and part of the object falls into the black hole at the right moment, the rest of the object can fly out of the ergosphere with more energy than the object initially had. The extra energy was gained from the rotation of the black hole, causing it to spin more slowly.

If a black hole were made to spin progressively faster, its inner event horizon would move outward and its outer event horizon would move inward. Eventually, a point on the outer event horizon's equator would travel almost exactly at the speed of light. The outer event horizon's radius would then be a bit more than half the Schwarzschild radius. If a black hole rotated faster than this, at the speed of light, the outer and inner event horizons would merge together and vanish. Unlike the case of a nonrotating black hole, for which the singularity is always unreachably hidden within the event horizon, in this case distant observers could receive signals from the singularity.

Such a point is called a **naked singularity**. Most theorists assume the existence of a law of "cosmic censorship," which requires all singularities to be "clothed" in event horizons – that is, not naked. Since so much energy might erupt from a naked singularity, we can conclude from the fact that we do not find signs of them that there are probably none in our Universe. Thus, each black hole cannot exceed its maximum rate of rotation (or a naked singularity could be seen).

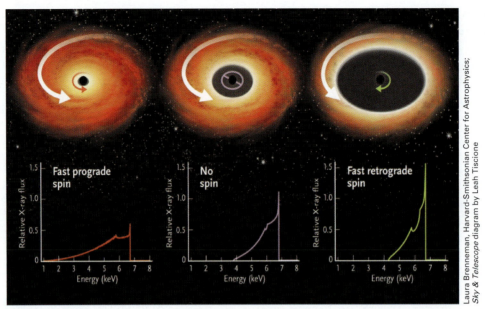

■ **FIGURE 14–9** Scientists can tell that a black hole is spinning (rotating), and whether it is spinning in the same direction as the accretion disk (prograde, the term used for rotation and revolution in our Solar System) or in the opposite direction (retrograde, again used the way in which we are familiar from our Solar System). A black hole's spin rate plays a crucial role in determining how close material can orbit before falling in. Frame-dragging, an effect predicted by general relativity, allows material to orbit very close to a fast-spinning, prograde black hole before falling in. In contrast, material orbits farther away from non-spinning and retrograde-spinning black holes before it meets its doom.

One method for measuring spin looks at an iron emission line known as K-α in the x-ray part of the spectrum. (The "K" does not stand for potassium in this case; rather, it refers to an electron in the innermost shell.) Its shape – its asymmetry and its width – reveals how far out from the black hole's event horizon we find the innermost stable circular orbit. Comparing that result with the theoretical model reveals whether the spin is prograde or retrograde. A second method, not shown here, looks at the blackbody (Planck) curve, whose peak is also in the x-rays because the gas is so hot. A third method, also not shown here, uses very-long-baseline-interferometry of millimeter radio waves to image the shadow of the event horizon, so far succeeding in two sources.

From our perspective, a nearby black hole can be thought of as being connected, through a wormhole, with a "white hole" at another location. A "white hole" is like the inverse of a black hole: Matter and radiation come out from it, but cannot go back in. We have no evidence for the existence of white holes in our Universe; they almost certainly do not exist.

14.5b Measuring Black-Hole Spin

Measuring the rate at which black holes rotate, or spin, is currently an active and challenging topic of research. There are two methods that have been successfully employed to date to measure spin (■ Fig. 14–9): thermal continuum fitting (that is, fitting the Planck curves; see Chapter 2) and emission-line modeling (that is, matching the detailed shapes of the x-ray emission lines when their brightnesses are graphed against wavelength). Both techniques involve examining spectra of the radiation emanating from gas rotating very close to the event horizon of a black hole, just before it falls in.

Suppose a black hole is orbited by a normal star that fills its "Roche lobe" (see Fig. 13–9 in Chapter 13) – that is, the region in which its gravitational pull dominates over that of the black hole. Gas crossing the intersection point in the "figure 8" will begin to accelerate toward the black hole; like all objects in the Universe, the black hole attracts matter. Some of the gas will be pulled directly into the black hole, never to be seen again. But other matter will go into high-speed orbit around the black hole. This added matter forms

an **accretion disk**; "accretion" is growth in size by the gradual addition of matter. We previously encountered accretion disks in Chapter 13, when we discussed novae (Section 13.1e) and Type Ia supernovae (Section 13.2b).

The accretion disk's inner edge is set by general relativity through a phenomenon known as "frame-dragging": the faster a black hole spins, the smaller is the radius of the innermost stable circular orbit in the accretion disk, assuming particles orbit in the same direction as the black hole spins (i.e., "prograde" rotation of the black hole). A NASA spacecraft known as Gravity Probe B recently confirmed the existence of frame-dragging, after a complicated analysis of the data. An Italian Space Agency satellite, the **La**ser **Re**lativity **S**atellite (LARES), was launched in 2012 to measure frame-dragging to, perhaps, within 1 per cent from its monitoring of the precession (gradual rotation, or shift) of the spacecraft's orbit.

Frame-dragging around rapidly spinning prograde black holes rotationally supports orbits in the disk that extend down nearly to the event horizon, while nonrotating black holes have an innermost stable circular orbit nearly six times farther out caused by the lack of frame-dragging. If the black hole and its accretion disk happen to be rapidly rotating in opposite directions (known as "retrograde" rotation of the black hole), frame-dragging would cause gas in the disk to plunge in toward the black hole at radii nearly nine times farther out than the innermost stable circular orbit for a rapidly rotating prograde black hole. Therefore, if we can measure the radius of the inner edge of the disk, assuming this is the innermost stable circular orbit, we can infer the spin of the black hole.

Gas in the accretion disk spiraling in toward a stellar-mass black hole releases gravitational potential energy, allowing it to be heated to a very high temperature by friction and by compression into a very small volume. The thermal-continuum emission from the inner disk thus takes the form of a blackbody (a Planck curve) in x-rays. Measuring the flux and temperature of this blackbody through spectral modeling (so-called thermal-continuum fitting) allows one to calculate the radius of the innermost edge of the accretion disk, where most of the emission originates. The black hole's spin can then be deduced by the monotonic relationship between the innermost stable circular orbit and the spin, provided that the mass of the black hole, its distance, and the inclination angle of the accretion disk to the line of sight are very well-known.

Similarly, many black-hole accretion disks also display spectral emission lines that have their shapes highly broadened and skewed (Fig. 14–9) by a combination of the disk's rapid rotation and "relativistic effects" (effects having to do with Einstein's special and general theories of relativity). Over the last ten years, fully relativistic spectral models have been developed to characterize these lines, enabling us to disentangle the effects that contribute to their broadening and to measure black-hole spin (by emission-line modeling). This method works for black holes of all masses, and is preferentially used for supermassive black holes in active galactic nuclei, which usually do not have precise measurements of their distances, masses, or disk inclination angles, and are therefore difficult to assess via thermal-continuum fitting. However, the spectra of active galactic nuclei are very complex, so isolating the disk's emission lines from the surrounding continuum can be difficult.

14.6 PASSAGEWAYS TO DISTANT LANDS?

In science-fiction movies (such as *Contact*, starring Jodi Foster and based on a novel by the astronomer Carl Sagan), it is often claimed that one can travel through a black hole to a distant part of our Universe in a very short amount of time, or perhaps even travel to other universes. This misconception arises in part from diagrams such as ■ Figure 14–10 for two nonrotating black holes. One black hole is connected to another black hole by a tunnel, or **wormhole**

■ **FIGURE 14–10** Wormholes hypothetically connect black holes (a) in distant parts of our Universe, or (b) perhaps even in different universes.

(officially called an "Einstein-Rosen bridge"), and it appears possible to traverse this shortcut.

However, this map is misleading; it does not adequately describe the structure of space-time inside a black hole. In particular, one would need to travel through space faster than the speed of light (which nobody can do) to avoid the singularity and end up in a different region. Thus, nonrotating black holes definitely seem to be excluded as passageways to distant lands.

In the case of a rotating black hole, on the other hand, travel through the wormhole at speeds slower than that of light, avoiding the singularity, initially seems feasible. In fact, it appears as though one could travel back to one's starting point in space, possibly arriving at a time prior to departure! This is quite disturbing, since "causality" could then be violated. For example, the traveller could affect history in such a way that he or she would not have been born and could not have made the journey!

More detailed analysis, however, shows that this favorable geometry of a rotating black hole is only valid for an idealized black hole into which no material or radiation is falling (or has previously fallen). As soon as anything actually tries to traverse the wormhole, the wormhole is destroyed! Moreover, at the inner event horizon, collisions between incoming and outgoing matter and radiation lead to a phenomenon known as the "mass inflation instability"; material effectively gets fried and destroyed by the intense radiation field.

In principle, one might try to keep the wormhole open by using a very exotic form of matter with antigravitating properties. But there is no evidence for the existence of such matter, at least not in the form where it can be gathered. (In Chapter 18, we will introduce the concept of "dark energy," which causes the expansion of the Universe to accelerate. However, this energy seems uniformly spread throughout space, and probably cannot be concentrated into a small volume.)

14.7 DETECTING A BLACK HOLE

A star collapsing to become a black hole would blink out in a fraction of a second, so the odds are unfavorable that we would actually see the crucial stage of star collapse as it approached the event horizon. And a black hole is too small to see directly. But all hope is not lost for detecting a black hole. Though the black hole disappears, it leaves its gravity behind. It is a bit like the Cheshire Cat from Alice in Wonderland, which fades away, leaving only its grin behind (■ Fig. 14–11).

14.7a Hot Accretion Disks

We have already discussed how a stellar-mass black hole in a binary system with a mass-donating star has a very hot accretion disk around it (■ Fig. 14–12a). The inner 200 km should reach ten million kelvins, giving off bursts of x-rays sporadically as clumps develop in

■ **FIGURE 14–11** Lewis Carroll's Cheshire Cat, from Alice in Wonderland, shown here in John Tenniel's drawing. The Cheshire Cat is analogous to a black hole in that the cat left its grin behind when it disappeared, just as a black hole leaves its gravity behind when it disappears from view. Alice thought that the Cheshire Cat's persisting grin was "the most curious thing I ever saw in my life!" We might say the same about a black hole and its persisting gravity.

the disk. Thus, though we cannot observe the black hole itself, we can hope to observe x-rays from the gas surrounding it.

A NASA satellite known as the Rossi X-ray Timing Explorer measured such rapid changes in x-rays, detecting a number of interesting flickering x-ray sources before it ceased science operations in 2012. The Chandra X-ray Observatory and the XMM-Newton Mission, both launched in 1999, are finding still more black-hole candidates.

In fact, a large number of x-ray sources are binary-star systems. Some may contain black holes, but the majority of them probably have neutron stars, around which hot accretion disks can also form. To be certain that you have a black hole, it is not enough to find an x-ray source that gives off sporadic pulses, since both black holes and neutron stars can produce such pulses. We must, in addition, show that a collapsed star of greater than 3 solar masses is present.

14.7b Cygnus X-1: The First Plausible Stellar-Mass Black Hole

We can determine masses only for certain binary stars. When we search the positions of x-ray sources, we look for a single-lined spectroscopic binary (that is, an apparently single star whose spectrum shows a periodically changing Doppler shift that indicates the presence of an invisible companion; we discussed such binary stars in Chapter 11). Then, if we can show that the companion is too faint to be a normal, main-sequence star, it must be a collapsed star.

If, further, the mass of the unobservable companion is greater than 3 solar masses, it is likely to be a black hole, assuming that the general theory of relativity is the correct theory of gravity and that our present understanding of matter is correct. To definitively show that the black-hole candidate is indeed a black hole, however, additional evidence is needed, as we will discuss below.

The first and most discussed case is Cygnus X-1 (Fig. 14–12), where X-1 means that it was the first x-ray source to be discovered in the constellation Cygnus. A 9th-magnitude star called HDE 226868 has been found at its location. This star has the spectrum of a blue supergiant; its mass is thought to be about 20 times that of the Sun. Its radial velocity varies with a period of 5.6 days, indicating that the supergiant and the invisible companion are orbiting each other with that period.

The most recent measurements indicate that Cygnus X-1 is definitely a very rapidly rotating black hole. The Very Long Baseline Array was used to make a parallax measurement of its distance, which turns out to be 6,070 light-years. With this accurate value in hand, x-ray measurements made over decades from the Chandra X-ray Observatory, the Rossi X-ray Timing Explorer, the Advanced Satellite for Cosmology and Astrophysics, Suzaku, and visible-light observations from telescopes on mountaintops show that Cygnus X-1's black hole has 15 times the Sun's mass. Moreover, the measurements indicate that its event horizon spins over 800 times per second, in the same direction as the accretion disk.

Radio astronomers have discovered a rapidly expanding 10 light-year-diameter bubble around Cygnus X-1, and optical

■ **FIGURE 14–12** The black hole Cygnus X-1. (a) An artist's conception of the binary system, where gas from the visible, blue supergiant star HDE 226868 is swirling over to and around the black hole. (The name comes from the Extension (E) of the Henry Draper (HD) catalog of stars compiled by Annie Cannon over 100 years ago.) The mutual orbits around each other take 5.6 days. See also the Chapter Opener and its caption. (b) The region of the source imaged with the Optical Monitoring Camera on the European Space Agency's Integral satellite, which was launched in 2002 to study gamma rays. (c) Gamma rays from the accretion disk surrounding the black hole in Cygnus X-1, viewed with the Spectrometer on Integral (SPI) – that is, on the European Space Agency's gamma-ray spacecraft named Integral. The binary x-ray source Cygnus X-3, which involves a neutron star rather than a black hole, is toward the upper left.

astronomers have seen it as well. Its boundary is a shock wave formed where a jet of high-speed charged particles coming from the vicinity of the black hole hits the interstellar medium. This discovery confirms that black holes affect the space around them in more ways than with just their gravity.

14.7c Other Black-Hole Candidates

Although Cygnus X-1 provides good evidence for the existence of stellar-mass black holes in our Galaxy, the large (and uncertain) mass of the visible star complicates the measurement of the mass of the dark object. More conclusive evidence for black holes comes from binary systems in which the visible star has a very small mass (for example, a K- or M-type main-sequence star); uncertainties in its

mass are nearly irrelevant (see *Figure It Out 14.1: Binary Stars and Kepler's Third Law*).

However, when such systems brighten suddenly at x-ray wavelengths, light from the accretion disk also dominates at other wavelengths, and the low-mass star becomes too difficult to see. Observers must wait until the outburst subsides, and the companion star once again becomes visible, before they are able to measure the orbital parameters.

The first well-studied system of this type was A0620–00, which is in the constellation Monoceros. During an x-ray outburst in 1975, the source was even brighter than Cygnus X-1, but it subsequently faded and the properties of the visible star were measured. The derived minimum mass of the invisible companion was 3.2 Suns, close to but above the limit at which a collapsed star could be a neutron star instead of a black hole.

FIGURE IT OUT 14.1   Binary Stars and Kepler's Third Law

Consider Kepler's third law, as generalized by Newton and discussed in Chapter 5:

$$(m_1 + m_2)P^2 = (4\pi^2/G)R^3.$$

If m_1 (the mass of the invisible object) is much larger than m_2 (the mass of the visible star), then we can safely ignore m_2 in the sum. Thus

$$m_1 P^2 = (4\pi^2/G)R^3, \text{ or } m_1 = (4\pi^2/GP^2)R^3,$$

regardless of the exact value of m_2.

If the visible star has a circular orbit of radius R, it travels the circumference ($2\pi R$) in one orbital period (P) at speed v, so $2\pi R = vP$ (distance = speed × time). Dividing by 2π, we have $R = vP/2\pi$. Substituting this expression for R into the previous equation for m_1, we find

$$m_1 = (4\pi^2/GP^2)(vP/2\pi)^3 = v^3P/2\pi G.$$

If v and P are respectively measured as the amplitude (height) and period of the radial-velocity curve (observed speed versus time) of the visible star, then m_1 is found from this equation. Actually, this value

is merely the *minimum* mass that the invisible object can have, since the measured radial velocity is less than the true orbital velocity if the orbit isn't exactly perpendicular to the line of sight. Thus we aren't actually able to find the true mass without more information.

For example, the figure below shows the radial-velocity curve for the visible star in an x-ray binary system. When the star is moving most directly toward us or away from us, we see the maximum blueshift or redshift of the absorption lines, +520 km/s and –520 km/s, respectively. At other times in the circular orbit, a smaller part of the velocity is radial, so intermediate values are measured, but we conclude that $v = 520$ km/s for this system. The orbital period is measured to be only 8.3 hours. Using these values in the equation $m_1 = v^3P/2\pi G$ (and being careful with units!), we find that $m_1 = 5.0$ solar masses. This is just the minimum mass of the dark object; further considerations suggest that the true mass is 8 to 9 Suns. Hence, the visible star's companion is almost certainly a black hole.

(a)

GS 2000+25

N

E

Alex Filippenko (UC Berkeley)

(b)

Alex Filippenko and Tom Matheson (UC Berkeley)

FIGURE FIO 14.1 (a) An image of the visible star in the x-ray binary system known as GS 2000+25, long after it had faded from its x-ray outburst in 1988. (b) Radial-velocity curve of the visible star, obtained by one of the authors (A.F.) with the Keck-I 10-m telescope.

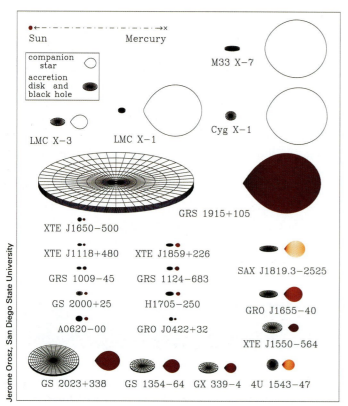

Jerome Orosz, San Diego State University

■ **FIGURE 14–13** The sizes of the accretion disks, companions, and their separations for some of the leading stellar-mass black-hole candidates.

An even more convincing case is the dark object in the x-ray binary star V404 Cygni. It has a mass of at least 6 Suns, but probably closer to 12 solar masses (■ Fig. 14–13). One of the authors (A.F.) has found four additional black-hole candidates with more than 5 times the Sun's mass and having a low-mass companion star, as well as two somewhat weaker cases for black holes. Over two dozen good or excellent stellar-mass black-hole candidates in binary systems have been identified and measured in our Galaxy.

The most massive stellar-mass black hole known at the time of writing (2013) is the variable x-ray source IC 10 X-1, in the nearby galaxy IC 10 (about 2.2 million light-years away). In 2008, the research of one of the authors (A.F.) and his graduate student at that time, Jeffrey M. Silverman, conclusively confirmed the tentative suggestion of A. Prestwich and collaborators that its mass is at least 23 Suns, and it may be as high as 33 solar masses or even slightly larger. The range stems largely from the uncertain mass of the companion star, an evolved, massive "Wolf-Rayet star."

Astronomers are now trying to conclusively demonstrate that the stellar-mass black-hole candidate systems discussed above do indeed contain black holes. After all, the argument thus far has been somewhat indirect: We think black holes are present partly because we don't know of any kind of object denser than a neutron star but not quite a black hole.

One of the most intriguing tests is to see whether the material from the accretion disk lands on a hard stellar surface, releasing its considerable kinetic energy in the form of electromagnetic radiation,

or instead passes quietly through an event horizon, never to be seen again. So far, the results seem to favor the presence of an event horizon rather than a stellar surface in several black-hole candidates, but more research is necessary in this area.

Recently, a few solitary black holes (not in binary systems) may have been detected as they gravitationally bent and magnified light from more distant stars along our line of sight. This phenomenon of "gravitational lensing" is discussed in Chapter 16.

14.7d The Strange Case of SS433

One of the oddest x-ray binaries is known as SS433, based on its number in a catalogue. It is in the constellation Aquila, the Eagle, and is about 16,000 light-years away. From measurements of Doppler shifts, we detect gas coming out of this x-ray binary at about 25 per cent of the speed of light, a huge speed for a source in our Galaxy.

Until recently, the most widely accepted model (■ Fig. 14–14) considered SS433 to be a neutron star surrounded by a disk of matter it has taken up from a companion star. Measurements of the Doppler shifts in optical light show us light coming toward us from one jet and going away from us from the other jet at the same time. The disk would wobble like a top (a precession, similar to the one we briefly discussed for the Earth's axis in Chapter 4). As it wobbles, the apparent to-and-fro velocities decrease and increase again, as we see the jets from different angles. We have even detected the jets in radio waves and x-rays (■ Fig. 14–15).

SS433 was an obvious candidate for imaging by the Chandra X-ray Observatory. With such observations, scientists found that the jets originated closer to the compact object than they had expected, only about 0.3 au, closer than Mercury is to the Sun. Spectroscopic information gleaned with Chandra showed temperatures (from the isotopes they observed) and velocities in the jets; they measured that the gas in the jet drops from 100 million kelvins to 10 million kelvins within about two million kilometers, only five times the Earth–Moon distance. From this information, and from details of how long and how often the companion star blocks the compact x-ray star, they

From a model of Modecai Milgrom, Bruce Margon, Jonathon Katz, and George Abell

■ **FIGURE 14–14** A model of SS433 in which the radiation emanates from two narrow beams of charged particles that are given off by the disk of matter orbiting the compact object (a neutron star or a black hole). When we discuss active galaxies in Chapter 17, we will see that high-speed jets are not unusual on the much larger galactic scale.

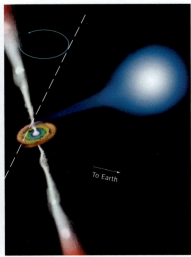

■ **FIGURE 14–15** An x-ray image of SS433 from the Chandra X-ray Observatory, accompanied by an artist's conception of how the lobes are produced. We see a binary system, which is thought to contain a massive star (bloated blue balloon-shaped object) and a black hole with its accretion disk. Jets emitted from the accretion disk produce the lobes. The detection of iron so highly ionized that only one of its electrons remains indicates that the temperature of the gas is 50 million kelvins. The jets wobble (blue circular arrow) between the positions shown and the dotted line. Finding the lobes so hot so far out ($1/4$ light-year) implies that the gas in them has been reheated. Apparently, blobs of gas ejected from near the black hole catch up and collide with earlier ejections after travelling outward for months at about $1/4$ the speed of light (as measured by the Doppler effect), thereby getting heated again.

NASA/CXC/U. Amsterdam/Simone Migliari and Rob Fender (University of Amsterdam) and Mariano Mendez (National Institute for Space Research, SRON)

could deduce that the probable mass of the compact object is about 16 solar masses. This result shows that the object is likely to be a black hole rather than a neutron star, solving a long-standing question.

SS433 and its jet make a relatively close-by analogue to the quasars that we will discuss in Chapter 17.

14.8 SUPERMASSIVE BLACK HOLES

Although the stellar-mass black holes discussed above provided the best initial evidence for the existence of black holes, we now have firm observational support for black holes containing millions or billions of solar masses. Let us discuss these **supermassive black holes** briefly in this black-hole chapter, and say more about them in Chapter 17, when we consider quasars and active galaxies.

The more mass involved, the lower the average density needed for a black hole to form. For a very massive black hole, one containing hundreds of millions or even billions of solar masses, the average density would be so low when the event horizon formed that it would be close to the density of water.

Thus, if we were travelling through the Universe in a spaceship, we couldn't count on detecting a black hole by noticing a volume of high density, or by measuring large tidal effects. We could pass through the event horizon of a high-mass black hole without being stretched tidally to oblivion, though visual effects would certainly be bizarre. We would never be able to get out, but it would be several minutes, or even over an hour, before we would notice that we were being drawn into the center at a rapidly accelerating rate.

Where could such a supermassive black hole be located? The center of our Milky Way Galaxy certainly contains a black hole of about 4 million solar masses, measured from stars that orbit near it, as we shall see in Chapter 16. Though we do not observe radiation

from the black hole itself, the gamma rays, x-rays, and infrared radiation we detect come from the gas surrounding the black hole.

The Hubble Space Telescope has been able to take spectra very close to the centers of certain galaxies. The Doppler shifts on opposite sides of the galaxies' cores, especially in disks of gas surrounding these cores, reveal how fast these points are revolving around the centers of the galaxies (■ Fig. 14–16). Similarly, the average orbital speeds of stars near the center of each galaxy can be measured from the spectra. These speeds, in turn, can be used to calculate how much mass is present in such a small volume. Only a giant black hole can

L. Ferrarese (Johns Hopkins University) and NASA

■ **FIGURE 14–16** A Hubble Space Telescope image of a spiral-shaped disk of gas and dust fueling a supermassive black hole in the center of the galaxy NGC 4261. The disk shown is 800 light-years wide and is 100 million light-years from us. Astronomers have measured the speed of the gas swirling around the central object. The object must have 1.2 billion times the mass of our Sun in order to have enough gravity to keep the orbiting gas in place, but it is concentrated into a region of space not much larger than our Solar System. Only a black hole easily meets both these requirements.

NASA/CXC/CfA/R. Kraft et al.

■ **FIGURE 14–17** (a) A Chandra X-ray Observatory view of jet of gas being emitted from the galaxy Centaurus A, no doubt from the vicinity of a central supermassive black hole. (The name came when it was the first galaxy, "A," to be discovered in radio waves in the constellation Centaurus.) The galaxy is about 11 million light-years away. The colors represent x-ray energies (lowest energy = red; middle = green; highest = blue). We see the jet extending 13,000 light-years toward upper left, and a smaller counterjet extending toward lower right. The x-ray emission comes from electrons spiraling around magnetic field lines in the jet. The dark green and blue bands crossing the jet are dust lanes absorbing the x-rays. Many bright points are x-ray binary stars.

NASA/CSC/SAO

■ **FIGURE 14–18** A composite image of the galaxy NGC 1068, known from Chandra X-ray Observatory images and spectra to contain a supermassive black hole. We see x-ray data from Chandra (red), optical data from the Hubble Space Telescope (green), and radio data from the Jansky Very Large Array in New Mexico (blue). The radio data show the jet; the x-ray and optical data show the spiral structure. The Chandra observations revealed that an equivalent of the Sun's mass is sent out to large distances by a strong wind from the vicinity of the black hole a few times a year, showing how a supermassive black hole can affect its environment.

have so much mass in such a small volume. It now seems that most relatively normal galaxies have a supermassive black hole at their center; over 5 dozen such cases have been found as of 2013. The Chandra X-ray Observatory has found independent evidence that the center of the Andromeda Galaxy contains a supermassive black hole (■ Fig. 14–17).

So-called "active galaxies" and quasars (see Chapter 17) are also probable locations for supermassive black holes. The Hubble Space Telescope and other instruments have been used to take images of galaxies with the highest possible resolution, and found extremely compact, bright cores at the centers of some of them. Most of these cores probably contain supermassive black holes that have an accretion disk of matter around them, making them emit exceptionally large amounts of electromagnetic radiation from very small volumes. The Chandra X-ray Observatory has found evidence of jets of gas emitted from the vicinity of supermassive black holes in several active galaxies (■ Figs. 14–18, 14–19).

Radio techniques that provide very high angular resolution can also be used to study such jets. The jet in the galaxy M87, the central galaxy in the Virgo Cluster of galaxies, has been imaged in various ways (■ Fig. 14–20). The jet has been traced so close to the central object that it shows a widening, which seems to indicate that the jet

NASA/SAO/CXC/M. Garcia et al.

■ **FIGURE 14–19** A Chandra X-ray Observatory view of a tiny region at the core of the Andromeda Galaxy. A hotter object 10 light-years north of (that is, slightly above) the relatively cool object shown in blue in this false-color image is at the very center of the galaxy and is deduced to be a black hole about 100 million times more massive than the Sun.

■ **FIGURE 14–20** A combination of observations in different parts of the spectrum reveals the region of formation of the jet at the center of the giant elliptical galaxy M87 (an object that we will discuss further in Chapter 17). A radio image made with the Jansky Very Large Array (VLA) shows the jet, which is imaged in visible light with the Hubble Space Telescope. The region of origin of that jet can be imaged in radio waves with the Very Long Baseline Array (VLBA), a network of radio telescopes spanning the globe, giving the angular resolution (clarity) of a single radio telescope close to the size of the Earth. The color coding shows the intensity of the radiation recorded.

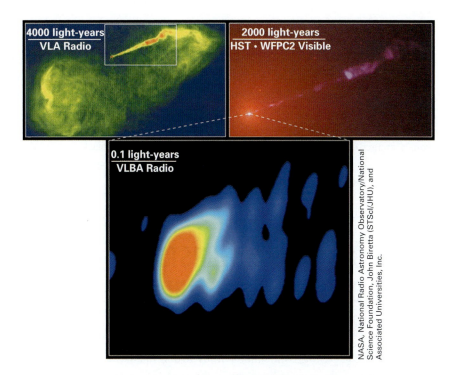

comes from the accretion disk rather than from the immediate vicinity of the central black hole itself.

The black hole in M87 has a mass of more than 6 billion Suns, recently revised upward from about 3 billion solar masses as a result of more detailed modeling of the data. As of 2013, the record mass of a supermassive black hole is 9.7 billion solar masses, in an elliptical galaxy 320 million light-years from us in the constellation Leo, with a contender in Coma Berenices perhaps even more massive. They are probably defunct quasars (Chapter 17).

Chandra observations of one x-ray source in the galaxy M74 (■ Fig. 14–21) discovered strong variations in its x-ray brightness that repeated nearly periodically about every two hours. Because they are almost, but not quite, periodic, the phenomenon is known as "quasi-periodic oscillations." The object is one of many known as "ultraluminous x-ray sources," which are 10 to 1,000 times stronger in x-rays than neutron stars or black holes of stellar mass.

The question has been whether this type of source was an intermediate-mass black hole or a stellar-mass black hole for which we are looking straight into the jet, increasing the x-ray intensity.

14.9 MODERATION IN ALL THINGS

Until recently, many scientists doubted that there were black holes with masses intermediate between those of stars (say, 3 to 40 solar masses) and those in the centers of all or most galaxies (millions to billions of times the Sun's mass). But one of the latest advances in black-hole astrophysics is the probable discovery of so-called **intermediate-mass black holes**, with masses of "only" 100–100,000 times the Sun's, bridging this gap. Chandra X-ray images reveal luminous, flickering sources in some galaxies, suggesting that intermediate-mass black holes are accreting material sporadically from their surroundings.

One possible place for such black holes to form is in the centers of dense clusters of stars, perhaps even young globular clusters, which retain the black holes after they form. Indeed, measurements of the motions of stars in the central regions of a few old globular clusters seemed to imply the presence of intermediate-mass black holes, although the interpretation of the data is still somewhat controversial.

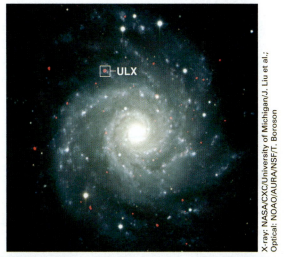

■ **FIGURE 14–21** A visible-light image of the galaxy M74, the 74th object in Charles Messier's catalogue, with the position of an ultraluminous x-ray source (ULX) marked. This source is thought to be a black hole having about 10,000 times the Sun's mass because of its quasi-periodic two-hour variation. The red spots are x-ray sources observed with the Chandra X-ray Observatory.

The existence of quasi-periodic oscillations tends to show that the intermediate-mass idea is the actual situation.

Observations of many black-hole x-ray sources of all types have shown that the quasi-period of oscillations is linked to the mass of the black hole. Based on this relation, the unusual x-ray source in M74 would have a mass about 10,000 times that of our Sun. Such intermediate-mass black holes, it is thought, would have resulted from a merger of perhaps hundreds or thousands of stellar-mass black holes in a star cluster, or are nuclei of small galaxies that are being incorporated into (one might say "eaten" by) a larger galaxy.

14.10 GAMMA-RAY BURSTS: BIRTH CRIES OF BLACK HOLES?

One of the most exciting astronomical fields in the middle of this second decade of the new millennium is the study of **gamma-ray bursts**. These events are powerful bursts of extremely short-wavelength radiation, lasting only about 20 seconds on average. Each gamma-ray photon, to speak in terms of energies instead of wavelength, carries a relatively large amount of energy. Gamma-ray bursts flicker substantially in gamma-ray brightness, and each of them has a unique light curve (■ Fig. 14–22). The joke so far has been "If you see one gamma-ray burst, you've seen one gamma-ray burst," a far cry from "you've seen them all."

14.10a How Far Away Are Gamma-ray Bursts?

Rarely have astronomers had ideas as wrong as the first ones about gamma-ray bursts. Indeed, in some previous editions of this textbook, we discussed gamma-ray bursts in the chapter on the Milky Way Galaxy, since that is where people first thought they were coming from. But now we know that most of them are in extremely distant galaxies, typically billions of light-years away. To appear as bright as they do to us (or to our gamma-ray satellites), they must come from exceptionally powerful explosions – and indeed, we now think they represent the formation of black holes during certain kinds of stellar collapses.

The gamma-ray-burst story started with the "Vela" military satellites used in the late 1960s to search for gamma rays that might come from surreptitious atomic-bomb tests. Some bursts of gamma rays were indeed detected by the orbiting satellites, but they came from above, not from below. Moreover, their properties seemed inconsistent with the explosions of nuclear bombs. However, since gamma-ray telescopes don't give accurate positions, astronomers were unable to link the bursts to specific astronomical objects.

As the data gradually accumulated, it became clear that gamma-ray bursts were not associated with the plane of our Milky Way Galaxy, unlike most bright stars or neutron stars (pulsars; see Figure 13–28 for the latter). This meant that gamma-ray bursts come from either a very large, spherical halo around our Galaxy or from far outside our Galaxy, perhaps from extremely distant galaxies; otherwise they would have shown some variation with respect to the plane of the Milky Way.

In the mid-1990s, an Italian satellite named BeppoSax ("Beppo" was the nickname of a distinguished Italian scientist, "Sa" was for satellite, and the "x" was for "x-ray") was able to swiftly reposition itself when it sensed a gamma-ray burst to point an x-ray telescope at that part of the sky. The x-ray telescope on board, in turn, gave a good enough position to allow optical telescopes on the ground and in space to look at that region. In some cases, an optical "afterglow" was visible at the same position. Additional studies with the largest telescopes, including the Keck telescopes in Hawaii and the Hubble Space Telescope (■ Fig. 14–23), revealed that these afterglows were coincident with faint galaxies billions of light-years away. Thus, gamma-ray bursts must be incredibly powerful to appear as bright as they do from such large distances.

In 2004, NASA launched a satellite, Swift, to discover gamma-ray bursts and rapidly turn toward them, obtaining accurate positions with x-ray and ultraviolet/optical telescopes. (It got its name because it can swiftly turn to look at gamma-ray bursts.) It had been used first with NASA's High-Energy Transient Explorer 2, since 2008 with the Fermi satellite, and with 7 other current spacecraft to study

Time (sec)

■ **FIGURE 14–22** A montage of six gamma-ray burst light curves from the Swift satellite. The gamma-ray brightness is plotted against time. Each of the light curves differs from the others.

■ **FIGURE 14–23** A Hubble Space Telescope image of the optical afterglow of gamma-ray burst 990123. At its peak, the optical afterglow was about four million times brighter and could be seen through a good pair of binoculars, despite being in a galaxy over 9 billion light-years away from Earth. The host galaxy shows signs of having recently formed a substantial number of very massive stars.

■ **FIGURE 14–24** (a) An image of the optical afterglow of gamma-ray burst 021211, obtained only a few minutes after the burst with a 0.76-meter robotic telescope at Lick Observatory. (b) An image of the same region of the sky, taken several hours later with the same telescope. Note how much the optical afterglow has faded.

The distances to gamma-ray bursts are determined by using Hubble's law (Chapter 16) with the afterglows' spectra.

both gamma-ray bursts and x-ray flashes, similar phenomena that give off most of their bursts of radiation in their respective parts of the spectrum.

The Swift satellite immediately relays the positions of gamma-ray bursts to ground-based observatories via the Internet. A network of small telescopes, the Robotic Optical Transient Experiment (ROTSE), spanning the world so that at least one is always in darkness, can study the associated "optical afterglow," which lasts longer than the gamma-ray burst itself. Sometimes even an optical burst that is *simultaneous* with the gamma rays is detected! Several other telescopes have also been automated to make rapid follow-up observations of gamma-ray bursts. For example, one of the authors (A.F.) operates a robotic telescope at Lick Observatory (see *A Closer Look 13.1: Search for Supernovae*) that is able to slew over to the position of a gamma-ray burst and take relatively deep images within a few tens of seconds after the outburst; it has detected several optical bursts and afterglows (■ Fig. 14–24).

On March 19, 2008, Swift discovered an incredibly powerful gamma-ray burst. A Polish team of astronomers using a small, very wide-field ($^1/_{12}$ of the sky at a time) robotic telescope at the Las Campanas Observatory in Chile imaged the associated optical burst while it was still brightening, and at its peak brightness it would have been barely visible to the naked eye on a dark, moonless night if someone had been looking at precisely the right location. Yet the burst was found to be about 7.5 billion light-years away, more than half the distance to the edge of the visible Universe! Had it been at a distance of 6,000 light-years, it would have appeared as bright as the daytime Sun, which is only 8.3 light-minutes away.

14.10b Models of Gamma-ray Bursts

Analysis of observations of gamma-ray bursts shows convincingly that their radiation is highly beamed, like that of pulsars (Section 13.3c) – it cannot be emitted uniformly over the sky. A leading

hypothesis is known as the "fireball model," in which some sort of compact engine releases blobs of material at speeds close to that of light along a narrow "jet." Rapidly moving blobs overtake more slowly moving ones, creating internal shocks within the jet that release the gamma rays. When the blobs subsequently collide with surrounding gas, additional shocks are produced and emit the afterglow visible at wavelengths other than gamma rays. The afterglow fades with time as the shock's energy dissipates.

How are these "relativistic" (rapidly moving) jets formed? A clue comes from the fact that at least some gamma-ray bursts, those with long duration (over 2 seconds) and characterized by lower-energy gamma rays, often appear in galaxies undergoing vigorous bursts of massive-star formation. Because massive stars live only short lives, such stars are also dying in those galaxies. The jet may have something to do with the rotation of the massive collapsing star known as a "collapsar."

If the massive star is rotating just before its core collapses, the collapsed part will spin even faster than before because of the conservation of angular momentum, a concept we have already encountered several times in this book. A disk will form, as in the formation of planetary systems. Since escaping material has a hard time going through the disk, it flows along the axis of rotation instead. If the star does not have a thick envelope, the material might burst all the way through the envelope (■ Fig. 14–25).

Recall (Section 13.2a) that massive stars without hydrogen envelopes, or even without helium envelopes, are thought to produce Type Ib and Type Ic supernovae, respectively (Fig. 13–15). Thus, perhaps long-duration gamma-ray bursts are essentially these kinds of supernovae (usually Type Ic), but with the core collapsing all the way to form a rapidly rotating black hole instead of just a neutron star, thereby releasing an even larger amount of gravitational energy. The rotation energy of the black hole and the surrounding disk might also be tapped, as discussed in Section 14.5. If the jet breaks through the star's surface and happens to be pointing at Earth, we will see an incredibly bright flash, especially because the brightness of the relativistic material moving along the line of sight is enhanced.

NASA/Dana Berry/SkyWorks Digital

■ **FIGURE 14–25** Artist's conception of the collapsar model of a gamma-ray burst. A rotating, massive star largely devoid of its hydrogen and helium envelopes collapses at the end of its life, forming a black hole and two oppositely directed jets along the star's rotation axis. The jets break out of the star, and if our line of sight is along one of them, we see a gamma-ray burst. If our line of sight is *not* along one of them, then the outburst is much fainter and might not be detected at all, or as a faint supernova generally of Type Ic (though only a small fraction of Type Ic supernovae form powerful jets that can produce a gamma-ray burst).

An expectation of the collapsar model is that a Type Ib or (more frequently) Type Ic supernova should become visible at the location of the gamma-ray burst when its afterglow has faded sufficiently. During the past few years, compelling evidence has indeed been found for Type Ic supernovae associated with several long-duration gamma-ray bursts. (However, most Type Ib and Ic supernovae do not produce gamma-ray bursts, regardless of the direction from which we view them.) A few of these supernovae show signs of being especially powerful, prompting some astronomers to call them "hypernovae."

There are a few other viable models for gamma-ray bursts, especially the short-duration bursts (less than 2 seconds) characterized by higher-energy gamma rays, which have not been studied as extensively as long-duration gamma-ray bursts. These objects are currently thought to come from collisions of collapsed objects, either two neutron stars (forming a black hole) or a neutron star and a black hole (increasing the mass of the black hole). For example, a tight (short separation) binary neutron-star system emits gravitational waves (Section 13.3f), with the neutron stars spiraling toward each other and eventually merging to form a black hole. The neutron stars would become torn apart near the final stages of merging, forming a disk with an axis of rotation. A jet of relativistic particles and radiation could then emerge along that axis.

Other possibilities remain for the short-period bursts, such as the formation of "magnetars," isolated neutron stars in our Galaxy with extremely high magnetic fields, or perhaps super-magnetar flares in distant galaxies. Swift finds a gamma-ray burst every few days. Maybe more similarities among different gamma-ray bursts will become clear. Robotic ground-based telescopes will conduct successful follow-up observations at other wavelengths, with progressively shorter time delays after the gamma-ray burst discoveries. Speeding up the rate at which astronomy is done is unusual and difficult. Gamma-ray bursts are but the latest of violent phenomena discovered in the Universe, which is by no means as placid as had been long thought.

14.11 MINI BLACK HOLES

We have discussed how black holes can form by the collapse of massive stars. But theoretically a black hole should result if a mass of any amount is sufficiently compressed. No object containing less than 2 or 3 solar masses will contract sufficiently under the force of its own gravity in the course of stellar evolution. However, the density of matter was so high at the time of the origin of the Universe (see Chapter 19) that smaller masses may have been sufficiently compressed to form **mini black holes**.

Stephen Hawking, an English astrophysicist, has suggested their existence. Mini black holes the size of pinheads would have masses equivalent to those of asteroids. There is no observational evidence for a mini black hole, but they are theoretically possible. Hawking has deduced that small black holes can emit energy in the form of elementary particles (electrons, neutrinos, and so forth). The mini black holes would thus evaporate through the emission of "Hawking radiation" and eventually disappear, with the final stage being an explosion of gamma rays.

Hawking's idea that black holes radiate may seem to be a contradiction to the concept that mass can't escape from a black hole. But when we consider effects of quantum physics, the simple picture of a black hole that we have discussed up to this point is not sufficient. Physicists already know that "virtual pairs" of particles and antiparticles can form simultaneously in empty space, though they disappear by destroying each other a short time later. Hawking suggests that a black hole so affects space near it that the particle or antiparticle disappears into the black hole, allowing its partner to escape (■ Fig. 14–26). Photons, which are their own antiparticles, appear as well.

As the mass of the mini black hole decreases, the evaporation rate increases, and the typical energy of emitted particles and photons increases. The final result is an explosion of gamma rays.

Although "gamma-ray bursts" have indeed been detected in the sky (see the preceding section), their observed properties are *not* consistent with the explosions of mini black holes. For example, gamma-ray bursts flicker substantially in gamma-ray brightness, with

■ **FIGURE 14–26** The formation of "virtual pairs" of particles and antiparticles (here represented as electrons and positrons) in and near a black hole. In some cases, the particle or antiparticle can escape, while its partner goes into the black hole with negative energy (as seen by an outside observer) and decreases the black hole's mass.

each one being different (looking back at Fig. 14–22). Such behavior is not expected of exploding black holes.

Instead, as we have seen, gamma-ray bursts probably result from neutron stars merging to form black holes, or from supernovae in which the rotating, collapsing core forms a black hole. Gamma-ray bursts are extremely powerful, and probably have something to do with the formation of black holes, but unfortunately for Hawking they are not evidence for exploding mini black holes.

Hawking's quantum process should apply to any black hole. However, the rate of evaporation is negligible for all but the smallest black holes. Indeed, in most cases it is easily exceeded by the rate at which a black hole swallows (accretes) surrounding gas.

CONCEPT REVIEW

» Black holes result when too much mass is present in a collapsed star to stop at the neutron-star stage (introductory section). The strong gravitational field of a **stellar-mass black hole** bends light, according to Einstein's general theory of relativity (Section 14.1). Only light within an **exit cone** escapes (Section 14.2). Light on the edge of the exit cone orbits in the **photon sphere**.

» When the radius of the star becomes equal to the **Schwarzschild radius**, it is so compact that the exit cone closes, and no light escapes; the star is entirely within its **event horizon** (Section 14.3), and it is called a black hole. The Schwarzschild radius, only 3 km for a black hole having the mass of the Sun, is directly proportional to the mass. In the Newtonian view, when a star's size decreases to the Schwarzschild radius, the **escape velocity** from the star's surface reaches the speed of light (Section 14.3a).

» Near a stellar-mass black hole, or on the surface of a star collapsing to form a black hole, the **tidal force** can be immense, pulling a local observer apart (Section 14.3b). This effect occurs because of the large difference in gravity between his head and feet: the force on the latter is much larger than that on the former. The tidal force on an observer would be smaller near a more massive black hole than near a less massive black hole.

» Once a collapsing star passes inside its event horizon, it continues to contract, according to the general theory of relativity. Nothing can ever stop its contraction. In fact, the classical mathematical theory predicts that it will contract to zero radius – it will reach a **singularity** (Section 14.3b). (But quantum effects probably prevent it from actually reaching exactly zero radius.) The event horizon, however, does not change in size as long as the amount of mass inside is constant.

» If you were far from a black hole, watching clocks fall toward it, you would see them run progressively more slowly, an effect known as **time dilation** (Section 14.4). Indeed, it would take an infinite amount of time (as measured by your clock) for them to reach the horizon. From the falling clocks' perspective, in contrast, no such time dilation occurs. On the other hand, if the falling clocks were to approach the event horizon and subsequently escape from the vicinity of the black hole, *less* time would have passed for them than for you; they would have aged less.

» In equilibrium, a black hole "has no hair" – in other words, its only measurable properties are mass, angular momentum (loosely, the rate at which it is spinning), and total electric charge (Section 14.5).

» The spherical event horizon of a rotating black hole is smaller than that of a nonrotating black hole of the same mass (Section 14.5a). Moreover, rotating black holes have a **stationary limit** within which no particles can remain at rest, even though they are outside the event horizon. The space between the stationary limit and the event horizon is the **ergosphere**, from which energy (work) can be extracted. If a black hole were to rotate faster than a certain value, its event horizon would vanish, revealing a **naked singularity** from which distant observers could receive signals. Most theorists assume that all singularities are clothed by an event horizon.

» Determining the spin of a black hole is difficult, but there are now two methods for doing so by measuring the properties of the **accretion disk** of gas surrounding the black hole (Section 14.5b). The accretion disk's inner edge is set by a relativistic phenomenon known as frame-dragging. Stable circular orbits in the same direction as the black-hole spin can exist closer to rapidly spinning black holes than to slowly spinning ones. However, the last stable orbit is farther out if the black hole's spin is opposite in direction to that of the accretion disk.

» At first glance, it appears as though black holes provide passageways to distant parts of our Universe, or even to other universes (Section 14.6). Such **wormholes** do not work, even in principle, for nonrotating black holes; one would need to travel faster than the speed of light to traverse them. In the case of a rotating black hole, on the other hand, travel through the wormhole at speeds slower than that of light, avoiding the singularity, initially seems feasible. But actually, the gravity of any object attempting to traverse a wormhole destroys the passage, precluding a safe journey.

» Black holes can be detected through their gravitational effects on their surroundings (Section 14.7). We look for black holes from collapsed stars in regions where flickering x-rays come from an accretion disk (Section 14.7a). We think a black hole is present where, as for Cygnus X-1, we find a visible object that is in an orbit (detected via a periodically varying Doppler shift in its spectrum) around an invisible object that is too massive and too faint to be anything other than a black hole (Section 14.7b). More than two dozen probable stellar-mass black holes in our Galaxy have been identified and measured (Section 14.7c). The most massive known stellar-mass black hole is in another nearby galaxy and has a mass of 23 to 33 Suns. Astronomers are now searching for signs of event horizons in these systems. A peculiar x-ray binary known as SS433, now thought to contain a black hole, shoots out jets of matter reaching 25 per cent the speed of light (Section 14.7d).

» Astronomers conclude that **supermassive black holes**, with millions or billions of times the mass of the Sun, exist in the centers of quasars and most galaxies (Section 14.8). Measurements of stars and rotating gas disks in the central regions of these objects reveal very high speeds, necessitating strong gravitational fields in very small volumes. Even our own Milky Way Galaxy contains a central black hole having several million solar masses. The largest known black holes have the equivalent of about 10 billion Suns.

» Evidence for **intermediate-mass black holes** (containing about 100 to 100,000 solar masses) has been found (Section 14.9). Such black holes may have resulted from a merger of perhaps hundreds or thousands of stellar-mass black holes in a star cluster,

or perhaps they are nuclei of small galaxies being consumed by a larger galaxy.

» One of the most exciting new astronomical fields is the study of **gamma-ray bursts**, brief flashes of extremely short-wavelength radiation (Section 14.10). Although their distances were for many years highly controversial, most of them are now known to come from galaxies billions of light-years away (Section 14.10a). Their radiation is highly beamed, like that of pulsars; it cannot be emitted uniformly over the sky (Section 14.10b). A leading hypothesis is known as the "fireball model," in which some sort of compact engine releases blobs of material at speeds close to that of light along a narrow "jet."

» Gamma-ray bursts may be the birth cries of certain types of stellar-mass black holes (Section 14.10b). Specifically, bursts of relatively long duration (typically a few tens of seconds) appear to be produced when the core of a very massive star deficient in hydrogen and usually even helium collapses to form a black hole, with the remaining material ejected as an unusually energetic Type Ib or (more frequently) Type Ic supernova. Those of short duration (typically less than 2 s) are less well understood, but probably result from either the merging of two neutron stars to form a black hole or the merging of a neutron star and a black hole.

» **Mini black holes** may have formed in the early Universe, but there is no direct evidence for them (Section 14.11). If they were indeed present, they should gradually evaporate by a quantum process, finally resulting in a violent burst of gamma rays. They are not, however, the origin of observed gamma-ray bursts.

QUESTIONS

1. Why doesn't the pressure from electrons or neutrons prevent a sufficiently massive star from becoming a black hole?

2. Why is a black hole blacker than a black piece of paper?

3. Discuss the escape of light from the surface of a star that is collapsing to form a black hole, at different stages of the collapse.

4. Explain the bending of light as a property of a warping of space, as discussed in Chapter 10.

†5. How would the gravitational force at the surface of a star change if the star contracted to one-fifth of its previous diameter, without losing any of its mass?

†6. What is the Schwarzschild radius of a nonrotating black hole of 10 solar masses? What is the radius of its photon sphere?

†7. What would be your Schwarzschild radius, if you were a black hole?

†8. What is the relation in size of the photon sphere and the event horizon of a nonrotating black hole?

9. If you were to throw a mortal enemy into a black hole, would you be able to actually see him cross the event horizon of the black hole? Explain.

10. Describe what we mean when we say that time slows down near a black hole.

11. Explain what we mean by the statement "A black hole has no hair."

12. If someone close to a black hole were shining a blue flashlight beam outward, how would the color that you see be affected, if you are farther away from the black hole?

13. If you were an astronaut in space, could you in principle escape (a) from within the photon sphere of a nonrotating black hole, (b) from within the ergosphere of a rotating black hole, or (c) from within the event horizon of any black hole? Explain each of these cases.

14. If the Sun suddenly became a black hole (it won't!), would the orbits of planets in our Solar System be affected?

15. How could the mass of a black hole that results from a collapsed star increase?

16. How could mini black holes, if they exist, lose mass? What can eventually happen to them?

17. Would we always know when we reached a black hole's event horizon by noticing its high density? Explain.

18. Is it possible to detect a black hole that is not part of a binary system, or is not surrounded by stars and gas in a galaxy? How?

19. Under what circumstances does the presence of an x-ray source associated with a spectroscopic binary suggest to astronomers the presence of a black hole?

20. Why is it generally easier to accurately determine the mass of a black hole orbited by a low-mass star than by a massive star?

21. Explain how the spin of a black hole may, in principle, be determined.

22. Why is SS433 so unusual?

23. When and how were mini black holes possibly formed?

24. Where are supermassive black holes thought to exist?

25. What is the link between quasi-periodic oscillations and intermediate-mass black holes?

26. **True or false?** The "ergosphere" is a region outside the event horizon of a rotating black hole from which the rotational energy of the black hole can sometimes be extracted.

27. **True or false?** The chemical composition of a black hole can be determined from measurements conducted outside the black hole.

28. **True or false?** Mathematically, it appears that simple, ideal, rotating black holes might connect our Universe to other universes, or to other distant parts of our Universe, but in reality matter probably cannot traverse the passage.

29. **True or false?** The radius of the event horizon of a nonrotating black hole is proportional to the square of the black hole's mass.

30. **True or false?** Any object trying to orbit a black hole will necessarily be swallowed after orbiting just a few times.

31. **True or false?** Light can orbit in circles around a nonrotating black hole, at a radius of 1.5 times the Schwarzschild radius of the black hole.

32. **True or false?** If we measure the distance and the gamma-ray apparent brightness of a "gamma-ray burst," we can accurately determine its actual gamma-ray luminosity by assuming that the gamma rays are emitted uniformly in all directions.

33. **Multiple choice:** Which one of the following statements about known "gamma-ray bursts" is *false*? (a) They have been observed most often in distant galaxies. (b) Some of them may be the result of two merging neutron stars. (c) We have shown that some of them are produced by the quantum evaporation of mini black holes. (d) They emit as much energy as, if not more than, a normal supernova. (e) Some of them may be caused by the collapse of the core of very massive, rotating stars at the end of their lives, forming a black hole.

34. **Multiple choice:** Which one of the following statements about detecting stellar-mass black holes in binary systems is *true*? (a) Absorption lines from both the black hole and the companion star can be seen in the spectrum. (b) Kepler's third law must be used to determine the product of the masses (i.e., $m_1 m_2$) of the two components of the binary. (c) The black hole periodically passes between us and the companion star, causing the latter to periodically disappear, and allowing us to deduce the black hole's presence. (d) Because of the generally unknown inclination of the orbit, spectroscopic observations only give us an upper limit (i.e., a maximum value) for the mass of a black hole. (e) The strongest candidates for black holes are generally found in binary systems with low-mass K and M main-sequence stars.

35. **Multiple choice:** The presence of a black hole in a galaxy core can be inferred from (a) the total mass of the galaxy; (b) the speeds of stars near the core; (c) the color of the galaxy; (d) the distance of the galaxy from the Milky Way Galaxy; or (e) the diminished brightness of starlight in the galaxy core, relative to surrounding areas.

36. **Multiple choice:** Which one of the following statements about black holes is *false*? (a) Inside a black hole, matter is thought to consist primarily of iron, the end point of nuclear fusion in massive

stars. **(b)** Photons escaping from the vicinity of (but not inside) a black hole lose energy, yet still travel at the speed of light. **(c)** Near the event horizon of a small black hole (mass = a few solar masses), tidal forces stretch objects apart. **(d)** A black hole that has reached an equilibrium configuration can be described entirely by its mass, electric charge, and amount of spin ("angular momentum"). **(e)** A black hole has an "event horizon" from which no light can escape, according to classical (i.e., non-quantum) ideas.

†**37.** **Multiple choice:** Which one of the following statements about stellar-mass black holes is *true*? **(a)** Black holes are generally more massive than 3 solar masses because neutron degeneracy pressure can hold up smaller masses. **(b)** Our Sun will become a black hole at the end of its life. **(c)** The Schwarzschild radius of a 100 solar mass black hole is 100 times that of a 10 solar mass black hole. **(d)** A stellar-mass black hole can suddenly brighten after swallowing a clump of gas in the surrounding accretion disk; the light comes out from inside the event horizon. **(e)** Most pulsars are thought to be rapidly rotating black holes.

38. **Fill in the blank:** Mass causes the surrounding space-time to _____.

39. **Fill in the blank:** According to Einstein's general theory of relativity, a black hole is completely described by its _____, _____, and _____.

40. **Fill in the blank:** The _____ of a black hole is its boundary, defining the region from within which nothing can escape (ignoring quantum effects).

41. **Fill in the blank:** All of the matter that falls into a black hole becomes concentrated in a very small volume called a _____.

42. **Fill in the blank:** The phenomenon of _____ is responsible for changing the radius of the last stable circular orbit around a spinning black hole.

†This question requires a numerical solution.

TOPICS FOR DISCUSSION

1. If you were given the chance, would you travel close to a black hole and then return to Earth, having aged very little compared with your friends and family? Would you, effectively, want to permanently jump to the future (and, if so, by what amount of time)? Explain.

2. What sorts of problems could be produced by the violation of causality – that is, if you could travel through a wormhole and return before your departure?

3. If "gamma-ray bursts" are beamed, are the energy requirements per gamma-ray burst (calculated from the observed brightness of each burst) smaller than if their energy were emitted uniformly in all directions? Is the total number of gamma-ray bursts we detect in the sky affected by the beaming?

X-ray (NASA/CXC/SAO/Emanuele Nardini, Junfent Wang, Pepi Fabbiano, Martin Elvis, Silvia Pellegrini, Guido Risalti, Margarita Karovska, and Andreas Zezas); Optical (NASA/STScl)

■ **FIGURE 14–27** At the midst of the optical image from the Hubble Space Telescope (shown in red, green, and blue, usually resulting in white) lie two supermassive black holes (Section 14.8) at the centers of each of a pair of galaxies in collision. A whitish tidal tail extends to lower right (see Figure 16–14). The purple shows a Chandra X-ray Observatory image of hot gas surrounding the system, NGC 6240, which is about 330 million light-years from us. The field of view is about 300,000 light-years across in the constellation Ophiuchus. Two spiral galaxies similar in size to our own Milky Way Galaxy are merging, eventually to form an elliptical galaxy (see Section 16.2b). The supermassive black holes at their centers are spiraling toward each other; eventually, they will probably merge to form a single, larger black hole. The collision of the galaxies has led to a burst of star formation that lasted over 200 million years, heating the gas to over 7 million kelvins, allowing Chandra's imaging.

light year

The 4-million-solar-mass supermassive black hole Sagittarius A* (Sgr A*, pronounced "Saj A-star") at the center of our Galaxy was "weighed" by following a star called S0–2 (marked with yellow circles) in a close elliptical orbit around it. From the star's period and orbit, Kepler's third law as modified by Newton gave the black hole's mass. Other stars near the center were a measured and analyzed in a similar manner. In 2011, it was announced that a 3-Earth-mass blob of dusty, ionized gas is on an orbit (lower-left ellipse) approaching Sgr A* and that the b should break up with pieces hitting the black hole sometime in 2013–14.

During 2013–14, as it reaches only 3,100 times the supermassive black hole's event horizon (36 light-hours), the cloud is expected to be further "spaghettified" – that is, drawn out by differential force of gravity from one side to the other (as described in Section 14.3b). It is simultaneously heating up through its interaction with the gas already being accreted by the bla hole. The radio and x-ray emission may increase a bit as a result. The next phase, according to simulations, could take years or decades, as gas captured from the incoming cloud and the orbiting the supermassive black hole falls in toward the black hole's event horizon. The quicker the infall, the more noticeable its effect on the x-ray and radio brightness of Sgr A* would Only observations will tell us if any additional luminosity will be boringly minor or interestingly substantial.

Reinhard Genzel, Max Planck Institute for Extraterrestrial Physics, and University of California, Berkeley

15

The Milky Way: Our Home in the Universe

ORIGINS

Our Sun is part of the Milky Way Galaxy, an enormous collection of billions of stars bound together by gravity. New stars forming from dense clouds of gas and dust in spiral arms provide clues to the Sun's birth.

We have already described the stars, which are important parts of any galaxy, and how they are born, live, and die. In this chapter, we discuss the gas and dust (small particles of matter) that are present to some extent throughout a galaxy. Substantial clouds of this gas and dust are called **nebulae** (pronounced "neb'yu-lee" or "neb'yu-lay"; singular: **nebula**); "nebula" is Latin for "fog" or "mist." New stars are born from such nebulae. We also discuss the overall structure of the Milky Way Galaxy and how, from our location inside it, we detect this structure.

Excitingly, new methods of adaptive optics and extended abilities to observe in the infrared have allowed detailed measurements to be made far into the center of our Galaxy, some 26,000 light-years from us. Two groups of scientists have actually seen stars orbiting close to the center of the Galaxy, so close and so fast that a supermassive black hole must be there with a mass of about 4 million times the mass of our Sun (see the Chapter Opener). Moreover, what appears to be a giant blob of gas is approaching the supermassive black hole at such a speed that it should be there around 2013–14. We await the possible fireworks!

AIMS

1. Understand the nature of the Milky Way, the faint band of light arching across the dark sky (Section 15.1).

2. Learn about our Galaxy and its constituents, and appreciate the position of our Sun and Earth, far from its center (Sections 15.2 to 15.6).

3. Discuss the shape of our Galaxy and why it has spiral arms (Sections 15.7, 15.8).

4. Explore the gas and dust between the stars – interstellar matter from which new stars form (Sections 15.9, 15.13).

5. See how radio observations are critical to studying interstellar matter and mapping our Galaxy (Sections 15.10 to 15.12, 15.14).

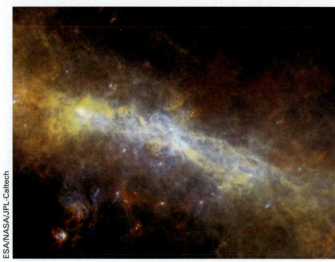

ESA/NASA/JPL-Caltech

■ **FIGURE 15–1** A twisted ribbon of gas 600 light-years across near the center of the Milky Way, imaged with the European Space Agency's Herschel Space Observatory.

15.1 OUR GALAXY: THE MILKY WAY

On the clearest moonless nights, when we are far from city lights, we can see a hazy band of light stretching across the sky (■ Fig. 15–1 and ■ Fig. 15–2). This band is the **Milky Way** – the gas, dust, nebulae, and stars that make up the Milky Way Galaxy in which our Sun is located. All this matter is our celestial neighborhood, typically within a few hundred or a thousand light-years from us. If we look a few thousand light-years in a direction away from that of the Milky Way, we see out of our Galaxy. But it is much, much farther to the other galaxies and beyond.

Don't be confused by the terminology: The Milky Way itself is the band of light that we can see from the Earth, and the **Milky Way Galaxy** is the whole galaxy in which we live. Also, note that when we refer to our Milky Way Galaxy, we say "our Galaxy" or "the Galaxy" with an uppercase "G." When we refer to other galaxies, we use a lowercase "g." This convention is similar to that used for the Solar System (our Solar System), so as not to confuse it with other planetary systems (other solar systems).

Like other large galaxies, our Milky Way Galaxy is composed of perhaps a few hundred billion stars plus many different types of gas, dust, planets, and so on. In the directions in which we see the Milky Way in the sky, we are looking through the relatively thin, pancake-like disk of matter that forms a major part of our Milky Way Galaxy. This disk is about 90,000 light-years across, an enormous, gravitationally bound system of stars.

The Milky Way appears very irregular when we see it stretched across the sky – there are spurs of luminous material that stick out in one direction or another, and there are dark lanes or patches in which much less can be seen (see *Star Party 15.1: Observing the Milky Way*). This patchiness is due to the splotchy distribution of nebulae and stars.

Here on Earth, we are inside our Galaxy together with all of the matter we see as the Milky Way (recall Figure 15–2). Because of our position, we see a lot of our own Galaxy's matter when we look along the plane of our Galaxy. On the other hand, when we look "upward" or "downward" out of this plane, our view is not obscured by matter, and we can see past the confines of our Galaxy.

15.2 THE ILLUSION THAT WE ARE AT THE CENTER

The gas in our Galaxy is more or less transparent to visible light, but the small solid particles that we call "dust" are opaque. So the distance we can see through our Galaxy depends mainly on the amount of dust that is present. This is not surprising: We can't always see far on a foggy day. Similarly, the dust between the stars in our Galaxy dims the starlight by absorbing it or by scattering (reflecting) it in different directions.

The dust in the plane of our Galaxy prevents us from seeing very far toward its center with the unaided eye and small telescopes. With visible light, on average we can see only one-tenth of the way in (about 2,000 light-years), regardless of the direction we look in the plane of the Milky Way. These direct optical observations fooled astronomers at the beginning of the 20th century into thinking that the Earth was near the center of the Universe (■ Fig. 15–3).

We shall see in this chapter how the American astronomer Harlow Shapley (pronounced to rhyme with "map lee," as in "road map") realized in 1917 that our Sun is not in the center of the Milky Way. This fundamental idea took humanity one step further away from thinking that we are at the center of the Universe. Copernicus, in 1543, had already made the first step in removing the Earth from the center of the Universe.

In the 20th century, astronomers began to use wavelengths other than optical ones to study the Milky Way Galaxy. In the 1950s and 1960s especially, radio astronomy gave us a new picture of our Galaxy. In the 1980s and 1990s, we began to benefit from space infrared observations at wavelengths too long to pass through the Earth's atmosphere. In 2003, NASA launched its Spitzer Space Telescope, which has now used up its coolant and so is operating in "warm mode," with only some of its channels. The European Space Agency launched its Herschel Space Observatory (named after the discoverer of the infrared, the British astronomer William Herschel, who also discovered the planet Uranus), in 2009; NASA participates in the project. The Herschel Space Observatory, with a mirror 3.5 m in diameter, is currently the largest telescope in space, and will remain so until the launch not before 2018 of NASA's James Webb Space Telescope, with its 6-m mirror. It used up its coolant in 2013. Infrared and radio radiation can pass through the Galaxy's dust and allow us to see our Galactic center and beyond.

■ **FIGURE 15–2** The Milky Way appears as the horizontal bar in these full-sky maps. The earlier far-infrared map released by NASA's Wilkinson Microwave Anisotropy Probe (WMAP) is at top, showing 7 years of results; WMAP collected data between 2001 and 2010. The later far-infrared map, from the first year of results from the European Space Agency's Planck spacecraft that was launched in 2009, was released in 2010 and appears at middle. Note that it is of higher spatial resolution; it also spans a wider range of radio frequencies. The map projection is such that the Milky Way defines the equator, and the north and south Galactic poles are at top and bottom, respectively. Around it, a smoothly varying background has been subtracted and we see primordial fluctuations in the temperature of the Universe's cosmic microwave background that we will discuss in Chapter 19. At the bottom, we see a visible-light composite photograph of our sky, in the same map projection. The full-resolution version was made from about 5,100 individual frames.

■ **FIGURE 15–3** A cross-sectional view of the structure of the Universe, as perceived by astronomers near the turn of the 20th century. The effects of interstellar gas and dust had not been taken into account, so they mistakenly concluded that the Sun was near the center of the Universe.

STAR PARTY 15.1   **Observing the Milky Way**

To see the Milky Way, go far away from city lights on a dark, clear, moonless night and look up at the sky. Depending on the season, you should be able to see a faint band of light somewhere in the sky. It will become more and more apparent as your eyes become adapted to the darkness, a process that usually takes about 15 minutes. From the northern hemisphere, the Milky Way is best seen in the summer sky, especially toward the constellations Scorpius and Sagittarius, but the winter Milky Way is also quite bright.

Notice how irregular some parts of the Milky Way appear, due to different amounts of gas, dust, and stars. There is an especially prominent, dark rift through Cygnus (the Swan), visible in the summer and early fall.

After viewing with your unaided eye, try using binoculars; you will probably see many open star clusters and nebulae sprinkled throughout the Milky Way. Refer to the Sky Maps on the inside covers of this book for their names and positions, and also consult Chapter 1 ("Observing the Constellations"). The Orion Nebula, in the sword of Orion, is an impressive emission nebula. The Pleiades (the Seven Sisters, in Taurus) form an obvious open star cluster surrounded by a very faint reflection nebula. If you have a small telescope, you can get reasonably detailed views of several other star clusters and nebulae that are marked on the Sky Maps.

A new generation of telescopes on high mountains enables us to see parts of the infrared and submillimeter spectrum. The Atacama Large Millimeter Array, now being built in Chile, is already, with its partial installation, giving us high-resolution views in the millimeter part of the spectrum. It is at an extremely high site, over 5,000 m (16,000 feet) of altitude, to get above so much of the infrared-absorbing water vapor in the Earth's atmosphere. Giant arrays of radio telescopes spanning not only local areas but also continents and the Earth itself, as well as with an antenna in space linking with those on Earth, enable us to get crisp views of what was formerly hidden from us.

15.3 NEBULAE: INTERSTELLAR CLOUDS

The original definition of "nebula" was a cloud of gas and dust that we see in visible light, though we now detect nebulae in a variety of ways. When we see the gas actually glowing in the visible part of the spectrum, we call it an **emission nebula** (■ Fig. 15–4).

Gas is ionized by ultraviolet light from very hot stars within the nebula; it then glows at optical (and other) wavelengths when electrons recombine with ions and cascade down to lower energy levels,

releasing photons. Additionally, free electrons can collide with atoms (neutral or ionized) and lose some of their energy of motion, kicking the bound electrons to higher energy levels. Photons are emitted when the excited bound electrons jump down to lower energy levels, so the gas glows even more. The spectrum of an emission nebula therefore consists of emission lines.

Emission nebulae often look red (on long-exposure images; the human eye doesn't see these colors directly), because the red light of hydrogen is strongest in them. Electrons are jumping from the third to the second energy levels of hydrogen, producing the Hα emission line in the red part of the spectrum (6,563 Å).

Other types of emission nebulae can appear green in photographs, because of green light from doubly ionized oxygen atoms. Additional colors occur as well. Don't be misled by the pretty, false-color images that you often see in the news. In them, color is assigned to some specific type of radiation and need not correspond to colors that the eye would see when viewing the objects through telescopes.

Sometimes a cloud of dust obscures our vision in some direction in the sky. When we see the dust appear as a dark silhouette, as we do across the center of the Lagoon Nebula, number 8 in Messier's catalogue (■ Fig. 15–5), we call it a **dark nebula** (or, often, an **absorption nebula**, since it absorbs visible light from stars behind it).

The Horsehead Nebula (■ Fig. 15–6) is another example of an object that is simultaneously an emission and an absorption nebula.

(a)

NASA, C.R. O'Dell and S.K. Wong (Rice University)

(b)

NASA/JPL-Caltech/T. Megeath (University of Toledo, Ohio)

(c)

Institute for Astronomy/University of Hawaii/PanSTARRS Consortium

■ **FIGURE 15–4** (a) The central part of the Orion Nebula, a large emission nebula about 1,500 light-years away. The gas is ionized by ultraviolet radiation from the cluster of hot stars. (b) A wider view of the Orion Nebula, which is in Orion's belt, from the Spitzer Space Telescope. This active nursery for stars is generating thousands of young stars and protostars. (c) A wide-field view of the Orion Nebula with the prototype Pan-STARRS telescope in Hawaii.

The reddish emission from glowing hydrogen gas spreads across the sky near the leftmost (eastern) star in Orion's belt. A bit of absorbing dust intrudes onto the emitting gas, outlining the shape of a horse's head. We can see in the picture that the horsehead is a continuation of a dark area in which very few stars are visible. In this region, dust is obscuring the stars that lie beyond.

Clouds of gas and dust surrounding relatively hot stars, like some of the stars in the star cluster known as the Pleiades (■ Fig. 15–7), are examples of **reflection nebulae**. Dust within them merely reflects the starlight toward us without emitting visible radiation of its own. Reflection nebulae usually look bluish for two reasons: (1) They reflect the light from relatively hot, luminous stars, which are bluish, and (2) dust reflects (scatters) blue light more efficiently than it does red light. (Similar scattering of sunlight in the Earth's atmosphere makes the sky blue; see *A Closer Look 4.1: Colors in the Sky*.)

Whereas an emission nebula has its own spectrum, as does a neon sign on Earth, a reflection nebula shows the spectral lines of the star or stars whose light is being reflected. Dust tends to be associated more with young, hot stars than with older stars, since the older stars would have had a chance to wander away from their dusty birthplaces.

The Orion Nebula (Figure 15–4) is an emission nebula. In the winter sky, we can readily observe it through even a small telescope or binoculars, and sometimes it has a tinge of color. We need longer photographic exposures or larger telescopes to study its structure in detail. Deep inside the Orion Nebula and the gas and dust alongside it, we see stars being born this very minute; many telescopes are able to observe in the infrared, which penetrates the dust. An example in a different region of the sky is shown in ■ Fig. 15–8.

We have already discussed (in Chapter 13) some of the most beautiful emission nebulae in the sky, composed of gas thrown off in the late stages of stellar evolution. They include planetary nebulae (■ Fig. 15–9a) and supernova remnants. Thus, nebulae are closely associated with both stellar birth (■ Fig. 15–9b,c) and stellar death. The chemically enriched gas blown off by unstable or exploding stars at the end of their lives becomes the raw material from which new stars and planets are born. As we emphasized in Chapter 13, we are made of the ashes of stars!

■ FIGURE 15–5 The Lagoon Nebula, M8. (a) The center of the nebula, imaged in high detail with the Hubble Space Telescope. Dark, absorbing dust appears as black. Emission from ionized sulfur atoms (red), from doubly ionized oxygen atoms (blue), and from electrons recombining with ionized hydrogen and cascading down to lower energy levels (green) are put together in this false-color view. The hot star at lower right appears as bright white, with red surroundings and reddish spikes that are an artifact caused by the telescope. This star provides the ultraviolet radiation that ionizes the brightest region of the nebula. (b) The whole Lagoon Nebula, in a huge image with great detail from the European Southern Observatory. (c) A cut across the Lagoon Nebula, from the Gemini Observatory.

15.4 THE PARTS OF OUR GALAXY

It was not until 1917 that the American astronomer Harlow Shapley realized that we are not in the center of our Milky Way Galaxy. He was studying the distribution of globular clusters and noticed that, as seen from Earth, most of them are in roughly the same general area of the sky. When he plotted their distances and directions, he noticed that they formed a spherical halo around a point thousands of light-years away from us (■ Fig. 15–10a).

Shapley's touch of genius was to realize that this point is likely to be the center of our Galaxy. After all, if we are at a party and discover that most people we see are off to our left, we soon figure out that we aren't at the party's center. ■ Figs. 15–10b,c show other spiral galaxies for comparison, illustrating what our Galaxy would look like if we could see it from high above; the Sun would be far from the center if it were in one of these galaxies.

Though Shapley correctly deduced that the Sun is far from our Galactic center, he actually overestimated the distance. The reason is that dust dims the starlight, making the stars look too far away, and he didn't know about this "interstellar extinction." The amount of dimming can be determined by measuring how much the starlight has been **reddened**: Blue light gets scattered and absorbed more easily than red light, so the star's color becomes redder than it should be for a star of a given spectral type. This is the same reason sunsets tend to look yellow, orange, or red, not white (again see *A Closer Look 4.1: Colors in the Sky*). But note that this "reddening" is *not* a

FIGURE 15–6 The Horsehead Nebula in Orion is dark dust superimposed on glowing gas.

Doppler redshift! Interstellar reddening suppresses blue light relative to red, but does not shift the wavelengths of absorption or emission lines, as the Doppler effect does.

Our Galaxy has several parts:

1. *The nuclear bulge*. Our Galaxy has the general shape of a pancake with a bulge at its center that contains billions of stars, primarily old ones. This **nuclear bulge** has the Galactic nucleus at its center. The nucleus itself is only about 10 light-years across.

2. *The disk*. The part of the pancake outside the bulge is called the Galactic **disk**. It extends 45,000 light-years or so out from the center of our Galaxy, though there is no well-defined edge. The Sun is located about one-half to two-thirds of the way out. The disk is very thin – roughly 2 per cent of its width – like a CD or DVD. It contains all the young stars and interstellar gas and dust, as well as some old stars. The disk is slightly warped at its ends, perhaps by interaction with our satellite galaxies, the Magellanic Clouds. Our Galaxy looks a bit like a hat with a turned-down brim.

It is very difficult for us to tell how the material in our Galaxy's disk is arranged, just as it would be difficult to tell how the streets of a city were laid out if we could only stand on one street corner without moving. Still, other galaxies have similar properties to our own, and their disks are filled with great **spiral arms** – regions of dust, gas, and stars in the shape of a pinwheel (Figure 15–10b). So, we assume that the disk of our Galaxy has spiral arms, too. Though the direct evidence is ambiguous in the visible part of the spectrum, radio observations have better traced the spiral arms.

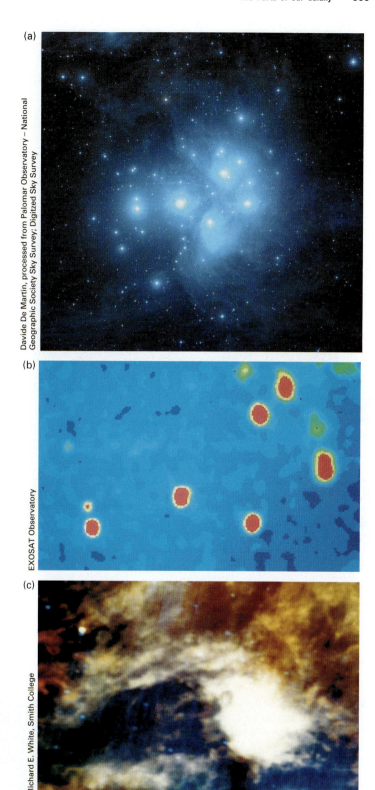

FIGURE 15–7 (a) The Pleiades in visible light, showing the blue reflection nebula around the open cluster. The spikes from bright stars, as well as the rings of light around them, are caused by the telescope optics. (b) This x-ray view of the Pleiades penetrates the dust to show the relatively hot stars. It was made from the EXOSAT in orbit. (c) The Pleiades (buried in the large region that appears white) have left a wake (dark region extending leftward) as they moved through the interstellar medium, as we see in this infrared image. Partly as a result of detecting this motion, scientists think that the Pleiades are undergoing a chance passage through the interstellar matter that forms the reflection nebula; the cluster did not actually form from this particular nebula.

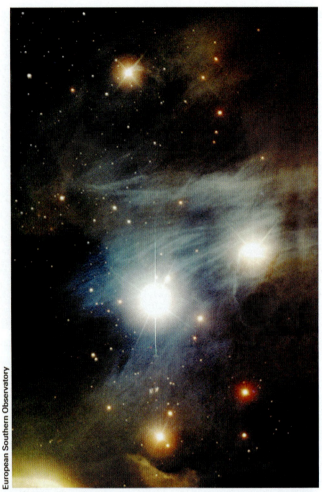

■ **FIGURE 15–8** An image of the Chameleon I region made with the Very Large Telescope of the European Southern Observatory, in Chile. The three colors used were visual (yellow), red, and infrared. The infrared radiation penetrates the dust especially well.

Table 15.1 Parts of Our Galaxy

Thin disk, within ± 0.35 kiloparsecs of the Galactic plane: Has the youngest stars, and stars are still forming from molecular clouds. Roughly called Population I stars. Type II and other core-collapse supernovae form in the thin disk.

Thick disk: Older stars (about 10 billion years old), and lower abundance of heavy elements. Roughly called Population II stars.

First generation of stars (as yet undiscovered): No heavy elements; known as Population III stars.

Galactic halo: Stars in the halo revolve as a unit, not differentially; $1/20$ to $1/10,000$ the abundances of elements heavier than hydrogen compared with the Sun. Globular clusters are in the halo. (Undoubtedly surrounded by a larger, dark-matter halo.)

The disk looks different when viewed in different parts of the spectrum (■ Fig. 15–11). Infrared and radio waves penetrate the dust that blocks our view in visible light, while x-rays show the hot objects best.

3. ***The halo***. Old stars (including the globular clusters) and very dilute interstellar matter form a roughly spherical Galactic **halo** around the disk. The inner part of the halo is at least as large across as the disk, perhaps 60,000 light-years in radius. The gas in the inner halo is hot, roughly 100,000 K, though it contains only about 2 per cent of the mass of the gas in the disk. As we discuss in Chapter 16, the outer part of the halo extends much farther, out to perhaps 200,000 or 300,000 light-years. Believe it or not, this Galactic outer halo apparently contains 5 or 10 times as much mass as the nucleus, disk, and inner halo together – but we don't know what it consists of! We shall see in Section 16.4 that such "dark matter" (invisible, and detectable only through its gravitational properties) is a very important constituent of the Universe.

■ **FIGURE 15–9** (a) The Butterfly Nebula, a planetary nebula imaged with the Very Large Telescope of the European Southern Observatory, in Chile. (b) A combination Herschel Space Observatory image (displayed as green for 70 µm and red for 160 µm radiation) and nearer-infrared National Optical Astronomy Observatory (Kitt Peak) image (displayed as blue) of a star-forming region in Orion, NGC 1999, the glowing green cloud. Multiple jets and blasts of radiation have carved out the dark hole in it. The nebula glows because of radiation from the star V380 Ori, a triple star, one of whose components launched the jet that helped clear the hole – apparently really an empty 3D region of space rather than a dark dust Cloud's silhouette. The orangish spots are three embryonic stars. Herbig-Haro objects HH $1/2$ appear at 2.2 µm (showing up in blue).

(a)

(b)

(a)

Halo

Globular clusters

Bulge

Nucleus   Disk

8 light minutes   Sun

Earth

Venus   Mercury

(b)

(background) NASA and the Hubble Heritage Team (STScI/AURA/N. Scoville (Caltech) and T. Rector (NOAO)

Spiral arm

Nucleus

Bulge

Open Clusters

(c)

NASA and the Hubble Heritage Team (STScI/AURA/ G. R. Meurer and T. M. Heckman (JHU), C. Leitherer, J. Harris, and D. Calzetti (STScI); and M. Sirianni (JHU)

■ **FIGURE 15–10** (a) The drawing shows our Milky Way Galaxy's nuclear bulge surrounded by the disk, which contains the spiral arms. The globular clusters are part of the halo, which extends above and below the disk. (From Earth, we see fewer clusters near the plane of our Galaxy than are actually there because interstellar dust gets in the way. This drawing was made from the actual observed distribution of globular clusters.) From the fact that most of the clusters appear in less than half of our sky, Harlow Shapley deduced that the Galactic center is in the direction indicated. (b) The center of the Whirlpool Galaxy, M51, imaged with the Hubble Space Telescope. Its spiral arms, with its dust clouds in which stars are being born, show clearly. Our Galaxy probably looks something like this one if seen from a comparable vantage point. (c) Another spiral galaxy, M100 in the constellation Coma Berenices, imaged with the European Southern Observatory's Very Large Telescope.

15.5 THE CENTER OF OUR GALAXY

We cannot see the center of our Galaxy in the visible part of the spectrum because our view is blocked by interstellar dust. Radio waves and infrared, on the other hand, penetrate the dust. The Hubble Space Telescope, with its superior resolution, has seen isolated stars where before we saw only a blur (■ Fig. 15–12).

In 2003, NASA launched a 0.85-m infrared telescope, the Spitzer Space Telescope (Section 3.8c, Figure 3–32a). Its infrared detectors were more sensitive than those on earlier infrared telescopes. Spitzer

NASA Goddard Space Flight Center Data Facility/David Friedlander

■ **FIGURE 15–11** Views of the Milky Way at a wide variety of wavelengths. In all cases, we see the concentration toward the disk of the Galaxy. In some wavelength/ frequency/energy bands, such as the infrared views, the radiation comes all the way from the Galactic center. We see data (a) in a radio continuum at 408 MHz = 73.5 cm; (b) the 21-cm radio-wavelength band from neutral hydrogen; (c) a higher-frequency (shorter wavelength) radio continuum at 2.5 GHz = 12 cm; (d) the 115 GHz = 2.6-mm radio-wavelength band from molecular hydrogen; (e) a combination of 12, 60, and 100 μm infrared bands imaged with the InfraRed Astronomy Satellite; (f) a combination of infrared bands imaged with the Spitzer Space Telescope; (g) 2 μm infrared imaged as part of the 2MASS survey; (h) optical hydrogen radiation's Balmer Hα line; (i) visible light; (j) a false-color soft x-ray image with observations at approximately 8–50 Å; (k) a false-color hard x-ray region with even shorter wavelengths; (l) a gamma-ray band at wavelengths shorter than 0.0001 Å, about 100,000 times shorter than the x-ray bands, from the Compton Gamma-Ray Observatory; and (m) another gamma-ray image, from the current Fermi Gamma-ray Space Telescope.

completed NASA's series of Great Observatories, including the Compton Gamma-Ray Observatory (now defunct), the Chandra X-ray Observatory, and the Hubble Space Telescope. As we previously saw, the European Space Agency's Herschel Space Observatory is even bigger and better.

Further mapping of our Galaxy has been carried out by the Wilkinson Microwave Anisotropy Probe and now the Planck spacecraft, both of whose major task is to study the cosmic background radiation. NASA's WISE (Wide-field Infrared Survey Explorer) made a map of the whole sky (covering it one-and-one-half times) in 2010

before it used up its coolant supply. As had been expected, the detectors then warmed up from their 17 K operating temperature, ending the most sensitive phase of the mission.

One of the brightest infrared sources in our sky is the nucleus of our Galaxy, only about 10 light-years across. This makes it a very small source for the prodigious amount of energy it emits: as much energy as radiated by 80 million Suns. It is also a radio source and a variable x-ray source.

High-resolution radio maps of our Galactic center (■ Figs. 15–13 and 15–14) show a small bright spot, known as Sgr A* (pronounced

Don F. Figer (STScI) and NASA

■ **FIGURE 15–12** A cluster of stars near the center of our Galaxy, imaged with the infrared camera aboard the Hubble Space Telescope.

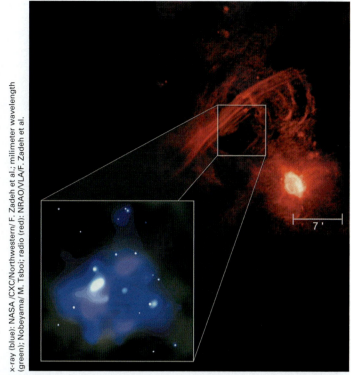

7 ′

x-ray (blue): NASA /CXC/Northwestern/ F. Zadeh et al.; milimeter wavelength (green): Nobeyama/ M. Tsboi; radio (red): NRAO/VLA/F. Zadeh et al.

■ **FIGURE 15–13** Radio measurements using the Jansky Very Large Array show (in red) that a vast Arc of parallel filaments stretches over 130 light-years perpendicularly to the plane of our Galaxy, which runs from upper left to lower right. The field of view is about 1° across; the wavelength used was in the continuous emission near 21 cm. (The Arc is also visible in Figure 15–13d.) A Chandra view in x-rays of part of the field is singled out and shown at lower left (blue). The current model for the x-ray emission is that it results from the collision of electrons from the filaments with a cloud containing a million solar masses of cold gas.

"Saj A-star"), in the middle of the bright radio source Sgr A. The radio radiation could well be from gas surrounding a central giant black hole (see the Chapter Opener). Extending somewhat farther out, a giant Arc of parallel filaments stretches perpendicularly to the plane of the Galaxy.

As we discuss further in Chapter 17, adaptive-optics techniques in the near-infrared have allowed very rapid motions of stars to be measured much nearer the Galactic center than was previously possible (■ Fig. 15–15). The observed orbits reveal the presence of a supermassive black hole that is about 4 million times the Sun's mass, with the main uncertainty coming from the uncertainty in the distance to the Galactic center, measured at 8.0 kiloparsecs (26,000 light-years). Two of the stars come within an astonishing 17 light-hours of Sgr A*, and have been followed through complete orbits. Two main research groups, one at UCLA in the U.S. and one in Germany, have been following this and the other stars that adaptive optics has revealed so close to the center of our Galaxy.

Sgr A*'s 4 million solar masses may seem like a lot, but we will see in Chapter 17 that many galaxies have black holes tens, hundreds, or even a thousand times more massive. So Sgr A*'s supermassive black hole is only a baby dragon compared with the dragons in the midst of other galaxies. Our Milky Way's black hole is apparently accreting less gas than the central black holes in other galaxies. Strangely, in 2011, a 3-Earth-mass blob of gas and dust was discovered to be heading toward Sgr A*, to arrive by the end of 2013 (■Fig. 15–15d). Though it is being disrupted by gravity so we cannot predict the trajectories of all its resulting parts, we will see at that time whether Sgr A* grows brighter, and even stays brighter, from the gas its supermassive black hole attracts inward and swallows.

Observations of the Galactic center with the Chandra X-ray Observatory and the European Space Agency's INTEGRAL (**INTE**rnational **G**amma-**R**ay **A**strophysics **L**aboratory) spacecraft (■ Fig. 15–16), launched in 2002 and still functioning, reveal the presence of hot, x-ray-luminous gas and stars there.

An audacious plan announced in 2012 was to put together a "world-wide telescope," linking submillimeter and radio telescopes in the mainland U.S. and Canada, Hawaii, Europe, and maybe also ALMA in Chile and even eventually the South Pole Telescope to get such long baselines that the event horizon of the Sgr A* supermassive black hole could be imaged. Only about 20 times the Sun's diameter in spite of its 4 million solar masses, much smaller than the size of Mercury's orbit, this black hole's shadow might be detected against the background of the gas behind it.

15.6 ALL-SKY MAPS OF OUR GALAXY

The study of our Galaxy provides us with a wide range of types of sources to study. Many of these have been known for decades from

(a)

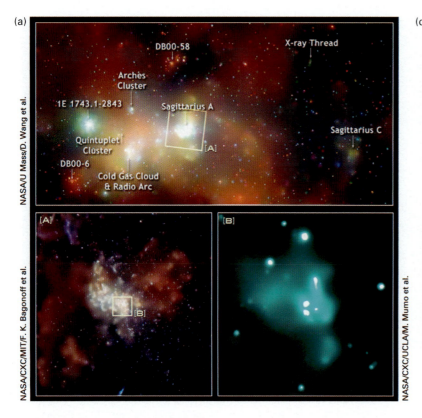

NASA/U Mass/D. Wang et al.

NASA/CXC/MIT/F. K. Bagonoff et al.

NASA/CXC/UCLA/M. Mumo et al.

(d)

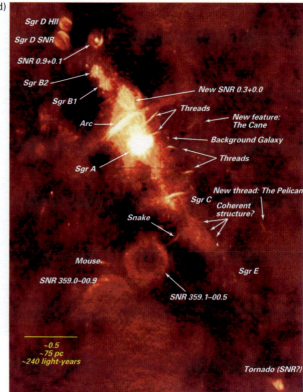

Image processing at the Naval Research Laboratory using DoD High Performance Computing Resources: produced by N. E. Kassim, D. S. Briggs, T. J. W. Lazio, T. N. LaRosa, J. Imamura, and S. D. Hyman; original data from the NARO Very Large Array, courtesy of A. Pedlar, K. Anantharamiah, M. Goss, and R. Ekers

■ FIGURE 15–14 (a) A Chandra X-ray Observatory view of a 400-light-year-across region of the center of our Galaxy, in the direction of the constellation Sagittarius. In the midst of the labeled "Sagittarius A" source is "Sagittarius A*," which marks the location of a supermassive black hole. The detail box at lower left, labeled A, includes about 2,000 individual x-ray sources, many of which are probably black holes several times more massive than our Sun in close orbits with normal stars. (b = sublabel A) Chandra's view of a small region, and (c = sublabel B) an even smaller region only 6 light-years across centered on Sagittarius A*. Since even such a small region contains several of the x-ray binaries, a dense swarm of 10,000 black holes and neutron stars may well have formed around Sagittarius A*. (d) A wide-field view of the center of our Galaxy, observed with the VLA (before it became the JVLA) at a wavelength of 90 cm. (e) A flare lasting several hours at our Galaxy's central black hole seen in the highest-energy x-rays from the NuSTAR (Nuclear Spectroscopic Telescope Array), launched in 2012. The region in the central inset is only 100 light-years across.

(e)

NASA/JPL–Caltech/NuSTAR Project

optical studies (■ Fig. 15–17, and also the seventh and eighth panels from the top in Figure 15–11). The infrared sky looks quite different (Figure 15–11, the fourth, fifth, and sixth panels from the top), with its appearance depending strongly on wavelength. The radio sky provides still different pictures, depending on the wavelength used (Figure 15–11, top three panels; see also Figure 15–2).

Maps of our Galaxy in the x-ray region of the spectrum (Figure 15–11, third and fourth panels from the bottom) show the hottest individual sources (such as x-ray binary stars) and diffuse gas that was heated to temperatures of a million degrees by supernova explosions. The Compton Gamma-Ray Observatory produced maps of the steady gamma rays (Figure 15–11, second panel from the bottom),

FIGURE 15–15 Using adaptive optics, usually at the infrared wavelenth of 2.2 μm, to study stars orbiting close to the center of our Galaxy, scientists have deduced that a giant black hole with about 4 million times the Sun's mass is lurking there. (a) The center of the Milky Way, with measurements made of the position of a star near it from the Very Large Telescope by a largely German research group led by Reinhard Genzel (Max Planck Institute for Extraterrestrial Physics; and the University of California, Berkeley). The image is about 2 light-years across (see the 6-light-month scale bar). At the right side of the image, enlarged (see the 2-light-day scale bar), we see the orbit of the star known as S0–2, which has been followed since 1992, more than a complete orbit. This star comes within only about 17 light-hours of Sgr A*. Following Kepler's second law, the star moves much more quickly when close to Sgr A*. From Kepler's third law, the mass of the central object was measured, given the size and duration of the orbit. (b) Images made at the Keck telescope by the research group of Andrea Ghez (University of California, Los Angeles), working at 3.8 μm, showing the orbits of several stars in the central 1 square arcsecond around Sgr A*. All of the stars are seen to move over time compared with the background image. Advancing time is shown with increasing saturation of the colors of the symbols. In 2012, she and her colleagues reported that the orbital period of star S0–102, imaged with adaptive optics with the Keck telescopes, around the black hole is only 11.5 years, even shorter than the 16-year period of S0–2 (where the "S" stands for Sagittarius, the constellation through which we look toward the galactic center, and the following 0 means that the star is within the innermost zone around Sgr A*, that is, within 1 arcsec). Watching the "tango" of these two stars as they interact can help us understand what other material might be present besides the central black hole and so clarify the conclusions about that central black hole. At the stars' closest approach to the supermassive black hole (which for S0–2, 15 times brighter than S0–102, will occur in 2018), the curvature of space-time following Einstein's general theory of relativity will most strongly affect their motion. (c) An image at 3.8 μm made by the Ghez group, using a "laser guide star," a technique having the adaptive optics use a laser spot projected on the sky as a known image to allow calculation of the feedback needed for adjustment of the mirror shape. (d) Adaptive-optics images with the Very Large Telescope show the cloud of gas approaching Sgr A*, discovered in 2011. (1) and (2) The cloud is visible in only one of the two near-infrared bands, showing that it is a dusty cloud instead of a star. (3) The cloud's proper motion: 2004.5 (red), 2008.3 (green), and 2011.3 (blue). (4) The highly eccentric ($e = 0.94$) orbit of the cloud goes close to Sgr A* in 2013. (5) The cloud's velocity over the next few years, quite different from its past velocity, which is also shown.

(a)

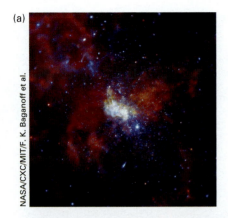

NASA/CXC/MIT/F. K. Baganoff et al.

(b)

NASA/MIT/Frederick K. Baganoff et al.

(c)

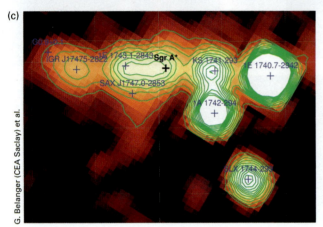

G. Belanger (CEA Saclay) et al.

■ **FIGURE 15–16** (a) A close-up view of the region of Sgr A* from the Chandra X-ray Observatory. This extended observation, to which Chandra was devoted for 1 million seconds (about 2 weeks), clearly shows the lobes of hot gas that erupted to over 10 light-years to the sides of the supermassive black hole. The long x-ray filaments are probably magnetic structures interacting with pulsar jets. Calculations show that the black hole eats only $1/10{,}000$ of the gas near it because a region just outside the event horizon provides outward pressure that turns the flow of gas away. (b) A mosaic from the Chandra X-ray Observatory showing x-ray emission from the Milky Way. (c) Emission from the Galactic center imaged with the European INTEGRAL observatory at energies near the boundary between x-rays and gamma rays. The grid marks are separated by 0.5°. The image shows an extended source around Sgr A*.

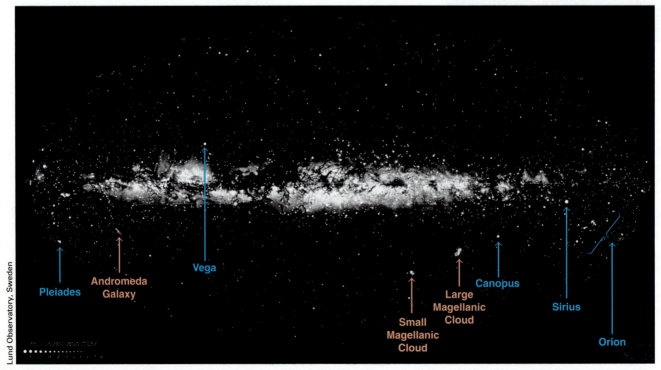

Lund Observatory, Sweden

■ **FIGURE 15–17** A drawing of the Milky Way as it appears to the eye, in visible light. Seven thousand stars plus the diffuse glow of the Milky Way are shown in this panorama.

most of which come from collisions between cosmic rays (see our discussion in Section 13.2f) and atomic nuclei in clouds of gas. More recently, the Fermi Gamma-ray Space Telescope has made additional gamma-ray maps of our Galaxy (Figure 15–11, bottom panel).

A different instrument on the Compton Gamma-Ray Observatory detected bursts of gamma rays that last only a few seconds or minutes (■ Fig. 15–18). These **gamma-ray bursts**, which were seen at random places in the sky roughly once per day, are especially intriguing. They were pinpointed with ESA's BeppoSax, which was able to turn rapidly toward the vague position from which a burst of gamma-rays was coming so that its x-ray telescope could image a higher-resolution view.) NASA's Swift satellite, mentioned in Sections 3.7a and 14.10a, was sent aloft in 2004 specifically to study them in detail.

Though some models suggested that the gamma-ray bursts were produced within our Galaxy (either very close to us or in a very extended halo), more recent observations have conclusively shown that most of them are actually in galaxies billions of light-years away. (It is possible that some of them are in our Galaxy, the result of sudden changes in the surface structure of extremely magnetic neutron stars known as "magnetars.") As we discussed in Chapter 14, these distant gamma-ray bursts may be produced when extremely massive stars collapse to form black holes, or when a neutron star merges with either another neutron star (forming a black hole) or with a stellar-mass black hole.

The Chandra X-ray Observatory is producing more detailed images of x-ray sources than had ever before been available. Studies of the highest-energy electromagnetic radiation, like x-rays and gamma rays, and of rapidly moving cosmic-ray particles (Section 13.2f) guided to some extent by the Galaxy's magnetic field, are part of the field of **high-energy astrophysics**. Riccardo Giacconi received a share of the 2002 Nobel Prize in Physics for his role in founding this field.

15.7 OUR PINWHEEL GALAXY

It is always difficult to tell the shape of a system from a position inside it. Think, for example, of being somewhere inside a maze of tall hedges; we would find it difficult to trace out the pattern. If we could fly overhead in a helicopter, though, the pattern would become very easy to see (■ Fig. 15–19). Similarly, we have difficulty tracing out the spiral pattern in our own Galaxy, even though the pattern would presumably be apparent from outside the Galaxy. Still, by noting the distances and directions to objects of various types, we can determine the Milky Way's spiral structure.

Young open clusters are good objects to use for this purpose, because they are always located in spiral arms. We think that they formed there and that they have not yet had time to move away (■ Fig. 15–20). We know their ages from the lengths of their main sequences on the temperature-luminosity diagram (Chapter 11). Also

useful are main-sequence O and B stars; the lives of such stars are so short we know they can't be old. But since our methods of determining the distances to open clusters, as well as to O and B stars, from their optical spectra and apparent brightnesses are uncertain to 10 per cent, they give a fuzzy picture of the distant parts of our Galaxy. Parallaxes measured from the Hipparcos spacecraft do not go far enough out into space to help in mapping our Galaxy. Gaia will soon do much better.

Another sign of young stars is the presence of emission nebulae. We know from studies of other galaxies that emission nebulae are preferentially located in spiral arms. In mapping the locations of emission nebulae, we are really again studying the locations of the O stars and the hottest of the B stars, since it is ultraviolet radiation from these hot stars that provides the energy for the nebulae to glow.

It is interesting to plot the directions to and distances of the open clusters, the O and B stars, and the clouds of ionized hydrogen as seen from Earth. When we do so, they appear to trace out bits of three spiral arms, which are relatively nearby.

Interstellar dust prevents us from using this technique to study parts of our Galaxy farther away from the Sun. However, another valuable method of mapping the spiral structure in our Galaxy involves spectral lines of hydrogen and of carbon monoxide in the radio part of the spectrum. Radio waves penetrate the interstellar dust, allowing us to study the distribution of matter throughout our Galaxy, though getting the third dimension (distance) that allows us to trace out spiral arms remains difficult. We will discuss the method later in this chapter.

15.8 WHY DOES OUR GALAXY HAVE SPIRAL ARMS?

The Sun revolves around the center of our Galaxy at a speed of approximately 200 kilometers per second. At this rate, it takes the Sun about 250 million years to travel once around the center, only 2 per cent of the Galaxy's current age. (Our Galaxy, after all, must be older than its globular clusters, whose age we discussed in Chapter 11.) But stars at different distances from the center of our Galaxy make a complete revolution in different lengths of time. (As we will see in Chapter 16, the Galaxy does not rotate like a solid disk.) For example, stars closer to the center revolve much more quickly than does the Sun. Thus the question arises: Why haven't the arms wound up very tightly, like the cream in a cup of coffee swirling as you stir it?

The leading current solution to this conundrum says, in effect, that the spiral arms we now see do not consist of the same stars that would previously have been visible in those arms. The spiral-arm pattern is caused by a **spiral density wave**, a wave of increased density that moves through the gas in the Galaxy. This density wave is a wave of compression, not of matter being transported. It rotates

Fermi team and NASA, courtesy of Lynn Cominsky, Sonoma State University

■ FIGURE 15–18 The distribution of 500 gamma-ray bursts recorded with NASA's Swift satellite. Surprisingly, the sources were found to be randomly distributed, so they seemed either very close to us or far outside our Galaxy. Studies with x-ray telescopes, the Hubble Space Telescope, ground-based telescopes, and now Swift and the Fermi Gamma-ray Observatory have revealed that most of the gamma-ray bursts are very distant, and hence are incredibly powerful. Fermi has recorded over 1,000 gamma-ray bursts. The background is an infrared sky image (2MASS survey), with our Galactic plane and the Large and Small Magellanic Clouds showing.

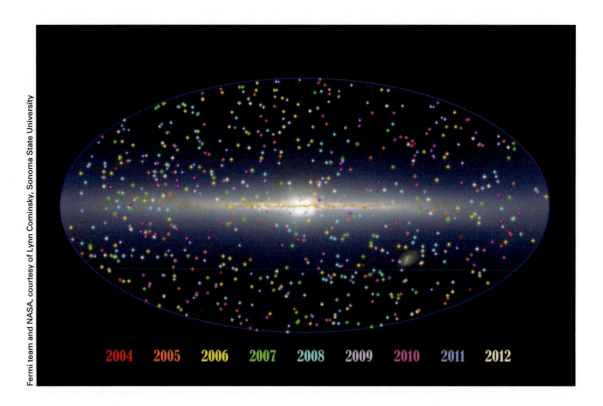

2004 2005 2006 2007 2008 2009 2010 2011 2012

more slowly than the actual material and causes the density of passing material to build up. Stars are born at those locations and appear to form a spiral pattern (■ Fig. 15–21), but the stars then move away from the compression wave.

Think of the analogy of a crew of workers fixing potholes in two lanes of a four-lane highway. A bottleneck occurs at the location of the workers; if we were in a traffic helicopter, we would see an increase in the number of cars at that place. As the workers continue

slowly down the road, fixing potholes in new sections, we would see what seemed to be the bottleneck moving slowly down the road. Cars merging from four lanes into the two open lanes need not slow down if the traffic is light, but they are compressed more than in other (fully open) sections of the highway. Thus the speed with which the bottleneck advances is much smaller than that of individual cars.

Similarly, in our Galaxy, we might be viewing only some galactic bottleneck at the spiral arms. The new, massive stars would ionize the

(a)

(b)

■ FIGURE 15–19 The view from the level of a maze, the "Galaxy Garden," located on the Big Island of Hawaii (a) makes it hard to figure out the pattern, which is readily visible from above (b). The plantings are an accurate representation of our Galaxy's shape, according to current models and calculations, with even a supermassive black hole (represented by a funnel-shaped fountain) at the center. The Sun and the nearest stars are marked on a single leaf.

■ **FIGURE 15–20** The Jewel Box, an open cluster of stars. The orange star is κ (the Greek letter "kappa") Crucis, a red supergiant (a cool star), while most of the other stars are bluish and therefore hot.

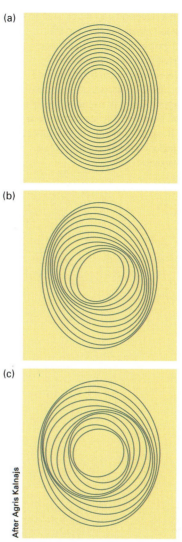

(a)

(b)

(c)

■ **FIGURE 15–21** Each part of the figure includes the same set of ellipses; the only difference is the relative alignment of their axes. Consider that the axes are rotating slowly and at different rates. The places where their orbits are close together take a spiral form, even though no actual spiral of physically connected points exists. The spiral structure of a galaxy may arise from an analogous effect.

interstellar gas so that it becomes visible. In fact, we do see young, hot stars and glowing gas outlining the spiral arms, providing a check of this prediction of the density-wave theory. This mechanism may work especially well in galaxies with a companion that gravitationally perturbs them (as seen in the opening image in Chapter 16).

15.9 MATTER BETWEEN THE STARS

The gas and dust between the stars is known as the **interstellar medium** or "interstellar matter." The nebulae represent regions of the interstellar medium in which the density of gas and dust is higher than average.

For many purposes, we may consider interstellar space as being filled with hydrogen at an average density of about 1 atom per cubic centimeter. (Individual regions may have densities departing greatly from this average.) Regions of higher density in which the atoms of hydrogen are predominantly neutral are called **H I regions** (pronounced "H one regions"; the Roman numeral "I" refers to the neutral, basic state). Where the density of an H I region is high enough, pairs of hydrogen atoms combine to form molecules (H_2). The densest part of the gas associated with the Orion Nebula might have a million or more hydrogen molecules per cubic centimeter. So hydrogen molecules (H_2) are often found in H I clouds.

A region of ionized hydrogen, with one electron missing from most of the atoms, is known as an **H II region** (from "H two," the second state – neutral is the first state and once ionized is the second). Since hydrogen, which makes up the overwhelming proportion of interstellar gas, contains only one proton and one electron, a gas of ionized hydrogen contains individual protons and electrons. Though all hydrogen is ionized in an H II region, only some of the helium is ionized. Heavier elements have sometimes lost 2 or even 3 electrons. Wherever a hot star provides photons having enough energy to ionize hydrogen, an H II region (emission nebula) results (■ Fig. 15–22 and ■ Fig. 15–23).

Studying the optical and radio spectra of H II regions and planetary nebulae tells us the abundances (proportions) of several of the

■ **FIGURE 15–22** (a) H I clouds are regions of neutral hydrogen of higher density than average, and H II regions are regions of ionized hydrogen. The protons and electrons that result from the ionization of hydrogen by ultraviolet radiation from a hot star, and the neutral hydrogen atoms, are shown schematically. The larger dots represent protons or neutrons, and the smaller dots represent electrons. The star that provides the energy for the H II region is shown. (b) The optical spectrum of an H II region in the galaxy NGC 4214. Emission lines are marked.

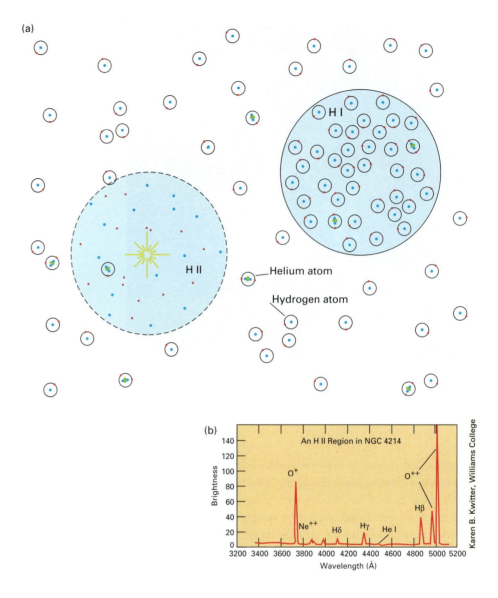

chemical elements (especially helium, nitrogen, and oxygen). How these abundances vary from place to place in our Galaxy and in other galaxies helps us choose between models of element formation and of galaxy evolution.

Tiny grains of solid particles are given off by the outer layers of red giants. They spread through interstellar space, and dim the light from distant stars. This "dust" never gets very hot, so most of its radiation is in the infrared. The radiation from dust scattered among the stars is faint and very difficult to detect, but the radiation coming from clouds of dust surrounding newly formed stars is easily observed from ground-based telescopes and from infrared spacecraft. They found infrared radiation from so many stars in our Galaxy that we think that about one star forms in our Galaxy each year.

Since the interstellar gas is often "invisible" in the visible part of the spectrum (except at the wavelengths of certain weak emission lines), different techniques are needed to observe the gas in addition to observing the dust. Radio astronomy is the most widely used technique, so we will now discuss its use for mapping our Galaxy.

15.10 RADIO OBSERVATIONS OF OUR GALAXY

The first radio astronomy observations were of continuous radiation; no spectral lines were known. If a radio spectral line is known, Doppler-shift measurements can be made, and we can determine motions in our Galaxy. But what is a radio spectral line? Remember that an optical spectral line corresponds to a wavelength of the optical spectrum that is more intense (for an emission line) or less intense (for an absorption line) than neighboring wavelengths. Similarly, a

(a)

ESO

(b)

Jeff Hester (Arizona State University) and NASA

■ **FIGURE 15–23** (a) The Trifid Nebula, M20, in Sagittarius, is the red H II region divided into (at least) three visible segments by absorbing dust lanes. The red Hα emission line is diluted with blue light scattered from the hot central stars. The blue reflection nebula at the top is unconnected to the Trifid. In the dark regions, the extinction (obscuration) is so high that we cannot see stars behind the dust causing the absorption. (b) A high-resolution Hubble Space Telescope view of about $1/4$ the diameter of the red nebula shown in part (a).

■ **FIGURE 15–24** An energy-level diagram showing that 21-cm radiation results from an energy difference between two sublevels in the lowest principal energy state of hydrogen. The energy difference is much smaller than the energy difference that leads to Lyman α (the Greek letter "alpha"), which was described in Chapter 2. So, because $E = hf = hc/\lambda$, the wavelength is much longer.

radio spectral line corresponds to a wavelength at which the radio radiation is slightly more, or slightly less, intense. A radio station is an emission line on a home radio.

Since hydrogen is by far the most abundant element in the Universe, the most-used radio spectral line is a line from the lowest energy levels of interstellar hydrogen atoms. This line has a wavelength of 21 cm. A hydrogen atom is basically an electron "orbiting" a proton. Both the electron and the proton have the property of

spin, as if each were spinning on its axis. The spin of the electron can be either in the same direction as the spin of the proton or in the opposite direction. The rules of quantum physics prohibit intermediate orientations. The energies of the two allowed conditions are slightly different.

If an atom is sitting alone in space in the upper of these two energy states, with its electron and proton spins aligned in the same direction, there is a certain small probability that the spinning electron will spontaneously flip over to the lower energy state and emit a bundle of energy – a photon (■ Fig. 15–24). We thus call this a **spin-flip transition** (■ Fig. 15–25). The photon of hydrogen's spin-flip transition corresponds to radiation at a wavelength of 21 cm – **the 21-cm line**. If the electron flips from the higher to the lower energy state, we have an emission line. If it absorbs energy from passing continuous radiation, it can flip from the lower to the higher energy state and we have an absorption line.

If we were to watch any particular group of hydrogen atoms in the slightly higher energy state, we would find that it would take about 11 million years before half of the electrons had undergone spin-flips; we say that the "half-life" is 11 million years for this transition. Thus, hydrogen atoms are generally quite content to sit in the upper state! But there are so many hydrogen atoms in space that enough 21-cm radiation is given off to be detected. The existence of the line was predicted in 1944 and discovered in 1951, marking the birth of spectral-line radio astronomy.

15.11 MAPPING OUR GALAXY

The 21-cm hydrogen line has proven to be a very important tool for studying our Galaxy (■ Fig. 15–26) because this radiation passes unimpeded through the dust that prevents optical observations very far into the plane of our Galaxy. It can even reach us from the

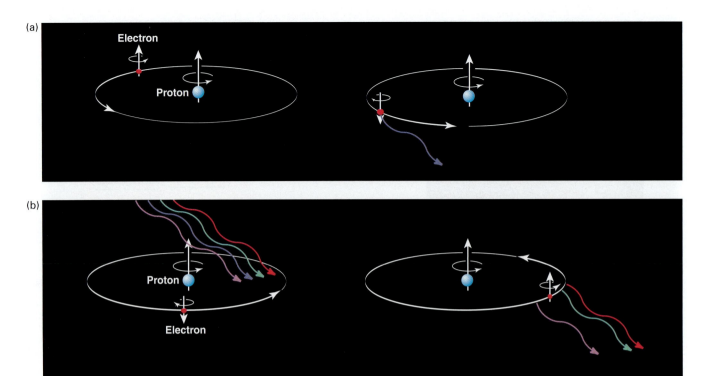

■ **FIGURE 15–25** In contrast to the energy-level diagram of the preceding figure, here we see what a spin-flip would look like if we could observe on such a submicroscopic level. (a) When the electron in a hydrogen atom flips over so that it is spinning in the opposite direction from the spin of the proton, an emission line at a wavelength of 21 cm results. To illustrate this effect, we show no rays coming in and the one ray emitted by the spontaneous spin-flip going outward (toward lower right). Note from the arrows the change in the direction of the electron's spin. (b) When an electron takes energy from a passing beam of radiation, causing it to flip from spinning in the direction opposite to that of the proton's spin to spinning in the same direction, then a 21-cm line in absorption results. We illustrate this effect schematically by showing four rays coming in from the upper left, but since only the one of them at the wavelength that matches the spontaneous spin-flip shown at the top causes the spin to flip back, only three rays go out. Actually, the spin-flip takes place only for a wavelength of 21 cm, so the ray has no actual color perceptible to our eyes; we arbitrarily use blue to show it. Physicists often show spin by wrapping their right hands around the spinning object in the direction of spin with their thumbs up. The direction of the thumb is called the direction (orientation) of the spin axis. We show the spin axes defined in this way as vertical lines with arrows.

opposite side of our Galaxy, whereas light waves penetrate the dust clouds in the Galactic plane only about 10 per cent of the way to the Galactic center, on average.

Astronomers have ingeniously been able to find out how far it is to the clouds of gas that emit the 21-cm radiation. They use the fact that gas closer to the center of our Galaxy rotates with a shorter period than the gas farther away from the center. Though there are substantial uncertainties in interpreting the Doppler shifts in terms of distance from the Galaxy's center, astronomers have succeeded in making some maps. These maps show many narrow arms but no

■ **FIGURE 15–26** A map of our Galaxy's 21-cm radiation, which comes from interstellar neutral hydrogen, with all gas at velocities between –400 km/s and +400 km/s included. To cover the whole sky, a Leiden/Dwingeloo Survey (The Netherlands) of the sky north of declination –30° was merged with the Instituto Argentino de Radioastronomia Survey of the southern sky. The Galactic center is in the middle of this Aitoff projection.

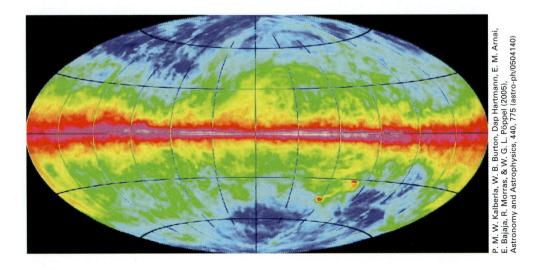

P. M. W. Kalberla, W. B. Burton, Dap Hartmann, E. M. Arnal, E. Bajaja, R. Morras, & W. G. L. Pöppel (2005), Astronomy and Astrophysics, 440, 775 (astro-ph/0504140)

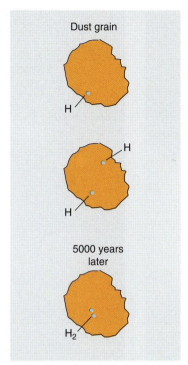

Dust grain

H

H

H

5000 years
later

H$_2$

■ **FIGURE 15–27** Hydrogen molecules are formed in space with the aid of dust grains at an intermediate stage. Terrestrial laboratory research reported in 2005 showed that irregularities on the dust grains appear to be needed to form the molecules.

Dan P. Clemens (Boston University), D. Sanders (University of Hawaii), and N. Scoville (Caltech)

■ **FIGURE 15–28** A map based on CO data does not show clear spiral arms. The point at (0, 0) on the graph marks the center of our Galaxy. The regions 4–6 kpc and near 7 kpc are more ring-like than spiral. Studies of other galaxies are showing that clear spiral arms are common in the outer regions of galaxies but not necessarily in the inner regions, which may be the case with our Galaxy as well.

clear pattern of a few broad spiral arms like those we see in some other galaxies (Chapter 16). The question emerged: Is our Galaxy really a spiral at all? With the additional information from studies of molecules in space that we describe in the next section, we finally made further progress.

15.12 RADIO SPECTRAL LINES FROM MOLECULES

Radio astronomers had only the hydrogen 21-cm spectral line to study for a dozen years, and then only the addition of one other group of lines for another five years. Then radio spectral lines of water (H$_2$O) and ammonia (NH$_3$) were found. The spectral lines of these molecules proved surprisingly strong, and were easily detected once they were looked for. Over 100 additional types of molecules have since been found.

The earlier notion that it would be difficult to form molecules in space was wrong. In some cases, atoms apparently stick to interstellar dust grains, perhaps for thousands of years, and molecules build up (■ Fig. 15–27). Though hydrogen molecules form on dust grains, many or most of the other molecules may be formed in the interstellar gas, or in the atmospheres of stars, without need for grains.

Studying the spectral lines provides information about physical conditions – temperature, densities, and motion, for example – in the

gas clouds that emit the lines. Studies of molecular spectral lines have been used together with 21-cm line observations to improve the maps of the spiral structure of our Galaxy (■ Fig. 15–28). Observations of carbon monoxide (CO), in particular, have provided better information about the parts of our Galaxy farther out from the Galaxy's center than our Sun. We use the carbon monoxide as a tracer of the more abundant hydrogen molecular gas, since the carbon monoxide produces a far stronger spectral line and is much easier to observe; molecular hydrogen emits extremely little.

15.13 THE FORMATION OF STARS

We have already discussed (in Chapter 12) some of the youngest stars known and how stars form. Here we will discuss star formation in terms of the gas and dust from which stars come. Astronomers have found that **giant molecular clouds** are fundamental building blocks of our Galaxy. Giant molecular clouds are 50 to 300 light-years across. There are a few thousand of them in our Galaxy. The largest giant molecular clouds contain about 100,000 to 1,000,000 times the mass of the Sun, though smaller ones go down to 10,000 solar

■ **FIGURE 15–29** (a) A visible-light image of the Orion Nebula, an H II region on the side of a molecular cloud. (b) A view of Orion taken with an infrared array. Three infrared wavelengths known as J, H, and K (1.25, 1.6, and 2.2 micrometers, respectively; out of alphabetical order for historical reasons) were used to make this false-color view.

(a) (b)

Ian Gatley (NOAO)

Ian Gatley (NOAO)

masses. Since giant molecular clouds break up to form stars, they only last 10 million to 100 million years.

Most radio spectral lines seem to come only from the molecular clouds. (Carbon monoxide is the major exception, for it is widely distributed across the sky.) Infrared and radio observations together have provided us with an understanding of how stars are formed from these dense regions of gas and dust. Carbon-monoxide observations reveal the giant molecular clouds, but it is molecular hydrogen (H_2) rather than carbon monoxide that contains a vast majority of the mass.

Many radio spectral lines have been detected only in a particular cloud of gas, the Orion Molecular Cloud. It is located close to a visible part, which we call the Orion Nebula, of a larger cloud of gas and dust. The Orion Molecular Cloud contains about 500 thousand times the mass of the Sun. It is relatively accessible to our study because it is only about 1,500 light-years away. Even though less than 1 per cent of the Cloud's mass is dust, that is still a sufficient amount of dust to prevent ultraviolet light from nearby stars from entering and breaking the molecules apart. Thus molecules can accumulate.

The properties of the molecular cloud can be deduced by comparing the radiation from its various molecules and by studying the radiation from each molecule individually. The average density is a few hundred to a thousand particles per cubic centimeter, but the cloud center may have up to a million particles per cubic centimeter. This central region is still billions of times less dense than our Earth's atmosphere, though it is much denser than the typical interstellar density of about 1 particle per cubic centimeter.

We know that young stars are found in the center of the Orion Nebula (■ Fig. 15–29). The Trapezium (■ Fig. 15–30; see also

C. R. O'Dell (Rice University) and NASA

■ **FIGURE 15–30** Part of the Orion Nebula only 1.6 light-years across diagonally, imaged with the Hubble Space Telescope with much higher detail than is possible from the ground. One of the Trapezium stars that provides energy for the nebula to glow is at top left. In the image, nitrogen emission is shown as red, hydrogen emission is shown as green, and oxygen emission is shown as blue. Astrometric measurements show the brightest object (see Figure 15–32) and two radio sources in the Trapezium moving away from the location where they were 500 years ago. They and other "runaway objects" may be young stars emerging from a multiple-star system that was disrupted by internal gravitational interactions.

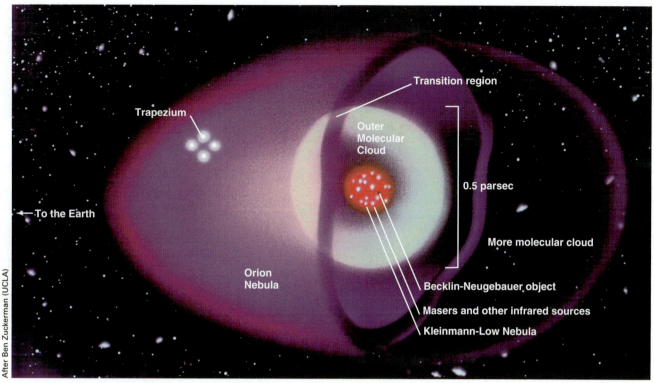

After Ben Zuckerman (UCLA)

FIGURE 15–31 A model of the structure of the Orion Nebula and the Orion Molecular Cloud.

Figure 15–4), a group of four hot stars readily visible in a small telescope, is the source of ionization and energy for the Orion Nebula. The Trapezium stars are relatively young, about 100,000 years old. The Orion Nebula, though prominent at visible wavelengths, is but an H II region located along the near side of the much more extensive molecular cloud (■ Fig. 15–31).

The Near-Infrared Camera and Multi-Object Spectrometer (NICMOS) on the Hubble Space Telescope is able to record infrared light that had penetrated the dust, bringing us images of newly formed stars within the Orion Molecular Cloud (■ Fig. 15–32). The Herschel Space Observatory was also used to observe this region.

WFPC2 Image: C. Robert O'Dell and Shui Kwan Wong (Rice University) and NASA; NICMOS image: Rodger Thompson, Susan Stolovy, and Marcia Rieke (University of Arizona), Glenn Schneider (Steward Obs., University of Arizona); Edwin Erickson (SETI Institute/ NASA's Ames Research Center); David Axon (STScI); and NASA

FIGURE 15–32 (a) Optical and (b) infrared views of the heart of the Orion Molecular Cloud, obtained with instruments (WFPC2 and NICMOS, respectively) on the Hubble Space Telescope. Most of the stars revealed in the infrared image were hidden from optical view; they are in a chaotic, active region of star birth. Stars and dust that has been heated by the intense starlight are shown as yellow-orange. Emission from excited hydrogen molecules is shown in blue. The brightest region is the "Becklin-Neugebauer object" (BN), a massive, young star. Outflowing gas from BN causes the bow shock above it. The view is about 0.4 light-years across the diagonal; details as small as our Solar System are visible.

FIGURE 15–33 The heart of the Jansky Very Large Array, near Socorro, New Mexico, with its dishes in its most compact configuration. One of us (J.M.P.) and colleagues used it to observe the Sun at high resolution during the 2012 annular solar eclipse. A similar Giant Meterwave Radio Telescope, a set of radio telescopes spaced over many kilometers, is in operation near Pune, India.

15.14 AT A RADIO OBSERVATORY

What is it like to go observing at a radio telescope? First, you decide just what you want to observe, and why. You have probably worked in the field before, and your reasons might tie in with other investigations under way. Then you decide with which telescope you want to observe, usually the most suitable one accessible to you; let us say it is the Jansky Very Large Array (VLA) of the National Radio Astronomy Observatory. You send in a written proposal describing what you want to observe, and why. You might be working with colleagues who are especially experienced in the field of research or with the telescope itself.

Your proposal is read by a panel of scientists. If the proposal is approved, it is placed in a queue to wait for observing time. You might be scheduled to observe for a five-day period to begin six months after you submitted your proposal.

At the same time, you might apply (usually to the National Science Foundation) for financial support to carry out the research. Your proposal possibly contains requests for some salary for yourself during the summer, and salary for a student or students to work on the project with you. You are not charged directly for the use of the telescope itself – that cost is covered in the observatory's overall budget.

You ordinarily carry out your observing at the JVLA headquarters at Socorro, New Mexico, though for the 2012 annular eclipse we were actually at the telescope site itself. A trained telescope operator runs the mechanical aspects of the telescope. A day or more before the actual observing you give the telescope operator a computer program that includes the coordinates of the points in the sky that you want to observe and how long to dwell at each location. The telescopes (■ Fig. 15–33) operate around the clock – one doesn't want to waste any observing time.

FIGURE 15–34 The sum of many partial maps shows the Arc near the center of our Galaxy. (See it also, at a shorter wavelength, in Figure 15–14.)

The electronics systems that are used to treat the incoming signals collected by the radio dishes are particularly advanced, especially after the upgrade that made the VLA into the JVLA. Computers combine the output from the 27 telescopes and may show you a color-coded image, with each color corresponding to a different brightness level (■ Fig. 15–34). Standard image-processing packages of programs are available for you to use back home, with the radio community generally using a different package from that used in the optical community.

You are expected to publish the results as soon as possible in one of the scientific journals, often after you have given a presentation about the results at a professional meeting, such as one of those held twice yearly by the American Astronomical Society.

Astronomy has become a very collaborative science. Many consortia of individual scientists, such as those studying distant supernovae,

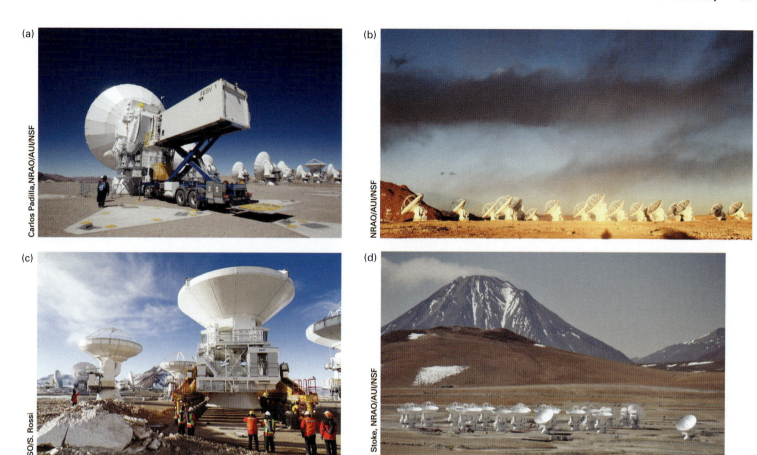

(a)

Carlos Padilla, NRAO/AUI/NSF

(b)

NRAO/AUI/NSF

(c)

ESO/S. Rossi

(d)

J. Stoke, NRAO/AUI/NSF

■ **FIGURE 15–35** A set of 66 linked telescopes, the Atacama Large Millimeter/submillimeter Array (ALMA), is being set up on the Chajnantor Plateau, a high, dry plain in Chile. It is a major international collaboration of countries in North America, Europe, and East Asia. ALMA's Technical Building, at 5,050 m (over 16,500 feet) of altitude, is the second highest building on Earth that is made of steel and concrete, the highest being a railway station on the China-Tibet railway. (a) A servicing vehicle installing a cryostat in an antenna. (b) Seventeen of the first antennas at the site, which is at about 5,000 m (16,500 feet) of altitude. (c) The first of the European antennas to be installed. (d) An overview.

have dozens of members. Telescope projects have also become so huge that collaboration is necessary. The Atacama Large Millimeter/submillimeter Array (ALMA), being completed in Chile on a high plain where it hasn't rained in decades (■ Fig. 15–35), will use 66 high-precision radio telescopes as an interferometer to examine our Galaxy and other celestial objects with high resolution. It is a joint project of the United States' National Science Foundation, the European Space Agency, Chile, and some East Asian countries.

Scientific observations are under way with ALMA (15–36). For example, the image shows the first direct observations of huge streams of gas flowing across a gap in the disk of gas and dust surrounding a young star.

(a)

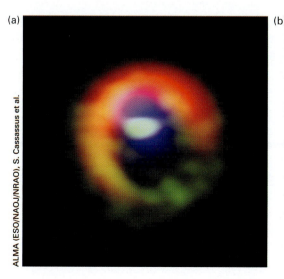

ALMA (ESO/NAOJ/NRAO), S. Cassassus et al.

(b)

■ **FIGURE 15–36** A sample of early observations from ALMA (Atacama Large Millimeter/submillimeter Array), released in 2013. (a) The ALMA data show a two-light-year-wide dust disk around the young star HD 142527 (in red) with dense gas (in green), HCO⁺, flowing across a gap in the dust disk. The streams may be a sign that giant planets are growing in this system. Diffuse gas, CO, in the central gap also shows (in blue). (b) An artist's conception of the disk and streams. Note that if this object were our own Solar System, the Voyager 1 spacecraft, the farthest object ever sent from Earth, would be at about the inner edge of the ring.

CONCEPT REVIEW

» Gas and dust (small particles of matter) are present to some extent throughout a galaxy, between the stars; a **nebula** (plural **nebulae**) is a substantial cloud of such gas and dust (introductory paragraph). At night, from a good observing location, we see a band of stars, nebulae, gas, and dust – the **Milky Way** – stretch across the sky (Section 15.1). It is our internal, edge-on view of our Galaxy, the **Milky Way Galaxy**. With the unaided eye, we cannot see more than a few thousand light-years through the Milky Way because of the large amount of dust; thus, it appears that we are close to the center of our Galaxy, but this is only an illusion (Section 15.2).

» **Emission nebulae** glow because ultraviolet radiation from hot stars ionizes the gas, and electrons jumping down to lower energy levels emit light (Section 15.3). **Absorption nebulae** (**dark nebulae**) block radiation that comes from behind them. **Reflection nebulae**, like those near the stars of the Pleiades, reflect radiation; they often look blue, for the same reason that the sky is blue.

» By studying the distribution of globular star clusters in the sky, Harlow Shapley deduced that we are not at the center of our Galaxy (Section 15.4). However, he overestimated our distance from the center because he didn't know about interstellar extinction, the dimming of starlight by dust. We now can measure this effect by noticing that it also **reddens** starlight: Dust preferentially scatters (reflects) or absorbs the violet and blue light, while the longer wavelengths pass through more easily.

» Our Galaxy has a **nuclear bulge** centered on the **nucleus**, and surrounded by a flat **disk** that contains **spiral arms** (Section 15.4). A spherical **halo** includes the globular clusters, and has a much greater diameter than the disk. We can detect the very center of our Galaxy in infrared light, radio waves, x-rays, or gamma rays that penetrate the dust between us and it (Section 15.5). The Galactic center is a bright infrared source and almost certainly contains a very massive black hole, about 4 million solar masses.

» Although mysterious **gamma-ray bursts** were at one time considered to be in our Galaxy, we now know that most of them originate billions of light-years away, probably from the formation of stellar-mass black holes (Section 15.6). NASA has sent the Swift satellite aloft to study them. The gamma rays and x-rays throughout our Galaxy and elsewhere are studied as part of **high-energy astrophysics**.

» Our Galaxy is a pinwheel-shaped spiral galaxy, with spiral arms marked by massive stars, open clusters, and nebulae (Section 15.7). However, from our vantage point inside the Galaxy, it is difficult to accurately trace out the arms. These spiral arms appear to be caused by a slowly rotating **spiral density wave** (Section 15.8).

» The matter between the stars, the **interstellar medium**, is mainly hydrogen gas (Section 15.9). Emission nebulae are mostly **H II regions**, regions of ionized hydrogen. Clouds of higher density in which the atoms of hydrogen are predominantly neutral are called **H I regions**. Hydrogen molecules (H_2) are very difficult to detect, but we think they are plentiful in regions where they are protected by dust from ultraviolet radiation.

» Observations of our Galaxy at radio wavelengths have been very important (Section 15.10). Specifically, the observed **21-cm line** comes from the **spin-flip transition** of hydrogen atoms, when the spin of the electron changes relative to that of the proton. Studies of the 21-cm line have enabled us to map our Galaxy by finding the distances to H I regions (Section 15.11). Observations of interstellar molecules, primarily carbon monoxide, have also been valuable in this regard (Section 15.12).

» **Giant molecular clouds**, containing a hundred thousand to a million times the mass of the Sun, are fundamental building blocks of our Galaxy; they are the locations at which new stars form (Section 15.13). Infrared satellites and radio telescopes have permitted mapping of the Orion Molecular Cloud and others. The Atacama Large Millimeter/submillimeter Array is a major international project to provide high-resolution images of giant molecular clouds and other objects at radio wavelengths (Section 15.14).

QUESTIONS

1. Why do we think our Galaxy is a spiral?

2. How would the Milky Way appear if the Sun were closer to the edge of our Galaxy?

3. Compare (a) absorption (dark) nebulae, (b) reflection nebulae, and (c) emission nebulae.

4. How can something be both an emission and an absorption nebula? Explain and give an example.

5. If you see a red nebula surrounding a blue star, is it an emission or a reflection nebula? Explain.

6. Discuss the key observations that led to the discovery of the Sun's location relative to the center of our Galaxy.

†7. If the Sun is 8 kpc from the center of our Galaxy and it orbits with a speed of 200 km/s, show that the Sun's orbital period is about 250 million years. (Assume the orbit is circular.)

8. Why may some infrared observations be made from mountain observatories while all x-ray observations must be made from space?

9. Describe infrared and radio results about the center of our Galaxy.

10. Discuss the possible explanations for gamma-ray bursts in the sky.

11. Why does the density-wave theory of spiral arms lead to the formation of stars?

12. (a) What are three tracers that we use for the spiral structure of our Galaxy? (b) What are two reasons why we expect them to trace spiral structure?

13. Discuss how infrared observations from space have added to our knowledge of our Galaxy.

14. While driving at night, you almost instinctively judge the distance of an oncoming car by looking at the apparent brightness of its headlights. Is your estimate correct if you don't account for fog along the way? Discuss your answer.

15. Describe the relation of hot stars to H I (neutral hydrogen) and H II (ionized hydrogen) regions.

16. Briefly define and distinguish between the redshift of a gas cloud and the reddening of that cloud.

17. What determines whether the 21-cm lines will be observed in emission or absorption?

18. Describe how a spin-flip transition can lead to a spectral line, using hydrogen as an example.

†19. Suppose the rest wavelength of the 21-cm line of hydrogen were exactly 21.0000 cm. (a) If this line from a particular cloud is observed at a wavelength of 21.0021 cm, is the cloud moving toward us or away from us? (b) How fast is it moving?

20. Why are dust grains important for the formation of some types of interstellar molecules?

21. Describe the relation of the Orion Nebula and the Orion Molecular Cloud.

22. Optical astronomers can observe only at night (except when studying the Sun). In what time period can radio astronomers observe? Explain any difference.

23. **True or false?** By measuring the wavelength of the peak of the reflected light received from a reflection nebula, we can determine the temperature of the nebula using Wien's law.

24. **True or false?** Globular clusters form a spherical "halo" around the center of the Milky Way Galaxy, and their observed positions suggest that our Solar System is far from the center of our Galaxy.

25. **True or false?** A dark (absorption) nebula blocks the light from background stars, and is sometimes so dense that new stars are forming within it.

26. **True or false?** An emission nebula can form when gas is ionized by a nearby young star, often of spectral type G, K, or M.

27. **True or false?** When the relative spin directions of the proton and electron in a hydrogen atom change, a photon having a wavelength of about 21 cm is either emitted or absorbed.

28. **True or false?** The spiral arms in a galaxy such as the Milky Way consist of the same groups of stars throughout the entire lives of these arms.

29. **True or false?** Direct observations of the trajectories of stars travelling near the center of our Milky Way Galaxy provide very strong evidence for the presence of a supermassive black hole, millions of times the mass of our Sun.

30. **True or false?** Interstellar dust and gas tend to absorb and scatter blue light more than red light, causing stars to appear redder than their true colors.

31. **Multiple choice:** Gaseous emission nebulae in the Milky Way Galaxy look red because **(a)** they are moving away from us, so that the light is redshifted; **(b)** many electrons are jumping from the third to the second energy levels of hydrogen, producing Hα emission; **(c)** they absorb red light from their surroundings; **(d)** they have temperatures of only about 100 K, and Wien's law tells us that the light they emit is therefore red; or **(e)** they are made mostly of iron compounds, like rust.

32. **Multiple choice:** In 1917, our perception of the Milky Way Galaxy changed when Harlow Shapley noticed **(a)** a high concentration of neutron stars in the halo of our Galaxy; **(b)** pulsars concentrated around the region of our Galaxy surrounding the Sun; **(c)** open star clusters located around the center of our Galaxy; **(d)** a concentration of other planetary systems near the edges of our Galaxy; or **(e)** globular star clusters centered around a point far from the Sun.

33. **Multiple choice:** At the present time, stars in our Galaxy tend to form most readily in **(a)** giant molecular clouds in spiral arms; **(b)** the Galactic halo; **(c)** the central supermassive black hole; **(d)** the Galactic bulge; or **(e)** globular clusters.

34. **Multiple choice:** Which one of the following statements about our Milky Way Galaxy is *false*? **(a)** New stars generally form in the spiral arms. **(b)** Nebulae such as the Orion Nebula are stellar nurseries – regions where new stars are forming, or recently formed. **(c)** It is difficult to see the central regions of our Galaxy in optical (visible) light because intervening dust absorbs and scatters the light. **(d)** Rapid motions of stars near the center suggest that it harbors a black hole, millions of times the mass of our Sun.

(e) Globular star clusters reside in the halo and contain main-sequence stars spanning all spectral types, from O through M.

35. **Multiple choice:** The 21-cm line observed by radio astronomers comes from **(a)** electrons in hydrogen atoms jumping from the third to the second energy levels; **(b)** the rotation of hydrogen molecules; **(c)** the atomic hydrogen spin-flip transition; **(d)** dust grains in molecular clouds; or **(e)** carbon monoxide (CO) molecules.

36. **Multiple choice:** Which one of the following statements concerning the Milky Way Galaxy is *false*? **(a)** It is flattened out into a disk because of its rotation. **(b)** It is a spiral galaxy, probably of the "barred" variety. **(c)** Old globular star clusters are found primarily in its halo. **(d)** Open star clusters and emission nebulae are found primarily in its spiral arms. **(e)** Most of it can be seen from Earth at optical wavelengths with a small telescope.

37. **Fill in the blank:** To form some types of molecules in interstellar space, grains of _____ appear to be necessary.

38. **Fill in the blank:** A(n) _____ nebula can form when gas is ionized by a nearby young star, usually of spectral type O.

39. **Fill in the blank:** The most distant parts of our Galaxy are most easily seen at _____ wavelengths.

40. **Fill in the blank:** A(n) _____ nebula reflects (scatters) light from stars that are near the gas.

41. **Fill in the blank:** Studies of the motions of stars near the center of our Galaxy suggest that a massive _____ is present there.

42. **Fill in the blank:** Clouds of ionized hydrogen are called _____.

†This question requires a numerical solution.

TOPICS FOR DISCUSSION

1. Can Harlow Shapley's conclusion regarding the Sun's location in our Galaxy be considered an extension of the Copernican revolution?

2. If the band of light called the Milky Way stretched only halfway around the sky, forming a semicircle rather than a circle, what would you conclude about the Sun's location in our Galaxy?

NGC 2071 NGC 2068

NGC 2024 NGC 2023 Horsehead Nebula

ar-infrared

  herschel

IC 434

Alnitak

NGC 2024 NGC 2023 Horsehead Nebula

ar-infrared

  hubble space telescope

Visible

  dss2

gion of the Horsehead Nebula in the constellation Orion, a NASA/ESA near-infrared Hubble Space Telescope image together with a wide-field view in the far-infrared taken with one of the Herschel's camera. A visible-light image from the Digitized Sky Survey 2 also appears. The star Altnitak that shows is the most easterly star in Orion's belt. Several reflection nebulas also

I: ESA/Herschel/PACS, SPIRE/N. Schneider, Ph. André, V. Könyves (CEA Saclay, France) for the "Gould Belt survey" Key Programme, N. Schneider, Ph. André, V. Könyves (CEA Saclay); Hubble: NASA, ESA, and the Hubble Heritage Team (AURA/ DSS2: ESO/Digitized Sky Survey 2, D. De Martin

160 µm   100 µm   70 µm

An infrared image of the Whirlpool Galaxy, M51, taken with the European Space Agency's Herschel Space Observatory. (See also an optical image from the Hubble Space Telescope as Figure 15–10b, a shorter-wavelength infrared image as Figure 16–1a, and an x-ray image as Figure 16–1b.) It uses false color to show radiation at three different infrared wavelengths, as indicated individually at bottom. These images are at much longer infrared wavelengths than had been imaged with the Spitzer Space Telescope (Figure 16–1b). The Whirlpool consists of a main spiral galaxy and a smaller companion with which the spiral is gradually merging. Note that the companion has little infrared emission, showing that its stars are older. The ongoing collision is thought to have caused (or influenced) the spiral structure and to be triggering star formation. This pair of galaxies is in the constellation Canes Venatici, the Hunting Dogs, and is 31 million light-years away from Earth.

16

A Universe of Galaxies

ORIGINS

We find that galaxies, of which many billions are known to exist, are the fundamental units of the Universe. Essentially all stars, including our Sun, are within them. Understanding the birth and lives of galaxies is necessary for a complete picture of our existence.

At the beginning of the 20th century, the nature of the faint, fuzzy "spiral nebulae" was unknown. In the mid-1920s, Edwin Hubble showed that they are distant galaxies like our own Milky Way Galaxy (see the Chapter Opener), and that the Universe is far larger than was previously thought. Galaxies are the fundamental units of the Universe, just as stars are the basic units of galaxies.

Like stars, many galaxies are found in clusters, and there are also superclusters of galaxies separated by enormous voids. By looking back in time at very distant galaxies and clusters, we can study how they formed and evolved. Surprisingly, we now know that all these enormous structures consist largely of "dark matter" that emits little or no electromagnetic radiation.

AIMS

1. Learn about the discovery of galaxies, and become familiar with the different types of galaxies and their habitats (Sections 16.1 to 16.3).

2. Investigate dark matter and how pervasive it is (Section 16.4).

3. Discover how gravitational lensing can tell us about both luminous and dark matter (Section 16.5).

4. Understand how astronomers can study the birth and lives of galaxies by looking far back in time (Section 16.6).

5. Introduce the expansion of the Universe (Section 16.7).

6. Explore the evolution of galaxies and large-scale structure (Sections 16.8 to 16.10).

NASA/JPL–Caltech/R. Kennicut (University of Arizona)/DSS

NASA/CXC/SAO/R. DiStefano et al.

■ **FIGURE 16–1** (a) M51, the Whirlpool Galaxy, also displayed in the Chapter Opener, in a Spitzer Space Telescope short-infrared set of four wavelengths: 3.6 μm (reproduced blue), 4.5 μm (reproduced green), 5.8 μm (reproduced orange), and 8.0 μm (reproduced red). The first two of these wavelengths show mainly stars, while the second two show mainly interstellar dust. Most of the dust is made of carbon-based organic molecules, and marks locations of future star formation. The thin, spoke-like filaments linking spiral arms in the infrared image were unexpected and are unexplained. (b) A Chandra X-ray Observatory view of M51. This false-color image shows the more energetic x-rays as blue, medium-energy x-rays as green, and lower-energy x-rays as red. A number of x-ray sources are visible, with temperatures between 1 and 4 million kelvins.

16.1 THE DISCOVERY OF GALAXIES

In the 1770s, the French astronomer Charles Messier was interested in discovering comets. To do so, he had to be able to recognize whenever a new fuzzy object (a candidate comet) appeared in the sky. To minimize possible confusion, he thus compiled a list of about 100 diffuse objects that could always be seen, as long as the appropriate constellation was above the horizon. Some of them are nebulae, and others are star clusters, which can appear fuzzy through a small telescope such as that used by Messier. To this day, the objects in Messier's list are commonly known by their Messier numbers (■ Figs. 16–1 and 16–2). They are among the brightest and most beautiful objects in the sky visible from mid-northern latitudes. A set of photographs of all the Messier objects appears as Appendix 7.

Other astronomers subsequently compiled additional lists of nebulae and star clusters. By the early part of the 20th century, several thousand nebulae and clusters were known. The nebulae were especially intriguing: Although some of them, such as the Orion Nebula (■ Fig. 16–3), seemed to be clouds of gas clearly associated with bright stars in our Milky Way Galaxy, the nature of others was more controversial. When examined with the largest telescopes then available, many of them showed traces of spiral structure, like pinwheels, but no obvious stars (■ Fig. 16–4).

16.1a The Shapley-Curtis Debate

Some astronomers thought that these so-called **spiral nebulae** were merely in our own Galaxy, while others suggested that they were very far away – "island universes" in their own right, so distant that the individual stars appeared blurred together. (The name "island universes" had originated with the philosopher Immanuel Kant in 1755, though at that time there was no evidence for the existence of galaxies outside our own.) The distance and nature of the spiral nebulae was the subject of the well-publicized "**Shapley-Curtis debate**," held in 1920 between the astronomers Harlow Shapley and Heber Curtis. Shapley argued that the Milky Way Galaxy is larger than had been thought, and could contain such spiral-shaped clouds of gas. Curtis, in contrast, believed that they are separate entities, far beyond the outskirts of our Galaxy. This famous debate is an interesting example of the scientific process at work.

Shapley's wrong conclusion was based on rather sound reasoning but erroneous measurements and assumptions. For example, one distinguished astronomer thought he had detected the slight angular rotation of a spiral nebula, and Shapley correctly argued that this change would require a preposterously high physical rotation speed if the nebula were very distant. It turns out, however, that the measurement was faulty. Shapley also argued that a bright nova that had appeared in 1885 in the Andromeda Nebula, M31, the largest and brightest spiral nebula in the sky (■ Fig. 16–5), would be far more

(a)

M82

Holmberg IX

M81

Ultraviolet
GALEX

Visible ▪ *HST*

(b)

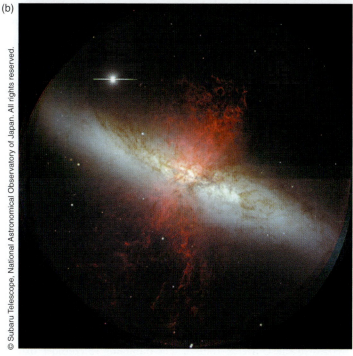

■ **FIGURE 16–2** (a) M81 and M82 in the constellation Ursa Major, two nearby galaxies in Charles Messier's catalogue of fuzzy, comet-like objects, imaged in the ultraviolet with NASA's Galaxy Explorer (GALEX) spacecraft. The blowup of the box shows some clusters at an unusual place, between the two galaxies and no doubt left over from their collision, given the clusters' age of 200 million years from their H–R diagram. These "blue blobs" are known as Arp's Loop. (b) A close-up of M82 from the 8-m Japanese national telescope, Subaru, on Mauna Kea. Extending outward from the galaxy's center are filaments of gas, seen in the red Hα light of hydrogen, tracing hydrogen that is forced outward by a superwind from numerous supernovae. The horizontal streak extending from the bright star is the result of an overloaded pixel in the CCD camera used to make the image.

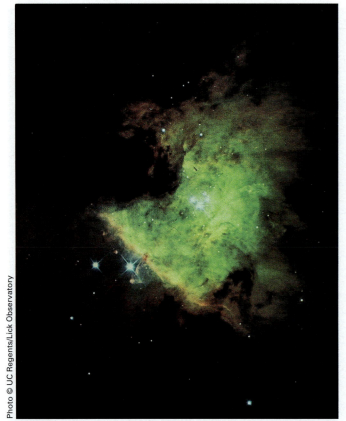

Photo © UC Regents/Lick Observatory

■ **FIGURE 16–3** The central region of the Orion Nebula in our Milky Way Galaxy, which is powered by the cluster of hot, massive stars. Green shows ionized oxygen.

NASA and The Hubble Heritage team (STScI/AURA); G. R. Meurer and T. M. Heckman (JHU), C. Leithere, J. harris, and D. Calzetti (STScI), and M. Sirianni (JHU)

■ **FIGURE 16–4** NGC 3310. Individual stars can be seen in this Hubble Space Telescope image, which is clearer than those from ground-based telescopes.

powerful than any previously known nova if it were very distant. Unfortunately, the existence of supernovae (this object turned out to be one), which are indeed more powerful than any known nova, was not yet known.

On the other hand, Curtis's conclusion that the spiral nebulae were external to our Galaxy was based largely on an incorrect notion of our Galaxy's size; his preferred value was much too small. Moreover, he treated the nova in Andromeda as an anomaly.

16.1b Galaxies: "Island Universes"

The matter was dramatically settled in 1923, when observations made by **Edwin Hubble** (■ Fig. 16–6) at the Mount Wilson Observatory in California proved that the spiral nebulae were indeed "island universes" (now called galaxies) well outside the Milky Way Galaxy. Using the 100-inch (2.5-m) telescope, Hubble discovered very faint Cepheid variable stars in several objects, including the Andromeda Nebula.

As we saw in Chapter 11, Cepheids are named after their prototype, the variable star δ (the Greek letter "delta") Cephei. Their light curves (brightness vs. time) have a distinctive, easily recognized shape. Cepheids are intrinsically very luminous stars, 500 to 10,000 times as

powerful as the Sun, so they can be seen at large distances, out to a few million light-years, with the 100-inch telescope used by Hubble.

Cepheids are very special to astronomers because measuring the period of a Cepheid's brightness variation (using what we are calling Leavitt's law, after Henrietta Leavitt) gives you its average luminosity. And comparing its average luminosity with its average apparent brightness tells you its distance, using the inverse-square law of light. The Cepheids in the spiral nebulae observed by Hubble turned out to be exceedingly distant. The Andromeda Nebula, for example, was found to be over 1 million light-years away (the more accurate value is now known to be about 2.4 million light-years) – far beyond the measured distance of any known stars in the Milky Way Galaxy.

From the distance and the measured angular size of the Andromeda Nebula, its physical size was found to be enormous. Clearly, the "spiral nebulae" were actually huge, gravitationally bound stellar systems like our own Milky Way Galaxy, not relatively small clouds of gas like the Orion Nebula (and so the Andromeda Nebula was renamed the Andromeda Galaxy).

The effective size of the Universe, as perceived by humans, increased enormously with this realization. In essence, Hubble brought the Copernican revolution to a new level; not only is the Earth just one planet orbiting a typical star among several hundred billion stars in the Milky Way Galaxy, but also ours is just one galaxy among the myriads in the Universe! Indeed, it is humbling to consider that the Milky Way is one of roughly 100 billion galaxies within the grasp of the world's best telescopes such as the Keck twins and the Hubble Space Telescope.

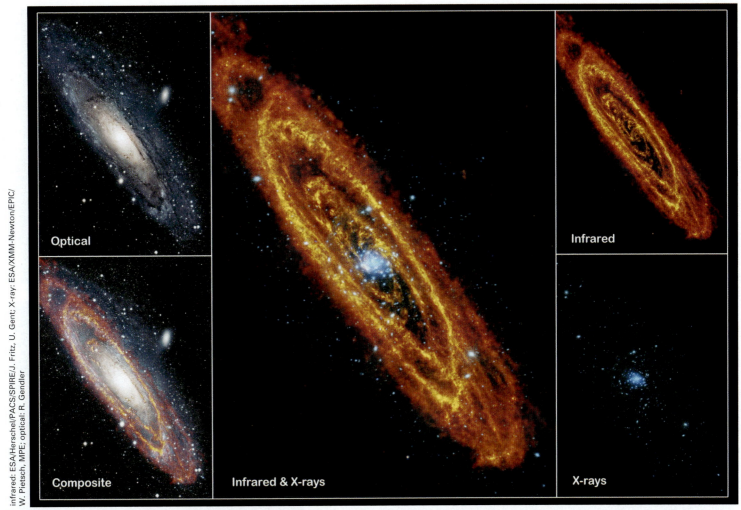

infrared: ESA/Herschel/PACS/SPIRE/J. Fritz, U. Gent; X-ray: ESA/XMM-Newton/EPIC/ W. Pietsch, MPE; optical: R. Gendler

Optical

Composite

Infrared & X-rays

Infrared

X-rays

■ **FIGURE 16–5** The Andromeda Galaxy, also known as M31 and NGC 224, is the nearest large spiral galaxy to the Milky Way. Its Hubble type is Sb, and it is accompanied by many much smaller galaxies. Two of these are the elliptical galaxies M32 (in projection, appearing closest to M31's arms) and NGC 205, sometimes called M110 (independently discovered by Charles Messier and Caroline Herschel in 1783). These galaxies are only 2.4 million light-years from Earth. In visible-light images, the older red and orange stars give the central regions of M31 a yellowish cast, in contrast to the blueness of the spiral arms from the younger stars there. At center, a Herschel Space Observatory view of Andromeda. Observing dust at different temperatures with its different wavelength bands shows the spiral arms continuing from the outside to their base inside the previously known star-forming ring. This ring is revealed to be separate from the arms, and to have a hole in it where the satellite galaxy M32 punched through. Views in other wavelength regions are included.

16.2 TYPES OF GALAXIES

Galaxies come in a variety of shapes. In 1925, Edwin Hubble set up a system of classification of galaxies, and we still use a modified form of it.

16.2a Spiral Galaxies

There are two main "Hubble types" of galaxies. We are already familiar with the first kind – the **spiral galaxies**. The Milky Way Galaxy and its near-twin, the Andromeda Galaxy (M31; look back at Figure 16–5), are relatively large examples containing several hundred billion stars. (Most spiral galaxies contain a billion to a trillion stars.) Another near-twin is NGC 7331 (■ Fig. 16–7).

Spiral galaxies consist of a bulge in the center, a roughly spherical halo around it, and a thin rotating disk with embedded spiral arms. There are usually two main arms, with considerable structure such as smaller appendages. Doppler shifts indicate that spiral galaxies rotate in the sense that the arms trail.

■ Fig.16–8, which is based on a very similar "tuning-fork diagram" drawn by Hubble, shows different types of spiral galaxies. Relative to the disk, the bulge is large in some spiral galaxies (known as "Sa"), which also tend to have more tightly wound spiral arms. The bulge is progressively smaller (relative to the disk) in spirals known as "Sb," "Sc," and "Sd," which also tend to have more loosely wound spiral arms. Moreover, spirals with smaller bulge-to-disk size ratios generally have more gas and dust, and larger amounts of active star formation within the arms at the present time. Spiral galaxies viewed along, or nearly along, the plane of the disk (that is, "edge-on") often exhibit

(a)

(b)

(c)

(d)

■ **FIGURE 16–6** (a) Edwin Hubble, who did his groundbreaking work at the Mt. Wilson Observatory above Los Angeles, California. Here he is shown with the 1.2-m Schmidt telescope (now, after upgrading, called the Oschin Schmidt) at the Palomar Observatory, on Palomar Mountain near San Diego, California. (b) Hubble's most famous photographic glass plate, from 1923. On the preceding night, he observed the same field including the Andromeda Galaxy. On this night, October 6, 1923, he observed again. In black ink, he noted several novae with the letter N. Subsequently, he noticed that one was a variable star; crossing out the N, he marked it "VAR!" in red. (The plate has been newly photographed in color at our request for this textbook.) It was a Cepheid variable, and from its period he found its absolute magnitude. By comparing its absolute magnitude with its apparent magnitude, he found out the distance of the star, and therefore of the Andromeda Galaxy as well. With the discovery, he solved the problem debated by Shapley and Curtis three years earlier: the Andromeda Galaxy and, no doubt, all other galaxies, were "island universes," external to our own Milky Way Galaxy. (c) Recent observations of the variable star discovered by Hubble in M31 ("V1"), which led to our current understanding of cosmic distances. (d) A light curve of the recent observations of V1; the new Hubble observations are marked with prominent star-shaped symbols.

■ **FIGURE 16–7** NGC 7331, a spiral galaxy that is thought to resemble our own Milky Way Galaxy so closely that it has been called our twin. It is 50 million light-years away, far beyond the stars in the constellation Pegasus. It was discovered in 1784 by William Herschel, who discovered not only this galaxy and the planet Uranus but also infrared light. This false-color infrared view, obtained with the Spitzer Space Telescope, reproduces wavelengths of 3.6 μm in blue, 4.5 μm in green, 5.8 μm in yellow, and 8.0 μm in red. At the shorter wavelengths, most of the radiation comes from stars, especially old, cool ones. At longer wavelengths, the starlight is less significant, and we see mainly radiation from glowing clouds of interstellar dust.

The longest of the wavelengths of this Spitzer image show a ring of active star formation about 20,000 light-years from the center of the galaxy. There is enough gas in the ring to produce about four billion Suns.

The three background galaxies in the image are about 10 times farther away than NGC 7331. The red dots are even more-distant galaxies. The blue dots are foreground stars in the Milky Way.

(a)

(b)

■ **FIGURE 16–8** (a) The Hubble "tuning fork" classification of galaxies from the 1930s. It is not meant to imply an evolutionary sequence among galaxies. (b) Images of all the Hubble types of galaxies, taken during the Sloan Digital Sky Survey. At top right we see a row of galaxies edge-on and a row face-on. The irregular (Irr) galaxy is not part of the tuning fork.

a dark dust lane that appears to divide the disk into two halves (■ Fig. 16–9).

In nearly one-half of all spirals, the arms unwind not from the nucleus but rather from a relatively straight bar of stars, gas, and dust that extends to both sides of the nucleus (■ Fig. 16–10). These "barred spirals" are similarly classified in the Hubble scheme from "SBa" to "SBc" (the "B" stands for "barred"), in order of decreasing size of the bulge and increasing openness of the arms (as we saw in Figure 16–8). In many cases, the distinction between a barred and nonbarred spiral is subtle. Studies show that our own Milky Way

■ **FIGURE 16–9** An edge-on view of NGC 4565, a type Sb spiral galaxy in the constellation Coma Berenices. The dust lane in the plane of the galaxy can be seen.

Galaxy is a barred spiral, probably of type SBbc (that is, intermediate between SBb and SBc).

Because they contain many massive young stars, the spiral arms appear bluish in color photographs. Between the spiral arms, the whitish-yellow disks of spiral galaxies contain both old and relatively young stars, but not the hot, massive, blue main-sequence stars, which have already died. Very old stars dominate the bulge, and especially the faint halo (which is difficult to see), and so the bulge is somewhat yellow/orange or even reddish in photographs (as we saw in Figure 16–5).

Since young, massive stars heat the dusty clouds from which they formed, resulting in the emission of much infrared radiation, the current rate of star formation in a galaxy can be estimated by measuring its infrared power. Space telescopes such as the Infrared Astronomical Satellite (IRAS, in the mid-1980s) and the Infrared Space Observatory (ISO, in the mid-1990s) were very useful for this kind of work, and it is being continued with the infrared camera on the Hubble Space Telescope, at even longer infrared wavelengths with Spitzer, and for a while at still longer infrared wavelengths with Herschel. We look forward to the James Webb Space Telescope, with an even larger mirror, being launched near the end of this decade; it will be especially sensitive at infrared wavelengths.

About half of the energy emitted by our own Milky Way Galaxy is in the infrared, indicating that a lot of stars are being formed. But we don't know why the Andromeda Galaxy, which in optical radiation resembles the Milky Way, emits only 3 per cent of its energy in the infrared. This galaxy and the Sombrero Galaxy emit infrared mostly in a ring rather than in spiral arms (■ Fig. 16–10).

A barred spiral is shown in Fig. 16–11.

(a)

European Southern Observatory

(b)

NASA/JPL-Caltech/R. Kennicutt (University of Arizona & the SINGS Team and NASA/ Hubble Heritage Team

■ **FIGURE 16–10** (a) The Sombrero galaxy (M104), named because it resembles a Mexican hat, is a type Sa spiral galaxy in the constellation Virgo. It is also seen close to edge-on, with its dust lane showing prominently. (b) A combination infrared/visible view of the Sombrero Galaxy. The visible-light view in this superposition of two views was taken with the Hubble Space Telescope's Advanced Camera for Surveys. The infrared view (in false color), taken with the Spitzer Space Telescope, shows how the dust is concentrated in a ring.

16.2b Elliptical Galaxies

Hubble recognized a second major galactic category: **elliptical galaxies** (■ Fig. 16–12). These objects have no disk and no arms, and generally very little gas and dust. Unlike spiral galaxies, they do not rotate very much. At the present time, nearly all of them consist almost entirely of old stars, so they appear yellow/orange in true-color photographs.

The dearth of gas and dust is consistent with this composition: There is insufficient raw material from which new stars can form. In many ways, then, an elliptical galaxy resembles the bulge of a spiral galaxy.

Elliptical galaxies can be roughly circular in shape (which Hubble called type E0), but are usually elongated (from E1 to E7, in order of increasing elongation). Since the classification depends on the observed appearance, rather than on the intrinsic shape, some E0

(a)

© European Southern Observatory

(b)

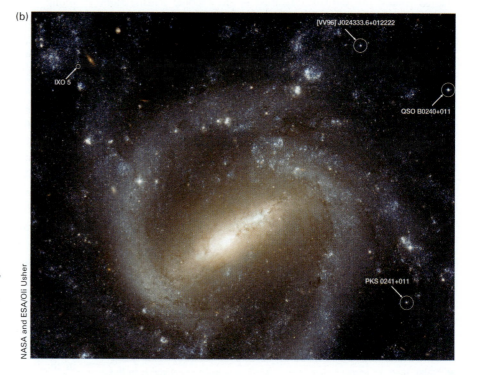

[VV96] J024333.6+012222

IXO 5

QSO B0240+011

PKS 0241+011

NASA and ESA/Oli Usher

■ **FIGURE 16–11** (a) A barred spiral galaxy (SBbc), NGC 1365, in the constellation Eridanus. (b) A barred spiral galaxy, NGC 1073, in the constellation Cetus. Three quasars (see Chapter 17) are circled, and the position of an x-ray source IXO 5, probably a binary system with a black hole and a normal star orbiting each other, is also shown, though which of two faint stars in the circle it is cannot yet be determined. Several more-distant galaxies are seen through the large galaxy in the foreground.

■ **FIGURE 16–12** (a) M87 (NGC 4486), a giant elliptical galaxy in the constellation Virgo. Two more-distant galaxies appear to the lower right. (b) The central region of the small elliptical galaxy M32 (see also Figure 16–5), imaged with the Hubble Space Telescope.

galaxies must actually be elongated, but are seen end-on (like a cigar viewed from one end).

Most elliptical galaxies are dwarfs, like the two main companions of the Andromeda Galaxy (look back at Figures 16–5 and 16–12b), containing only a few million solar masses – a few per cent of the mass of our Milky Way Galaxy. Some, however, are enormous, consisting of a few trillion stars in a volume that is several hundred thousand light-years in diameter (Fig. 16–12a). Many ellipticals may have resulted from two or more spiral galaxies colliding and merging, as we will discuss later.

16.2c Other Galaxy Types

"Lenticular" galaxies (also known as "S0" [pronounced "ess-zero"] galaxies) have a shape resembling an optical lens; they combine some of the features of spiral and elliptical galaxies. They have a disk, like

spiral galaxies. On the other hand, they lack spiral arms, and generally contain very little gas and dust, like elliptical galaxies. Hubble put them at the intersection between spiral and elliptical galaxies in his classification diagram (which we showed as Figure 16–8). Though they are sometimes called "transition galaxies," this designation should not be taken literally: The diagram is not meant to imply that spiral galaxies evolve with time into ellipticals (or vice versa) in a simple manner, contrary to the belief of some astronomers decades ago.

A few per cent of galaxies at the present time in the Universe show no clear regularity. Examples of these "irregular galaxies" include the Small and Large Magellanic Clouds, small satellite galaxies that orbit the much larger Milky Way Galaxy (■ Fig. 16–13). Sometimes traces of regularity – perhaps a bar – can be seen. Irregular galaxies generally have lots of gas and dust, and are forming stars rapidly. Indeed, some of them emit 10 to 100 times as much infrared as optical energy, probably because the rate of star formation is greatly elevated.

Some galaxies are called "peculiar." These often look roughly like spiral or elliptical galaxies, but have one or more abnormalities (■ Fig. 16–14), or like spirals without a nucleus (that is, like rings, ■ Fig. 16–15), or like ellipticals with a dark lane of dust and gas. The ring galaxies are the result of collisions of galaxies.

16.3 HABITATS OF GALAXIES

Most galaxies are not solitary; instead, they are generally found in gravitationally bound binary pairs, small groups, or larger **clusters of galaxies**. Binary and multiple galaxies consist of several members. An example is the Milky Way Galaxy with its two main companions (the Magellanic Clouds, ■ Fig. 16–16), or Andromeda and its two main satellites (as we saw in Figure 16–5). Both Andromeda and the Milky Way have many even smaller companions.

Galaxies and clusters of galaxies all over the Universe are studied with the Hubble Space Telescope in the ultraviolet, visible, and near-infrared, the Spitzer Space Telescope in the infrared, and the Chandra X-ray Observatory in x-rays. NASA's Galaxy Evolution Explorer (GALEX), launched in 2003, was a small satellite that studied galaxies and surveyed the sky in the ultraviolet. When its NASA support ended in 2012, it was turned over to Caltech, under whose control it studied our Galaxy instead of distant galaxies for a year. NASA's Nuclear Spectroscopic Telescope Array (NuSTAR), also controlled by Caltech, was launched in 2012 for high-resolution imaging of the sky with the highest-energy x-rays, higher than those observed with the Chandra X-ray Observatory. It is providing an increase in resolution by a factor of 10 and an increase in sensitivity by a factor of 100 compared with previous observations at those frequencies. NuSTAR's capabilities will allow it to make a census of black holes and to look closer toward the event horizons of a variety of black holes, probably even measuring their spin.

■ **FIGURE 16–13** (a) The Large Magellanic Cloud (LMC) and (b) the Small Magellanic Cloud (SMC) – small, irregular galaxies that orbit around our Milky Way Galaxy. Their respective distances from us are about 170,000 light-years and 210,000 light-years. The globular cluster omega Centauri appears near the LMC in this image. The Clouds were first reported to Europe by 16th-century Portuguese navigators who had travelled to the Cape of Good Hope at the southern tip of Africa. They were named a few decades later in honor of Ferdinand Magellan, who was then circumnavigating the globe by that route (although Magellan himself died before completing the journey). (c) A very wide angle view of Cerro Paranal, Chile, with the European Southern Observatory's Very Large Telescope array of four telescopes at right and the Milky Way arcing overhead, with the Large and Small Magellanic Clouds appearing below it. The reddish disk of the 2010 totally eclipsed moon appears directly over one of the VLT's Unit Telescopes, known as Kueyen (the Moon) in one of Chile's native languages. The zodiacal light surrounds Venus at lower left. One of the 1.8-m auxiliary telescopes used as part of using the VLT as an interferometer is dimly visible at center.

16.3a Clusters of Galaxies

The Local Group is a small cluster of nearly 60 galaxies, some of which are binary or multiple galaxies. Its two dominant members are the Andromeda (M31) and Milky Way Galaxies. M33, the Triangulum Galaxy (■ Fig. 16–17), is a smaller spiral. M31 and M33, at respective distances of 2.4 and 2.6 million light-years, are the farthest objects you can see with your unaided eye (see *Star Party 16.1: Observing Galaxies*). The Local Group also contains four irregular galaxies, at least a dozen dwarf irregulars (■ Fig. 16–18), four regular ellipticals, and the rest are dwarf ellipticals or the related "dwarf spheroidals." The diameter of the Local Group is about 3 million light-years.

The Virgo Cluster (in the direction of the constellation Virgo, but far beyond the stars that make up the constellation), at a distance of about 50 million light-years, is the largest relatively nearby cluster (■ Fig. 16–19). It consists of at least 2,000 galaxies spanning the full range of Hubble types, covering a region in the sky over 15° in

diameter – about 15 million light-years. The Coma Cluster of galaxies (in the direction of the constellation Coma Berenices) is very rich, consisting of over 10,000 galaxies at a distance of about 300 million light-years (■ Fig. 16–20).

A majority of the galaxies in rich clusters are ellipticals, not spirals. There is often a single, very large central elliptical galaxy (sometimes two) that is cannibalizing other galaxies in its vicinity, growing bigger with time (■ Fig. 16–21).

X-ray observations of rich clusters reveal a hot intergalactic gas (10 million to 100 million K) within them, containing as much (or more) mass as the galaxies themselves (■ Fig. 16–22). The gas is clumped in some clusters, while in others it is spread out more smoothly with a concentration near the center. This may be an evolutionary effect; the clumps occur in clusters that only recently formed from the gravitational attraction of their constituent galaxies and groups of galaxies. As clusters age, the gas within them becomes more smoothly distributed and partly settles toward the center.

■ **FIGURE 16–14** The peculiar, interacting spiral galaxies NGC 4038 and 4039, known as The Antennae. They are in the constellation Corvus, the Crow. We see the tidal tails as two galaxies move past each other, as they have been doing for a few hundred million years, with gravity pulling out the tails. Compare with the simulations in Figure 16–40. (The stars in a galaxy are spaced far enough apart that they don't actually collide with each other.) (a) Hubble Space Telescope images with the Advanced Camera for Surveys showed that the galaxies are 45 million light-years from us, 20 million light-years closer than had been thought, deduced from measurements of the colors and brightnesses of the red giants. (The luminosity of the brightest red giants on an H–R diagram is known.) Our own galaxy and the Andromeda Galaxy may put on such a show of interaction and merging in a few billion years. (b) Hubble Space Telescope close-up of The Antennae that shows details of gas and dust in the two central regions. (c) A Chandra X-ray Observatory view. The dozens of bright points are neutron stars or black holes pulling gas off nearby stars. The bright patches come from the accumulated power of thousands of supernovae. (d) A Spitzer Space Telescope view, showing glowing dust.

Curt Struck and Philip Appleton (Iowa State University), Kirk Borne (Hughes STX Corp.), Ray Lucas (STScI), and NASA

■ **FIGURE 16–15** A Hubble Space Telescope image of the Cartwheel Galaxy, in the constellation Sculptor, a ring galaxy probably caused by one of its satellite galaxies passing through it.

16.3b Superclusters of Galaxies

Clusters are seen to vast distances, in a few cases up to 8 billion light-years away. When we survey their spatial distribution, we find that they form clusters of clusters of galaxies, appropriately called **superclusters**. These vary in size, but a typical diameter is about 100 million light-years. The Local Group, dozens of similar groupings nearby, and the Virgo Cluster form the Local Supercluster.

Superclusters often appear to be elongated and flattened. The thickness of the Local Supercluster, for example, is only about 10 million light-years, or one-tenth of its diameter. Superclusters tend to form a network of bubbles, like the suds in a kitchen sink (■ Fig. 16–23). Large concentrations of galaxies (that is, several adjacent superclusters) surround relatively empty regions of the Universe, called **voids**, that have typical diameters of about 100 million light-years (but sometimes up to 300 million light-years). They formed as a consequence of matter gravitationally accumulating into superclusters; the regions surrounding the superclusters were left with little matter.

Does the clustering continue in scope? Are there clusters of clusters of clusters, and so on? The present evidence suggests that this is not so. Surveys of the Universe to very large distances do not reveal many obvious super-superclusters. There are, however, a few giant structures such as the "Great Wall" that crosses the center of the

■ **FIGURE 16–16** The Milky Way Galaxy, in which we can see the Milky Way as a band across the sky, has the Magellanic Clouds as satellite galaxies. In this view of the Andes Mountains in Chile, we see Comet McNaught during its wonderful 2007 apparition.

Miloslav Druckmüller

Black and white image from Palomar Observatory. Caltech; color HST image from Hui Yang (University of Illinois) and NASA

■ **FIGURE 16–17** The Triangulum Galaxy, M33, a member of the Local Group. It is about 2.6 million light-years away, in nearly the same direction as the Andromeda Galaxy but considerably smaller. At lower right we see a nebula in M33, imaged with the Hubble Space Telescope.

This program marks an extensive collaboration between Italian (Yatzan Momany, Enrico V. Held, and Marco Gullieuszik), E.S.O. (Ivo Saviane and Luigi Bedin), U.S.A. (Michael Rich, Luca Rizzi, and Dutch (Konrad Kuijen) institutions. This image was produced by the Hubble Heritage Team (STScI/AURA). Credit: NASA, ESA, and The Hubble Heritage Team (STScI/AURA). Acknowledgment: Y. Momany (University of Padua)

■ **FIGURE 16–18** This Hubble view of the Sagittarius Dwarf Irregular Galaxy shows its hot, bluish stars at a distance of 3.5 million light-years. Hubble's Advanced Camera for Surveys is able to resolve individual stars in such nearby members of the Local Group of galaxies. Studying them reveals how stars are still forming out of the gas that is present. Background objects, usually reddish/brown, often show spiral arms; they are galaxies at much larger distances.

slices shown in Figure 16–23. We will discuss in Chapter 19 how the "seeds" from which these objects formed were visible within 400,000 years after the birth of the Universe.

A useful analogy for the distribution of galaxies discussed above is to consider isolated houses (galaxies), small villages (loose groups like the Local Group), towns and cities (small and large clusters), and metropolitan areas (superclusters). Even isolated houses (say, in the countryside) are usually accompanied by smaller buildings such as barns and tool sheds, just as large galaxies often have smaller companions.

STAR PARTY 16.1   Observing Galaxies

The stars that make up the outlines of constellations, and essentially all other stars that you see in the night sky (even faint ones that require binoculars or a small telescope), are part of our Milky Way Galaxy. But a separate "island universe," the Andromeda Galaxy (M31; see Figure 16–5), is visible to the unaided eye if you know where to look.

On a clear, dark, moonless night you can see it as a faint, fuzzy patch of light between the brightest stars of the constellations Andromeda and Cassiopeia (consult the Sky Maps inside the back cover of this text). Through binoculars, it can be traced over an arc length of several degrees. (The width of your thumb at the end of your outstretched arm covers an arc length of about 2 degrees.) Ponder that you are seeing light that left the Andromeda Galaxy about 2.4 million years ago, around the time when early hominids were developing on Earth!

The smaller and fainter galaxy M33 (Figure 16–17), 2.6 million light-years away, is also sometimes visible in the constellation Triangulum near Andromeda, but it is considerably more difficult to see than M31.

With a good pair of binoculars, it is possible to just barely see the galaxies M81 and M82 (Figure 16–2). They are about 12 million light-years away, in the direction of the constellation Ursa Major. M81 is a type Sb spiral galaxy, while M82 is an irregular galaxy with a very large amount of gas and dust. Both galaxies will be much easier to see with your college telescope, if one is available.

If you have access to a telescope, try looking for galaxies in the Virgo Cluster. In clear, dark skies, many galaxies will be visible in a relatively small area of the sky. The light that you see actually left those galaxies about 50 million years ago, not long after the time when dinosaurs became extinct on Earth and long before people walked the Earth.

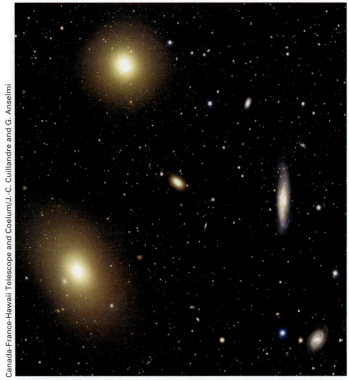

Canada-France-Hawaii Telescope and Coelum/J.-C. Cuillandre and G. Anselmi

■ **FIGURE 16–19** A visible-light image of the central part of the extensive Virgo Cluster of galaxies (in the direction of the constellation Virgo); with a distance of about 50 million light-years, it is the nearest large cluster to the Milky Way Galaxy. Fewer than a dozen of the roughly 2,000 cluster galaxies are shown here.

NOAO/Nigel Sharp

■ **FIGURE 16–20** The extremely rich Coma Cluster (in the direction of the constellation Coma Berenices), about 300 million light-years away and containing over 10,000 galaxies, only a fraction of which can be seen in this visible-light image. There are two dominant, large elliptical galaxies.

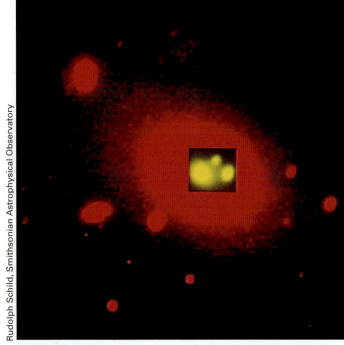

Rudolph Schild, Smithsonian Astrophysical Observatory

■ **FIGURE 16–21** An example of galactic cannibalism. Computer processing displays the central region of the large galaxy in yellow, and we also see two smaller galaxies in the process of being swallowed. The large galaxy, NGC 6166, is the central one in the rich cluster Abell 2199 (the 2199th object in a catalogue of clusters of galaxies compiled by UCLA astronomer George Abell).

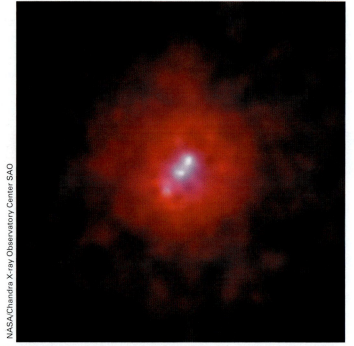

NASA/Chandra X-ray Observatory Center SAO

■ **FIGURE 16–22** A false-color x-ray image of 3C 295, a cluster of galaxies, taken with the Chandra X-ray Observatory. We see more irregularities than had been expected from earlier observations of clusters of galaxies, which didn't show such faint detail. It appears that the cluster has been built up from smaller elements like the constituent galaxies or small groups of galaxies.

(a)

(b)

John Huchra, Margaret Geller, and V. de Lapparent, © 1989 Smithsonian Astrophysical Observatory

John Huchra and Margaret Geller, Smithsonian Astrophysical Observatory

(c)

Messier 33

NGC 604

M. Blanton and the SDSS-III collaboration

SDSSIII

Southern Galactic Cap

Northern Galactic Cap

■ **FIGURE 16–23** "Slices of the Universe," illustrating the distribution of galaxies, a pioneering study by John Huchra and Margaret Geller with their student Valerie de Lapparent at the Harvard-Smithsonian Center for Astrophysics that later led to the large-scale mapping by the 2dF ("two-degree field)" and Sloan Digital Sky Survey projects that we will discuss later in this chapter. Each slice shows distance from the Earth (going along straight lines outward from the Earth, which is at the bottom point) versus position on the sky (in right ascension) for all galaxies in a strip of declination 6° in size. (Right ascension corresponds to longitude and declination to latitude on the sky; see Chapter 4.) Each wedge extends out to a distance of about one billion light-years. (a) Distance vs. position of galaxies in two adjacent slices centered on declinations of 29.5° (pink dots) and 35.5° (white dots). (b) Positions of galaxies in four adjacent slices. Note how the structures apparent in one slice continue to the next, indicating that the galaxies define giant bubbles, like the suds in a kitchen sink, or perhaps long, sponge-like structures in space. The Coma Cluster of galaxies is in the center, apparently at the intersection of several bubbles. (c) The universe as mapped by the third data release (2011) of the Sloan Digital Sky Survey. The channels and structure originally revealed by Huchra, Geller, and de Lapparent in their slices of the Universe still show in this image assembled by a team led by Michael Blanton of NYU. At top left, we see the SDSS field of M33, and the other images at top zoom in on M33. At bottom, the crab-like images (some remnants of the telescope's scanning across the sky are visible at the edges) show the clusters and walls that are the Universe's largest structures. The full resolution of the image is over a trillion pixels, which would take over half a million high-definition televisions to display.

16.4 THE DARK SIDE OF MATTER

There are now strong indications that much of the matter in the Universe does not emit any detectable electromagnetic radiation, but nevertheless has a gravitational influence on its surroundings. One of the first clues for such "**dark matter**" was provided by the flat (nearly constant) **rotation curves** of spiral galaxies.

16.4a The Rotation Curve of the Milky Way Galaxy

The rotation curve of any spinning galaxy is a plot of its orbital speed as a function of distance from its center. For example, the rotation curve of the Milky Way Galaxy has been determined through studies of the motions of stars and clouds of gas (■ Fig. 16–24). It rises from zero in the center to a value of about 200 km/s at a radial distance of about 5,000 light-years. The rotation curve farther out is rather "flat"; the orbital speed stays constant, all the way out to distances well beyond that of the Sun.

The speed of a star at any given distance from the center is determined by the gravitational field of matter enclosed within the orbit of that star – that is, by the matter closer to the center. (It can be shown that matter at larger distances, outside the star's orbit, does not affect the star's speed as long as the galaxy's disk has a smooth, symmetric distribution of matter.) So, we can use the rotation curve to map out the distribution of mass within our Galaxy. The speeds and distances of stars near our Galaxy's edge, for example, are used to measure the mass in the entire Galaxy.

Specifically, Kepler's third law can be manipulated to give an expression for the mass (M) enclosed within an orbit of radius R from the center (see *Figure It Out 16.1: Calculating the Mass from the Rotation Curve*). A similar method is used to find the amount of mass in the Sun by studying the orbits of the planets, or the amount of a planet's mass by observing the orbits of its moons. In the 17th century, Newton developed this technique as part of his derivation and elaboration of Kepler's third law of orbital motion (see Chapter 5).

If we insert the Sun's distance from the Galactic center (26,000 light-years) and the Sun's orbital speed (200 km/s) into the formula, we find that the matter within the Sun's orbit has a mass of about 100 billion (10^{11}) solar masses! But the mass of a typical star is about half that of the Sun. Thus, if most of the matter in our Galaxy is in the form of stars (rather than interstellar gas and dust, black holes, etc.), we conclude that there are about *200 billion stars* within the Sun's orbit, closer to the Galaxy's center than the Sun is. (However, as we shall see below, part of the matter within the Sun's galactic orbit is actually dark matter, so a more accurate estimate of the number of stars is probably about 100 billion.)

The next thing to notice is that the flat rotation curve of the Milky Way Galaxy (Figure 16–24) is quite different from the rotation curve of the Solar System. As discussed in Chapter 5, the orbital

■ **FIGURE 16–24** The rotation curve of our Milky Way Galaxy. The vertical axis shows the speed with which stars travel as they orbit the center, and the horizontal axis is their distance from the center. Note how constant ("flat") the speed is beyond distances of a few thousand light-years; it certainly doesn't decline, and might even rise a little.

speeds of distant planets are slower than those of planets near the Sun (see *Figure It Out 5.3: Orbital Speed of Planets*), instead of being roughly independent of distance. In the Solar System, the Sun's mass greatly dominates all other masses; the masses of the planets are essentially negligible in comparison with the Sun. But in the Milky Way Galaxy, the flat rotation curve implies that, except in the central region (where the rotation curve isn't flat), the mass grows with increasing distance from the center.

The growth in mass of our Galaxy continues to large distances beyond the Sun's orbit. (The rotation curve way out from the center is usually determined from the measured speeds of clouds of hydrogen gas, which can be easily seen at radio wavelengths.) This is very puzzling because few stars are found in those regions: The number of stars falls far short of accounting for the derived mass.

For example, at a distance of 130,000 light-years from the center, the enclosed mass is about 5×10^{11} solar masses, and the corresponding number of typical stars (each having half a solar mass) would be about a trillion – yet there are too few stars visible, by a large margin. Indeed, studies of the outer parts of the Milky Way Galaxy throughout the electromagnetic spectrum do not reveal sufficient quantities of material to account for the derived mass.

16.4b Dark Matter Everywhere

We conclude that the Milky Way Galaxy contains large quantities of "dark matter" – material that exerts a gravitational force but is invisible or at least very difficult to see!

This material has sometimes been called the "missing mass," especially in older texts, but the term is not appropriate because the mass is present. Instead, it is the light that's missing.

Many other spiral galaxies also have flat (speed roughly constant) rotation curves, as was first shown by Vera Rubin of the Carnegie Institution for Science (■ Fig. 16–25). Estimates suggest that 80 to 90 per cent of the mass of a typical spiral galaxy consists of dark matter. However, it has been shown that the amount of matter in the disk itself cannot exceed what is visible by more than a factor of 2. Instead, the dark matter is probably concentrated in an extended,

FIGURE IT OUT 16.1　　　　Calculating the Mass from the Rotation Curve

The rotation curve, a plot of orbital speed versus distance from the orbit's center, can be used to calculate the mass enclosed within an object's orbit. We shall first apply it to our Solar System; see also *Figure It Out 5.3: Orbital Speed of Planets.*

Recall from Chapter 5 that Newton's generalized version of Kepler's third law is

$$P^2 = \frac{4\pi^2}{G(m_1 + m_2)} R^3,$$

where m_1 and m_2 are the masses of the two bodies (1 and 2). Suppose m_2 (a planet) is negligible relative to m_1 (the Sun). Then we can ignore m_2 in the sum $(m_1 + m_2)$, and Kepler's third law becomes

$$P^2 = \frac{4\pi^2}{Gm_1} R^3.$$

If the planet's orbit is circular (which is roughly true), then the circumference of the orbit $(2\pi R)$ must equal the planet's speed multiplied by the period: $2\pi R = vP$. This is merely an application of *distance = speed × time*, which is correct for constant speed. Thus, $P = 2\pi R/v$. If we now substitute this into the above equation for P^2 and rearrange terms, we find that

$$m_1 = v^2 R / G.$$

So, knowing Earth's distance from the Sun $(R = 1.5 \times 10^8$ km) and its orbital speed $(v = 30$ km/s), we can deduce that the Sun's mass is $m_1 = 2 \times 10^{33}$ grams. This is about 330,000 times the Earth's mass, thereby justifying our assumption that Earth's mass (m_2) is negligible.

The formula $m_1 = v^2R/G$ also turns out to give the mass of a galaxy (m_1) within a radius R from its center. Let's use $R = 26,000$ light-years and $v = 200$ km/s, as is the case for the Sun in the Milky Way Galaxy. Being careful to keep track of units, we find that the mass within the Sun's orbit is

$$m_1 = (200 \text{ km/s})^2 (10^5 \text{ cm/km})^2 (26{,}000 \text{ lt-yr})$$
$$\times (9.5 \times 10^{17} \text{ cm/lt-yr})/(6.673 \times 10^{-8} \text{ cm}^3/\text{g/s}^2)$$
$$= 1.5 \times 10^{44} \text{ g}.$$

But the mass of the Sun is about 2×10^{33} g, so $m_1 = (1.5 \times 10^{44}$ g)/ $(2 \times 10^{33}$ g/solar mass) $\approx 7.4 \times 10^{10}$ solar masses, or roughly 10^{11} solar masses, as stated in the text.

■ **FIGURE FIO 16.1** The rotation curve of the Solar System, showing the orbital speeds for planets (or particles) at different distances from the Sun. Earth, with a distance of 1 au, has an orbital speed of 30 km/s, as marked. Particles at distances of 4 au and 9 au would have orbital speeds of 15 km/s and 10 km/s, respectively. See also *Figure It Out 5.3: Orbital Speed of Planets.*

spherical, outer halo of material that extends to perhaps 200,000 light-years from the galactic center.

Similar studies show that elliptical galaxies also contain large amounts of dark matter. Gas and stars are moving so quickly that they would escape if the visible matter alone produced the gravitational field. There must be another, more dominant, contribution to gravity in these galaxies.

The orbital speeds of galaxies in binary pairs, groups, and clusters can be used to determine the masses of these systems. Astronomers find that, in essentially all cases, the amount of mass required to produce the observed orbital speeds is larger than that estimated from the visible light (which is assumed to come from stars and gas). A related technique is to measure the typical speeds of particles of gas bound to a cluster of galaxies – and, once again, the particles could not be gravitationally bound to the cluster if its mass consisted only of that provided by the visible matter.

Decades ago, the Caltech astronomer Fritz Zwicky was the first to point out that clusters of galaxies could not remain gravitationally

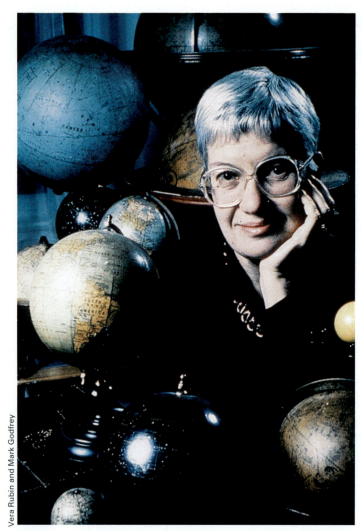

Vera Rubin and Mark Godfrey

■ **FIGURE 16–25** Vera Rubin, who was the first to show that many spiral galaxies have flat rotation curves and thus contain substantial dark matter.

bound if they contain only visible matter. He postulated the existence of some form of dark matter. However, this idea was largely ignored or dismissed – it was too far ahead of its time.

16.4c What Is Dark Matter?

What is the physical nature of the dark matter in single and binary galaxies, groups, and clusters? We just don't know – this is one of the outstanding unsolved problems in astrophysics. A tremendous number of very faint normal stars (or even brown dwarfs) is a possibility, though it seems unlikely, extrapolating from the numbers of the faintest stars that we can study.

There is some evidence (see Section 16.5 below) that part of the dark matter consists of old white dwarfs. If these and other corpses of dead stars (neutron stars, black holes) accounted for most of the dark matter, however, then where is the chemically enriched gas that the stars must have ejected near the ends of their lives? Other candidates for the dark matter are small black holes, massive planets ("Jupiters"), and neutrinos.

Actually, we will see in Chapter 19 that certain kinds of measurements indicate that only a small fraction of the dark matter can consist of "normal" particles such as protons, neutrons, and electrons; the rest must be exotic particles. Most of the normal dark matter consists of tenuous, million-degree gas in galactic halos. This gas was recently detected by the absorption spectra it produced in the radiation from background objects, and also from its emission at relatively long x-ray wavelengths. Though no longer technically "dark" (because we have seen it!), such matter is still generally considered to be part of the "dark matter" that pervades the Universe; it is difficult to detect.

Probably the most likely candidate for a majority of the dark matter (the "abnormal" part) is undiscovered subatomic particles with unusual properties, left over from the big bang, such as **WIMPs** – "weakly interacting massive particles." Physicists studying the fundamental forces of nature suggest that many WIMPs exist, though it is disconcerting that none has been unambiguously detected in a laboratory experiment.

If it is unsatisfactory to you that most of the mass in our Galaxy (and, indeed, most of the mass in the Universe!) is in some unknown form, you may feel better by knowing that astronomers also find the situation unsatisfactory. But all we can do is go out and conduct our research, and try to find out more. Clever new techniques, such as one described below, may provide the crucial clues that we seek.

16.5 GRAVITATIONAL LENSING

The phenomenon of **gravitational lensing** of light provides a powerful probe of the amount (and in some cases the nature) of dark matter. Hundreds of cases of gravitational lensing have been found.

If the light from a distant object passes through a gravitational field, the light is bent – that is, it follows a curved path through the warped space-time. This is analogous to (but differs in detail from) the effect that a simple glass lens has on light. It is, perhaps, more akin to the warping of images we get from a fun-house mirror, but the terms "lens" and "lensing" have caught on.

We have already encountered lensing in Chapter 10: Recall that Einstein's general theory of relativity predicts that the Sun should bend the light of stars beyond it by an amount twice that predicted with Newtonian theory, and that this effect was first measured in 1919 by Arthur Eddington and others during a total solar eclipse.

If an observer, a galaxy acting as a gravitational lens, and a more distant, very compact object are nearly perfectly aligned (collinear), the distant object will look like a circle (known as an "Einstein ring") centered on the lens (■ Fig. 16–26). More usually, deviations from symmetry (for example, slight misalignment of the lens) lead to the formation of several discrete, well-separated images (■ Fig. 16–27). The apparent brightness of the object is magnified, in some cases by large amounts.

When a massive cluster of galaxies lenses many distant galaxies, a collection of arcs tends to be seen (■ Fig. 16–28). The number of

■ **FIGURE 16–26** (a) Simulation of the formation of an "Einstein ring" from the gravitational lensing of a compact object by a lensing object, a galaxy or cluster of galaxies, that is along a straight line between the observer and the compact object. The apparent brightness of the lensed object is increased due to the magnification of the gravitational lens. Einstein would be very pleased with the actual discovery of such objects; he had thought that the effect – predicted not only by himself but also by one earlier theorist – would never be observed. (b) This cosmic mirage, mostly of a circular Einstein ring, popped into view when observed with one of the 8.2-m telescopes of the Very Large Telescope Array in Chile. The distances of the lensing galaxy and the lensed galaxy are nearly 8 and 12 billion light-years, respectively. The magnification factor of at least 13 helps us see the lensed galaxy at a time when the Universe was only about 12 per cent of its current age.

arcs, their magnification factors, and their distorted pattern depend on the mass of the cluster, as well as on the distribution of mass within the cluster. This method measures the total mass (visible and dark) in the cluster, and gives results consistent with those obtained from other techniques. Again, we conclude that dark matter dominates most clusters of galaxies.

Projects in which the apparent brightness of millions of stars in the Large and Small Magellanic Clouds are systematically monitored have revealed that occasionally a star brightens and fades over the course of a few weeks. (These galaxies provide a nice background field of stars that are out of the Galactic plane.) The light curve has exactly the shape expected if a compact lens such as another star were to pass between the background star and us, temporarily focusing the background star's light toward us (■ Fig. 16–29). Moreover, the shape and height (amplitude) of the light curve are independent of the filter through which it was obtained, as predicted for gravitational lensing and unlike the case for intrinsically variable stars.

Note that if our angular resolution were sufficiently high, each gravitationally lensed star would look like a tiny Einstein ring (if the alignment were perfectly collinear), or more generally as several discrete images with very small angular separations. But such high

resolution is not achievable with current technology, so all we detect is an overall brightening of the lensed star – and sometimes even a brief secondary brightening, when an exoplanet bound to the lens also magnifies the lensed star (recall the discussion of such "gravitational microlensing" in Chapter 9).

Some of these searches for lensed stars in the Magellanic Clouds were motivated by the opportunity of finding "**ma**ssive **c**ompact **h**alo **o**bjects," or **MACHO**s. In principle, such objects would be too massive to be brown dwarfs, yet could not be normal main-sequence stars because we would see them. White dwarfs are a possible candidate, but in this case the population of stars that produced them must have been devoid of very low-mass stars, since we do not see enough stars in the halo, where astronomers expect that MACHOs would be found.

The searches have detected hundreds of lensed stars in the Magellanic Clouds, but there is considerable controversy about the interpretation of these detections. In a few especially favorable cases, the distance of the lens has been determined. In two cases, the lens turns out to be in the Large Magellanic Cloud itself, rather than in the halo of the Milky Way Galaxy. Another Hubble Space Telescope discovery of a lens in the direction of the Large Magellanic Cloud

■ **FIGURE 16–27** (a) A slight misalignment of the lensing object leads instead to the formation of several discrete images of the compact lensed object. (b) A visible-light image of a lensed object (a quasar, as described in the next chapter), PG1115+080, that was not directly in line with the lensing object.

(a)

(b)

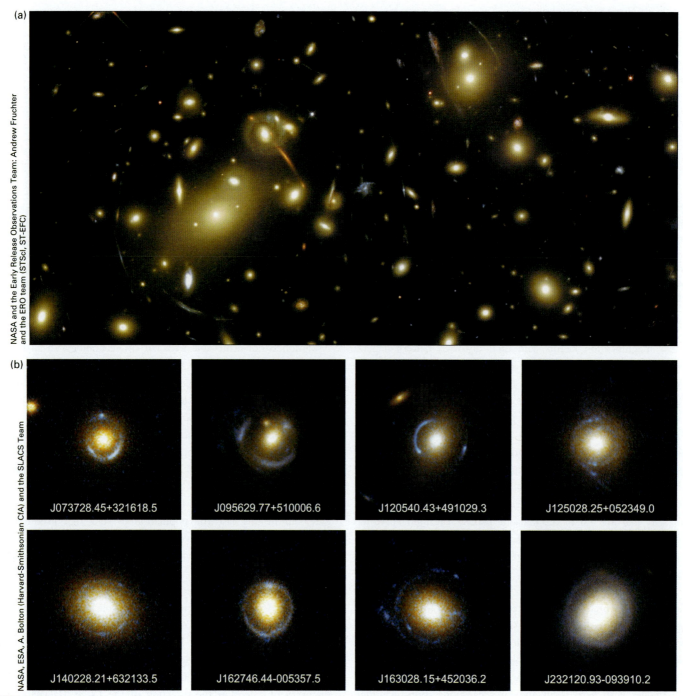

NASA and the Early Release Observations Team: Andrew Fruchter and the ERO team (STScI, ST-EFC)

NASA, ESA, A. Bolton (Harvard-Smithsonian CfA) and the SLACS Team

J073728.45+321618.5 J095629.77+510006.6 J120540.43+491029.3 J125028.25+052349.0

J140228.21+632133.5 J162746.44-005357.5 J163028.15+452036.2 J232120.93-093910.2

■ **FIGURE 16–28** If the lensed object is an extended (rather than a compact) galaxy, the lensed images often appear as arcs. (a) Distant galaxies gravitationally lensed by the cluster Abell 2218 appear as arcs centered on the cluster and imaged with the Hubble Space Telescope. The bluish color of many of the lensed galaxies indicates that their rate of star formation is high. (b) A set of rings, arcs, and other gravitationally lensed patterns detected by the Hubble Space Telescope.

turns out to be an ordinary star in the disk of the Milky Way Galaxy. So far, the searches have not revealed any definitive lenses in the halo of our Galaxy.

If most of the lenses are not in our Galaxy's halo, then the evidence for dark, compact objects greatly decreases. Indeed, estimates suggest that no more than about 10 or 20 per cent of the halo's mass consists of MACHOs. Thus, it appears that much of the dark matter in the halo may consist of subatomic particles like the WIMPs mentioned above; these particles almost never interact with normal matter, except through gravity. It is exciting to think that astronomical observations may end up providing the crucial evidence for the existence of tiny, otherwise undetectable particles.

■ FIGURE 16–29 Light curves, through two different filters, of a star that has been gravitationally lensed by a dark, compact object. Although several discrete images are presumably present, even through a telescope the star appears point-like and its brightness can be measured. The star's light is temporarily focused toward us, with the same magnification at all wavelengths. (The important point is that, as we show here, the shape and height of the light curves as seen through the different color filters are identical.) Unlike the case for variable stars, which brighten and fade periodically, a star is generally lensed only once while we are watching, since the chances of a lens lining up perfectly are exceedingly small.

Exceptionally strong evidence that most of the dark matter consists of WIMPs comes from studies of a few special galaxy clusters, especially the "Bullet Cluster," that show a clear separation between hot x-ray-emitting gas and the dominant dark matter (■ Fig. 16–30). Two clusters are in the process of passing through each other, and the hot gas in them collides and slows down. But maps of the distribution of mass in the clusters, made through gravitational lensing, show that *most* of the mass in each cluster continues to move without slowing down; there is a clear offset in the distribution of the main mass (which is centered on the visible-light image of the cluster) and the x-ray-emitting gas. This implies that most of the dark matter consists of particles that *don't* collide, unlike normal gas but exactly as expected for stars and WIMPs. (The stars generally rarely collide because they are so small compared with the vast spaces between them, and WIMPs hardly ever interact with normal matter.)

Gravitational lensing also helps us find out about dark matter by revealing how mass is distributed in galaxies. Just how centralized the mass is in galaxy cores can, in principle, be revealed by the study of gravitational lenses with multiple images.

16.6 THE BIRTH AND LIFE OF GALAXIES

It is difficult or impossible to determine what any given nearby galaxy (or our own Milky Way Galaxy) used to look like, since we can't view it as it was long ago. However, as we discussed in Chapter 1, the finite speed of light effectively allows us to view the past history

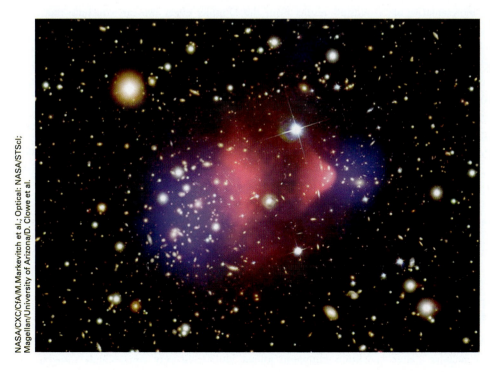

■ FIGURE 16–30 The "Bullet Cluster," in which two clusters of galaxies are colliding with and passing through each other. We see the distribution of hot x-ray-emitting gas (in red) and of dark matter (in blue); there is a clear spatial offset between the two. The hot gas in the two clusters collides and slows down, but the dark matter and the visible galaxies continue unimpeded (except for some slowing down due to gravity). Most of the dark matter probably consists of WIMPs that don't collide the way gas does.

■ **FIGURE 16–31** (top) The optical spectrum of a very nearby galaxy, having essentially zero redshift. Various emission and absorption lines are marked. (bottom) The spectrum of the same galaxy, if it were at a redshift $z = 0.1$ (in other words, a recession speed of about 30,000 km/s; see *Figure It Out 16.2: Redshifts and Hubble's Law*). The same patterns of emission and absorption lines are visible, but their wavelengths are shifted toward the red by 10 per cent.

of the Universe: We see different objects at different times in the past, depending on how long the light has been travelling toward us. At least in a statistical manner we can explore galactic evolution by examining galaxies at progressively larger distances or **lookback times**, and hence progressively farther back in the past.

An important but likely valid assumption is that we live in a typical part of the Universe, so that nearby galaxies are representative of galaxies everywhere. Hence, galaxies several billion light-years away, viewed as they were billions of years ago, probably resemble what today's nearby galaxies used to look like. The refurbished Hubble Space Telescope has led to the most progress in this field, since it provides detailed images of faint, distant galaxies. Also, the Chandra X-ray Observatory has revealed objects that might be very primitive, distant galaxies; the seemingly uniform x-ray glow that previous x-ray telescopes had detected is actually produced by many individual discrete sources.

It is crucial to know the distances of the very distant galaxies, but they are so far away that no Cepheid variables or other normal stars can be seen and compared with nearby examples. So, an indirect technique is used: Hubble's law, as described below.

16.7 THE EXPANDING UNIVERSE

Early in the 20th century, Vesto Slipher of the Lowell Observatory in Arizona noticed that the optical spectra of "spiral nebulae" (later

recognized by Edwin Hubble to be separate galaxies) almost always show a **redshift** (recall our discussion of redshifts and blueshifts in Chapter 11). The absorption or emission lines seen in the spectra have the same *patterns* as in the spectra of normal stars or emission nebulae, but these patterns are displaced (that is, shifted) to longer (redder) wavelengths (■ Fig. 16–31). Under the assumption that the redshift results from the Doppler effect, we can conclude that most galaxies are moving away from us, regardless of their direction in the sky. (In Chapter 18, we will see that the redshift is *actually* caused by the stretching of space itself, not by the motion of objects *through* space, but the equation is the same as that for the Doppler effect, at least at low redshifts.)

In 1929, using newly derived distances to some of these galaxies (from Cepheid variable stars), Hubble discovered that the displacement of a given line (that is, the redshift) is *proportional* to the galaxy's distance. (In other words, when the redshift we observe is greater by a certain factor, the distance is greater by the same factor.)

Thus, under the Doppler-shift interpretation, the recession speed, v, of a given galaxy must be proportional to its current distance, d (■ Fig. 16–32a). This relation is known as **Hubble's law**, $v = H_0 d$ (see *Figure It Out 16.2: Redshifts and Hubble's Law*), where H_0 (pronounced "H nought") is the *present-day* value of the constant of proportionality, H (the factor by which you multiply d to get v). H_0 is known as **Hubble's constant**.

For various reasons, Edwin Hubble's original data were suggestive but not conclusive. Subsequently, Hubble's assistant and disciple Milton Humason joined Hubble in very convincingly showing the

(a)

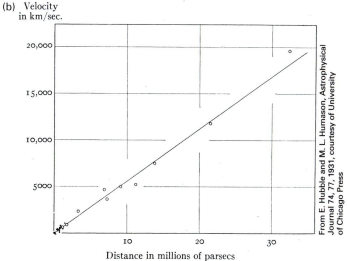

(b) Velocity in km/sec.

From E. Hubble and M. L. Humason, Astrophysical Journal 74, 77, 1931, courtesy of University of Chicago Press

National Academy of Sciences

Distance in millions of parsecs

■ **FIGURE 16–32** (a) The speed-distance relation, in Hubble's original diagram from 1929. Note that the units of velocity (speed) on the vertical axis should be km/s, not km; Hubble, or the artist who labeled the graph, made a mistake in the units. One million parsecs equals 3.26 million light-years. Dots are individual galaxies; open circles represent groups of galaxies. The scatter to one side of the line or the other is substantial. (b) By 1931, Hubble and Humason had extended the measurements to greater distances, and Hubble's law was well established. All the points shown in the 1929 work appear bunched in the lower-left corner of this graph. The relation $v = H_0 d$ represents a straight line of slope H_0. These graphs use older distance measurements than we now use, and so give different values for H_0 than we now derive.

relationship (Fig. 16–32b). (Interestingly, Humason had first come to Mt. Wilson as a mule-team driver, helping to bring telescope parts up the mountain. He worked his way up within the organization.)

This behavior is similar to that produced by an explosion: Bits of shrapnel are given a wide range of speeds, and those that are moving fastest travel the largest distance in a given amount of time. The implication of Hubble's law is that the Universe is *expanding*! (Interestingly, Edwin Hubble himself resisted this idea, and never quite signed on to the expanding Universe whose concept he had fostered.) As we shall see in Chapter 18, however, there is no unique center to the expansion, so in this way it is *not* like an explosion. Moreover, the expansion of the Universe marks the creation of *space itself*, unlike the explosion of a bomb in a preexisting space.

Library research in 2011 dealt with a historical matter long pondered: Did the Belgian abbé Georges Lemaître deserve credit

for what we call Hubble's law? Indeed, he put together Slipher's velocities and some of Hubble's distances in 1927, two years before Hubble did, and published it in French in an obscure Belgian journal. After Hubble's 1929 widely read paper, the editor of a major British journal asked Lemaître to translate his paper. In the translation, key paragraphs about the velocity-distance relation disappeared. A few conspiracy-prone astronomers speculated that the journal editor left them out on purpose, or even that Hubble had exerted influence to omit this discovery, which preceded his own. But the original correspondence, examined in 2011 by Mario Livio of the Space Telescope Science Institute, showed that Lemaître himself modestly omitted the paragraphs in his translation, saying they had "no actual interest" – perhaps because Hubble's work had already been published. Decades later, Lemaître did entitle an article (in French) "The Expanding Universe," perhaps to retrospectively establish a claim to priority.

One of the greatest debates in 20th century astronomy has been over the numerical value of Hubble's constant. In Chapter 18 we will explain in detail how it is determined. Measurements of the recession speeds of galaxies at known distances showed that $H_0 = 50$ to 80 km/s/Mpc. (An "Mpc" is a megaparsec, a million parsecs. Parsecs are units of distance preferred by astronomers because they come from direct measurements, though only for the nearest stars; 1 parsec (1 pc) is equal to 3.26 light-years (see Chapter 11). A capital M is the prefix meaning a million. As of 2013, $H_0 = (67.8 \pm 1.2)$ km s^{-1} Mpc^{-1} from ESA's Planck spacecraft (see Chapter 19) and $74.3 \pm 1.5 \pm 2.1$ (the first $\pm$ is statistical; the second $\pm$ is systematic) from Hubble measurements linked with ground-based observations of supernovae. We will use 71 km s^{-1} Mpc^{-1} as a compromise. For example, a galaxy 10 Mpc (32.6 million light-years) away recedes from us with a speed of about 710 km/s, and a galaxy 20 Mpc away recedes with a speed of about 1420 km/s.

Hubble's constant is always quoted in the strange units of km/s/Mpc. The value $H_0 = 71$ km/s/Mpc simply means that for each megaparsec (3.26 million light-years) of distance, galaxies are receding 71 km/s faster.

The expansion of the Universe will be the central theme in Chapters 18 and 19; we will discuss its implications and associated phenomena. For now, however, let us simply use Hubble's law to determine the distances of very distant galaxies and other objects. Knowing the value of H_0, a measurement of a galaxy's recession speed v then gives the distance d, since $d = v/H_0$; see *Figure It Out 16.3: Using Hubble's Law to Determine Distances* and *Figure It Out 16.4: Relativistic Effects.*

Note that Hubble's law cannot be used to find the distances of stars in our own Galaxy, or of galaxies in our Local Group; these objects are gravitationally bound to us, and hence the expansion of the intervening space is overcome. Moreover, Hubble's law does not imply that objects in the Solar System or in our Galaxy are themselves expanding; they are bound together by forces strong enough to overcome the tendency for empty space to expand. Hubble's law applies to distant galaxies and clusters of galaxies; the space between us and them is expanding.

FIGURE IT OUT 16.2   Redshifts and Hubble's Law

For a given absorption or emission line in a spectrum of a galaxy, define $\Delta\lambda$ (the Greek letters uppercase "delta" and lowercase "lambda") to be the displacement (shift) in the wavelength – that is, the observed wavelength (λ) minus the wavelength measured at rest in a laboratory gas (λ_0). Although different lines have different displacements $\Delta\lambda$, the ratio of the displacement to the rest wavelength ($\Delta\lambda/\lambda_0$) is the same for all lines in a galaxy's spectrum.

This quantity $z = \Delta\lambda/\lambda_0$ is defined as the redshift (z) of the galaxy. When multiplied by 100 per cent, the redshift z can be considered to be the percentage by which lines are shifted toward longer wavelengths. For example, if $z = 0.5$, then a line with $\lambda_0 = 6,000$ Å is shifted by $\Delta\lambda = 3,000$ Å (50 per cent of 6,000 Å) to an observed wavelength of $\lambda = 9,000$ Å.

Empirically, Hubble found that the redshift is larger in the spectra of distant galaxies than in the spectra of nearby galaxies. Thus, redshift is proportional to distance (or, $z \propto d$). If this redshift z is produced by motion away from us, then we can use the Doppler formula to derive v, the speed of recession (but keeping in mind that actually, the redshift is caused by an expansion of *space itself*, not by motion *through* space; see Chapter 18). Since $z = \Delta\lambda/\lambda_0 \approx v/c$ (the Doppler formula; see Chapter 11), we have

$$z \approx v/c, \text{ and therefore } v \approx cz,$$

where c is the speed of light (3×10^5 km/s). The approximation $v \approx cz$ is quite accurate as long as the redshift z is not larger than about 0.2, and then it becomes progressively less accurate (but the definition $z = \Delta\lambda/\lambda_0$ is exact for all redshifts). For example, if $z = 0.1$, then $0.1 = \Delta\lambda/\lambda_0 \approx v/c$, and so $v \approx 0.1c = 0.1 \times (3 \times 10^5$ km/s$) = 3 \times 10^4$ km/s.

Since speed v is proportional to redshift z, which Hubble found to be proportional to distance d, a galaxy's speed of recession must be proportional to its distance. This relation is known as Hubble's law,

$$v = H_0 d,$$

where H_0 is the constant of proportionality, Hubble's constant. Thus, a galaxy twice as far away from us as another galaxy recedes twice as fast; if 10 times as distant, it recedes 10 times as fast.

Note that $v = H_0 d$ does *not* imply that the speed of a given galaxy increases with time as its distance increases. We have defined H_0 to be the present-day value of Hubble's constant H, but the value of H actually changes with time. (Indeed, in most simple models of the Universe, $H \propto 1/t$.) Since the "Hubble constant" changes with time, some astronomers prefer to call it the "Hubble parameter." It is a constant only in space, having the same value throughout the Universe at any given time.

FIGURE IT OUT 16.3   Using Hubble's Law to Determine Distances

Suppose you find a galaxy that is so faint and far away that conventional techniques for measuring distances, such as the use of Cepheid variable stars, cannot be used, or give inaccurate results. You can instead do the following. (1) Obtain a spectrum of the galaxy. (2) Measure the redshift, $z = \Delta\lambda/\lambda_0$, of an absorption or emission line that you know has rest wavelength λ_0. (3) Compute the approximate recession speed v from $z \approx v/c$ – that is, solve for the recession speed $v \approx cz$. (4) Compute the distance d from Hubble's law, $v = H_0 d$, using your preferred value for Hubble's constant H_0 – in other words, solve for the distance $d = v/H_0$.

For example, suppose an absorption line with a known rest wavelength $\lambda_0 = 5,000$ Å is at an observed wavelength $\lambda = 5,100$ Å in the spectrum of a galaxy. Then

$$\Delta\lambda = \lambda - \lambda_0 = 5,100 \text{ Å} - 5,000 \text{ Å} = 100 \text{ Å, so}$$
$$z = \Delta\lambda/\lambda_0 = (100 \text{ Å})/(5,000 \text{ Å}) = 0.02.$$

Therefore the recession speed is $v \approx cz = (3 \times 10^5$ km/s$) \times 0.02 = 6,000$ km/s. If we assume that $H_0 = 71$ km/s/Mpc, we find that the galaxy's distance is

$$d = v/H_0 = (6,000 \text{ km/s})/(71 \text{ km/s/Mpc}) \approx 85 \text{ Mpc},$$

or about 280 million light-years, since 1 Mpc = 3.26 million light-years.

Several large "redshift surveys" of galaxies have each produced many thousands of distances, allowing study of the large-scale, three-dimensional distribution of galaxies. These are the studies that led to the conclusion that galaxies tend to be found in superclusters separated by large voids (see Figures 16–23 and 16–43).

One of the largest published sets of distances (there are over a million of them) is the "Two Degree Field Galaxy Redshift Survey" conducted with the Anglo-Australian 4-m telescope in Australia and named because the field of view is unusually wide for a telescope that size: 2°. The overall project is known as 2dF. (A 6dF project, with a field three times wider, is under way.) The Sloan Digital Sky Survey, which uses a telescope in New Mexico, surpassed that record and continues to measure positions and distances to tens of millions of galaxies and quasars.

16.8 THE SEARCH FOR THE MOST DISTANT GALAXIES

With the Hubble Space Telescope, we obtained relatively clear images of faint galaxies that are suspected to be very distant. The main imaging camera of the time (the Wide Field/Planetary Camera 2) exposed on a small area of the northern sky for 10 days in December 1995. Though it covers only about one 30-millionth of the area of the sky (roughly the apparent size of a grain of sand held at arm's length), this **Hubble Deep Field** contains several thousand extremely faint galaxies (■ Fig. 16–33a). If we could photograph the entire sky with such depth and clarity, we would see about 50 to 100 billion galaxies, each of which contains billions of stars!

Three years after the Hubble Deep Field, which was in the northern sky (and thus inaccessible to ground-based telescopes far south of the equator), the Hubble Space Telescope got very deep images of another region, this time in the southern celestial hemisphere: the

---

FIGURE IT OUT 16.4   Relativistic Effects

The Doppler-formula conversion between redshift and recession speed ($z \approx v/c$) used in *Figure It Out 16.2* and *16.3* loses significant accuracy when z is larger than about 0.2, and it appears to fail badly at redshifts approaching or exceeding 1 because the recession speed becomes close to (or exceeds) the speed of light. Thus, at high redshifts, some authors instead suggest using the "relativistic Doppler formula" from Einstein's special theory of relativity,

$$z = \sqrt{\frac{(1+v/c)}{(1-v/c)}} - 1 = \left[\frac{(1+v/c)}{(1-v/c)}\right]^{1/2} - 1$$

which leads to $v < c$ regardless of the redshift z. But even this equation is not strictly correct, since the redshift is actually due to the expansion of *space itself* rather than to motion *through* space; Einstein's *general* (rather than special) theory of relativity must be used for a correct solution.

An object at high redshift is currently at a greater distance than when it emitted the light you now see, because at each moment the space through which the light had already travelled subsequently expanded while the light was still on its way to us. (Indeed, objects that emitted their light soon after the origin of the Universe 13.7 billion years ago are now about 45 billion years away from us!) Hence, the high-redshift object's formally calculated speed (as defined by the total distance divided by the time it took light to travel) can easily *exceed* the speed of light! This is *not* a violation of Einstein's special theory of relativity: The object did *not* move *through* pre-existing space at a "superluminal" speed; instead, space itself expanded, and no information was transmitted at $v > c$.

So, at high redshifts, when the "distance" of an object derived from Hubble's law is very large (say, billions of light-years), it is better to think of this distance as a "lookback time" – that is, how long the light that we are currently receiving has been travelling from the object. Since the Universe expanded during the light's long journey, it had to cover a greater distance than if the Universe were static.

The derived "distance" corresponds neither to the object's distance when the light was emitted (which was smaller) nor to the object's distance now (which is larger), but rather to the distance at some intermediate time. This is much more confusing than simply saying that the light was emitted a certain time ago – the lookback time. Such lookback times can be compared with the age of the Universe for various models; see the example in Table 16–1. So, remember – if a newspaper article says that astronomers found a galaxy 8 billion light-years away, what it *really* means is that the light we are seeing was emitted 8 billion years ago!

Table 16–1 Lookback Times	
Redshift	**Lookback time**
0	0 years
0.1	1.3 billion years
0.2	2.4 billion years
0.3	3.4 billion years
0.4	4.3 billion years
0.6	5.7 billion years
0.8	6.8 billion years
1.0	7.7 billion years
1.5	9.3 billion years
2.0	10.3 billion years
2.5	11.0 billion years
3.0	11.5 billion years
4.0	12.1 billion years
5.0	12.5 billion years
6.0	12.7 billion years
7.0	12.9 billion years
8.0	13.0 billion years
9.0	13.1 billion years
10.0	13.2 billion years
12.0	13.3 billion years
15.0	13.4 billion years
∞	13.7 billion years

Note: Here we assume that the Universe's matter density is 27 per cent of the critical density, and that the energy density associated with the cosmological constant is 73 per cent of the critical density; see Chapter 18 for definitions and details. We also adopt a Hubble constant of 71 km/s/Mpc. The Universe is 13.7 billion years old in this model.

FIGURE 16–33 (a) Part of the original Hubble Deep Field, a tiny section of the sky in the northern celestial hemisphere, near the Big Dipper. About a thousand galaxies at a range of distances are visible in this exposure. In fact, only a few foreground stars in our own Milky Way Galaxy can be seen. (b) Part of the Hubble Ultra Deep Field, a region in the constellation Fornax that appears almost blank to ground-based telescopes. Overall, it looks similar to the original Hubble Deep Field but with finer resolution. This exposure, a million seconds (about 11 days) long, shows even fainter galaxies and was made more efficiently, thanks to the capabilities of Hubble's Advanced Camera for Surveys. In this image, blue and green are their true colors but red shows an infrared image. The data were collected over 400 Hubble orbits.

Hubble Deep Field – South. It looks similar to the northern field, even though it is nearly in the opposite direction in the sky, providing some justification for our assumption that the Universe is reasonably uniform over large scales. Later, after the Advanced Camera for Surveys (ACS) was installed on Hubble in 2002, it was used to make a **Hubble Ultra Deep Field** (■ Fig. 16–33b). These regions of the deep fields and ultra deep field have since been observed by many other telescopes on the ground and in space, notably the Chandra X-ray Observatory. (Although the ACS broke because of an electrical short-circuit in 2007, two of its three cameras were repaired by astronauts in 2009. They also installed the Wide Field Camera 3. It makes

the high-resolution images needed in the visible, and also extends the sensitivity into the ultraviolet on the shorter-wavelength end and the infrared on the longer-wavelength end.) ACS and WFC3 were used to make the Hubble eXtreme Deep Field (XDF), released in 2012.

Other deep surveys further strengthen our conclusion that we live in a rather typical place in the Universe. Spectra obtained with large telescopes, especially the two Keck telescopes in Hawaii, confirm that many galaxies in the Hubble Deep/Ultra Deep Fields and other deep surveys have large redshifts and hence are very far away. Though a few of the galaxies have relatively low redshifts, typical redshifts of the faintest objects are between 1 and 4 (■ Fig. 16–34). Light that

FIGURE 16–34 Part of the original Hubble Deep Field, with measured redshifts (mostly from the Keck telescopes) indicated for many of the galaxies.

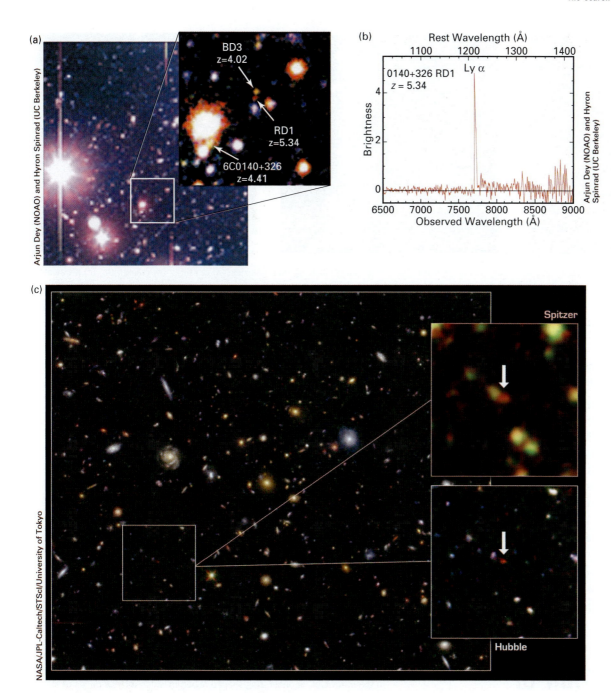

FIGURE 16–35 (a) The image and (b) the spectrum of a distant galaxy, obtained with the Keck telescopes. The measured redshift is 5.34, until recently one of the highest known. The spectrum shows ultraviolet light that was redshifted to optical wavelengths. (c) The galaxy GN-108036 at redshift 7.2, which puts it 12.9 billion light-years from us, only 750 million years after the big bang. Observations with the Hubble and Spitzer Space Telescopes discovered that stars are forming in it at a surprising rate – 100 suns per year, which is 30 times the rate of star formation in our own Milky Way Galaxy. The Hubble observations are part of the Great Observatories Origins Deep Survey (GOODS); this image shows Advanced Camera for Surveys orange light at 0.6 μm (reproduced as blue) and near-infrared light at 0.9 μm (reproduced as green), with a Wide Field Camera 3 infrared image at 1.6 μm (reproduced as red). The Spitzer image shows wavelengths 3.6 μm (in green) and 4.5 μm (in red). This extremely red galaxy does not appear in most of the visible-light Hubble images.

we observe at visible wavelengths actually corresponds to ultraviolet radiation emitted by the galaxy, but shifted redward by 100 per cent to 400 per cent! If we convert these redshifts to "distances" (or, more precisely, lookback times – see Table 16–1), we find that the galaxies are billions of light-years away. We see them as they were billions of years in the past, when the Universe was much younger than it is now. Note that when astronomers say that light from a high-redshift galaxy comes from "the distant universe," what they really mean is "distant parts of our Universe." We do not receive light from other universes, even if they exist!

In the past few years, many galaxies with redshifts exceeding 5 (that is, all the spectral lines are shifted by over 500 per cent, putting

them at more than 600 per cent of their original values) were discovered (■ Fig. 16–35). Identifying the spectral lines at such redshifts is difficult, and is done by fitting templates. At least one galaxy has been reported with a likely redshift of 10. The lookback time corresponding to redshift 6 is about 12.7 billion years (Table 16–1) and at redshift 10 over 13 billion years; we are seeing denizens of an era shortly (less than a billion years) after the birth of the Universe. They are probably newly formed galaxies.

Hubble Space Telescope and Spitzer Space Telescope observations have shown that a galaxy at redshift 7.2 (Fig. 16–35c), with a lookback time of 12.9 billion years, is astonishingly bright; it has a rate of star formation 30 times that of our own Galaxy and is called a

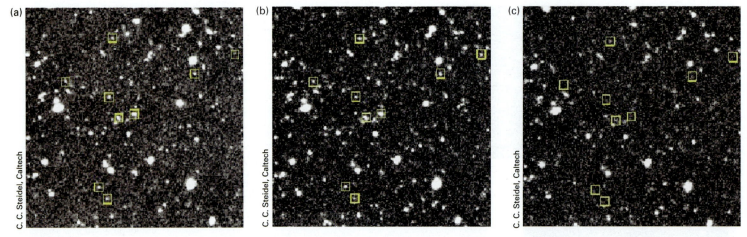

■ **FIGURE 16–36** Deep images through several different filters, obtained at the Keck Observatory, showing faint galaxies: (a) red filter, (b) green filter, (c) ultraviolet filter. Galaxies marked with boxes are visible in the red and green images, but are extremely faint or invisible in the ultraviolet image, suggesting that they are extremely far away; their intrinsic ultraviolet light has been redshifted, as well as absorbed by intervening clouds of gas.

"starburst galaxy." Only two galaxies have confirmed redshifts greater than this one's, though a few tentative redshift identifications are also greater.

Out of the thousands of galaxies found in images such as the Hubble Deep and Ultra Deep Fields, how do we go about choosing those that are likely to be at high redshift? (After all, with limited time for spectroscopy using large telescopes such as Keck, it is important to improve the odds if the goal is to find the highest redshifts.) One very effective technique is to first measure the "color" of each galaxy (■ Fig. 16–36). For two reasons, those that are likely to be very far away look very red, and have little if any of the blue or ultraviolet light normally emitted by stars. First, their redshift moves light to redder (longer) wavelengths. Second, there is often another galaxy or large cloud of gas along the way, and the hydrogen gas within it completely absorbs the ultraviolet light.

16.9 THE EVOLUTION OF GALAXIES

Comparisons of the appearance of distant galaxies and nearby galaxies provide clues to the way in which galaxies evolve. These comparisons must be done carefully to avoid wrong conclusions. For example, a visible-light image of a high-redshift galaxy actually corresponds to ultraviolet radiation emitted by the galaxy and shifted into the visible band, so it would not be fair to compare the image with the visible-light appearance of a nearby (and hence essentially unshifted) galaxy.

One way around this problem is to compare the visible-light images of high-redshift galaxies with ultraviolet images of nearby galaxies, as obtained with the Hubble Space Telescope. These images emphasize regions containing hot, massive, young stars that glow brightly in the ultraviolet (■ Fig. 16–37).

■ **FIGURE 16–37** Galaxies can appear dramatically different at different wavelengths. The galaxy NGC 6782 looks very different in ultraviolet and visible light, as shown in these images obtained with the Hubble Space Telescope. (a) In the ultraviolet, an inner, ring-shaped region in which intense bursts of star formation have occurred is obvious. An outer ring is faintly visible. The nucleus consists mostly of old stars, and is relatively faint in the ultraviolet. (b) The visible-light/infrared image, on the other hand, shows the pinwheel shape that is typical of spiral galaxies. The image reveals that the ring of star formation appears at the outer edge of the galaxy's bar. The image is made from three exposures: one in the blue, one in the yellow, and one in the near-infrared.

Another technique is to obtain infrared images of the high-redshift galaxies and compare them with the visible-light images of nearby galaxies. Such images tend to be dominated by light from older, less massive stars that more accurately reflect the overall shape of the galaxy rather than pockets of recent, intense star formation (■ Fig. 16–38). So far, the clearest infrared images have been made with the Hubble Space Telescope. (Its infrared camera ran out of solid nitrogen coolant sooner than anticipated. A new, electronic

Mark Dickinson (STScl) and NASA

■ **FIGURE 16–38** An infrared image of a high-redshift galaxy (z = 1), obtained with the Hubble Space Telescope. This can be compared with visible-light (optical) images of nearby galaxies that have only a small redshift, since the intrinsic wavelengths viewed are then the same in both cases.

method of cooling the camera was used in equipment installed during a servicing mission, and it is still working even better than before.)

These data, together with various types of analyses, such as computer simulations (for example, of what happens when two galaxies collide and merge), provide many interesting results. One spectacular conclusion is that most spiral galaxies used to look quite peculiar; there were essentially no large galaxies with distinct, well-formed spiral arms beyond redshift 2. By redshift 1 there were quite a few of them, but many took on their current, mature shapes more recently, in the past 5 billion years (that is, at redshifts below about 0.5).

Another conclusion is that there used to be a large number of small, blue, irregular galaxies that formed stars at an unusually high rate (■ Fig. 16–39). Their strange shape might be partly caused by an irregular distribution of young star clusters. However, they appear peculiar even at infrared wavelengths, which are more sensitive to older stars, so they must be structurally disturbed. Some of them probably later merged together to form larger galaxies, including disturbed spirals. Perhaps others faded and are now difficult to find because they are so dim.

It appears that most elliptical galaxies formed early in the Universe, beyond redshift 2 (lookback time 10 billion years); there are many old-looking, well-formed ellipticals between redshifts of 1 and 2. On the other hand, we also think that an elliptical galaxy can be produced by the collision and **merging** of two spiral galaxies. (Note that as in the case of the Bullet Cluster mentioned above, few if any *stars* actually collide during the merging process, because stars occupy only a tiny fraction of the volume of galaxies.) Many new stars are created from the interstellar gas in the spirals, especially when one cloud of gas collides with another.

Computer models of the merging process also show that long "tails" of material are sometimes temporarily formed (■ Fig. 16–40), just as we see in nearby examples of interacting galaxies (■ Fig. 16–41; also see the example we showed in Figure 16–14). Later

Rogier Windhorst and Sam Pascarelle (Arizona State University) and NASA

■ **FIGURE 16–39** Small, generally blue, irregular galaxies that have unusually high rates of star formation, imaged with the Hubble Space Telescope. These objects have lookback times of about 10 billion years. Some of them probably merged to form larger galaxies; others may have faded away with time.

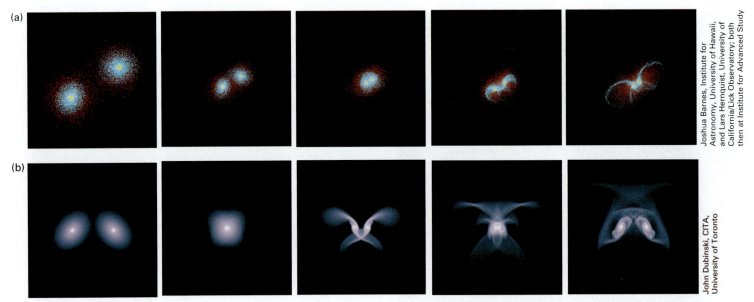

Joshua Barnes, Institute for Astronomy, University of Hawaii, and Lars Hernquist, University of California/Lick Observatory; both then at Institute for Advanced Study

John Dubinski, CITA, University of Toronto

■ **FIGURE 16–40** Calculations made with supercomputers, showing time sequences of the very close encounter of two identical model galaxies. (a) A time interval of 250 million years separates successive frames. About 10,000 particles represent each galaxy. (b) Advances in computing power allow even more particles as well as giant molecular clouds to be considered. The formation of tidal tails is obvious. Extension of such calculations to even later times shows the formation of an elliptical galaxy from the merged pair. The Antennae (images of which we saw in Figure 16–14) are thought to have taken their shape in the manner shown above.

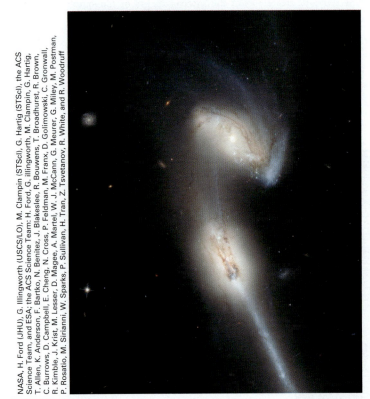

NASA, H. Ford (JHU), G. Illingworth (USCS/LO), M. Clampin (STScI), G. Hartig (STScI), the ACS Science Team, and ESA; the ACS Science Team: H. Ford, G. Illingworth, M. Clampin, G. Hartig, T. Allen, K. Anderson, F. Bartko, N. Benítez, J. Blakeslee, R. Bouwens, T. Broadhurst, R. Brown, C. Burrows, D. Campbell, E. Cheng, N. Cross, P. Feldman, M. Franx, D. Golimowski, C. Gronwall, R. Kimble, J. Krist, M. Lesser, D. Magee, A. Martel, W. J. McCann, G. Meurer, G. Miley, M. Postman, P. Rosatio, M. Sirianni, W. Sparks, P. Sullivan, H. Tran, Z. Tsvetanov, R. White, and R. Woodruff

■ **FIGURE 16–41** The Mice, NGC 4676, with long tails of stars and gas drawn out as two galaxies collided, imaged with the Advanced Camera for Surveys on the Hubble Space Telescope. Supercomputer simulations such as those in Figure 16–40 show that we are seeing two spiral galaxies approximately 160 million years after their closest encounter.

these tails disappear, leaving a more normal looking elliptical galaxy, but with a population of stars younger than in the really ancient ellipticals. In fact, the Milky Way Galaxy and the Andromeda Galaxy are approaching each other and may collide (or barely miss each other) in about 5 or 6 billion years. Subsequently, they are likely to merge and become an elliptical galaxy over the course of billions of years.

The total rate at which stars currently form in the Universe is rather small compared with what it was billions of years ago. We see that the star-formation rate has decreased since a redshift of 1 (8 billion years ago) to the present time. The rate may have been constant at still larger redshifts, up to about 4 or 5 (12 billion years ago), though we are unsure because many high-redshift galaxies are cloaked with dust. The dust seems to have been produced by the first few generations of stars, making it difficult to detect high-redshift galaxies at visible wavelengths. But infrared and submillimeter telescopes are finding them in progressively larger numbers, so we can expect a more accurate census in the near future.

As galaxies age, they evolve chemically, primarily because of supernovae that create many of the heavy elements through nuclear reactions and disperse them into the cosmos. Large, massive galaxies, whose gravitational fields don't allow much of the gas to escape, tend to become more chemically enriched than small galaxies that are not able to retain the hot gas. So, we don't expect to find many rocky, Earth-like planets in small (dwarf) galaxies like the Magellanic Clouds. The formation of massive galaxies like the Milky Way seems to be a critical step for the existence of humans.

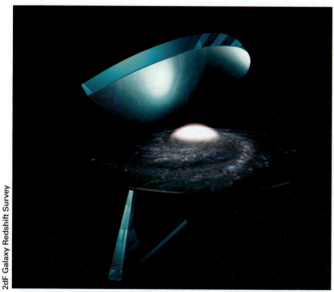

FIGURE 16–42 The two slices of space viewed by the Two Degree Field (2dF) project, in an artist's conception. The vertex of each wedge is the Earth, and the wedges were chosen to show some galaxies in each celestial hemisphere. The observations were made with a telescope in Australia.

16.10 EVOLUTION OF LARGE-SCALE STRUCTURE

We have also studied the evolution of large-scale structure in the Universe. By getting the redshifts of hundreds of thousands of galaxies over large regions of the sky (■ Fig. 16–42), the growth of clusters, superclusters, and voids can be traced (■ Fig. 16–43). As mentioned earlier, it appears that superclusters and giant walls of galaxies are the largest structures in the Universe; we have no clear evidence for super-superclusters of galaxies.

Galaxies generally preceded clusters, and then gravitationally assembled themselves into clusters. Many clusters formed relatively recently, within the past 5 billion years (that is, at redshifts below 0.5), and in fact are still growing now. However, cluster formation did begin earlier. Some very large, well-formed clusters have been found at redshift 1 (8 billion light-years away), and evidence exists for substantial concentrations of matter (which later formed clusters) at a redshift of 4, corresponding to a lookback time of about 12 billion years.

In 2012, the Third Sloan Digital Sky Survey (SDSS) team released BOSS, the Baryon Oscillation Spectroscopic Survey, with information about galaxies with redshifts up to $z = 0.8$ (see Figure 16–43b), corresponding to a distance of about 7 billion light-years (actually, a lookback time of 7 billion years), and quasars (discussed in the next chapter) with redshifts between 2.1 and 3.5, from 10 to 11.5 billion light-years away.

The observed distribution of superclusters and voids (as the 2dF and BOSS mapping projects showed in Figure 16–43) can be compared with the predictions of various theoretical models using computer simulations (■ Fig. 16–44). One important conclusion

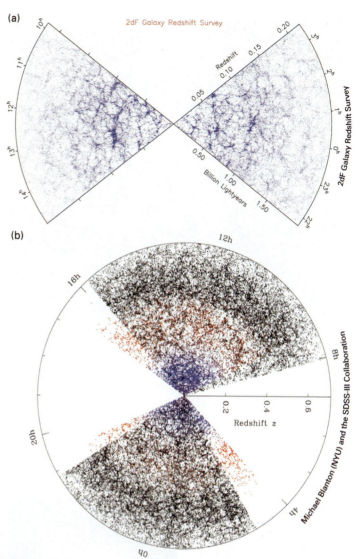

FIGURE 16–43 (a) A map of space, made by the Two Degree Field Galaxy Redshift Survey (2dF GRS), showing a slice whose vertex is at the Earth, as in Figures 16–23 and 16–42. The distances come from the measured redshifts. We see over two hundred thousand galaxies and the structure they reveal in the Universe. Many clusters of galaxies, superclusters, filaments, and voids can be traced. This plot shows the completed survey. (b) A map of space from 2012, made by the Sloan Digital Sky Survey 3 (SDSS-III). The black dots are massive galaxies from the Baryon Oscillation Spectroscopic Survey (BOSS). The red dots are very similar massive galaxies (Luminous Red Galaxies) from the SDSS-II survey. The blue dots are mostly galaxies closer to the Milky Way's mass (the "Main sample"). We see the Main-sample galaxies only close up, because they are intrinsically fainter so harder to see far away.

is that dark matter pervades the Universe; otherwise, it is difficult to produce very large structures. Galaxies and clusters seem to form at unusually dense regions ("peaks") in the somewhat more uniform dark matter distribution, like snow on the peaks of mountains.

Agreement with observations seems best when most of the dark matter used in the simulations is "cold" – that is, moving relatively slowly compared with the speed of light. Work is in progress to determine what specific type of **cold dark matter** is likely to account for most of the material. Simulations that use primarily **hot dark matter**

Virgo Consortium, University of Durham

■ **FIGURE 16–44** This supercomputer calculation of the formation of clusters of galaxies shows the existence of filaments and voids. The details of the results depend on whether cold dark matter, hot dark matter, or a mixture of the two fills the Universe. The ability of astrophysicists to model the formation of galaxies in the Universe increases as computing power increases.

(such as neutrinos, with speeds close to that of light) do not produce galaxy distributions that resemble those observed. Specifically, hot dark matter has a hard time clustering on small scales, like those of individual galaxies and small clusters.

Results announced from NASA's Wilkinson Microwave Anisotropy Probe (WMAP; see Chapter 19), which operated through 2010, show that about 16 per cent of the *matter* in the Universe consists of protons, neutrons, and electrons – although only about one-tenth of this normal matter resides in stars and other objects that we can easily see at visible wavelengths, the rest being in the form of relatively hot x-ray-emitting gas, MACHOs, and other constituents. The remaining 84 per cent is matter that does *not* consist of normal particles; moreover, it is dark. This is the cold dark matter discussed above. It may be mostly WIMPs (see Section 16.4c), but we don't yet know this for sure, and no actual WIMPs have ever been directly detected in a laboratory. Again, we emphasize that the nature of dark matter is one of the outstanding mysteries of modern astrophysics!

But the situation is actually even stranger! When we say that about 16 per cent of the matter in the Universe consists of protons, neutrons, and electrons, we are really talking about just 4–5 per cent of the *total* matter plus energy. This is because matter (most of which is dark) constitutes only 27 per cent of the Universe; the rest, as we shall see in Chapter 18, is dark energy, which is very different from dark (or luminous) matter (■ Fig. 16–45).

The Planck mission is now aloft, and in Chapter 19 we discuss its 2013 data release that shows its results about the curvature of the Universe and of density variations that seem to have

■ **FIGURE 16–45** The constituents of the Universe, as shown by Adam Riess in his 2011 Nobel Prize lecture in Stockholm. The numbers are approximate; more accurately, 23 per cent of the Universe consists of dark matter, and 73 per cent is dark energy. Also, the vast majority of the gas (marked here as 4 per cent) is actually not visible at optical wavelengths but instead radiates x-rays because it is quite hot.

Adam Reiss, StScI

a characteristic scale of angular distance in the Universe as a whole. A next big step might be WFIRST, the Wide-Field Infrared Survey Telescope, the surprise top priority in the 2010 decadal survey of recommendations by the astronomical community. With its 1.5-m mirror, it would search for signs of dark matter in three different ways starting in the year 2020. Shortly before that, the European Space Agency plans to launch its Euclid mission, which will search for dark energy by surveying galaxies and clusters of galaxies out to a redshift of 2 with a 1.2-m telescope from a "Lagrange," gravity-balancing point in the direction of the Sun (near the location of the Solar and Heliospheric Observatory). We can hope to get data from at least one of these two telescopes by early in the next decade.

The overlapping recommended ground-based survey, now under construction for use in the 2015 to 2025 time range, is the LSST (Large Synoptic Survey Telescope), to be built on the mountain Cerro Pachón in Chile. With its 8.4-m mirror, it is to photograph the whole sky visible from its location every three nights. To provide redshifts, a new spectrograph is to make use of 500 nights on the Cerro Tololo Inter-American Observatory's existing 4-m telescope to carry out the BigBOSS survey (where, again, BOSS stands for Baryon Oscillation Spectroscopic Survey). (This bunching up, or oscillation, of galaxies every 500 million light-years or so and at other intervals links back to pressure waves moving through the early universe, as surveyed in the cosmic microwave background by Planck and earlier spacecraft, which we will discuss in Chapter 18.) BigBOSS is to map 20 million galaxies and quasars, extending the size of the Sloan Digital Sky Survey's map by a factor of about 10 in distance and therefore 1,000 in volume, going back 10 billion years.

CONCEPT REVIEW

» Around the turn of the 20th century, some astronomers believed that the **spiral nebulae** were clouds of gas within our Milky Way star system, while others thought they might be huge, completely separate systems of stars bound together by gravity (Section 16.1a). Their distance and nature was the subject of the famous **Shapley-Curtis debate**. In the mid-1920s, **Edwin Hubble** recognized Cepheid variable stars within a few of these objects, and their observed faintness implied an enormous distance (Section 16.1b). Thus, the true nature of the spiral nebulae was finally unveiled: They are distant galaxies like our own Milky Way Galaxy, and the Universe is far larger than previously thought. Galaxies are the fundamental units of the Universe; each one contains billions of stars spread over a diameter exceeding 10,000 light-years.

» There are various types of galaxies. **Spiral galaxies** contain a thin, rotating disk of stars and gas; new stars tend to form in the spiral arms (Section 16.2a). They also have a spherical bulge of old stars. **Elliptical galaxies** have no disk or arms, and generally very little gas and dust; they consist almost entirely of old stars (Section 16.2b). They are roughly spherical or elongated in shape, and don't rotate much. Lenticular (S0) galaxies are a cross between spirals and ellipticals (Section 16.2c). The shapes of irregular and peculiar galaxies are strange. Galaxies frequently occur in **clusters of galaxies** (Section 16.3a), and these in turn congregate in **superclusters** separated by large **voids** (Section 16.3b).

» We now have strong indications that much of the matter in the Universe emits little if any detectable electromagnetic radiation, but it has a gravitational influence on its surroundings. One of the first clues was provided by the flat (nearly constant) **rotation curves** of spiral galaxies, including our own Milky Way Galaxy (Section 16.4a). These are in stark contrast with the orbital speeds of planets in the Solar System, which decrease with increasing distance from the Sun. There is also evidence for such **dark matter** in pairs, groups, and clusters of galaxies (Section 16.4b).

» The exact composition of the dark matter is unknown (Section 16.4c); possibilities include white dwarfs, neutron stars, black holes, brown dwarfs, large "planets," and neutrinos. Most of the dark matter is probably exotic subatomic particles such as **WIMPs** (weakly interacting massive particles). Much of what we call the "normal" dark matter (that is, consisting of protons, neutrons, and electrons) is gas at a temperature of about a million degrees; it has now barely been detected, and therefore isn't really "dark." Studies of **gravitational lensing** can help reveal the presence of dark matter (Section 16.5). They suggest that a small fraction of the dark matter in our Milky Way Galaxy may consist of **MACHOs** (massive compact halo objects), though the vast majority of it seems to be WIMPs.

» Clues to the formation and evolution of galaxies are found from observations of very distant galaxies, at large **lookback times**; we are viewing them as they were when the Universe was far younger than it is today (Section 16.6). Their distances are obtained from the **redshifts** measured in their spectra and interpreted according to the Doppler formula and **Hubble's law** (Section 16.7). The Universe is expanding, with the speeds of different galaxies being proportional to their current separations from us, and the constant of proportionality is called **Hubble's constant**. (We will see in Chapter 18 that the expansion is currently accelerating.)

» Galaxies at a variety of lookback times are seen in the **Hubble Deep Fields** and other long-exposure photographs of the Universe (Section 16.8). Recently, galaxies with redshifts exceeding 5, 6, and even 7 have been found; we see them as they were only about one billion years after the birth of the Universe. The Hubble Space Telescope's Advanced Camera for Surveys and Wide Field Camera 3 are more efficient and cover a wider range of wavelengths than the earlier cameras.

» Though most elliptical galaxies formed a very long ago, some were produced relatively recently by the **merging** of two spiral galaxies (Section 16.9). Early in the history of the Universe, there were many small, irregular galaxies that may have coalesced to form large spirals. Massive galaxies can retain the heavy elements synthesized and blown out by supernovae, while small galaxies tend not to become as chemically enriched.

» Computer models of the **large-scale structure** of the Universe suggest that clusters and superclusters of galaxies formed at the peaks of a somewhat more uniform distribution of dark matter (Section 16.10). **Cold dark matter**, which moves slowly compared with light, provides a better match to the data than does rapidly moving **hot dark matter**. Cold dark matter constitutes most of the matter in the Universe, but it cannot be normal matter consisting of protons, neutrons, and electrons. Though cold dark matter may be mostly WIMPS, its exact nature is one of the outstanding mysteries of modern astrophysics. All matter (of which about 84 per cent is cold dark matter and 16 per cent is normal matter) constitutes only 27 per cent of the Universe; the rest (73 per cent) is gravitationally repulsive "dark energy," to be discussed in Chapter 18.

QUESTIONS

1. Summarize the topic (and its significance) of the Shapley-Curtis debate.

2. What was the physical principle used by Edwin Hubble to demonstrate that the "spiral nebulae" are very distant and large?

3. Explain why Cepheid variable stars are so useful for determining the distances of galaxies.

4. What are the main morphological types of galaxies, and their most important physical characteristics?

5. In which types of structures are many or most galaxies found?

6. Why can't we determine the distances to galaxies using the geometric method of trigonometric parallax (triangulation), as we do for stars?

7. How is the discovery of other galaxies an extension of the Copernican revolution?

8. Describe the rotation curve of the Milky Way Galaxy and contrast its shape with that of the Solar System.

†9. Show that the mass of the Milky Way Galaxy within a radius of 10,000 light-years from the center is about 3×10^{10} solar masses. (Assume a rotation speed of 200 km/s at that radius.)

10. How does the shape of the Milky Way Galaxy's rotation curve imply the presence of dark matter, given the distribution of visible light in the Galaxy?

11. The sense of rotation of galaxies is determined spectroscopically. How might this be done in practice?

12. What is the evidence for dark matter in larger structures such as clusters of galaxies?

13. Describe the phenomenon of gravitational lensing.

14. Define MACHOs, state the tentative evidence for their presence, and discuss their possible nature.

15. How can we statistically study the evolution of galaxies many billions of years ago?

16. What if the speed of light were infinite? Would there still be a way for astronomers to look back at galaxies as they appeared long ago? Explain.

17. What is meant by the redshift of a galaxy?

18. State Hubble's law, explain its meaning, and argue that the Universe is expanding.

19. Is your body expanding according to Hubble's law? Is the Earth expanding?

20. Can Hubble's law be used to determine the distance of **(a)** α Centauri and other stars relatively near the Sun, **(b)** the center of our Galaxy, **(c)** the Large Magellanic Cloud, a satellite of our Galaxy, and **(d)** a cluster of galaxies very far from the Local Group?

†21. What is the recession speed of a galaxy having a distance of 50 Mpc, if Hubble's constant is 71 km/s/Mpc?

†22. At what speed is a galaxy 100 million light-years away receding from us, if Hubble's constant is 71 km/s/Mpc? (Be careful with units!)

†23. What is the redshift z of a galaxy whose Hα absorption line (rest wavelength 6,563 Å) is measured to be at 6,707 Å?

†24. Calculate the distance of a galaxy with a measured redshift $z = 0.03$, using 71 km/s/Mpc for Hubble's constant.

25. Of what importance are spectra and clear images of distant galaxies, in studies of galactic evolution?

26. Summarize some of the tentative conclusions drawn about the formation and evolution of galaxies.

27. True or false? Spiral galaxies contain only relatively young stars; there are no ancient stars anywhere within them.

28. True or false? The nearly "flat" rotation curve of our Milky Way Galaxy implies that the mass of our Galaxy interior to a given radius grows linearly with increasing radius (that is, $M \propto r$).

29. True or false? Giant "voids" in the Universe are the result of enormous explosions that blew matter out of those regions.

30. True or false? Some elliptical galaxies appear to have formed as a result of the merging of two spiral galaxies.

31. True or false? If you measure the redshift of a certain galaxy to be 4 right now, then an observer living in that galaxy would measure the Milky Way Galaxy to have a redshift of 4 right now.

32. Multiple choice: Which one of the following has never been used by astronomers measuring distances? **(a)** Cepheid variable stars. **(b)** Trigonometric parallax of nearby galaxies. **(c)** Hubble's law. **(d)** Main-sequence stars. **(e)** Inverse-square law of light.

33. Multiple choice: Which one of the following has *not* been used to suggest that there is "dark matter" in the Universe? **(a)** The orbital speeds of planets around the Sun. **(b)** The orbital speeds of stars in our Milky Way Galaxy and in other spiral galaxies. **(c)** The orbital speeds of galaxies in binary pairs. **(d)** The orbital speeds of galaxies in clusters. **(e)** The gravitational lensing of background galaxies by foreground clusters of galaxies.

34. Multiple choice: How do we try to detect MACHOs? **(a)** By looking at a dimming of stars when MACHOs pass in front of them. **(b)** By observing the Large and Small Magellanic Clouds night after night. **(c)** Through their interactions with WIMPs. **(d)** By looking at the Doppler shift of stars. **(e)** From their occasional emission of x-rays.

†35. Multiple choice: If an absorption line of calcium is normally found at a wavelength of about 4,000 Å in a laboratory gas, and you see it at 4,400 Å in the spectrum of a galaxy, what is the approximate distance to the galaxy? (For simplicity, you may assume that Hubble's constant is 100 km/s/Mpc, instead of the more correct value of 71 km/s/Mpc.) **(a)** 300 Mpc. **(b)** 40,000 light-years. **(c)** 3 Mpc. **(d)** 3 million light-years. **(e)** 3,000 Mpc.

36. Multiple choice: Spiral arms are usually the most prominent features in the disk of a spiral galaxy. Which one of these statements about spiral arms is *false*? **(a)** Clouds of gas and dust are mostly found in spiral arms. **(b)** Emission nebulae are mostly found in spiral arms. **(c)** Spiral arms contain most of the hot, young, massive stars. **(d)** Spiral arms are usually the most prominent features in the disk. **(e)** Spiral arms consist mostly of dark matter.

37. Multiple choice: Which one of the following statements about the expanding Universe and Hubble's law ($v = H_0 d$) is *false*? **(a)** If Galaxies Vesto and Slipher both satisfy Hubble's law, and Galaxy Vesto's distance is three times Galaxy Slipher's distance, then Galaxy Vesto is moving away from us three times as fast as Galaxy Slipher. **(b)** The distance of a galaxy whose recession velocity is 7,100 km/s is about 100 Mpc. **(c)** Despite the expansion of the Universe, pairs of galaxies can still sometimes move toward each other and even merge together. **(d)** Hubble's law can be used to determine the lookback times of distant galaxies – that is, how far back in time we are seeing them. **(e)** Hubble's law alone implies that as the distance of a particular galaxy increases with time, it moves away from us progressively faster.

38. Multiple choice: Which one of the following statements about gravitational lensing is *false*? **(a)** Spectra taken of different images of a single gravitationally lensed object will be essentially identical to each other. **(b)** An asymmetric distribution of matter in the lensing or lensed object leads us to observe discrete points or arcs instead of full Einstein rings. **(c)** Gravitational lensing bends blue light more than red light, since blue light is more energetic than red light and the warping of space-time is therefore greater. **(d)** It provides strong support for Einstein's general theory of relativity, which predicts that mass and energy can warp space-time. **(e)** By using the method of gravitational lensing, we can measure the total mass (luminous plus dark matter) of the lensing object.

39. Multiple choice: Which one of the following statements about galaxy evolution is *false*? **(a)** At least some galaxies formed from the joining together (merging) of small groups of stars. **(b)** In the past, there existed a population of small, blue, irregular galaxies that have either merged together or faded from view. **(c)** Our Milky Way Galaxy and the Andromeda Galaxy are likely to form an elliptical galaxy at some time in the future. **(d)** Astronomers can learn about the evolution of galaxies by comparing recently obtained images of galaxies with images of the same galaxies taken by Edwin Hubble in the early 20th century. **(e)** Most elliptical galaxies formed within the first few billion years after the big bang.

40. **Fill in the blank:** High-redshift arcs around clusters of galaxies at lower redshifts are now thought to be examples of _____.

41. **Fill in the blank:** Most of the dark matter is thought to consist of _____; it is not the "normal" dark matter composed of neutrons, protons, and electrons.

42. **Fill in the blank:** Edwin Hubble realized that some of the "spiral nebulae" visible in photographs must be outside our own Milky Way Galaxy because he saw very faint _____ within them, yet such stars are known to be intrinsically luminous.

43. **Fill in the blank:** Distant galaxies have large _____ times, allowing us to study the evolution of galaxies.

†44. **Fill in the blank:** If Hubble's constant is 71 km/s/Mpc, then a galaxy whose redshift $z = 0.05$ has a distance of _____.

†45. **Fill in the blank:** If Hubble's constant is 71 km/s/Mpc, then a galaxy 400 Mpc away has a redshift z of _____.

†This question requires a numerical solution.

TOPICS FOR DISCUSSION

1. How is the discovery of other galaxies an extension of the Copernican revolution?

2. Are you bothered by the notion that most of the matter in the Universe might be dark, detectable only through its gravitational influence? Can you think of any alternative explanations for the data? Explain.

3. What might be some possible explanations for the redshifts of galaxies that do not involve the expansion of the Universe? How could you test these hypotheses?

A combination of x-ray data (shown in blue) from NASA's Hubble Space Telescope and visual-light imaging (shown in yellow) from the Baade Telescope in Chile (one of the Magellan p... of telescopes) reveals a pair of quasars formed by the merger of two galaxies. This system, SDSS 1254+0846 (named from the Sloan Digital Sky Survey and the object's position in rig... ascension and declination), is the first time such a merger of galaxies, which have supermassive black holes at their cores, has been seen. The pair is about 4.6 billion light-years from... We see a region about 1.3 million light-years across in the direction of the constellation Virgo (though well past the stars in our Galaxy that make up the constellation).

X-ray (NASA/CXC/SAO/P. Green et al.), Optical (Carnegie Obs./Magellan/W. Baade Telescope/J.S. Mulchaey et al.)

17

Quasars and Active Galaxies

ORIGINS

We explore the significance of quasars as an early phase in the origin and evolution of large galaxies, such as our Milky Way Galaxy. We find that quasars allow us to study the distribution of matter, both visible and dark, in the vast space between us and them.

Quasars, and the way in which they became understood, have been one of the most exciting stories of the last half century of astronomy. First noticed as seemingly peculiar stars, quasars turned out to be some of the most powerful objects in the Universe, and represent violent forces at work (see the Chapter Opener). We think that giant black holes, millions or even billions of times the Sun's mass, lurk at their centers. A quasar shines so brightly because its black hole is pulling in the surrounding gas, causing the gas to glow vividly before being swallowed.

Our interest in quasars is further piqued because many of them are among the most distant objects we have ever detected in the Universe. Since, as we look out, we are seeing light that was emitted farther and farther back in time, observing quasars is like using a time machine that enables us to see the Universe when it was very young. We find that quasars were an early stage in the evolution of large galaxies. As time passed, gas in the central regions was used up, and the quasars faded, becoming less active. Indeed, we see examples of active galaxies relatively near us, and in some of these the presence of a massive black hole has been all but proven.

■ **FIGURE 17–1** Hubble Space Telescope images of the central regions of (a) the normal galaxy NGC 7626 and (b) the active galaxy NGC 5548. Both galaxies have the same general appearance (Hubble type), but note the bright, star-like nucleus in the active galaxy. The spikes of light from this star-like nucleus are a telescope artifact ("diffraction spikes," like those in images of bright stars).

17.1 ACTIVE GALACTIC NUCLEI

The central regions of normal galaxies tend to have large concentrations of stars. For example, at infrared wavelengths we can see through our Milky Way Galaxy's dust and penetrate to the center. When we do so, we see that the bulge of our Galaxy becomes more densely packed with stars as we look closer to the nucleus. With so many stars confined there in a small volume, the nucleus itself is relatively bright. This concentrated brightness appears to be a natural consequence of galaxy formation; gas settles in the central region because of gravity, and subsequently forms stars.

In a minority of galaxies, however, the nucleus is far brighter than usual at optical and infrared wavelengths, when compared with other galaxies at the same distance (■ Fig. 17–1). Indeed, when we compute the optical luminosity (power) of the nucleus from its apparent brightness and distance, we have trouble explaining the result in terms of normal stars: It is difficult to cram so many stars into so small a volume.

Such nuclei are also often very powerful at other wavelengths, such as x-rays, ultraviolet, and radio. These galaxies are called "active" to distinguish them from normal galaxies, and their luminous centers are known as **active galactic nuclei**. Clusters of ordinary stars rarely, if ever, produce so much x-ray and radio radiation.

Active galaxies that are extraordinarily bright at radio wavelengths often exhibit two enormous regions (known as "lobes") of radio emission far from the nucleus, up to a million light-years away. The first "radio galaxy" of this type to be detected, Cygnus A (■ Fig. 17–2), emits about a million times more energy in the radio region of the spectrum than does the Milky Way Galaxy.

Close scrutiny of such radio galaxies sometimes reveals two long, narrow, oppositely directed "jets" joining their nuclei and lobes (■ Fig. 17–3). The jets are thought to consist of charged particles moving at close to the speed of light and emitting radio waves. Sometimes radio galaxies appear rather peculiar when we look at visible wavelengths, and the jet is visible in x-rays, as in the case of Centaurus A (■ Fig. 17–4).

■ **FIGURE 17–2** A radio map of Cygnus A, with shading indicating the intensity of the radio emission in the two giant lobes. A Hubble Space Telescope optical image, with the blue component provided by one of the Keck telescopes, shows detailed structure. Particles feeding the large radio lobes come from the very central part.

William C. Keel (University of Alabama), Karl-Heinz Mack, Dayton Jones (JPL), and Rick Perley (NRAO)

NGC 6251 radio structure

30 arc min

1 arc min

1 arc min

0.1 arc sec

■ **FIGURE 17–3** Radio maps of the active galaxy NGC 6251 shown on four scales. (At the distance of NGC 6251, 30 arc minutes corresponds to about 3 million light-years, 1 arc minute is about 100,000 light-years, and 0.1 arc second is about 170 light-years.) One of the radio jets is pointing roughly in our direction, and thus appears much brighter than the other jet, which points away from us.

(a) Centaurus A
NGC 5128
HST WFC3/UVIS

F225W+F336W+F438W
F547M *y*
F657N Hα+[N II]
F673N [S II]
F814W *I*

3000 light-years
1400 parsecs
56″

NASA, ESA, and the Hubble Heritage (STScI/AURA)-ESA/Hubble Collaboration; Acknowledgment: R. O'Connell (University of Virginia) and the WFC3 Scientific Oversight Committee

(b)

NASA/Chandra X-ray Observatory Center/SAO

(c)

NASA, ESA, and J. Madrid (McMaster University)

STIS
May 17, 1999

STIS
Feb. 27, 2002

STIS
Jul. 17, 2002

ACS
Apr. 17, 2003

ACS
May 9, 2005

ACS
Nov. 28, 2006

■ **FIGURE 17–4** (a) Centaurus A (NGC 5128) looks like an elliptical or S0 galaxy seen through an extensive dust lane, which is shown in this Hubble Space Telescope image with its new Wide Field Camera 3. The approximate wavelengths of the filters used appear as color labels. (b) A Chandra Observatory x-ray view of the jet in Centaurus A. The orientation is the same but the image scale is 2.5 times greater (in other words, a more close-up view). (c) Hubble Space Telescope views of the changes in the jet of M87 (Figure 16–12a) over a few years.

■ **FIGURE 17–5** A very small part of the optical spectra of the nuclei of two galaxies, plotted with the redshifts removed (so that we see light at the emitted wavelengths rather than the observed, redshifted wavelengths). (a) A normal galaxy in which hot, massive stars have recently formed. Note the two narrow emission lines, whose width indicates that gas moves with a speed of only a few hundred km/s. There is also some Hβ absorption, formed by relatively massive stars. (b) A typical active galactic nucleus, in this case the Seyfert galaxy NGC 5548 (an image of which is shown in Fig. 17–1b). Note the broad Hβ emission line along with narrower lines (in this case, O++ and the narrow component of Hβ). The broad line is formed by gas, close to the center, that is moving very rapidly (thousands of km/s). The narrow lines come from gas, farther out from the nucleus, having a speed of just a few hundred km/s.

Optical spectra of the active nuclei often show the presence of gas moving with speeds in excess of 10,000 km/s, far higher than in normal galactic nuclei. We measure these speeds from the spectra, which have broad emission lines (■ Fig. 17–5). Atoms that are moving toward us emit photons that are then blueshifted, while those that are moving away from us emit photons that are then redshifted, thereby broadening the line by the Doppler effect. Early in the 20th century, Carl Seyfert was the first to systematically study galaxies with unusually bright optical nuclei and peculiar spectra, and in his honor they are often called "Seyfert galaxies."

Although spectra show that gas has very high speeds in supernovae as well, the overall observed properties of active galactic nuclei generally differ a lot from those of supernovae, making it unlikely that stellar explosions are responsible for such nuclei. Indeed, it is difficult to see how stars of any kind could produce the unusual activity. However, for many years active galaxies were largely ignored, and the nature of their central powerhouse was unknown.

17.2 QUASARS: DENIZENS OF THE DISTANT PAST

Interest in active galactic nuclei was renewed with the discovery of **quasars** (shortened form of "quasi-stellar radio sources"), the recognition that quasars appear similar to active galactic nuclei, and the realization that both kinds of objects must be powered by a strange process that is unrelated to stars.

17.2a The Discovery of Quasars

In the late 1950s, as radio astronomy developed, astronomers found that some celestial objects emit strongly at radio wavelengths.

Catalogues of them were compiled, largely at Cambridge University in England, where the method of pinpointing radio sources was developed. For example, the third such Cambridge catalogue is known as "3C," and objects in it are given numerical designations like 3C 48.

Although the precise locations of these objects were difficult to determine with single-dish radio telescopes (since they had poor angular resolution; recall our discussion of radio telescopes in Chapter 3), sometimes within the fuzzy radio image there was an obvious probable optical counterpart such as a supernova remnant or a very peculiar galaxy. More often, there seemed to be only a bunch of stars in the field – yet which of them might be special could not be identified, and in any case there was no known mechanism by which stars could produce so much radio radiation.

Special techniques were developed to pinpoint the source of the radio waves in a few instances. Specifically, the occultation (hiding) of 3C 273 by the Moon provided an unambiguous identification with an optical star-like object. When the radio source winked out, we knew that the Moon had just covered it while moving slowly across the background of stars. Thus, we knew that 3C 273 was somewhere on a curved line marking the front edge of the Moon. When the radio source reappeared, we knew that the Moon had just uncovered it, so it was somewhere on a curved line marking the Moon's trailing edge at that time. These two curves intersected at two points, and hence 3C 273 must be at one of those points. Though one point seemed to show nothing at all, the other point was coincident with a bluish, star-like object about 600 times fainter than the naked-eye limit.

When the positions of other radio sources were determined accurately enough, it was found that they, too, often coincided with faint, bluish-looking stars (■ Fig. 17–6). These objects were dubbed "quasi-stellar radio sources," or "quasars" for short. Optically they looked like stars, but stars were known to be faint at radio wavelengths, so they had to be something else. Object 3C 273 seemed to be especially interesting: A jet-like feature, visible at optical wavelengths (■ Fig. 17–7) and radio wavelengths (■ Fig. 17–8), stuck out from it.

(a)   3C 48

(b)   3C 147

(c)   3C 273

(d)   3C 196

■ **FIGURE 17–6** Original optical photographs of four of the first known quasars, from the photographic era before the current era of making electronic images. The objects appear star-like, but measurements with radio telescopes show that they are much brighter than normal stars at radio wavelengths.

■ **FIGURE 17–7** (a) A Hubble Space Telescope visible-light image of 3C 273, the first quasar to be unambiguously identified, shows a star-like central brightness. The diffraction spikes going off in several directions are artifacts formed in the telescope. The image also shows, to the lower right, a faint jet. (b) Hubble's Advanced Camera for Surveys includes a dark spot that artificially eclipses the central quasar 3C 273, greatly dimming its light. This provides a much better view of the quasar's host galaxy because it removes the glare from the ultra-bright central source.

17.2b Puzzling Spectra

Several astronomers, including Maarten Schmidt of Caltech, photographed the optical spectra of some quasars with the 5-m (200-inch) Hale telescope at the Palomar Observatory. These spectra turned out to be bizarre, unlike the spectra of normal stars. They showed bright, broad emission lines, at wavelengths that did not correspond to lines emitted by laboratory gases at rest. Moreover, different quasars had emission lines at different wavelengths.

Schmidt made a breakthrough in 1963, when he noticed that several of the emission lines visible in the spectrum of 3C 273 had the pattern of hydrogen – a series of lines with their spacing getting closer together toward shorter wavelengths – though not at the normal hydrogen wavelengths (■ Fig. 17–9). He realized that he could simply be observing hot hydrogen gas (with some contaminants to produce the other lines) that was Doppler shifted. The required redshift would

be huge, about 16 per cent (that is, $z = \Delta\lambda/\lambda_0 \approx 0.16$), corresponding to 16 per cent of the speed of light (since $z \approx v/c$, or $v \approx cz$, valid for z less than about 0.2).

This possibility had not been recognized because nobody expected stars to have such large redshifts. Also, the spectral range then available to astronomers, who took spectra on photographic film, did not include the bright Balmer-α line of hydrogen (that is, Hα), which is normally found at 6,563 Å but was shifted over to 7,600 Å in 3C 273. As soon as Schmidt announced his insight, the spectra of other quasars were interpreted in the same manner. Indeed, one of Schmidt's Caltech colleagues, Jesse Greenstein, immediately realized that the spectrum of quasar 3C 48 looked like that of hydrogen redshifted by an even more astounding amount, 37 per cent.

Subsequent searches for blue star-like objects revealed a class of "radio-quiet" quasars – their optical spectra are similar to those of quasars, yet their radio emission is weak or absent. These are often

■ **FIGURE 17–8** The jet of the quasar 3C 273 stems from the nucleus. Left to right, we see the jet in visible light from Hubble, x-rays from Chandra, and radio waves from MERLIN, a British set of radio telescopes.

called QSOs ("**q**uasi-**s**tellar **o**bjects"), and they are about ten times more numerous than "radio-loud" quasars. Consistent with the common practice of using the terms interchangeably, here we will simply use "quasar" to mean either the radio-loud or radio-quiet variety, unless we explicitly mention the radio properties.

17.2c The Nature of the Redshift

How were the high redshifts produced? The Doppler effect is the most obvious possibility. But it seemed implausible that quasars were discrete objects ejected like cannonballs from the center of the Milky

Way Galaxy (■ Fig. 17–10); their speeds were very high, and no good ejection mechanism was known. Also, we would then expect some quasars to move slightly across the sky relative to the stars, since the Sun is not at the center of the Galaxy, but such motions were not seen. Even if these problems could be overcome, we would then have to conclude that only the Milky Way Galaxy (and not other galaxies) ejects quasars – otherwise, we would have seen "quasars" with blueshifted spectra, corresponding to those objects emitted toward us from other galaxies.

Similarly, there were solid arguments against a "gravitational redshift" interpretation (recall our discussion of this effect in Chapter 14), one in which a very strong gravitational field causes the emitted light to lose energy on its way out. This possibility was completely ruled out later, as we shall see.

If, instead, the redshifts of quasars are due to the expansion of the Universe (as is the case for normal galaxies), then quasars are receding with enormous speeds and hence must be very distant. Quasar 3C 273, for example, has $z = 0.16$, so $v \approx 0.16c \approx 48,000$ km/s. According to Hubble's law, $v = H_0 d$, so if $H_0 = 71$ km/s/Mpc, then

$$d = v/H_0 \approx (48,000 \text{ km/s})/(71 \text{ km/s/Mpc})$$
$$\approx 680 \text{ Mpc} \approx 2.2 \text{ billion light-years},$$

nearly a sixth of the way back to the origin of the Universe! A few galaxies with comparably high redshifts (and therefore distances) had previously been found, but they were fainter than 3C 273 by a factor of 10 to 1,000, and they looked fuzzy (extended) rather than star-like.

Quasar 3C 273 turns out to be one of the closest quasars. Other quasars found during the 1960s had redshifts of 0.2 to 1, and hence are billions of light-years away. While we could use the relativistic Doppler formula given in *Figure It Out 16.4: Relativistic Effects* to calculate their recession speeds, this would be misleading and, strictly speaking, incorrect: The redshift is produced by the *expansion* of space itself, not by motion *through* space (and thus not actually the Doppler effect). The concept of "speed" then takes on a somewhat different meaning.

■ **FIGURE 17–9** (a) The upper spectrum is a photographic optical spectrum of the quasar 3C 273. Dark lines represent emission, not absorption (this is a "negative" image). The hydrogen Balmer emission lines (Hβ, Hγ, and Hδ) appear as fuzzy (broad) bands. When properly scanned, the spectrum would look much like the modern, well-calibrated spectrum of 3C 273 shown in (b). The comparison spectrum in (a) is of a hot lamp in the telescope dome; it consists of hydrogen, helium, neon, and other elements that emit lines at known wavelengths. This "comparison spectrum" establishes the wavelength scale. A color bar shows the colors of the different wavelengths of the comparison spectrum. The hydrogen Balmer lines (Hβ, Hγ, and Hδ) in the quasar (upper) spectrum in (a) are at longer wavelengths (labels in red) than in the comparison spectrum (labels in green). Similarly, in (b), the observed hydrogen emission lines appear redshifted to wavelengths longer than their laboratory values (indicated at the top). The redshift of 16 per cent corresponds, according to Hubble's law (with $H_0 = 71$ km/s/Mpc), to a distance of about 2.2 billion light-years.

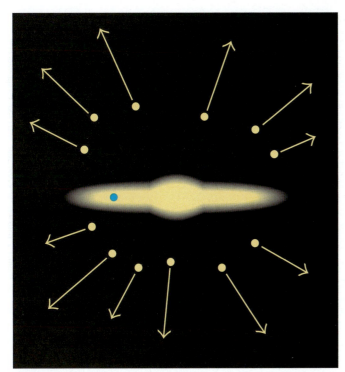

FIGURE 17–10 The idea that quasars were local, and were ejected from our Galaxy, could explain why they all have high redshifts. (The blue dot shows the position of the Sun.) But this hypothesis has a number of problems and was quickly rejected by most astronomers.

As discussed in Chapter 16 for galaxies, it makes more sense to refer to the "lookback time" of a given quasar (the time it has taken for light to reach us) than to its distance: $v = H_0 d$ is inaccurate at large redshifts for a number of reasons. The lookback time formula is complicated, but some representative values are given in Table 16–1.

A few dozen quasars with redshifts exceeding 6 have been discovered. (■ Fig. 17–11). The highest redshift known for a quasar as of 2013 is $z = 7.085$, which means that a feature whose laboratory (rest) wavelength is 1,000 Å is observed to be at a wavelength 708.5 per cent larger, or 1,000 Å + 7,085 Å = 8,085 Å. (Recall that $z = \Delta\lambda/\lambda_0$.) According to Table 16–1, the quasar's lookback time is roughly 12.9 billion years. Comparing this with the Universe's age of 13.7 billion years, we see the quasar as it was when the Universe was about 800 million years old, or just 6 per cent of its current age!

For a long time, quasars held the record for the highest redshifts ever identified for discrete objects (that is, not the cosmic background radiation that we discuss in Chapter 19). However, in the past few years normal galaxies at somewhat higher redshifts, possibly even surpassing $z = 10$, have been found. Also, a gamma-ray burst (see Chapter 14) having $z = 8.2$ was discovered in 2009, as was another such object at a probable redshift of 9.4.

How do we detect quasars? Many of them are found by looking for faint objects with unusual colors – that is, the relative amounts of blue, green, and red light differ from those of normal stars.

(a)

(b)

(c)

FIGURE 17–11 (a) A false-color image of the most distant known quasar, ULAS J1120+0641, at the center of the image, discovered in 2010. The blue-green-red colors of the displayed image translate from the near-infrared filters used (*i* from about 0.7 to 0.8 μm; *z* from about 0.8 to 0.9 μm; *Y* from about 1.0 to 1.1 μm); since the quasar appears only in the *Y* filter, it reproduces here as red. (b) Its spectrum, from the Gemini North telescope in Hawaii, shows that all the emission lines are shifted by 7.085 times their original wavelengths. It is currently the only known quasar with a redshift greater than 7. The high redshift puts the ultraviolet line Lyman-α, normally at 1,216 Å, into the near-infrared region of the spectrum. At these redshifts, we are looking back to only about 800 million years after the birth of the Universe. (c) An artist's conception of the region of the quasar as it would be seen from a few megaparsecs away.

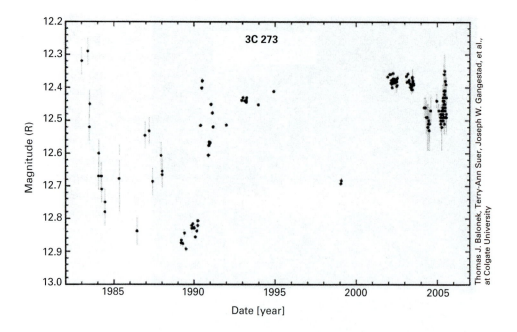

FIGURE 17–12 The apparent brightness of optical radiation from the quasar 3C 273 varies on a timescale of months. Note that a difference from magnitude 12.3 to 12.9, a fading by 0.6 magnitudes, is a fading by a factor of about 2.

Low-redshift quasars tend to look bluish, because they emit more blue light than typical stars. But the light from high-redshift quasars is shifted so much toward longer wavelengths that these objects appear very red, especially since intergalactic clouds of gas absorb much of the blue light. Quasars have also been found in maps of the sky made with x-ray satellites, and of course with ground-based radio surveys.

After finding a quasar candidate with any technique, however, it is necessary to take a spectrum in order to verify that it is really a quasar and to measure its redshift. As we have seen, the spectra of quasars are quite distinctive, and are rarely confused with other types of objects. Hundreds of thousands of quasars are now known, and more are being discovered very rapidly, especially by the Sloan Digital Sky Survey.

17.3 HOW ARE QUASARS POWERED?

Astronomers who conducted early studies of quasars (mid-1960s) recognized that quasars are very powerful, 10 to 1,000 times brighter than a galaxy at the same redshift. But while galaxies looked extended in photographs, quasars with redshifts comparable to those of galaxies appeared to be mere points of light, like stars. Their diameters were therefore smaller than those of galaxies, so their energy-production efficiency must have been higher, already making them unusual and intriguing.

17.3a A Big Punch from a Tiny Volume

However, these astronomers were in for a big surprise when they figured out just how compact quasars really are. They noticed that

some quasars vary in apparent brightness over short timescales – days, weeks, months, or years (Fig. 17–12). This implies that the light-emitting region is probably smaller than a few light-days, light-weeks, light-months, or light-years in diameter, in all cases a far cry from the tens of thousands of light-years for a typical galaxy.

The argument goes as follows. Suppose we have a glowing, spherical, opaque object that is 1 light-month in radius (Fig. 17–13). Even if all parts of the object brightened instantaneously by an intrinsic factor of two, an outside observer would see the object brighten gradually over a timescale of 1 month, because light from the near side of the object would reach the observer 1 month earlier than light from the edge. Thus, the timescale of an observed variation sets an upper limit (that is, a maximum value) to the size of the emitting region: The actual size must be smaller than this upper limit.

Although this conclusion can be violated under certain conditions (such as when different regions of a transparent object brighten in response to light reaching them from other regions, creating a "domino effect"), such models generally seem unnatural. Proper use of Einstein's special theory of relativity (in case the light-emitting material is moving very fast) can also change the derived upper limit to some extent, but the basic conclusion still holds: Quasars are very small, yet they release tremendous amounts of energy. For example, a quasar only 1 light-month across can be 100 times more powerful than an entire galaxy of stars 100,000 light-years in diameter!

17.3b What Is the Energy Source?

The nature of the prodigious (yet physically small) power source of quasars was initially a mystery. How does such a small region give off so much energy? After all, we don't expect huge explosions from tiny firecrackers. There was some indication that these objects might be related to active galactic nuclei: They have similar optical spectra and are bright at radio wavelengths. So, perhaps the same mechanism

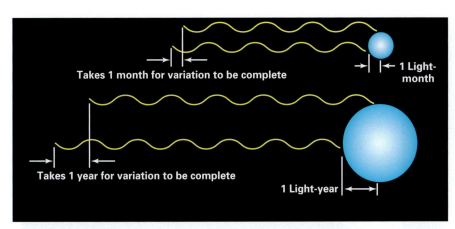

■ **FIGURE 17–13** Why a large object cannot appear to fluctuate in brightness as rapidly as a smaller object. Suppose that each object abruptly brightens at one instant. The wave emitted from the edge of the object takes longer to reach the observer than light from the near side of the object, because it has to travel farther. We don't see the full variation until waves from all parts of the object reach us. The same is true for both opaque and transparent objects. For the latter, light from the far side takes longer to reach us than light from the near side.

might be used to explain the unusual properties of both kinds of objects. In fact, maybe active galactic nuclei are just low-power versions of quasars! If so, quasars should be located in the centers of galaxies. Later we will see that this is indeed the case.

The fact that the incredible power source of quasars is very small immediately rules out some possibilities. Such a process of elimination is often useful in astronomy; recall, for instance, how we deduced that pulsars are rapidly spinning neutron stars. It turns out that, for quasars, chemical energy is woefully inadequate: They cannot be wood on fire, or even chemical explosives, because the most powerful of these is insufficient to produce so much energy within such a small volume.

Even nuclear energy, which works well for stars, is not possible for the most powerful quasars. They cannot be radiation from otherwise-unknown supermassive stars or chains of supernovae going off almost all the time, or other more exotic stellar processes, because once again the efficiency of nuclear energy production is not high enough. To produce that much nuclear energy, a larger volume of material would be needed.

The annihilation of matter and antimatter is energetically feasible, since it is 100 per cent efficient. That is, all of the mass in a matter-antimatter collision gets turned into photons (radiation), and in principle a very small volume can therefore be tremendously powerful. However, the observed properties of quasars do not support this hypothesis. Specifically, matter-antimatter collisions tend to emit excess amounts of radiation at certain wavelengths, and this is not the case for quasars.

The release of gravitational energy, on the other hand, can in some cases be very efficient, and seemed most promising to several theorists studying quasars in the mid-1960s. We have already discussed how the gravitational contraction of a ball of gas (a protostar), for example, both heats the gas and radiates energy. But to produce the prodigious power of quasars, a very strong gravitational field is needed.

The conclusion was that a quasar is a **supermassive black hole**, perhaps 10 million to a billion times the mass of the Sun, in the process of swallowing ("accreting") gas. The black hole is in the center of a galaxy. The rate at which matter can be swallowed, and hence the power of the quasar, is proportional to the mass of the black hole, but it is typically a few solar masses per year. Although the Schwarzschild radius of, say, a 50 million solar-mass black hole is

150 million km, this is just 1 au (i.e., 8.3 light-minutes, the distance between the Earth and the Sun), and hence is minuscule compared with the radius of a galaxy (many thousands of light-years).

17.3c Accretion Disks and Jets

The matter generally swirls around the black hole, forming a rotating disk called an accretion disk (■ Fig. 17–14), a few hundred to a thousand times larger than the Schwarzschild radius of the black hole (and hence up to a few light-days to a light-week in size). As the matter falls toward the black hole, it gains speed (kinetic energy) at the expense of its gravitational energy, just as a ball falling toward the ground accelerates. Compression of the gas particles in the accretion disk to a small volume, and the resulting friction between the particles, causes them to heat up; thus, they emit electromagnetic radiation, thereby converting part of their kinetic energy into light.

Note that energy is radiated *before* the matter is swallowed by the black hole – nothing escapes from within the black hole itself. This process can convert the equivalent of about 10 per cent of the rest-mass energy of matter into radiation, more than 10 times more efficiently than nuclear energy. (Recall from Chapter 11 that the fusion of hydrogen to helium converts only 0.7 per cent of the mass into energy.)

A spinning, very massive black hole is also consistent with the well-focused "jets" of matter and radiation that emerge from some quasars, typically reaching distances of a few hundred thousand light-years and ending in wider lobes of emitting particles. Again, no material actually comes from *within* the black hole; instead, its origin is the accretion disk. The charged particles in the jets are believed to shoot out in a direction perpendicular to the accretion disk, along the black hole's axis of rotation (recall Figure 17–14). They emit radiation as they are accelerated. In addition to the radio radiation, high-energy photons such as x-rays can be produced (■ Fig. 17–15). The impressive focusing might be provided by a magnetic field, as in the case of pulsars, or by the central cavity in the disk.

Recall that jets and lobes are also seen in some types of active galaxies, which appear to be closely related to quasars (see Figure 17–3 again). As discussed in more detail later in this chapter, we know that the particles move with very high speeds because a jet can

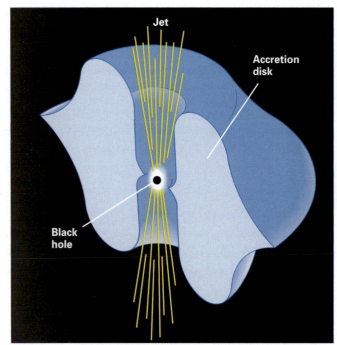

■ **FIGURE 17–14** Cross-sectional view of an accretion disk (doughnut) surrounding a black hole, with high-speed jets of particles emerging from the disk's nozzle. The nozzle points along the black hole's rotation axis.

■ **FIGURE 17–15** An image of the quasar PKS 0637–752 ($z = 0.65$) obtained with the Chandra X-ray Observatory. We see it as it was about 6 billion years ago. Note the "jet" that extends about 200,000 light-years out from the quasar; its power at x-ray energies exceeds that at radio energies, an important constraint that a successful theory of jet formation will have to explain.

sometimes appear to travel faster than the speed of light – an effect that occurs only when an object travels nearly along our line of sight, nearly at the speed of light.

Recently, indirect evidence for accretion disks surrounding a central, supermassive black hole has been found in several active galaxies from observations with various x-ray telescopes (Japan's ASCA and Suzaku, the European Space Agency's XMM-Newton Mission, and NASA's Chandra X-ray Observatory and NuSTAR). The specific shape of emission lines from highly ionized iron atoms that must reside very close to the galaxy center resembles that expected if the light is coming from a rotating accretion disk. Moreover, these lines exhibit a "gravitational redshift" – they appear at a somewhat longer wavelength than expected from the recession speed of the galaxy, because the photons lose some energy (and hence get shifted to longer wavelengths) as they climb out of the strong gravitational field near the black hole (see Chapter 14).

Similar emission lines have been seen in x-ray binary systems in which the compact object is likely to be a black hole (see the discussion in Section 14.7). Such lines, in both active galaxies and x-ray binaries, are now being analyzed in detail to detect and study predicted relativistic effects such as the strong bending of light and the "dragging" of space-time around a rotating black hole.

17.4 WHAT ARE QUASARS?

The idea that quasars are energetic phenomena at the centers of galaxies is now strongly supported by observational evidence. First of all,

the observed properties of quasars and active galactic nuclei are strikingly similar. In some cases, the active nucleus of a galaxy is so bright that the rest of the galaxy is difficult to detect because of contrast problems, making the object look like a quasar; special techniques are needed to reveal the galaxy (■ Fig. 17–16). This is especially true if the galaxy is very distant: We see the bright nucleus as a point-like object, while the spatially extended outer parts (known as "fuzz" in this context) are hard to detect because of their faintness and because of blending with the nucleus.

In the 1970s, a statistical test was carried out with quasars. A selection of quasars, sorted by redshift, was carefully examined. Faint fuzz (presumably a galaxy) was discovered around most of the quasars with the smallest redshifts (the nearest ones), a few of the quasars with intermediate redshifts, and none of the quasars with the largest redshifts (the most distant ones). Astronomers concluded that the extended light was too faint and too close to the nucleus in the distant quasars, as expected. In the 1980s, optical spectra of the fuzz in a few nearby quasars revealed absorption lines due to stars, but the vast majority of objects were too faint for such observations. In any case, the data strongly suggested that quasars could indeed be extreme examples of galaxies with bright nuclei. (The quasar absorption lines are narrow because they are formed by cold gas in intergalactic space between us and the quasar. Spectral lines formed in the quasar itself, typically emission lines, are broad because the gas moves rapidly in the gravitational field of the supermassive black hole.)

More recently, images obtained with the Hubble Space Telescope demonstrate conclusively that quasars live in galaxies, always at their centers. With a clear view of the skies above the

Credit (rotated, left side): John N. Bahcall and Sofia Kirhakos (Institute for Advanced Study), and Donald P. Schneider (Pennsylvania State University), and NASA

■ **FIGURE 17–16** A Hubble Space Telescope image of the quasar PKS 2349–014, whose spectral lines are redshifted by 17.3 per cent – that is z = 0.173. The loops of gas around this quasar suggest that it is being fueled by gas from the merging of two galaxies. The galaxies above the quasar are probably interacting with it.

Earth's atmosphere, and equipped with CCDs, the Hubble Space Telescope easily separates the extended galaxy light from the point-like quasar itself at low redshifts. In some cases the "host galaxy" is obvious (■ Fig. 17–17), but in others it is barely visible, and special techniques are used to reveal it; recall, for example, 3C 273 in Figure 17–7.

Adaptive optics is now allowing high-resolution imaging from mountaintop observatories in addition to the Hubble Space Telescope. An image with adaptive optics on the Gemini North telescope has enabled the central quasar peak of brightness to be subtracted from the overall image. A flat edge-on disk, interpreted to be the host galaxy, was revealed.

Further solidifying the association of quasars with galaxies, ground-based optical spectra of many relatively nearby quasars (z = 0.2–0.3) show unambiguous stellar absorption lines at the same redshift as that given by the quasar emission lines (■ Fig. 17–18). If there were some exotic reason other than the expansion of the Universe producing the high redshifts of quasars, the same process would also have to affect the normal host galaxy of stars, yet none is known.

Quasars exist almost exclusively at high redshifts and hence large distances. The peak of the distribution is at $z \approx 2$ (■ Fig. 17–19), though new studies at x-ray wavelengths suggest that it might be at an even higher redshift. With lookback times of about 10 billion years, quasars must be denizens of the *young* Universe. What happened to them? Quasars probably faded with time, as the central black hole gobbled up most of the surrounding gas; the quasar shines only while it is pulling in material.

Credit (rotated, panels a & b): John N. Bahcall and Sofia Kirhakos (Institute for Advanced Study), Donald P. Schneider (Pennsylvania State University), and ANSA

Credit (rotated, panels c & d): John B. Hutchings (Dominion Astrophysical Obs.)/CFHT, and NASA/STScI

■ **FIGURE 17–17** (a) The quasar PG 0052+251, at redshift z = 0.155, is in the center of a normal spiral galaxy. (b) The quasar PHL 909, at z = 0.171, is in the core of a normal elliptical galaxy. Both of these images were obtained with the Hubble Space Telescope.
(c) A ground-based image of QSO 1229+204 made with the Canada–France–Hawaii Telescope on Mauna Kea. The quasar is in the core of a spiral galaxy (the "fuzz") that is colliding with a dwarf galaxy (not seen in this image). The resolution is about 0.5 arc sec.
(d) The high resolution of the Hubble Space Telescope reveals structure in the fuzz of QSO 1229+204. On one side of the galaxy we see a string of knots, which are probably massive young star clusters. They may have been formed as a result of the collision. The resolution is about 0.1 arc sec.

■ FIGURE 17–18 Keck telescope spectra of (top) the quasar 4C 31.63 ($z = 0.296$) and (below) of "fuzz" surrounding the quasar. The spectrum of the fuzz shows absorption lines typical of those that are produced by the relatively cool outer parts of normal stars. Thus the fuzz seen around nearby quasars is really starlight from galaxies surrounding the bright central object. In this double graph, the brightness of the quasar was reduced substantially to fit it on the same part of the vertical axis as the fuzz spectrum.

Thus, some of the nearby active and normal galaxies may have been luminous quasars in the distant past, but they now exhibit much less activity because of a slower accretion rate. On the other hand, many of these objects were probably never luminous enough to be considered genuine quasars; either their central black hole wasn't sufficiently massive to pull in much material, or there was little gas available to be swallowed.

Perhaps the nucleus of the Milky Way Galaxy, which is now only slightly active, was much more powerful in the past, when the putative black hole had plenty of material to accrete. In fact, recent evidence suggests that our Galaxy did indeed have an episode of activity in the past 10 million years, resulting from the relatively sudden accretion of gas: Two large lobes of gas emitting gamma-rays and other radiation extend from the center, perpendicular to the disk (■ Fig. 17–20). They are reminiscent of those seen in some active galaxies and quasars, but it is also possible that they were produced by a central cluster of massive stars formed in the past few million years.

Though most quasars are very far away, some have relatively low redshifts (like 0.1). If quasars were formed early in the Universe, how can these quasars still be shining? Why hasn't all of the gas in the central region been used up? High-resolution images (■ Fig. 17–21) show that, in many cases, the galaxy containing the quasar is interacting or merging with another galaxy. This result suggests that gravitational tugs end up directing a fresh supply of gas from the outer part of the galaxy (or from the intruder galaxy) toward its central black hole, thereby fueling the quasar and allowing it to continue radiating so strongly. Some quasars may have even faded for a while, and then the interaction with another galaxy rejuvenated the activity in the nucleus.

17.5 ARE WE BEING FOOLED?

A few astronomers have disputed the conclusion that the redshifts of quasars indicate large distances, partly because of the implied enormously high luminosity produced in a small volume. If Hubble's law doesn't apply to quasars, maybe they are actually quite nearby.

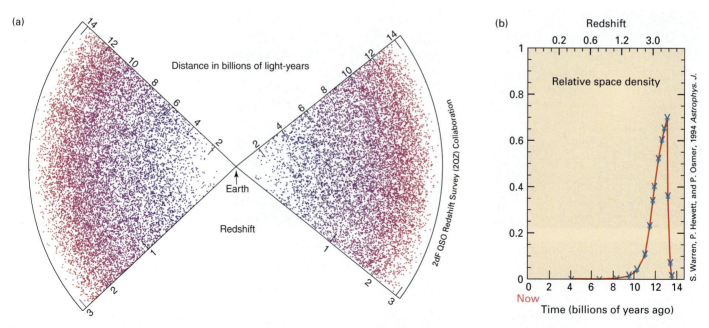

■ FIGURE 17–19 (a) The Sloan Digital Sky Survey and the 2dF Quasar Survey (as mentioned in the preceding chapter, 2dF stands for "Two degree field") have confirmed and extended the earlier results graphed in part (b). The new surveys measured many more quasars at high redshift, in particular. Here we see the quasars mapped in the 2dF Quasar Survey. (b) The number density of quasars (number per billion cubic parsecs) is plotted versus cosmic time, for an assumed Universe age of 14 billion years. There was a bright, spectacular era of quasars billions of years ago, and essentially none now remain.

(a)

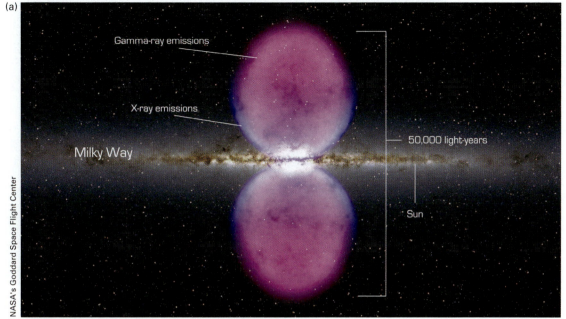

Gamma-ray emissions

X-ray emissions

Milky Way

50,000 light-years

Sun

NASA's Goddard Space Flight Center

(b)

NASA/DOE/Fermi LAT/
D. Finkbeiner et al.

■ **FIGURE 17–20** A gamma-ray image of the region near the center of the Milky Way Galaxy obtained with the Fermi satellite, showing two lobes of emission emerging from the nucleus. A large amount of energy must have been emitted by the central region within the past 10 million years. It is possible that a star or cloud of gas was accreted by the supermassive black hole in the Galactic center, releasing a burst of energy. Alternatively, the lobes might have been produced by the energy emitted by massive stars and supernovae formed in the central region in the past few million years. (a) An artist's conception of the bubbles drawn onto an image of the center of our galaxy. (b) Actual observations from the Fermi Gamma-Ray Space Telescope.

Specifically, Halton Arp has found some cases where a quasar seems associated with an object of a different, lower redshift (■ Fig. 17–22). However, the vast majority of astronomers blame the association on chance superposition. There could also be some amplification of the brightnesses of distant quasars, along the line of sight, by the gravitational field of the low-redshift object; this would produce an apparent excess of quasars around such objects.

We now have little reason to doubt the conventional interpretation of quasar redshifts (though of course as scientists we should keep an open mind). Quasars clearly reside in the centers of galaxies having the same redshift. They are simply the more luminous cousins of active galactic nuclei, and a plausible energy source has been found. In addition, gravitational lensing (see below) shows that quasars are indeed very distant.

(a)　　　　(b)　　　　(c)　　　　(d)

John N. Bahcall and Sofia Kirhakos (Institute for Advanced Study), Donald P. Schneider (Pennsylvania State University), Mike Disney (University of Wales), and NASA

■ **FIGURE 17–21** Quasars in interacting or merging galaxies. (a) A Hubble Space Telescope image showing debris from a collision between two galaxies fueling a quasar. A ring galaxy left by the collision is at bottom; a foreground star in our Galaxy is at top. (b) A Hubble view showing a tidal tail above a quasar, perhaps drawn out by a galaxy that is no longer there. (c) A quasar merging with the bright galaxy that appears just below it. The swirling wisps of dust and gas surrounding them indicate that an interaction is taking place. (d) A pair of merged galaxies have left loops of gas around this quasar.

■ FIGURE 17–22 Two galaxies with relatively low redshifts are seen close in the sky to high-redshift quasars. Most astronomers think these are just chance projections along nearly the same lines of sight; the quasars and nearby galaxies are not physically associated with each other.

European Southern Observatory; from an ASP slide set by William C. Keel, U. Alabama

17.6 FINDING SUPERMASSIVE BLACK HOLES

We argued above, essentially by the process of elimination, that the central engine of a quasar or active galaxy consists of a supermassive black hole swallowing material from its surroundings, generally from an accretion disk. Is there any more direct evidence for this?

Well, the high speed of gas in quasars and active galactic nuclei, as measured from the widths of emission lines, suggests the presence of a supermassive black hole. A strong gravitational field causes the gas particles to move very quickly, and the different emitted photons are Doppler shifted by different amounts, resulting in a broad line. On the other hand, alternative explanations such as supernovae might conceivably be possible; they, too, produce high-speed gas, but without having to use a supermassive black hole.

Recently, however, very rapidly rotating disks of gas have been found in the centers of several mildly active galaxies. Their motion is almost certainly produced by the gravitational attraction of a compact central object, because we see the expected decrease of orbital speed with increasing distance from the center, as in Kepler's laws for the Solar System. The galaxy NGC 4258 (■ Fig. 17–23) presents the most convincing case, one in which radio observations were used to obtain very accurate measurements. The typical speed is $v = 1120$ km/s at a distance of only 0.4 light-year from the center. The data imply a mass of about 3.6×10^7 solar masses in the nucleus (see *Figure It Out 17.1: The Central Mass in a Galaxy*).

The corresponding density is over 100 million solar masses per cubic light-year, a truly astonishing number. If the mass consisted of stars, there would be no way to pack them into such a small volume, at least not for a reasonable amount of time: They would rapidly collide and destroy themselves, or undergo catastrophic collapse. The natural conclusion is that a supermassive black hole lurks in the center. Indeed, this is now regarded as the most conservative

Lincoln J. Greenhill (Harvard University), and Holland Ford (STScI/Johns Hopkins University)

■ FIGURE 17–23 An optical image of the spiral galaxy NGC 4258 that emphasizes star-forming regions. Radio observations of the nucleus have revealed gas orbiting a central, exceedingly small, very massive object – almost certainly a black hole.

explanation for the data: If it's not a black hole, it's something even stranger!

One of the most massive black holes ever found is that of M87, an active galaxy in the Virgo Cluster that sports a bright radio and optical jet (■ Fig. 17–24a,b). Spectra of the gas disk surrounding the nucleus were obtained with the Hubble Space Telescope (■ Fig. 17–24c), and the derived mass in the nucleus was about 3 billion

FIGURE IT OUT 17.1

The Central Mass in a Galaxy

We saw in Chapter 16 that the mass enclosed within a circular orbit of radius R is $M = v^2R/G$, where v is the orbital speed and G is Newton's constant of gravitation. For NGC 4258, we measure $v =$ 1120 km/s at a distance of 0.4 light-year from the center. Converting units, we find that $M = 7.1 \times 10^{40}$ g $= 3.6 \times 10^7$ solar masses. Most of this mass must be in a black hole rather than in stars.

■ **FIGURE 17–24** (a) The nucleus (bright point at top) and jet of M87, in a computer-processed view from the Faint Object Camera aboard the Hubble Space Telescope. In the jet of high-speed charged particles, which extends 8000 light-years, the image reveals detail as small as 10 light-years across. (b) This later Hubble image of M87's nucleus and jet also shows (enlarged) an unusual spiral disk in the galaxy's center. See Figure 17–28 for a radio view of the jet. (c) Spectra of the regions shown on the image of the center of M87, taken with the Faint Object Spectrograph aboard the Hubble Space Telescope, reveal Doppler shifts of the gas. (The single emission line appears at different wavelengths.) The orbital speed of 550 km/s at this distance of 60 light-years from the nucleus allows astronomers to calculate how much mass must be inside those locations to keep the gas in orbit. The result is about 3 billion solar masses, after various effects like the inclination of the disk are taken into account. More recent analysis and modeling of the data suggests that the mass might be as high as 6 or 7 billion Suns.

solar masses. More recent analysis of these and other data now suggests that the actual mass might be as high as 6 to 7 billion solar masses, depending on the assumptions used in theoretical models.

If some nearby, relatively normal-looking galaxies were luminous quasars in the past, and a significant fraction even show some activity now, we suspect that supermassive black holes are likely to exist in the centers of many large galaxies today. Sure enough, when detailed spectra of the nuclear regions of a few galaxies were obtained (especially with the Hubble Space Telescope), strong evidence was found for rapidly moving stars. The masses derived from Kepler's third law were once again in the range of a million to a billion Suns. By 2013, the central regions of nearly 100 galaxies had been observed in this manner, revealing the presence of supermassive black holes, in a few cases (reported very recently, late 2011) having masses up to about 10 billion Suns.

Probably the most impressive and compelling case is our own Milky Way Galaxy. As we discussed in Chapter 15, stars in the highly obscured nucleus were seen from Earth at infrared wavelengths, and their motions were measured over the course of a few years; see Figure 15–15. The data are consistent with stars orbiting a single, massive, central dark object (■ Fig. 17–25). The implied mass of this object is 4 million solar masses, and it is confined to a volume only 0.03 light-year in diameter! The only known explanation is a black hole. Thus, our Galaxy could certainly have been more active in the past, though never as powerful as the most luminous quasars, which require a black hole of 10^8 to 10^9 solar masses.

In the past decade or so, it has been found that the mass of the central black hole is proportional to the mass of the bulge in a spiral galaxy, or to the total mass of an elliptical galaxy (■ Fig. 17–26). But recall from Chapter 16 that the bulges of spiral galaxies are old, as are elliptical galaxies (which resemble the bulges of spiral galaxies). Thus, there is evidence that the formation of the supermassive

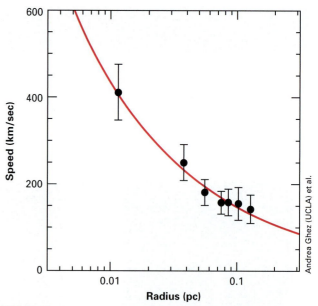

■ **FIGURE 17–25** The observed speeds of stars vs. their distance from the center of our Galaxy, a radio source known as Sgr A* ("Sagittarius A-star"). Recent data suggest that the mass is close to 4 million solar masses. Note that 1 pc = 3.26 light-years. See the information about stars orbiting close to our Galaxy's center in Figure 15–15.

black hole is related to the earliest stages of formation of galaxies. We don't yet understand this relation, but clearly it offers a clue to physical processes long ago, when most galaxies were being born. Recent studies also show that, for a given bulge mass, the more compact the bulge, the more massive the black hole, suggesting an even closer link between bulge formation and black-hole formation.

Yet more evidence for the presence of supermassive black holes in galactic nuclei comes from what are known as "tidal disruption

■ **FIGURE 17–26** The mass of the supermassive black hole in the center of a galaxy appears to be larger in spiral galaxies with larger bulges. Similarly, the black-hole mass scales with the mass of the elliptical galaxy in which it resides.

NASA/CXC/M.Weis

■ **FIGURE 17–27** Artist's conception of the tidal disruption of a star that happens to wander too close to the supermassive black hole at the center of a galaxy, brightening suddenly. Some of the gas is ejected from the system, while other material is gradually swallowed by the black hole. During this accretion process, the gas continues to glow.

flares." Astronomers have recently found that such black holes occasionally tear apart stars that approach them too closely. The stars are disrupted because of exceptionally strong tidal forces, as discussed in Chapter 14. A flash of light can be emitted from the nucleus, and the galaxy becomes "active" while gas from the stellar disruption accretes onto it (■ Fig. 17–27). Several tidal disruption flares have been found in the past few years, and we seem to have viewed at least one object nearly along the black hole's axis, the direction in which matter can be propelled at speeds close to that of light.

17.7 THE EFFECTS OF BEAMING

Radio observations with extremely high angular resolution, generally obtained with the technique of very-long-baseline interferometry, have shown that the jets seen in some quasars contain bright "blogs" (or "knots"), many of which move down the jet. In many cases, observations over a few years reveal that the components are apparently separating very fast (■ Fig. 17–28), given the conversion from the angular change in position we measure across the sky to the actual physical speed in km/s at the distance of the quasar. Indeed, some of the components appear to be separating at **superluminal speeds** – that is, at speeds greater than that of light! But Einstein's special theory of relativity says that no objects can travel through space faster than light, an apparent contradiction.

Astronomers can explain how the components only *appear* to be separating at speeds greater than that of light even though they are actually physically moving at allowable speeds (less than that of light). If one of the components is a jet approaching us almost along our line of sight, and nearly at the speed of light, then according to our perspective the jet is nearly keeping up with the radiation it emits (■ Fig. 17–29).

If the jet moves a certain distance in our direction in (say) 5 years, the radiation it emits at the end of that period gets to us sooner than it would have if the jet were not moving toward us. So, in fewer than 5 years, we see the jet's motion over 5 full years. In the interval between our observations, the jet had several times longer to move than we would naively think it had. So it could, without exceeding the speed of light, appear to move several times as far.

Whether a given object looks like a quasar or a less-active galaxy with broad emission lines probably depends in part on the orientation of the jet relative to our line of sight: Jets pointing at us appear far brighter than those that are misaligned. Thus, quasars are probably often beamed very roughly toward us, a conclusion supported by the fact that many radio-loud quasars show superluminal motion.

However, if the jet is pointing *straight* at us, it can greatly outshine the emission lines, and the object's optical spectrum looks rather featureless, unlike that of a normal quasar. It is then called a "BL Lac object," after the prototype in the constellation Lacerta, the Lizard; sometimes the term "blazar" is used instead. Many such objects have recently been

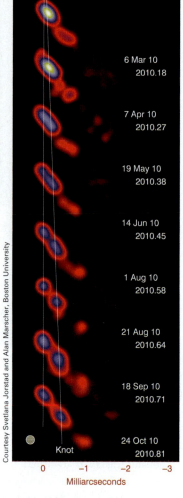

■ FIGURE 17–28 (a) A series of views of the quasar 3C 279 with radio interferometry at a wavelength of 1.3 cm. The apparent speed translates (for $H_0 = 71$ km/s/Mpc) to a speed of 8.5c, though such apparent "superluminal speeds" can be explained in conventional terms. (b) Superluminal speeds also appear in the active galaxy M87. The inset shows features that seem to move at 2.3c. Red shows the brightest and blue shows the faintest emission. (c) Changes in the jet of 3C 273, the first quasar recognized, in less than a year. The images were taken with the Very Long Baseline Array, with a resolution (shown with a cross-hatched circle) of only 0.0001 arcsec. The "knot" that appears to be moving away from the core of 3C 273 has an apparent speed of 9.5 times the speed of light, though we know that that value is not the real speed. A gamma-ray flash occurred when the knot moved through the core in March 2010.

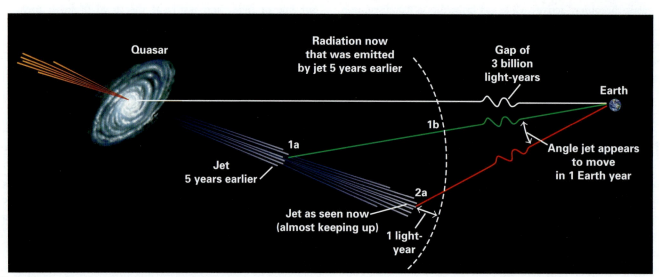

■ FIGURE 17–29 The leading model for explaining how a jet of high-energy particles emitted from a quasar, shown at left with its accretion disk surrounding a central black hole, can appear to exceed the speed of light when seen from Earth (shown at right), in this case 3 billion light-years away. Point 1a marks the position of the jet 5 years earlier than the jet we are seeing now. (It actually takes another 3 billion years for the radiation to reach us, an extra duration that we can ignore for the purposes of this example because we are interested in *differences* in the arrival times of the light from two jet positions.) After an interval of 5 years, the light emitted from point 1a has reached point 1b, while the jet itself has reached point 2a. The jet has been moving so fast that points 2a and 1b are separated by only 1 light-year in this example. The light emitted then from the jet at point 2a is thus only 1 year behind the light emitted 5 years earlier at point 1a. So, within a 1-year interval we receive light from two positions that the jet has taken 5 years to go between. The jet actually therefore had 5 years to move across the sky to make the angular change in position that we see over a 1-year interval. We therefore think that the jet is moving faster than it actually moves.

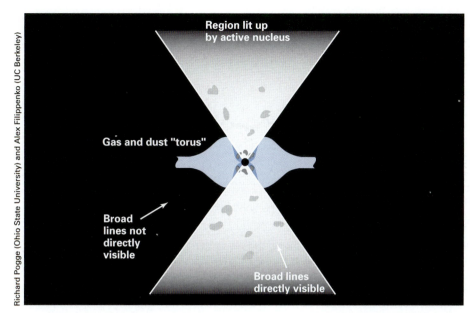

FIGURE 17–30 The appearance from Earth of an active galaxy can depend on the orientation relative to our line of sight of the torus (doughnut) that surrounds it. If we view the nucleus from a direction within the indicated cones, the nucleus appears bright, and we can see the broad emission lines produced by gas (shown as dark clouds) in rapid motion near the supermassive black hole. If instead we view from outside the cones, then the active nucleus and central, rapidly moving clouds can be hidden at optical wavelengths by gas and dust in the torus. Narrow emission lines from gas farther from the nucleus (and within the cones; lighter clouds), however, should still be visible.

Richard Pogge (Ohio State University) and Alex Filippenko (UC Berkeley)

Zlatan Tsvetanov (Johns Hopkins University), and Clive Tadhunter (University of Sheffield)

FIGURE 17–31 Radiation from the core of this active galaxy, NGC 5252, illuminates matter only within two oppositely directed cones. Presumably the cones are aligned along the axis of a rotating black hole surrounded by a torus of gas and dust that blocks our direct view of the active nucleus, as shown in Figure 17–30.

found with a variety of telescopes, including the Fermi Gamma-ray Space Telescope and the Wide-field Infrared Survey Explorer.

At the other extreme, if the jet is close to the plane of the sky, dust and gas in a torus (doughnut) surrounding the central region may hide the active nucleus from us (recall Figure 17–14). The galaxy nucleus itself may then appear relatively normal, although the active nature of the galaxy could still be deduced from the presence of extended radio emission from the jet.

This general idea of beamed, or directed, radiation probably accounts for many of the differences seen among active galactic nuclei. For example, in one type of Seyfert galaxy, the very broad emission lines are not easily visible, despite other evidence that indicates considerable

activity in the nucleus. (For example, bright, narrow emission lines can be seen.) We think that, in some cases, the broad emission lines are present, but simply can't be directly seen because they are being blocked by an obscuring torus of material (■ Fig. 17–30). But light from the broad lines can still escape along the axis of this torus and reflect off of clouds of gas elsewhere in the galaxy. Observations of these clouds then reveal the broad lines, but faintly.

Similarly, some galaxies hardly show any sort of active nucleus directly – it is too heavily blocked from view by gas and dust along our line of sight, in the central torus. However, radiation escaping along the axis of this torus can still light up exposed parts of the galaxy, indirectly revealing the active nucleus (■ Fig. 17–31).

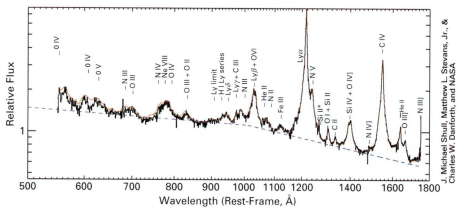

■ **FIGURE 17–32** Spectra of two quasars, plotted with the redshift removed (so that we see light at the emitted wavelengths rather than the observed, redshifted wavelengths). A high-redshift quasar (bottom) shows a large number of absorption lines produced by clouds of gas and galaxies along our line of sight to the quasar. Most of these lines, especially the ones in the blue (left) part of the spectrum, correspond to the hydrogen Lyman-α transition produced by intergalactic clouds at many different (and generally large) distances from us. We see the lines at many different wavelengths because of the different redshifts of the clouds. The spectrum of 3C 273 (top), a low-redshift quasar, has far fewer absorption lines, indicating that intergalactic clouds of gas are scarce at the present time in the Universe. (c) a spectrum in the extreme ultraviolet of 3C 279 with the Cosmic Origins Spacecraft that was added to the Hubble Space Telescope on its final servicing mission.

17.8 PROBES OF THE UNIVERSE

Quasars are powerful beacons, allowing us to probe the amount and nature of intervening material at high redshifts. For example, numerous narrow absorption lines are seen in the spectra of high-redshift quasars (■ Fig. 17–32). These spectral lines are produced by clouds of gas at different redshifts between the quasar and us. The lines can be identified with hydrogen, carbon, magnesium, and other elements.

Analysis of the line strengths and redshifts allows us to explore the chemical evolution of galaxies, the distribution and physical properties of intergalactic clouds of gas, and other interesting problems. The lines are produced by objects that are generally too faint to be detected in other ways. One surprising conclusion is that all of the clouds have at least a small quantity of elements heavier than helium. Since stars and supernovae produced these heavy elements, the implication is that an early episode of star formation preceded the formation of galaxies.

Another way in which quasars are probes of the Universe is the phenomenon of **gravitational lensing** of light (Chapter 16). In fact, such lensing was first confirmed through studies of quasars. In 1979, two quasars were discovered close together in the sky, only a few seconds of arc apart (■ Fig. 17–33a). They had the same redshift, yet their spectra were essentially identical, arguing against a possible binary quasar. A cluster of galaxies with one main galaxy (Fig. 17–33b,c) was subsequently found along the same line of sight, but at a smaller redshift. The most probable explanation is that light from the quasar is bent by the gravity of the cluster (warped space-time), leading to the formation of two distinct images (■ Fig. 17–34). The cluster is acting like a gravitational lens.

Since then, dozens of gravitationally lensed quasars have been found. For a point lens and an exactly aligned object, we can get an image that is a ring centered on the lensing object. Such a case is called an "Einstein ring," and quite a few are now known (■ Fig. 17–35a). Some gravitationally lensed quasars have quadruple quasar

■ **FIGURE 17–33** (a) A Hubble Space Telescope wide-angle view of two quasars, 0957+561 A and B, that are only about 6 seconds of arc apart in the sky. They have the same redshifts (z = 1.4136) and essentially identical spectra. (b) Narrow-angle view of the two quasars, obtained with a ground-based telescope; they appear slightly oblong because of imperfect tracking by the telescope during the exposure. An intervening galaxy can be seen only 1 second of arc from the quasar on the right, in the direction of the quasar on the left. (c) The image of the quasar on the left has been subtracted from that on the right to reveal just the intervening galaxy.

Path of light ray

Image B
Quasar

Galaxy

Observer
on Earth

Image A

■ FIGURE 17–34 The geometry of gravitational lensing. Light from a distant quasar can take two different paths on its way to Earth; each is bent because of the gravitational field of the intervening galaxy or cluster of galaxies. The two paths produce two distinct images of the same quasar close together in the sky. Note that the lengths of the two paths generally differ.

(a)   (b)   (c)

NASA, ESA, A. Bolton (Harvard-Smithsonian CfA) and the SLACS Team

Jacqueline N. Hewitt (MIT) et al.

National Radio Astronomy Observatory

■ FIGURE 17–35 (a) This particular quasar is so exactly aligned with the intervening gravitational lens that its light is spread out into an "Einstein ring" (Figure 16–26). The Hubble Space Telescope has imaged several dozen Einstein rings, in most cases produced by galaxies rather than quasars. (b) A Hubble Space Telescope image of an "Einstein cross" – a quasar that is gravitationally lensed into four images by a galaxy along the line of sight. In this case, the lensing galaxy is at a much lower redshift than the quasar; the galaxy's reddish nucleus is visible in the center of the image. The four bluish images of comparable brightness are separated by less than 2 seconds of arc and have essentially identical spectra. The rare configuration and identical spectra show that we are indeed seeing gravitational lensing rather than a cluster of quasars. (c) With complicated foreground mass distributions, one can get strange image configurations. In this system there are 6 images of a quasar at redshift 3.235 produced by a compact triple of foreground galaxies.

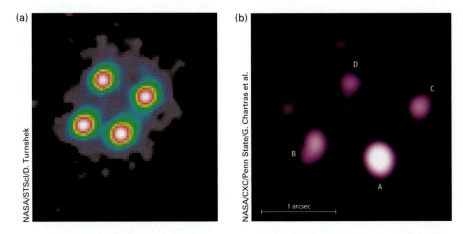

(a)   (b)

NASA/STScI/D. Turnshek

NASA/CXC/Penn State/G. Chartras et al.

D

C

B

A

1 arcsec

■ FIGURE 17–36 The "clover leaf" (H1413+117), a quadruply imaged quasar ("Einstein cross") at redshift 2.55. (a) A Hubble Space Telescope view. (b) A Chandra X-ray Observatory view. The four images are only about 1 second of arc apart. Note that a different one of the four images, marked "A," is brighter in the x-ray band. This difference results from "microlensing," when a single or binary star in one of the galaxies between us and the quasar passes right in front of the small, x-ray-producing region around the quasar's supermassive black hole. Since the optical light comes from a much larger region, it is not similarly magnified.

images that resemble a cross (■ Figs. 17–35b, 17–36), or even more complicated configurations (Fig. 17–35c). Only gravitational lensing seems to be a reasonable explanation of these objects, the redshifts of whose components are identical. In 2012, quasars themselves lensing even more distant galaxies were reported in Hubble Space Telescope images (■ Fig. 17–37).

Moreover, in some cases continual monitoring of the brightness of each quasar image has revealed the same pattern of light variability, but with a time delay between the different quasar images. This

delay occurs because the light travels along two different paths of unequal length to form the two quasar images; recall Figure 17–34. The variability pattern is not expected to be identical in two entirely different quasars that happen to be bound in a physical pair.

The multiple imaging of quasars is an exciting verification of a prediction of Einstein's general theory of relativity. The lensing details are sensitive to the total amount and distribution of matter (both visible and dark) in the intervening cluster. Thus, gravitationally lensed quasars provide a powerful way to study dark matter.

FIGURE 17–37 Careful analysis of 23,000 SDSS spectra of quasars has revealed that they themselves are in line with even more distant galaxies; spectral features of the galaxies were detected in the quasar spectra. The search led to the Hubble Space Telescope making images that show that these quasars are gravitationally lensing galaxies beyond them. The arcs and rings of these amplified and distorted galaxies are marked with arrows. Also, the host galaxy of each quasar is readily visible in these high-resolution images.

NASA, ESA, and F. Courbin (EPFL, Switzerland)

CONCEPT REVIEW

» The central regions of some galaxies are unusually luminous – they are very powerful emitters of all kinds of electromagnetic radiation, and they sometimes spew out enormous jets of particles (Section 17.1). It is unlikely that normal stellar processes can account for these objects, called **active galactic nuclei**. The mysterious **quasars** are apparently related: They are star-like objects that in some cases shine very brightly at radio wavelengths (Section 17.2a). Their optical spectra were initially difficult to interpret, but eventually astronomers realized that these spectra were substantially redshifted (Section 17.2b). The discovery that quasars have high redshifts implied that they are at vast distances and have astonishingly large luminosities (Section 17.2c).

» Rapid variations in the observed brightnesses of quasars were used to deduce that they are small, in some cases less than one light-week across – yet they are more powerful than entire galaxies 100,000 light-years in diameter (Section 17.3a). The energy of a quasar probably comes from matter falling into a **supermassive black hole**, 10 million to several billion times the mass of the Sun, in the central region of a galaxy (Section 17.3b). The matter forms a rotating **accretion disk** around the black hole (Section 17.3c). In some cases, well-focused "jets" of matter and radiation emerge along the disk's axis of rotation.

» We now know that quasars are indeed phenomena in the centers of a minority of galaxies (Section 17.4). Hubble Space Telescope images of low-redshift quasars reveal nebulous structures that closely resemble galaxies, and spectra confirm that the extended light comes from stars. Quasars generally lived in the distant past; as the material surrounding the supermassive black holes was used up, the quasars faded, becoming less-active galaxies and finally normal-looking galaxies. Although a few quasars appear close in the sky to galaxies with much lower redshifts, these are thought by most astronomers to be chance projections (Section 17.5).

» There is strong evidence for the existence of gigantic black holes in the centers of many relatively normal, nearby galaxies; our own Milky Way Galaxy harbors such a beast (Section 17.6). The more compact and massive the bulge of a spiral galaxy, the more massive is the black hole within it. Some quasars appear to be revived through gravitational interactions and collisions with galaxies, which direct gas toward the central black hole.

» Apparent **superluminal speeds** of components seen in some quasars are produced by material moving near (but below) the speed of light, close in direction to our line of sight; they are not violations of Einstein's special theory of relativity (Section 17.7). The spectra of quasars often exhibit absorption lines due to material at intermediate distances, allowing astronomers to study galaxies and clouds of gas that are otherwise too dim to see (Section 17.8). Some quasars look multiple because of **gravitational lensing** by intervening galaxies or clusters.

QUESTIONS

1. Summarize the main observed characteristics of active galactic nuclei.

2. Describe the historical development of the study of quasars, and list their peculiar properties.

3. Why is it useful to find the optical objects that correspond in position with radio sources?

4. Why did the optical spectra of quasars initially seem so mysterious, before the correct interpretation was made?

5. What was the key breakthrough in the interpretation of quasar spectra? What was its significance?

6. Why do you think it was initially difficult for astronomers to entertain the possibility that the spectra of quasars are highly redshifted?

7. It was suggested in the 1960s that quasars might be compact objects ejected at high speeds from the centers of nearby ordinary galaxies. Why does the absence of blueshifted quasars provide strong evidence against this hypothesis?

†8. We observe a quasar with a spectral line at 5,000 Å that we know is normally emitted at 4,000 Å. **(a)** By what percentage is the line redshifted? **(b)** By approximately what percentage of the speed of light is the quasar receding? (A nonrelativistic formula is acceptable.) **(c)** At what speed in km/s is the quasar receding? **(d)** Using Hubble's law, to what distance does this speed correspond?

†9. Suppose you observe a quasar with a redshift of 3.6 – that is, all lines are shifted to longer wavelengths by 360 per cent. At what wavelength would you observe the ultraviolet Lyman-α line of hydrogen, whose laboratory (rest) wavelength is 1,216 Å? What region of the spectrum is this? (Just give the general name, such as "ultraviolet.")

10. Explain what is meant by the "lookback time" of a quasar.

†11. A certain quasar has redshift $z = 2.5$. Roughly what is its lookback time?

12. Outline the argument used to infer that the physical size of quasars is very small.

13. Why does the rapid variability of the apparent brightness of some quasars make "the energy problem" even more difficult to solve using conventional energy sources such as chemical burning or nuclear fusion?

14. What is the most probable physical mechanism that produces a quasar's energy? Is anything actually escaping from within the black hole's event horizon?

15. What are three differences between quasars and pulsars?

16. What is the evidence suggesting that quasars live in the centers of galaxies?

17. Summarize what generally happens to a quasar as it ages.

18. Why are relatively nearby quasars so rare, compared with very distant quasars?

19. Why do we expect many or most large, nearby galaxies to have a supermassive black hole in their center?

20. Why do we think that some relatively nearby quasars are rejuvenated?

21. Explain how we deduce the presence and measure the mass of a supermassive black hole in the center of a galaxy.

22. Qualitatively explain how parts of a quasar could appear to be moving at greater than the speed of light without actually violating the special theory of relativity, which states that no physical object can exceed the speed of light.

23. Describe how quasars can be used as probes of matter between them and us.

24. What do we mean by a "gravitationally lensed quasar"?

25. How would the presence or absence of dark matter affect the gravitational lensing of quasars?

26. **True or false?** Most astronomers think that the enormous amount of energy emitted by a quasar is produced by nuclear reactions in gas confined to a radius of a light-year or less.

27. **True or false?** The term "active galaxy" refers to a galaxy undergoing rapid evolutionary change in shape, such as spiral to elliptical.

28. **True or false?** The energy source that produces jets of particles and radiation in the centers of radio galaxies is thought to be associated with a supermassive black hole.

29. **True or false?** Quasars are the only known objects that emit light from within the event horizon of a supermassive black hole.

30. **True or false?** The observed variability of the light emitted from a quasar places an approximate maximum (upper limit) on the physical size of the quasar.

31. **True or false?** Quasars can be far more luminous than galaxies at comparable redshifts, but they are also physically far larger than galaxies.

32. **True or false?** Multiple quasars with identical spectra and the same redshift, very near each other in the sky, generally are examples of gravitational lensing of light.

33. **Multiple choice:** After carefully measuring the brightness of an astronomical object for one month, you find that the brightness changes significantly from one night to the next. The safest conclusion is that this object is (a) a perfectly normal galaxy; (b) no more than one light-month away from the Sun; (c) not within our Milky Way Galaxy; (d) not larger than about one light-day across; or (e) a perfectly normal main-sequence star.

†34. **Multiple choice:** If two quasars have the same apparent brightness, but Quasar R2D2 has a redshift of 0.1 while Quasar C3PO has a redshift of 0.2, the luminosity of Quasar C3PO is _____ that of Quasar R2D2. (a) one-quarter; (b) one-half; (c) equal to; (d) twice; or (e) four times.

35. **Multiple choice:** Quasars are probably powered by (a) extremely hot, young, massive stars, since they are in very young galaxies; (b) colliding neutron stars or black holes; (c) material falling into a central supermassive black hole; (d) supernovae; or (e) chemical reactions.

†36. **Multiple choice:** Suppose you obtain a spectrum of a quasar. You find that a hydrogen emission line (Lyman-α; rest wavelength $\lambda_0 \approx 1{,}200$ Å) is observed to be at a wavelength of $\lambda \approx 6{,}000$ Å. You also measure two Lyman-α absorption lines at 3,600 Å and 4,800 Å. Which one of the following statements is definitely *true*? (a) The quasar is at a redshift $z = 5$. (b) Hydrogen must be present along this line of sight at a redshift $z = 3$. (c) The quasar is lensed into multiple quasar images having redshifts $z = 2$ and 3. (d) There are no hydrogen clouds along this line of sight at a redshift $z = 6$. (e) The absorption lines are produced by hydrogen in the galaxy containing the quasar.

37. **Multiple choice:** Which one of the following statements about quasars is *true*? (a) Quasars appear blue because they are highly blueshifted. (b) The spectrum of a quasar appears to be so unusual because it is composed of exotic molecules. (c) Quasars are most likely powered by supermassive stars undergoing nuclear reactions at an unusually fast rate. (d) Quasars probably faded with time, and have now evolved into relatively normal galaxies. (e) Quasars are powered by particle-antiparticle annihilation, which has an energy conversion efficiency of 100 per cent.

38. **Fill in the blank:** Quasars were first detected in the _____ region of the electromagnetic spectrum.

39. **Fill in the blank:** The "fuzz" seen around some quasars is evidence for a _____ in which the quasar lives.

†40. **Fill in the blank:** Light that we are now seeing from a redshift $z = 2$ quasar actually left that quasar roughly _____ billion years ago.

41. **Fill in the blank:** Broad _____ provide evidence that the gas in a quasar is moving around at very high speeds.

42. **Fill in the blank:** Quasars whose relativistic jets are pointing nearly along our line of sight sometimes exhibit the phenomenon of _____ motion.

†This question requires a numerical solution.

TOPICS FOR DISCUSSION

1. Does it seem paradoxical that black holes, from which nothing can escape, account for the incredible power of quasars? Explain.

2. Suppose *no* nearby galaxies exhibited evidence for the presence of supermassive black holes in their centers. Would this be a problem for the hypothesis of what powers quasars? Why or why not?

3. What do you think of the hypothesis, favored by very few astronomers, that quasars are ejected from relatively nearby galaxies, and hence are not at the distances implied by their redshifts?

The dark night sky with 2011's Comet Lovejoy next to one of the domes of the Very Large Telescope in Chile. The Moon appears at left. The puzzle of why the entire sky isn't bright is knc Olbers's paradox and is fundamental to cosmology.

ESO/Gabriel Brammer, Williams College '02 and Ph.D. Yale

18

Cosmology: The Birth and Life of the Cosmos

ORIGINS

Knowledge of the age, geometry, structure, and evolution of the Universe is fundamental to our overall understanding of our place in the Cosmos. An amazing discovery is that the expansion of the Universe is now speeding up, not slowing down as expected if normal attractive gravity dominates throughout.

Cosmology is the study of the structure and evolution of the Universe on its grandest scales (see the Chapter Opener). Some of the major issues studied by cosmologists include the Universe's birth, age, size, geometry, and ultimate fate. We are also interested in the birth and evolution of galaxies, topics already discussed in Chapter 16.

As in the rest of astronomy, we are trying to discover the fundamental laws of physics, and use them to understand how the Universe works. We do not claim to determine the purpose of the Universe or why humans, in particular, exist; these questions are more in the domain of theology, philosophy, and metaphysics. However, in the end we will see that some of our conclusions, which are solidly based on the methods of science, nonetheless seem to be untestable with our present knowledge; they are at the boundary of what is generally considered to be science.

AIMS

1. See how the simple observation that the night sky is dark has profound implications for the origin or extent of the Universe (Section 18.1).

2. Understand the expansion of the Universe, and how it has no center (Section 18.2).

3. Learn how astronomers determine the age of the Universe (Section 18.3).

4. Explore various possibilities for the overall geometry, expansion history, and fate of the Universe (Section 18.4).

5. Learn how the expansion of the Universe now appears to be accelerating, propelled by mysterious "dark energy" (Section 18.5). This discovery was honored with the 2011 Nobel Prize in Physics.

6. Discuss major stages in the future of the Universe, should it expand forever (Section 18.6).

■ FIGURE 18–1 If we look far enough in any direction in an infinite universe, our line of sight should eventually hit the surface of a star. This leads to Olbers's paradox.

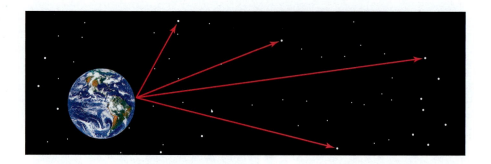

18.1 OLBERS'S PARADOX

We begin our exploration of cosmology by considering a deceptively simple question: "Why is the sky dark at night?" The answer seems so obvious ("The sky is dark because the Sun is down, dummy!") that the question may be considered absurd.

Actually, though, it is very profound. If the Universe is static (that is, neither expanding nor contracting), and infinite in size and age, with stars spread uniformly throughout it, then every line of sight should intersect a shining star (■ Fig. 18–1) – just as in a hypothetical infinite forest, every line of sight eventually intersects a tree. So, the sky should be bright everywhere, even at night. But it clearly isn't, thereby making a paradox – a conflict of a reasonable deduction with our common experience.

One might argue that distant stars appear dim according to the inverse-square law of light, so they won't contribute much to the brightness of the night sky. But the apparent size of a star also decreases with increasing distance. (This can be difficult to comprehend: Stars are so far away that they are usually approximated as points, or as

the blur circle produced by turbulence in Earth's atmosphere, but intrinsically they really do have a nonzero angular area.) So, a star's brightness per unit area remains the same regardless of distance (■ Fig. 18–2). If there were indeed an infinite number of stars, and if we could see all of them, then every point on the sky would be covered with a star, making the entire sky blazingly bright!

Another way to think about Olbers's paradox is to consider an infinite number of spherical shells centered on Earth (■ Fig. 18–3). Each shell has the same thickness, but the volume occupied by progressively more distant shells grows. Although individual stars in the distant shells appear dimmer than in the nearby shells, there are more stars in the distant shells (because of the growing volume).

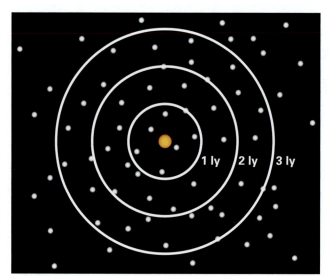

■ FIGURE 18–3 A model non-expanding universe in which the stars are uniformly distributed in spherical shells, all of the same thickness, centered on the Earth. Individual stars in the distant shells appear dim, but there are more stars (because of the larger volume of progressively more-distant shells). The total brightness contributed by each shell is the same, as long as the universe is infinitely old and stars have always been present. With an infinite number of shells, the sky should be very bright, even at night.

Now suppose the stars were suddenly created everywhere at the same instant. In the first year after creation, the observer would see only those stars within 1 light-year of him or her (that is, all stars within the innermost sphere), and most of the sky would appear not to have stars; thus, it would be dark. With each additional year, he or she would see all stars within another light-year farther away (progressively larger spheres). The total number of visible stars would steadily increase with time, thereby brightening the sky. Stated another way, the size of dark gaps between the stars would steadily decrease as time passed and more stars became visible.

Jay M. Pasachoff

■ FIGURE 18–2 The lower halves of trees are sometimes painted white, and mimic seeing a uniform expanse of starlight. Note that the "surface brightness" (the brightness per unit area) of a tree is the same whether the tree is close to us or far away.

These two effects exactly cancel each other, so each shell contributes the same amount to the brightness of the sky. With an infinite number of shells, the sky should be infinitely bright – or at least as bright as the surface of a star (since distant stars will tend to be blocked by closer stars along the same line of sight).

This dark-sky paradox has been debated for hundreds of years. It is known as Olbers's paradox, though the 19th-century astronomer Wilhelm Olbers wasn't the first to realize the problem. Kepler and others considered it, but not until the 20th century was it solved. (Note that the name "Olbers" has an "s" on the end before we make it possessive, so we write of "Olbers's paradox," or, alternatively, "Olbers' paradox." It is incorrect to write "Olber's paradox," with an apostrophe before a sole "s," since his name wasn't "Olber.")

Actually, there are several conceivable resolutions of Olbers's paradox, each of which has profound implications. For example, the Universe might have finite size. It is as though the whole Universe were a forest, but the forest has an edge – and if there are sufficiently few trees, one can see to the edge along some lines of sight. Or, the Universe might have infinite size, but with few or no stars far away. This is like a forest that stops at some point, or thins out quickly, with an open field (the rest of the Universe) beyond it.

Another possible solution is that the Universe has a finite age, so that light from most of the stars has not yet had time to reach us. If the forest suddenly came into existence, an observer would initially see only the most nearby trees (due to the finite speed of light), and there would be gaps between them along some lines of sight.

There are other possibilities as well. Most of them are easily ruled out by observations or violate the Copernican principle (that we are at a typical, non-special place in the Universe), and some are fundamentally similar to the three main suggestions discussed above. One idea is that dust blocks the light of distant stars, but this doesn't work because, if the Universe were infinitely old, the dust would have time to heat up and either glow brightly or be destroyed.

It turns out that the primary true solution to Olbers's paradox is that the Universe has a *finite age*, about 14 billion years. There has been far too little time for the light from enough stars to reach us to make the sky bright. Effectively, we see "gaps" in the sky where there are no stars, because light from the distant stars still hasn't been detected.

For example, consider the static (non-expanding) universe in Figure 18–3. If stars were suddenly created as shown, then in the first year the observer would see only those stars within 1 light-year of him or her, because the light from more distant stars would still be on its way. After 2 years, the observer would see more stars – all those within 2 light-years; the gaps between stars would be smaller, and the sky would be brighter. After 10 years, the observer would see all stars within 10 light-years, so the sky would be even brighter. But it would be a long time before enough very distant stars became visible and filled the gaps.

Our own Universe, of course, is not so simple – it is expanding. To some degree, the expansion of the Universe also helps solve Olbers's paradox: As a galaxy moves away from us, its light is redshifted from visible wavelengths (which we can see) to longer wavelengths (which we can't see). Indeed, the energy of each photon actually decreases because of the expansion. But the effects of expansion are minor in resolving Olbers's paradox, compared with the finite, relatively short age of the Universe.

Regardless of the actual resolution of Olbers's paradox, the main point is that such a simple observation and such a silly-sounding question lead to incredibly interesting possible conclusions regarding the nature of the Universe. So the next time your friends are in awe of the beauty of the stars, point out the profound implications of the darkness, too!

18.2 AN EXPANDING UNIVERSE

To see that the Universe has a finite age, we must consider the expansion of the Universe. In Chapter 16, we described how spectra of galaxies studied by Edwin Hubble led to this amazing concept, one of the pillars of cosmology.

18.2a Hubble's Law

Hubble found that, in every direction, all but the closest galaxies have spectra that are shifted to longer wavelengths. Moreover, the measured **redshift**, z, of a galaxy is *proportional* to its current distance from us, d (■ Fig. 18–4). If this redshift is produced by motion away from us, then we can use the Doppler formula to derive the speed of recession, v. (Recall from Chapter 16 that the redshift z of an object is defined by $z = \Delta\lambda/\lambda_0$, where $\Delta\lambda = \lambda - \lambda_0$ is the observed wavelength of an absorption or emission line in the object's spectrum minus its laboratory (rest) wavelength. If $z < 0.2$, a reasonable approximation is the Doppler formula, $z = \Delta\lambda/\lambda_0 \approx v/c$, where v is the object's speed of recession.)

The recession speed can be plotted against distance for many galaxies in a **Hubble diagram** (■ Fig. 18–5), and a straight line nicely represents the data. The final result is **Hubble's law**, $v = H_0 d$, where H_0 is the *present-day value* of **Hubble's constant**, H. (Note that H is constant throughout the Universe at a given time, but its value decreases with time.) The Wilkinson Microwave Anisotropy Probe (WMAP), to be discussed in Chapter 19, in 2010 provided a precise estimate of Hubble's constant, $H_0 = 71.0 \pm 2.5$ km/s/Mpc, within the uncertainty of the Cepheid/supernova-determined value of 73.8 ± 2.4 km/s/Mpc that we will soon discuss. We adopt the WMAP value, $H_0 = 71$ km/s/Mpc, throughout this book.

An explosion can give rise to Hubble's law. If we kick a pile of balls, for example, some of them are hit directly and given a large speed, while others fly off more slowly. After a while, we see that the most distant ones are moving fastest, while the ones closest to the original pile are moving slowly (■ Fig. 18–6). The reason is obvious: To have reached a particular distance in a given amount of time, the distant balls must have been moving quickly, while the nearby ones must have been moving slowly. Speed is indeed proportional to distance, which is the same formula as Hubble's law.

Hubble's law with $H_0 = 71$ km s^{-1} Mpc^{-1}

Stars in our galaxy
z=0 (non-cosmological)
v=~0 (non-cosmological)
d=<10 kpc

Spiral galaxy
z=~5%
v=15,000 km s^{-1}
d=v/71 = ~210 Mpc

More distant galaxy
z=~25%
v=75,000 km s^{-1}
d=v/71 = ~1.1 Gpc

Wavelength (Å)

SWEEPS FIELD

NGC 5584

Primeval Galaxy
HUDF WFC3/IR

Karen Teramura, Institute for Astronomy, University of Hawaii

NASA images. Teramura art @kthonhi

■ **FIGURE 18–4** An illustration of Hubble's law, based on $H_0 = 71$ km/s/Mpc. Our own Galaxy is gravitationally bound and thus not expanding; stars have relative speeds close to 0. But other galaxies are moving away from ours. On the whole, the farther away a galaxy is, the smaller it looks. The arrows on the spectra show how far the "H and K" lines of singly ionized calcium are redshifted, with the arrow corresponding to the K line. (Fraunhofer named the dark H-line; the K-line was named a few decades later.)

Based on Hubble's law, we conclude that the Universe is expanding. Like an expanding gas, its density and temperature are decreasing with time. Extrapolating the expansion backward in time, we can reason that the Universe began at a specific instant when all of the material was in a "singularity" at essentially infinite density and infinite temperature. (This is not the only possible conclusion, but other evidence, to be discussed in Chapter 19, strongly supports it.)

We call this instant the **big bang** – the initial event that set the Universe in motion – though we will see below that it is not like a conventional explosion. "Big bang" is both the technical and popular name for the current class of theories that deal with the birth and evolution of the Universe. In Chapter 19, we will discuss in more detail the reasons astronomers think that the Universe began its life in a hot, dense, expanding state.

18.2b Expansion Without a Center

Does the observed motion of galaxies away from us imply that we are the center of expansion, and hence in a very special position? Such a conclusion would be highly anti-Copernican: Looking at the billions

of other galaxies, we see no scientifically based reason for considering our Galaxy to be special, in terms of the expansion of the Universe. Historically, too, we have encountered this several times. The Earth is not the only planet, and it isn't the center of the Solar System, just as Copernicus postulated. The Sun is not the only star, and it isn't the center of the Milky Way Galaxy. The Milky Way Galaxy is not the only galaxy ... and it probably isn't the center of the Universe.

If our Galaxy *were* the center of expansion, we would expect the number of other galaxies per unit volume to decrease with increasing distance, as shown by the balls in Figure 18–6. In fact, however, galaxies are *not* observed to thin out at large distances – thus providing direct evidence that we are not at the unique center.

A different conclusion that is consistent with the data, and also satisfies the Copernican principle, is that there is *no* center – or, alternatively, that all places can claim to be the center. Consider a loaf of raisin bread about to go into the oven. The raisins are spaced at various distances from each other. Then, as the uniformly distributed yeast causes the dough to expand, the raisins start spreading apart from each other (■ Fig. 18–7). If we were able to sit on any one of those raisins, we would see our neighboring raisins move away from us at a certain speed. Note that raisins far away from us recede faster

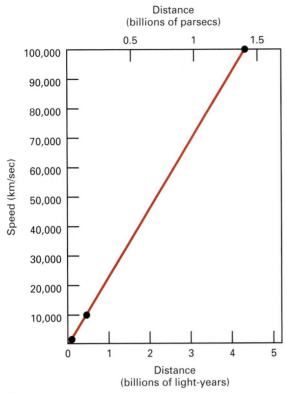

FIGURE 18–5 The Hubble diagram for the galaxies shown in Figure 18–4.

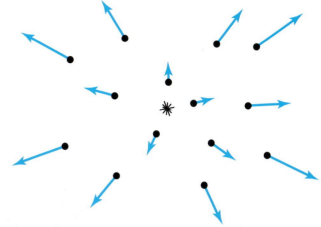

FIGURE 18–6 After receiving a kick, balls go flying away from their common point of origin at different speeds, depending on how directly each one was hit. After any given amount of time, it is clear that the most distant balls are those with the highest speeds; to have travelled farthest, they must be moving fastest.

than nearby raisins because there is more dough between them and us, and all of the dough is expanding uniformly.

For example, suppose that, after 1 second, each original centimeter of dough occupies 2 cm (Figure 18–7). From our raisin, we will see that another raisin initially 1 cm away has a distance of 2 cm after 1 second. It therefore moved with an average speed of 1 cm/s. A different raisin initially 2 cm away has a distance of 4 cm after 1 second, and moved with an average speed of 2 cm/s. Yet another raisin initially 3 cm away has a distance of 6 cm after 1 second, so its average speed was 3 cm/s. We see that speed is proportional to distance ($v \propto d$), as in Hubble's law.

It is important to realize that it doesn't make any difference which raisin we sit on; all of the other raisins would seem to be receding, regardless of which one was chosen. (Of course, any real loaf of raisin bread is finite in size, while the Universe may have no limit so that we would never see an edge.) The fact that all the galaxies are receding from us does not put us in a unique spot in the Universe; there is no center to the Universe. Each observer at each location would observe the same effect.

A convenient one-dimensional analogue is an infinitely long rubber band with balls attached to it (Fig. 18–8). Imagine that we are on one of the balls. As the rubber band is stretched, we would see all other balls moving away from us, with a speed of recession proportional to distance. But again, it doesn't matter which ball we chose as our home.

Another very useful analogy is an expanding spherical balloon (Fig. 18–9). Suppose we define this hypothetical universe to be only the *surface* of the balloon. It has just two spatial dimensions, not three like our real Universe. We can travel forward or backward, left or right, or any combination of these directions – but "up" and "down" (that is, out of, or into, the balloon) are not allowed. All of the laws of physics are constrained to operate along these two directions; even light travels only along the surface of the balloon in this hypothetical universe.

FIGURE 18–7 From every raisin in a rising loaf of raisin bread, every other raisin seems to be moving away from you at a speed that depends on its distance from you. This leads to a relation like Hubble's law between the speed and the distance. Note also that each raisin would be at the center of the expansion measured from its own position, yet the bread is expanding uniformly. The raisins themselves do not expand, just as galaxies do not expand. For a better analogy with the Universe, consider an infinite loaf; unlike the finite analogy pictured here, there is then no overall center to its expansion.

■ **FIGURE 18–8** A one-dimensional representation of the expanding Universe in which balls are attached to an infinite, expanding rubber band. No matter which ball you choose to sit on, all others move away from it. Distant balls recede faster than nearby ones, since each bit of rubber expands uniformly and there is more rubber between you and distant balls than between you and nearby ones.

■ **FIGURE 18–9** The surface of a spherical balloon provides a good example of a finite, expanding universe. The space inside and outside the balloon is forbidden in this hypothetical universe. Stickers on the balloon represent galaxies, and their distances from each other increase according to Hubble's law as the balloon expands. The center of expansion is the center of the balloon, but this is not part of the surface itself, and hence is outside this hypothetical universe. Note also that while the stickers don't expand, light waves *do* expand, since they stretch with the rubber – and this fundamentally gives rise to the redshift.

If we put flat stickers on the balloon, they recede from each other as the balloon expands, and flat creatures on them would deduce Hubble's law. A clever observer could also reason that the surface is curved: By walking in one direction, for example, the starting point would eventually be reached. With enough data, the observer might even derive an equation for the surface of the balloon, but it would reveal an *unreachable third spatial dimension*. The observer could conclude that the center of expansion is in this "extra" dimension, which exists only mathematically, not physically!

It is possible that we live in an analogue of such a spherical universe, but with three spatial dimensions that are physically accessible, and an additional, inaccessible spatial dimension around which space mathematically curves. We will discuss this idea in more detail later.

18.2c What Is Actually Expanding?

Besides illustrating Hubble's law and the absence of a unique center, the above analogies accurately reproduce two additional aspects of the Universe. First, according to Einstein's general theory of relativity, which is used to quantitatively study the expansion and geometry of the Universe, it is *space itself* that expands.

The dough or the rubber expands, making the raisins, balls, and stickers recede from each other. They do not travel *through* the dough or rubber. Similarly, in our Universe, galaxies do not travel through a preexisting space; instead, space itself is expanding. (We sometimes

say that the "fabric of space-time is expanding.") In this way, the expansion of the Universe differs from a conventional explosion, which propels material through a preexisting space.

Second, note that the raisins, balls, and stickers themselves don't expand – only the space *between* them expands. (We intentionally didn't draw ink dots on the balloon, because they would expand, unlike stickers.) Similarly, galaxies and other gravitationally bound objects such as stars and planets do not expand: The gravitational force is strong enough to overcome the tendency of space within them to expand. Nor do people expand, because electrical forces (between atoms and molecules) strongly hold us together.

Strictly speaking, even most clusters of galaxies are sufficiently well bound to resist the expansion. Only the space between the *clusters* expands, and even in these cases the expansion is sometimes diminished by the gravitational pull between clusters (as in superclusters).

However, electromagnetic waves or photons do expand with space; they are not tightly bound objects. Thus, for example, blue photons turn into red photons as the Universe expands (Figure 18–9). In fact, this is what actually produces the observed redshift of galaxies. Technically, the redshift is *not* a Doppler effect, since nothing is moving *through* the Universe, and the Doppler effect was defined in terms of the motion of an object relative to the waves it emits. The Doppler formula remains valid at low speeds, though, and it is convenient to think about the redshift as a Doppler effect, so we will continue to do so – but you should be aware of the deeper meaning of redshifts.

18.3 THE AGE OF THE UNIVERSE

The discovery that our Universe had a definite beginning in time, the big bang, is of fundamental importance. The Universe isn't infinitely old. But humans generally have a fascination with the ages of things, from the Dead Sea Scrolls to movie stars. Naturally, then, we would like to know how old the Universe itself is!

18.3a Finding Out How Old

There are at least two ways in which to determine the age of the Universe. First, the Universe must be at least as old as the oldest objects within it. Thus, we can set a minimum value to the age of the Universe by measuring the ages of progressively older objects within it. For example, the Universe must be at least as old as you – admittedly, not a very meaningful lower limit! More interestingly, it must be at least 200 million years old, since there are dinosaur fossils of that age. Indeed, it must be at least 4.6 billion years old, since Moon rocks and meteorites of that age exist.

The oldest discrete objects whose ages have reliably been determined are globular star clusters in our Milky Way Galaxy (■ Fig. 18–10). The oldest ones are now thought to be 12–13 billion years old, though the exact values are still controversial. (Globular clusters used to be thought to be about 14–17 billion years old.) Theoretically, the formation of globular clusters could have taken place only a few hundred million years after the big bang; if so, the age of the Universe would be about 14 billion years. Since no discrete objects have been found that appear to be much older than the oldest globular clusters, a reasonable conclusion is that the age of the Universe is indeed at least 13 billion years, but not much older than 14 billion years.

A different method for finding the age of the Universe is to determine the time elapsed since its birth, the big bang. At that instant, all the material of which any observed galaxies consist was essentially at the same location. Thus, by measuring the distance between our Milky Way Galaxy and any other galaxy, we can calculate how long the two have been separating from each other if we know the current recession speed of that galaxy.

At this stage, we have made the simplifying assumption that the recession speed has always been *constant*. So, the relevant expression is distance equals speed multiplied by time: $d = vt$. For example, if we measure a friend's car to be approaching us with a speed of 60 miles/hour, and the distance from his home to ours is 180 miles, we calculate that the journey took 3 hours if the speed was always constant and there were no rest breaks. In the case of the Universe, the amount of time since the big bang, assuming a constant speed for any given galaxy, is called the **Hubble time** (see *Figure It Out 18.1: The Hubble Time*).

If, however, the recession speed was faster in the past, then the true age is less than the Hubble time. Not as much time had to elapse for a galaxy to reach a given distance from us, compared to the time needed with a constant recession speed. Using the previous example, if our friend started his journey with a speed of 90 miles/

Michael Rich, Kenneth Mighell, and James D. Neill (Columbia University); Wendy Freedman (Carnegie Observatories); and NASA

■ **FIGURE 18–10** A Hubble image of a globular cluster in the Large Magellanic Cloud, seen as clearly as Earth-bound telescopes image clusters within our own Galaxy. Two bright foreground stars also appear.

hour, and gradually slowed down to 60 miles/hour by the end of the trip, then the average speed was clearly higher than 60 miles/hour, and the trip took fewer than 3 hours. (This is why people often break the posted speed limit!)

Astronomers have generally expected such a decrease in speed because all galaxies are gravitationally pulling on all others, thereby presumably slowing down the expansion rate. In fact, many cosmologists have believed that the deceleration in the expansion rate is such that it gives a true age of exactly two-thirds of the Hubble time. This is, in part, a theoretical bias; it rests on an especially pleasing cosmological model. Later in this chapter we will see how attempts have been made to actually measure the expected deceleration.

18.3b The Quest for Hubble's Constant

To determine the Hubble time, we must measure Hubble's constant, H_0. This can be done if we know the distance (d) and recession speed (v) of another galaxy, since rearrangement of Hubble's law tells us that $H_0 = v/d$. Many galaxies at different distances should be used, so that an average can be taken.

The recession speed is easy to measure from a spectrum of the galaxy and the Doppler formula. We can't use galaxies within our own Local Group (like the Andromeda galaxy, M31), however, since they are bound to the group by gravity and are not expanding away.

Galaxy distances are notoriously hard to determine, and historically this has led to large uncertainties in the derived value of Hubble's constant. We can't use triangulation because even the nearest galaxies are much too far away. In principle, the distance of a galaxy can be determined by measuring the apparent angular size of an object (such as a nebula) within it, and comparing it with an

FIGURE IT OUT 18.1   The Hubble Time

There is a simple relationship between the value of Hubble's constant and the time elapsed since the big bang, assuming that the recession speed of any given galaxy has always been constant. For galaxies, we know that Hubble's law holds: $v = H_0 d$. Rearranging this relation, we have

$$d = v/H_0 = vT_0,$$

where we define T_0 (the "Hubble time") to be $1/H_0$, the reciprocal or inverse of Hubble's constant. Note that this equation looks just like the relation $d = vt$ for the distance travelled by an object at constant speed.

Consider Hubble's constant, 71 km/s/Mpc. What is the corresponding Hubble time? This is mostly an exercise in keeping track of units. First, 1 pc is approximately equal to 3.26 light-years (see Chapter 11), or 3.09×10^{18} cm, so 1 Mpc $\approx 3.09 \times 10^{24}$ cm =

3.09×10^{19} km. Recall that 1 Mpc = 1 megaparsec, a million parsecs, which is 3.26 million light-years.

We then convert $H_0 = 71$ km/s/Mpc = 71 km/(Mpc s) = 71 km Mpc^{-1} s^{-1} to units of inverse seconds:

$$H_0 = (71 \, \text{km}/(\text{Mpc s})) / (3.09 \times 10^{19} \, \text{km}/\text{Mpc}) \approx 2.3 \times 10^{-18}/\text{s}.$$

Thus, $T_0 = 1/H_0 \approx 4.3 \times 10^{17}$ s. But one year (that is, 365.25 days at 86,400 seconds per day) is about 3.16×10^7 s, so

$$T_0 \approx (4.3 \times 10^{17} \, \text{s}) / (3.16 \times 10^7 \, \text{s} / \text{year}) \approx 1.4 \times 10^{10} \, \text{years}.$$

This is roughly 14 billion years!

Note that, as the Universe ages, T_0 increases, and so H_0 decreases – the value of Hubble's constant decreases with time. However, at a given time, its value is thought to be the same everywhere in the Universe – hence the name "Hubble's constant."

assumed physical size. But this method generally gives only a crude estimate of distance, because the physical sizes of different objects in a given class are not uniform enough.

More frequently, we measure the apparent brightness of a star, and compare this with its intrinsic brightness (luminosity, or power) to determine the distance. This is similar to how we judge the distance of an oncoming car at night: We intuitively use the inverse-square law of light (discussed in Chapter 11) when we see how bright a headlight appears to be. We must be able to recognize that particular type of star in the galaxy, and we assume that all stars of that type are "standard candles" (a term left over from the 19th century, when sets of actual candles were manufactured to a standard brightness) – that is, they all have the same luminosity.

Historically, the best such candidates have been the Cepheid variables, at least in relatively nearby galaxies. Though not all of uniform luminosity, they do obey a "period-luminosity relation" (■ Fig. 18–11), as shown by Henrietta Leavitt in 1912. Thus, if the variability period of a Cepheid is measured, its average luminosity can be read directly off the graph. This means that Cepheid variables should actually be referred to as "standardizable candles" instead of "standard candles," though many astronomers still use the latter term, with the understanding that the period-luminosity relation must actually be used to make the Cepheids "standard."

We emphasize that the period-luminosity relation, now often called Leavitt's law, is valid only for the specific class of variable star known as Cepheid variables, after its prototype, δ (delta) Cephei.

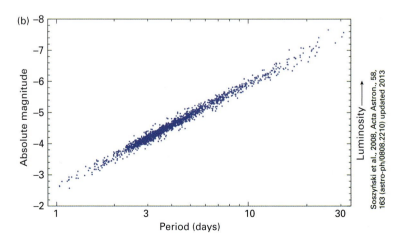

Soszyński et al., 2008, Acta Astron., 58, 163 (astro-ph/0808.2210) updated 2013

■ **FIGURE 18–11** (a) The light curve of a typical Cepheid variable star. (b) The period-luminosity relation of Cepheid variable stars, which the authors like to refer to as "Leavitt's law" in honor of Henrietta Leavitt, who discovered it (see our discussion in Chapter 11). Longer-period Cepheids have higher average luminosities than shorter-period Cepheids. Though they have the same light-curve shape, with sharp rises and slower falls, do not confuse the Cepheids, whose periods are over a day, with a different class of variable star, known as RR Lyrae stars, whose periods are a day or shorter. These recent data are from the OGLE (Optical Gravitational Lensing Experiment) survey, a Polish project based at the University of Warsaw. The Cepheids with periods 30–100 days saturated OGLE's detector so are not shown. These infrared data are corrected for the effect of the dust between the stars and us, so-called reddening.

And let us further emphasize that you can identify Cepheid variables by plotting their light curves – their brightnesses over time – and seeing that they simultaneously meet two criteria: (1) they must have a sharp rise with a slower fall, and (2) their periods must be days or weeks, longer than those of another class of variable star with a similar periodic shape, the so-called RR Lyrae variables.

But individual stars are difficult to see in distant galaxies: They merge with other stars when viewed with ground-based telescopes. Other objects that have been used include luminous nebulae, globular star clusters, and novae – though all of them have substantial uncertainties. They aren't excellent standard candles, or are difficult to see in ground-based images, or depend on the assumed distances of some other galaxies.

Even entire *galaxies* can be used, if we determine their luminosity from other properties. The luminosity of a spiral galaxy is correlated with how rapidly it rotates, for example. Also, the brightest galaxy in a large cluster has a roughly standard luminosity. Again, however, significant uncertainties are associated with these techniques, or they depend on the proper calibration of a few key galaxies.

Throughout the 20th century, many astronomers have attempted to measure the value of Hubble's constant. Edwin Hubble himself initially came up with 550 km/s/Mpc, but several effects that were at that time unknown to him conspired to make this much too large. From the 1960s to the early 1990s, the most frequently quoted values were between 70 and 50 km/s/Mpc, due largely to the painstaking work of Allan Sandage (■ Fig. 18–12), a disciple of Edwin Hubble himself.

Such values yielded a Hubble time of about 14–20 billion years, the probable maximum age of the Universe. The true expansion age could therefore have been 9–13 billion years, if it were only two-thirds of the Hubble time, as many cosmologists have been prone to believe. These lower numbers may give rise to a discrepancy if the globular clusters are 12–13 billion years old. Despite uncertainties in the measurements, an "age crisis" would exist if the clusters were 14–17 billion years old, as was thought until the mid-1990s.

But at that time, some astronomers obtained considerably larger values for H_0, up to 100 km/s/Mpc. This value gave a Hubble time of about 10 billion years, and two-thirds of it is only 6.7 billion years. Even the recently revised ages of globular star clusters are substantially greater (12–13 billion years), leading to a sharp age crisis. The various teams of astronomers who got different answers all claimed to be doing careful work, but there are many potential hidden sources of error, and the assumptions might not be completely accurate. The debate over the value of Hubble's constant was often heated, and sessions of scientific meetings at which the subject was discussed were well attended.

Note that the value of Hubble's constant also has a broad effect on the perceived size of the observable Universe, not just its age. As mentioned above, Hubble's constant is now well determined to be 71.0 ± 2.5 km/s/Mpc. Thus, a galaxy whose recession speed is measured to be 7,100 km/s would be at a distance

$$d = v/H_0 = (7{,}100 \text{ km/s}) / (71 \text{ km/s /Mpc}) = 100 \text{ Mpc}.$$

Andrew Fraknoi, Astronomical Society of the Pacific

■ **FIGURE 18–12** Allan Sandage (*right*) of the Carnegie Observatories in Pasadena, California, who devoted much of his career to the measurement of Hubble's constant. He is seen here with Maarten Schmidt of Caltech, the first person to realize that the spectra of quasars are highly redshifted.

18.3c A Key Project of the Hubble Space Telescope

The aptly named Hubble Space Telescope was expected to provide a major breakthrough in the field. It was to obtain distances to many important galaxies, mostly through the use of Cepheid variable stars (■ Fig. 18–13). Indeed, very large amounts of telescope time were to be dedicated to this "Key Project" of measuring galaxy distances and deriving Hubble's constant. But astronomers had to wait a long time, even after the launch of the Hubble Space Telescope in 1990, because the primary mirror's spherical aberration (see our discussion in Chapter 4) made it too difficult to detect and reliably measure Cepheids in the chosen galaxies.

Finally, in 1994, the Hubble Key Project team announced their first results, based on Cepheids in only one galaxy (■ Fig. 18–14). Their value of H_0 was about 80 km/s/Mpc, higher than many astronomers had previously thought. This implied that the Hubble time was 12 billion years; the Universe could be no older, but perhaps significantly younger (down to 8 billion years) if the expansion decelerates with time.

Because these values are less than 14–17 billion years (the ages preferred for globular star clusters at that time), this disagreement brought the age crisis to great prominence among astronomers, who shared it with the public. How could the Universe, as measured with the mighty Hubble Space Telescope, be younger than its oldest contents? There were several dramatic headlines in the news.

Admittedly, the Hubble team's quoted value of H_0 had an uncertainty of 17 km/s/Mpc, meaning that the actual value could have been between about 63 and 97 km/s/Mpc. Thus, the Universe could have been as old as 15–16 billion years, especially if there has been

| April 23 | May 4 | May 9 | May 16 | May 20 | May 31 |

Wendy L. Freedman, Observatories of the Carnegie Institution of Washington, and NASA

■ **FIGURE 18–13** Cepheid variable stars in the galaxy M100, as observed with the Hubble Space Telescope. They have been used to calibrate the cosmic distance scale through the measurement of Hubble's constant.

little deceleration. The ages of globular clusters were uncertain as well, so it was not entirely clear that the age crisis was severe. But, as is often the case with newspaper and popular magazine articles, these subtleties are ignored or barely mentioned; only the "bottom line" gets reported, especially if it's exciting.

In 2001, the Hubble team announced a final answer, which was based on several methods of finding distances, with Cepheid variables as far out as possible and Type Ia supernovae (to be discussed later in this chapter) pinning down the greatest distances. Their preferred value of H_0 was 72 km/s/Mpc, with an uncertainty of about 8 km/s/Mpc (■ Fig. 18–15). But the Hubble team was not the only game in town, and other groups of scientists measured slightly different values. The most accurate and precise existing value directly measured with Cepheids and supernovae, reported in 2011, is 73.8 ± 2.4 km/s/Mpc, consistent (within the uncertainties) with our adopted WMAP value of 71.0 ± 2.5 km/s/Mpc.

A value of 72 km/s/Mpc for Hubble's constant, as reported by the Hubble team in 2001, would mean that the Universe has been expanding for 13.6 billion years, if there is no deceleration. By assuming only a small amount of deceleration (not as much as many theorists would have preferred), the Hubble team announced a best-estimate expansion age of 12 billion years for the Universe. Moreover, around 2000, the preferred ages of globular clusters had shifted from 14–17 billion years to only 11–14 billion years, based on accurate new parallaxes of stars from the Hipparcos satellite and on some new theoretical work. (We are now awaiting still more accurate parallaxes, as well as parallaxes farther out, from the Gaia satellite in a few years.) This age value meant that the age discrepancy had subsided to some extent, but did not fully disappear if the globular clusters are actually as old as 13–14 billion years.

But on what basis was the amount of deceleration estimated? We will discuss this more fully in Section 18.5, with the surprising result

■ **FIGURE 18–14** A Hubble Space Telescope image of the galaxy M100 in the Virgo Cluster, the first galaxy whose distance was determined by the Hubble Space Telescope's Key Project team that used Cepheid variable stars. The zigzag square on Hubble's Wide Field and Planetary Camera 2 had twice the resolution of the larger squares (it has since been replaced with Wide Field Camera 3), so the galaxy's center can be enlarged as at upper right without loss of detail. The Cepheids, though, are in the outer spiral arms.

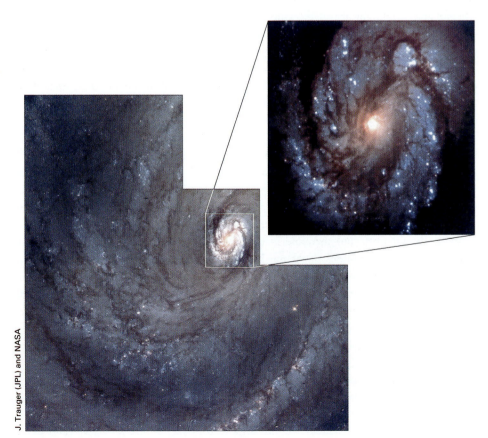

J. Trauger (JPL) and NASA

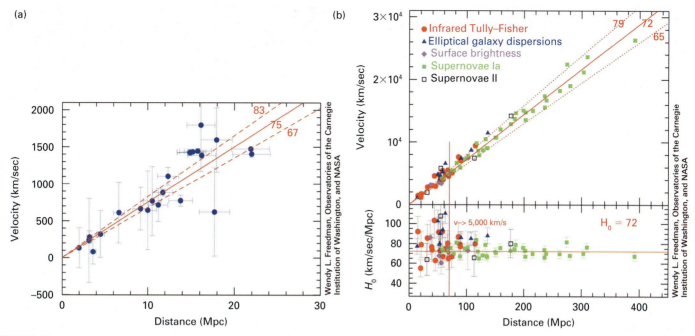

■ FIGURE 18–15 Hubble's law as determined by the Key Project on the Cosmic Distance Scale of the Hubble Space Telescope. (a) Hubble's law as determined from Cepheid variable stars out to a distance of about 20 Mpc (about 65 million light-years), our "local" region. (b) The Hubble diagram made by evaluating and averaging several distance indicators (which are plotted with different symbols). Including supernovae brings the graph out to 400 Mpc (about 1.3 billion light-years). At the bottom, we see deviations of the individual values of Hubble's constant from the average. In Chapter 19, we will see how the value of Hubble's constant reported from the Wilkinson Microwave Anisotropy Probe (WMAP) is in close agreement with this Hubble Key Project value. In 2013, results from the Planck spacecraft suggest a somewhat lower value, but it must still be checked by other astronomers.

that the assumed deceleration may have been erroneous. Instead, the expansion rate of the Universe appears to actually be *increasing* with time! This exciting discovery, known as the "accelerating universe," is now accepted by most astronomers and physicists, contrary to the situation when it was initially announced in 1998 based on supernovae found since 1995 (■ Fig. 18–16a). As we shall see later in this chapter, recent evidence makes it quite convincing. In fact, the evidence is convincing enough that the three leaders of the two teams that nearly simultaneously made the discovery were given the 2011 Nobel Prize in Physics (■ Fig. 18–16b). In recognition of the team effort, the prize winners invited their 48 collaborators to join them in Stockholm for the Nobel festivities (■ Fig. 18–16c,d), and in speeches there they emphasized everyone's contributions.

The discovery of acceleration (not mere expansion, which has been known since the time of Lemaître and Hubble in the 1920s) implies some very intriguing, but also troubling, new aspects to the nature and evolution of the Universe. It fully resolves the age crisis: Using WMAP and other data, the expansion age of the Universe is found to be 13.75 ± 0.11 billion years, consistent with the 12–13 billion year ages of globular clusters estimated most recently.

18.3d Deviations from Uniform Expansion

A major problem with using relatively nearby galaxies for measurements of Hubble's constant is that proper corrections must be made

for deviations from the "Hubble flow" (the assumed uniform expansion of the Universe). As we discussed in Chapter 16, there are concentrations of mass (clusters and superclusters) in certain regions, and large voids in others, so a specific galaxy may feel a greater pull in one direction than in another direction. It will therefore be pulled *through* space (relative to the Hubble flow), and its apparent recession speed may be affected. Though the galaxy's recession speed is easy to measure from a spectrum, it might not represent the true expansion of space.

For example, the Virgo Cluster of galaxies (■ Fig. 18–17) is receding from us more slowly than it would if it had no mass: The Milky Way Galaxy is "falling" toward the Virgo Cluster, thereby counteracting part of the expansion of space. Such gravitationally induced **peculiar motions** are typically a few hundred kilometers per second, but can occasionally reach as high as 1,000 km/s. Their exact size is difficult to determine without detailed knowledge of the distribution of matter in the Universe. In the case of the Virgo Cluster, the average observed recession speed is about 1,100 km/s, and the peculiar motion is thought to be about 300 km/s, but this is uncertain. Errors in the adopted "true" recession speed directly affect the derived value of Hubble's constant.

A surprising discovery was that even the Virgo Cluster is moving with respect to the average expansion of the Universe. Some otherwise unseen "Great Attractor" is pulling the Local Group, the Virgo Cluster, and even the much larger Hydra-Centaurus Supercluster toward it. Redshift measurements by a team of astronomers informally known as the "Seven Samurai" showed the location of the

(a)

Circular No. 6270

Central Bureau for Astronomical Telegrams
INTERNATIONAL ASTRONOMICAL UNION
Postal Address: Central Bureau for Astronomical Telegrams
Smithsonian Astrophysical Observatory, Cambridge, MA 02138, U.S.A.
IAUSUBS@CFA.HARVARD.EDU or FAX 617-495-7231 (subscriptions)
BMARSDEN@CFA.HARVARD.EDU or DGREEN@CFA.HARVARD.EDU (science)
Phone 617-495-7244/7440/7444 (for emergency use only)

SUPERNOVAE

The Supernova Cosmology Project [S. Perlmutter, S. Deustua, G. Goldhaber, D. Groom, I. Hook, A. Kim, M. Kim, J. Lee, J. Melbourne, C. Pennypacker, and I. Small, Lawrence Berkeley Lab. and the Center for Particle Astrophysics; A. Goobar, Univ. of Stockholm; R. Pain, CNRS, Paris; R. Ellis and R. McMahon, Inst. of Astronomy, Cambridge; and B. Boyle, P. Bunclark, D. Carter, and M. Irwin, Royal Greenwich Obs.; with A. V. Filippenko and A. Barth (Univ. of California, Berkeley) at the Keck telescope; W. Couch (Univ. of N.S.W.) and M. Dopita and J. Mould (Mt. Stromlo and Siding Spring Obs.) at the Siding Spring 2.3-m telescope; H. Newberg (Fermi National Accelerator Lab.) and D. York (Univ. of Chicago) at the ARC telescope] report eleven supernovae found with the Cerro Tololo (CTIO) 4-m telescope in their 1995 High Redshift Supernova Search:

SN	1995 UT	R.A.	(2000)	Decl.	R	Offset
1995aq	Nov. 19	0 29 04.26	+ 7 51 20.0		22.4	0".6 W, 1".4 S
1995ar	Nov. 19	1 01 20.41	+ 4 18 33.8		23.1	2".9 W, 0".5 S
1995as	Nov. 19	1 01 35.30	+ 4 26 14.8		23.3	0".7 W, 0".7 N
1995at	Nov. 20	1 04 50.94	+ 4 33 53.0		22.7	0".3 W, 0".4 S
1995au	Oct. 29	1 18 32.60	+ 7 54 03.5		20.7	1".4 E, 3".3 N
1995av	Nov. 20	2 01 36.75	+ 3 38 55.2		20.1	0".2 W, 0".0 N
1995aw	Nov. 19	2 24 55.54	+ 0 53 07.5		22.5	0".2 W, 0".2 S
1995ax	Nov. 19	2 26 25.80	+ 0 48 44.2		22.6	0".3 W, 0".2 S
1995ay	Nov. 20	3 01 07.52	+ 0 21 19.4		22.7	0".9 W, 1".4 S
1995az	Nov. 20	4 40 33.59	− 5 30 03.6		24.0	1".6 W, 1".7 N
1995ba	Nov. 20	8 19 06.46	+ 7 43 21.2		22.6	0".1 E, 0".2 N

The spectra (Keck, Nov. 26-28) are consistent with type-I supernovae (except SN 1995av, a probable type II) at the redshift of the host galaxy: z = 0.45, 0.46, 0.49 (preliminary type-I identification), 0.65, 0.16, 0.30, 0.4 (supernova redshift only), 0.61, 0.48, 0.45, 0.39. Photometry obtained on Nov. 21-23 at CTIO (A. Walker) and Nov. 23-27 at WIYN (D. Harmer, D. Willmarth) indicates that SNe

IAUC, Smithsonian Astrophysical Observatory

(b)

Alex Filippenko

(c)

Nicholas Suntzeff

(d)

Ellie Newberg

■ **FIGURE 18–16** (a) The International Astronomical Union Circular (IAUC) sent out by U.S. Mail (officially an International Astronomical Union card from their Central Bureau of Astronomical Telegrams, a term left over from pre-internet days) on which one of the first sets of distant supernovae was announced. This image was shown during Saul Perlmutter's Nobel Prize talk in Stockholm in 2011. (b) The 2011 Nobel Laureates in Physics (left to right): Adam G. Riess, Brian P. Schmidt, and Saul Perlmutter. (c) Most of the members of the High-*z* Supernova Search Team, including Physics Laureates Schmidt and Riess, and team member A.F., immediately after the Nobel award ceremony in Stockholm, Sweden, on December 10, 2011. (d) Most of the team members of the Supernova Cosmology Project, including Physics Laureate Perlmutter, at a Swedish Christmas smorgasbord on an island near Stockholm, after the Nobel lectures on December 8, 2011.

■ **FIGURE 18–17** The central part of the Virgo Cluster of galaxies. This cluster is not receding from the Milky Way Galaxy as quickly as it would if galaxies had no mass and didn't gravitationally attract each other.

18.3e Type Ia Supernovae as Cosmological Yardsticks

In the 1990s, a remarkable reliable method was developed for measuring the distances of very distant galaxies. It is based on Type Ia supernovae ("white-dwarf supernovae"), which are exploding stars that result from a nuclear runaway in a white dwarf (see our discussion in Chapter 13). When they reach their peak power, these objects shine with the luminosity (intrinsic brightness) of about 4 billion Suns, or about a million times more than Cepheid variables. So, they can be seen at very large distances, 1,000 times greater than Cepheid variables (■ Fig. 18–18).

Most observed Type Ia supernovae are found to have nearly the same peak luminosity, as would be expected since the exploding white dwarf is thought to always have the same mass (nearly the Chandrasekhar limit). Type Ia supernovae are therefore very good "standard candles" for measuring distances. (They do show small variations in peak luminosity, but we have ways of taking this into account – essentially like reading the wattage label on a light bulb. Thus, we "standardize" them, as with Cepheid variables.) By comparing the apparent brightness of a faint Type Ia supernova in a distant galaxy with the supernova's known luminosity, and by using the inverse-square law of light, we obtain the distance of the supernova, and hence of the galaxy in which it exploded (see Figure 18–18a).

Of course, to apply this method successfully, we need to know the peak luminosity of a Type Ia supernova. But this can be found by measuring the peak apparent brightness of a supernova in a relatively nearby galaxy – one whose distance can be measured by other techniques, such as Cepheid variable stars. So, an important part of

giant mass that must be involved. It is about three times farther from us than the Virgo Cluster, and includes tens of thousands of galaxies or their equivalent mass.

Measurements of still more distant galaxies avoid the problem of peculiar motions when trying to determine Hubble's constant. For example, compared with galaxies having recession speeds of 15,000–30,000 km/s, the peculiar motions are negligible. So, measurements of their distances, when combined with their recession speeds, can yield an accurate value of H_0. The trick is to find their distances – and this can't be done directly with Cepheid variable stars because they aren't intrinsically bright enough.

■ **FIGURE 18–18** (a) Type Ia supernovae, marked with cross-hairs, in four different galaxies. Near its peak, a Type Ia supernova can shine as brightly as several billion Suns. (b) Light curves of four recent Type Ia supernovae obtained through various optical filters.

the Hubble Key Project was to find the distances of galaxies in which Type Ia supernovae had previously been seen, and in that way to calibrate the peak luminosity of Type Ia supernovae. By 2011, reliable distances to over a dozen such galaxies had been measured. Indeed, our adopted value of $H_0 = 71$ km/s/Mpc is partly dependent on this work. The August 2011 discovery of a Type Ia supernova only 21 million light-years from us in the "Pinwheel Galaxy," M101, only 11 hours after its explosion and long before its light curve reached its peak (Figure 18–18b), is enabling us to refine our understanding of this low step on the cosmic distance ladder.

18.4 THE GEOMETRY AND FATE OF THE UNIVERSE

We have seen that to determine the age of the Universe, its expansion history (in addition to Hubble's constant) must be known. It turns out that, under certain assumptions, the expansion history is closely linked to the eventual fate of the Universe as well as to its overall (large-scale) geometry.

18.4a The Cosmological Principle: Uniformity

Mathematically, we use Einstein's general theory of relativity to study the expansion and overall geometry of the Universe. Since matter produces space-time curvature (as we have seen when studying black holes in Chapter 14), we expect the average density to affect the overall geometry of the Universe. The average density should also affect the way in which the expansion changes with time: High densities are able to slow down the expansion more than low densities, due to the gravitational pull of matter. Thus, the average density appears to be the most important parameter governing the Universe as a whole.

To simplify the equations and achieve reasonable progress, we assume the **cosmological principle**: On the largest size scales, the Universe is very *uniform* – it is **homogeneous** and **isotropic**. Homogeneous means that it has the same average density everywhere at a given time (though the density *can* change with time). Isotropic means that it looks the same in all directions – there is no preferred axis along which most of the galaxies are lined up, for example (■ Fig. 18–19). Note that we can check for isotropy only from our own position in space. However, for even greater simplicity we could suppose that the Universe looks isotropic from *all* points. (In this case of isotropy everywhere, the Universe is also necessarily homogeneous.)

The cosmological principle is basic to most big-bang theories. But it is clearly incorrect on small scales: A human, the Earth, the Solar System, the Milky Way Galaxy, and our Local Group of galaxies have a far higher density than average. Even the supercluster of galaxies to which the Milky Way belongs is somewhat denser than average.

(a)

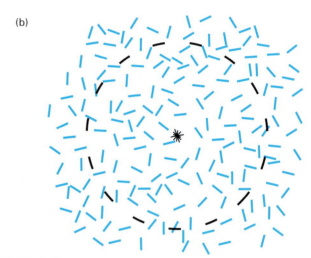
(b)

■ **FIGURE 18–19** Illustration of (a) a homogeneous universe that is not isotropic, and (b) a universe that is both homogeneous and isotropic. The average density of galaxies in a sufficiently large volume is the same in both universes, but in (a) the galaxies are all lined up in the same direction. The dashed circle is the observable part of the universe.

However, averaged over volumes about a billion light-years in diameter, the cosmological principle does appear to hold. The largest structures in the Universe seem to be superclusters and huge voids, but these are only a few hundred million light-years in diameter. Moreover, as we will see in Chapter 19, the strongest evidence comes from the "cosmic background radiation" that pervades the Universe: It looks the same in all directions, and it comes to us from a distance of about 14 billion light-years. Thus, over large distances, the Universe is indeed uniform.

18.4b No "Cosmological Constant"?

Another assumption we will make, at least temporarily, is that there are no long-range forces other than gravity, and that only "normal" matter and energy (with an attractive gravitational force) play a significant role – there is no "dark energy" having a repulsive effect.

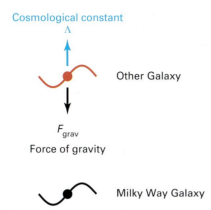

Cosmological constant
Λ

Other Galaxy

F_{grav}
Force of gravity

Milky Way Galaxy

■ **FIGURE 18–20** Albert Einstein introduced the cosmological constant, Λ (Lambda), in order to produce a static universe. He knew that normal gravity always pulls, so there had to be a counteracting repulsive force to prevent collapse. In this diagram, another galaxy is pulled toward the Milky Way Galaxy by their mutual gravitational attraction, but pushed apart with equal force provided by the cosmological constant.

Prior to Edwin Hubble's discovery that the Universe is expanding, most people thought the Universe is static (neither expanding nor contracting), which in some ways is aesthetically pleasing. Einstein knew that normal gravity should make the Universe contract, so in 1917 he postulated a long-range repulsive force, sort of a "cosmic antigravity," with a specific value that made the Universe static (■ Fig. 18–20). This "fudge factor" became known as the **cosmological constant**, denoted by the Greek capital letter Λ (lambda).

Though not mathematically incorrect, the cosmological constant is aesthetically displeasing, and it implies that the vacuum has a nonzero energy. Einstein was never fond of it, and reluctantly introduced it only because of the existing evidence for a static universe.

In 1929, when Hubble discovered the expansion of the Universe (actually, he found the redshift-distance relation, which others starting from Lemaître interpreted as an expansion of the Universe), the entire physical and philosophical motivation for the cosmological constant vanished. The Universe wasn't static, and no forces are needed to make it expand. After all, the Universe could have simply begun its existence in an expanding state, and is still coasting. Einstein renounced the cosmological constant and was unhappy that he had apparently erred; after all, he could have predicted that the Universe is dynamic rather than static. (Einstein is often quoted as saying that he made the "biggest blunder" of his life (or career) with the cosmological constant, but these exact words are not found in any of his written work. They were written only in a memoir by a respected physicist, George Gamow, who was known for exaggerating and joking. So it is possible, and maybe even probable, that Einstein never used the term "blunder" at all.)

Interestingly, the concept of the cosmological constant itself (or, more generally, repulsive "dark energy"; see Section 18.5d) should perhaps not be considered erroneous. In a sense, it is just a generalization of Einstein's relativistic equations for the Universe. The mistake was in supposing that the cosmological constant has the precise value needed to achieve a static universe – especially since this turns out

to be an unstable mathematical solution (slightly perturbing the Universe leads to expansion or collapse).

Nevertheless, it isn't clear what could physically produce a nonzero cosmological constant, and the simplest possibility is that the cosmological constant is zero (Λ = 0). Since there has generally been no strong observational evidence for a nonzero cosmological constant, astronomers have long assumed that its value is indeed zero. This is what we will initially assume here, too – but later in this chapter we will discuss exciting evidence that the cosmological constant (or some kind of "dark energy" that behaves in a similar way) isn't zero after all.

18.4c Three Kinds of Possible Universes

Given the assumptions of the cosmological principle and no long-range antigravity, and also that no new matter or energy are created after the birth of the Universe, the general theory of relativity seemed to allow only three possibilities. These are known as "Friedmann universes" in honor of the Russian scientist Alexander Friedmann, who, in the 1920s, was the first to derive them mathematically.

In each case the expansion decelerates with time, but the ultimate fate (that is, whether the expansion ever stops and reverses) depends on the overall *average density of matter* relative to a specific **critical density**. If we define the average matter density divided by the critical density to be Ω_M, where Ω is the Greek capital letter "Omega" and the subscript M stands for "matter," then the three possible universes correspond to the cases where this ratio is greater than one, equal to one, and less than one. (Details are explored in *Figure It Out 18.2: The Critical Density and Ω_M*.)

The separation between any two galaxies versus time is shown in ■ Figure 18–21 for the three types of universes. It is best to choose galaxies in different clusters (or even different superclusters, to be absolutely safe), since we don't want them to be bound together by gravity. This galaxy separation is often called the "scale factor" of the Universe; it tells us about the expansion of the Universe itself.

If $\Omega_M > 1$ (that is, the average density is above the critical density), galaxies separate progressively more slowly with time, but they eventually turn around and approach each other (in other words, the recession speed becomes negative), ending in a hot "big crunch." (Some astronomers also jokingly call it a "gnab gib," which is "big bang" backwards!) A good analogy is a ball thrown upward with a speed less than Earth's escape speed; it eventually falls back down. It is conceivable that another big bang subsequently occurs, resulting in an "oscillating universe," but we have little confidence in this hypothesis since the laws of physics as currently stated cannot be traced through the big crunch.

If $\Omega_M = 1$ (that is, the average density is exactly equal to the critical density), galaxies separate more and more slowly with time, but as time approaches infinity, the recession speed approaches zero. Thus, the Universe will expand forever, though just barely. The relevant analogy is a ball thrown upward with a speed equal to Earth's escape speed; it continues to recede from Earth ever more slowly, and it stops when time reaches infinity. This turns out to be the type of

The "critical density," ρ_{crit}, is defined to be that density of matter that would allow the Universe to expand forever, but only just barely. Its formula is

$$\rho_{crit} = 3H_0^2/(8\pi G),$$

where H_0 is Hubble's constant and G is Newton's universal constant of gravitation. If the average matter density of the Universe (ρ_{ave}) is greater than, equal to, or less than ρ_{crit}, then Ω_M (Omega of the matter) is greater than, equal to, or less than 1, respectively, since we have defined $\Omega_M = \rho_{ave}/\rho_{crit}$.

For $H_0 = 71$ km/s/Mpc, we find that $\rho_{crit} = 9.4 \times 10^{-30}$ g/cm³, which is the equivalent of only about 5.6 hydrogen atoms per cubic meter of space! Clearly, our local surroundings are much denser, but this isn't relevant. Only the average density of the *Universe as a whole*

should be compared with ρ_{crit} to determine the value of Ω_M. In fact, we find that Ω_M is almost certainly less than 1. This means that on large scales the Universe is remarkably empty.

Note that, as the Universe ages, the average density of matter in the Universe decreases. However, if the value of Ω_M initially exceeds 1, it does not later drop to 1 or less. This is because the value of H_0 decreases with time, and hence the critical density also decreases. The ratio of the average density to the critical density, Ω_M, remains either above 1 or below 1 (or is exactly equal to 1) forever.

As we will discuss in Section 18.5, we now know that in addition to matter, there is mysterious "dark energy" that contributes to the total density of the Universe.

universe predicted by most "inflation theories" (which we will study in Chapter 19).

If $\Omega_M < 1$ (that is, the average density is below the critical density), galaxies separate more and more slowly with time, but as time approaches infinity, the recession speed (for a given pair of galaxies) approaches a constant, nonzero value. Thus, the Universe will easily expand forever. Once again using our ball analogy, it is like a ball thrown upward with a speed greater than Earth's escape speed; it continues to recede from Earth ever more slowly, but it never stops receding.

These three kinds of universes have different overall geometries. The $\Omega_M = 1$ case is known as a **flat universe** or a **critical universe**. It is described by "Euclidean geometry" – that is, the geometry worked out first by the Greek mathematician Euclid in the third century BCE. In particular, Euclid's "fifth postulate" is satisfied: Given a line and a point not on that line, only one unique parallel line can be drawn through the point (■ Fig. 18–22a). Such a universe is spatially flat, formally infinite in volume (but see the caveat at the end of Section 18.4c), and barely expands forever. Its age is exactly two-thirds of the Hubble time, $(^2/_3)/H_0 = (^2/_3)T_0$.

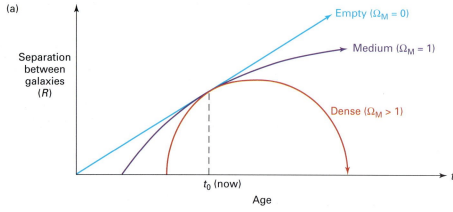

(a)

Separation between galaxies (*R*)

Empty ($\Omega_M = 0$)

Medium ($\Omega_M = 1$)

Dense ($\Omega_M > 1$)

t_0 (now)

t

Age

(b)

Jay M. Pasachoff

■ **FIGURE 18–21** (a) To trace back the growth of our Universe, we would like to know the rate at which its expansion rate is changing with time. Big-bang models of the Universe are shown; the vertical axis represents the separation between any two galaxies (preferably in different superclusters), and the horizontal axis is time. If the Universe were empty, then the rate of expansion would be constant forever, since there would be no galaxies and hence no gravitational forces. In a more realistic case of a non-empty Universe, gravity tends to decelerate the expansion. However, if the ratio of the average density of matter to the critical density is less than 1 (i.e., $\Omega_M < 1$), then the Universe is still able to expand forever. If $\Omega_M > 1$, on the other hand, gravity causes the expansion to eventually halt, and the Universe subsequently collapses. The dividing line, where the ratio of the matter density is equal to the critical density (that is, $\Omega_M = 1$), corresponds to expansion that continues forever, but just barely: Galaxies recede from each other with a speed that approaches zero as time approaches infinity. (b) A two-dimensional version of the big crunch; a reversal of the balloon-surface expansion shown in Figure 18–9.

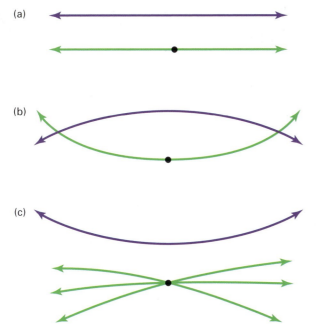

(a)

(b)

(c)

■ **FIGURE 18–22** (a) Euclid's "fifth postulate" for spatially flat geometry states that given a line and a point not on that line, only one unique parallel line can be drawn through the point. (b) In a positively curved geometry, no parallel lines can be drawn through the point. (c) In a negatively curved geometry, infinitely many parallel lines can be drawn through the point. (Of course, we cannot accurately draw curved geometries on flat sheets of paper, so these diagrams are only meant to convey the general idea.)

In the $\Omega_M > 1$ universe, Euclid's fifth postulate fails in the following way: Given a line and a point not on that line, *no* parallel lines can be drawn through the point (Fig. 18–22b). Such a universe has positive spatial curvature, is finite ("closed") in volume, but has no boundaries (edges) like those of a box. Its fate is a hot "big crunch." Generally known as a closed universe, it is also sometimes called a "spherical" ("hyperspherical") or "positively curved" universe. Its age is less than two-thirds of the Hubble time.

Finally, in the $\Omega_M < 1$ universe, Euclid's fifth postulate fails in the following way: Given a line and a point not on that line, many (indeed, infinitely many) parallel lines can be drawn through the point (Fig. 18–22c). Such a universe has negative spatial curvature, is formally infinite ("open") in volume (but see the caveat at the end of Section 18.4c), and easily expands forever. Generally known as an **open universe**, it is also sometimes called a "hyperbolic" or "negatively curved" universe. Its age is between $(2/3)T_0$ and T_0 (the latter extreme only if $\Omega_M = 0$).

Note that, in some texts and magazine articles, the $\Omega_M = 1$ universe is called "closed," but only because it is *almost* closed. It actually represents the *dividing line* between "open" and "closed."

Under certain conditions, flat or negatively curved universes might have exotic shapes with finite volume (see *A Closer Look 18.1: Finite Flat and Hyperbolic Universes*). Even positively curved universes might not be simple hyperspheres. It is difficult, but not impossible, to distinguish such universes from the "standard" ones discussed above, and so far no clear observational evidence for them has been found.

Though quite intriguing, in this book we will not further consider this possibility. Keep in mind, though, that convincing support for a finite, strangely shaped universe might be found in the future; we should always be open to potential surprises.

18.4d Two-Dimensional Analogues

It is useful to consider analogues to the above universes, but with only two spatial dimensions (■ Fig. 18–23). The flat universe is like an infinite sheet of paper. One property is that the sum of the interior angles of a triangle is always 180°, regardless of the shape and size of the triangle. Moreover, the area A of a circle of radius R is proportional to R^2 (that is, $A = \pi R^2$). This relation can be measured by scattering dots uniformly (homogeneously) across a sheet of paper, and seeing that the number of dots enclosed by a circle grows in proportion to R^2.

The positively curved universe is like the surface of a sphere: The sum of the interior angles of a triangle is always greater than 180°. For example, a triangle consisting of a segment along the equator of the Earth, and two segments going up to the north pole at right angles from the ends of the equatorial segment, clearly has a sum greater than 180°. Moreover, the area of a circle of radius R falls short of being proportional to R^2. If the sphere is uniformly covered with dots, the number of dots enclosed by a circle grows more slowly with R than in flat space because a flattened version of the sphere contains missing slices.

The negatively curved universe is somewhat like the surface of an infinite horse's saddle or potato chip. These analogies are not perfect because a horse's saddle (or potato chip) embedded in a universe with three spatial dimensions is not isotropic; the saddle point, for example, can be distinguished from other points. The sum of the interior angles of a triangle is always less than 180°. The area of a circle of radius R is more than proportional to R^2. If the saddle is homogeneously covered with dots, the number of dots enclosed by a circle grows more quickly with R than in flat space because a flattened version of the saddle contains extra wrinkles.

With three spatial dimensions, we can generalize to the growth of volumes V with radius R. In a flat universe, the volume of a sphere is proportional to R^3 (that is, $V = {}^4/_3 \pi R^3$). In a positively curved universe, the volume of a sphere is not quite proportional to R^3. In a negatively curved universe, the volume of a sphere is more than proportional to R^3.

18.4e What Kind of Universe Do We Live In?

How do we go about determining to which of the above possibilities our Universe corresponds? There are a number of different methods. Perhaps most obvious, we can measure the average density of matter, and compare it with the critical density. The value of Ω_M (again, the ratio of the average matter density to the critical density) is greater than 1 if the Universe is closed, equal to 1 if the Universe is flat (critical), and less than 1 if the Universe is open.

Or, we can measure the expansion rate in the distant past (preferably at several different epochs), compare it with the current expansion rate, and calculate how fast the Universe is decelerating. This can be done by looking at very distant galaxies, which are seen as they were long ago, when the Universe was younger.

We can also examine geometrical properties of the Universe to determine its overall curvature. For example, we can see whether the sum of the interior angles of an enormous triangle is greater than, equal to, or less than 180°. (One application of this method has been successfully used with WMAP data; as we discuss below, the Universe is found to be essentially flat.) Or, we can see whether "parallel lines" ever meet – but this is not practical, since we cannot reach sufficiently large distances.

Another geometrical method is to measure the angular sizes of galaxies as a function of distance. High-redshift galaxies of fixed physical size will appear larger in angular size if space has positive curvature than if it has zero or negative curvature, because light rays diverge

more slowly in a closed universe than in a flat universe or in an open universe (■ Fig. 18–24a). Or, we could instead look at the apparent brightness of objects as a function of distance. High-redshift objects of fixed luminosity (intrinsic brightness) will appear brighter if space has positive curvature than if it has zero or negative curvature; again, light rays diverge more slowly in a closed universe (Fig. 18–24b).

We might also count the number of galaxies as a function of distance to see how volume grows with radius (if galaxies don't evolve with time, something known to be untrue). This is analogous to the measurement of area in two-dimensional universes, as explained in Section 18.4d. If space is flat, volume is exactly proportional to R^3; thus, doubling the surveyed distance (R) should increase the number of galaxies by a factor of 8. On the other hand, if space has positive curvature, the factor will be smaller than 8, while if space has negative curvature, the factor will be larger than 8.

A completely different technique is to measure the relative abundances (proportions) of the lightest elements and their isotopes, which were produced shortly after the big bang. As we will discuss in Chapter 19, these depend on the value of Ω_M.

■ **FIGURE 18–23** (*Top row*) Two-dimensional analogues to three-dimensional space. (a) A flat universe is like an infinite sheet of paper; the laws of Euclidean geometry are obeyed. For example, the sum of the interior angles of a triangle is equal to 180°. (b) A positively curved universe is like the surface of a sphere; the sum of the interior angles of a triangle is larger than 180°. (c) A negatively curved universe is like the surface of an infinitely large saddle or potato chip in some respects; the sum of the interior angles of a triangle is smaller than 180°. (*Second row*) Consider dots uniformly distributed on each two-dimensional surface. If we count the total number of dots within circles of larger and larger radius R in flat space, we find that the total number of dots grows exactly in proportion to the area, $A = \pi R^2$. But in positively curved space there are fewer dots within a given radius, and in a negatively curved space there are more dots within a given radius. (*Third row*) It is easiest to visualize these relationships by trying to flatten out the curvature in each case. The positively curved surface, when flattened, gives rise to empty sections, while the flattened negatively curved surface has folds with extra dots. (*Bottom row*) The resulting distribution of dots in the flattened versions shows a relative deficit of dots at large radii in the positively curved surface and a relative excess of dots in the negatively curved surface.

Mathematicians have pointed out that the local isotropy and homogeneity of a flat or negatively curved (hyperbolic) space might not extend globally, to every part of the universe. (After all, we observe our Universe to be isotropic only from our own vantage point; we don't really know that it must be isotropic everywhere.) If so, one can avoid the conclusion that universes with these geometries are infinite in size. Indeed, there is an infinite number of examples of finite flat and hyperbolic spaces! The shape might be like that of a torus (doughnut) or a horn, for instance. One could even have unusual positively curved (spherical) spaces that are dodecahedral, or some other exotic shape.

These finite universes evolve in exactly the same way as their infinite counterparts, making them difficult to tell apart observationally. One distinguishing feature is that in a finite universe, light could in principle circumnavigate space one or more times. This possibility was first suggested in 1900 by the German physicist Karl Schwarzschild (after whom the radius of the event horizon of a nonrotating black hole is named; see our discussion in Chapter 14).

Schwarzschild reasoned that if space were small enough, there would be a "hall-of-mirrors" effect caused by the light wrapping around the universe, and we would see multiple images of a single astronomical object. Conversely, the absence of multiple images could be used to determine the minimum size of the universe. Based on observations available at the time, Schwarzschild concluded that the volume of the universe must be much greater than the volume ascribed to the Milky Way.

Over the years, people have used ever-improving observations to extend this minimum size, and most cosmologists now think that the scale over which a spatially flat universe wraps around is probably much larger than the distance to which we can see.

However, some physicists have recently suggested that the image of the edge of the visible Universe, the "cosmic background radiation" as measured by the Wilkinson Microwave Anisotropy Probe (WMAP; see our extensive discussion in Chapter 19), shows slight deviations from isotropy like those one would expect in a finite universe. Others point to the lack of large-scale features in the WMAP image as another possible sign that the Universe might be finite, in much the same way that water in a small bathtub cannot have waves as large as those found in the open ocean.

A group of scientists applied a new version of Schwarzschild's test that looks for multiple images of the same hot and cold spots in the WMAP data. If the Universe were finite on a scale smaller than the diameter of visible space, then we would see the same portion of the early universe when looking in different directions in the sky. The multiple images would give rise to matching circles of hot and cold spots in the WMAP image. Computers were used to look for such patterns, but none was found. This implies that the Universe has a diameter of at least 78 billion light-years.

When the analysis of data from the European Space Agency's Planck spacecraft is released, all these investigations will be redone at Planck's higher spatial resolution.

Neil Cornish and Jeff Weeks

Repeating images of the Earth give the illusion that space is infinite in this "torus universe." Actually, though, a single Earth is viewed at many times in the past, as light repeatedly circumnavigates (that is, completes a full path around) the finite universe. The illusion is created by gluing together opposite faces of a cubic cell, the edges of which are indicated with wooden struts.

Some astronomers measure the motions of galaxies and clusters of galaxies through the Universe (that is, relative to the smooth "Hubble flow"). These are produced by the gravitational tug of other clusters, and hence provide a measure of the mass and distribution of clumped matter (both visible and dark).

There are many other, related techniques. For a number of reasons, almost all of them (including most of those listed here) are difficult and uncertain.

18.4f Obstacles Along the Way

One problem is that local, dense objects (planets, stars, galaxies, etc.) produce spatial curvature larger than the gradual, global effect that we seek. Moreover, we know that the Universe is nearly flat, so to detect any slight overall curvature one needs to look very far, and this is difficult. Another problem is that when counting galaxies or determining the average density, how does one know that a representative

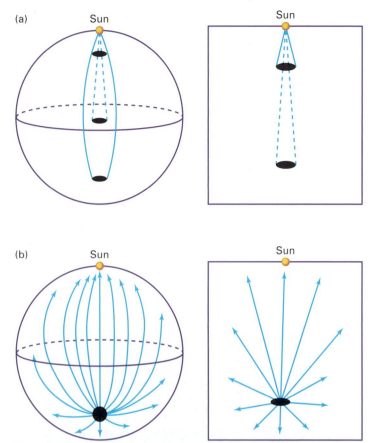

■ **FIGURE 18–24** (a) In flat space (*right*), distant galaxies look much smaller than a nearby galaxy; in fact, the angular size is inversely proportional to distance. In positively curved space (*left*), on the other hand, the angular size is somewhat larger than expected when compared with flat space. Extremely distant galaxies can even start looking larger again, because of the curved trajectory of light rays. (b) In flat space (*right*), distant galaxies look very faint because of the inverse-square law of light. But in positively curved space (*left*), the light rays don't diverge as fast, and even begin to converge again at sufficiently large distances; thus, distant objects look brighter than expected.

■ **FIGURE 18–25** Typical galaxies photographed by the Hubble Space Telescope, at redshifts between 0.3 and 0.7 (that is, roughly 3.4 to 6.3 billion light-years away). They look peculiar relative to nearby galaxies, suggesting that galaxies evolve over time in shape, size, and luminosity.

volume was chosen? After all, there are deviations from uniformity (inhomogeneities) on large scales, such as superclusters of galaxies.

A major difficulty is that galaxies evolve with time. So, we cannot assume that a known type of galaxy has a constant physical size (when measuring angular sizes as a function of distance), and the quantitative evolution of physical size is difficult to predict. We also do not know how the luminosity of a typical galaxy evolves with time, yet galaxies certainly do evolve (■ Fig. 18–25). Counts of galaxies to various distances might be dominated by intrinsic luminosity differences, rather than by the volume of space surveyed.

Clusters of galaxies also evolve with time, and are therefore subject to similar uncertainties. Determining the total amount of matter in a cluster is difficult; much of the matter is dark, and can be detected only through its gravitational effects. In addition, we do not know how much dark matter is spread somewhat uniformly, rather than clumped in clusters and superclusters, and this affects the calculated average density. Of course, an observational problem is that distant objects appear small and faint, and are therefore subject to considerable measurement uncertainties.

At the time of writing, in 2013, there is consensus that $\Omega_M \approx 0.3$. If true, the Universe is spatially infinite (but see the caveat at the end of Section 18.4c) and will expand forever.

Yet there are theoretical reasons (described in Chapter 19) for believing that Ω_M (or Ω_{total}, which might include some new kind of energy) is exactly 1.0, if it is known to be at least relatively close to 1.0, such as the value of about 0.3 currently favored. If it were not *exactly* 1.0 initially, it should have deviated very far from 1.0 (for example, 10^{-7} or 10^{15}) by the present time. This behavior is like nudging a pencil balanced on its tip: It quickly falls to the surface.

18.5 MEASURING THE EXPECTED DECELERATION

Perhaps the most direct way of determining the expected deceleration of the Universe is to measure the expansion rate as a function of time by looking at very distant objects. As discussed earlier, the separation between two clusters of galaxies varies with time in different ways, depending on the deceleration rate of the Universe (recall Figure 18–21). For any of the curves, the slope at a given time is the expansion rate at that time. (Hubble's constant itself is the slope divided by the separation between galaxies at that time.)

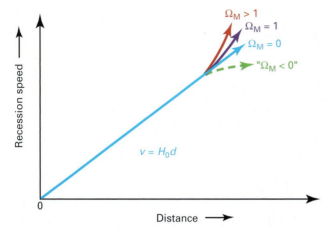

FIGURE 18–26 A plot of the measured recession speeds of objects (as given by their redshifts) vs. their distances. Relatively nearby galaxies all satisfy the same relationship, Hubble's law. But at large recession speeds there are differences that depend on how rapidly the Universe's expansion rate is decelerating, and hence on Ω_M, the ratio of the average density of matter to the critical density. For a given redshift, a galaxy is closer if the Universe is denser than if it is less dense. If the density of matter could be negative (that is, "$\Omega_M < 0$"), distances larger than those in an empty (non-decelerating) universe would be measured.

18.5a The High-Redshift Hubble Diagram

Observationally, we construct the Hubble diagram by plotting the measured recession speed (from the redshift) vs. the distance (from the inverse-square law) for a set of objects (■ Fig. 18–26). At small distances, Hubble's law holds: Speed is directly proportional to distance ($v = H_0 d$). At very large distances, however, there should be deviations from this. For a given distance, the speed should be higher if Ω_M is greater than 1 than if Ω_M is less than 1, because the Universe used to be expanding much more quickly if Ω_M is greater than 1.

Or, for a given speed, the distance should be larger if Ω_M is less than 1 than if Ω_M is greater than 1, because the expansion didn't slow down as much if Ω_M is less than 1. The distance should be even larger if Ω_M is less than 0 (which is physically impossible, since matter is gravitationally attractive), because the expansion of the Universe will have accelerated, pushing objects even farther away.

What kinds of objects can be seen at sufficiently large distances to accomplish this task? We could try galaxies – but they evolve with time because of mergers, bursts of star formation, and other processes that vary from one galaxy to another and are not well understood. Clusters of galaxies also evolve with time in ways that are difficult to predict accurately, although recent progress in the use of clusters looks promising (see below).

Some astronomers had hoped that quasars would serve well – but they exhibit a tremendous range in luminosity (intrinsic brightness), and they evolve quickly with time. Gamma-ray bursts, being so luminous, are obvious candidates as well, but they too span a wide range in luminosity. On the other hand, very recent attempts to calibrate gamma-ray bursts have shown signs of success; they may one day

provide important complementary information on the history of the expansion rate.

18.5b Type Ia (White-Dwarf) Supernovae

Starting in the 1990s, the most progress on this front was made with Type Ia supernovae, whose utility for finding the Hubble constant has already been discussed (Section 18.3e). Type Ia supernovae are nearly ideal objects for such studies. They are very luminous, and although not exactly standard candles, accurate corrections can be made for the differences in luminosity by measuring their light curves. (The width of the peak of brightness is correlated with the peak brightness, allowing scientists to specify their peak brightness more accurately than had previously been possible; Type Ia supernovae are therefore standardizable candles.) Their intrinsic properties should not depend very much on redshift: Long ago (that is, at high redshift), white dwarfs should have exploded in essentially the way they do now.

Starting in the 1990s, two teams found and measured high-redshift ($z = 0.3$ to $z = 0.7$) Type Ia supernovae. The first was led by Saul Perlmutter (Lawrence Berkeley National Laboratory, and the University of California, Berkeley), and the second by Brian Schmidt (Australian National University) and Adam Riess (then at UC Berkeley, now at the Space Telescope Science Institute and the Johns Hopkins University). These three scientists were the recipients of the 2011 Nobel Prize in Physics for the discoveries they made as a result of these supernova studies. One of us (A.F.) was the only scientist to have worked with both groups, although his primary association since 1996 was with the Schmidt/Riess team.

A Type Ia supernova occurs somewhere in the observable universe every few tens of seconds. Thus, if very deep photographs are made of thousands of distant galaxies (■ Fig. 18–27a), the chances of catching a supernova are reasonably good. The same regions of the sky are photographed about a month apart with large telescopes. Comparison of the photographs with sophisticated computer software reveals the faint supernova candidates ((■ Fig. 18–27b).

A spectrum of each candidate is obtained, often with the Keck telescopes in Hawaii. This reveals whether the object is a Type Ia supernova, and provides its redshift. Follow-up observations of the Type Ia supernovae with many telescopes, including the Hubble Space Telescope (■ Fig. 18–27c,d), provide their light curves. The peak apparent brightness of each supernova, together with the known (appropriately standardized) luminosity, gives the distance using the inverse-square law.

Incidentally, the light curves of high-redshift supernovae appear broader than those of low-redshift examples; that is, the former take longer to brighten and fade than the latter. This results from the expansion of the Universe: Each successive photon has farther to travel than the previous one. Indeed, this observed "time dilation" effect (Figure 18–27c,d), and especially its spectroscopic counterpart (the spectrum of a supernova evolves in a manner consistent with time dilation), currently provides the best evidence that redshifts really are produced by the expansion of the Universe, rather than by

(a)

(b)

(c) (d)

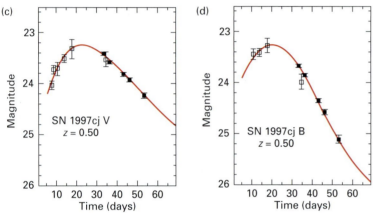

■ **FIGURE 18–27** (a) A deep, wide-angle CCD image of the sky showing thousands of faint, distant galaxies. Almost every visible object is another galaxy, not a star in the Milky Way Galaxy. (b) Discovery of SN 1997cj (redshift $z = 0.5$). The top two images are small subsets of much larger CCD frames obtained with a 3-week gap, with a 4-m telescope in Chile. The difference between the two images, at bottom, reveals a supernova candidate that was subsequently confirmed to be a Type Ia supernova with a spectrum obtained by one of the Keck 10-m telescopes. At left is an image obtained with the Hubble Space Telescope, showing the supernova well separated from the center of the host galaxy. (c and d) Light curves of SN 1997cj (redshift $z = 0.5$), obtained with ground-based telescopes and the Hubble Space Telescope. The data were obtained through filters that pass light emitted by the supernova at (c) "visual" (roughly yellow/green; labeled "V") and (d) blue (labeled "B") wavelengths. These light curves are broader than those of low-redshift supernovae because the Universe is expanding, producing a kind of "time dilation."

some other mechanism (such as light becoming "tired," losing energy during its long journey). The time dilation factor by which the light curves are broader is $1 + z$.

18.5c An Accelerating Universe!

Several dozen supernovae at typical lookback times of 4 to 5 billion years were measured in this manner by early 1998, and many additional ones have been observed since then (■ Fig. 18–28). The results are astonishing: Both teams found that the high-redshift supernovae are fainter, and hence more distant, than expected in a standard universe with attractive gravity.

A simple analogy illustrates the essence of what the data imply. Suppose you toss an apple up from your hand and measure its distance after 1 second. If the Earth's mass were less, the gravitational force on the apple would be less, producing less deceleration, so the traversed distance after 1 second would be greater. If Earth's mass were zero, the apple wouldn't slow down at all, and the traversed distance after 1 second would be still greater. But there is no way to get an even larger distance unless the apple *accelerates*, as it would

if it had a rocket attached to it. Applied to supernovae at different redshifts, the argument turns out to be independent of the assumed age of the Universe: in all cases, the supernovae appear too dim, and hence are farther away than is possible unless the expansion of space accelerates with time.

The data agree best with the "$\Omega_M < 0$" curve in Figure 18–26, yet we know that the matter density is greater than 0 since we (and planets, stars, and galaxies) exist and there are no known forms of negative matter! Note that the possible presence of antimatter would not resolve this problem: Both matter and antimatter have positive energy, and exert an *attractive* gravitational force. (As we briefly mentioned in Chapters 12 and 14, for every subatomic particle, there is a corresponding antiparticle. This antiparticle has the same mass as the particle, but is opposite in all other properties, such as charge and the direction it is spinning. If a particle and its antiparticle meet, they annihilate each other; their total mass is 100 per cent transformed into the energy of electromagnetic radiation in the amount $E = mc^2$. Antiparticles together make up antimatter.)

We conclude that there is a previously unknown component of the Universe with a long-range *repulsive* effect – essentially a cosmic "antigravity." Its presence is felt only over exceedingly large

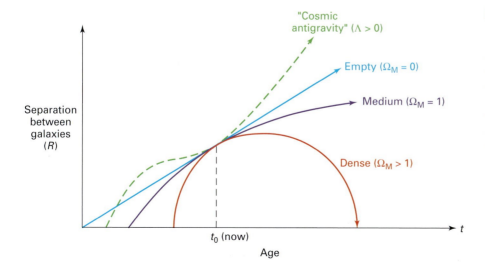

NASA, ESA, and A. Riess (STScI)

■ **FIGURE 18–28** Five supernovae (*top row*), together with images of the same galaxies before or long after the supernova explosion (*bottom row*). These supernovae are between 3.5 and 10 billion light-years away; they were imaged by Adam Riess and his collaborators in 2006 with the Advanced Camera for Surveys on the Hubble Space Telescope.

distances. One cannot use this substance to make "antigravity boots," for example.

The extra stretching of space makes the supernovae more distant than they would have been had the Universe's expansion decelerated throughout its history. Thus, we now live in an **accelerating universe**, one whose expansion rate is *increasing* rather than decreasing with time (■ Fig. 18–29)! Unless the repulsive effect changes sign and becomes attractive in the future (which it might, since we do not know what causes the effect), the Universe will expand forever, even if it is closed (finite in volume). Distant galaxies are receding from each other faster and faster (a "runaway Universe"), so eventually they will fade away, as their light becomes redshifted to essentially zero energy, and as time dilation causes photons to arrive extremely rarely.

The two teams calculate the age of the Universe to be about 13.7 billion years, if $H_0 = 71$ km/s/Mpc. This is the same as the value of 13.7 billion years expected if the Universe were empty ($\Omega_M = 0$), but larger than the value of 9.2 billion years expected with $\Omega_M = 1$ (Figure 18–29). (Coincidentally, the recent acceleration of the Universe is almost exactly balanced by an early era of deceleration, to be discussed below, yielding the same age as would have

been obtained with constant-speed expansion.) An expansion age of 13.7 billion years is quite consistent with the recently revised ages of globular clusters: 12–13 billion years. The pesky age crisis seems finally to be over!

Although astronomers and physicists were at first skeptical of the initial results announced by the two teams in February 1998, nobody has since found any obvious flaws in their data, analysis methods, or conclusions. The discovery soon gained international prominence (■ Fig. 18–30). The quest for the cause of the acceleration (and the physical nature of the responsible agent) became one of the hottest topics in all of physics; indeed, many now consider it to be the top observationally motivated, unsolved problem in all of physics.

18.5d Einstein's Biggest Blunder?

Perhaps the simplest explanation for the observed acceleration is that Einstein's cosmological constant "Lambda" (Λ), the "fudge factor" he introduced to make a static universe, is nonzero. In essence, space appears to have a repulsive aspect to it, a cosmic "antigravity." But instead of exactly negating the attractive force of gravity, the

■ **FIGURE 18–29** Similar to Figure 18–21, but with an additional (*dashed*) curve corresponding to what our Universe appears to be doing. The ratio of the average density of matter to the critical density, Ω_M, is about 0.3 – but recent data seem to show that the cosmological constant (Λ, Lambda) is positive. The expansion of the Universe is now accelerating, and will continue to do so forever (unless the "dark energy" driving it someday becomes gravitationally attractive). The derived current age of the Universe turns out to be about 14 billion years.

■ FIGURE 18–30 The cover of *Science* magazine from December 18, 1998; the discovery of the accelerating expansion of the Universe, made by two teams both of which included one of the authors (A.F.), was named the most important scientific breakthrough of 1998. The caricature of Einstein is surprised because the "bubble universe" that he blew out of his pipe appears to be accelerating in its expansion, rather than decelerating as expected. He is doubly surprised because he predicted (but for the wrong reason) the possibility of large-scale cosmic repulsion in 1917, and later renounced the idea. Thirteen years later, the scientists who headed the discovery teams received the 2011 Nobel Prize in Physics. The fact that the same astonishing result has been verified by other astronomers using several different methods no doubt helped persuade the Nobel Committee that the discovery was solid enough to award the Prize.

cosmological constant slightly dominates gravity over very large distances, producing a net acceleration.

It is ironic that Einstein introduced the cosmological constant and later completely rejected it. The idea itself may not have been wrong; rather, only the exact value that Einstein gave to the cosmological constant was slightly erroneous. Even if the acceleration turns out to be caused by something other than Λ (see below), one could say that Einstein was right after all, because there is indeed a new, previously unanticipated, repulsive effect present in the Universe. The idea may have been his greatest triumph rather than his biggest blunder!

Many theorists find a positive cosmological constant to be very disconcerting. In the context of quantum theories, it suggests that the vacuum has a nonzero, positive energy density (a "vacuum energy") due to **quantum fluctuations** – the spontaneous formation (and then rapid destruction) of virtual pairs of particles and antiparticles.

(They are called "virtual" because they form out of nothing and last for only a very short time, unlike "real matter.") For subtle reasons in general relativity having to do with its "negative pressure," a positive energy density of this type causes space to expand faster and faster with time.

Expressed in the same units as Ω_M, essentially as a ratio of densities, the current value of Λ that the two teams measure is $\Omega_\Lambda \approx 0.69$ (making use of the cosmic background radiation as well – see Chapter 19). But theorists generally expected that $\Omega_\Lambda \approx 0$ because of an exact cancellation of all the quantum fluctuations; otherwise, they predicted the exceedingly large value $\Omega_\Lambda \approx 10^{120}$, or possibly down to "only" about 10^{50}, clearly neither of which is actually observed. (We would definitely not exist if the cosmological constant were so large; the Universe would have expanded much too quickly for galaxies and stars to form.) The discrepancy between the observed and expected values of Ω_Λ has been named the greatest error (or embarrassment) ever in theoretical physics!

Although Λ itself is constant with time, the density ratio Ω_M decreases as the Universe ages, while Ω_Λ increases. This, however, leads to another problem: Why should these two density ratios be roughly equal right now? They are measured to be $\Omega_\Lambda \approx 0.7$ and $\Omega_M \approx 0.3$, but they could have been 0.00001 and 0.99999, for example, or any other two numbers (between 0 and 1) whose sum is 1.0000. Do we live in a cosmically "special" time?

18.5e Dark Energy

Because of these and other problems associated with the cosmological constant, physicists have eagerly sought alternative explanations for the observed acceleration of the Universe. Some of the hypotheses seem rather wild, to say the least. For example, gravity might be "leaking out" of our Universe and into extra dimensions, or perhaps "other universes" in some larger "hyperspace" (see Chapter 19 for more details) are "pulling out" on our Universe! Perhaps more realistically, Einstein's general theory of relativity might be wrong on large scales, thereby potentially explaining the supernova data without invoking the cosmological constant or anything similar to it.

Most of the alternatives, however, invoke a new kind of particle or field within our Universe, similar to the cosmological constant but having different specific properties. The most general term for the responsible substance (including also the cosmological constant) is **dark energy** – in some ways an unfortunate choice of words because of the possible confusion with "dark matter." Despite Einstein's famous equation, $E = mc^2$, it is important to remember that although "dark energy" has a mass equivalent, it is *not* the same thing as "dark matter." Dark energy causes space (over the largest distances) to expand more and more quickly, whereas dark matter is gravitationally attractive and produces deceleration.

Dark energy is thought to be present throughout space, even here on Earth. However, only on the largest scales (exceeding roughly 100 million light-years) does its density exceed that of luminous and dark matter, making its effects noticeable. Dark energy causes superclusters of galaxies, and especially the space between them, to expand progressively faster. Space does not expand on Earth, in our Solar System, in our Galaxy, or even in our Local Group of galaxies.

One popular set of dark-energy models, having hundreds of possibilities, is called "quintessence" – named after the Aristotelian "fifth essence" that complements Earth, air, fire, and water. In the quintessence models, the value of "Ω_X" (associated with the new energy "X") decreases with time in a manner similar to that of Ω_M, so it is not surprising that the two values are now roughly comparable. But these hypotheses have their own set of problems. Moreover, detailed observations (including those of Type Ia supernovae) are beginning to rule out entire classes of quintessence models.

In one interesting form of quintessence, the amount of dark energy per unit volume actually grows with time. In the future, dark energy could thus become strong enough to rip apart clusters of galaxies, followed successively by the disruption of galaxies, planetary systems, stars, planets, and (regrettably) people. Eventually, even atoms would get ripped apart! Fortunately, this model has substantial theoretical difficulties, and current observations do not appear to strongly support it.

It turns out that, regardless of the nature of the dark energy, its value of Ω plus that of matter add up to 1 (that is, $\Omega_X + \Omega_M = \Omega_{total} = 1$). This total value means that the Universe is *spatially flat* on large scales, or nearly so (given that we don't know the exact value of Ω_{total}). Some theorists predicted that the Universe is flat (see our discussion in Chapter 19), but without dark energy. Perhaps the most natural model is one in which the Universe is formally closed (like the three-dimensional version of a sphere), but so incredibly large that it appears flat; this way, we don't need to deal with a formally infinite universe. On the other hand, theorists have shown that an infinite universe is also a very reasonable possibility, strange as it may sound.

18.5f The Cosmic Jerk

Given how bizarre most of the above dark-energy hypotheses sound, perhaps we should question more carefully the data on which the accelerating-universe conclusion rests. Is it possible that the high-redshift supernovae appear fainter than expected not because of their excessively large distances, but for a different reason? For example, maybe long ago they were intrinsically less luminous than now. Keck-telescope spectra of high-redshift supernovae, however, look very similar to those of nearby supernovae (■ Fig. 18–31); there is no clear observational evidence for an intrinsic difference.

Or, perhaps there is dust between us and the high-redshift supernovae, making them look too faint. If this is normal dust, it would redden the light (that is, preferentially absorb and scatter blue photons) – but our procedure already accounts for this kind of dust by measurements of the reddening. If the dust grains are larger than normal, on the other hand, then the reddening is less (that is, blue light is not as preferentially extinguished), making such dust more difficult to detect. Nevertheless, some observable consequences are predicted, yet none has been seen.

If the dust grains are very large, then there would be essentially no difference between the amounts by which blue and red light are extinguished, making such dust extremely difficult to detect directly. However, there are theoretical reasons against the formation of such large dust grains, and they would have adverse observable effects on other aspects of cosmology.

A convincing test of the accelerating universe hypothesis is provided by Type Ia supernovae at redshifts exceeding 1. If the unanticipated faintness measured for the $z = 0.3$–0.7 supernovae (discussed above) were caused by evolution of the intrinsic luminosity of supernovae or by the presence of dust, then we would expect supernovae at even higher redshifts to experience a still larger effect, and thus to appear even fainter than expected. On the other hand, if the cosmological constant (or some similar effect) were causing the observed faintness of the $z = 0.3$–0.7 supernovae, then we would expect supernovae at higher redshifts to not appear as dim.

The reason for the latter effect is that at redshifts exceeding 1, corresponding to lookback times of at least 8 billion years, the Universe was so young that it should have been decelerating, even though it has been accelerating the past 4 or 5 billion years. Galaxies were closer together back then, so their gravitational attraction for

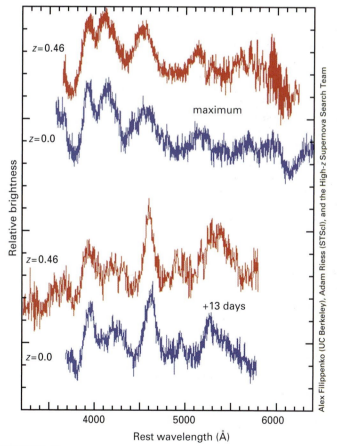

FIGURE 18–31 The spectrum of a Type Ia supernova at redshift $z = 0.46$, compared with that of a nearby (redshift about 0) example, at two different times: (*top*) near peak brightness, and (*bottom*) 13 days after peak brightness. There are no clear differences, suggesting that the two objects are physically very similar.

Alex Filippenko (UC Berkeley), Adam Riess (STScI), and the High-z Supernova Search Team

■ **FIGURE 18–32** Three extremely distant Type Ia supernovae, at redshifts close to or exceeding 1, imaged with the Advanced Camera for Surveys on the Hubble Space Telescope. These stars exploded when the Universe was roughly half its present age. Analysis of the data suggests that the expansion of the Universe was decelerating for its first 9 billion years, but then switched to acceleration about 4 or 5 billion years ago.

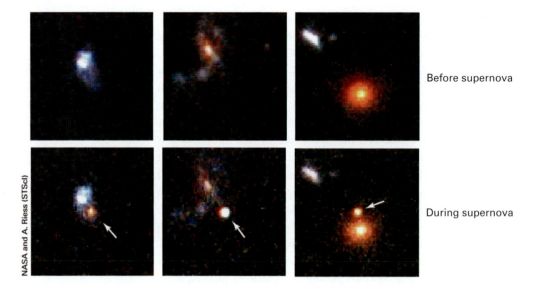

Before supernova

During supernova

NASA and A. Riess (STScI)

each other was stronger. Moreover, if the density of the dark energy does not decrease very much as the Universe ages (for example, the density stays constant if dark energy is a property of space itself), then its cumulative effect was smaller when galaxies were closer together. For both reasons, attractive gravity should have dominated over antigravity, thereby producing deceleration of the expansion rate when the Universe was young.

In 2004, a team led by Adam Riess (Space Telescope Science Institute, and the Johns Hopkins University), of which one of the authors (A.F.) was a member, announced measurements of about 10 supernovae with redshifts close to or exceeding 1 (■ Fig. 18–32). They were brighter than would have been expected if luminosity evolution or dust had affected the results for the previous, redshift 0.3–0.7 supernovae. Instead, the data show that for roughly the first 9 billion years of its existence, the Universe was decelerating. Then, 4 or 5 billion years ago, the expansion rate began to accelerate. This conclusion was supported by additional data that the team published in 2007.

Mathematically, the rate of change of acceleration is called "jerk." (So, to list the relevant terminology, the rate of change of position is called velocity; the rate of change of velocity is called acceleration; and the rate of change of acceleration is called jerk.) A nonzero jerk clearly occurred when the expansion of the Universe went from a state of deceleration to acceleration about 4 to 5 billion years ago. So, it can be said that the Universe experienced a "cosmic jerk" at that time!

No known effect other than distance has been found to account for the measured brightness of supernovae as a function of redshift. Although we should remain vigilant for subtle problems that could invalidate the conclusions based on supernovae, for the time being they seem reliable.

Moreover, completely independent techniques that give the same results have increased our confidence in the conclusion that the expansion rate of the Universe has been accelerating the past 4–5 billion years. The high-redshift Hubble diagrams constructed from observations of clusters of galaxies and gamma-ray bursts, for example, are consistent with those obtained from Type Ia supernovae, though one

should view these new Hubble diagrams with caution because they depend on certain unproven assumptions.

The most convincing new support for the existence of dark energy and the accelerating Universe comes from the cosmic background radiation (discussed in Chapter 19), the "afterglow" of the big bang. We will see that the data imply a flat universe, which means that $\Omega_{total} = 1$. Yet, other compelling observations yield $\Omega_M = 0.27$ (that is, the contribution of luminous and dark matter is about 27 per cent of the total density, averaged over large scales). The difference, 0.73, is consistent with the contribution of dark energy suggested by the measurements of Type Ia supernovae together with the large-scale distribution of matter (galaxies, clusters of galaxies) in the Universe.

Detailed studies of the cosmic background radiation (Chapter 19) even confirm a subtle effect expected from dark energy: superclusters of galaxies, being only loosely bound by gravity, expand more quickly than would have been the case without dark energy. We can tell that the superclusters expand this way because cosmic background photons gain some energy from gravity (that is, they become blueshifted) as they travel toward the center of a supercluster, but they lose less energy from gravity (becoming redshifted, though by a smaller amount) as they travel out from the center of the now more extended cluster. The blueshift slightly exceeds the redshift because of the extra expansion of the supercluster during the time taken for the photons to traverse it. Astronomers recently found a small net blueshift of cosmic background photons in the directions of known superclusters, thus detecting their extra expansion, propelled by dark energy.

18.6 THE FUTURE OF THE UNIVERSE

It seems that no matter what kind of universe we live in, the end is somewhat depressing.

Even for the cases now ruled out, our distant future wouldn't be good: If the Universe is closed and the "big crunch" would occur in the reasonably near future (say, within 100 billion or a trillion years), some stars would be present throughout much of the time. During the final collapse, however, all matter and energy would be squeezed into a tiny volume, at blazingly high temperatures (since compressed gases heat up). If a rebirth occurred (but there is no good reason to suggest this), it would not contain any traces of complexity from the former universe.

If, on the other hand, the Universe is open (or critical) and expands forever, or if dark energy exists (as seems to be the case) and continues to stretch space progressively faster in the future, a number of interesting physical stages will be encountered – though the end result will still be dreary. Freeman Dyson (at the Institute for Advanced Study in Princeton) was among the first to discuss these stages, which we will now consider in turn. The specific timescales mentioned here were calculated in 1998 by Fred Adams (at the University of Michigan) and Greg Laughlin (now at the University of California, Santa Cruz), who assumed a universe with low matter density and zero dark energy (since its existence was still quite controversial at the time). However, the overall picture remains similar even for the Universe that we now know to be accelerating.

We currently live in the **stelliferous era** – there are lots of stars! The Sun will become a white dwarf in about 6 or 7 billion years. Current M-type main-sequence stars will die as white dwarfs by the time the Universe is about 10^{13} years old. Gas supplies will be exhausted by $t \approx 10^{14}$ years, and the last M-type stars will become white dwarfs shortly thereafter (that is, within another 10^{13} years). L-type main-sequence stars won't last too much longer. Certainly by $t \approx 10^{15}$ years, normal stars will be gone.

The Universe will then enter the **degenerate era** filled with cold brown dwarfs, old white dwarfs ("black dwarfs," since they will be so cold and dim), and neutron stars. There will also be black holes and planets. Most stars and planets will be ejected from galaxies by $t \approx 10^{20}$ years. Black holes will swallow a majority of the remaining objects in galaxies by $t \approx 10^{30}$ years. All remaining objects (that is, outside of galaxies) except black holes will disintegrate by $t \approx 10^{38}$ years, due to proton decay. The timescale for this is very uncertain since the lifetime of a proton has not yet been measured.

Next, the Universe will enter the **black-hole era**: The only discrete objects will be black holes. Stellar-mass black holes will evaporate by $t \approx 10^{65}$ years, because of the Hawking quantum process (see Chapter 14). Supermassive black holes having a mass of a million Suns (such as those commonly thought to be in the centers of today's galaxies) will evaporate by $t \approx 10^{83}$ years. The largest galaxy-mass black holes will evaporate by $t \approx 10^{100}$ years.

■ **FIGURE 18–33** Schematic of a "positronium atom": an electron bound to an anti-electron (positron), rather than to a proton as in a normal hydrogen atom. Weakly bound pairs of this type, millions of light-years in size, might form during the "dark era" of the Universe, when space is nearly empty of matter.

Finally, the Universe will enter the **dark era**: There will be only low-energy photons (with a characteristic temperature of nearly absolute zero), neutrinos, and some elementary particles that did not find a partner to annihilate. The electron–positron pairs may form slightly bound "positronium" atoms (■ Fig. 18–33) millions of light-years in size. (They will be disturbed very little, because of the very low space density of objects.) The positronium atoms will decay (combine to form photons) by $t \approx 10^{110}$ years.

If the Universe is open or critical, it seems unlikely that "life" of any sort will be possible beyond $t \approx 10^{14}$ to 10^{15} years, except perhaps in rare cases (such as when brown dwarfs collide and temporarily form a normal star). Certainly beyond $t \approx 10^{20}$ years, the prospects for life appear grim. The Universe will be very cold, and all physical processes (interactions) will be very slow.

On the other hand, Freeman Dyson and others have pointed out that the available timescale for interactions will become very long. Hence, perhaps something akin to life will be possible. This is the "Copernican time principle" – we do not live at a special time, the only one that permits life. Instead, life of unknown forms might be possible far into the future.

However, the situation is now thought to be much bleaker, given that the Universe is accelerating and will continue to do so: It will become exceedingly cold and nearly empty much faster. Within a hundred billion (10^{11}) years, there won't even be any galaxies visible in the sky, because they will have accelerated away to exceedingly large distances, making them too faint even for the biggest conceivable telescopes.

In Chapter 19 we will see that other universes might be born from within our Universe, or in other regions of a larger "hyperspace." Thus, in some form the "universe" (more correctly, the "multiverse") might continue to exist, and to support life, essentially forever. Such ideas are intriguing and fun, but perhaps they do not currently belong to the realm of science, for we don't know any observational or experimental ways to test them. Indeed, they are very speculative, much more so than most of the other ideas described in this book.

CONCEPT REVIEW

» **Cosmology** is the study of the Universe on its grandest scale – its birth, evolution, and ultimate fate (introductory paragraph). **Olbers's paradox** – the darkness of the night sky – has profound implications, and is resolved primarily by the finite age of the Universe (Section 18.1).

» Historically, the first piece of evidence for this definite beginning, the **big bang**, was Edwin Hubble's discovery of the expansion of the Universe (Section 18.2a): Most galaxies are moving away from us, and the speed of any galaxy, determined from the **redshift** of its spectrum, is proportional to its distance. The plot of redshift (or recession speed) vs. distance is the **Hubble diagram**, the proportionality is known as **Hubble's law**, and the constant of proportionality is called **Hubble's constant**, H_0. The expansion of the Universe has no definite center within any physically accessible dimensions (Section 18.2b). Note that *space itself* expands; galaxies are *not* flying apart from each other through a preexisting space (Section 18.2c).

» There are at least two ways in which to determine the age of the Universe (Section 18.3a). First, the Universe must be at least as old as the oldest objects within it. Second, the age of the Universe is the elapsed time since the big bang, when all the material that formed our Galaxy and any other was essentially in the same place. The **Hubble time** is the expansion age of the Universe under the assumption that gravity has not been slowing down the expansion; its value depends on Hubble's constant. If the true expansion age of the Universe is only two-thirds of the Hubble time, as expected by some theorists, then its value is smaller than the ages of the oldest globular star clusters, leading to an age crisis.

» The quest for an accurate measurement of Hubble's constant has been long and arduous, with many conflicting results (Section 18.3b). Generally, astronomers have used the inverse-square law of light to determine the distances of galaxies from the luminosities and measured brightnesses of certain kinds of stars within them. Cepheid variables have played a very important role in this regard. A major project of the Hubble Space Telescope was to determine the Hubble constant by measuring the distances of galaxies with Cepheid variables, which can be identified by looking at their light curves to see that they (1) have a sharp rise and slower fall, and (2) have periods of days or weeks rather than shorter than a day (Section 18.3c). One difficulty is that the recession speeds of relatively nearby galaxies can deviate from the **Hubble flow**, the assumed smooth expansion of the Universe (Section 18.3d). Such **peculiar motions** are produced by the gravitational tug of the Local Group and other mass concentrations. Nevertheless, several techniques were combined in the late 1990s and in the subsequent decade to achieve a convincing value for Hubble's constant, $H_0 = 71$ km/s/Mpc (Section 18.3e).

» Mathematically, studies of the expansion history and overall geometry of the Universe are conducted with Einstein's general theory of relativity (Section 18.4a). To simplify the equations and achieve reasonable progress, we assume the **cosmological principle**: On the largest size scales, the Universe is **homogeneous** (has the same average density everywhere) and **isotropic** (looks the same in all directions). Observationally, this assumption appears to be valid. In addition, we initially assume that Einstein's **cosmological constant** (Λ, Lambda), or any other source of long-range repulsion, is zero (Section 18.4b). However, as we discuss later in the chapter, this constant (or something that behaves like it) actually appears to have a nonzero value.

» The result (ignoring Λ) is that the ultimate fate of the Universe depends only on the ratio of the average matter density of the Universe to a specific **critical density** (Section 18.4c). If the average density exceeds the critical density, then this ratio, known as Ω_M, would be greater than 1 and gravity will eventually cause the Universe to collapse to a fiery death informally called the "big crunch." On the other hand, if the average density is low, such that Ω_M is less than or equal to 1, the Universe will expand forever, though more and more slowly with time; indeed, in the case that Ω_M is equal to 1, the expansion would halt as time approaches infinity.

» In the absence of Λ, the value of Ω_M also determines the overall geometry of the Universe (Section 18.4c). If the average density equals the critical density, we live in a **flat (critical) universe** whose volume is formally infinite, and whose age is two-thirds ($2/3$) of the Hubble time. A **closed universe** has $\Omega_M > 1$; it has positive curvature and finite volume, resembling a sphere (technically, a "hypersphere"). Its age is less than $2/3$ of the Hubble time. The last alternative, an **open universe**, occurs if $\Omega_M < 1$; it has negative curvature ("hyperbolic"), and its volume is formally infinite. Its age is $2/3$ to 1 times the Hubble time. Two-dimensional analogues for such flat, positively curved, and negatively curved universes are an infinite sheet of paper, a sphere, and an infinite horse's saddle or potato chip (Section 18.4d). We can attempt to predict the ultimate fate of the Universe from a number of methods (Section 18.4e), all of which are difficult and have substantial uncertainties (Section 18.4f).

» One can determine the expansion history of the Universe by measuring the distance and redshift of objects at a very wide range of distances (Section 18.5a). The most successful recent technique for determining accurate distances at high redshifts has been to measure the apparent brightness of Type Ia (incinerating-white-dwarf) supernovae (Section 18.5b). Since the very distant ones are fainter than expected given our past understanding of an expanding Universe, the result is that the Universe appears to be expanding faster now than in the past: We currently live in an **accelerating universe** (Section 18.5c)! This observation resolves the age crisis of the Universe and shows that the Universe will expand forever if the acceleration continues.

» Apparently, a long-range "antigravity" effect exists in the Universe, like the cosmological constant that Einstein had previously postulated with a specific value to obtain a static universe (Section 18.5d). Physically, the cosmological constant may correspond to a vacuum energy consisting of **quantum fluctuations**, the spontaneous formation (followed by rapid destruction) of virtual pairs of particles and antiparticles. However, there are many alternatives to the cosmological constant; we don't know the nature of this **dark energy** that fills the Universe (Section 18.5e). Dark energy is ubiquitous, and it constitutes about 73 per cent of the Universe. All together, there is enough luminous matter, dark matter, and dark energy to make the Universe spatially flat (Euclidean) over the largest scales.

» The above conclusions are so astonishing that alternative explanations have been sought, but they have not been compelling (Section 18.5f). One recent triumph is that measurements of supernovae from a time when the Universe was half its present age show that, as expected, the expansion rate was initially decelerating. The change from deceleration to acceleration, a "cosmic jerk," occurred about 4 to 5 billion years ago. Moreover, there are now additional observations, independent of supernovae, which imply the presence of dark energy and cosmic acceleration (or, alternatively, to a failure of the general theory of relativity over very large distances).

» If the Universe lasts forever, many interesting physical processes will occur (Section 18.6). We now live in the **stelliferous era**, filled with stars, but eventually all normal stars will burn out, and the Universe will enter the **degenerate era**, dominated by objects like white dwarfs and neutron stars. After a very long time, these degenerate objects will disintegrate, leaving the Universe in the **black-hole era**. But even black holes decay due to the Hawking evaporation process after exceedingly long times; the Universe will enter its last stage, the **dark era**, as a cold, nearly empty space consisting of extremely low-energy photons and a few elementary particles.

QUESTIONS

1. Describe Olbers's paradox – the darkness of the night sky – and several possible resolutions.

2. Summarize the first observational evidence found for the expansion of the Universe and a possible beginning of time.

3. Why does the recession of galaxies not necessarily imply that the Milky Way Galaxy is at the center of the Universe?

4. Explain how the effective center of expansion can be in an unobservable spatial dimension.

5. (a) What do we mean when we say that the Universe is expanding? (b) Are galaxies moving through some preexisting space? (c) Do humans, planets, stars, and galaxies themselves expand?

6. What are two ways of estimating the age of the Universe?

7. Describe how the current value of Hubble's constant is measured.

8. Why can relatively nearby galaxies give erroneous values for Hubble's constant?

†9. Calculate the Hubble time (that is, the expansion age of the Universe assuming no deceleration) if Hubble's constant is 71 km/s/Mpc.

10. Suppose we assume that the recession speed of a given galaxy never changes with time. Why is the derived age of the Universe likely to be an overestimate of its true age?

11. If the expansion rate of the Universe were increasing (rather than decreasing) with time, would the true expansion age be less than or greater than that derived by assuming a constant expansion rate?

12. What is the "age crisis" that rocked cosmology until the late 1990s? Discuss two developments that helped alleviate this crisis.

13. Why are exploding white dwarf stars so useful for measuring the distances of galaxies?

14. State the cosmological principle. Why is it a reasonable assumption?

15. Define what we mean by the cosmological constant, Λ.

16. Summarize the possible types of overall geometry for a homogeneous, isotropic universe with a cosmological constant (or dark energy) equal to zero.

17. Describe how the geometry of the Universe is intimately connected to the ultimate fate of the Universe, assuming the cosmological constant (or dark energy) is zero.

18. Explain what we mean by the critical density of the Universe.

19. Discuss why Ω_M, the ratio of the average matter density to the critical density of the Universe, is such an important parameter.

20. (a) Summarize different methods for determining the value of Ω_M. (b) What are the difficulties in actually implementing these methods?

21. Does the average matter density appear to be large enough to close the Universe?

22. Discuss the main conclusion of the distant supernova studies, and its implications.

23. Explain how Einstein's cosmological constant is relevant to the conclusion in question 22.

24. What do we mean by "dark energy"? How might it affect the expansion of the Universe?

25. If the expansion of the Universe is currently accelerating, can we conclude that the Universe will necessarily expand forever? Why?

26. What is the overall shape of the Universe, as implied by the total amount of luminous matter, dark matter, and dark energy?

27. Why might high-redshift supernovae appear fainter than expected, other than an acceleration of the expansion rate of the Universe?

28. (a) Has the expansion rate of the Universe always been accelerating? (b) What could cause a change from deceleration to acceleration?

29. What are some of the processes that will eventually occur if the Universe expands forever?

30. **True or false?** The curvature of three-dimensional space cannot be detected because we live within the three spatial dimensions.

31. **True or false?** Current observational evidence suggests that the Universe will eventually collapse; the expansion rate is decelerating rapidly.

32. **True or false?** Most astronomers believe that "Olbers's paradox," the darkness of the night sky, is resolved by the fact that the Universe has a finite age.

33. **True or false?** According to the "cosmological principle," the Universe is homogeneous and isotropic on the largest size scales, but its properties can change with time.

34. **True or false?** The cosmological constant was a concept introduced by Albert Einstein to support a static universe, one that is neither expanding nor contracting.

35. **True or false?** A value of Ω_{total} that is greater than 1 indicates that the Universe has positive curvature, while a value less than 1 indicates that the Universe has negative curvature.

36. **True or false?** According to Einstein's equation, $E = mc^2$, dark energy and dark matter are essentially the same thing.

37. **Multiple choice:** Which one of the following is an assumption made as part of the "cosmological principle"? (a) The Universe is homogeneous on scales smaller than galaxies. (b) The Universe has had a constant average density over time. (c) The Universe is isotropic on scales of billions of light-years. (d) The age of the Universe is finite. (e) The Universe is expanding at an accelerating rate.

†38. **Multiple choice:** Using incomplete data, suppose you find that the value of Hubble's constant in our Universe is about 700 km/s/Mpc instead of about 70 km/s/Mpc. How old would you calculate the Universe to be? (a) 137 billion years. (b) $^1/_{1700}$ billion years. (c) 1.37 billion years. (d) 700 billion years. (e) 13.7 billion years.

†39. **Multiple choice:** Suppose from the top of your palace you notice a circular patch of land of radius 10 meters that needs landscaping. You hire Edwin, who charges $5 per square meter. Edwin completes the job, and comes back with a bill of $3,000. What type of homogeneous, isotropic space can you deduce that you live on? (a) Spherical. (b) Hyperbolic. (c) Euclidean. (d) Shaped like a donut. (e) A space devoid of logic.

40. **Multiple choice:** Which one of the following statements about Hubble's constant, H_0, is *false*? (a) Hubble's constant changes with time. (b) The value of Hubble's constant is best determined with the very nearest galaxies, whose distances can be measured accurately. (c) Hubble's constant is believed to be constant throughout the Universe at a given time. (d) If the expansion of the Universe has always been constant (neither speeding up nor slowing down), then the age of the Universe is $1/H_0$. (e) H_0 is the constant of proportionality between the observed recession speeds (v) and distances (d) of galaxies.

41. **Multiple choice:** Which one of the following statements about the big bang theory is *false*? (a) The redshifts of distant galaxies are a consequence of the wavelength of light stretching during its journey to us. (b) All, or almost all, clusters of galaxies move away from each other. (c) Everything used to be much closer together

and hotter than at the present time. **(d)** The matter density of the Universe is decreasing with time. **(e)** There is a unique center within the Universe that coincides with where the big bang happened.

42. **Multiple choice:** Observations of what kind of objects were the first to strongly imply that the expansion of the Universe is currently accelerating? **(a)** The cosmic background radiation left over from the big bang. **(b)** Supernovae. **(c)** Gamma-ray bursts. **(d)** Clusters of galaxies. **(e)** Gravitational lenses.

43. **Multiple choice:** Which one of the following distributions is isotropic but is *not* homogeneous? **(a)** The distribution of stars in a globular cluster, as measured from its center. **(b)** The distribution of globular clusters in our Galaxy, as measured from Earth. **(c)** The distribution of stars in our Galaxy, as measured from Earth. **(d)** The distribution of galaxy clusters in the Universe averaged over large scales, as measured from Earth. **(e)** The distribution of galaxy clusters in the Universe averaged over large scales, as measured from anywhere.

44. **Multiple choice:** Which one of the following is *not* a possible homogeneous, isotropic, expanding universe according to Einstein's general theory of relativity? (Assume there is no dark energy.) **(a)** A "flat" universe in which the laws of Euclidean geometry are satisfied. **(b)** A "closed" universe in which there is

no such thing as parallel lines. **(c)** A "closed" universe that will expand forever. **(d)** An "open" universe whose volume is infinite, and has been infinite since the beginning of time. **(e)** An "open" universe in which the volume of a large sphere of radius R is greater than $(^4/_3)\pi R^3$.

45. **Fill in the blank:** Two types of stars that have played central roles as cosmological yardsticks are _____ variables and _____ supernovae.

46. **Fill in the blank:** A time far in the future of the Universe, when the density of photons and elementary particles will be very low, is known as the _____.

47. **Fill in the blank:** The expansion of the Universe appears to be accelerating, driven by the repulsive effect of the cosmological constant or, more generally, _____.

48. **Fill in the blank:** A universe with _____ curvature is said to be "hyperbolic," shaped somewhat like a potato chip.

†49. **Fill in the blank:** In a universe where the Hubble constant is 0.05 Gyr^{-1} and in which $\Omega_M = 0$ and there is no dark energy, you can determine that the universe has an age of _____.

†This question requires a numerical solution.

TOPICS FOR DISCUSSION

1. Are you convinced that Olbers's paradox really does present a problem whose main possible solutions are profound? Can you think of any other solutions besides those discussed here? Explain.

2. Can you visualize what our Universe might look like if it's the three-dimensional analogue of the two-dimensional surface of a balloon? Explain.

3. Which prospect do you find more depressing: eternal expansion, or the fiery "big crunch"? Or perhaps you find neither one depressing – all things must die, and the Universe is no exception. Explain.

4. What do you think Einstein's reaction would be to the discovery that the cosmological constant might actually be nonzero, after he had rejected it (anecdotally as his "biggest blunder")? Explain.

5. Why was the 2011 Nobel Prize in Physics awarded to a set of astronomers?

A cluster of galaxies 5 billion light-years away, with faint, blue arcs representing gravitational lensing of a still more distant galaxy, nearly 10 billion light-years away. This image from 2011 was obtained with the Wide Field Camera 3 on the Hubble Space Telescope. Because of the great mass of the foreground cluster and the special alignment, the background galaxy appears about 20 times larger and more than 3 times brighter than typical lensed galaxies.

NASA, ESA, J. Rigby (NASA Goddard Space Flight Center), K. Sharon (Kavli Institute for Cosmological Physics, University of Chicago), and M. Gladders and Wuyts (University of Chicago)

19

In the Beginning

ORIGINS

The birth of the Universe itself is probably the most fundamental of all questions about origins. The behavior of the Universe at very early times greatly affected its contents, producing the matter of which we are made and the relative proportions of the light elements that dominate all others. It is possible that many universes exist, yet only a small fraction might have conditions suitable for life as we know it.

Distant galaxies (see the Chapter Opener) are moving away from us; the Universe is expanding. But, by itself, the expansion of the Universe does not prove that there was a big bang; indeed, one could postulate that the Universe had no beginning in time and will have no end. The fatal blow to this "steady-state theory" was the discovery of a faint radio glow that pervades all of space and was produced when the Universe was very young. The existence of a nearly uniform amount of helium and deuterium (heavy hydrogen) throughout the Universe provides additional evidence for a very hot, dense phase in its early history.

But how did the Universe achieve a uniform temperature, and why is space so nearly flat? These troubling issues led to a magnificent but still unproven hypothesis: The early Universe apparently went through a stage of phenomenally rapid expansion, doubling its size many times in just a tiny fraction of a second. Such a process may have created the Universe from essentially nothing! And if it happened once, it might well happen again; perhaps there exist multiple "universes" physically disconnected from ours.

We emphasize from the start that these last few ideas are quite speculative, much more so than the other material presented in this book. Nevertheless, based on what we now know, they are physically reasonable possibilities.

AIMS

1. Learn about radio (microwave) radiation that now uniformly permeates the Universe and provides compelling evidence for the big bang, ruling out a competing theory (Sections 19.1 and 19.2).

2. Consider tiny ripples in the cosmic microwave background radiation, and how they tell us many properties of the Universe (Section 19.3).

3. Discuss what happened during the first few minutes after the birth of the Universe (Section 19.4).

4. See how a proposed modification of the original big-bang theory in the first wink of the Universe's existence explains some otherwise puzzling observations (Section 19.5).

5. Consider the very speculative possibility of multiple universes, possibly having different properties (Sections 19.6 and 19.7).

Hubble Space Telescope WFPC Team and NASA, courtesy of William A. Baum, University of Washington

■ **FIGURE 19–1** A Hubble Space Telescope image of galaxies in the direction of the Coma Cluster of galaxies. The smaller, fainter galaxies are generally more distant than the brighter, larger galaxies. Essentially all galaxies are receding from us, with the distant galaxies moving away faster than the nearby ones.

19.1 THE STEADY-STATE THEORY

In the previous chapter, we considered the big-bang theory of the Universe, which is motivated in part by the observed recession of galaxies away from our own Milky Way Galaxy (■ Fig. 19–1). We saw that the matter density of the Universe dictates the history, as well as the future, of its expansion rate. Moreover, there is widespread acceptance that a cosmic "antigravity" effect over large scales accelerates the expansion of the Universe, causing it to grow larger (and therefore less dense) at an ever-increasing rate.

Although the expansion of the Universe suggests that there was a definite beginning of time when matter was in a very compressed and hot state, this is not the only logically consistent possibility. Here we consider a reasonable alternative that turned out to be incorrect, but that nevertheless served science well because it forced cosmologists to question critically their assumptions and conclusions.

In 1948, Fred Hoyle of Cambridge University (■ Fig. 19–2) and two of his associates (Hermann Bondi and Thomas Gold) proposed the **steady-state theory** as an alternative to the hot big bang. It is based on a modification of the cosmological principle called the **perfect cosmological principle**: The Universe is homogeneous and isotropic on large scales, and *its average properties never change with time*. Therefore, there was no well-defined beginning (the Universe is infinitely old), and there will be no end; the Universe is simply expanding throughout all time. To keep the average density constant as the Universe expands, new matter must be created continuously, and new galaxies form out of this material.

We might argue against the steady-state theory on the grounds that it requires something to be produced from nothing, which is thought to be impossible over long timescales. That is, we believe in the law of "conservation of energy" – the combination of mass plus energy is neither created nor destroyed.

But this objection turns out to be observationally weak. Only one hydrogen atom must be created per cubic meter, per billion

Clemson University and Donald D. Clayton (Clemson University)

■ **FIGURE 19–2** Fred Hoyle, one of the main advocates of the steady-state theory of the Universe. Though now known to be incorrect, this theory helped accelerate progress in cosmology because it stimulated new observational and theoretical advances.

years, to satisfy the requirements of the steady-state theory. This rate is well below the current detection limit of laboratory experiments. In other words, we have not yet verified the law of conservation of energy to this degree of accuracy. Moreover, the original big-bang theory postulates that the Universe was created out of nothing in a single instant ($t = 0$), so it too appears to violate conservation of energy (although we will later see that no significant violation might occur).

Despite its aesthetic appeal to some astronomers, the steady-state theory gradually lost support as observational evidence showed that the Universe has a finite age (about 14 billion years, not much greater than the age of the oldest globular clusters) and that its properties change with time. One of the first arguments for evolution of the Universe was that radio galaxies seem to be more numerous and luminous at large distances.

An even more compelling case was made with quasars (see Chapter 17): They are clearly denizens of the distant past; all have high redshifts, so we see them at large lookback times. There are no nearby quasars, seen as they are in the present-day Universe, but only lower-luminosity active galaxies. In recent years, Hubble Space Telescope images of distant galaxies have shown that they differ from nearby galaxies, providing additional evidence for the evolution of the Universe.

■ **FIGURE 19–3** Arno Penzias (*left*) and Robert W. Wilson (*right*) with their horn-shaped antenna in the background. After they removed all possible sources of noise (by fixing faulty connections and loose antenna joints, and by removing "sticky white contributions" from pigeons), a certain amount of radiation remained. It was the 3 K background radiation.

Bell Labs–Alcatel-Lucent

19.2 THE COSMIC MICROWAVE RADIATION

The fatal blow to the steady-state theory, however, was the discovery of the **cosmic microwave radiation** – a faint afterglow from the Universe's hot past. The steady-state theory has no explanation for it, but big-bang models require it, as we will now discuss.

19.2a A Faint Hiss from All Directions

From 1964 through 1965, two young researchers (Arno A. Penzias and Robert W. Wilson) were observing with a 20-foot horn-shaped radio antenna owned by Bell Labs near Holmdel, New Jersey (■ Fig. 19–3). The antenna had previously been used to detect radio signals from communications satellites, but Penzias and Wilson were allowed to use it for astronomical observations. They initially wanted to accurately measure the radio brightness of Cas A, a prominent supernova remnant that is used to calibrate radio data. (It got its name from being the first radio source to be found in the constellation Cassiopeia.) Then they wanted to map the 21-cm radio emission in the Milky Way Galaxy.

For the most accurate measurements, they needed to make their microwave antenna as sensitive as possible. After removing all possible sources of noise, they found a very faint but persistent hiss that seemed to defy explanation. It was independent of location in the sky, and also independent of the time of day. If produced by a "blackbody" (Chapter 2), it corresponded to a temperature of about 3 K, which is just 3°C above absolute zero and is equal to –270°C. The researchers were not aware that it might be the afterglow of the big bang!

Meanwhile, at nearby Princeton University in New Jersey, a team led by Robert Dicke was building a radio telescope with which to

■ **FIGURE 19–4** The COBE spacecraft (NASA's Cosmic Background Explorer), remade smaller to fit on a Delta rocket after the explosion of the space shuttle Challenger in 1986. COBE was launched in 1989 and sent back data for four years. It fantastically improved the accuracy of measurements of the cosmic background radiation.

NASA COBE Science Team, courtesy of Nancy Boggess

detect this expected big-bang afterglow. They had predicted that it should exist if the Universe began in a very hot and compressed state.

Actually, this prediction had already been made in the 1940s, by the Russian astrophysicist George Gamow (who was then working in the United States), but the Princeton group was unaware of it. Gamow and his students, Robert Herman and especially Ralph Alpher, had based their conclusion on the idea that all of the chemical elements were created shortly after the big bang. Although incorrect in detail, this is correct in spirit; only the lightest elements were produced early in the Universe, as we shall see later. The Universe must have been very hot to do this, but expansion would have cooled it, so the radiation should now be that of a cold blackbody.

Another astronomer alerted the Princeton and Bell Labs teams to each other, and that's how Penzias and Wilson found out what they had discovered! The Princeton group made a measurement at another wavelength, and it agreed with the prediction of a blackbody spectrum. Other subsequent measurements were also consistent with blackbody radiation from a gas at $T \approx 3$ K, a very low temperature indeed. In 1990, astronomers announced results from NASA's Cosmic Background Explorer (COBE, pronounced "koh´ bee"; ■ Fig. 19–4): The spectrum was that of a perfect, cold blackbody (■ Fig. 19–5). Later in this chapter, we will discuss more recent

■ **FIGURE 19–5** This first spectrum based on COBE observations drew a standing ovation from astronomers when it was initially shown at an American Astronomical Society meeting. It represents astonishingly good evidence that the cosmic background radiation comes from a blackbody. The data points and their error bars (*black*) were fit precisely by a blackbody curve (*red*) for 2.735 ± 0.06 K, a value that has since been slightly improved, based on the final data set, to 2.728 ± 0.004 K. Current error bars are thinner than the thickness of the curve drawn! Astrophysicist John Mather received a share of the 2006 Nobel Prize in Physics for his leadership in this discovery.

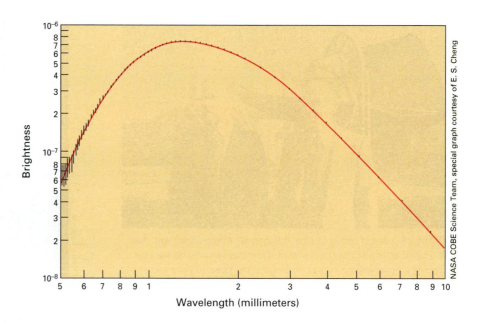

satellite results, from the Wilkinson Microwave Anisotropy Probe and, in 2013, from Planck.

If you have a TV that gets its signal from an aerial rather than from cable or from the internet, about 1 per cent of the "snow" on the screen when you are tuned to non-broadcasting channels is from this cosmic microwave background radiation.

19.2b Origin of the Microwave Radiation

From where, exactly, does a photon of the cosmic microwave background radiation come? The early Universe was very hot and ionized: There were no bound atoms, but rather free electrons and atomic nuclei. Because photons easily scatter (bounce, or reflect) off of free electrons, they cannot travel in a straight line from one object to another, and instead jump randomly from one direction to another. So instead of being transparent, the Universe looked opaque, like thick fog. Even if there were discrete objects at this time (but there weren't), one would not have been able to see them. The Universe was very bright and filled with photons, however, since it was hot – and the photons simply scattered around randomly in a sea of hot particles.

As the Universe expanded, it cooled. Eventually, about 400,000 years (more accurately, 379,000 years) after the big bang, it reached a temperature of about 3,000 K, and electrons were able to combine with protons to form neutral hydrogen atoms. The process is called "recombination" – even though in this case, electrons were combining with protons and forming neutral atoms for the first time in the history of the Universe.

Unlike free electrons, electrons bound in atoms are able to interact with photons of only certain specific energies, absorbing them (Chapter 2). They are not effective at scattering photons at other energies. Thus, at this time most photons stopped bouncing around, and became free to travel unhindered. The Universe therefore became

transparent over a very broad range of electromagnetic waves. Astronomers say that matter and radiation "decoupled" from each other at $t \approx 400{,}000$ years.

Before recombination, the Universe was very bright, because there were many optical photons travelling through it. As space expanded, however, the wavelengths of photons increased, just as a wavy line drawn on an expanding balloon expands; they lost energy, and the Universe became progressively dimmer. They maintained the spectrum of a blackbody, but of lower and lower temperature (■ Fig. 19–6). Over time, typical photons gradually went from being optical/infrared to radio (microwave). The cosmic microwave radiation now corresponds to such a low temperature, about 3 K, because the Universe has expanded a great deal over its 14-billion-year life. Since it started off as radiation filling the entire Universe, it is easy to understand why it now comes equally from every direction (that is, it appears isotropic).

In essence, the cosmic background photons come from an opaque "wall" at redshift about 1,000 (the ratio of 3,000 K to 3 K). We cannot see electromagnetic radiation from beyond this wall because the Universe was opaque. This marks the "boundary" of the observable Universe: We cannot see electromagnetic radiation from times prior to about 400,000 years after the big bang. In principle, we could see neutrinos from $t < 400{,}000$ years, since the Universe was not opaque to them. But, unfortunately, neutrinos are very difficult to detect.

The cosmic microwave radiation is left over from when the Universe was very hot, dense, and opaque. Being almost perfectly isotropic, it provides the best-known evidence in support of the cosmological principle that the Universe is homogeneous and isotropic. The steady-state theory has no reasonable explanation for its presence.

Penzias and Wilson shared the 1978 Nobel Prize in Physics for their important discovery: They had been very careful, and had not ignored the faint (almost nonexistent) hiss. Gamow could not receive the Nobel Prize, as he had already died, but it seems somewhat unfair

■ **FIGURE 19–6** The cosmic background radiation was set free at a temperature of 3,000 K (*blue curve*) when the Universe became transparent at the time of recombination, about 400,000 years after the big bang. As the Universe expanded, the spectrum was transformed into that of a lower-temperature gas (*red curve at lower right*). The rainbow shows the wavelengths of visible light; the cosmic background radiation is detected near its peak at microwave wavelengths, and not in visible light or other wavelengths far from its peak.

that Alpher was omitted. (According to the official rules, each Nobel Prize, other than the Peace Prize, can be shared among a maximum of three people.)

19.3 DEVIATIONS FROM ISOTROPY

The cosmic microwave background radiation shows a slight deviation from isotropy of about two parts per thousand. It looks slightly hotter in one direction of the sky, and slightly cooler in the opposite direction. Such an "anisotropy" is caused by a combination of motions: the Sun's motion around the center of the Milky Way Galaxy, the Milky Way's motion within the Local Group, and, most important, the Local Group's motion relative to the "Hubble flow" (that is, relative to the smoothly expanding Universe).

These gravitational perturbations are all known. The sum of the motions produces a net motion in a particular direction of space, and the slight temperature anisotropy is attributed to the Doppler effect. The radiation coming from the direction of motion is slightly blueshifted ("hotter"), while that coming from the opposite direction is slightly redshifted ("cooler"). Maps made with COBE very clearly show this anisotropy (■ Fig. 19–7), as do maps with later satellites.

19.3a Ripples in the Cosmic Microwave Background

A major puzzle for many years was the apparent absence of small spatial temperature variations (often called "fluctuations," and

(a)
3.3 mm

(b)
5.7 mm

(c)
9.6 mm

■ **FIGURE 19–7** Maps of the sky from all four years of COBE data at radio wavelengths of (a) 3.3 mm, (b) 5.7 mm, and (c) 9.6 mm. At the longest wavelength, we see a discrete source in Cygnus (*green dot at middle left*) and a sign of the Galactic plane (*horizontal red at right*). The rest of the signal is from the cosmic microwave background radiation. The asymmetry from bottom left to top right is produced by the Sun's motion relative to the smoothly expanding Universe. The range shown is 63 millikelvins.

Updated in 2001 from Max Tegmark, Angelico de Oliveira-Costa, Marc Devlin, Barth Netterfield, Lyman Page, and Ed Wollack, in Astrophys J. 474, L77-80, 1996

■ FIGURE 19–8 A full-sky map of the variations in the 3 K cosmic background radiation, as analyzed from all four years of COBE data, after removal of the Sun's motion relative to the Universe. The temperature variations are typically only 30 microkelvins (0.001 per cent). The red band across the center of this map results from the microwave emission of the Milky Way Galaxy. The variations above and below the Milky Way are real fluctuations in the background radiation. The results are also shown for a balloon experiment known as QMAP and for a ground-based experiment carried out in Saskatoon, Saskatchewan, Canada. The latter observed only a small area around the north celestial pole but obtained higher angular resolution within that region.

■ FIGURE 19–9 (a) A map of the tiny variations in the 3 K cosmic background radiation over part of the sky, based on data from the BOOMERANG balloon project. The full moon is shown (*as a black dot at lower right*) for comparison. Note that the temperature differences, as shown in the scale bar, vary from the orange spots to the black spots by only about 600 microkelvins – that is, 0.0006 K. (b) The BOOMERANG mission's observed variations match the predictions of the flat model for the Universe (*bottom center*) rather than the open or closed models, for which simulations are shown at *bottom left* and *bottom right*.

associated with differences in matter density) in the background radiation corresponding to the size scales of clusters and superclusters of galaxies. If these structures formed by the gravitational contraction of matter, there should have been slight ripples in the density of matter early in the Universe. Photons escaping from higher-density clumps would have a slightly different redshift than those coming from regions of average or below-average density.

There are several relevant processes. One effect, dominant on large scales, is that photons lose more energy when they come out of stronger gravitational fields, and the different redshifts translate into slight variations in the associated temperature of the radiation. Another effect, more important at small scales, is that compressed pockets of gas get hotter.

No such ripples were found until the early 1990s. Some specific models for the formation of large-scale structure were consequently eliminated from further consideration because they predicted variations larger than the observed upper limits.

In 1992, a breakthrough was made with the COBE satellite: Tiny spatial temperature variations (fluctuations; about one part per hundred thousand, or 30 microkelvins, 30 μK) were finally found, at least in a statistical sense. (Ripples were present, but no great significance could be given to any particular one.) After a few more years of data collection, individual variations became clearly visible (■ Fig. 19–8). The angular resolution of COBE was low, so even the smallest detected variations correspond to structures billions of light-years across. The 2006 Nobel Prize in Physics was awarded to John Mather (NASA/Goddard Space Flight Center) and George Smoot (University of California, Berkeley) for their primary COBE discoveries: the blackbody nature of the cosmic background radiation to high precision, and the tiny temperature variations.

Subsequently, several balloon-based experiments found similar temperature (and hence density) variations, but on smaller angular scales (■ Fig. 19–9a), corresponding to superclusters of galaxies. These are the "seeds" from which large-scale structure grew in the Universe. They are the imprints of minuscule ripples in the distribution of matter established shortly after the big bang. One of the missions was BOOMERANG: **B**alloon **O**bservations of **M**illimetric **E**xtragalactic **R**adiation **an**d **G**eophysics. Their results and others showed that the Universe is globally very close to being flat (■ Fig. 19–9b), as we will now discuss.

(a)

(b)

(c)

■ **FIGURE 19–10** The magnitude (amplitude) of temperature variations in the sky versus their angular size. Though variations appear over a wide range of scales, the strongest ones appear to be about 1° in diameter, or twice the angular size of the full moon. (The angular scale is 360° divided by the term (known as the "multipole moment") shown on the lower x-axis. In essence, the multipole moment is the number of waves of a given wavelength that would fit within a 360° circle.) Several peaks can be seen. The sizes and positions of the peaks in the graph indicate that the Universe is flat and allows the percentage of nuclear particles in the Universe to be calculated. The curves illustrate theoretical fits to the data shown with error bars. Do not confuse such diagrams of variation peaks with Planck curves, which graph intensity versus wavelength for a given temperature. Note that the theoretical curves shown in colors other than red represent models for the Universe that had been under serious consideration before the cosmic background radiation measurements were made that clearly supported the standard model, which is shown in red. (a) Experimental results showing a selection of the different telescopes devoted to the topic, illustrating its scientific importance and widespread interest. (b) Experimental results based on 7 years of observations (2002–2009) with the **W**ilkinson **M**icrowave **A**nisotropy **P**robe (WMAP), with a solid line drawn from the best theoretical model; uncertainties in the model are shown in gray. (c) The WMAP points define the curve from the left side of the graph through the major peak. Points from the ACBAR experiment (**A**rcminute **C**osmology **B**olometer **A**rray **R**eceiver, an Antarctic experiment at the South Pole led by scientists from UC Berkeley and Case Western Reserve University) cover the second peak. Planck's analyzed data are even better (see *A Closer Look 19.1: Planck Maps the Cosmic Background Radiation*). The standard model matches the observational points much better than alternative cosmological models match.

19.3b The Overall Geometry of the Universe

By 2001, several missions had shown that the angular size of the most obvious temperature variations in the cosmic background radiation is typically about one degree (1°). This statement means that the typical angular size of the highest-magnitude ("amplitude") ripples is around 1°, and it tells us something about the overall geometry of the Universe. Until the 2013 release of the maps from the Planck spacecraft, the best measurement was from the 9-year set of data from NASA's Wilkinson Microwave Anisotropy Probe (■ Fig. 19–10), discussed further below. The data are consistent with the presence of considerable amounts of dark matter and other energy in the Universe – enough to make the Universe spatially flat, or very

nearly so! (Looking at the data showed that the Universe is flat, but analysis of the sizes of the variations gives us much more detail, as well as some explanation regarding why and how.)

A spatially flat geometry means that Ω, the ratio of average total density (of luminous and dark matter, normal energy, and other energy) to the critical density, is 1 (see Chapter 18). This "other energy" may well be the "vacuum energy" or "dark energy" associated with the cosmological constant, whose nonzero value was first suggested by observations of distant supernovae (see Chapter 18). Galaxies and clusters of galaxies seem to have formed at peaks in the overall distribution of dark matter.

From the peaks in the graph, which show compressions and rarefactions in the early Universe, scientists infer an amount of dark energy matching that determined from supernova observations that led to the accelerating-universe conclusion.

19.3c The Wilkinson Microwave Anisotropy Probe (WMAP)

A NASA spacecraft, the Wilkinson Microwave Anisotropy Probe (WMAP), covered the whole sky at 30 times the resolution of COBE. It was launched in 2001 into an orbit a million miles from us, on the side of the Earth opposite to the direction to the Sun, a so-called Lagrangian point. It started taking scientific data the next year, and was shut off in 2010. What follows are results from 9 years of observations, published in December 2012. The probe was named after David Wilkinson, an important member of the COBE and WMAP teams and a member of the original Princeton team that was among the first to understand the cosmic microwave background radiation.

WMAP's map of the entire sky (■ Fig. 19–11) is spectacularly detailed compared with that from COBE, with an angular resolution of only 0.2°. It shows temperature variations measured in microkelvins – millionths of degrees. COBE also had mapped the whole sky, but at lower resolution, and though several recent experiments had mapped at high resolution, they had covered only tiny portions of the sky. Thus the results from WMAP, taken at five wavelengths between 2.3 mm and 13 mm, represented a major advance. In *A Closer Look 19.1: Planck Maps the Cosmic Background Radiation*, we will see how the 2013 release of a map from the European Space Agency's Planck spacecraft was even more detailed.

Comparing the WMAP maps at the different wavelengths allowed the effects of the Milky Way to be subtracted. The results for a variety of cosmological parameters are so precise that they have greatly helped transform cosmology into a precision science.

From studying various features (especially the first peak) in the graph showing the distribution of the sizes of the variations, and making use of additional information from other astronomical studies, we now know the age of the Universe. WMAP's value is 13.74 billion years, with a formal statistical uncertainty of only 0.08 billion years. This value fits with our current understanding of the main-sequence turnoff in globular clusters of 12 billion years and the ages of the oldest white dwarfs in clusters of 12.7 billion years, both with uncertainties of about 1 billion years.

The WMAP data analysis found that Hubble's constant is 70.0 ± 2.2 km/s/Mpc; in conjunction with other studies, the data imply that $H_0 = 69.3 ± 0.8$ km/s/Mpc, an uncertainty of only 1 per cent. These values are consistent with the direct measurement of 73.8 ± 2.4 km/s/Mpc obtained from Cepheid variable stars with the Hubble Space Telescope and Type Ia supernovae (Chapter 18). The long battle over the value of Hubble's constant is finally over, or at least reduced from a factor of 2 to only about 2 per cent.

Other observations had shown that the cosmic microwave background is polarized – that is, radiation coming from a given region of the sky has more of its electric field oscillating in one direction than in the perpendicular direction. Such polarization is typically produced when light scatters off of a surface. For example, light reflected from water or pavement is polarized, and polarizing sunglasses reduce the glare by blocking light having the dominant direction of oscillation.

In the Universe, the level of polarization is set by the density of normal, ordinary matter ("baryonic matter," in most cases consisting of protons and neutrons; the associated electrons are also included, though strictly speaking they are not baryons). This matter is attracted by the regions of greater-than-average density, which are largely set by the distribution of dark matter. This distribution also corresponds to the distribution of the temperature anisotropy.

The Wilkinson Microwave Anisotropy Probe's measurements (■ Fig. 19–11c) show that the theories of polarization correctly predict what is seen. The results tell us that the cosmic background radiation is indeed coming from a surface of last scattering, a moment before recombination occurred and the Universe became transparent. The Planck spacecraft should provide even more observational evidence about the polarization of the cosmic microwave background, whose analysis will be a key beneficiary of the higher resolution. The cosmic microwave background is more strongly polarized than the foreground, which will help scientists pick it out from the data.

The WMAP data, when linked with measurements of Cepheids and distant supernovae, and also with maps of the distribution of galaxies in the sky, show that the Universe is essentially flat on global scales: The total amount of matter and energy in the Universe, Ω_{total}, is 1.0023 ± 0.0055, an astonishing degree of precision and consistent with 1 (the condition of flatness, according to Einstein's general theory of relativity). In 2013, analysis of data from the Planck satellite gave a consistent value of 1.0005 ± 0.0065.

WMAP's coverage allows the distribution of the sizes of temperature variations to be measured much more accurately than had previously been done. The distribution of its peaks reveals other cosmological parameters.

In particular, using values from the 9-year results of WMAP and complementary data, the density of normal, baryonic types of matter comprises only 4.63 per cent of the total, to within 0.10 per cent. And of that same total, only about 0.5 per cent is luminous matter (stars, nebulae, planets that reflect light, etc.), the type that we can easily detect. Hence, the stuff of which we are made is just a small minority of all that exists; we are like the debris of the Universe. The remaining 4 per cent of normal matter contributes to the dark matter that scientists have been detecting for some decades by its gravitational effects in the halos of galaxies and in clusters of galaxies. MACHOs are an example of this type of dark matter, but recent results suggest that much of it actually consists of tenuous, million-degree gas in the halos of galaxies.

Together with other data, WMAP also showed that another 22.7 per cent (to within 1.7 per cent) is other types of "cold dark matter." In this context, "cold" means that the matter is moving slowly, not close to the speed of light. We do not yet know what this matter consists of, but the "best bet" is exotic particles left over from the big bang.

We can tell that most of the dark matter cannot be "hot" (moving at very high speeds, as neutrinos do) because such light particles do not cluster enough under the influence of gravity to form galaxies. If a substantial amount had been present in the early Universe, too few galaxies would have formed to account for the young, high-redshift galaxies seen with the Hubble Space Telescope.

(a)

(b)

(c)

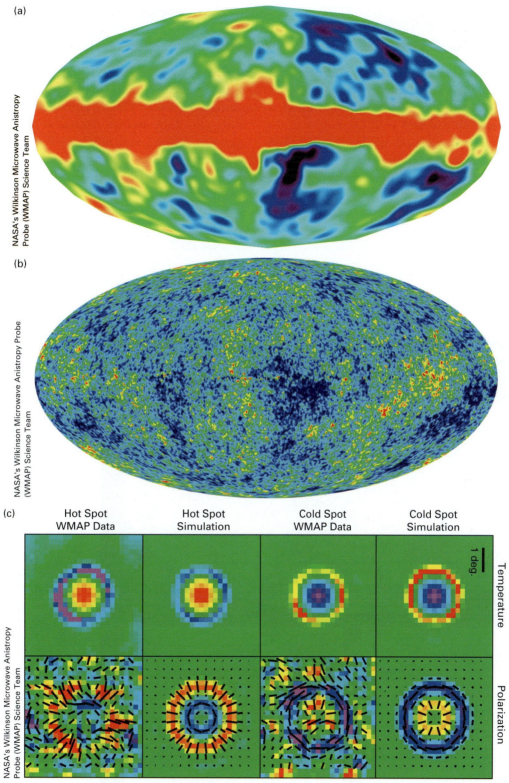

Hot Spot
WMAP Data

Hot Spot
Simulation

Cold Spot
WMAP Data

Cold Spot
Simulation

1 deg.

Temperature

Polarization

NASA's Wilkinson Microwave Anistropy Probe (WMAP) Science Team

NASA's Wilkinson Microwave Anistropy Probe (WMAP) Science Team

NASA's Wilkinson Microwave Anistropy Probe (WMAP) Science Team

■ **FIGURE 19–11** Temperature variations on the microkelvin level displayed on an all-sky map. (a) The calibration of all four years of COBE data. These variations, about 5 parts per million, show the seeds from which structure in the Universe evolved nearly 14 billion years ago. Astrophysicist George Smoot received a share of the 2006 Nobel Prize in Physics for his leadership in this discovery. (b) The higher resolution provided from NASA's WMAP. We see a weighted combination of the five WMAP maps at different frequencies, with weights chosen to minimize the contributions of our Galaxy. The image, released in late 2012, shows the result of all 9 years of observations. (c) This set of comparisons shows that the polarization pattern measured by WMAP around hot and cold spots matches what was expected based on the standard model. Improved polarization measurements are expected to be a major benefit of the Planck mission.

A CLOSER LOOK 19.1 — Planck Maps the Cosmic Background Radiation

The wonderful mapping of the cosmic background radiation from NASA's COBE spacecraft, launched in 1990, was superseded by the even better mapping from NASA's WMAP spacecraft, launched in 2001. Now the fantastic map released in 2013, made with the European Space Agency's Planck spacecraft, which was launched in 2009 and is named in honor of Max Planck (one of the fathers of quantum physics), reigns supreme. As with the previous spacecraft's telemetry, scientists from many countries poured over the data in secret for several years, not letting out the results until they agreed on them.

Planck's map (Figure ACL19.1-a) has higher resolution and sensitivity than those of the previous spacecraft, and covers the whole sky with that high resolution and sensitivity.

The all-sky map (Figure ACL19.1-b) shows minuscule differences in the temperature (corresponding to variations in matter density) as measured by one of Planck's instruments in the submillimeter/millimeter part of the spectrum. Because of its unprecedently high spatial resolution, it gives a much better determination of the extent of temperature variations at the time that the cosmic background radiation was released, about 380,000 years after the big bang, so looking back in time more than 13.7 billion years. We are looking at the Universe as it was when "recombination" took place: nuclei (chiefly hydrogen) and electrons came together to make atoms, which absorb and emit radiation chiefly in spectral lines rather than as continuous rainbows of radiation. The temperature of the Universe has cooled, as it expanded, from about 3000 K at the time of its release to 2.7 K now.

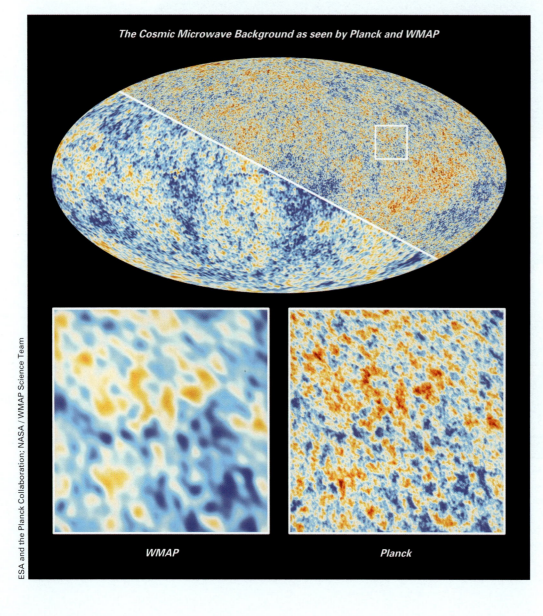

The Cosmic Microwave Background as seen by Planck and WMAP

WMAP

Planck

ESA and the Planck Collaboration; NASA / WMAP Science Team

■ **Figure ACL19.1-a** The Planck spacecraft's resolution was higher than that of previous spacecraft, including that of the Wilkinson Microwave Anisotropy Probe (WMAP).

ESA and the Planck Collaboration

■ Figure ACL19.1-b The 2013 release of the all-sky map of minuscule spatial temperature variations (known as "fluctuations") measured in the early Universe with the European Space Agency's Planck spacecraft. Teams of scientists from dozens of countries worked to remove the effects of the Milky Way and other foreground objects to obtain this background map.

■ Figure ACL19.1-c The "power spectrum" of the square of the magnitude (amplitude) of temperature fluctuations vs. their angular size, as measured with Planck and released in 2013.

Having the fine resolution allowed the angular sizes of spatial temperature variations (fluctuations) to be measured so accurately that the graph of the square of the magnitude (often called "amplitude") of the variations vs. their angular sizes is much more precise than was previously possible to make. The uncertainties shown in WMAP's similar graph (Figure 19–10) are thus reduced (Figure ACL19.1-c).

Studies of the fluctuations allow such important parameters to be determined as the relative amounts of ordinary matter, dark matter, and dark energy, and to make even more definite that the overall density of the Universe is just on the borderline between open and closed. The numbers that came out were a bit surprising, since they indicated a larger change from the WMAP numbers than had been expected (Figure ACL19.1-d). But the changes were merely quantitative adjustments rather than requiring any qualitative changes in scientists' conclusions. Normal matter makes up only 4.9 per cent of whatever is in the Universe and dark matter makes up 25.8 per cent. Dark energy takes up the remaining 69.3 per cent.

The results come out of a calculation that simultaneously fits many variables, including also Hubble's constant. The value from Planck data, 67.8 km/s/Mpc, is about 4 per cent lower than the previously accepted value, outside the "standard error" (a so-called 1-sigma error bar that means that repeating a measurement would give a similar value about $^2/_3$ of the time, still leaving a $^1/_3$ chance that the result would be outside this error bar) but within 2-sigma, not an unreasonable variation for a scientific experiment. This value of Hubble's constant corresponds to an age of the Universe of 13.80 billion years (with an uncertainty of only 0.04 billion years), after taking into account the proportions on the right side of Figure ACL19.1-d.

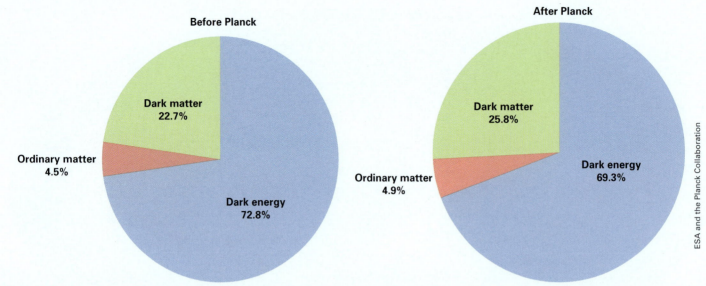

ESA and the Planck Collaboration

■ **Figure ACL19.1-d** The relative proportions of ordinary matter, dark matter, and dark energy, as determined first by scientists using WMAP data and then by scientists using Planck data.

The first release of Planck data reported some minor "anomalies" in which the map differs from the "standard model" of the early Universe, but as of this writing it is too early to say that they are real. There appears to be a minor difference in the magnitudes of temperature fluctuations in opposite hemispheres of the sky, while we expect the Universe to be the same in all directions on a large scale. Also, a slightly colder than average spot seems larger in extent than expected. Similar anomalies had been reported on the basis of WMAP data; they will be studied even more carefully now that they have shown up in the Planck results.

Post-Planck, normal matter and cold dark matter contribute about 31 per cent to the total matter plus energy content of the Universe. In other words, the matter content, Ω_M, is about 0.31 of the total density ratio of the Universe, Ω_{total}. Since $\Omega_{total} = 1$, this leaves roughly 69 per cent as "dark energy" (■ Fig. 19–12), which we discussed in Chapter 18. We have even less understanding of dark energy than we do of cold dark matter. We appear to live in a Universe in which we can relate to only about 4.9 per cent of its constituents, if even that (i.e., only 0.5 per cent is easily visible)! As John Bahcall of the Institute for Advanced Study summarized, "We are in an implausibly crazy Universe, but one whose characteristics we know."

The spacecraft findings represent strong endorsement of the big-bang model of the Universe. They also support some versions of inflationary cosmology ("inflation" will be discussed later in this chapter), though other specific inflationary scenarios are now ruled out. Also, the "quintessence" model that had rivaled Einstein's cosmological constant as a contender for explaining the dark energy now seems somewhat less likely than it had previously. ("Quintessence," named after the Aristotelian fifth essence – in addition to earth, air, fire, and water – refers to possible new forces or fields that might be causing the expansion of the Universe to accelerate.)

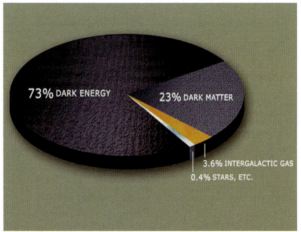

■ **FIGURE 19–12** Pie diagram showing WMAP's conclusions about the composition of the Universe. Normal matter (made primarily of protons, neutrons, and electrons) constitutes only about 4 per cent of the total, and only about one-tenth of the normal matter is easily visible in the form of stars, nebulae, and planets. Gravitationally attractive dark matter of an exotic nature (but slowly moving, so it is called "cold") is 23 per cent, and mysterious, gravitationally repulsive dark energy is 73 per cent.

19.3d Ground-Based Telescopes for the Cosmic Background Radiation

Studies of the cosmic microwave background continue with telescopes on the Earth's surface, not only in space (■ Fig. 19–13). A 10-m telescope has operated at the South Pole since 2007, and a "Polarbear Collaboration" of scientists from many institutions are studying its data. The 6-m Atacama Cosmology Telescope has been operating since 2008 in the high Atacama desert in Chile.

These telescopes have lots of astronomy to study besides the background radiation. They can image distant clusters of galaxies, for example, out beyond $z = 1$.

19.3e The Planck spacecraft

Since Max Planck was a German scientist who, in 1900, came up with the formula that describes the blackbody curve, it is fitting that

(a)

(b)

■ FIGURE 19–13 (a) The 10-m South Pole Telescope at the Amundsen-Scott South Pole Station, a project of the National Science Foundation being done by a multi-institution collaboration. Detectors at these millimeter wavelengths were once single but now contain over 1,000 elements. The various halos and arcs around the telescope are caused by the interaction of sunlight with ice crystals in the air. (b) El Gordo is the most massive known cluster of galaxies in the distant universe. The cluster was discovered by the Atacama Cosmology Telescope using a subtle effect in the cosmic microwave background. This image is shown on the left. To determine the cluster's distance and mass, follow-up observations in the optical and x-rays were made with ESO's Very Large Telescope and NASA's Chandra X-ray Observatory, respectively. A composite of these observations is displayed at right, showing the x-ray emission in light blue on top of the optical image. Note that the cosmic background radiation is not centered on the x-ray emission, because they arise from different physical processes within the cluster. El Gordo is one of the most amazing known clusters, because it is massive, distant, and in the process of doing something that looks like merging. It is one of the Atacama Cosmology Telescope's best-known finds. (c) The 6-m Atacama Cosmology Telescope is run by a collaboration of many universities and institutions.

(c)

■ **FIGURE 19–14** One of the superclusters of galaxies discovered with the Planck spacecraft compared with a view from the XMM-Newton x-ray spacecraft that shows three component clusters (and whose x-ray contours are also superimposed on the Planck image). The supercluster was discovered by its silhouette against the cosmic background radiation. This effect, named after the scientists Sunyaev and Zel'dovich, provides the temperature and density of the hot gas in which the cluster is embedded.

the European Space Agency gave his name to its spacecraft meant to map the Universe's blackbody curve. The Planck spacecraft was launched on the same rocket as the Herschel Space Observatory.

Planck mapped the whole sky in the infrared and submillimeter parts of the spectrum, as we saw in Chapter 15. It has twice the spatial resolution and 100 times the sensitivity of WMAP. In 2012, Planck's High Frequency Instrument, HFI, ran out of coolant having completed five full-sky surveys over 30 months. (The Low Frequency Instrument, LFI, worked longer.) Planck's first results about the cosmic microwave background radiation were released in 2013 based on the first 15 months of data, with the full data-release expected about a year later.

Planck scientists have compiled a catalogue of thousands of very cold objects, which are being explored by a variety of ground-based and space-based telescopes. For example, it found a previously unknown population of galaxies billions of light-years away that are so covered in dust that we couldn't see them in the visible part of the spectrum. Yet they formed stars at a rate 10 to 1,000 times higher than the star formation rate in our Milky Way Galaxy. Planck has also located dozens of previously unknown distant clusters of galaxies (■ Fig. 19–14).

19.4   THE EARLY UNIVERSE

The evolution of the Universe can be studied nearly back to the moment of creation, "$t = 0$," with an increasing amount of uncertainty as we go to progressively earlier times. Just how close to $t = 0$ we can get with reasonable accuracy is debated.

19.4a   Going Back in Time

Almost all astronomers and physicists agree that we can go quite far back in time because the early Universe was relatively simple: It consisted of uniformly distributed particles and photons in thermal equilibrium (that is, at the same temperature), and we seem to know

the few fundamental laws that governed their behavior. Compare this with a complex problem like Earth's weather, which is affected by factors such as local heating, ocean currents, cloud formation, continents, mountain ranges, and so on. It is easier to predict the properties of the early Universe than it is to predict the weather!

Certainly we can reach something like $t \approx 1$ s, because conditions were similar to those in the central regions of stars, and the nuclear physics is well understood. We also have substantial confidence in our conclusions down to $t \approx 10^{-12}$ s, since the corresponding temperatures and energy densities have been reproduced in high-energy particle accelerators like the **L**arge **H**adron **C**ollider (LHC), which started operating in 2008 at CERN, the European Laboratory for Particle Physics near Geneva, Switzerland (■ Fig. 19–15). (Scientists working at CERN reported, in 2012, the discovery of the "Higgs boson," an elementary particle that acts as a kind of molasses that other particles have to swim through, giving them their mass and explaining why particles have mass. An analogy to the Higgs boson is the presence of people in a room when the president enters; if people swarm around the president, his progress across the room is slowed, similar to the way more massive particles go slower when given the same energy as less massive particles.)

It can also be argued that we can extrapolate all the way to about 10^{-35} s, or perhaps somewhat earlier; once again, hot gases in equilibrium are relatively easy to understand. However, we do not yet know everything down to such small timescales; our physical theories are incomplete, and no conceivable particle accelerators can achieve temperatures that are high enough to test them. Thus, some of our current conclusions about this era are rather speculative. Turning this problem around, we can use celestial observations to test the physical theories: The Universe is our great "accelerator in the sky."

Although times such as 10^{-35} s may seem ridiculously short, with no real "action," it is important to understand that lots of things could have happened. This is because the interaction timescales and distances were far smaller than they are today.

19.4b   A Brief History of the Early Universe

Concerning times before 10^{-43} s, we know essentially nothing, except that the temperature exceeded about 10^{32} K. This is the **Planck time:** It is thought that time itself might be packaged in small units ("quantized") at about this interval, or at least it becomes unpredictable. Some physicists have the notion of "space-time foam," where packets of time and space themselves flit into and out of existence. We must develop a self-consistent quantum theory of gravity to better understand this era.

If the standard big-bang theory is correct, the temperature at the Planck time was about 10^{32} K, and the Universe was expanding and gradually cooling down.

When the Universe was between 10^{-35} and 10^{-6} s old, there was equilibrium among particles, antiparticles, and photons. Particle-antiparticle

(a)

CERN

■ **FIGURE 19–15** (a) Inside the CERN **L**arge **H**adron **C**ollider (LHC), at its site outside Geneva. The ring of the LHC is drawn on the aerial view. (b) Some scenes of one of the major experiments, ALICE (**A** **L**arge **I**on **C**ollider **E**xperiment), deep underground at the LHC to receive elementary particles that result from the LHC's collisions. In July 2012, a team using ALICE (and another team using a different detector at the LHC) announced the probable discovery of the Higgs boson, a long-sought particle whose presence completes the "Standard Model" of particle physics. Sometimes called the "God particle" by the popular press, the Higgs boson (and its associated "Higgs field") is thought to be responsible for giving mass to other fundamental particles.

(b)

pairs annihilated each other and produced photons. Photons spontaneously formed particle-antiparticle pairs. **Quarks**, particles that are normally bound together by "gluons" to form protons and neutrons, were plentiful and unbound during most of this era.

Quarks come in six types or "flavors": up, down, strange, charmed, top (or truth), and bottom (or beauty). Each flavor also comes in three "colors": blue, green, and red. (These names are whimsical; they don't reflect the real "character" of each quark.) Normal matter consists of the "up" and "down" quarks, together with electrons and electron neutrinos (a type of neutrino associated with electrons). For example, a proton consists of two "up" quarks (each with a charge of $+2/3e$, where e is the unit of electric charge) and one "down" quark (with a charge of $-1/3e$). A neutron is one "up" quark and two "down" quarks (■ Fig. 19–16). Each quark has a corresponding antiquark.

At some stage, probably early in this era of the quark-gluon-photon mixture, a slight imbalance of matter (quarks) over antimatter (antiquarks) was formed. It amounts to about one part per billion. It is

Proton

+1 Charge unit

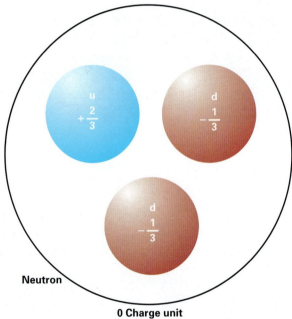

Neutron

0 Charge unit

■ **FIGURE 19–16** The most common quarks are *up* (*u*) and *down* (*d*); ordinary matter is made of them. The up quark has an electric charge of $+2/3e$, and the down quark has a charge of $-1/3e$, where *e* is the unit of electric charge. This fractional charge is one of the unusual things about quarks; prior to their invention, it had been thought that all electric charges come in whole numbers. Note how the charge of the proton and of the neutron is the sum of the charges of their respective quarks.

Great discoveries sometimes have humble beginnings, especially when the discoverers are not aware that they are making a great discovery. Because of this fact we do not have an attractive picture of the neutral kaon apparatus. I am sorry about this.

Sincerely,

James W. Cronin

James W. Cronin

■ **FIGURE 19–17** James Cronin and Val Fitch received the 1980 Nobel Prize in Physics for their experiment that showed an asymmetry in the decay of a certain kind of elementary particle. This result may point the way to the explanation of why there is now more matter than antimatter in the Universe. When one of us (J.M.P.) asked Professor Cronin for a picture to illustrate the Nobel Prize-winning apparatus, he replied with the letter that is excerpted in the figure.

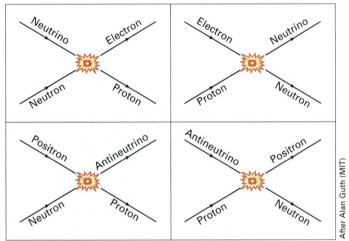

■ **FIGURE 19–18** Methods by which protons and neutrons are converted back and forth into each other, early in the history of the Universe when temperatures were very high.

not known which specific reactions took place, but laboratory experiments have demonstrated the existence of a matter–antimatter asymmetry for some processes (■ Fig. 19–17). We owe our existence to this tiny asymmetry.

The annihilation of protons, neutrons, and their antiparticles occurred at $t \approx 10^{-6}$ s, one microsecond, when the temperature was about 10^{13} K, leaving a sea of photons. The annihilation was not complete, because of the slight imbalance of matter over antimatter.

Then there was rapid conversion of protons and neutrons back and forth into each other (■ Fig. 19–18). However, since neutrons are slightly more massive than protons, they are harder to make, and the number of neutrons settled to only about one-quarter that of protons. By the time the Universe was about 1 second old, its temperature had dropped to about 10^{10} K (10 GK, ten billion degrees), and the typical photon energies became too low to spontaneously produce electron-positron pairs. Electrons and positrons therefore annihilated each other for the last time, producing more photons. The remaining (excess) electrons balanced the positive charge of the protons. Moreover, at this time, neutrons began to decay into protons, electrons, and antineutrinos; it was not possible to rapidly replenish the supply of neutrons. This further increased the imbalance between protons and neutrons.

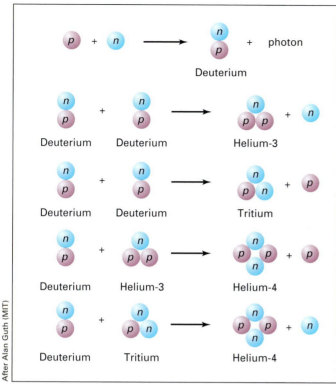

After Alan Guth (MIT)

■ FIGURE 19–19 Reactions during the first few minutes of the Universe's existence that produced various isotopes of the lightest elements, the process of primordial nucleosynthesis. In addition, a very small amount of lithium-7 was produced by other reactions.

Based on Kepler's *Rudolphine Tables* (1627), collection of Jay M. Pasachoff

■ FIGURE 19–20 Pillars of cosmology: expansion (from Hubble's law), background radiation (and the deductions from studying ripples in it), and deuterium abundance. The evolution of the Universe is an additional pillar.

19.4c Primordial Nucleosynthesis

The temperature of the universe had dropped to only a billion degrees by an age of about 100 s, allowing the formation of heavy isotopes of hydrogen, two isotopes of helium, and a little bit of lithium. Before this time, collisions between particles would sometimes produce bound states, but subsequent collisions destroyed them.

At the temperature and density of matter when $t \approx 100$ s, collisions tended to produce a surplus of bound particles; they were not immediately destroyed (■ Fig. 19–19). Elements heavier than lithium were not formed at this time: The temperature and density dropped too rapidly. By an age of about ten minutes, the Universe had completed its **primordial nucleosynthesis** – the formation of the light elements shortly after the big bang.

Primordial nucleosynthesis is predicted to have happened when the proton to neutron ratio was about 7:1. For every 16 nuclear particles, 2 were neutrons, and these generally combined with 2 protons to form helium, thereby using up 4 of the 16 particles. The other 12 particles remained as protons (simple hydrogen nuclei). So, about 25 per cent of the matter (by mass) should have turned into helium, while most of the rest was hydrogen. (A little bit of the hydrogen was in the form of deuterium and tritium, but the latter is unstable and quickly decays.)

The fact that helium is observed to be about 25 per cent by mass throughout the Universe (as far as we can tell) strongly supports the big-bang theory: The Universe had to be hot and dense at early times. The rather uniform relative proportion (abundance) of helium suggests a primordial origin, before stars were born. Heavier elements, on the other hand, exhibit rather large variations in their relative proportions: In some stars they are rare, and in others they are common, implying an origin that is not primordial or uniform throughout the Universe.

The verification of the predictions of primordial nucleosynthesis is one of the four major pillars on which the big-bang theory rests (■ Fig. 19–20). (The other three are the observed expansion of the Universe, the existence of the cosmic microwave background radiation, and the observed evolution of the Universe from its beginning about 14 billion years ago.) Being inconsistent with the competing steady-state theory, it is more compelling than the mere expansion of the Universe.

George Gamow and his students had the right idea when they concluded that the Universe must have been hot at an early stage. Their mistake was in thinking that *all* of the elements (instead of just the lightest ones) were synthesized shortly after the big bang. They didn't know about nucleosynthesis in the cores of stars.

Physicists at CERN are making tiny amounts of the "primordial soup" of free quarks and gluons by smashing nuclei of lead into

After Robert V. Wagoner (Stanford University)

■ **FIGURE 19–21** The horizontal axis shows the current cosmic density of matter. From our knowledge of the approximate rate of expansion of the Universe, we can deduce what the density was long ago. The abundance of deuterium is very sensitive to the density; thus, present-day observations of the relative amount of deuterium tell us the cosmic density, by following the arrow on the graph. (The nucleus of a deuterium atom is called a deuteron. There is a joke that the study of deuterons, and the cosmological importance of studying their formation in the first few minutes after the big bang, is known as "deuteronomy.")

each other at very high energies to make a quark-gluon plasma. They analyze the results with the giant ALICE detector (**A** **L**arge **I**on **C**ollider **E**xperiment; aliceinfo.cern.ch) at the LHC. The experiment tests the idea that quarks and gluons were once freely interacting in such a plasma instead of being trapped in composite particles as they are today.

The approximate observed abundances of the light elements show that the general ideas of primordial nucleosynthesis are correct. Detailed measurements provide some additional constraints; the exact current proportions reflect conditions in the early Universe. For example, deuterium is rapidly destroyed, forming helium. Thus, if the density remained high for a long time (the expansion of the Universe was relatively slow), the current deuterium abundance would be very low. The fact that deuterium is reasonably abundant right now (a few times 10^{-5} of the amount of ordinary hydrogen) tells us that the baryonic component of Ω_M, the ratio of the average matter density to the critical density, is only about 0.05 (5 per cent) (■ Fig. 19–21).

In spite of years of trying, including by one of us (J.M.P.), nobody succeeded in observing the fundamental spin-flip line of

deuterium, the analogue to the 21-cm line of ordinary hydrogen (see our discussion in Chapter 15), until the report of MIT's Alan Rogers and colleagues in 2005. Observing in Massachusetts, they used a field of broadband antennas, linked electronically, to detect deuterium in the direction opposite to that of our Galaxy's center. The background noise in this anticenter direction is so much lower than that toward the Galactic center that the deuterium was detectable, in spite of its being weaker. They found that deuterium nuclei are fewer than hydrogen nuclei by a ratio of 2.3×10^{-5}. The ultraviolet spectral line of deuterium can now be observed in the distant universe, seen in absorption against a background of quasar light.

As mentioned above, the derived value $\Omega_M = 0.05$ from primordial nucleosynthesis considerations refers only to normal, baryonic matter (mostly protons and neutrons). This is in remarkable agreement with values measured with WMAP and Planck.

There might be additional contributions to Ω_M, but they must be "exotic" matter such as neutrinos and "weakly interacting massive particles" (WIMPs). If we believe that $\Omega_M \approx 0.27$, as suggested by studies of clusters of galaxies and the peculiar motions of galaxies (Chapter 18), then a large fraction of the matter in the Universe is not only dark but must also be composed of some kind of exotic matter! This conclusion is consistent with the findings of WMAP and Planck discussed above (Section 19.3c).

19.5 THE INFLATIONARY UNIVERSE

The original, "standard" hot-big-bang theory described in Section 19.4 is remarkably successful. Its four main observational foundations are the existence of the cosmic microwave background radiation, the relative proportions of various isotopes of the lightest elements (hydrogen, helium, and lithium), the expansion of the Universe, and the evolution of the Universe over a finite amount of time.

However, there are some puzzling aspects of the Universe that the original big-bang theory cannot explain, at least not without imposing "initial conditions" that seem unlikely and unjustified. Here we will discuss two of the main problems. We will also present a brilliant possible refinement to the big-bang theory that appears to resolve them by affecting just the first blink-of-an-eye of the Universe's existence.

19.5a Problems with the Original Big-Bang Model

The first apparent difficulty is the great uniformity of the Universe – its observed homogeneity and isotropy. This uniformity is known as the **horizon problem**.

For example, the temperature indicated by the cosmic background radiation is essentially identical (2.728 K) in all directions, excluding the two known effects that produce deviations from isotropy (our Galaxy's motion through the Universe, and early variations in density from which clusters and superclusters of galaxies formed).

How can the Universe be so uniform? Widely separated regions of the Universe could never have been in thermal equilibrium with each other (that is, at the same temperature): No physical signals, not even those travelling at the speed of light, could have crossed such vast distances over the age of the Universe. Thus, distant regions had no way of "telling" each other which temperature to choose – they are beyond each other's "horizon" (and hence the name "horizon problem").

For example, assume the Universe is 14 billion years old, and consider two points on opposite sides of the sky. Cosmic background photons are just now reaching us, after travelling almost the entire age of the Universe (minus the first 400,000 years, when the Universe was still opaque). Clearly, photons from one side have not yet reached the other side; these regions should therefore have different temperatures.

One might suggest that the big bang started as a "point," and hence could have achieved a uniform temperature. But actually, the big bang was never a true "point"; we need a quantum theory of gravity to figure out its initial size. Running the original big-bang equations backward in time, we find that the Universe was always larger than the distance light (or any other signal) could have travelled by that age, even if the Universe was always transparent. Indeed, the number of "causally disconnected" regions was larger in the past, making the observed homogeneity and isotropy (uniformity) even more perplexing. For many years, cosmologists simply assumed, as an initial condition without physical justification, that the Universe originated with a uniform temperature.

The second major problem with the original big-bang theory is that it has no natural way of explaining why the overall geometry of the Universe is so close to being flat (Euclidean). This difficulty is known as the **flatness problem**. Observations tell us that the ratio of matter density to the critical density, Ω_M, is certainly larger than 0.01, just from the luminous matter and a small correction for dark matter in galaxies. But in fact $\Omega_M \approx 0.27$ if dark matter (both baryonic and exotic) is included.

Moreover, as discussed in Chapter 18, observations of distant supernovae show that there is a significant contribution to the total density from the "dark energy" ("vacuum energy") associated with the cosmological constant, so the total Ω is apparently 1. And, as we saw above, recent measurements of the cosmic microwave background radiation show that the overall geometry of the Universe is essentially flat, meaning that Ω_{total} is indeed 1. (Recall that the latest WMAP, Planck, and other data show that $\Omega_{total} = 1.0005 \pm 0.0065$.)

Why should we be surprised that the Universe is so flat right now ($\Omega_{total} \approx 1$)? Well, according to the equations describing the evolution of the Universe in the original big-bang theory, Ω must have been exceedingly close to 1 (that is, exceedingly flat) in the distant past (■ Fig. 19–22). Otherwise, as the Universe aged, Ω would have rapidly deviated from 1 by a very large extent, so we should not be measuring $\Omega \approx 1$ at present.

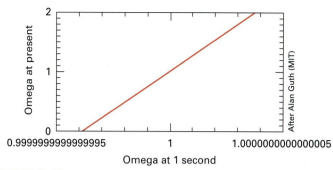

■ **FIGURE 19–22** To achieve a current value close to 1 for the density of the Universe divided by the critical density (that is, $\Omega \approx 1$), the ratio had to be almost exactly 1 to an exceedingly high precision when the Universe was very young (say, 1 second old, as shown here).

It turns out that $\Omega = 1$ is an equilibrium value, but it is unstable in the sense that the slightest deviation from 1 quickly grows into a huge deviation. It is like a pencil standing on its point: The slightest push makes it fall over. You are unlikely to walk into a room to find a pencil standing on its point, unless the pencil was balanced very precisely from the start – and this is unlikely to happen by chance.

On the other hand, if there were some physical reason why the pencil was forced to stand on its point, then you might expect to find it in that position. The original big-bang theory provides no suitable explanation for why the Universe started out with $\Omega = 1$, so for many years cosmologists simply assumed this initial condition; they had no reasonable explanation for it.

19.5b Inflation to the Rescue

In 1980, Alan Guth, now at the Massachusetts Institute of Technology, came up with a brilliant possible solution to these problems. His idea, known as **inflation**, is a modification of the original big-bang theory. Guth's initial proposal had important flaws, which were partially resolved in 1981 by Andrei Linde (then in the Soviet Union, but now at Stanford University), and later also by Paul Steinhardt (now at Princeton University) and Andreas Albrecht (now at the University of California, Davis).

The final improvement was made by Linde in 1983, in his theory of "chaotic inflation." He demonstrated that inflation is a generic process, which may occur even if one abandons the usual assumption about thermal equilibrium in the very early universe. (In 2012, Guth and Linde each received a Fundamental Physics Prize for their extraordinary research; this new award, worth $3 million per recipient, was established by Yuri Milner, a former physics student who later earned billions investing in Internet companies.) Most physicists and astronomers accept some form of inflation, although the proposed details and specific mechanisms currently have much greater variety than (and generally differ from) the theories of a few decades ago. But the earliest example, as formulated by Guth, Linde, Steinhardt, and Albrecht, illuminates many of the essential principles, so we discuss it here. Most physicists and astronomers accept some form of inflation.

■ **FIGURE 19–23** The rate at which the Universe expands, for both inflationary and noninflationary models. The vertical axis shows the radius of the region that expands to become the presently observable part of the Universe, and the horizontal axis shows time. The inflationary period is marked on the graph as a vertical yellow band. Before inflation, the size of the Universe in the inflationary model is much smaller than in the original big-bang model, allowing the observed Universe to come to a uniform temperature in the time available.

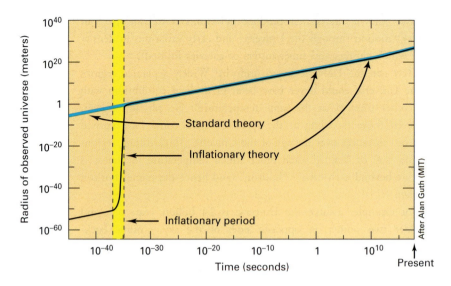

Suppose the Universe started out *much smaller* than the size implied by an extrapolation of the equations of the original big-bang theory. Guth proposed that it might have subsequently "inflated" – expanded *extremely fast* and by an immense factor (■ Fig. 19–23), perhaps 10^{50} or even 10^{100} (a "googol"). ("Googol" is the name of the number 10^{100}, a 1 with 100 zeroes after it. The popular search engine "Google" was named after this number, but the creators of the search engine intentionally misspelled it because they believed that most people would naturally spell it this way rather than the correct way.) The Universe achieved this huge expansion by doubling in size every tiny fraction of a second for a significant amount of time, contrary to the constant or decelerating expansion predicted in the original big-bang theory.

An intuitive feel for such "exponential" growth can be obtained by envisioning a chessboard. If there is one grain of wheat on the first square, and the number of grains in each successive square is doubled, the wheat on the 64th square (2^{63} grains) would be the size of a mountain that covers Manhattan!

Notice that, before inflating, the tiny initial Universe could have achieved a uniform temperature (thermal equilibrium). Unlike the case in the original big-bang theory, it was sufficiently small that signals could travel across it. Specifically, collisions between particles and photons would have distributed the heat uniformly. But the Universe would have retained its uniformity as it inflated, because there is no physical mechanism by which it could have lost this property. With inflation, we expect the currently observable Universe to be homogeneous and isotropic, eliminating the horizon (uniformity) problem. (For numerical examples, see *Figure It Out 19.1: Inflation of the Early Universe.*)

Moreover, inflation would have flattened out any initial curvature in the Universe, like a balloon expanding to immense size. In effect, the Universe became so large that the entire region we can see with our most powerful telescopes (which is only a minuscule fraction, perhaps 10^{-40} in radius, of what really exists) appears flat or nearly flat, just as the Earth's surface appears flat when viewed over small distances. Inflation therefore resolves the flatness problem.

After the end of inflation, the Universe would have continued expanding at a rate consistent with the original big-bang theory.

So, inflation affected only the first brief moments in the life of the Universe. It is not a replacement for the big-bang theory, but rather a modification (or refinement) of it. In fact, inflation can be thought of as the mechanism that "launched" the big bang!

In the inflationary universe, space itself expands far faster than the speed of light. After all, the Universe was initially much smaller than a subatomic particle, yet it quickly inflated to an almost uncountable number of light-years across! Einstein's theory of relativity permits space to expand faster than the speed of light; it only restricts the transmission of information *through* space to a speed no greater than that of light. But the expansion of space cannot be used to transmit *information* from one object to another elsewhere. Put another way, relativity states that two objects cannot *pass* each other at speeds exceeding that of light. But the expansion of space itself does not allow particles to pass each other.

19.5c   Forces in the Universe

Why should the Universe have experienced a period of inflationary expansion? Physicists are now considering a variety of possible reasons, some quite far removed from those advocated by the early pioneers of inflation. But the originally favored reason remains instructive to study. It stems from theories that attempt to unify the known fundamental forces of nature.

For much of the 20th century, physicists thought that there are four fundamental forces in nature. The first force, **gravity**, is extremely weak, but it acts over large distances, and all types of matter (and energy) are affected in the same way, regardless of charge or other properties. We are already very familiar with gravity, since it has played a prominent role in many topics discussed in this book.

Electromagnetism, the second force, is about 10^{39} times stronger than gravity, and it acts over large distances, but only charged or magnetized objects are affected. (This particular strength comparison uses an electron and a proton in a hydrogen atom. The exact numbers are different if two protons or two electrons are used.) Since most large

Let us consider a numerical example to clarify the concept of inflation in the early Universe. At time $t = 10^{-37}$ s, the radius of the *currently* observable part of the Universe (at $t \approx 14$ billion years) was about 0.01 m, according to the original big-bang theory. Even if the early Universe were transparent, light could have travelled only

$$d = ct = (3 \times 10^8 \text{ m/s})(10^{-37} \text{ s}) = 3 \times 10^{-29} \text{ m}$$

in that amount of time. (The actual distance is somewhat greater, because of the expansion of space itself. Typically it might be a factor of 3 larger – but in this case still only about 10^{-28} m.) This distance is far too small, compared with the radius of the Universe at that time. Thermal equilibrium could not have been established, even with light signals.

Now suppose the radius of the currently observable Universe was actually only 10^{-32} m at $t = 10^{-37}$ s. This radius was sufficiently small that the transmission of signals (even slow ones such as particle collisions) *could* have easily produced thermal equilibrium.

If the Universe subsequently inflated by a factor of 10^{31} by $t = 10^{-35}$ s, then its new radius would be about 0.1 m, in agreement with the original big-bang theory; see Figure 19–21. Thereafter, the

radius of the currently observable part of the Universe grew at a rate consistent with the original big-bang theory. Inflation affects *only* the first roughly 10^{-35} s of the Universe's existence.

Here is another example, this time focusing on the entire Universe (not just the currently observable part). Suppose the Universe had a radius of 10^{-30} m at $t = 10^{-37}$ s. If it inflated by a factor of 10^{100} by $t = 10^{-35}$ s, its new radius at the end of inflation would be

$$10^{-30} \text{ m} \times 10^{100} = 10^{70} \text{ m},$$

or about 10^{54} light-years! But the currently observable part of the Universe had a radius of about 0.1 m at $t = 10^{-35}$ s, and the radius is now about 10^{26} m (or 10^{10} light-years). Both of these radii are vastly smaller than the total size of the Universe after inflation.

The times discussed here, and especially the initial sizes and the inflation factors, are somewhat arbitrary; they depend on the details of the specific inflationary model under consideration. (In some versions of inflation, the Universe inflates by a factor of 10 to the *trillionth* power!) The important point is that inflation produces a truly *enormous* Universe, far larger than we can see. This means that the geometry of the Universe should appear flat ($\Omega = 1$), as observed.

objects are electrically neutral and not magnetized, electromagnetic forces are generally not felt over large distances. Electromagnetism is critical, however, to the structure of liquids and solids such as your body, and of course to atoms and molecules themselves.

The third force, known as the **strong nuclear force**, is about 100 times stronger than electromagnetism, but it operates only over an extremely short range. The protons and neutrons in an atomic nucleus are held together by the strong force. Actually, we now know that the "strong force" is just the residue of a more fundamental (and even stronger) "color force" that binds quarks together. However, for historical purposes we will still refer to it as the strong force.

Finally, the **weak nuclear force** is about a million times weaker than electromagnetism, and it operates only over an extremely short range. It is the force that governs certain types of radioactive decay, such as the neutron becoming a proton, an electron, and an antineutrino. Any time neutrinos or antineutrinos interact with matter, the weak force is involved (■ Fig. 19–24).

In the late 1970s, theorists suggested that the electromagnetic and weak forces are different manifestations of a single more fundamental force, the **electroweak force**. The electromagnetic and weak forces seem different at the low energies characteristic of the Universe today. But at higher energies and temperatures (such as when the Universe was younger than about 10^{-12} s, and hotter than about 10^{16} K), they behave in a similar manner. We say that the "symmetry is broken" at low energies. In high-energy interactions, the two forces cannot be distinguished and hence appear **symmetric** (unified). A useful analogy is a coin flicked along the surface of a table: When it is

spinning rapidly at high energies, it appears symmetric, but we see either "heads" or "tails" when it comes to rest, a "broken symmetry."

High-energy experiments at the CERN particle accelerator in Switzerland, at the Stanford Linear Accelerator Center in California, and elsewhere subsequently confirmed the predictions of the electroweak theory. Specifically, for example, physicists found the expected "W" and "Z" particles in 1983. Thus, we now often say that there are only three fundamental forces. James Clerk Maxwell achieved a similar "unification" in the 1860s, when he showed that electricity and magnetism are different manifestations of a more fundamental interaction, electromagnetism. (He was arguably one of the three greatest physicists of all time, the list commonly including Newton, Einstein, and Maxwell, not necessarily in that order. Incidentally, the original family name was "Clerk," pronounced "Clark," but his father took on "Maxwell" as an additional surname after inheriting a country estate from his Maxwell ancestors.)

Physicists are now trying to see whether the three known forces are different low-energy manifestations of only one or two fundamental forces. **Grand unified theories** (GUTs) attempt to unify the electroweak and strong nuclear forces. In principle, these forces behave in a similar ("symmetric") manner at temperatures above about 10^{29} K (which characterized the Universe at times less than 10^{-37} s).

GUTs predict that the big bang produced slightly more matter than antimatter; all of the antimatter subsequently annihilated with matter, leaving behind the matter we see today. GUTs also predict that protons are unstable, though proton decay has not yet been observed (see Figure 19–24 again). The experimental minimum value of about 10^{33}

■ **FIGURE 19–24** This cavity, deep underground in a salt mine near Cleveland, was filled with 10,000 tons of water to make the Irvine-Michigan-Brookhaven (IMB) detector. The primary purpose was to confirm the hypothesized decay of protons to more elementary particles. To detect flashes of light that result as particles pass through it, 2,400 light-detectors were installed around the sides. Its detection of a few neutrinos from Supernova 1987A confirmed that massive stars can undergo core collapse to form neutron stars. No supernova neutrinos have been detected since then.

Joe Stancampiano/National Geographic Stock

years for the proton lifetime already rules out some candidate GUTs, but many still remain, and we don't know which (if any) is correct.

"**T**heories **of e**verything" (TOEs), though still rather speculative, attempt to unify the grand unified force and gravity in a single super-unified force. Physicists conjecture that these forces are symmetric at temperatures above about 10^{32} K, which characterized the Universe at times less than about 10^{-43} s (the "Planck time").

To understand the physics of this very early era in cosmic history, we need a unification of quantum mechanics and general relativity, the two pillars of 20th century physics – that is, a quantum theory of gravity. No fully self-consistent theory has yet emerged, but many theorists are focusing their attention on **superstring theories** (often called simply string theories). Modern string theory was born with the work of John Schwarz (Caltech) and Michael Green (then of Queen Mary College and now Lucasian Professor of Mathematics at Cambridge University, the academic chair once held by Newton), and one of the current leaders in this quest is Edward Witten (Institute for Advanced Study, Princeton).

In superstring theories, each fundamental particle (quark, electron, neutrino) is postulated to be a different vibration mode of a tiny, one-dimensional energy packet called a "string" (or perhaps a tiny "membrane," having two or more dimensions). A string is an elongated, one-dimensional analogue to the zero-dimensional points that have often been used in theories of matter. Strings are so small (about 10^{-33} cm) that they seem almost like points. Superstring theories thus far appear to be a promising way to get around theoretical difficulties on the extremely small scale where a quantum theory of gravity is needed.

An informal summary of the situation has been put into verse:

A Brief History of Gravity (A Series of Limericks by Bruce Elliot)

It filled Galileo with mirth
To watch his two rocks fall to Earth.
He gladly proclaimed,
"Their rates are the same,

And quite independent of girth!"
Then Newton announced in due course
His own law of gravity's force:
"It goes, I declare,
As the inverted square
Of the distance from object to source."
But remarkably, Einstein's equation
Succeeds to describe gravitation
As space-time that's curved,
And it's this that will serve
As the planets' unique motivation.
Yet the end of the story's not written;
By a new way of thinking we're smitten.
We twist and we turn,
Attempting to learn
The Superstring Theory of Witten!

Recent superstring theories tend to favor an 11-dimensional Universe (revised up from 10 dimensions), including the time dimension, but 7 of the dimensions are too small to see; they are curled up on scales of about 10^{-33} cm, the Planck length (■ Fig. 19–25). A useful analogy is to view a sheet of sandpaper from far above: it looks only two-dimensional, but is really three-dimensional because of the granularity. Another example is a garden hose: From far away, it appears one-dimensional (like a string) instead of three-dimensional. Nobody knows why the Universe has only four macroscopically large dimensions (usually called x, y, z, and t; three spatial dimensions and one time dimension).

Steinhardt, one of the originators of inflation, now has a different take on the subject. He envisions a periodic type of cosmology, an alternative to the big bang and inflation, that he calls "ekpyrotic." In this model, quite unpopular among cosmologists, our Universe "remembers" some events that happened before the big bang, but there is no known mechanism for this. Steinhardt tries to formulate some quantum relation to a theory of gravity. He needs the

■ FIGURE 19–25 Theorists calculate that superstrings require an 11-dimensional space. We are familiar with only a 4-dimensional space, having 3 spatial dimensions and 1 time dimension.

existence of extra dimensions, and perhaps the collision of "branes," a multidimensional generalization of what we are familiar with as a two-dimensional membrane (perhaps curved into a third dimension). The branes, our visible one and another, would collide in a non-visible fifth dimension. (The name "ekpyrotic" comes from the Greek stoic philosophers' model in which the Universe is consumed by fire many times over and regenerated.) Steinhardt's model makes predictions for gravitational waves and the polarization of the background radiation that differ from those of standard big-bang cosmology, so it will be tested when gravitational wave detectors find gravitational waves and Planck scientists report on the polarization they measure.

19.5d Supercooling the Universe

Right after the big bang, when temperatures were exceedingly high, superstring theories suggest that there should have been just one symmetric, superunified force.

A sphere of liquid water in a vacuum, floating in space, looks the same when viewed from different directions, so physicists say it is "rotationally symmetric." In contrast, an ice crystal has a preferred axis (the axis of symmetry), and hence does not appear the same when viewed from different directions. Physicists say that ice is less symmetric than liquid water when they use the word "symmetry" in this manner.

Given the uncertainty over what happens before the Planck time (10^{-43} s), we will instead consider the situation when the Universe was about 10^{-37} s old, focusing on GUTs. The grand unified theory (GUT) force and gravity would have existed as distinguishable forces.

At $t \approx 10^{-37}$ s, the temperature would have dropped to 10^{29} K, below which the GUT force should have started to break symmetry, becoming the strong nuclear force and the electroweak force. This

■ FIGURE 19–26 Ice (solid water) and liquid water are different states of matter. The phase change between them is, in some ways, analogous to the phase change that must have taken place early in the history of the Universe.

phase transition is similar to what happens when liquid water is cooled below a temperature of 0°C: It generally turns to ice (■ Fig. 19–26). When water turns to ice, the freezing process liberates excess energy so that the water molecules can line up to form crystals.

Suppose, however, that the Universe cooled below 10^{29} K *without* breaking symmetry – that is, the strong and electroweak forces remained unified as the GUT force.

This **supercooled** Universe would have contained too much energy relative to what it should have had at that temperature. For example, water can sometimes be supercooled below 0°C without turning into ice, if one is very careful not to disturb the water in any way. Its excess energy keeps it in the liquid state. The excess energy is released when the water freezes.

The energy associated with the supercooled, symmetric (GUT) Universe would have caused space to double its size every tiny fraction of a second – that is, the Universe would have entered an epoch of accelerated, inflationary expansion. (This is where the analogy with water breaks down: The excess energy associated with the supercooled phase does not make the water expand exponentially!) For reasons that are not intuitively obvious, the Universe's excess energy had a stretching effect on space, like a cosmic "antigravity." Physicists call this excess energy the "energy of the false vacuum" because a conventional vacuum should have zero energy. It is essentially Einstein's cosmological constant Lambda (Λ; see Chapter 18), but in this case very early in the history of the Universe. The inflationary expansion made the Universe truly *enormous*; it is much, much larger than the regions that we can actually see.

Eventually, perhaps at an age of 10^{-35} s, the Universe experienced a phase transition, with the electroweak and strong forces separating (that is, the symmetry was broken). At this time all of the excess energy of the false vacuum was released, thereby forming matter, antimatter, and photons. The energy of the false vacuum thus provided all of the energy in the Universe.

With no more "antigravity" force, the inflationary expansion of the Universe subsided, and the regular Hubble expansion began. The Universe started "coasting," but gravity caused the expansion to decelerate. In essence, inflation is what gave the Universe its "big bang" – its initial kick up to a time of about 10^{-35} s, after which all of the events discussed previously in this chapter unfolded.

19.5e Successes of Inflation

Despite its abstract and speculative nature, the achievements of the inflationary theory are magnificent. First, it explains the horizon problem – the great uniformity (homogeneity and isotropy) of the Universe. The Universe was sufficiently small to have achieved thermal equilibrium (the same temperature) at very early times, and it maintained this as it inflated.

Second, the Universe is very close to having a spatially flat geometry. After inflation, it was so large that any observable region would appear flat (that is, $\Omega_{total} = 1$; the average total density is equal to the critical density).

Third, the observable Universe has few, if any, "magnetic monopoles" (isolated north or south magnetic poles, without the opposite pole being present as well); none has been found despite intensive searches for them. There are also no apparent "cosmic strings" – long, skinny regions that contain huge amounts of mass. Many such "defects" should have been produced by phase transitions soon after the big bang, but the rapid expansion of the Universe during the inflationary era would have whisked them away, thereby accounting for their scarcity.

Fourth, the predictions of inflation are consistent with the WMAP and Planck measurements of tiny temperature variations in the young universe (Figures 19–10 and ACL 19.1–c).

Perhaps the most spectacular success of inflation is its prediction of the minuscule spatial temperature variations measured by WMAP and Planck. This prediction, which is shown by the green line in Figure ACL 12.1–c, is in excellent agreement with Planck results shown there by red dots. These temperature variations represent tiny quantum fluctuations, which were stretched by inflation to exponentially large size. According to inflationary theory, these quantum fluctuations are ultimately responsible for the creation of all galaxies in our Universe.

Advanced versions of inflationary theory, such as the theory of "chaotic inflation," do not require supercooling or phase transitions. Inflation may occur during a slow evolution of various "scalar fields," similar to the famous Higgs field that has an associated Higgs particle, discovered in 2012 at the LHC.

19.5f The Ultimate Free Lunch?

In the inflationary theory, matter, antimatter, and photons were produced by the energy of the false vacuum, which was released following the phase transition. All of these particles consist of positive energy. This energy, however, is exactly balanced by the *negative* gravitational energy of everything pulling on everything else. In other words, the *total* energy of the Universe is zero! It is remarkable that

the Universe might consist of essentially nothing, but (fortunately for us!) in positive and negative parts.

You can easily see that gravity is associated with negative energy: If you drop a ball from rest (defined to be a state of zero energy), it gains energy of motion (kinetic energy) as it falls. But this gain is exactly balanced by a larger negative gravitational energy as the ball comes closer to the Earth's center, so the sum of the two energies remains zero.

The idea of a zero-energy Universe, together with inflation, suggests that all one needs is just a tiny bit of energy to get the whole thing started (that is, a tiny volume of energy in which inflation can begin). The Universe then experiences inflationary expansion, but without creating net energy.

What produced the energy before inflation? This is perhaps the ultimate question. As crazy as it might seem, the energy may have come out of *nothing*! The meaning of "nothing" is somewhat ambiguous here. It might be the vacuum in some preexisting space and time, or it could be nothing at all – that is, all concepts of space and time were created with the Universe itself.

Quantum theory provides a natural explanation for how that energy may have come out of nothing (see *Figure It Out 19.2: Heisenberg's Uncertainty Principle*). As mentioned in Chapter 18, it turns out that particles and antiparticles spontaneously form and quickly annihilate each other, microscopically violating the law of energy conservation (■ Fig. 19–27). (Alternatively, if the energy of the particles and antiparticles is "borrowed" from the vacuum, briefly creating a negative-energy "hole," then energy is conserved at all times.)

These spontaneous births and deaths of so-called virtual-particle pairs are known as "quantum fluctuations." Indeed, laboratory experiments have proven that quantum fluctuations occur everywhere, all the time. Virtual particle pairs (such as electrons and positrons) directly affect the energy levels of atoms, and the predicted energy levels disagree with the experimentally measured levels unless quantum fluctuations are taken into account.

Perhaps many quantum fluctuations occurred before the birth of our Universe. Most of them quickly disappeared. But one of them lived sufficiently long and had the right conditions for inflation to have been initiated. Thereafter, the original tiny volume inflated by an enormous factor, and our macroscopic Universe was born. The original particle-antiparticle pair (or pairs) may have subsequently

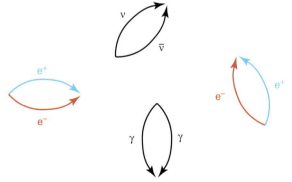

■ **FIGURE 19–27** Formation of virtual pairs of particles and antiparticles, quickly followed by their mutual destruction. According to quantum theory, such spontaneous births and deaths ("quantum fluctuations") can and do exist, even in a vacuum. Their effects have been experimentally verified.

FIGURE IT OUT 19.2   Heisenberg's Uncertainty Principle

A key feature of quantum theory is the Heisenberg Uncertainty Principle, which implies that one cannot know the exact energy at each point in space in a finite amount of time. The energy at each point must therefore be randomly fluctuating; otherwise, we would know that its value is zero. The creation and annihilation of "virtual" pairs of particles and antiparticles, even in a complete vacuum, is the manifestation of such quantum energy fluctuations.

Quantitatively, virtual particles of energy ΔE spontaneously form and live for a time Δt if the product $\Delta E \Delta t$ is no larger than about h/2π, where h is Planck's constant:

$$\Delta E \Delta t \leq h/2\pi.$$

In a sense, space itself is uncertain of its energy by an amount up to ΔE, if the timescale over which it is measured is less than or equal to Δt, where the product of the two does not exceed h/2π.

Thus, the smaller the energy (ΔE) of the virtual particles that are spontaneously created, the larger the time interval (Δt) over which they can exist. For example, a virtual proton–antiproton pair, which has a mass of about 3.3×10^{-24} g and a corresponding energy given by $E = mc^2$, can live up to about 3×10^{-25} s. But a virtual electron–positron pair can exist much longer, since its mass is nearly 2000 times less than that of a virtual proton–antiproton pair.

annihilated each other, disappearing from the Universe. But even if they didn't, the violation of energy conservation would be minuscule, not large enough to be measurable (and perhaps even zero, in the case of a pair borrowing energy from the vacuum).

If this admittedly speculative hypothesis is indeed correct, then the answer to the ultimate question is, in the words of Alan Guth, that the Universe is "the ultimate free lunch"! (The Nobel Prize–winning economist Milton Friedman, who asserted that "there is no such thing as a free lunch," would surely be surprised. Friedman was academically noted for Friedman and Anna Schwartz's *A Monetary History of the United States, 1867–1960*; he published a book in 1975 called *There's No Such Thing as a Free Lunch*. Previously the science-fiction author Robert Heinlein had popularized a similar phrase, "There ain't no such thing as a free lunch," often referred to by the acronym "TANSTAAFL," in his 1966 novel, *The Moon Is a Harsh Mistress*. But apparently, the phrase was in common usage before either Friedman or Heinlein.)

The Universe came from nothing, and its total energy is zero, but it nevertheless has incredible structure and complexity. Physicists are attempting to come up with ways to test this idea in principle, but no method has emerged yet; thus, in some respects it is not currently within the realm of science, which requires hypotheses to be experimentally or observationally testable.

19.6 A UNIVERSE OF UNIVERSES

Is our Universe the only universe, or could it be one of many? The entire history of our Universe might be just one episode in the much grander **multiverse** consisting of many (perhaps infinitely many) universes. This idea is the most extreme extension of the Copernican principle. Some scientists, notably the celebrated British cosmologist Lord Rees of Ludlow (Martin Rees), the English Astronomer Royal, find the multiverse idea quite attractive. Others, however, remain skeptical. The concept is certainly quite controversial at this time,

and more speculative than even the "free lunch" hypothesis discussed above.

The concept of another "universe" is difficult to define. It could be a disjoint region that occupies dimensions separate from (and physically inaccessible to) those of our Universe. This other universe might, however, be connected with ours via a wormhole – a passage between two black holes (see our discussion in Chapter 14). Another possibility is a region in our physical Universe, or arising from our Universe, in which inflation occurred or ended in a different way.

In some or all cases, the physical constants (or, more speculatively, perhaps even the laws of physics) in other universes may differ from those in our Universe. By "physical constants" we mean the relative strengths of the various forces, the values of quantities such as the speed of light, and the masses of fundamental particles such as electrons and neutrinos.

As summarized below, there are scientifically valid reasons to think that other universes might exist. But by its very nature, this interesting conclusion removes itself from the realm of science, at least temporarily, because we don't know of a way to contact other universes and perform experiments on them. In other words, the hypothesis generally seems to be untestable.

One exception would be a different region in our Universe that eventually becomes observable when photons reach us, far in the future. Another possibility is the collision of a different universe with ours, which could produce a recognizable imprint on the cosmic microwave background radiation. (A few astrophysicists claimed to see possible evidence of this sort in the WMAP map, but most are very skeptical. Just how some "anomalies" apparent in the Planck map pan out remains to be seen.) Also, consider sending light signals through a wormhole: It is conceivable that a future generation will discover matter with antigravity properties (on small scales) and will transport it into the wormhole, in this way perhaps keeping it open. Finally, there may be properties of our Universe that depend on the existence of other universes; thus, our descendants might eventually

■ **FIGURE 19–28** Many separate universes hypothetically arising from quantum fluctuations in a grander hyperspace, the "multiverse." The different colors of these "bubble universes" are meant to illustrate that the values of their physical constants (or perhaps even their physical laws) could differ.

discover some method of testing for their presence. But currently, at least, we can only speculate about the multiverse.

There seem to be at least three or four ways to produce other universes. For example, distinct quantum fluctuations could arise out of "nothing," some of which give rise to a universe (■ Fig. 19–28). This "nothing" might be the vacuum of our Universe, or the "nothing" outside our Universe, in some sort of a larger "hyperspace." If a quantum fluctuation out of "nothing" created our Universe, it seems reasonable that such a process may have occurred many times (thereby making the hyperspace a multiverse), perhaps even infinitely many. We do not yet know whether this is a viable process. To find out, we need a fully self-consistent quantum theory of gravity that unites relativity with quantum physics, and theorists are currently quite excited by the potential of superstring theories.

Other universes could also be regions of different broken symmetry in an inflating volume. As the false vacuum inflates, the phase transition (symmetry breaking) may occur in different ways in different regions (■ Fig. 19–29). This is analogous to water freezing along different axes in a jar: A jumble of ice crystals forms, with various orientations. The values of the physical constants could differ in those regions where symmetry breaking had occurred differently. (More speculatively, the physical laws themselves might differ in them.) These regions could remain separated from each other because of the presence of a still-inflating false vacuum between them.

A popular related scenario, promoted largely by Andrei Linde, is that quantum effects in an inflating volume spawn new, distinct, inflating regions within it. Although most of the volume will have undergone a phase transition (symmetry breaking) and stopped inflating, some regions will continue to inflate. These inflating regions will subsequently spawn other regions that continue to inflate, even as symmetry is broken throughout most of their volume (■ Fig. 19–30). The process never ends, so it is called "eternal inflation." Again, the values of the physical constants may differ in the different regions of inflation. As Linde said more than 30 years ago, extending the "ultimate free lunch" sentence by Guth, "The inflationary universe is the only lunch at which all possible dishes are available." This idea became especially popular after 2003, when it was found that string theory may describe a multiverse consisting of perhaps 10^{500} different types of inflationary universes permanently producing each other.

Yet another idea is that a new universe can sprout from within any black hole in our Universe. However, this hypothesis is very speculative, and we know of no way to test for the presence of the other universe.

19.7 A UNIVERSE FINELY TUNED FOR LIFE?

We will end this chapter on an intriguing but very controversial note, one with which some scientists vigorously disagree. It is at the extreme boundary of what can currently be called science, and may be more suitable to theology, philosophy, or metaphysics. But it is worth bringing up for completeness; it is based on the theory of the inflationary multiverse and it addresses our most basic origins.

Our existence may allow us to draw some interesting conclusions about the Universe – for example, that it contains elements heavier

■ **FIGURE 19–29** Regions of different broken symmetry in an inflating volume; each might have a different set of values for the physical constants.

■ **FIGURE 19–30** A series of inflating universes, each giving rise to new regions of inflation. The different colors represent changes in the values of physical constants (or, more speculatively, perhaps even different physical laws) among the universes.

Courtesy of Andre Linde (Stanford University)

■ **FIGURE 19–31** A straight flush, which in the game of poker is the best hand possible (the Royal flush, Ace high, being the top, special case of the straight flush).

than hydrogen and helium. Such deductions utilize the **anthropic principle**: We exist, hence the Universe must have certain properties or we wouldn't be here to see it.

In a few cases, apparently "mysterious" relationships between numbers were explained with anthropic reasoning. A previously unknown property of the energy levels of the carbon nucleus was actually *predicted* by Fred Hoyle in this manner. He argued that it must exist – otherwise there would not be enough carbon in the Universe.

The values of the physical constants (not to mention the laws of physics) seem to be spectacularly "fine-tuned" for life as we know it – indeed, almost "tailor-made" for humans. In many cases, if things were altered just a tiny amount, the results would be disastrous for life, and even for the production of heavy elements or molecules. The Universe would be "stillborn" in terms of its complexity. Here are a few examples.

If the weak nuclear force were weaker, most neutrons would not have turned into protons shortly after the big bang. The protons and neutrons would have then fused to form helium, leaving little hydrogen. A similar conclusion is reached if the weak nuclear force were much stronger: Protons would quickly combine with each other to form helium nuclei, converting two of the four protons to neutrons.

Now suppose the strong nuclear force were weaker. It turns out that the elements heavier than hydrogen would be unstable. The periodic table of the elements would be very boring, and stars would not produce nuclear energy. Conversely, if the strong nuclear force were stronger, protons would bind together so easily that hydrogen would not exist.

Or, if the mass of a neutron and a proton were more nearly equal (they are the same to within about 0.14 per cent), most protons would be converted to neutrons. Without hydrogen, stars would live only a short time, and there would be no hydrogen-rich compounds such as water and hydrocarbons.

Additional examples abound. If gravity were considerably stronger relative to electromagnetism, stars would be more massive, and they would burn out too quickly for very complex life (as we know it) to develop. Also, if certain relationships between the energy levels of different kinds of atomic nuclei were not present, the heavy elements

would not have formed; the case of carbon, predicted by Hoyle, was mentioned above. If electrons were not much less massive than protons, complex chemistry would not be possible. Large molecules such as DNA can have well-defined shapes only if the nuclei of atoms are much heavier than electrons.

Some people use these and other "cosmic coincidences" to argue for a divine Creator who specifically designed the Universe in this manner. But this theological conclusion is not testable by the methods of science. Therefore, here we will not consider it further; we are sticking to science in this book.

Or, one could say that there is only a single Universe, and it has these properties by chance. This conclusion is difficult to accept; the odds are astronomically low (pun intended), if most other possible combinations of physical constants lead to conditions unsuitable for life. However, it may be true.

Finally, one could say that there is an ensemble of universes, perhaps even infinitely many, which span a very wide range of properties (values of the physical constants, particle masses, etc.). We, of course, were dealt a "winning hand" – indeed, something like a straight flush (■ Fig. 19–31): We live in a universe that allowed life and sentient beings to develop. Such universes might be exceedingly rare; most random hands are not winners under the rules of poker.

Note that ours is not necessarily the *only* interesting universe, or the *most* interesting possible universe. But we argue that a far larger number of *uninteresting* possibilities exist.

Though an excellent example of the Copernican principle applied to the grandest imaginable scale, the multiverse idea may not be directly testable (but some strong indirect tests might be found). Thus, some people argue that it is more in the realm of philosophy and theology than of science. It seems that we are still left with the ultimate question: How and why did the Universe (or the hyperspace that contains it) occur? Perhaps there is no way, even in principle, for humans to answer this question.

CONCEPT REVIEW

» The expansion of the Universe alone suggests, but does not demand, that there was a big bang (Section 19.1). According to the alternative **steady-state theory**, the Universe had no beginning in time and will have no end. This idea is based on the **perfect cosmological principle**: The average properties of the Universe do not change at all with time.

» Historically, there were several observational arguments against the steady-state theory, but the fatal blow was the discovery of the **cosmic microwave radiation**, a faint radio glow that uniformly pervades the Universe (Section 19.2a). It was produced when the Universe was hot, before an age of about 400,000 years, and now corresponds to a very low temperature because of the cooling produced by expansion (Section 19.2b).

» In the 1990s, tiny ripples were detected in the cosmic background radiation, corresponding to the seeds from which superclusters and clusters of galaxies were formed (Section 19.3a). The observed angular size of the most obvious variations indicates that the Universe is spatially flat on large scales (Section 19.3b). These ripples were recently mapped in great detail (Section 19.3c and ACL 19.1). They were used to show that Hubble's constant is 71 km/s/Mpc, and that normal matter, exotic "cold dark matter," and "dark energy" constitute about 5 per cent, 26 per cent, and 69 per cent of the total energy density of the Universe, respectively.

» One can consider the Universe when it was very young; it consisted of uniformly distributed gas and radiation at the same temperature, governed by just a few fundamental laws (Section 19.4a). Specifically, we discussed the first several minutes after the big bang, starting from an almost inconceivable age of 10^{-43} s, the **Planck time** (Section 19.4b). Up to an age of about 1 microsecond, the Universe consisted of a primordial soup of **quarks**, gluons (which normally bind quarks together), and photons. A slight dominance of matter over antimatter was established, making possible our existence.

» The lightest elements were formed in the first 10 minutes through the process of **primordial nucleosynthesis** (Section 19.4c), and their currently observed proportions show striking agreement with the predictions of the big-bang theory. The relative amounts of different isotopes, combined with other observations, tell us that much of the dark matter in the Universe must consist of exotic particles such as neutrinos.

» Some observed aspects of the Universe, such as its incredible uniformity (the **horizon problem**) and nearly flat spatial geometry (the **flatness problem**), are difficult to understand in the context of the original big-bang theory (Section 19.5a). It has been conjectured that the Universe began much smaller than we had thought, was able to achieve a uniform temperature, and subsequently went through an era of extremely rapid expansion known as **inflation** (Section 19.5b). The resulting Universe would then be truly enormous, much larger than what we can see — and our relatively small, observed region would naturally appear to be homogeneous and flat. Inflation affects just the first tiny fraction of a second of the Universe's existence, yet it removes some of the arbitrary assumptions of the original big-bang theory.

» Although many alternative possibilities are now being considered, a clue to why inflation occurred may be provided by theories that postulate a high-energy merging of at least three of the four fundamental forces of nature: **gravity**, **electromagnetism**, **strong nuclear force**, and **weak nuclear force** (Section 19.5c). Already physicists have shown that electromagnetism and the weak nuclear force are unified at high energies to form the **electroweak force**; the two forces cannot be distinguished and hence appear **symmetric** (unified). At still higher energies, the electroweak and strong nuclear forces might merge, according to **grand unified theories (GUTs)**. It is possible that even gravity can be unified with the grand unified force. Perhaps the best candidates for such speculative "theories of everything" (TOEs) are **superstring** (string) **theories**, which postulate that all fundamental particles are different forms of vibration of a tiny string or membrane.

» As the Universe expanded and cooled, the various forces should have become apparent as different manifestations of the same force — that is, **phase transitions** should have occurred, like water turning to ice. But if the Universe **supercooled** without undergoing a phase transition, it would have contained excess energy that caused inflation for a short time (Section 19.5d). Inflation solves the horizon and flatness problems; moreover, it explains the absence of "magnetic monopoles," and its predictions are consistent with the observed distribution of matter on various scales (Section 19.5e). It is speculated that, before inflation, the Universe could have begun as a quantum fluctuation out of nothing, the "ultimate free lunch" (Section 19.5f).

» We end with the still more speculative possibility that there are many, perhaps even infinitely many, universes in a **multiverse** (Section 19.6). The values of the physical constants, or even the physical laws, might be different in universes other than our own. According to the **anthropic principle**, our existence allows us to deduce some properties of the Universe (Section 19.7). Many calculations suggest that the consequences of modified constants would be disastrous for life as we know it, or even for the production of heavy elements or molecules. Perhaps our Universe is simply one of few in the multiverse with conditions suitable for the formation of complexity, culminating in life.

QUESTIONS

1. Explain how, in the absence of other information, the expansion of the Universe could be consistent with no beginning and no end in time: a "steady-state universe."

2. Summarize some observational arguments against the steady-state theory.

3. Describe the origin of the cosmic background radiation and the reason it now corresponds to a very low temperature.

4. Would you believe the announcement of the discovery of a galaxy at redshift $z = 10{,}000$? Why or why not?

5. Explain why the cosmic background radiation looks slightly hotter in one direction of the sky, and slightly cooler in the opposite direction.

6. Discuss the significance of tiny variations ("ripples") in the cosmic background radiation.

7. Explain why we can consider with some confidence the behavior of the Universe early in its history, even before it was 1 second old.

8. Discuss the significance of the matter–antimatter asymmetry in the Universe.

9. Summarize the main physical events that occurred in the Universe's first 10 minutes of existence.

10. Describe how the currently observed proportions of the lightest chemical elements reflect conditions in the Universe shortly after its birth.

11. What are two observational problems with the original big-bang theory of the Universe?

12. Explain how the main problems with the original big-bang theory can be resolved if, at very early times, the Universe was much smaller than we had thought and subsequently inflated by an enormous factor.

13. Explain how space itself can expand faster than light without violating Einstein's theory of relativity.

14. Describe the fundamental forces of nature and attempts to unify them.

15. Discuss what should have happened to the "grand unified force" as the Universe expanded and cooled at an age of about 10^{-37} s.

16. Outline how the Universe may have supercooled, leading to inflationary expansion by an almost arbitrarily large factor when the Universe was only 10^{-35} s old.

†17. Suppose that, during inflation, the size-doubling timescale was only 10^{-38} s. If inflation started at $t = 10^{-37}$ s and continued until $t = 10^{-35}$ s, show that about 1000 doublings occurred.

18. Explain how the total energy of the Universe might be zero, or nearly zero.

19. Why has it been suggested that the Universe may be the "ultimate free lunch"?

20. Define what is meant by the term "universe."

21. Describe several ways in which there might be other universes besides our own.

22. Outline the basic idea behind superstring theories and the possible existence of dimensions far too small to detect.

23. Discuss how the properties of the Universe depend on the values of the physical constants.

24. Summarize various possibilities for why our Universe seems so finely tuned for life.

25. State the anthropic principle and its implications.

26. **True or false?** Sensitive observations have not revealed any variations in the temperature of the cosmic microwave background radiation, leading to major conflicts with theories of galaxy formation.

27. **True or false?** The expansion of the Universe is consistent with Hoyle's "steady-state theory" of the Universe.

28. **True or false?** It is quite possible that the Universe started from essentially zero energy and has remained at essentially zero total energy ever since.

29. **True or false?** The most highly redshifted electromagnetic radiation we have detected so far is from quasars or very distant galaxies.

30. **True or false?** The force involved in chemical reactions is the strong nuclear force.

31. **True or false?** Right after photons formed as a result of the final annihilation of protons, neutrons, and their antiparticles, the Universe became transparent to light.

32. **True or false?** According to quantum theory, particle-antiparticle pairs having positive total energy can spontaneously appear out of nothing, if they exist for only a very short time before destroying themselves.

33. **Multiple choice:** Which one of the following is *not* a success of the original, pre-inflation big-bang theory? **(a)** It gives a good estimate of the age of the Universe. **(b)** It is consistent with the observed evolution of galaxies. **(c)** It explains the large-scale uniformity of the temperature of the Universe. **(d)** It explains the existence of the cosmic microwave background radiation. **(e)** It is consistent with the expansion of the Universe.

34. **Multiple choice:** Which one of the following statements about the cosmic background radiation (CBR) is *false*? **(a)** The CBR shows that the early Universe was highly (but not perfectly) homogeneous and isotropic. **(b)** There are small variations in the CBR that correspond to slight differences in density from which galaxies, superclusters, and voids eventually formed. **(c)** We can deduce the overall geometry of the Universe from studies of the CBR. **(d)** The CBR was produced by the first few generations of stars in the early Universe. **(e)** The CBR is inconsistent with the steady-state theory of the Universe.

35. **Multiple choice:** Which one of the following was *not* a result derived in part from the Wilkinson Microwave Anisotropy Probe (WMAP) satellite? **(a)** We live in a spatially flat, or nearly flat, universe. **(b)** Most of the dark matter consists of massive compact halo objects (MACHOs). **(c)** The current age of the Universe is about 13.7 billion years. **(d)** Only about 4.5 per cent of the contents of the Universe is in the form of normal matter consisting of protons, neutrons, and electrons. **(e)** The contents of the Universe are currently dominated by dark energy.

36. **Multiple choice:** There are good reasons to believe that at one time the Universe expanded extremely rapidly. This "inflation" occurred primarily **(a)** within the first 10^{-30} seconds after the Universe was created; **(b)** when the Universe was about 10^{-6} seconds old; **(c)** when the Universe was about 1 second old; **(d)** when the Universe was about 1 million years old; or **(e)** when the Universe was about 10 billion years old.

37. **Multiple choice:** Which one of the following prevented photons from travelling long distances in the early Universe? **(a)** Before inflation, the Universe was small enough for the photon to wrap back around to its source. **(b)** Free electrons easily interacted with the photons, scattering them. **(c)** Neutron decay absorbed most photons. **(d)** The large expansion of the early Universe redshifted the photons to slow speeds. **(e)** The dark energy absorbed the photons.

38. **Multiple choice:** Which one of the following statements about the inflationary theory is *false*? **(a)** The Universe may have inflated because of a strange vacuum energy associated with a "supercooled" state of the Universe. **(b)** The Universe is uniform (i.e., homogeneous and isotropic) on large scales because long ago, before inflation, parts of the Universe that are currently very distant were able to come into equilibrium. **(c)** The Universe started off in a very hot, dense state and has inflated exponentially ever since. **(d)** According to the inflationary theory, the Universe is very much larger than the parts that we currently see. **(e)** During inflation, the speed of the expansion of space can exceed the speed of light.

39. **Multiple choice:** Which one of the following statements about the early universe is *true*? **(a)** There were many forces in the past, which eventually unified, leaving four fundamental forces. **(b)** The Universe started off very large and has remained the same average density because of the random creation of small amounts of matter over time. **(c)** When matter and antimatter combined, a small excess of matter was left over, which later formed stars and galaxies. **(d)** The creation of matter left very little room in the Universe, and space was forced to expand as a result. **(e)** Small amounts of heavy elements, including carbon and oxygen, were formed within a few minutes after the big bang.

40. **Multiple choice:** The first helium nuclei in the Universe were synthesized **(a)** during the inflationary epoch; **(b)** by the first generation of stars; **(c)** immediately after the Universe became transparent to radiation; **(d)** when the Universe was roughly 10^{-6} seconds old; or **(e)** when the Universe was a few minutes old.

41. **Fill in the blank:** According to the _____, we can deduce certain properties of the Universe based on our mere existence.

42. **Fill in the blank:** The cosmic microwave background radiation was released when _____ and _____ combined together for the first time.

43. **Fill in the blank:** The lightest elements formed shortly after the big bang in a fusion process known as _____.

44. **Fill in the blank:** Recent measurements have shown that the ratio of the actual density of the Universe to the critical density is _____, meaning that over very large distances the shape of the Universe is ____.

45. **Fill in the blank:** Rather than being fundamental, protons and neutrons consist of smaller particles called _____, which themselves might be tiny, vibrating packets of energy known as _____.

†This question requires a numerical solution.

TOPICS FOR DISCUSSION

1. Does it even make sense to include times before $t = 0$ in a scientific discussion, given that science is supposed to deal with predictions that are testable in principle? Explain.

2. What do you think of the argument that the Universe has a total energy of zero, and that it arose from a quantum fluctuation out of "nothing"?

3. Do you think some forms of "life" could arise in most universes, almost regardless of the specific values of the physical laws and constants? Explain.

Curiosity, part of Mars Science Laboratory, on the surface of Mars, where it landed in 2012. It has instruments to look for signs that the conditions for life on Mars were met eons ago.
self-portrait was stitched together from 130 individual images, with enough selection that the ones showing the boom holding the cameras that made the portrait could be omitted. Th
image was taken in the Yellowknife Bay region of Gale Crater on Mars, where Curiosity first drilled into a martian rock.

NASA/JPL-Caltech/MSSS, panorama by Andrew Bodrov

20

Life in the Universe

ORIGINS

One of the most important questions we can pursue in our quest to determine our origin is whether life, and especially intelligent, communicating life, has arisen independently elsewhere in the Universe. Are creatures like humans a common phenomenon, or are we rare or even alone?

We have discussed the eight planets, Pluto and the other plutoids, asteroids, and some of the moons in the Solar System, and have found most of them to be places that seem hostile to terrestrial life forms. Yet a few locations besides the Earth – most notably Mars, with its signs of ancient running water, and Europa and Enceladus (see Figure 7–25), with liquid water below their icy crusts – have characteristics that suggest life may have existed there in the past, or might even be present now or develop in the future. **Astrobiology** or **exobiology** is the study of life elsewhere than Earth. It has also been suggested that "astrobiology" or "exobiology" is merely a NASA term, and that the field should more correctly be called **bioastronomy** (a part of astronomy) rather than "astrobiology," which would arguably be part of biology.

In our first real attempt to search for life on another planet, in the 1970s, NASA's Viking landers carried out biological and chemical experiments with martian soil. The results seemed to show that there is probably no life on Mars (■ Fig. 20–1). A small, inadequately funded British probe, Beagle 2 (the first Beagle having been Darwin's ship), failed as it approached Mars in 2003. (One reason it became so popular among the public may have been that it was supposed to send back tones from the musical group Blur when it landed, though those tones never came.)

AIMS

1. Understand the nature of life and conditions necessary for its emergence (Section 20.1).

2. Identify the kinds of planets and stars that might be most suitable for the development of life and intelligence (Sections 20.2 and 20.3).

3. Explore ways of searching for, and communicating with, extraterrestrial life (Sections 20.4 and 20.5).

4. Assess the probability that intelligent, communicating life exists elsewhere in our Galaxy besides the Earth (Section 20.6).

5. Discuss UFOs and how they relate to the scientific method (Section 20.7).

NASA/JPL/Caltech

■ **FIGURE 20–1** The surface of Mars from a Viking Lander, which carried a small biology laboratory to search for signs of life. It probably didn't find any. We now still look for life on Mars but are also considering Europa and Enceladus as other possible abodes.

Jay M. Pasachoff

■ **FIGURE 20–2** The Mars Science Laboratory, nicknamed Curiosity, landed on Mars in 2012 for the first major search for life there since the 1976 arrivals of the Viking Landers. A full-size model appears at left for comparison with models of earlier Mars landers at a meeting of the Division of Planetary Sciences of the American Astronomical Society.

NASA continues to explore Mars with robotic spacecraft and one day should have a more sophisticated biology lab landing on it. In the meantime, its Mars Exploration Rovers (Spirit and Opportunity) and the European Space Agency's Mars Express have found clear signs that water flowed on Mars in the distant past, raising hope that life could have formed there at that time and might even have survived. NASA's Mars Science Laboratory's rover, Curiosity, is extending the search for conditions favorable for life on Mars (■ Fig. 20–2).

Although studies of a martian meteorite in 1996 gave some indications of ancient, primitive life on Mars (see Chapter 6), this idea has not been generally accepted, though it is still the subject of much discussion. Jupiter's moon Europa and Saturn's moons Titan and Enceladus (see Chapter 7) are also intriguing places where scientists think it is possible that life has begun.

Since it seems reasonable that life as we know it would be on planetary bodies, we first discuss the chances of life arising elsewhere in the Solar System, as well as the kinds of stars most likely to have planets suitable for the emergence and development of intelligent life. Next we explore attempts to receive communication signals from intelligent extraterrestrials.

We also consider a way to estimate the number of communicating civilizations elsewhere in our Milky Way Galaxy, or at least to see which factors most seriously limit our ability to do so. It is possible that humans are the only technologically advanced civilization in our Galaxy, or one of very few. Finally, we explain why most scientists do not consider reported sightings of UFOs to be good evidence for extraterrestrial visitations to Earth.

NASA has formed an institute of "astrobiology," with scientists at institutions across the country and around the world, and is making a major push to investigate matters of biology that can be important to understanding the origin of life or to space exploration. The institute is "virtual," in that it has no actual buildings but rather is made of individuals who communicate by e-mail and by occasional meetings.

20.1 THE ORIGIN OF LIFE

It would be very helpful if we could state a clear, concise definition of life, but unfortunately that is not yet possible. Biologists state several criteria that are ordinarily satisfied by life forms – reproduction, for example. Still, there exist forms on the fringes of life, such as viruses, that need a host organism in order to reproduce. Scientists cannot always agree whether some of these things are "alive."

In science fiction, authors sometimes conceive of beings that show such signs of life as the capability for intelligent thought, even though the being may share few of the other criteria that we ordinarily recognize. In Fred Hoyle's novel *The Black Cloud*, for instance, an interstellar cloud of gas and dust is as alive as (and smarter than!) human beings. But we can make no concrete deductions if we allow such wild possibilities, so exobiologists prefer to limit the definition of life to forms that are more like "life as we know it."

Life on Earth is based on **amino acids** – chains of carbon, in which each carbon atom is bonded to hydrogen and sometimes to oxygen and nitrogen. Chemically, carbon is essentially unique in its ability to form such long chains; indeed, we speak of compounds that contain carbon as being **organic**. But life is selective, incorporating only about 20 of all the possible amino acids. Similarly, long chains of amino acids form **proteins**, though life utilizes only a small fraction of the multitude of possible combinations of amino acids.

The genetic code of any living creature is contained in one extremely long and complex structure: DNA (**d**eoxyribo**n**ucleic **a**cid), the famous "double helix." This magnificent molecule, consisting of two chains of nucleotides joined by base pairs (adenine and thymine; cytosine and guanine), is the key to life as we know it.

How hard is it to build up long organic chains? To the surprise of many, an experiment performed in the 1950s showed that making organic molecules is easier than had been supposed. Stanley Miller and Harold Urey, at the University of Chicago, filled a glass jar with simple molecules like water vapor (H_2O), methane (CH_4), and ammonia (NH_3), along with hydrogen gas (H_2). They exposed it to electric sparks, simulating the vigorous lightning that may have existed in the stages soon after Earth's formation.

After a few days, long chains of atoms formed in the jar, in some cases complex enough to include simple amino acids, the building blocks of life. Later versions of these experiments created even more complex organic molecules from a wide variety of simple actions on simple molecules. Such molecules may have mixed in the oceans to become a "primordial soup" of organic molecules (■ Fig. 20–3).

Since the original experiment of Miller and Urey, most scientists have come to think that the Earth's primitive atmosphere was not made of methane and ammonia, which would have disappeared soon after the Earth's formation. Instead, it may have consisted mostly of carbon dioxide, carbon monoxide, and water, and such a mixture does not generally lead to a large abundance of amino acids.

On the other hand, complex molecules may have formed near geothermal sources of energy under the oceans or on Earth's surface, or perhaps in vents deep underground, where the right raw materials

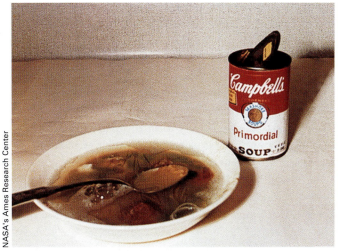

■ FIGURE 20–3 The simple solution of organic material in the oceans, from which life may have arisen, is informally known as "primordial soup."

■ FIGURE 20–4 Close-up view of a part of the Murchison meteorite, in which simple amino acids have been found. Analysis shows that they are truly extraterrestrial and the sample had not simply been contaminated on Earth.

existed. Also, extraterrestrial amino acids have been found in several meteorites that had been long frozen in Antarctic ice, as well as in some other meteorites (■ Fig. 20–4). In any case, some exoplanets or moons in our Galaxy may have had primitive atmospheres similar to the mixture used by Miller and Urey, so the results of their experiments are interesting.

Also relevant are laboratory observations with conditions mimicking the low density (strong vacuum) of space. Amino acids have been observed to form.

However, mere amino acids or even DNA molecules are not life itself. A jar containing a mixture of all the atoms that are in a human being is not the same as a human being. This is a vital gap in the chain; astronomers certainly are not yet qualified to say what supplies the "spark" of life.

Still, many astronomers think that since it is not difficult to form complex molecules, primitive life may well have arisen not only on the Earth but also in other locations. The appearance of very simple organisms in Earth rocks that are 3.5 billion years old, and indirect evidence for life as far back as 3.8 billion years (not long after the end of the bombardment suffered by the newly formed Earth), suggests that primitive life arises quite easily. Similarly, the presence of life in what appear to be very harsh environments on Earth (water that is highly acidic or near its boiling point, for example) shows that life can exist in extreme conditions.

The discovery of indigenous life on at least one other planet or moon in the Solar System would provide much support for the hypothesis that simple organisms such as microbes and bacteria form readily. But even if life is not found elsewhere in the Solar System, there are so many other stars and planets in space – hundreds of billions in our Galaxy alone – that it would seem that life might have arisen at some location.

■ **FIGURE 20–5** The martian meteorite ALH 84001, in which it has been controversially suggested that tiny "nano-fossils" have been found, as shown in Figures 6–56 and 6–57. The conclusion is not widely accepted by other scientists working in this field.

20.2 LIFE IN THE SOLAR SYSTEM

Life elsewhere in our planetary system, if present at all, is primitive at best (single cells, or perhaps very simple multi-cellular organisms). Among all of the planets and moons, only a few have nonzero odds for life.

Mars has provided the best evidence thus far, but it is still very controversial (■ Fig. 20–5). Moreover, even if real, life on Mars might not have been independent of Earth – a meteorite from Mars containing simple life may have contaminated the young Earth, "seeding" it with life, though this idea is of course quite speculative.

Europa, one of Jupiter's large moons, looks like a promising environment: Below its icy surface there is almost certainly water slush or even an ocean. NASA is contributing an instrument and other hardware to ESA's Jupiter Icy Moons Explorer (JUICE), scheduled for a 2022 launch to study Ganymede, Callisto, and Europa.

Titan, Saturn's largest moon, has a thick atmosphere of nitrogen molecules. As we saw, in 2004, the Cassini spacecraft sent the Huygens probe into this atmosphere. There is evidence for substantial methane and ethane on Titan, perhaps in the form of lakes; methane can be solid, liquid, or gas at temperatures reasonable for Titan, just as water can have solid, liquid, or gaseous phases on Earth. Huygens, as it approached Titan's surface, imaged what appeared to be a lake shore. The Cassini orbiter's radar has found a reflection that seems lake-like, presumably with the liquid being a tarry substance. The absence of liquid water, however, makes it more difficult for life as we know it to form.

Io, another Galilean satellite, has conditions that may be suitable for life resembling that found near volcanic vents on Earth. Again, however, the apparent absence of water is a major problem.

A surprise major candidate for life in our Solar System is Saturn's moon Enceladus (see Figure 7–25). Its water jets reveal the presence of an ocean below; the Cassini spacecraft was even able to fly through the jets and measure their constituents in detail.

Note that an intelligent alien who obtains and correctly interprets a spectrum of Earth could deduce the probable presence of life here. The large amount of free oxygen suggests the continuous production by a process like photosynthesis; otherwise, oxygen would rapidly be depleted because it is so reactive. In addition, methane quickly reacts with oxygen, so the significant amount of methane (largely from cows – "bovine flatulence," to quote the late Carl Sagan) in our atmosphere implies a steady production mechanism – the decay of organic compounds. Such signs have been picked up from spacecraft, such as Galileo and Cassini, while they were near the Earth and headed for the outer planets.

20.3 SUITABLE STARS FOR INTELLIGENT LIFE

If we seek indigenous (that is, originating locally, not from elsewhere), intelligent life on planets orbiting other stars, what kinds of stars have the best odds? Stars that are either near the beginning or end of their lives are not very good bets. For example, intelligent life may take a long time to form, so very young stars are less suitable than older stars. Stars heading to the red giant or white dwarf stages go through rapid changes, making it difficult for complex life to survive. White dwarfs and neutron stars certainly have passed through stages that would have destroyed life.

Some types of main-sequence stars are also not especially suitable. The lives of O-type and B-type stars are probably too short for the development of life of any kind. Planets around A-type stars might have life, but probably not intelligent life. M-type and L-type stars have a small **habitable zone** (also known as the ecosphere) – the range of distances in which conditions suitable for life might be found. There

Single stars, stars in very wide binary systems (with planets orbiting close to one star), or closely spaced binary stars (which planets could orbit at large distances) are most suitable. Planets that move from one star to another in a binary system (for example, like a "figure eight") tend to have unstable orbits and are ejected.

Over 1,000 exoplanets have already been found, with more than 2,000 additional good candidates (see our discussion in Chapter 9). In most cases they orbit F, G, and K stars because the searches specifically targeted those stars. Generally these planets are gas giants, in many cases very close to the star or in highly eccentric orbits and hence probably inhospitable to intelligent life. But increasingly, smaller and rocky ones are being found. A few of these planets appear to have orbital properties potentially suitable for the emergence of life; perhaps at least simple life exists within their atmosphere or on moons orbiting them. We are finding closer and closer analogues to Earth, which encourages those trying to discover signs of life on exoplanets.

■ **FIGURE 20–6** Goldilocks, at the three bears' house (seen here as gingerbread covered with icing and candies), found some porridge too hot, some porridge too cold, and some porridge just right. We now look for just-right conditions for life ("Goldilocks conditions") on planets in habitable zones around exoplanets.

is unlikely to be a planet in this narrow region around any specific low-mass star. On the other hand, the enormous number of such stars in our Galaxy suggests that there are many planets in the habitable zones around low-mass stars. The origin of life on their surfaces might be impeded because of frequent flares of high-energy radiation, though one side would be shielded by their synchronous rotation.

Main-sequence stars of types F, G, and K are the most likely candidates to have suitable planets in habitable zones, on a "per star" basis. These stars live a long time, and models of their habitable zone lead to reasonably large sizes. Such models must explain the "Goldilocks effect" (Fig. 20–6) – in the case of the Solar System, why Venus is too hot, Mars is too cold, but the Earth is just right. The Kepler spacecraft is finding suitable stars with exoplanets in or close to the habitable zones, providing excitement for those interested in evaluating the chances for life on exoplanets.

20.4 THE SEARCH FOR EXTRATERRESTRIAL INTELLIGENCE

How should we look for intelligent extraterrestrial life? Perhaps we can find evidence for such life here on Earth, in the form of alien spacecraft that have landed here. After all, Pioneers 10 and 11 and Voyagers 1 and 2 are even now carrying messages out of the Solar System in case an alien interstellar traveller should happen to encounter these spacecraft (■ Fig. 20–7).

The odds, however, seem very small, given the vastness of space. Although spaceships can, in principle, travel between the stars, even

■ **FIGURE 20–7** (a) The gold-plated copper record bearing two hours' worth of Earth sounds and a coded form of photographs, carried by Voyagers 1 and 2. The sounds include a car, a steamboat, a train, a rainstorm, a rocket blastoff, a baby crying, animals in the jungle, and greetings in various languages. Musical selections include Bach, Beethoven, rock, jazz, and folk music. (b) The record includes 116 photographs. Among them is this view, which one of the authors (J.M.P.) took when in Australia for a solar eclipse. It shows Heron Island on the Great Barrier Reef in Australia, in order to illustrate an island, an ocean, waves, a beach, and signs of life.

FIGURE IT OUT 20.1 — Interstellar Travel and Einstein's Relativity

Interstellar travel is, in principle, possible (but extremely difficult!) for humans to achieve. If the speed is constant, distance is equal to speed multiplied by time,

$$d = vt,$$

and we can calculate that a rocket moving with the Earth's escape velocity (11 km/s) would reach the bright star Sirius (8.7 light-years away) in about 240,000 years. (Recall from Chapter 14 that the escape velocity is the speed needed at liftoff to escape from an object's surface.)

One can imagine a huge spaceship containing a community of explorers that has permanently left Earth. They would have to be self-sufficient (for example, grow their own food), and the journey to Sirius would span roughly 10,000 generations. To be sure, humans would need to overcome enormous obstacles to successfully complete such a trip, but it is not categorically impossible.

But what if you want to make the journey within a *single* human lifetime? It is possible, according to Einstein's special theory of relativity. The key fact is that time slows down if you move rapidly: A given time interval t_{Earth} (according to a clock on Earth) is reduced to

$$t_{\text{Travel}} = t_{\text{Earth}} \times [1 - (v/c)^2]^{1/2}$$

for the traveller, where c is the speed of light. For example, at a speed of $v = 0.995c$, the round-trip travel time to Sirius according to an observer on Earth is 17.5 years, but only 1.75 years according to the traveller. (Here we ignore the additional time it takes to accelerate up to that speed, but the general idea still holds.)

At higher speeds, the traveller ages even less: If $v = 0.9999c$, 17.4 years will have passed on Earth, but the journey is only 3 months long according to the traveller's clock! Thus, by moving fast one ages less and effectively "jumps" into the future. However, the actual life span according to the traveller's clock is unchanged: In 3 months of elapsed time, for example, the traveller reads far fewer books than the earthling does in the corresponding 17.4 years.

How can this be, according to the traveller's frame? Well, he sees the Earth zooming away from him, and Sirius is rapidly approaching – but special relativity tells us that lengths along the direction of motion contract. The distance between Earth and Sirius is reduced to

$$L_{\text{Travel}} = L_{\text{Earth}} \times [1 - (v/c)^2]^{1/2}$$

for the traveller, where L_{Earth} is the length measured by a person on Earth. At $v = 0.995c$, the round-trip distance is only

$$(17.4 \text{ lt-yr})(0.1) = 1.74 \text{ lt-yr}$$

according to the traveller, and this distance is traversed in only 1.75 years at $v = 0.995c$. At $v = 0.9999c$, the round-trip distance is only 3 light-months, which is traversed in just 3 months. Thus, at high speeds, times dilate and lengths contract. Everything is self-consistent, and rapid interstellar travel is possible in principle.

The overwhelming *practical* problem is that travel close to the speed of light requires astronomically large (pun intended) amounts of energy, if the rocket's mass is non-negligible. According to special relativity, the total energy E of the rocket is given by

$$E = m_0 c^2 / [1 - (v/c)^2]^{1/2},$$

where m_0 is the rest mass. For $v = 0.995c$, $E \approx 10 m_0 c^2$ – that is, 10 times its rest mass; for $v = 0.9999c$, $E \approx 71 m_0 c^2$. These are the energies that must be expended to accelerate to such speeds, assuming 100 per cent efficiency! To give you some feeling for this quantity of energy, note that the most powerful bombs ever built by humans release the energy equivalent of only about 2 kilograms of material – that is, $E = m_0 c^2$, where $m_0 = 2$ kg. Rocket ships are far more massive, and we require 10 or 71 times their rest-mass energy in the above two examples. But the situation is actually much worse: If the efficiency of energy conversion is only 1 per cent (the maximum possible for nuclear energy), then in reality the necessary values are $E \approx 1,000 m_0 c^2$ or $7,100 m_0 c^2$. These are truly *staggering* amounts of energy!

When one also considers that the rockets sent to the Moon had about 1,000 times as much mass as their useful payload, the energy requirements become even *more* prohibitive, since the rest mass m_0 is so large. It would probably be best to gather the fuel along the way, from interstellar gas. Maybe some highly advanced aliens have overcome these and other formidable problems, such as reaching very high speeds in relatively small amounts of time (which requires large accelerations – and these are harmful to life).

---

rather quickly according to Einstein's special theory of relativity (see *Figure It Out 20.1: Interstellar Travel and Einstein's Relativity*), such journeys are difficult, expensive, dangerous, and probably rare compared to communication with electromagnetic waves.

A potentially more fruitful approach is to search for electromagnetic signals, but some waves are more suitable than others. For example, x-ray and gamma-ray photons have high energy and are

therefore expensive to produce, and typical atmospheric gases block them. Ultraviolet photons are absorbed by interstellar dust in the plane of our Milky Way Galaxy. And, at infrared wavelengths, Earth's warm atmosphere makes the sky bright. At optical wavelengths, the signal from a planet orbiting a star is very difficult to distinguish from the bright light of the star itself, unless a large amount of energy is concentrated into a narrow range of wavelengths (as in a laser) or into a short pulse. During the past few years, searches for special, narrow optical emission lines have been made in high-resolution optical spectra of many nearby stars, but without success.

For many years, radio waves have seemed to be the best choice: They are easy and cheap to produce, and are not generally absorbed by interstellar matter. Also, there are few sources of contamination – although the increasing number and strength of radio and television stations is a threat to radio astronomers in the same manner that city lights brighten the sky at optical wavelengths.

At least initially, it seems too overwhelming a task to listen for signals at all radio frequencies in all directions and at all times. One must make some reasonable guesses on how to proceed.

A few frequencies in the radio spectrum seem especially fundamental, such as the 21-cm line of neutral hydrogen (as described in Chapter 15). This wavelength corresponds to 1,420 MHz, a frequency over ten times higher than stations at the high end of the normal FM band. We might conclude that creatures on a far-off planet would decide that we would be most likely to listen near this frequency because it is so fundamental. The "water hole," the wavelength range between the radio spectral lines of H and OH, has a minimum of radio noise from background celestial sources, the telescope's receiver, and the Earth's atmosphere, and so is another favored possibility.

Humans have conducted several searches for extraterrestrial radio signals. No unambiguous evidence has been found, but the quest is worthwhile and continues. A telltale signal might be an "unnatural" set of repeating digits, such as the first 100 digits of the irrational number π (the Greek letter "pi"). We must verify that the signal could not be produced by a natural, non-intelligent source (such as a pulsar), and also that it is not of human origin (either unintentionally or intentionally).

Though there is agreement to verify the veracity of the supposition that a received signal is indeed a message or a sign of extraterrestrial life, and to do so before making a public announcement, the discovery of such an extraterrestrial signal would be so momentous that it would surely be released to the public without major delay. A protocol has been accepted by researchers in the field as to how to announce any believable signal from extraterrestrials, so as not to cause fear and panic.

In 1960, Frank Drake conducted the first serious search for extraterrestrial signals. He used a telescope at the National Radio Astronomy Observatory (in West Virginia) for a few months to listen for signals from two of the nearest stars – tau Ceti and epsilon Eridani. He was searching for any abnormal kind of signal – a sharp burst of energy, for example. He called this investigation Project Ozma after the queen of the land of Oz (■ Fig. 20–8) in L. Frank Baum's stories.

A NASA-sponsored group based at the Search for Extraterrestrial Intelligence (SETI) Institute in California began an ambitious effort on October 12, 1992, the 500th anniversary of Columbus's landing in the New World. It made use of sophisticated signal-processing capabilities of powerful computers to search millions of radio channels simultaneously in the microwave region of the spectrum (■ Fig. 20–9). Consisting of both a sky survey and a targeted study of

■ **FIGURE 20–8** Dorothy and Ozma visit the Nome King in *Ozma of Oz* (1907).

■ **FIGURE 20–9** This "waterfall" display from Project Phoenix shows the instantaneous power in 462 narrow channels, each only 2 Hz wide. Each dot represents a channel, and the size of the dot is proportional to the strength of the signal. As a new spectrum is plotted in the top row across the screen, the older spectra move downward. Here, the signal changes steadily in frequency, so it slants on the screen. It is readily distinguishable from the background noise.

Photo by Seth Shostak/SETI Institute

■ **FIGURE 20–10** Jill Tarter, Director of Project Phoenix at the SETI Institute.

Jay M. Pasachoff

■ **FIGURE 20–11** The Arecibo radio telescope in Puerto Rico, used once to send a message into space. The telescope is 305 meters across, the largest telescope on Earth; you can see it in the movie *Contact*. It is sensitive enough to pick up a hypothetical cellular telephone signal on Venus.

individual stars, in its first fraction of a second it surpassed the entire Project Ozma. But Congress cut off funds anyway, after about a year.

The targeted search of the project, Project Phoenix, lasted a decade, backed by private funds. It was led by astrophysicist Jill Tarter (■ Fig. 20–10), a real-life model for actor Jodie Foster's character Ellie Arroway in the movie *Contact*. It examined about 1,000 stars; no unexplained signals were found, though there were some exciting false alarms. The Institute is now using the Allen Telescope Array (Figure 20–17) to survey so-called "habitable star systems," the planet candidate systems found so far by NASA's Kepler spacecraft, and a small region near the Galactic center.

A well-known project is "SETI@home," whose operation is based at the University of California, Berkeley (see http://seti.berkeley.edu). It has established a way for the general public to contribute: During otherwise unused time on your home computer, a special program can automatically analyze data from the giant Arecibo radio telescope in Puerto Rico (■ Fig. 20–11) for signs of extraterrestrial signals (■ Fig. 20–12). Thus, there is a very small but nonzero chance that the first unambiguous evidence of intelligent life elsewhere in the Universe would be found by your computer, should you choose to participate!

The SETI@home effort involves 155,000 active participants and 235,000 computers in 200 countries, effectively forming the Earth's largest supercomputer; millions of computer-years have been contributed. The scientists conducting the project have taken the most suspicious signals from the widespread data analysis and gone back to Arecibo to observe those sources in detail. Obviously, no confirmation was obtained.

The same Berkeley team runs SERENDIP, an acronym for "Search for Extraterrestrial Radio Emissions from Nearby Developed Intelligent Populations." It uses an antenna at the giant Arecibo radio telescope to scan for SETI signals while the telescope is doing its other tasks. It has lower sensitivity but a wider range of radio frequencies than SETI@home. ("Serendipity" means, essentially, finding something by accident while looking for something else. It comes from a 16th-century Italian story, perhaps after an earlier Persian original, about three princes from a place called Serendippo. There is a related Talmudic story in Jewish literature.)

Although most searches for extraterrestrials have been conducted at radio wavelengths, some optical and near-infrared searches are also under way. Because the plane of the Milky Way Galaxy absorbs visible light, we can't expect to survey as many stars as at radio wavelengths,

SETI@home, UC Berkeley

■ **FIGURE 20–12** Output on the screen of a computer that has been used to analyze some of the radio data collected by the "SETI@home" project. There have been no clear signs of intelligent life among the data processed so far.

but there are still plenty of them. If we search for short, very intense pulses of light emitted by lasers, we can actually see quite far in the plane of our Galaxy, increasing the number of available stars. More importantly, the laser pulses outshine the light from the star that a planet is orbiting. Finally, such laser pulses are very difficult or impossible to produce by any natural phenomena other than life; detection of them could provide strong evidence for the existence of extraterrestrials. One such optical search was with a Harvard telescope in interior Massachusetts, sponsored by The Planetary Society. Another such search is the Berkeley group's SEVENDIP, "Search for Extraterrestrial Visible Emissions from Nearby Developed Intelligent Populations."

Only some scientists think that there is a reasonable chance of detecting signals from extraterrestrial beings, at least in the near future. But it would be a shame if there were abundant signals that we missed just because we weren't looking. Of course, now that we have looked and listened for a while, that possibility seems to be somewhat reduced. There are other reasons to think that there may not be any extraterrestrials at all in our neighborhood within the Milky Way Galaxy, or perhaps even in our entire Galaxy. For example, if intelligent life is reasonably common, the odds are that it would have arisen elsewhere than on Earth first, so why aren't the aliens already here? Maybe they aren't here because they are very rare or don't exist.

Even if we don't get messages from outer space, there are many scientific spinoffs of the search. Investigating thousands of stars in detail, and mapping the sky in different parts of the spectrum, gives us bits of information that can lead to scientific breakthroughs of other kinds.

20.5 COMMUNICATING WITH EXTRATERRESTRIALS

All of the searches described above are passive – astronomers are simply looking for signals from intelligent, communicating extraterrestrials. If such a signal is ever found and confirmed by several cross-checks, we might choose to "reply" – but not until some global consensus has been reached about who will speak for Earth and what they will say. Nevertheless, we have already intentionally sent our own signals toward very distant stars that may or may not be orbited by planets containing intelligent life.

For example, on November 16, 1974, astronomers used the giant Arecibo radio telescope in Puerto Rico to send a coded message about people on Earth (■ Fig. 20–13) toward the globular star cluster M13 (■ Fig. 20–14) in the constellation Hercules. The idea was that the presence of over 200,000 closely packed stars in that location would increase the chances of our signal being received by a civilization in the planetary system of one of them.

But the travel time of the message (at the speed of light) to M13 is about 24,000 years, so we certainly cannot expect to have

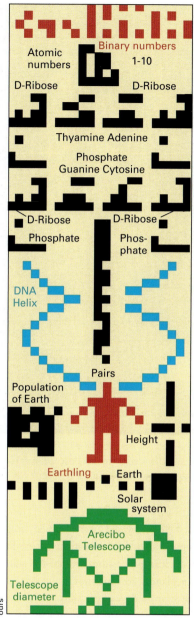

■ **FIGURE 20–13** The message sent to the globular cluster M13, plotted out and with a translation into English added. The basic binary-system count at upper right is provided with a position-marking square below each number. The message was sent as a string of 1679 consecutive characters, in 73 groups of 23 characters each (two prime numbers). There were two kinds of characters (0 and 1), each represented by a frequency; they are shown here as yellow background or other colors, respectively. If interpreted correctly, the diagram provides information about humans (DNA structure, population, etc.).

an answer before twice 24,000, or 48,000 years, have passed. If any creature is observing the Sun at the right frequency (2,380 MHz) when the signal arrives, the radio brightness of the Sun (which is at essentially the same location as Earth, when seen from a great distance) will increase by 10 million times over a 3-minute period. A similar signal, if received from a distant star, could be the giveaway that there is intelligent life there.

In retrospect, M13 may not have been the best choice as a target, despite its large number of stars. The problem is that the stars in

Globular Cluster M13
HST ACS/WFC WFPC2

20 light-years
6.1 parsecs   2′44″
WFPC2 F814W *I*
WFPC2 F785LP *I*
WFPC2 F555W *V*
ACS/WFC F814W *I*
ACS/WFC F606W *V*
ACS/WFC F435W *B*

N

E

Ground · NOAO

NASA, ESA, Z. Levay (STScI), and the Hubble Heritage Team (STScI/AURA) and NOAO

■ **FIGURE 20–14** The globular star cluster M13, toward which a 3-minute message was sent with the Arecibo radio telescope in 1974. We see both a ground-based image and a Hubble Space Telescope clearer view of the globular cluster's central portion.

globular clusters, being very old, were formed from gas that did not have a large proportion of heavy elements; it had not gone through many stages of nuclear processing by massive stars and supernovae. Rocky, Earth-like planets are thus not as likely to have formed there as they would have around younger stars.

A quarter century later, in 1999, a new message was sent by Canadian astronomers toward several relatively nearby (50 to 70 light-years away) Sun-like stars, including 16 Cygni B, which is known through Doppler measurements to have at least one planet orbiting it (see our discussion in Chapter 9). The complete message, which is much larger in size, duration, and scope than the one sent in 1974, was transmitted three times over a 3-hour period in the direction of each star. Still later, starting on July 5, 2003, some much more complicated messages were sent out toward five nearby stars (32 to 46 light-years away) from a 70-m radio telescope in Evpatoriya, Ukraine. One would hope to get answers in fewer than 100 years instead of tens of thousands, but nobody is betting on a return message.

Even had we not sent these few directed messages, during much of the 20th century (and so far in the 21st century) we have been unintentionally transmitting radio signals into space on the normal broadcast channels. Aliens within a few tens of light-years could listen to our radio and television programs. A wave bearing the voice

of Winston Churchill is expanding into space, and at present is about 70 light-years from Earth. And once a week or so a new episode of *American Idol* is carried into the depths of the Universe. Do you think aliens would get a favorable impression of us from most of what they hear?

20.6 THE STATISTICS OF INTELLIGENT EXTRATERRESTRIAL LIFE

Is it reasonable to expect that there are any signals out there that we can hope to detect with projects such as that at the SETI Institute? What if no intelligent creatures exist elsewhere in our Milky Way Galaxy or even in the observable parts of the Universe?

20.6a The Drake Equation

Instead of phrasing an all-or-nothing question about extraterrestrial life, we can use a procedure developed by Frank Drake, then of Cornell University, and extended by Carl Sagan and Joseph Shklovskii, among others. The overall problem is broken down into a chain of simpler questions, the results of which are multiplied together in what is called the **Drake equation** to give the final answer. The formulation is unusual for an equation because it is really a guide to thought rather than a real way of calculating; there is no right answer or unique solution to the Drake "equation."

In the standard formulation of the Drake equation, we first estimate the rate at which stars form in our Galaxy. If many stars form, the chances for life are higher than if there are few stars.

Second, we consider the probability that stars are suitable to allow intelligent life to evolve on at least one body orbiting them. For example, as we have already discussed, the most massive stars evolve rather quickly, remaining stable for too short a time to allow intelligent life a good chance to evolve.

Third, we ask what the probability is that a suitable star has planets. With the multitude of new detections of planets orbiting other stars, astronomers know that the chances are very high.

Fourth, we need planets with suitable conditions for the origin of life, so we multiply by the average number of such planets per suitable star. A planet like Jupiter might be ruled out, for example, because it lacks a solid surface and because its surface gravity is high. (Alternatively, though, one could consider a liquid region, if it were at a suitable temperature, to be as advantageous as the oceans on Earth may have been to the development of life here. Or a moon of the planet could provide the solid surface.) Also, planets probably must be in orbits in which the temperature does not vary too much.

Fifth, we have to consider the fraction of the suitable planets on which life actually begins. This is a very large uncertainty, for if this

(a)

E. Imre Friedmann and NASA

(b)

Todd Stevens, photo from Pacific Northwest Regional Laboratory, operated by Batelle Memorial Institute

■ FIGURE 20–15 (a) We see the inside of an Antarctic rock, with a lichen growing safely insulated from the external cold. The image shows the outer 1 cm of the rock. (b) Bacteria that live over 1 km below the Earth's surface and that fluoresce as red patches on this microscope image. The bacteria, discovered in deep wells, survive on hydrogen generated by a chemical reaction between water and iron silicates in the surrounding volcanic rock. Thus they do not need sunlight (or oxygen). Do similar types of microbes exist on Mars?

fraction is zero (with the Earth being the only exception), then we get nowhere with this entire line of reasoning. Still, the discovery that amino acids can be formed in laboratory simulations of at least some kinds of potential primitive atmospheres, and the discovery of complex molecules (such as ammonia, formic acid, and vinegar) in interstellar space, indicate to many astronomers that it is relatively easy to form complicated molecules. As mentioned previously, amino acids, much less complex than DNA but also basic to life as we know it, have even been found in some meteorites.

Moreover, life is found in a wide range of extremes on Earth, including oxygen-free environments near geothermal sulfur vents on the ocean bottom or surface hot springs, under rocks in the Antarctic, and deep underground in some other parts of the Earth (■ Fig. 20–15). These bacteria do not survive in the presence of oxygen; their existence supports the idea that life evolved before oxygen appeared in Earth's atmosphere. So environments on other planets may not be as hostile to life as we had thought, even though they couldn't support the types of animal and plant life with which we are most familiar. It is these hardy examples of life on Earth that give us hope that the possible discovery of primitive life on Mars will someday be confirmed.

Sixth, if we want to hear meaningful signals from aliens, we must have a situation where not just life but *intelligent* life has evolved. We cannot converse with algae or paramecia, and certainly not with the organic compounds reported on Mars. Furthermore, the life must have developed a technological civilization capable of interstellar communication. So we consider the fraction of stars with life on which intelligence and communication actually develop.

Finally, such a civilization must also live for a fairly long time; otherwise, its existence would be just like a flashbulb going off in an otherwise perpetually dark room. Humans now have the capability of destroying themselves either dramatically in a flurry of hydrogen bombs or more slowly by, for example, dramatically altering our climate, destroying our ozone shield (which keeps harmful ultraviolet radiation out), or greatly increasing the level of atmospheric pollution. Natural disasters must also be avoided. That an Earth-crossing asteroid or comet will eventually impact the Earth with major consequences seems statistically guaranteed on a timescale of a few hundred million years or less, unless we take preventive action.

It is a sobering question to ask whether the typical lifetime of a technological civilization is measured in decades and centuries, or whether all the problems that we have – political, environmental, and otherwise – can be overcome, leaving our civilization to last for millions or even billions of years. So, to complete our calculation, we must multiply by the average lifetime of a civilization.

We can try to estimate answers for each of these simpler questions within our chain of reasoning, though some of them are actually more like wild guesses. (Consequently, some scientists don't find the exercise useful, though we disagree.) We can then use these answers together to figure out the larger question of the probability that communicating extraterrestrial life exists (see *Figure It Out 20.2: The Drake Equation*).

Fairly reasonable assumptions can lead to the conclusion that there may be thousands, or even tens of thousands, of planets in our Galaxy on which technologically advanced life has evolved (handy "Drake equation calculators" can be found at several websites; search with Google to find some). Perhaps overly optimistically, Carl Sagan estimated that there might be a million such planets.

On the other hand, adopting more pessimistic (but possibly more realistic) values for the probabilities of intelligent and technologically advanced life, and for the typical lifetime of such a civilization, gives a much bleaker picture. Indeed, humans may be at the pinnacle of intelligence and technological capability in our Galaxy, with few (if any) comparably advanced civilizations.

20.6b Where Is Everyone?

One interesting argument is as follows: If there are many (say, a million) intelligent, communicating civilizations in our Galaxy, why haven't we detected any of them? Where are they all? Surely most must be far more advanced than we are, and have sent signals if not

(a) NASA/JPL/Caltech

(b) NASA/JPL/Malin Space Science Systems

(c) © ESA/DLR/FU Berlin (G. Neukum), MOC (Malin Space Science Systems)

■ **FIGURE 20–16** The "face" on Mars, seen at low resolution from an earlier spacecraft (a), turns out to be an ordinary set of landforms when imaged from spacecraft capable of higher resolution, such as in this Mars Global Surveyor image (b) or the more recent Mars Express image (c).

spacecraft that have reached Earth, yet there is no evidence for them. This reasoning, promoted among others by the late Enrico Fermi of the University of Chicago, suggests that advanced creatures such as us might be very rare, even if primitive life (such as microbes and bacteria) is abundant in our Galaxy. (Note that various surface features claimed by some people to have been built by extraterrestrials are either nonexistent, such as the "face" on Mars – ■ Figure 20–16 – or have more conventional explanations, like the huge drawings on the Peruvian desert known as the Nazca lines.)

Indeed, a more extreme version of this argument points out that there could be self-sustaining colonies voyaging through space for generations. They need not travel close to the speed of light if families are aboard. Even if colonization of space took place at a rate of only 1 light-year per century, our entire Galaxy would still have been colonized during a span of just a hundred million years, less than one per cent of its lifetime. The fact that we do not find extraterrestrial life here indicates that the Solar System has probably not been colonized, which in turn may imply that technologically advanced life has not arisen elsewhere in our Galaxy.

Another possibly relevant fact suggesting that life having our capabilities is very rare in our Galaxy is that at least a few billion species have ever lived on Earth, yet apparently we are the only species to have developed space communication or even acquired technology. Similarly, it is sobering to realize that a communicating civilization developed on Earth only in the last century, despite evidence for primitive life on Earth for the past 3.5 to 3.8 billion years.

Recently, astronomers and other scientists have carefully considered the many factors that affect complex life on Earth. For example, the stability of the Earth's axis of rotation depends on the presence of our rather massive Moon. If we had no moon, or only a small one, then Earth's axis would undergo rather rapid, random changes in its orientation, typically several tens of millions of years apart, causing large variations in the seasons and climate, and presumably making it more difficult for complex life to develop. Similarly, the presence of the very massive planet Jupiter in the Solar System, yet fairly far from Earth and having a nearly circular orbit, is a blessing: Jupiter's gravitational tugs have cleared out the Solar System, making collisions between Earth and large "killer meteoroids" infrequent.

There are many other relevant factors. The presence of plate tectonics on Earth is very important, for instance, because it allows carbon to be recycled in a steady manner through Earth's atmosphere, oceans, interior, and surface. This "carbon cycle" is important for maintaining global climate stability and for certain aspects of life itself. Having an abundant supply of heavy elements during the formation of the Solar System was also crucial; many regions of our Milky Way Galaxy were quite deficient in such elements billions of years ago, when the Sun formed. These arguments, and others, have led many astronomers to conclude that Earth-like planets capable of developing complex, technologically advanced, communicating civilizations really are rare in our Milky Way Galaxy, though primitive life such as bacteria might be very common indeed. However, other astronomers (such as those at the SETI Institute) argue that there are major potential flaws in the reasoning for this "rare Earth" hypothesis, and that communicating civilizations may be common.

It is not clear, for example, whether the apparently "special" conditions of Earth are essential to the development of intelligent life; maybe our view has been too highly skewed by the single example

FIGURE IT OUT 20.2   The Drake Equation

We can make an educated guess regarding the present number of communicating civilizations (those that should be able to contact each other) in our Milky Way Galaxy by using the "Drake equation," first written by Frank Drake in 1961:

$$N = R\, f_s f_p n_e f_l f_i f_c L.$$

Let us examine each of the terms in this equation.

R is the star-formation rate in our Galaxy. It is roughly equal to the total number of stars divided by the age of our Galaxy. A reasonable estimate is $R = 10$ stars per year. The true rate is lower now, but was higher in the past.

f_s is the fraction of stars that are suitable. Various criteria can be applied, such as F, G, and K main-sequence stars, and a sufficiently high abundance of heavy elements to make rocky planets possible. Perhaps 0.1 is a good guess.

f_p is the fraction of stars with planetary systems. This fraction could be as little as 0.1 (many binary systems, and even some solitary stars, might not have planets) or as large as 1.

n_e is the number of Earth-like planets or moons per planetary system. We could guess that this number is typically between 0.1 and 1. It might be a little larger than 1, if a planetary system has several suitable moons. For example, in the Solar System, it is not unreasonable to suppose that Mars, Europa, Enceladus, and Titan (besides Earth) had or will have conditions suitable for the formation of life at some time in their existence.

f_l is the fraction of Earth-like planets on which life (even if primitive) arises. This fraction, though quite speculative, might be in the range 10^{-3} to 1.

f_i is the fraction of life-bearing planets on which intelligent life actually arises. Though perhaps even more speculative, it might be 10^{-6} to 1.

f_c is the fraction of intelligent life that has the ability and desire to communicate with aliens. Once again, we can only speculate; perhaps 10^{-3} to 1 is reasonable.

L is the average lifetime of a communicating civilization, or the cumulative lifetime of such civilizations on a given planet (if there are more than one at any given time, or if the annihilation of one is followed at some later time by the emergence of another). Humans have sent radio signals for only about 100 years – and we hope that we have many years ahead of us! However, perhaps advanced civilizations tend to destroy themselves after a short time. The possible range for L is enormous: Let's say 100 to 10^9 years.

The overall result is the product of all these factors, most of which are highly uncertain. The pessimistic estimates give a value of 10^{-12}, suggesting that creatures like humans are exceedingly rare – only one case per 10^{12} galaxies. There are about 10^{11} galaxies in the observable part of the Universe, so we might be the only communicating civilization within this volume, or even within a volume 10 times larger!

On the other hand, the most optimistic estimates give a value of 10^9 – essentially the lifetime L, since all of the other factors are either about 1 or balance each other out (note that $R \times f_s \approx 1$). This means that one star out of several hundred might now have a communicating civilization. The fraction is even larger if some civilizations achieve interstellar travel and colonize more than one star.

Although use of the Drake equation doesn't answer our question very precisely at all, the approach is systematic, and it gives us a range of reasonable values. It also focuses our attention on which factors are highly uncertain and need the most improvement. For example, we can see that our uncertainty in the rate at which stars form is just a minor fraction of the total uncertainty in the final answer; far more important are the uncertainties in the fraction of life with intelligence and the lifetime of a communicating civilization.

---

we know – ourselves. Moreover, is it necessarily true that if other intelligent creatures evolved, they would choose to colonize space, or have the ability to do so? The rather late appearance of technologically advanced life on Earth may also be a statistically unlikely fluke. Finally, some of the arguments relied on the unproven assumption that life elsewhere is quite similar to that on Earth in its properties and evolutionary path.

In any case, unless we make the effort to actually look for signs of intelligent extraterrestrial life, we might never know whether humans are indeed alone in our Galaxy, or simply one of many such creatures. Thus, many astronomers support the search for extraterrestrial intelligence, especially using telescopes that are simultaneously doing other, more conventional types of research projects.

Some major radio-telescope projects have been built for SETI purposes, with more ordinary astronomy carried out as a bonus

(■ Fig. 20–17). Perhaps someday we will be scanning the skies with radio telescopes on the far side of the Moon, shielded from Earth's radio interference by the Moon's bulk. The chances for success might be slim, but nearly all agree that the actual detection of extraterrestrial signals would be one of the most important and mind-blowing discoveries ever, if not the greatest discovery of all time.

20.7 UFOS AND THE SCIENTIFIC METHOD

If some or many astronomers believe that technologically advanced life exists elsewhere in our Galaxy, why do they not accept the idea

■ **FIGURE 20–17** The Allen Telescope Array, built in California for the SETI Institute and the University of California, Berkeley, though now managed by SRI International. The goal was to have about 350 antennas, though only 42 were completed because of insufficient funding.

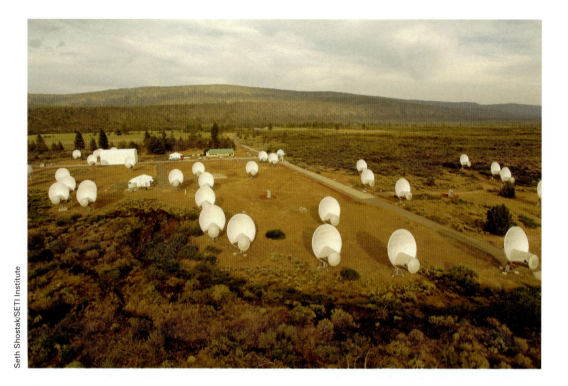

Seth Shostak/SETI Institute

that unidentified flying objects (UFOs) represent visitations from these other civilizations? The answer to this question leads us not only to explore the nature of UFOs but also to consider the nature of knowledge and truth. The discussion that follows is a personal view of the authors, but one that is shared to a greater or lesser extent by most scientists.

20.7a UFOs

The most common UFO is a light that appears in the sky that seems to the observer unlike anything made by humans or any commonly understood natural phenomenon. UFOs may appear as a point or extended, or may seem to vary in size or brightness. But the observations are usually anecdotal, are not controlled as in a scientific experiment, and are not accessible to study by sophisticated instruments.

Most of the sightings of UFOs that are reported can actually be explained in terms of natural phenomena. The Earth's atmosphere can display a variety of strange effects, and these can be used to explain many apparent UFOs. When Venus shines brightly near the horizon, for example, we sometimes get telephone calls from people asking us about the "UFO" – especially if the crescent moon happens to also be in that general direction. It is not well known that a bright star or planet low on the horizon can seem to flash red or green because of atmospheric refraction. Atmospheric effects can affect radar (radio) waves as well as visible light.

Sometimes other natural phenomena (flocks of birds, for example) are mistakenly reported as UFOs. Even the more mysterious sightings lack convincing evidence that rules out alternatives. One should not

accept explanations that UFOs are flying saucers from other planets before more mundane explanations – including hoaxes, exaggeration, and fraud – are exhausted (■ Figs. 20–18, 20–19).

For many of the effects that have been reported, the UFOs would have been defying well-established laws of physics. Why haven't we heard the expected sonic booms, for example, from rapidly moving UFOs? Scientists treat challenges to laws of physics very seriously, since our science and technology are based on these laws and they seem to work extremely well.

Most professional astronomers feel that UFOs can be so completely explained by natural phenomena that they are not worthy of more of our time. Although most of us do not categorically deny the possibility that UFOs exist (after all, the Voyager spacecraft might someday pass by a planet orbiting another star), the standard of evidence expected of all claims in science has not yet been met. Some individuals may ask why we reject the identification of UFOs with flying saucers when – they may say – that explanation is "just as good an explanation as any other." Let us go on to discuss what scientists mean by "truth" and how that applies to the above question.

20.7b Of Truth and Theories

At every instant, we can explain what is happening in a variety of ways. When we flip a light switch, for example, we assume that the switch closes an electric circuit in the wall and allows the electricity to flow. But it is certainly possible, although not very likely, that the switch activates a relay that turns on a radio that broadcasts a message to the president of the United States. The president then might send back a telepathic message to the electricity to flow, making the light go on. The latter

■ **FIGURE 20–18** A Christmas-light UFO seen on a rooftop.

■ **FIGURE 20–19** Though they may look like giant UFOs, these objects are really the planetarium and the Arts Palace in Valencia, Spain, projects of the architect Santiago Calatrava.

explanation sounds so unlikely that we don't seriously consider it. We would even call the former explanation "true," without qualification.

We usually regard as "true" the simplest explanation that satisfies all the data we have about any given thing. This "Principle of Simplicity" is known as **Occam's Razor**, since it cuts like a razor through a variety of options; it is named after the 14th-century British philosopher who originally proposed it. Without this rule, we would always be subject to such complicated doubts that we would accept nothing as known to be true.

Science is based on Occam's Razor, though we don't usually bother to think about it. Sometimes something we call "true" might be more accurately described as a theory (see Chapter 1). An example of a theory is the Newtonian theory of gravitation, which for many years explained almost all of the planetary motions. Albert Einstein's 1916 theory of gravity, known as the general theory of relativity, provided an explanation for a nagging discrepancy in the orbit of Mercury, as we described in Chapter 10.

Is Newton's theory "true"? Though we know it is actually "false," it is a good approximation of the truth in most regions of space; it is generally an accurate model for what we observe. Is Einstein's theory "true"? We may say so, although we know it is not compatible with quantum physics. One day a newer theory may come along that is more general than Einstein's in the same way that Einstein's is more general than Newton's. Indeed, as we discussed in Chapter 19, superstring theory is a leading candidate for the unification of general relativity and quantum physics, and hence may someday be a more complete theory than Einstein's.

How does this view of truth tie in with the previous discussion of UFOs? Scientists have assessed the probability of UFOs being flying saucers from other worlds, and most have decided that the probability is so low that we have better things to do with our time and with our national resources. We have so many other, simpler explanations of the phenomena that are reported as UFOs that, when we apply Occam's Razor, we call the identifications of UFOs with extraterrestrial visitation "false." However, it is conceivable that we will someday have compelling evidence for real UFOs associated with intelligent extraterrestrials, though at this time we do not.

20.8 CONCLUSION

You have covered a lot of material in this book, and learned much about the Universe. We, the authors, hope that this new knowledge increases your awe and fascination for nature. The understanding of a phenomenon should enhance its beauty, not detract from it. The magnificence of the Universe comes in part from its logical structure – and the foundation is perhaps unexpectedly simple. Indeed, Einstein remarked that "The most incomprehensible thing about the Universe is that it's comprehensible."

The actual consequences of the basic laws can be extraordinarily complicated and varied. The best example is life itself: It is the most complex known structure. Even the simplest cell is more difficult to understand than the formation of galaxies or the structure of stars. Our highly advanced brains and dexterity are what sets humans apart from other forms of life. We are able to question, explore, and ultimately understand the inner workings of nature through a process of observation, experimentation, and logical thought.

It is almost as if the Universe has found a way to know itself, through us. We are the observers and explorers of the Universe; we are its brains and its conscience. This makes each one of us special to the Universe as a whole. Perhaps it need not have been this way: Alter the physical constants ever so slightly and the Universe might have been stillborn, with no such complexity. But here we are, living in this beautiful, amazing, mind-blowing Universe. May you continue to enjoy the wonders of the cosmos!

CONCEPT REVIEW

» The study of life elsewhere than on Earth is called **astrobiology**, **exobiology**, or **bioastronomy** (introductory paragraph). One difficulty in the search for life is that we don't have a clear definition of what life is, and we generally must restrict ourselves to life as we know it, which is based on **organic** compounds – those containing carbon (Section 20.1). Specifically, life on Earth is based on chains of carbon called **amino acids**, which follow the genetic code transmitted by DNA to link together many at a time to form **proteins**. Experiments in which sparks were passed through certain mixtures of gases showed that amino acids form easily, but it is now thought that these mixtures are not representative of Earth's primitive atmosphere. Nevertheless, amino acids seem to form in a variety of environments, including harsh ones.

» Primitive life arose very quickly on Earth, suggesting that it is not highly improbable, though the discovery of indigenous life in at least one other location would provide much support for this hypothesis. Within the Solar System, Mars and Europa are the most likely candidates for life (Section 20.2). Extraterrestrial life outside our planetary system, especially intelligent life, is most likely to have formed on moons or planets orbiting Sun-like main-sequence stars; such stars have a long life and a relatively large **habitable zone** (ecosphere), or range of distances in which conditions suitable for life might be found (Section 20.3).

» In searching for signs of extraterrestrial life, we can wait for their spacecraft to reach us, but the odds of success are low, even though Einstein's special theory of relativity in principle allows interstellar voyages to be completed in a reasonably short time (Section 20.4). A much more likely method of detection is by searching for patterns among electromagnetic signals, especially radio waves. Several such projects are being conducted, most notably the "search for extraterrestrial intelligence, at home" (SETI@home), in which the general public can participate with personal computers. Humans have sent radio signals into space, intentionally and unintentionally, which might someday be detected by extraterrestrials (Section 20.5).

» Although we do not know how many communicating civilizations exist in the Milky Way Galaxy, an estimate can be attempted by using the **Drake equation** (Section 20.6a). Among the most uncertain factors are the probability that intelligence arises after the formation of primitive life and the typical lifetime of a communicating civilization. Some scientists think that intelligent life might be common in our Galaxy, but many others are concluding that intelligence and technological capability at the level of humans is rather rare (Section 20.6b). Unless we actually search for signals, however, the answer may always elude us.

» Evidence that UFOs visit us is not convincing; most of the reports can be easily explained by atmospheric and other natural phenomena (Section 20.7a). Although it is possible that some of the sightings are of actual UFOs, **Occam's Razor** (the Principle of Simplicity) suggests otherwise, until the evidence becomes much stronger (Section 20.7b). Having finished this textbook, you should now have a much deeper and broader perspective of the Universe and our place within it (Section 20.8). We, the authors, hope that this new knowledge increases your awe and fascination for nature.

QUESTIONS

1. Discuss the most likely places in the Solar System, other than Earth, where there might be primitive (microbial) life.

2. What is the significance of the Miller-Urey experiments?

3. What types of stars are most likely to have planets on which indigenous life developed? Explain.

4. Why do we think that indigenous intelligent life wouldn't evolve on planets orbiting stars much more massive than the Sun?

5. List two means by which we might detect extraterrestrial intelligence.

6. Why have astronomers generally used radio wavelengths in their search for life elsewhere?

7. In what way can optical (visible) signals be used in a search for extraterrestrial life?

8. Describe the radio signals humans have sent to outer space, both intentionally and unintentionally.

†9. Work out the binary value given for the population of Earth on the message we sent to M13, and compare it with the current actual value. (This question assumes knowledge of the binary system.)

†10. Work out the binary value given for the size of the Arecibo telescope on the message we sent to M13, and compare it with the actual value. (This question assumes knowledge of the binary system.)

11. Why can't we hope to carry on a conversation at a normal rate with extraterrestrials on a planet orbiting a distant star?

12. State the Drake equation and its significance. Discuss the importance of the value of L, the lifetime of the civilization.

†13. In the Drake equation, use your own preferences for each quantity to derive the number of intelligent, communicating civilizations in our Galaxy.

†14. If 10 per cent of all stars are of suitable type for life to develop, 30 per cent of all stars have planets, and 20 per cent of planetary systems have a planet or moon at a suitable distance from the star, what fraction of stars have a planet suitable for life? Roughly how many such stars would there be in our Milky Way Galaxy?

†15. How long would it take a rocket ship travelling at 100 km/s to reach a star that is 20 light-years away?

16. Explain how it is possible, in principle, to travel many light-years in a short time interval as measured by the traveller.

†17. Suppose you are in a rocket ship that is moving at 97 per cent of the speed of light. If your journey to another star appeared to take 30 years from the perspective of an Earth-bound observer, how long did it take in your frame of reference?

†18. Referring back to question 17, how far did you travel, in your frame of reference?

19. Describe how Einstein's general theory of relativity serves as an example of the scientific method.

20. **True or false?** Unambiguous bacteria fossils have been found in at least one martian meteorite.

21. **True or false?** Biological processes on Earth are largely responsible for the free oxygen in its atmosphere.

22. **True or false?** Organic compounds can be created by sending electric discharges through mixtures of gases, some of which might be similar to the early atmospheres of Earth or some exoplanets.

23. **True or false?** Planets orbiting the most massive stars have the best chance of developing intelligent life, since these stars have long lives.

24. **True or false?** One of the most uncertain factors in estimating the number of communicating civilizations in our Galaxy is the lifetime of such civilizations.

25. **True or false?** Astronomers currently have no compelling evidence for other life in the Solar System, but the most promising places to look are Mars and a few specific moons of Jupiter and Saturn.

26. **True or false?** Intelligent aliens could find significant evidence for life on Earth by detecting nitrogen molecules and ammonia in the spectrum of Earth's atmosphere.

27. **True or false?** Despite the highly probable language barriers in communicating with intelligent alien life, we could signal our own existence by transmitting fundamental mathematical constants (such as π) using radio waves.

28. **Multiple choice:** Around which one of the following objects is there thought to be the *best* chance of finding life, outside the Solar System? **(a)** A neutron star. **(b)** A main-sequence F-type star. **(c)** A red supergiant star. **(d)** A main-sequence O-type star. **(e)** A planetary nebula.

29. **Multiple choice:** Which one of the following is *not* a conclusion from Einstein's special theory of relativity? **(a)** It is possible to travel into the future, but not into the past. **(b)** A moving clock runs more slowly according to a person at rest. **(c)** A moving ruler looks shorter according to a person at rest. **(d)** The speed of light can be exceeded, in principle, if an object is given a truly gargantuan amount of energy. **(e)** Interstellar travel is possible within a single human lifetime, in principle, if one travels sufficiently close to the speed of light.

30. **Multiple choice:** Which one of the following is *not* used in the Drake equation? **(a)** The fraction of Earth-like planets or moons on which life actually arises. **(b)** The fraction of suitable stars with planetary systems. **(c)** The average rate of star formation in the Milky Way Galaxy. **(d)** The fraction of stars that are on the main sequence. **(e)** The fraction of intelligent civilizations that develop electromagnetic communication techniques.

31. **Multiple choice:** Which one of the following is a reason to believe there might be life elsewhere in the Universe, outside our Solar System? **(a)** The presence of a large number of O-type and B-type stars in the Milky Way Galaxy. **(b)** Most of the recently discovered exoplanets are a lot like the Earth. **(c)** The fact that life exists in many extreme environments on Earth, and seems to have arisen shortly after Earth formed. **(d)** The substantial quantities

of free methane and oxygen (together) detected in spectra of some exoplanets, suggesting that the methane might have been produced by decaying organisms. **(e)** The many UFO sightings reported in the U.S. and elsewhere in the world.

32. **Multiple choice:** Which one of the following is the *least* likely place in our Solar System to find life? **(a)** Mars. **(b)** Mercury. **(c)** Europa. **(d)** Enceladus. **(e)** Titan.

†33. **Multiple choice:** According to the most recent data and the Drake equation, the odds that there is intelligent, communicating life outside our Solar System, but somewhere within 10,000 light-years of it, are roughly which one of the following? **(a)** 1 in 10,000. **(b)** 1 in 10^8 (i.e., $1/10,000^2$, according to the inverse-square law). **(c)** Virtually certain, since there are so many stars within 10,000 light-years. **(d)** 0, since intelligent life on Earth has been scientifically proven to be the one-time-only work of a special Creator. **(e)** Nobody really knows.

34. **Multiple choice:** Which one of the following is *not* a problem with interstellar travel? **(a)** The distance to even the closest star is so great that it would take many thousands of years to get there with current technology. **(b)** The amount of energy required to get to the nearest stars within a single lifetime vastly exceeds what we can currently produce. **(c)** The mass of the fuel required for interstellar travel can be far greater than the mass of the rocket that would carry you. **(d)** The acceleration of the rocket would have to be limited to prevent injury to human beings. **(e)** There would be no way for a human being to survive a trip to a star more than 100 light-years away, since nothing travels faster than the speed of light.

35. **Fill in the blank:** Life on Earth is based on _____ consisting of chains of carbon, and in turn they form long chains called _____.

36. **Fill in the blank:** The experiments done by _____ were the first to show that long organic compounds could be made from some types of primitive atmospheres containing simple molecules.

37. **Fill in the blank:** Jupiter's moon _____ probably has liquid water beneath its icy surface, and hence could be an abode for primitive life.

38. **Fill in the blank:** Searches for signals from extraterrestrial life have been conducted mainly at _____ wavelengths.

39. **Fill in the blank:** The volume around a star in which conditions may be suitable for life is called the _____.

40. **Fill in the blank:** According to Einstein's special theory of relativity, a traveller moving very fast relative to people on Earth ages more _____ than they do.

41. **Fill in the blank:** The principle that the simplest explanation that satisfies all the data regarding a phenomenon is "true" is known as _____.

42. **Fill in the blank:** A well-known effort to find signals from intelligent life elsewhere in our Galaxy is called _____; it uses the computers of millions of people to process data.

†This question requires a numerical solution.

TOPICS FOR DISCUSSION

1. How restrictive do you think we are being when we consider only "life as we know it?" Explain.

2. How useful, in your opinion, is the Drake equation for estimating the number of intelligent, communicating civilizations in our

Galaxy? Is it perhaps more useful as a tool for identifying the most uncertain aspects of the issue?

3. Do you think humans or their successors (machines?) will ever overcome the enormous barriers to rapid interstellar travel? Explain.

EPILOGUE

We have completed our grand tour of the Universe. We have seen stars and planets, matter between the stars, giant collections of stars called galaxies and clusters of galaxies, and very distant objects such as quasars with curved space lensing even farther objects. We have witnessed the evolution of stars, in some cases ending with spectacular explosions leaving compact remnants the size of a city but half a million times more massive than Earth. We have pondered the properties of still more bizarre objects, black holes. We have learned how our Universe began in a hot, compressed state and has been expanding ever since – seemingly faster and faster during the past five billion years, perhaps driven by a cosmic antigravity effect. We have explored the origins of the Universe, galaxies, stars, the chemical elements, planets, and ultimately life itself. If you are thirsty for more information, as the authors would like you to be, you can consult the books listed in the Selected Readings.

Moreover, we have seen the vitality of contemporary science in general and astronomy in particular. The individual scientists who call themselves astronomers are engaged in fascinating studies, often pushing modern technologies to their limits. New telescopes on the ground and in space, new types of detectors, new computer capabilities for studying data and carrying out calculations, and new theoretical ideas are linked in research about the Universe.

Our views at the cutting edge of astronomy are changing so rapidly that within a few years some of what you read here will have been revised. Science is a dynamic process: New ideas are developed and tested, and modified when necessary. We, the authors, hope that, over the years, you will keep up by following astronomical articles and stories in newspapers, magazines, and books, and on television. We would like you to consider the role of scientific research as you vote. And we hope you will remember the methods of science – the mixture of logic and standards of evidence by which scientists operate – that you have seen illustrated in this book.

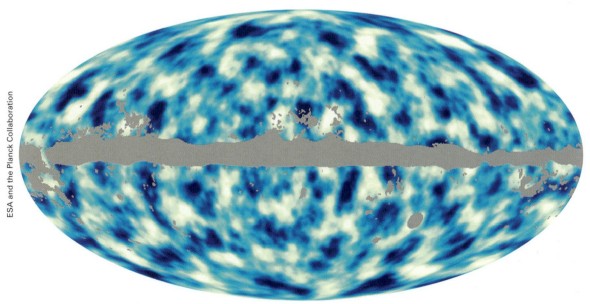

ESA and the Planck Collaboration

An all-sky map (with the Milky Way grayed out) showing the distribution of dark matter in the Universe, calculated as the effects of gravitational lensing by massive cosmic structures. It was derived from the European Space Agency's Planck spacecraft's measurements of the cosmic microwave background radiation, released in 2013.

Appendix 1A International System of Units (Système International d'Unités)

	SI Units	SI Abbrev.	Other Abbrev.
length	meter	m	
volume	liter	L	ℓ
mass	kilogram	kg	kgm
time	second	s	sec
temperature	kelvin	K	°K

Appendix 1B Conversion Factors

1 erg = 10^{-7} joule = 10^{-7} kg · m²/s²
1 joule = 6.2419×10^{18} eV
1 in = 25.4 mm = 2.54 cm
1 yd = 0.9144 m
1 mi = 1.6093 km $\simeq$ 8/5 km
1 oz = 28.3 g
1 lb = 0.4536 kg

Appendix 2A Physical Constants (2010 CODATA Recommended Values)

Speed of light*	c	= 299 792 458 m/sec (exactly)
Constant of gravitation	G	= $(6.673\ 84 \pm 0.000\ 80) \times 10^{-11}$ m³/kg · s²
Planck's constant	h	= $(6.626\ 069\ 57 \pm 0.000\ 000\ 29) \times 10^{-34}$ J · s
Boltzmann's constant	k	= $(1.380\ 6488 \pm 0.000\ 0013) \times 10^{-23}$ J/K
Stefan-Boltzmann constant	σ	= $(5.670\ 373 \pm 0.000\ 021) \times 10^{-8}$ W/m² · K⁴
Wien displacement constant	$\lambda_{max} T$	= $(2.897\ 7721 \pm -0.000\ 0026) \times 10^{7}$ Å · K

See http://physics.nist.gov/constants as of 2012. From 2010 CODATA recommended values.

Appendix 2B Astronomical Constants

Astronomical Unit†		1 au = 149 597 870 700 m exactly
		1 pc = 3.086×10^{16} m
		= 206 264.806 au
		= 3.261 633 1t yr
Light-year		1 1t-yr = $(9.460\ 530) \times 10^{15}$ m
		= 6.324×10^{4} au
Mass of Sun*	$M_\odot$	= $(1.988\ 843 \pm 0.000\ 03) \times 10^{30}$ kg
Radius of Sun*	$R_\odot$	= 696 000 km
Luminosity of Sun	$L_\odot$	= 3.827×10^{26} J/sec
Mass of Earth*	M_E	= $(5.972\ 23 \pm 0.000\ 08) \times 10^{24}$ kg
Equatorial radius of Earth*	R_E	= 6 378.140 km

†This definition of the astronomical unit was adopted by the General Assembly of the International Astronomical Union in Beijing in 2012.

*Adopted as "IAU (2009) system of astronomical constants" at the General Assembly of the International Astronomical Union in Rio de Janeiro. The meter was redefined in 1983 to be the distance travelled by light in a vacuum in 1/299,792,458 second.

See http://maia.usno.navy.mil/NSFA.

APPENDIX 3   Planets and Dwarf Planets

Appendix 3A Our Solar System: Intrinsic and Rotational Properties of Planets

Name	Equatorial Radius km	Equatorial Radius ÷ Earth's	Mass ÷ Earth's	Mean Density (g/cm³)	Oblateness	Surface Gravity (Earth = 1)	Sidereal Rotation Period	Inclination of Equator to Orbit	Apparent Magnitude During 2013
Mercury	2,439.7	0.3824	0.0553	5.43	0	0.378	58.646^d	0.0°	–2.2 to +5.5
Venus	6,051.8	0.9489	0.8150	5.24	0	0.894	243.02^{d}R	177.3°	–4.9 to –3.8
Earth	6,378.14	1	1	5.513	0.0034	1	23^{h}56^{m}04.1^s	23.45°	–
Mars	3,397	0.5326	0.1074	3.93	0.006	0.379	24^{h}37^{m}22.662^s	25.19°	–0.8 to +1.6
Jupiter	71,492	11.194	317.894	1.33	0.065	2.54	9^{h}50^m to > 9^{h}55^m	3.12°	–2.7 to –1.9
Saturn	60,268	9.41	95.185	0.69	0.098	1.07	10^{h}39.9^m	26.73°	+0.1 to 0.7
Uranus	25,559	4.0	14.537	1.27	0.022	0.8	17^{h}14^{m}R	97.86°	+5.7 to +5.9
Neptune	24,764	3.9	17.151	1.64	0.017	1.2	16^{h}7^m	29.56°	+7.8 to +8.0

The table was verified and updated by E. Myles Standish, JPL, 2012. See also *The Astronomical Almanac* for the year 2013.
See http://ssd.jpl.nasa.gov/?planet_phys_par. See also http://nssdc.gsfc.nasa.gov/planetary/factsheet/Marsfact.html, etc.

Appendix 3B Our Solar System: Intrinsic and Rotational Properties of Dwarf Planets (Ceres, Pluto, Eris, Makemake, Haumea)

Name	Equatorial Radius km	Equatorial Radius ÷ Earth's	Mass ÷ Earth's	Mean Density (g/cm³)	Oblateness	Surface Gravity (Earth = 1)	Sidereal Rotation Period	Inclination of Equator to Orbit	Apparent Magnitude at Opposition
1 Ceres	476.2	0.07					9.07 h		6.7
134340 Pluto	1150–1200	0.18					153.29 h		14.0
136108 Haumea	~500–980	0.10	0.0007	2.6–3.3	50%		3.915 h		17.3
136472 Makemake	~730	0.11	0.0005				7.77 h		16.7
136199 Eris	1163	0.18					25.9 h		18.7

Appendix 3C Our Solar System: Orbital Properties of Planets

Name	Semimajor Axis		Sidereal Period		Synodic Period (Days)	Eccentricity	Inclination to Ecliptic
	au	10⁵ km	Years	Days			
Mercury	0.387	57.9	0.241	87.97	115.9	0.205 63	7.005°
Venus	0.723	108.2	0.615	224.70	583.9	0.006 77	3.395°
Earth	1	149.60	1.000	365.26	–	0.016 71	0.000°
Mars	1.524	227.94	1.881	686.98	779.9	0.093 41	1.851°
Jupiter	5.203	778.30	11.862	4,332	398.9	0.048 39	1.305°
Saturn	9.537	1429.39	29.457	10,759	378.1	0.054 15	2.484°
Uranus	19.191	2875.04	84.01	30,685	369.7	0.047 168	0.770°
Neptune	30.069	4504.45	164.79	60,189	367.5	0.008 59	0.769°

Elements of planetary orbits for 2012, referred to the mean ecliptic and equinox of J2000 (E. M. Standish, X. X. Newhall, J. G. Williams, and D. K. Yeomans, *Explanatory Supplement to the Astronomical Almanac*, P. K. Seidelmann, ed., 1992). Periods are from http://nssdc.gsfc.nasa.gov/planetary/factsheet/Marsfact.html.

For planetary satellites, see http://ssd.jpl.nasa.gov/sat_props.html; http://ssd.jpl.nasa.gov/sat_elem.html, from NASA's JPL.

See also Sean E. Urban and P. Kenneth Seidelmann, *Explanatory Supplement to the Astronomical Almanac* (Mill Valley, CA: University Science Books, 2013).

Appendix 3D Our Solar System: Orbital Properties of Dwarf Planets

Name	Semimajor Axis		Sidereal Period		Discovery Date	Eccentricity	Inclination to Ecliptic
	au	10⁶ km	Years	Days			
1 Ceres	2.765	413.64	4.60	1680	January 1, 1801	0.079	10.587°
134340 Pluto	39.452	5901.94	247.74		February 18, 1930	0.250	17.089°
136108 Haumea	42.985	6430.46	281.83		March 7, 2003	0.198	28.202°
136472 Makemake	45.436	6797.13	306.28		March 31, 2005	0.163	29.012°
136199 Eris	68.047	10,179.69	561.34		January 5, 2005	0.435	43.824°

From NASA's JPL at http://ssd.jpl.nasa.gov.

APPENDIX 4

Appendix 4 The Brightest Stars

Rank	Star Name	RA (2012.5) Dec				m_V	B–V	MK Type	M_V	D lt-yr	
		h	m	°	′						
	Sun	–		–		−26.75	0.63	G2 V	4.8	[8.3 lt-min]	Sun
1	α CMa A	6	45.7	−16	44	−1.44	0.01	A0 V	1.5	9	Sirius
2	α Car	6	24.2	−52	42	−0.62	0.16	A9 Ib	−5.4	310	Canopus
3	α Boo	14	6.2	+19	07	−0.05	1.24	K1.5 III	−0.6	37	Arcturus
4	α Cen A	14	40.5	−60	53	−0.01	0.71	G2 V	4.2	4	Rigil Kentaurus A[1]
5	α Lyr	18	37.3	+38	48	0.03	0.00	A0 V	0.6	25	Vega
6	α Aur Aa+Ab	5	17.6	+46	01	0.08	0.80	G6+G2:III	−0.8	42	Capella
7	β Ori A	5	15.1	−8	11	0.18	−0.03	B8 Ia	−6.6	800	Rigel
8	α CMi A	7	40.0	+5	12	0.40	0.43	F5 IV–V	2.8	11	Procyon
9	α Eri	1	38.2	−57	10	0.45	−0.16	B3 V	−2.9	144	Achernar
9	α Ori	5	55.8	+7	24	0.45v	1.50	M2 Iab	−5.0:	520	Betelgeuse
11	β Cen AB	14	04.7	−60	26	0.58v	−0.23	B1 III	−5.5	500	Hadar
12	α Aql	19	51.4	+8	54	0.76	0.22	A7 V	2.1	17	Altair
13	α Tau A	4	36.6	+16	32	0.87v	1.54	K5 III	−0.8	65	Aldebaran
14	α Vir	13	25.9	−11	1	0.98v	−0.24	B1 V	−3.6	260	Spica
15	α Sco A	16	30.2	−26	28	1.06v	1.86	M1.5 Iab	−5.8	600	Antares
16	β Gem	7	46.1	+28	00	1.16	0.99	K0 IIIb	1.1	34	Pollux
17	α PsA	22	58.3	−29	33	1.17	0.14	A3 V	1.6	25	Fomalhaut
18	α Cru A	12	27.3	−63	10	1.25	−0.20	B0.5 IV	−4.0	320	Acrux A[2]
18	β Cru	12	48.5	−59	45	1.25v	−0.24	B0.5 III	−4.0	350	Becrux
18	α Cyg	20	41.9	+45	20	1.25	0.09	A2 Ia	−7.5:	1500	Deneb
21	α Cen B	14	40.4	−60	53	1.35	0.90	K1 V	6.2	4	Rigil Kentaurus B[1]
22	α Leo A	10	09.0	+11	54	1.36	−0.09	B7 V	−0.6	78	Regulus
23	ε CMa A	6	59.1	−28	59	1.50	−0.21	B2 II	−4.1	400	Adhara
24	γ Cru	12	31.9	−57	11	1.59v	1.60	M3.5 III	−0.7	88	Gacrux
25	λ Sco	17	34.5	−37	07	1.62v	−0.23	B1.5 IV	−3.6:	360	Shaula
26	γ Ori	5	25.8	+6	22	1.64	−0.22	B2 III	−2.8	240	Bellatrix
26	α Cru B	12	27.3	−63	10	1.64	−0.18	B1 V	−3.6	320	Acrux B[2]
28	β Tau	5	27.1	+28	37	1.65	−0.13	B7 III	−1.3	130	El Nath
29	β Car	9	13.3	−69	46	1.67	0.07	A1 III	−1.1	111	Miaplacidus
30	ε Ori	5	36.8	−1	12	1.69	−0.18	B0 Ia	−6.6	1000	Alnilam
31	α Gru	22	09.0	−46	54	1.73	−0.07	B7 V	−0.9	101	Alnair
32	ζ Ori A	5	41.4	−1	56	1.74	−0.20	O9.5 Ib	−5.5	800	Alnitak
33	γ2 Vel	8	09.9	−47	22	1.75v	−0.14	WC8+O9	−5.8	800	Suhail
34	ε UMa	12	54.6	+55	54	1.76v	−0.02	A0 IV	−0.2	81	Alioth
35	α Per	3	25.2	+49	54	1.79	0.48	F5 Ib	−4.9	600	Mirfak

35	ε	Sgr	18	25.0	−34	23	1.79	−0.03	A0 II	−1.4	140	Kaus Australis
37	α	UMa AB	11	04.5	+61	41	1.81	1.06	K0 IIIa	−1.3	124	Dubhe
38	δ	CMa	7	08.9	−26	25	1.83	0.67	F8 Ia	−7.2	2000	Wezen
39	η	UMa	13	48.0	+49	15	1.85	−0.10	B3 V	−1.8	101	Alkaid
40	ε	Car	8	22.8	−59	33	1.86v	1.20	K3:III+B2:V	−4.8	600	Avior
40	θ	Sco	17	38.2	−43	00	1.86	0.41	F1 III	−3.0	270	Sargas
42	β	Aur	6	00.4	+44	57	1.90v	0.08	A1 IV	−0.2	82	Menkalinan
43	α	TrA	16	50.0	−69	03	1.91	1.45	K2 IIb–IIIa	−5.0	400	Atria
44	γ	Gem	6	38.4	+16	23	1.93	0.00	A1 IV	−0.6	100	Alhena
44	α	Gem A	7	35.4	+31	52	1.93	0.03	A1 V	0.6	52	Castor A
44	δ	Vel AB	8	45.0	−54	45	1.93	0.04	A1 V	0.0	80	
47	α	Pav	20	26.6	−56	42	1.94	−0.12	B2.5 V	−2.1	180	Peacock
48	α	UMi A	2	46.9	+89	19	1.97v	0.64	F5–8 Ib	−4.1	400	Polaris
49	β	CMa	6	23.2	−17	58	1.98v	−0.24	B1 II–III	−4.0	500	Mirzam
50	α	Hya	9	28.2	−8	43	1.99	1.44	K3 II–III	−2.1	180	Alphard

[1]To the unaided eye, the combined magnitude of Rigil Kentaurus A and B is −0.28.
[2]To the unaided eye, the combined magnitude of Acrux A and B is 0.68.
© 2012 Royal Astronomical Society of Canada.

The components of visual binary stars are listed separately; their combined magnitudes appear in footnotes. Where more than one star has the same apparent magnitude, they have been sorted by right ascension.

Apparent Visual Magnitude ($m_V = V$): Apparent magnitudes, with "v" appended for variables. The (yellow) V filter corresponds roughly to the response of the eye.

Color Index ($B − V$): Since B on this system is the brightness of a star through a blue filter, the difference $B − V$ measures apparent color (as possibly reddened by interstellar dust, although, in general, $B − V$ and spectral type are well correlated).

Absolute Visual Magnitude (M_V) and *Distance in Light-Years* (D): Absolute magnitudes and distances are determined from parallaxes, except where a colon follows the absolute magnitude; in these cases, both quantities are determined from a calibration of the spectral classification.

APPENDIX 5

Appendix 5A The Nearest Stars

Name			RA (2000) Dec				parallax		D lt-yr	motion mas/yr	θ °	V_{rad} km/s	Sp.T.	m_V	M_V
			h	m	°	′	mas	±mas							
	Sun												G2V	−26.72	4.83
1	Proxima Cen	(C)	14	30	−62	41	768.85	0.29	4.24	3853	281.5	−16	M5.0V	11.05	15.48
	α Cen	A	14	40	−60	50	747.23	1.17	4.36	3710	277.5	−26	G2V	0.01	4.38
		B	14	40	−60	50				3724	284.8	−18	K0.0V	1.34	5.71
2	Barnard's		17	58	+04	42	545.51	0.29	5.98	10358	355.6	−111	M3.5V	9.57	13.25
3	Wolf 359		10	56	+07	01	419.10	2.10	7.78	4696	234.6	+19	M5.5V	13.53	16.64
4	Lalande 21185		11	03	+35	58	393.25	0.57	8.29	4802	186.9	−85	M2.0V	7.47	10.44
5	Sirius	A	06	45	−16	43	380.02	1.28	8.58	1339	204.1	−9	A1V	−1.43	1.47
		B	06	45	−16	43							DA2	8.44	11.34
6	BL Cet	(A)	01	39	−17	57	373.70	2.70	8.73	3368	080.4	+29	M5.5V	12.61	15.47
	UV Cet	(B)	01	39	−17	57							+32	M6.0V	13.06	15.92
7	Ross 154		18	50	−23	50	337.22	1.97	9.67	666	106.8	−10	M3.5V	10.44	13.08
8	Ross 248		23	42	+44	11	316.37	0.55	10.31	1617	177.0	−78	M5.5V	12.29	14.79
9	ε Eri		03	33	−09	27	311.22	0.09	10.48	977	271.1	+16	K2.0V	3.73	6.20
10	CD −36 15693		23	06	−35	51	305.08	0.70	10.69	6896	078.9	+9	M1.0V	7.34	9.76
11	Ross 128		11	48	+00	48	298.14	1.37	10.94	1361	153.6	−31	M4.0V	11.16	13.53
12	EZ Aqr	A	22	39	−15	18	289.50	4.40	11.27	3254	046.6	−50	M5.0VJ	13.03	15.34
		B	22	39	−15	18								13.27	15.58
		C	22	39	−15	18								15.07	17.38
13	61 Cyg	A	21	07	+38	45	286.08	0.48	11.40	5281	051.9	−66	K5.0V	5.20	7.48
		B	21	07	+38	45				5172	052.6	−64	K7.0V	6.03	8.31
14	Procyon	A	07	39	+05	13	285.17	0.64	11.44	1259	214.7	−4	F5 IV–V	0.37	2.65
		B	07	39	+05	13							DQZ	10.70	12.98
15	BD +59 1915	A	18	43	+59	38	283.83	1.46	11.49	2238	323.6	−1	M3.0V	8.90	11.17
		B	18	43	+59	38				2313	323.0	+1	M3.5V	9.69	11.96
16	GX And	(A)	00	18	+44	01	279.87	0.60	11.65	2918	081.9	+12	M1.5V	8.08	10.31
	GQ And	(B)	00	18	+44	01						+11	M3.5V	11.06	13.29
17	ε Ind	A	22	03	−56	47	276.07	0.28	11.81	4704	122.7	−40	K3.0V	4.68	6.89
		B	22	04	−56	47				4823	121.1		T1.0V		
		C	22	04	−56	46							T6.0V		
18	DX Can		08	30	+26	47	275.80	3.00	11.83	1290	242.2	−5	M6.0V	14.90	17.10
19	τ Cet		01	44	−15	56	273.97	0.17	11.91	1922	296.4	−17	G8.5V	3.49	5.68
20	GJ 1061		03	36	−44	31	272.01	1.30	11.99	831	118.8	−20	M5.0V	13.09	15.26
21	YZ Cet		01	13	−17	00	269.08	2.99	12.12	1372	061.9	+28	M4.0V	12.10	14.25
22	Luyten's		07	27	+05	14	266.23	0.66	12.25	3738	171.2	+18	M3.5V	9.85	11.98
23	SCR 1845–6357	A	18	45	−63	58	259.50	1.11	12.57	2558	074.7		M8.5V	17.40	19.47
		B	18	45	−63	58							T6.0V		
24	SO 0253+1652		02	53	+16	53	259.41	0.89	12.57	5050	137.9		M6.5V	15.14	17.21
25	Kapteyn's star		05	12	−45	01	255.67	0.91	12.76	8670	131.4	+245	M2.0VI	8.85	10.89

By Todd Henry, © 2012 Royal Astronomical Society of Canada. For explanatory text, see the *Observer's Handbook* of the Royal Astronomical Society of Canada.

Appendix 5B Selected Nearby Stars. The binary stars are ordered by magnitude of the primary component

Name		RA (2000) Dec				parallax mas	D lt-yr	motion		Sp.T.	m_V
		h	m	°	′			″/yr	°		
Sirius[A]	A	6	45	−16	43	380	8.6	1.3	204	A1	−1.4
α Cen[B]	A	14	40	−60	50	747	4.4	3.7	278	G2	0.0
	B	14	40	−60	50			3.7	285	K0.0	1.3
Procyon[C]	A	7	39	5	13	285	11.4	1.3	215	F5	0.4
τ Cet[D]		1	44	−15	56	274	11.9	1.9	296	G8.5	3.5
ε Eri[E]	+1 Planet	3	33	−09	27	311	10.5	1.0	271	K2.0	3.7
o Eri[F]	A	4	15	−07	39	201	16.3	4.1	213	K1.0	4.4
	B	4	15	−07	39			4.1	212	DA4	9.5
ε Ind	A	22	3	−56	47	276	11.8	4.7	123	K3.0	4.7
61 Cyg[G]	A	21	7	38	45	286	11.4	5.3	52	K5.0	5.2
	B	21	7	38	45			5.2	53	K7.0	6.0
Kapteyn's		5	12	−45	01	256	12.8	8.7	131	M2.0	8.9
Barnard's[H]		17	58	4	42	546	6.0	10.4	356	M3.5	9.6
Luyten's[I]		7	27	5	14	266	12.3	3.7	171	M3.5	9.9

[A] α CMa.
[B] Rigil Kentaurus, similar to Sun.
[C] α CMi.
[D] Similar to Sun, the nearest such star visible to observers at north-temperate latitudes.
[E] The star nearest to Sun known to have at least one planet.
[F] B is the nearest white-dwarf star that is easily observable.
[G] First star to have its stellar parallax measured (Bessel, 1838).
[H] Star with the largest proper motion; second-closest star system to Sun (Barnard, 1916).
[I] Only 1.2 lt-yr from Procyon.
The symbol for milliarcseconds is mas.

1 *Crab Nebula* (supernova)   2 Globular cluster   3 Globular cluster   4 Globular cluster   5 Globular cluster   6 *Butterfly Cluster* (open)   7 Open cluster

15 Globular cluster   16 *Eagle Nebula* (diffuse)   17 *Omega Nebula* (diffuse)   18 Open cluster   19 Globular cluster   20 *Trifid Nebula* (diffuse)   21 Open cluster

29 Open cluster   30 Globular cluster   31 *Andromeda Galaxy* (spiral)   32 Elliptical galaxy   33 *Pinwheel Galaxy* (spiral)   34 Open cluster   35 Open cluster

43 Diffuse nebula   44 *Beehive Cluster* (open)   45 *Pleiades* (open cluster)   46 Open cl. & plan. neb.   47 Open cluster   48 Open cluster   49 Elliptical galaxy

57 *Ring Nebula* (planetary)   58 Spiral galaxy   59 Elliptical galaxy   60 Elliptical galaxy   61 Spiral galaxy   62 Globular cluster   63 *Sunflower Galaxy* (spiral)

71 Globular cluster   72 Globular cluster   73 Asterism   74 Spiral galaxy   75 Globular cluster   76 *Little Dumbbell* (planetary)   77 Spiral galaxy

85 Elliptical galaxy   86 Elliptical galaxy   87 Elliptical galaxy   88 Spiral galaxy   89 Elliptical galaxy   90 Spiral galaxy   91 Spiral galaxy

99 Spiral galaxy   100 Spiral galaxy   101 Spiral galaxy   102 Messier object number 102 is a duplicate of M101, so there are really only 109 Messier objects.   103 Open cluster   104 *Sombrero Galaxy* (spiral)   105 Elliptical galaxy

8 *Lagoon Nebula* (diffuse) 9 Globular cluster 10 Globular cluster 11 *Wild Duck Cluster* (open) 12 Globular cluster 13 *Hercules Cluster* (globular) 14 Globular cluster

22 Globular cluster 23 Open cluster 24 *Sagittarius Star Cloud* 25 Open cluster 26 Open cluster 27 *Dumbbell Nebula* (planetary) 28 Globular cluster

36 Open cluster 37 Open cluster 38 Open cluster 39 Open cluster 40 Double star 41 Open cluster 42 *Orion Nebula* (diffuse)

50 Open cluster 51 *Whirlpool Galaxy* (spiral) 52 Open cluster 53 Globular cluster 54 Globular cluster 55 Globular cluster 56 Globular cluster

64 *Black-Eye Galaxy* (spiral) 65 Spiral galaxy 66 Spiral galaxy 67 Open cluster 68 Globular cluster 69 Globular cluster 70 Globular cluster

78 Diffuse nebula 79 Globular cluster 80 Globular cluster 81 Spiral galaxy 82 Irregular galaxy 83 Spiral galaxy 84 Elliptical galaxy

92 Globular cluster 93 Open cluster 94 Spiral galaxy 95 Spiral galaxy 96 Spiral galaxy 97 *Owl Nebula* (planetary) 98 Spiral galaxy

106 Spiral galaxy 107 Globular cluster 108 Spiral galaxy 109 Spiral galaxy 110 Elliptical galaxy

Photos by Tim Hunter and James McGaha, Grasslands Observatory, http://www.3towers.com

APPENDIX 7

Appendix 7 The Constellations

Latin Name	Genitive	Abbreviation	Translation	Latin Name	Genitive	Abbreviation	Translation
Andromeda	Andromedae	And	Andromeda*	Lacerta	Lacertae	Lac	Lizard
Antlia	Antliae	Ant	Pump	Leo	Leonis	Leo	Lion
Apus	Apodis	Aps	Bird of Paradise	Leo Minor	Leonis Minoris	LMi	Little Lion
Aquarius	Aquarii	Aqr	Water Bearer	Lepus	Leporis	Lep	Hare
Aquila	Aquilae	Aql	Eagle	Libra	Librae	Lib	Scales
Ara	Arae	Ara	Altar	Lupus	Lupi	Lup	Wolf
Aries	Arietis	Ari	Ram	Lynx	Lyncis	Lyn	Lynx
Auriga	Aurigae	Aur	Charioteer	Lyra	Lyrae	Lyr	Harp
Boötes	Boötis	Boo	Herdsman	Mensa	Mensae	Men	Table (mountain)
Caelum	Caeli	Cae	Chisel	Microscopium	Microscopii	Mic	Microscope
Camelopardalis	Camelopardalis	Cam	Giraffe	Monoceros	Monocerotis	Mon	Unicorn
Cancer	Cancri	Cnc	Crab	Musca	Muscae	Mus	Fly
Canes Venatici	Canum Venaticorum	CVn	Hunting Dogs	Norma	Normae	Nor	Level (square)
Canis Major	Canis Majoris	CMa	Big Dog	Octans	Octantis	Oct	Octant
Canis Minor	Canis Minoris	CMi	Little Dog	Ophiuchus	Ophiuchi	Oph	Ophiuchus* (serpent bearer)
Capricornus	Capricorni	Cap	Goat	Orion	Orionis	Ori	Orion*
Carina	Carinae	Car	Ship's Keel**	Pavo	Pavonis	Pav	Peacock
Cassiopeia	Cassiopeiae	Cas	Cassiopeia*	Pegasus	Pegasi	Peg	Pegasus* (winged horse)
Centaurus	Centauri	Cen	Centaur*	Perseus	Persei	Per	Perseus*
Cepheus	Cephei	Cep	Cepheus*	Phoenix	Phoenicis	Phe	Phoenix
Cetus	Ceti	Cet	Whale	Pictor	Pictoris	Pic	Easel
Chamaeleon	Chamaeleonis	Cha	Chameleon	Pisces	Piscium	Psc	Fish
Circinus	Circini	Cir	Compass	Piscis Austrinus	Piscis Austrini	PsA	Southern Fish
Columba	Columbae	Col	Dove	Puppis	Puppis	Pup	Ship's Stern**
Coma Berenices	Comae Berenices	Com	Berenice's Hair*	Pyxis	Pyxidis	Pyx	Ship's Compass**
Corona Australis	Coronae Australis	CrA	Southern Crown	Reticulum	Reticuli	Ret	Net
Corona Borealis	Coronae Borealis	CrB	Northern Crown	Sagitta	Sagittae	Sge	Arrow
Corvus	Corvi	Crv	Crow	Sagittarius	Sagittarii	Sgr	Archer
Crater	Crateris	Crt	Cup	Scorpius	Scorpii	Sco	Scorpion
Crux	Crucis	Cru	Southern Cross	Sculptor	Sculptoris	Scl	Sculptor
Cygnus	Cygni	Cyg	Swan	Scutum	Scuti	Sct	Shield
Delphinus	Delphini	Del	Dolphin	Serpens	Serpentis	Ser	Serpent
Dorado	Doradus	Dor	Swordfish	Sextans	Sextantis	Sex	Sextant
Draco	Draconis	Dra	Dragon	Taurus	Tauri	Tau	Bull
Equuleus	Equulei	Equ	Little Horse	Telescopium	Telescopii	Tel	Telescope
Eridanus	Eridani	Eri	River Eridanus*	Triangulum	Trianguli	Tri	Triangle
Fornax	Fornacis	For	Furnace	Triangulum Australe	Trianguli Australis	TrA	Southern Triangle
Gemini	Geminorum	Gen	Twins	Tucana	Tucanae	Tuc	Toucan
Grus	Gruis	Gru	Crane	Ursa Major	Ursae Majoris	UMa	Big Bear
Hercules	Herculis	Her	Hercules*	Ursa Minor	Ursae Minoris	UMi	Little Bear
Horologium	Horologii	Hor	Clock	Vela	Velorum	Vel	Ship's Sails**
Hydra	Hydrae	Hya	Hydra* (water monster)	Virgo	Virginis	Vir	Virgin
Hydrus	Hydri	Hyi	Sea serpent	Volans	Volantis	Vol	Flying Fish
Indus	Indi	Ind	Indian	Vulpecula	Vulpeculae	Vul	Little Fox

*Proper names
**Formerly formed the constellation Argo Navis, the Argonauts' ship

SELECTED READINGS

Magazines

Astronomy, 21027 Crossroads Circle, P.O. Box 1612, Waukesha, WI 53187, 800–533–6644, astronomy.com.

Discover, P.O. Box 42105, Palm Coast, FL 34142–0105, 800–829–9132; discovermagazine.com.

The Griffith Observer, 2800 East Observatory Road, Los Angeles, CA 90027, www.griffithobs.org.

Journal of the Royal Astronomical Society of Canada, rasc.ca/jrasc.

Mercury, Astronomical Society of the Pacific, 390 Ashton Avenue, San Francisco, CA 94112, 800–962–3412, www.astrosociety.org.

National Geographic, P.O. Box 96583, Washington, DC 20078–9973, 800-NGC-LINE, www.nationalgeographic.com.

Natural History, P.O. Box 5000, Harlan, IA 51593–5000, 800–234–5252, www.naturalhistorymag.com.

New Scientist, 151 Wardour Street, London W1V 4BN, U.K., 888–822–4342, rbi.subscriptions@rbi.co.uk; www.newscientist.com.

Physics Today, American Institute of Physics, 2 Huntington Quadrangle, Melville, NY 11747, 800–344–6902, www.physicstoday.org.

The Planetary Report, The Planetary Society, 85 South Grand Avenue, Pasadena, CA 91105, 818–793–5100, www.planetary.org/explore/the-planetary-report/.

Science News, P.O. Box 1925, Marion, OH 43305, 800–552–4412, www.sciserv.org, mainly online.

Science Year, World Book Encyclopedia, Inc., P.O. Box 11207, Des Moines, IA 50340–1207, 800–504–4425, www.worldbook.com.

Scientific American, P.O. Box 3186, Harlan, IA 51593–2377, 800–333–1199, www.scientificamerican.com.

Sky and Telescope, P.O. Box 9111, Belmont, MA 02138, 800–253–0245, www.skyandtelescope.com.

Smithsonian, P.O. Box 420311, Palm Coast, FL 32142–0311, Washington, DC 20560, 800–766–2149; www.smithsonianmag.si.edu.

StarDate, McDonald Observatory, University of Texas, One University Station A2100, Austin, TX 78712, 800-STARDATE, www.stardate.org.

Observing Reference Books

The Astronomical Almanac (Washington, DC 20402: U.S. Government Printing Office, published yearly).

David S. Chandler, *The Night Sky Planisphere* (Springville, CA: The David Chandler Company, 1992). Planispheres showing constellations.

Guy Consolmagno and Dan M. Davis, *Turn Left at Orion: Hundreds of Night Sky Objects to See in a Home Telescope – and How to Find Them* (Cambridge: Cambridge University Press, 2011). Covers both northern and southern hemispheres.

Michael Covington, *Digital SLR Astrophotography* (Cambridge: Cambridge University Press, 2007).

Terence Dickinson, *Nightwatch: A Practical Guide to Viewing the Universe*, 3rd ed. (Willowdale, Ontario: Firefly Books, 1998).

R. Gendler, L. Lindberg Christensen, and D. Malin, *Treasures of the Southern Sky* (New York: Springer, 2012).

Philip S. Harrington, *Cosmic Challenge: The Ultimate Observing List for Amateurs* (Cambridge: Cambridge University Press, 2010).

Alan Hirshfeld, Roger W. Sinnott, and François Ochsenbein, 1991 (vol. 1, 2nd ed.); Alan Hirshfeld and Roger W. Sinnott, 1985 (vol. 2, 1st ed.). *Sky Catalogue 2000.0*, (Cambridge, MA: Sky Publishing Corp.). Volume 1 is stars, and Volume 2 is full of tables of all other objects.

Chris Kitchen and Robert W. Forrest, *Seeing Stars: The Night Sky through Small Telescopes* (Berlin/New York: Springer, 1997).

David H. Levy, *David Levy's Guide to Variable Stars* (Cambridge: Cambridge University Press, 2006).

David H. Levy, *David Levy's Guide to Eclipses, Transits, and Occultations* (Cambridge: Cambridge University Press, 2010).

Axel Mellinger and Ronald Stoyan, *The Cambridge Photographic Star Atlas* (Cambridge: Cambridge University Press, 2011).

Jim Mullaney, *Celestial Harvest: 3000-Plus Showpieces of the Heavens for Telescope Viewing & Contemplation* (self-published, P.O. Box 1146, Exton, PA 19341, 1998); jimmullaneysm@nospam.com; store.yahoo.com/doverpublications.

Observer's Handbook (yearly), Royal Astronomical Society of Canada, 203–4920 Dundas Street W., Toronto, Ontario M9A 1B7, Canada.

Stephen James O'Meara, *Exploring the Solar System with Binoculars: A Beginner's Guide to the Sun, Moon, and Planets* (Cambridge: Cambridge University Press, 2010).

Stephen James O'Meara and Mario Motta, *Deep Sky Companions: The Secret Deep* (Cambridge: Cambridge University Press, 2011).

Guy Ottewell, *Astronomical Calendar* (yearly) and *The Astronomical Companion* (Raynham, MA 02767–0102 : Universal Workshop), 800–533–5083; www.UniversalWorkshop.com.

Jay M. Pasachoff, *A Field Guide to the Stars and Planets*, 4th ed. (Boston: Houghton Mifflin Co., 2000; updated 2010). All kinds of observing information, including monthly maps and the 2000.0 sky atlas by Wil Tirion, and Graphic Timetables to locate planets and special objects like clusters and galaxies.

Jay M. Pasachoff, *Peterson's First Guide to Astronomy* (Boston: Houghton Mifflin Co., 1997; updated printing, 2012). A brief, beautifully illustrated introduction to observing the sky. Tirion monthly maps.

Jay M. Pasachoff, *Peterson's First Guide to the Solar System* (Boston: Houghton Mifflin Co., 1997). Color illustrations and simple descriptions mark this elementary introduction, now out of print.

H. A. Rey, *Find the Constellations*, 2nd ed. (Boston: Houghton Mifflin Harcourt, 2008). An excellent guide to finding and recognizing the constellations.

Ian Ridpath, ed., *Norton's Star Atlas and Reference Handbook (epoch 2000.0)*, 20th ed. (New York: Dutton, 2004). An updated old standard.

Roger W. Sinnott, ed., *NGC 2000.0: The Complete New General Catalogue and Index Catalogues of Nebulae and Star Clusters* (Cambridge, MA: Sky Publishing Corp. and New York: Cambridge University Press, 1998). A centennial reissue of Dreyer's work with updated data.

Roger W. Sinnott and Michael A. C. Perryman, *Millennium Star Atlas* (Cambridge, MA: Sky Publishing Corp., 1997). Uses the Hipparcos data.

Wil Tirion, *Cambridge Star Atlas*, 4th ed. (Cambridge: Cambridge University Press, 2011). A naked-eye star atlas in full color. A moon map, 24 monthly sky maps, 20 detailed star charts, and 6 all-sky maps.

Wil Tirion, Barry Rappaport, and Will Remaklus, *Uranometria 2000.0*, 2nd ed. (Richmond, VA: Willmann-Bell, 2001). Volume 1 covers +90° to –6°. Volume 2 covers +6° to –90°. Star maps to magnitude 9.7.

Wil Tirion and Roger W. Sinnott, *Sky Atlas 2000.0*, 2nd ed. (Cambridge, MA: Sky Publishing Corp. and New York: Cambridge University Press, 1998). Large-scale star charts to magnitude 8.5.

Sean E. Urban and P. Kenneth Seidelmann, *Explanatory Supplement to the Astronomical Almanac* (Mill Valley, CA: University Science Books, 2013).

Willmann-Bell, *Astronomy Books* catalogue (P.O. Box 35025, Richmond, VA 23235), 800–825–7827, www.willbell.com. They publish many observing guides and distribute all kinds of astronomy books.

For Information About Amateur Societies

American Association of Variable Star Observers (AAVSO), 49 Bay State Road, Cambridge, MA 02138, www.aavso.org.

American Meteor Society, Dept. of Astronomy and Astrophysics, Pennsylvania State University, 121 EEE, University Park, PA 16802, www.amsmeteors.org.

Astronomical League, the umbrella group of amateur societies. For their newsletter, *The Reflector*, see www.astroleague.org/al/reflectr/reflmain.html, www.astroleague.org.

Astronomical Society of the Pacific, 390 Ashton Avenue, San Francisco, CA 94112, www.astrosociety.org.

British Astronomical Association, Burlington House, Piccadilly, London W1V 0NL, U.K., www.britastro.org.

International Dark Sky Association, 3223 N. First Avenue, Tucson, AZ 85719, 520–293–3198, www.darksky.org.

The Planetary Society, 85 South Grand Avenue, Pasadena, CA 91105, 818–793–5100, www.planetary.org.

Royal Astronomical Society of Canada, 203–4920 Dundas Street W, Toronto ON M9A 1B7, Canada, 416–924–7973, www.rasc.ca.

Careers in Astronomy

AAS American Astronomical Society Education Office, 2000 Florida Avenue, N.W., Suite 400, Washington, DC 20009, aased@aas.org; aas.org/education. Information on careers in astronomy is available at aas.org/education/careers.php.

Teaching

Lucienne Gouguenheim, Derek McNally, and John R. Percy, *New Trends in Astronomy Teaching* (Cambridge: Cambridge University Press, 1997). The proceedings of International Astronomical Union Colloquium #162, held in London in 1996.

George Greenstein, *Understanding the Universe: An Inquiry Approach to Astronomy and the Nature of Scientific Research* (New York: Cambridge University Press, 2012). A student-active introduction to astronomy.

Jay M. Pasachoff and John R. Percy, *The Teaching of Astronomy* (Cambridge: Cambridge University Press, 1990; paperbound, 1993). The proceedings of International Astronomical Union Colloquium #105, held in Williamstown, MA, in 1988. Available at adswww.harvard.edu.

Jay M. Pasachoff and John R. Percy, *Teaching and Learning Astronomy: Effective Strategies for Educators Worldwide* (Cambridge: Cambridge University Press, 2005). The proceedings of International Astronomical Union Special Session, held at the Sydney, Australia, General Assembly in 2003.

Jay M. Pasachoff, Rosa M. Ros, and Naomi Pasachoff, *Innovation in Astronomy Education* (Cambridge: Cambridge University Press, 2008). The proceedings of International Astronomical Union Special Session, held at the Prague, Czech Republic, General Assembly in 2006.

John R. Percy, *Astronomy Education: Current Developments, Future Coordination* (San Francisco: Astronomical Society of the Pacific, 1996). Proceedings of an ASP conference, held in College Park, MD, in 1995.

Stephen M. Pompea, ed., *Great Ideas for Teaching Astronomy*, 3rd ed. (Pacific Grove, CA: Brooks Cole, 2000). Lots of interesting suggestions for effective astronomy teaching.

General Reading and Reference

Geoff Andersen, *The Telescope: Its History, Technology, and Future* (Princeton: Princeton University Press, 2007).

Jean Audouze and Guy Israel, eds. *The Cambridge Atlas of Astronomy*, 3rd ed. (Cambridge: Cambridge University Press, 1994). Coffee-table size with fantastic photos and authoritative text.

Marcia Bartusiak, *Archives of the Universe: A Treasury of Astronomy's Historic Works of Discovery* (New York: Pantheon, 2004; paperback, Vintage, 2006).

Christopher De Pree and Alan Axelrod, *The Complete Idiot's Guide to Astronomy*, 4th ed. (New York: Alpha/Penguin, 2008). *The Complete Idiot's Guide* series has as a theme "You're no idiot, of course," at what you do best, but on this subject….

David DeVorkin and Robert W. Smith, *Hubble: Imaging Space and Time* (Washington, DC: National Geographic Society, 2008).

Terence Dickinson, *The Universe and Beyond*, 4th ed. (Willowdale, Ontario: Firefly, 2004).

Ken M. Harrison, *Astronomical Spectroscopy for Amateurs* (New York: Springer, 2011).

Robin Kerrod and Carole Stott, *Hubble: The Mirror on the Universe*, revised and updated (Willowdale, Ontario: Firefly, 2007).

Charles H. Langmuir and Wally Broecker, *How to Build a Habitable Planet: The Story of Earth from the Big Bang to Humankind* (Princeton: Princeton University Press, 2012).

Stephen P. Maran, *Astronomy for Dummies*, 3rd ed. (New York: Wiley, 2013).

Stephen P. Maran and Laurence A. Marschall, *Galileo's New Universe: The Revolution in Our Understanding of the Cosmos* (Dallas: BenBella Books, 2009).

W. Patrick McCray, *Giant Telescopes: Astronomical Ambition and the Promise of Technology* (Cambridge, MA: Harvard University Press, 2004).

Paul Murdin, ed., *Encyclopedia of Astronomy and Astrophysics* (Bristol, U.K.: Macmillan and Institute of Physics, 2000). See www.ency-astro.com.

Robert Nemiroff and Jerry Bonnell, *The Universe: 365 Days* (New York: Abrams, 2003). Selections from Astronomy Picture of the Day; see antwrp.gsfc.nasa.gov.

Sybil P. Parker and Jay M. Pasachoff, *McGraw-Hill Encyclopedia of Astronomy* (New York: McGraw-Hill, 1993). Continual updates and new articles are available at www.accessscience.com/, for which author J.M.P. is astronomy consultant.

Philip C. Plait, *Bad Astronomy: Misconceptions and Misuses Revealed, from Astrology to the Moon Landing "Hoax,"* www.badastronomy.com.

Martin Rees, *Illustrated Encyclopedia of the Universe* (New York: DK Publishing, 2011). The ultimate guided tour of the Universe from the Astronomer Royal.

Martin Rees, *From Here to Infinity: A Vision for the Future of Science* (New York: W. W. Norton & Company, 2012).

George H. Rieke, *The Last of the Great Observatories: Spitzer and the Era of Faster, Better, Cheaper at NASA* (Tucson: University of Arizona Press, 2006).

Neil deGrasse Tyson and Donald Goldsmith, *Origins: Fourteen Billion Years of Cosmic Evolution* (New York: W. W. Norton, 2004).

Donald G. York, Owen Gingerich, and Shuang-Nan Zhang, *The Astronomy Revolution: 400 Years of Exploring the Cosmos* (Boca Raton, FL, 2012, CRC Press/Taylor & Francis).

Jack B. Zirker, *An Acre of Glass: A History and Forecast of the Telescope* (Baltimore: Johns Hopkins University Press, 2005).

History

James A. Connor, *Kepler's Witch* (San Francisco: Harper San Francisco, 2004).

Stillman Drake, *Galileo: A Very Short Introduction* (New York: Sterling Publishing Co., 2010, by arrangement with Oxford University Press, 1980).

Kitty Ferguson, *The Nobleman and His Housedog: Tycho Brahe and Johannes Kepler: The Strange Partnership that Revolutionized Science* (New York: Walker & Co, 2002), tychoandkepler.com.

Owen Gingerich and James MacLachlan, *Nicolaus Copernicus: Making the Earth a Planet* (New York: Oxford University Press, 2005).

Owen Gingerich, *The Eye of Heaven: Ptolemy, Copernicus, and Kepler* (New York: American Institute of Physics, 1993).

Owen Gingerich, *The Book Nobody Read: Chasing the Revolutions of Nicolaus Copernicus* (New York: Walker and Co., 2004, paperback, 2005). About the hunt for copies of Copernicus's book.

James Gleick, *Isaac Newton* (New York: Vintage, 2004).

John L. Heilbron, *Galileo* (Oxford: Oxford University Press, 2010). The soft-cover version of this masterful biography was released in 2012.

Michael Hoskin, ed., *The Cambridge Concise History of Astronomy* (Cambridge: Cambridge University Press, 1999).

Michael Hoskin, *Discoverers of the Universe, William and Caroline Herschel* (Princeton: Princeton University Press, 2011).

George Johnson, *Miss Leavitt's Stars* (New York: Atlas Books/W. W. Norton, 2005).

James MacLachlan, *Galileo Galilei: First Physicist* (New York: Oxford University Press, 1999).

Giulio Magli, *Mysteries and Discoveries of Archaeoastronomy: From Giza to Easter Island* (New York: Springer, 2009).

Arthur I. Miller, *Empire of the Stars: Obsession, Friendship, and Betrayal in the Quest for Black Holes* (Boston: Houghton Mifflin, 2005).

Simon Mitton, *Fred Hoyle: A Life in Science* (Cambridge: Cambridge University Press, 2011).

Andrew Norton, ed., *Observing the Universe* (Cambridge: Cambridge University Press, 2004). How professional astronomers work.

Naomi Pasachoff, *World Book Biographies in Context: Newton, Einstein, and Oppenheimer* (Chicago: World Book, 2006).

Dava Sobel, *A More Perfect Heaven: How Copernicus Revolutionized the Cosmos* (New York: Walker Publishing Company, Inc., 2011).

James Voelkel, *Johannes Kepler and the New Astronomy* (New York: Oxford University Press, 1999). A short biography in the Oxford Portraits series.

Fred Watson, *Stargazer: The Life and Times of the Telescope* (Boston and New York: Da Capo Press, 2004).

Robert Zimmerman, *The Universe in a Mirror: The Saga of the Hubble Space Telescope and the Visionaries Who Built It* (Princeton: Princeton University Press, 2010).

Art and Astronomy

Roberta J. M. Olson and Jay M. Pasachoff, *Fire in the Sky: Comets and Meteors, the Decisive Centuries, in British Art and Science* (Cambridge: Cambridge University Press, 1998, 1999). From the time of Newton and Halley to the present.

Solar System

David Baker and Todd Ratcliff, *The 50 Most Extreme Places in Our Solar System* (London and Cambridge, MA: The Belknap Press of Harvard University Press, 2010). Beautifully illustrated with many images.

J. Kelly Beatty, Caroline Collins Petersen, and Andrew Chaikin, *The New Solar System*, 4th ed. (Cambridge, MA: Sky Publishing Corp. and Cambridge, U.K.: Cambridge University Press, 1999). Each chapter was written by a different expert.

Jim Bell and Jacqueline Mitton, eds., *Asteroid Rendezvous: NEAR Shoemaker's Adventures at Eros* (Cambridge: Cambridge University Press, 2002).

Joseph M. Boyce, *The Smithsonian Book of Mars* (Washington, DC: Smithsonian Institution Press, 2002).

Alan Boyle, *The Case for Pluto: How a Little Planet Made a Big Difference* (New York: Wiley, 2009).

Mike Brown, *How I Killed Pluto and Why It Had It Coming* (New York: Spiegel & Grau, 2010).

Arlin Crotts, *The New Moon: Water, Exploration, and Future Habitation* (Cambridge: Cambridge University Press, 2012).

Imke de Pater and Jack J. Lissauer, *Planetary Sciences*, 2nd ed. (Cambridge: Cambridge University Press, 2010).

Monica Grady, Giovanni Pratesi, and Vanni Moggi Cecchi, eds., *Atlas of Meteorites* (Cambridge: Cambridge University Press, 2012).

Chris Impey, ed., *Talking about Life: Conversations on Astrobiology* (Cambridge: Cambridge University Press, 2010). Candid interviews with dozens of astronomers, geologists, biologists, and writers about the origin of life on Earth and beyond.

Barrie W. Jones, *Life in the Solar System and Beyond* (New York: Springer-Praxis Books, 2004).

Barrie W. Jones, *Pluto: Sentinel of the Outer Solar System* (Cambridge: Cambridge University Press, 2010).

Jeffrey S. Kargel, *Mars: A Warmer, Wetter Planet* (New York: Springer-Praxis Books, 2004).

Jeffrey Kluger, *Moon Hunters: NASA's Remarkable Expeditions to the Ends of the Solar System* (New York: Simon and Schuster, 2001).

Kenneth Lang, *The Cambridge Guide to the Solar System*, 2nd ed. (Cambridge: Cambridge University Press, 2011).

Nick Lomb, *Transit of Venus: 1631 to the Present* (New York: The Experiment Publishing, 2011). Includes many illustrations, artwork, and historical and contemporary photographs.

Ralph Lorenz and Jacqueline Mitton, *Titan Unveiled: Saturn's Mysterious Moon Explored* (Princeton: Princeton University Press, 2010).

Eli Maor, *June 8, 2004: Venus in Transit* (Princeton: Princeton University Press, 2000).

Laurence A. Marschall and Stephen P. Maran, *Pluto Confidential: An Insider Account of the Ongoing Battles over the Status of Pluto* (Dallas: BenBella Books, 2009).

W. Patrick McCray, *The Visioners: How a Group of Elite Scientists Pursued Space Colonies, Nanotechnologies, and a Limitless Future* (Princeton: Princeton University Press, 2012).

Rod Pyle, *Destination Mars: New Explorations of the Red Planet* (Amherst, NY: Prometheus Books, 2012).

Stephen J. Pyne, *Voyager: Seeking Newer Worlds in the Third Great Age of Discovery* (New York: Viking, 2010).

David A. Rothery, Neil McBride, and Iain Gilmour, eds., *An Introduction to the Solar System*, 2nd ed. (Cambridge: Cambridge University Press, 2011).

C. T. Russell, *New Horizons: Reconnaissance of the Pluto-Charon System and the Kuiper Belt* (New York: Springer, 2009).

Carl Sagan, *Pale Blue Dot* (New York: Random House, 1994). One of astronomy's most eloquent spokesperson's last works.

Paul Schenk, *Atlas of the Galilean Satellites* (Cambridge: Cambridge University Press, 2010).

Govert Schilling, *The Hunt for Planet X: New Worlds and the Fate of Pluto* (New York: Copernicus Books/Springer Science, 2009).

William Sheehan and John Westfall, *The Transits of Venus* (Amherst, NY: Prometheus Books, 2004).

Steven Squyres, *Roving Mars: Spirit, Opportunity, and the Exploration of the Red Planet* (New York: Hyperion, 2005). By the principal investigator of the Mars Exploration Rovers Spirit and Opportunity.

S. Alan Stern, *Our Worlds: The Magnetism and Thrill of Planetary Exploration: As Described by Leading Planetary Scientists* (Cambridge: Cambridge University Press, 1999).

S. Alan Stern, ed., *Worlds Beyond* (Cambridge: Cambridge University Press, 2003).

S. Alan Stern and Jacqueline Mitton, *Pluto & Charon: Ice Worlds on the Ragged Edge of the Solar System*, 2nd ed. (New York: Wiley, 2005).

Neil DeGrasse Tyson, *The Pluto Files: The Rise and Fall of America's Favorite Planet* (New York: W. W. Norton & Company, 2009).

David A. Weintraub, *Is Pluto a Planet?* (Princeton: Princeton University Press, 2009). Perspective on the long-term battles over the definition of the term "planet."

Andrea Wulf, *Chasing Venus: The Race to Measure the Heavens* (New York: Alfred A. Knopf, 2012). Chronicles the international scientific endeavor to observe and perform calculations of the 1761 transit of Venus.

Donald K. Yeomans, *Near-Earth Objects: Finding Them Before They Find Us* (Princeton: Princeton University Press, 2012).

Ben Zuckerman and Michael Hart, eds., *Extraterrestrials: Where Are They?* 2nd ed. (Cambridge: Cambridge University Press, 1995).

Exoplanets

Carole A. Haswell, *Transiting Exoplanets* (Cambridge: Cambridge University Press, 2010).

Ray Jayawardhana, *Strange New Worlds: The Search for Alien Planets and Life Beyond Our Solar System* (Princeton: Princeton University Press, 2011).

Chris Kitchin, *Exoplanets: Finding, Exploring, and Understanding Alien Worlds* (New York: Springer, 2011).

Michael Perryman, *The Exoplanet Handbook* (Cambridge: Cambridge University Press, 2011).

Sara Seager, *Exoplanet Atmospheres* (Princeton: Princeton University Press, 2010).

The Sun

David Alexander, *The Sun* (Santa Barbara: ABC-CLIO, LLC, 2009). One of the Greenwood Guides to the Universe.

Bob Berman, *The Sun's Heartbeat: And Other Stories from the Life of the Star That Powers Our Planet* (New York: Little, Brown and Company, 2011).

Arvind Bhatnagar and William C. Livingston, *Fundamentals of Solar Astronomy* (Singapore: World Scientific, 2005). Comprehensive and phenomenological but relatively non-mathematical.

Michael J. Carlowicz and Ramon E. Lopez, *Storms from the Sun: The Emerging Science of Space Weather* (Washington, DC: Joseph Henry Press, 2000).

Fred Espenak, *Fifty Year Canon of Solar Eclipses* (NASA Ref. Pub. 1178, Rev. 1987). Maps and tables.

Leon Golub and Jay M. Pasachoff, *Nearest Star: The Surprising Science of Our Sun*, 2nd ed. (New York: Cambridge University Press, 2013). A non-technical trade book.

Leon Golub and Jay M. Pasachoff, *The Solar Corona*, 2nd ed. (Cambridge: Cambridge University Press, 2010). An advanced text.

Kenneth R. Lang, *The Sun from Space* (New York: Springer, 2009).

Jay M. Pasachoff, *The Complete Idiot's Guide to the Sun* (Indianapolis: Alpha Books, 2003). Downloadable.

Jack B. Zirker, *Journey from the Center of the Sun* (Princeton: Princeton University Press, 2001, paperback, 2004).

Jack B. Zirker, *Sunquakes: Probing the Interior of the Sun* (Princeton: Princeton University Press, 2003).

Jack B. Zirker, *The Magnetic Universe: The Elusive Traces of an Invisible Force* (Baltimore: Johns Hopkins University Press, 2009).

Stars and the Milky Way Galaxy

Mitchell Begelman and Martin Rees, *Gravity's Fatal Attraction: Black Holes in the Universe* (New York: Scientific American Library, 1996).

Joshua S. Bloom, *What Are Gamma-Ray Bursts?* (Princeton: Princeton U. Press, 2011).

Timothy Ferris, *Coming of Age in the Milky Way* (New York: Harper Perennial, 2003).

Heather Couper and Nigel Henbest, *Encyclopedia of Space* (New York: Dorling Kindersley, 2009).

Martin Griffiths, *Planetary Nebulae and How to Observe Them* (New York: Springer, 2012).

Alan W. Hershfeld, *Parallax* (New York: W. H. Freeman, 2001).

James B. Kaler, *The Hundred Greatest Stars* (New York: Copernicus, 2002).

James B. Kaler, *Heaven's Touch: From Killer Stars to the Seeds of Life, How We Are Connected to the Universe* (Princeton: Princeton University Press, 2008).

James B. Kaler, *Stars and Their Spectra* (New York: Cambridge University Press, 2011). Revised and expanded 2nd edition.

Jonathan I. Katz, *The Biggest Bangs: The Mystery of Gamma-Ray Bursts: The Most Violent Explosions in the Universe* (New York: Oxford University Press, 2002).

Sun Kwok, *Stardust: The Cosmic Seeds of Life* (New York: Springer, 2013). Interplanetary and interstellar dust, and the origins of life.

NASA Astrobiology Program, *Astrobiology: The Story of Our Search for Life in the Universe*; Graphic Novels. With original artwork and storyline, chronicles the origin and evolution of astrobiology. http://www.astrobio.net/index.php?option=com_content&id=7 and http://missionscience.nasa.gov/astrobiology_graphic_novel/.

Govert Schilling, *Flash! The Hunt for the Biggest Explosions in the Universe* (Cambridge: Cambridge University Press, 2002).

Kip Thorne, *Black Holes and Time Warps: Einstein's Outrageous Legacy* (New York: Norton, 1994). An excellent account of general relativity.

Wallace H. Tucker and Karen Tucker, *Revealing the Universe: The Making of the Chandra X-ray Observatory* (Cambridge, MA: Harvard University Press, 2001).

William H. Waller, *The Milky Way: An Insider's Guide* (Princeton: Princeton U. Press, 2013).

J. Craig Wheeler, *Cosmic Catastrophes: Supernovae, Gamma-Ray Bursts, and Adventures in Hyperspace* (Cambridge: Cambridge University Press, 2000).

Galaxies and Cosmology

Fred Adams and Greg Laughlin, *The Five Ages of the Universe* (New York: Free Press, 1999). Tells the story of what will happen as the Universe expands forever.

Richard Berendzen, Richard Hart, and Daniel Seeley, *Man Discovers the Galaxies* (New York: Neale Watson Academic Publications, 1976). A historical review.

Dennis Danielson, *The Book of the Cosmos* (New York: Helix Books, 2000). A compendium from millennia of writings about the cosmos, with useful commentaries.

Timothy Ferris, *The Whole Shebang: A State of the Universe(s) Report* (New York: Simon & Schuster, 1997, 1998). Cosmology, from a gifted journalist.

Ann Finkbeiner, *A Grand and Bold Thing: An Extraordinary New Map of the Universe Ushering in a New Era of Discovery* (New York: Free Press, 2010).

Michael Friedlander, *A Thin Cosmic Rain: Particles from Outer Space* (Cambridge, MA: Harvard University Press, 2000).

Evalyn Gates, *Einstein's Telescope: The Hunt for Dark Matter and Dark Energy in the Universe* (New York: W. W. Norton & Company, 2010).

Riccardo Giacconi, *Secrets of the Hoary Deep, A Personal History of Modern Astronomy* (Baltimore: Johns Hopkins University Press, 2008). The story of x-ray astronomy from a recipient of the 2002 Nobel Prize in Physics.

Donald Goldsmith, *The Astronomers* (New York: St. Martin's Press, 1991). The companion to a PBS television series.

Donald Goldsmith, *The Runaway Universe* (Cambridge, MA: Perseus Books, 2000). Describes the discovery of the accelerating expansion of the Universe.

Brian Greene, *The Hidden Reality: Parallel Universes and the Deep Laws of the Cosmos* (New York: Knopf, 2011).

Brian Greene and Erik Davies, *The Elegant Universe: Superstrings, Hidden Dimensions, and the Quest for the Ultimate Theory* (New York: Norton, 1999). Superstrings and the Universe; updated from a Phi Beta Kappa book-award winner.

Alan H. Guth, *The Inflationary Universe: The Quest for a New Theory of Cosmic Origins* (New York: Helix Books/Addison Wesley, 1997). By one of the originators of the inflationary theory.

Edward Harrison, *Darkness at Night: A Riddle of the Universe* (Cambridge, MA: Harvard University Press, 1989). Discusses Olbers's paradox, the darkness of the night sky.

Stephen W. Hawking, *A Brief History of Time, Updated and Expanded* (New York: Bantam Books, 1998). A best-selling discussion of fundamental topics. *The Illustrated Brief History of Time* (1997) is also available.

Stephen W. Hawking and Leonard Mlodinow, *The Grand Design* (New York: Bantam Books, 2010).

Michio Kaku, *Hyperspace* (New York: Doubleday, 1994). A very readable introduction to string theory and other dimensions.

Robert P. Kirshner, *The Extravagant Universe: Exploding Stars, Dark Energy, and the Accelerating Cosmos* (Princeton: Princeton University Press, 2002, paperback 2004).

Lawrence Krauss, *A Universe from Nothing: Why There Is Something Rather than Nothing* (New York: Free Press, 2012). Discusses forms of repulsive dark energy that may be accelerating the Universe.

Leon M. Lederman and David N. Schramm, *From Quarks to the Cosmos*, 2nd ed. (New York: Scientific American Library, 1995).

Michael Lemonick, *Echo of the Big Bang* (Princeton: Princeton University Press, 2003). About the WMAP satellite and its results.

Alan Lightman and Roberta Brawer, *Origins: The Lives and Worlds of Modern Cosmologists* (Cambridge, MA: Harvard University Press, 1990). Interviews with 27 cosmologists.

Mario Livio, *The Accelerating Universe: Infinite Expansion, the Cosmological Constant, and the Beauty of the Cosmos* (New York: John Wiley & Sons, 2000). Contemporary cosmology by an expert at the Space Telescope Science Institute.

Abraham Loeb, *How Did the First Stars and Galaxies Form?* (Princeton: Princeton University Press, 2010).

David Malin, *The Invisible Universe* (Boston: Bulfinch Press, 1999). By the master of ground-based astronomical color photography.

John Mather and John Boslough, *The Very First Light: The True Inside Story of the Scientific Journey Back to the Dawn of the Universe*, 2nd ed. (New York: Basic Books, 2008). A personal account of the discovery of ripples in the cosmic background radiation.

Alain Mazure and Vincent LeBrun, *Matter, Dark Matter, and Anti-Matter: In Search of the Hidden Universe* (New York: Springer-Praxis Books, 2011).

Fulvio Melia, *The Edge of Infinity: Supermassive Black Holes in the Universe* (Cambridge: Cambridge University Press, 2003).

Fulvio Melia, *The Black Hole at the Center of Our Galaxy* (Princeton: Princeton University Press, 2003).

Fulvio Melia, *The Galactic Supermassive Black Hole* (Princeton: Princeton University Press, 2007).

Fulvio Melia and Roy Kerr, *Cracking the Einstein Code: Relativity and the Birth of Black Hole Physics* (Chicago: University of Chicago Press, 2009).

Jeremiah Ostriker and Simon Mitton, *Heart of Darkness: Unraveling the Mysteries of the Invisible Cosmos* (Princeton: Princeton University Press, 2012).

Dennis Overbye, *Lonely Hearts of the Cosmos: The Scientific Quest for the Secrets of the Universe* (New York: HarperCollins, 1991). Humanizing the cosmologists.

Jay M. Pasachoff, Hyron Spinrad, Patrick Osmer, and Edward S. Cheng, *The Farthest Things in the Universe* (Cambridge: Cambridge University Press, 1995). The cosmic background of radiation, quasars, and distant galaxies.

Carolyn Collins Petersen and John C. Brandt, *Visions of the Cosmos* (Cambridge: Cambridge University Press, 2003). Highly illustrated with images from Hubble, Chandra, and other telescopes on the ground and in space.

Martin Rees, *Our Cosmic Habitat* (Princeton: Princeton University Press, 2001). By the Astronomer Royal, a major researcher in the field.

Martin Rees, *Universe: The Definitive Visual Guide* (New York: Dorling Kindersley, 2008).

George H. Rieke, *Measuring the Universe: A Multiwavelength Perspective* (New York: Cambridge University Press, 2012). Guide to astronomical instrumentation and data acquisition.

Michael Rowan-Robinson, *The Nine Numbers of the Cosmos* (Oxford: Oxford University Press, 1999).

Vera Rubin, *Bright Galaxies, Dark Matters* (Woodbury, NY: American Institute of Physics, 1997). Science and autobiography.

Allan Sandage and John Bedke, *The Carnegie Atlas of Galaxies* (Washington, DC: Carnegie Institution of Washington, 1994). Images to stare at and pore over.

David A. Schultz, *The Andromeda Galaxy and the Rise of Modern Astronomy* (New York: Springer, 2012).

Joseph Silk, *The Infinite Cosmos: Questions from the Frontiers of Cosmology* (Oxford: Oxford University Press, 2006).

George Smoot and Keay Davidson, *Wrinkles in Time* (New York: Morrow, 1993). A personal account of the discovery of ripples in the cosmic background radiation.

Ian Stewart, *Flatterland: Like Flatland, Only More So* (New York: Perseus Books, 2001).

William H. Waller and Paul W. Hodge, *Galaxies and the Cosmic Frontier* (Cambridge, MA: Harvard University Press, 2004).

Steven Weinberg, *The First Three Minutes*, 2nd ed. (New York: Basic Books, 1993). A readable discussion of the first minutes after the big bang, including a discussion of the cosmic background radiation.

David A. Weintraub, *How Old Is the Universe?* (Princeton: Princeton University Press, 2011).

John Noble Wilford, ed., *Cosmic Dispatches: The New York Times Reports on Astronomy and Cosmology* (New York: W. W. Norton, 2002).

Richard Wolfson, *Simply Einstein: Relativity Demystified* (New York: W. W. Norton, 2003). An excellent account of relativity.

Ben Zuckerman and Matthew A. Malkan, *The Origin and Evolution of the Universe* (Boston: Jones and Bartlett, 1996).

Books and Other Resources for the Vision-Impaired or Blind

See *youcandoastronomy.com*.

Noreen Grice, *Everyone's Universe: A Guide to Accessible Astronomy Places*, 2nd ed. (You Can Do Astronomy LLC, 2012; youcandoastronomy.com). Large print/Kindle/EPub/iBook. They detail educator strategies for making astronomy accessible to students with mobility, visual, neurological, non-verbal, and hearing disabilities. Second half of the book is a state-by-state travel guide to astronomy destinations (planetariums, observatories, star parties) offering accessible astronomy experiences.

Noreen Grice, *Everyone's Universe: A Guide to Accessible Astronomy Places* (You Can Do Astronomy LLC, 2011; youcandoastronomy.com). Large print/Kindle/EPub.

Amy Hansen (with tactile graphics by Noreen Grice), *Touch the Earth: A Multimedia Book About Earth's Biomes* (NASA, 2010). Large print/Braille text with tactile color images.

Noreen Grice, Simon Steel, and Doris Daou, *Touch the Invisible Sky: A Multi-wavelength Braille Book Featuring Tactile NASA Images* (San Juan, PR: Ozone Publishing, 2007). Large print/Braille text with tactile color images.

Noreen Grice, *Touch the Sun: A NASA Braille Book* (Washington, DC: Joseph Henry Press, 2005). Large print/Braille text with tactile color images.

Noreen Grice, *The Little Moon Phase Book* (San Juan, PR: Ozone Publishing, 2005). Large print/Braille text with tactile images.

Noreen Grice, *Touch the Universe: A NASA Braille Book of Astronomy* (Washington, DC: Joseph Henry Press, 2002). Large print/Braille text with tactile color images.

Noreen Grice (with tactile graphics by Shirley Keller and Irma Goldberg), *Touch the Stars*, revised edition (Boston: National Braille Press, 2002). Large print/Braille text with tactile images. See also http://chandra.harvard.edu/fettss/.

See also Katherine M. Watson, *www.starsatyerkes.net/teacher-resources/images/tactile-images*.

For data set to music: http://www.cfa.harvard.edu/news/2013/pr201313.html.

GLOSSARY

absolute magnitude The magnitude that a star would appear to have if it were at a distance of ten parsecs (32.6 lt-yr) from us.

absorption line Wavelengths at which the intensity of radiation is less than it is at neighboring wavelengths.

absorption nebula Gas and dust seen in silhouette.

accelerating universe The model for the Universe based on observations late in the 1990s that the expansion of the Universe is speeding up over time, rather than slowing down in the way that gravity alone would modify its expansion.

accretion disk Matter that an object has taken up and that has formed a disk around the object.

active galactic nucleus (AGN) A galaxy nucleus that is exceptionally bright in some part of the spectrum; includes radio galaxies, Seyfert galaxies, quasars, and QSOs.

active galaxy A galaxy whose nucleus radiates much more than average in some part of the spectrum, revealing high-energy processes; radio galaxies and Seyfert galaxies are examples.

active optics Optical systems providing relatively slow adjustments to components to keep them lined up properly.

active region One of the regions on the Sun where sunspots, plages, flares, etc., are found.

active Sun The group of solar phenomena that vary with time, such as active regions and their phenomena.

adaptive optics Optical systems providing relatively rapid corrections to counteract atmospheric blurring.

AGN Active galactic nucleus.

albedo The fraction of incident light reflected by a body.

allowed states The energy values that atoms can have by laws of quantum theory.

ALMA The Atacama Large Millimeter/submillimeter Array, an array of 66 radio telescopes being built by an international consortium on a very high plateau in northern Chile. It was formally opened in 2013, providing excellent angular resolution in the millimeter and submillimeter bands.

alpha particle A helium nucleus; consists of two protons and two neutrons.

altitude (a) Height above the surface of a planet; (b) for a telescope mounting, elevation in angular measure above the horizon.

amino acid A type of molecule containing a chain of carbon atoms and the group NH_2 (the amino group). Amino acids are fundamental building blocks of life.

angstrom A unit of length equal to 10^{-8} cm.

angular momentum An intrinsic property of a system corresponding to the amount of its revolution or spin. The amount of angular momentum of a body orbiting around a point is the mass of the orbiting body times its (linear) velocity of revolution times its distance from the point. The amount of angular momentum of a spinning object is the amount of inertia, an intrinsic property of the distribution of mass, times the angular velocity of spin. The conservation of angular momentum is a law that states that the total amount of angular momentum remains constant in a system that is undisturbed from outside itself.

angular resolution See **resolution**.

angular velocity The rate at which a body rotates or revolves expressed as the angle covered in a given time (for example, in degrees per hour).

anisotropy Deviation from isotropy; changing with direction.

annular eclipse A type of solar eclipse in which a ring (annulus) of solar photosphere remains visible.

anthropic principle The idea that since we exist, the Universe must have certain properties or it would not have evolved so that life formed and humans evolved.

antimatter A type of matter in which each particle (antiproton, antielectron, etc.) is opposite in charge and certain other properties to a corresponding particle (proton, electron, etc.) of the same mass of the ordinary type of matter from which the Solar System is made.

antiparticle The antimatter corresponding to a given particle.

aperture The diameter of the lens or mirror that defines the amount of light focused by an optical system.

aphelion For an orbit around the Sun, the farthest point from the Sun.

Apollo asteroids A group of asteroids, with semimajor axes greater than Earth's and less than 1.017 au, whose orbits overlap the Earth's.

apparent magnitude The brightness of a star as seen by an observer, given in a specific system in which a difference of five magnitudes corresponds to a brightness ratio of one hundred times; the scale is fixed by correspondence with a historical background.

archaebacteria A primitive type of organism, different from either plants or animals, perhaps surviving from billions of years ago.

association A physical grouping of stars; in particular, we talk of O and B associations.

asterism A special apparent grouping of stars, part of a constellation.

asteroid A "minor planet," a non-luminous chunk of rock smaller than planet-size but larger than a meteoroid, in orbit around a star.

asteroid belt A region of the Solar System, between the orbits of Mars and Jupiter, in which most of the asteroids orbit.

astrobiology See **exobiology**.

astrometric binary A system of two stars in which the existence of one star can be deduced by study of its gravitational effect on the proper motion of the other star.

astrometric method A method for detecting exoplanets that uses astrometry, precise measurements of the positions of stars over time.

astrometry The branch of astronomy that involves the detailed measurement of the positions and motions of stars and other celestial bodies.

astronomical unit (au) 149 597 870 700 m exactly. This definition of the astronomical unit was adopted by the General Assembly of the International Astronomical Union in Beijing in 2012. It corresponds to some measure of the average distance from the Earth to the Sun.

astrophysics The science, now essentially identical with astronomy, applying the laws of physics to the Universe.

Atacama Large Millimeter/submillimeter Array See **ALMA**.

atom The smallest possible unit of a chemical element. When an atom is subdivided, the parts no longer have properties of any chemical element.

atomic clock A system that uses atomic properties to provide a measure of time.

atomic number The number of protons in an atom.

atomic weight The number of protons and neutrons in an atom, averaged over the abundances of the different isotopes.

aurora Glowing lights visible in the sky, resulting from processes in the Earth's upper atmosphere involving charged particles from the Sun and linked with the Earth's magnetic field.

aurora australis The southern aurora.

aurora borealis The northern aurora.

autumnal equinox Of the two locations in the sky where the ecliptic crosses the celestial equator, the one that the Sun passes each year when moving from northern to southern declinations.

au Astronomical unit.

azimuth The angular distance around the horizon from the northern direction, usually expressed in angular measure from 0° for an object in the northern direction, to 180° for an object in the southern direction, back around to 360°.

background radiation See **cosmic background radiation**.

Baily's beads Beads of light visible around the rim of the Moon at the beginning and end of a total solar eclipse. They result from the solar photosphere shining through valleys at the edge of the Moon.

Balmer series The set of spectral absorption or emission lines resulting from a transition up from or down to (respectively) the second energy level (first excited level) of hydrogen.

bar The straight structure across the center of some spiral galaxies, from which the arms unwind.

baryons Nuclear particles (protons, neutrons, etc.) subject to the strong nuclear force; made of quarks.

basalt A type of rock resulting from the cooling of lava.

baseline The distance between points of observation when it determines the accuracy of some measurement.

beam The cone within which a radio telescope is sensitive to radiation.

belts   Dark bands around certain planets, notably Jupiter.

big-bang theory   A cosmological model, based on Einstein's general theory of relativity, in which the Universe was once extremely hot and dense, and has been expanding ever since.

binary pulsar   A pulsar in a binary system.

binary star   Two stars revolving around each other.

bioastronomy   See **exobiology**.

bipolar flow   A phenomenon in young or forming stars in which streams of matter are ejected from the poles.

blackbody   An object that absorbs all radiation that hits it and emits radiation that exactly follows Planck's law.

blackbody curve   A graph of brightness vs. wavelength that follows Planck's law; each such curve corresponds to a given temperature. Also called a **Planck curve**.

blackbody radiation   Radiation whose distribution in wavelength follows Planck's law, the blackbody curve.

black dwarf   A feebly radiating ball of gas that results when a white dwarf has radiated nearly all its energy.

black hole   A region of space from which, according to the general theory of relativity, neither radiation nor matter can escape.

black-hole era   The future era, following the degenerate era, when the only objects (besides photons and subatomic particles) the Universe will contain will be black holes.

blink method   See **transit method**.

blueshift   A shift of optical wavelengths toward the blue or in all cases towards shorter wavelengths; when the shift is caused by motion, from a velocity of approach.

Bohr atom   Niels Bohr's model of the hydrogen atom, in which the energy levels are depicted as concentric circles of radii that increase as (level number)^2.

bound-free transition   An atomic transition in which an electron starts bound to the atom and winds up free from it.

brown dwarf   A self-gravitating, self-luminous object, insufficiently massive for sustained nuclear fusion (less than 0.075 solar mass).

carbonaceous   Containing a lot of carbon.

carbon-nitrogen-oxygen (CNO) cycle   A chain of nuclear reactions, involving carbon as a catalyst and including nitrogen and oxygen as intermediaries, that transforms four hydrogen nuclei into one helium nucleus with a resulting release in energy. The CNO cycle is important only in stars hotter than the Sun.

Cassegrain (telescope)   A type of reflecting telescope in which the light focused by the primary mirror is intercepted short of its focal point and refocused and reflected by a secondary mirror through a hole in the center of the primary mirror.

Cassini   (a) Jean Dominique Cassini (1625–1712); (b) the NASA mission that began to orbit Saturn and sent a probe into Titan upon its arrival there in 2004.

Cassini's division   The major division in the rings of Saturn.

catalyst   A substance that enables or speeds up a reaction but that is left over in its original form at the end.

CCD   Charge-coupled device, a solid-state imaging device.

celestial equator   The intersection of the celestial sphere with the plane that passes through the Earth's equator.

celestial poles   The intersection of the celestial sphere with the axis of rotation of the Earth.

celestial sphere   The hypothetical sphere centered at the center of the Earth to which it appears that the stars are affixed.

centaur   A Solar-System object, with 2060 Chiron as the first example, intermediate between comets and icy planets or satellites and orbiting the Sun between the orbits of Jupiter and Neptune.

center of mass   The "average" location of mass; the point in a body or system of bodies at which we may consider all the mass to be located for the purpose of calculating the gravitational effect of that mass or its mean motion when a force is applied.

central star   The hot object at the center of a planetary nebula, which is the remaining core of the original star.

Cepheid variable   A type of supergiant star that oscillates in brightness in a manner similar to the star δ Cephei. The periods of Cepheid variables, which are between 1 and 100 days, are linked to the average luminosity of the stars by known relationships; this allows the distances to Cepheids to be found.

Chandra X-ray Observatory   The major NASA x-ray observatory, the former Advanced X-ray Astrophysics Facility, launched in 1999 to make high-resolution observations in the x-ray part of the spectrum.

Chandrasekhar limit   The limit in mass, about 1.4 solar masses, above which electron degeneracy (a quantum-mechanical pressure) cannot support a star, and so the limit above which white dwarfs cannot exist.

chaos   The condition in which very slight differences in initial conditions typically lead to very different results. Also, specifically, examples of orbits or rotation directions that change after apparently random intervals of time because of such sensitivity to initial conditions.

charge-coupled device   See **CCD**.

chromatic aberration   A defect of lens systems in which different colors are focused at different points.

chromosphere   The part of the atmosphere of the Sun (or another star) between the photosphere and the corona. It is probably entirely composed of spicules and probably roughly corresponds to the region in which mechanical energy is deposited.

circle   A conic section formed by cutting a cone perpendicularly to its axis.

circumpolar stars   For a given observing location, stars that are close enough to the celestial pole that they never set.

classical   When discussing atoms, not taking account of quantum-mechanical effects.

closed universe   A big-bang universe with positive curvature; it has finite volume and will eventually contract, assuming there is no gravitationally repulsive dark energy.

cluster   (a) Of stars, a physical grouping of many stars; (b) of galaxies, a physical grouping of at least a few galaxies.

CME   See **coronal mass ejection.**

CNO cycle   See **carbon-nitrogen-oxygen cycle**.

COBE (Cosmic Background Explorer)   A spacecraft launched in 1989 to study the cosmic background radiation.

cold dark matter   Non-luminous matter that moves slowly through the expanding Universe, such as mini black holes and exotic nuclear particles.

color   (a) Of an object, a visual property that depends on wavelength; (b) an arbitrary name assigned to a property that distinguishes three kinds of quarks.

coma   (a) Of a comet, the region surrounding the head; (b) of an optical system, an off-axis aberration in which the images of points appear with comet-like asymmetries.

comet   A type of object orbiting the Sun, often in a very elongated orbit, that when relatively near to the Sun shows a coma and may show a tail.

comparative planetology   Studying the properties of Solar-System bodies by comparing them.

comparison spectrum   A spectrum of known elements on Earth photographed in order to provide a known set of wavelengths for zero Doppler shift.

composite spectrum   The spectrum that reveals a star is a binary system, since spectra of more than one object are apparent.

condensation   A region of unusually high mass or brightness.

conic sections   Geometric shapes obtained by slicing a cone.

conjunction   A planet is said to be in conjunction when it appears in the same (or nearly the same) direction as the Sun in the sky.

conservation law   A statement that the total amount of some property (angular momentum, energy, etc.) of a body or set of bodies does not change.

constellation   One of 88 areas into which the sky has been divided for convenience in referring to the stars or other objects therein.

continental drift   The slow motion of the continents across the Earth's surface, explained in the theory of plate tectonics as a set of shifting regions called plates.

continuous spectrum   A spectrum with radiation at all wavelengths but with neither absorption nor emission lines.

continuum (plural: **continua**)   The continuous spectrum that we would measure from a body if no spectral lines were present.

convection   The method of energy transport in which the rising motion of masses from below carries energy upward in a gravitational field. Boiling is an example.

convection zone   The subsurface zone in certain types of stars in which convection dominates energy transfer.

coordinate systems   Methods of assigning positions with respect to suitable axes.

core   The central region of a star or planet.

core-collapse supernova   A stellar explosion in which the iron core of a massive star collapses, forming a neutron star or a black hole, and the surrounding layers of gas are violently ejected.

corona, solar or stellar   The outermost region of the Sun (or of other stars), characterized by temperatures of millions of kelvins.

coronagraph   A type of telescope with which the corona (and, sometimes, merely the chromosphere) can be seen in visible light at times other than that of a total solar eclipse by occulting (hiding) the bright photosphere. Also, any telescope in which a bright central object is occulted to reveal fainter things.

coronal holes   Relatively dark regions of the corona having low density; they result from open field lines.

coronal mass ejection (CME)   An energetic puff of gas ejected from the Sun, generally spread out over a larger area than a solar flare and coming from the corona rather than from the photosphere.

correcting plate   A thin lens of complicated shape at the front of a Schmidt camera that compensates for spherical aberration.

cosmic abundances   The overall abundances of elements in the Universe.

cosmic background radiation   Isotropic millimeter and submillimeter radiation following a blackbody curve for about 3 K; interpreted as a remnant of the big bang.

cosmic microwave radiation   See **cosmic background radiation**.

cosmic rays   Nuclear particles or nuclei travelling through space at high velocity.

cosmogony   The study of the origin of the Universe, usually applied in particular to the origin of the Solar System.

cosmological constant A constant arbitrarily added by Einstein to an equation in his general theory of relativity in order to provide a solution in which the Universe did not expand or contract. It was only subsequently discovered that the Universe does expand after all, leading Einstein to renounce the cosmological constant. Recent evidence suggests, however, that its value may be nonzero, and is somewhat larger than originally postulated by Einstein, causing the Universe's expansion to currently accelerate. Alternatively, some other form of gravitationally repulsive "dark energy" may be causing the observed acceleration, or perhaps Einstein's general theory of relativity is wrong on large scales.

cosmological principle The principle that on the whole the Universe looks the same in all directions and in all regions.

cosmology The study of the Universe as a whole.

critical density The density of matter that would allow the Universe to expand forever, but at a rate that would decrease to 0 at infinite time. The density of the Universe with respect to the critical density defines whether the Universe will expand forever or eventually contract, assuming there is no gravitationally repulsive dark energy.

crust The outermost solid layer of some objects, including neutron stars and some planets.

dark energy A generic name for the gravitationally repulsive energy that is thought by many to be the cause of the observed acceleration of the expansion rate of the Universe.

dark era The final era of the Universe, when only low-energy photons, neutrinos, and some elementary particles (the ones that did not find a partner to annihilate) remain.

dark matter Non-luminous matter. See **hot dark matter** and **cold dark matter**.

dark nebula Dust and gas seen in silhouette.

declination Celestial latitude, measured in degrees north or south of the celestial equator.

deferent In the Ptolemaic system of the Universe, the larger circle, centered at the Earth, on which the centers of the epicycles revolve.

degenerate era The future era when the Universe will be filled with cold brown dwarfs, white dwarfs, and neutron stars. (Black holes will also exist.)

density Mass divided by volume.

density wave A circulating region of relatively high density, important, for example, in models of spiral arms of galaxies.

density-wave theory The explanation of spiral structure of galaxies as the effect of a wave of compression that rotates around the center of the galaxy and causes the formation of stars in the compressed region.

deuterium An isotope of hydrogen that contains 1 proton and 1 neutron.

deuteron A deuterium nucleus, containing 1 proton and 1 neutron.

diamond-ring effect The last bit of photosphere glowing brightly just before the beginning of the total phase of a solar eclipse, or its counterpart just after the end of totality.

differential force A net force resulting from the difference of two other forces; a tidal force.

differential rotation Rotation of a body in which different parts have different angular velocities (and thus different periods of rotation).

differentiation For a planet, the formation of layers of different structure or composition while it is molten.

diffraction A phenomenon affecting light as it passes any obstacle, spreading it out in a complicated fashion.

diffraction grating A very closely ruled series of lines that, through diffraction of light, provides a spectrum of radiation that falls on it.

dirty snowball A theory explaining comets as amalgams of ices, dust, and rocks.

discrete Separated; isolated.

disk (a) Of a galaxy, the disk-like flat portion, as opposed to the nucleus or the halo; (b) of a star or planet, the two-dimensional projection of its surface.

dispersion Of light, the effect that different colors are bent by different amounts when passing from one substance to another.

D lines A pair of lines from sodium that appear in the yellow part of the spectrum.

DNA Deoxyribonucleic acid, a long chain of molecules that contains the genetic information of life.

Dobsonian An inexpensive type of large-aperture amateur telescope characterized by a thin mirror, lightweight tube, and Teflon bearings on an altitude-azimuth mount.

Doppler effect A change in wavelength that results when a source of waves and the observer are moving relative to each other.

Doppler shift The change in wavelength that arises from the Doppler effect, caused by relative motion toward or away from the observer.

Doppler-wobble method A method for detecting exoplanets. Periodic changes in the Doppler shift (radial velocity) of a star are measured with high-quality spectrographs, revealing the subtle "reflex motion" caused by the orbiting exoplanet. Also called the **radial-velocity** method.

double star A binary star; two or more stars orbiting each other.

double-lobed structure An object in which radio emission comes from a pair of regions on opposite sides.

Drake equation An equation advanced by Frank Drake and popularized by Carl Sagan that attempts to estimate the number of communicating civilizations by breaking the calculation down into a series of steps that can be assessed individually, such as the rate of star formation, the fraction of stars with planets, and the average lifetime of a communicating civilization.

dust tail The dust left behind a comet, reflecting sunlight.

dwarf ellipticals Small, low-mass elliptical galaxies.

dwarf planet An object that satisfies all of the criteria for being a planet, except its gravitational field does not clearly dominate in its neighborhood; also, it is not a planetary satellite. Pluto is now called a dwarf planet rather than a planet because it has not cleared out other objects in the Kuiper belt. At its 2006 General Assembly, the International Astronomical Union resolved "that planets and other bodies, except satellites, in our Solar System be defined into three distinct categories in the following way. (1) A planet is a celestial body that (a) is in orbit around the Sun, (b) has sufficient mass for its self-gravity to overcome rigid body forces so that it assumes a hydrostatic equilibrium (nearly round) shape, and (c) has cleared the neighborhood around its orbit. (2) A "dwarf planet" is a celestial body that (a) is in orbit around the Sun, (b) has sufficient mass for its self-gravity to overcome rigid body forces so that it assumes a hydrostatic equilibrium (nearly round) shape, (c) has not cleared the neighborhood around its orbit, and (d) is not a satellite. (3) All other objects, except satellites, orbiting the Sun shall be referred to collectively as "Small Solar System Bodies.""

dwarf stars Main-sequence stars.

dwarfs Dwarf stars.

$E = mc^2$ Einstein's formula (special theory of relativity) for the equivalence of mass and energy.

early universe The Universe during its first minutes.

earthshine Sunlight illuminating the Moon after having been reflected by the Earth.

eccentric Deviating from a circle.

eccentricity A measure of the flatness of an ellipse, defined as the distance between the foci divided by the major axis.

eclipse The passage of all or part of one astronomical body into the shadow of another. Also often used to denote the passage of one body directly behind another, as in the Moon eclipsing the Sun.

eclipsing binary A binary star in which one member periodically hides the other.

ecliptic The path followed by the Sun across the celestial sphere in the course of a year.

ecliptic plane The plane of the Earth's orbit around the Sun.

ecosphere See **habitable zone**.

Edwin Hubble An American astronomer who, in the 1920s, showed that "spiral nebulae" were actually galaxies external to the Milky Way. He also classified galaxies into different categories or "Hubble types." In 1929, he discovered the linear relationship between the recession speeds and distances of galaxies (Hubble's law), later interpreted as the expansion of the Universe.

electric field A force set up by an electric charge.

electromagnetic force One of the four fundamental forces of nature, giving rise to electromagnetic radiation.

electromagnetic radiation Radiation resulting from changing electric and magnetic fields.

electromagnetic spectrum Energy in the form of electromagnetic waves, in order of wavelength.

electromagnetic waves Waves of changing electric and magnetic fields, travelling through space at the speed of light.

electromagnetism The combined force of electricity and magnetism, which follows the formulae unified by Maxwell.

electron A particle of one negative charge, $1/1830$ the mass of a proton, that is not affected by the strong force. It is a lepton.

electron volt (eV) The energy necessary to raise an electron through a potential of one volt.

electroweak force The unified electromagnetic and weak forces, according to a relatively recent theory.

element A kind of atom, characterized by a certain number of protons in its nucleus. All atoms of a given element have similar chemical properties.

elementary particle One of the constituents of an atom.

ellipse A curve with the property that the sum of the distances from any point on the curve to two given points, called the foci, is constant.

elliptical galaxy A type of galaxy characterized by elliptical appearance.

emission line Wavelengths (or frequencies) at which the brightness of radiation is greater than it is at neighboring wavelengths (or frequencies).

emission nebula A glowing cloud of interstellar gas.

energy A fundamental quantity usually defined in terms of the ability of a system to do something that is technically called "work"; that is, the ability to move an object by application of force, where the work is the force times the displacement.

energy, law of conservation of Energy is neither created nor destroyed, but may be changed in form.

energy level A state corresponding to an amount of energy that an atom is allowed to have by the laws of quantum mechanics.

energy problem For quasars, how to produce so much energy in such a small emitting volume.

ephemeris A listing of astronomical positions and other data that change with time. From the same root as *ephemeral*.

epicycle In the Ptolemaic theory, a small circle, riding on a larger circle called the deferent, on which a planet moves. The epicycle is used to account for retrograde motion.

equal areas, law of Kepler's second law.

equant In Ptolemaic theory, the point equally distant from the center of the deferent as the Earth but on the opposite side, around which the epicycle moves at a uniform angular rate.

equator (a) Of the Earth, a great circle on the Earth, midway between the poles; (b) celestial, the projection of the Earth's equator onto the celestial spheres; (c) Galactic, the plane of the Milky Way Galaxy's disk as projected onto a map.

equatorial coordinates The celestial coordinates based on the celestial equator, the extension of the Earth's equator into space, with right ascension measured around the celestial equator in hours, minutes, and seconds, and declination measured north and south of it in degrees.

equinox An intersection of the ecliptic and the celestial equator. Neglecting atmospheric refraction, the center of the Sun is geometrically above and below the horizon for equal lengths of time on the two days of the year when the Sun passes the equinoxes. If the Sun were a point and atmospheric refraction were absent, then day and night would be of equal length on those days.

erg A unit of energy in the metric system, corresponding to the work done by a force of one dyne (the force that is required to accelerate one gram by one cm/s²) producing a displacement of one centimeter.

ergosphere A region surrounding a rotating black hole (or other system satisfying Kerr's solution) from which work can be extracted.

escape velocity The initial velocity that an object must have to escape the gravitational pull of a mass, neglecting effects such as air resistance.

event horizon The sphere around a black hole from within which nothing can escape; the place at which the exit cones close.

evolutionary track The set of points on a temperature–luminosity diagram showing the changes of a star's temperature and luminosity with time.

excitation The raising of atoms to energy states higher than the lowest possible.

excited level An energy level of an atom above the ground level.

exit cone The cone that, for each point within the photon sphere of a black hole, defines the directions of rays of radiation that escape.

exobiology The study of life located elsewhere than Earth. Also called **astrobiology** or **bioastronomy.**

exoplanet A planet orbiting a star other than the Sun. Also called an **extra-solar planet.**

exponent The "power" representing the number of times a number is multiplied by itself.

exponential notation The writing of numbers as a power of 10 times a number with one digit before the decimal point.

extended objects Objects with detectable angular size.

extinction The dimming of starlight by scattering and absorption as the light traverses interstellar space.

extragalactic Exterior to the Milky Way Galaxy.

extra-solar planet See exoplanet.

eyepiece The small combination of lenses at the eye end of a telescope, used to examine the image formed by the objective.

field of view The angular expanse viewable.

filament A feature of the solar surface seen in Hα as a dark wavy line; a prominence projected on the solar disk.

fireball An exceptionally bright meteor.

fission, nuclear The splitting of an atomic nucleus.

flare An extremely rapid brightening of a small area of the surface of the Sun, usually observed in Hα and other strong spectral lines and accompanied by x-ray and radio emission.

flash spectrum The solar chromospheric spectrum seen in the few seconds before or after totality at a solar eclipse.

flat (critical) universe A universe in which Euclid's parallel postulate holds. It barely expands forever, assuming there is no gravitationally repulsive dark energy. It has infinite volume and its age is ⅔ the Hubble time (assuming no dark energy).

flatness problem One of the problems solved by the inflationary theory, that the Universe is exceedingly close to being globally flat for no obvious reason.

flavors A way of distinguishing quarks; up, down, strange, charmed, truth (top), beauty (bottom).

flux The amount of something (such as energy) passing through a surface per unit time.

focal length The distance from a lens or mirror to the point to which rays from an object at infinity are focused.

focus (plural: **foci**) (a) A point to which radiation is made to converge; (b) of an ellipse, one of the two points the sum of the distances to which remains constant.

force In physics, something that can or does cause change of momentum, measured by the rate of change of momentum with time.

Fraunhofer lines The absorption lines of a solar or other stellar spectrum.

frequency The rate at which waves pass a given point.

frost line Also known as the snow line, the distance from the Sun (roughly 3 au) beyond which icy clumps of material could form in the early Solar System.

full moon The phase of the Moon when the side facing the Earth is fully illuminated by sunlight.

fusion, nuclear The amalgamation of nuclei into heavier nuclei.

fuzz The faint light detectable around nearby quasars.

galactic cannibalism The incorporation of one galaxy into another.

galactic year The length of time the Sun takes to complete an orbit of our Galactic Center.

Galilean satellites Io, Europa, Ganymede, and Callisto: the four major satellites of Jupiter, discovered by Galileo in 1610.

Galileo (a) The Italian scientist Galileo Galilei (1564–1642); (b) the NASA spacecraft that orbited the planet Jupiter 1995–2003 and that sent a probe into Jupiter's clouds at its arrival on December 7, 1995.

gamma rays Electromagnetic radiation with wavelengths shorter than approximately 0.1 Å.

gamma-ray burst A brief burst of gamma rays coming from specific locations uniformly distributed in the sky; discovered to come, at least in most cases, from exceedingly distant and thus extremely powerful sources of energy.

gas tail The puffs of ionized gas trailing a comet.

general theory of relativity Einstein's 1916 theory of gravity, in which space and time are warped in the presence of mass or energy.

geocentric Earth-centered.

geology The study of the Earth, or of other solid bodies.

geothermal energy Energy from under the Earth's surface.

giant A star that is larger and brighter than main sequence stars of the same color.

giant ellipticals Elliptical galaxies that are very large.

giant molecular cloud A basic building block of our Galaxy, containing dust, which shields the molecules present.

giant planets Jupiter, Saturn, Uranus, and Neptune; or even larger extra-solar planets.

giant star A star more luminous than a main-sequence star of its spectral type; a late stage in stellar evolution.

gibbous moon The phases between quarter moon and full moon.

globular cluster A spherically symmetric, gravitationally bound collection of typically hundreds of thousands of stars that formed from the same cloud of gas. Generally found in the halo of our Galaxy and other galaxies.

Goldilocks planets Exoplanets in the habitable zone around a star, in which water can exist in the liquid state, being neither too hot nor too cold.

gluon The particle that carries the color force (and thus the strong nuclear force).

grand unified theories (GUTs) Theories unifying the electro-weak force and the strong force.

granulation Convection cells on the Sun about 1 arc sec across.

grating A surface ruled with closely spaced lines that, through diffraction, breaks up light into its spectrum.

gravitational force One of the four fundamental forces of nature, the force by which two masses attract each other.

gravitational instability A situation that tends to break up under the force of gravity.

gravitational lens In the gravitational lens phenomenon, a massive body changes the path of electromagnetic radiation passing near it so as to make more than one image (or a brightening) of an object. A double quasar was the first example to be discovered.

gravitational microlensing The gravitational focusing of the light of a background star by a foreground exoplanet. It causes the background star to additionally brighten for a day or two during a longer overall brightening and fading produced by gravitational lensing by the exoplanet's parent star.

gravitationally Controlled by the force of gravity.

gravitational redshift A redshift of light caused by the presence of mass, according to the general theory of relativity.

gravitational waves Waves that most scientists consider to be a consequence, according to the general theory of relativity, of changing distributions of mass.

gravity The tendency for all masses to attract each other; described in a formula by Newton and more recently described by Einstein as a result of a warping of space and time by the presence of a mass.

gravity assist Using the gravity of one celestial body to change a spacecraft's trajectory.

grazing incidence Striking at a low angle.

great circle The intersection of a plane that passes through the center of a sphere with the surface of that sphere; the largest possible circle that can be drawn on the surface of a sphere.

Great Dark Spot A giant circulating region on Neptune seen by Voyager 2 in 1989; it has since disappeared.

Great Red Spot   A giant circulating region on Jupiter.

greenhouse effect   The effect by which the atmosphere of a planet heats up above its expected equilibrium temperature because it is transparent to incoming visible radiation but opaque to the infrared radiation that is emitted by the surface of the planet.

Gregorian calendar   The calendar in current use, with normal years that are 365 days long, with leap years every fourth year except for years that are divisible by 100 but not by 400.

ground level   An atom's lowest possible energy level.

ground state   See **ground level**.

GUTs   See **grand unified theories**.

H₀   The Hubble constant (or Hubble parameter) at present; its value changes with time.

H I region   An interstellar region of neutral hydrogen.

H II region   An interstellar region of ionized hydrogen.

H line   The spectral line of singly ionized calcium at 3,968 Å.

Hα   The first line of the Balmer series of hydrogen, at 6,563 Å.

habitable zone   The region around a star in which conditions are suitable for life as we know it. Also called the **ecosphere**.

half-life   The length of time for half a set of particles to decay through radioactivity or instability.

halo   Of a galaxy, the region of the galaxy that extends far above and below the plane of the galaxy, containing globular clusters and much dark matter.

head   Of a comet, the nucleus and coma together.

heat flow   The flow of energy from one location to another.

heavyweight stars   Stars of more than about 8 or 10 solar masses.

heliocentric   Sun-centered; using the Sun rather than the Earth as the point to which we refer. A heliocentric measurement, for example, omits the effect of the Doppler shift caused by the Earth's orbital motion.

helioseismology   The study of oscillations (vibrations) of the Sun's surface, revealing conditions in the solar interior.

helium flash   The rapid onset of fusion of helium into carbon through the triple-alpha process that eventually takes place in most red-giant stars.

Herbig-Haro objects   Blobs of gas ejected in star formation.

Herschel Space Observatory   A telescope 1.5 million km from Earth, launched in 2009 by the European Space Agency. Its 3.5-m diameter mirror is the largest in space, and it was used for astronomical studies at far-infrared and submillimeter wavelengths until it ran out of coolant in 2013.

hertz   The measure of frequency, with units of /s (per second); formerly called cycles per second.

Hertzsprung–Russell diagram   A graph of temperature (or equivalent) vs. luminosity (or equivalent) for a group of stars.

high-energy astrophysics   The study of x-rays, gamma rays, and cosmic rays, and of the processes that make them.

highlands   Regions on the Moon or elsewhere that are above the level that may have been smoothed by flowing lava.

Hipparcos   A European satellite (1989–1993) that precisely measured the positions of a very large number of stars.

homogeneous   Uniform throughout.

horizon problem   One of the problems of cosmology solved by the inflationary theory: why the Universe has the same average temperature in all directions, even though widely separated regions could never have been in thermal equilibrium with each other since they are beyond each other's horizons.

hot dark matter   Non-luminous matter with large speeds, like neutrinos.

H–R diagram   Hertzsprung–Russell diagram.

Hubble Deep Field   A small part of the sky in Ursa Major extensively studied by the Hubble Space Telescope in December 1995 and thereafter also studied by a wide variety of telescopes on the ground and in space; the long exposures allowed astronomers to see "deep" into space, that is, to great distances and therefore far back in time. See hubblesite.org/gallery/album/pr1996001c/. A Hubble Deep Field–South was observed subsequently (/pr1998041b/), as was a Hubble Ultra Deep Field (/pr2006012b/) and a Hubble eXtreme Deep Field (/pr2012037a/).

Hubble diagram   The plot of redshift (or recession speed) versus distance, graphically representing Hubble's law.

Hubble flow   The assumed uniform expansion of the Universe, on which any peculiar motions of galaxies or clusters of galaxies is superimposed.

Hubble Space Telescope (HST)   One of NASA's Great Observatories, a 2.4-m telescope launched in 1990. It orbits a few hundred miles above Earth's surface, above most of the atmosphere, and has been upgraded by space-shuttle astronauts several times. A very large number of studies have been conducted with it at ultraviolet, optical, and infrared wavelengths.

Hubble time   The amount of time since the big bang, assuming a constant speed for any given galaxy; the Hubble time is calculated by tracing the Universe backward in time using the current Hubble's constant.

Hubble type   Hubble's galaxy classification scheme; E0, E7, Sa, SBa, etc.

Hubble's constant (H₀)   The constant of proportionality in Hubble's law linking the speed of recession of a distant object and its current distance from us.

Hubble's law   The linear relation between the speed of recession of a distant object and its current distance from us, $v = H_0\, d$.

hydrostatic equilibrium   The condition achieved when the inward pull of gravity is balanced by the outward pressure gradient.

hyperfine level   A subdivision of an energy level caused by such relatively minor effects as changes resulting from the interactions among spinning particles in an atom or molecule.

hypothesis   The first step in the traditional formulation of the scientific method; a tentative explanation of a set of facts that is to be tested experimentally or observationally.

igneous   Rock cooled from lava.

inclination   Of an orbit, the angle of the plane of the orbit with respect to the ecliptic plane.

inclined   Tilted with respect to some other body, usually describing the axis of rotation or the plane of an orbit.

inferior planet   A planet whose orbit around the Sun is within the Earth's, namely, Mercury and Venus.

inflation   The theory that the Universe expanded exponentially, by perhaps 10 to the 100th power, in the first fraction of a second after the big bang. The concept of inflation solves several problems in cosmology, such as the horizon problem.

inflationary universe   A model of the expanding Universe involving a brief period of extremely rapid, exponential expansion.

infrared   Radiation beyond the red, about 7,000 Å to 1 mm.

interference   The property of radiation, explainable by the wave theory, in which waves in phase can add (constructive interference) and waves out of phase can subtract (destructive interference); for light, this gives alternate light and dark bands.

interferometer   A device that uses the property of interference to measure such properties of objects as their positions or structure.

interferometry   Observations using an interferometer.

intergalactic medium   Material between galaxies in a cluster.

intermediate-mass black hole   A black hole whose mass is roughly in the range 100 to 100,000 solar masses.

interior   The inside of an object.

international date line   A crooked imaginary line on the Earth's surface, roughly corresponding to 180° longitude, at which, when crossed from east to west, the date jumps forward by one day.

interplanetary medium   Gas and dust between the planets.

interstellar medium   Gas and dust between the stars.

interstellar reddening   The relatively greater extinction of blue light than of red light by interstellar matter.

inverse-square law   Decreasing with the square of increasing distance.

ion   An atom that has lost one or more electrons.

ionized   Having lost one or more electrons.

ionosphere   The highest region of the Earth's atmosphere.

ion tail   See **gas tail**.

IRAS   The Infrared Astronomical Satellite (1983).

iron meteorites   Meteorites with a high iron content (about 90%); most of the rest is nickel.

irregular galaxy   A type of galaxy showing no regular shape or symmetry.

ISO (Infrared Space Observatory)   A European Space Agency project, 1995–1998.

isotope   A form of chemical element with a specific number of neutrons.

isotropic   Being the same in all directions.

Jansky Very Large Array   The National Radio Astronomy Observatory's set of radio telescopes in New Mexico, used together for interferometry, upgraded through 2012 with new electronics, fiber optics, and a supercomputer, and renamed in 2012 after Karl G. Jansky, the "father" of radio astronomy.

joule   The SI unit of energy, 1 kg · m²/s².

jovian planets   Same as **giant planets**.

JPL   The Jet Propulsion Laboratory in Pasadena, California, funded by NASA and administered by Caltech; a major space contractor.

JUICE   Jupiter Icy Moon Explorer; a flagship mission of the European Space Agency approved in 2012 for launch in 2022 and arrival at the Jupiter system in 2030.

Julian calendar   The calendar with 365-day years and leap years every fourth year without exception; the predecessor to the Gregorian calendar.

Julian day   The number of days since noon on January 1, 713 BCE; used for keeping track of variable stars or other astronomical events. January 1, 2014, noon, begins Julian day 2,456,659.

JVLA   See **Jansky Very Large Array**.

K line   The spectral line of singly ionized calcium at 3,933 Å.

Keplerian   Following Kepler's law.

Kepler mission   A spacecraft launched in 2009 that continually monitors the brightnesses of more than 150,000 stars, searching for transiting exoplanets. As of 2012, it had found more than 2,300 exoplanet candidates, over 90 per cent of which are thought to be genuine exoplanets.

Kerr's solution Equations describing rotating black holes.

Kuiper belt A reservoir of perhaps hundreds of thousands of Solar-System objects, each tens or hundreds of kilometers in diameter, orbiting the Sun outside the orbit of Neptune. It is the source of many short-period comets.

Kuiper-belt object An object in the Kuiper belt; many comets and probably even the dwarf planet Pluto are examples.

large-scale structure The network of filaments and voids or other shapes distinguished when studying the Universe on the largest scales of distance.

laser An acronym for "light amplification by stimulated emission of radiation," a device by which certain energy levels are populated by more electrons than normal, resulting in an especially intense emission of light at a certain frequency when the electrons drop to a lower energy level.

latitude Number of degrees north or south of the equator measured from the center of a coordinate system.

law of equal areas Kepler's second law.

leap year A year in which a 366th day is added.

lens A device that focuses waves by refraction.

lenticular galaxy A galaxy of type S0.

light Electromagnetic radiation between about 4,000 and 7,000 Å. Often called visible light or optical light, to explicitly distinguish it from other forms of electromagnetic radiation.

light curve The graph of the brightness of an object vs. time.

light pollution Excess light in the sky.

light-year The distance that light travels in a year, about 10 trillion km or 6 trillion miles.

lighthouse model The explanation of a pulsar as a spinning neutron star whose beam we see as it comes around.

lightweight stars Stars between about 0.075 and 8 or 10 solar masses.

limb The edge of a star or planet.

line profile The graph of the intensity of radiation vs. wavelength for a spectral line.

lithosphere The crust and upper mantle of a planet.

lobes Of a radio source, the regions to the sides of the center from which high-energy particles are radiating.

local In our region of the Universe.

Local Group The five dozen or so galaxies, including the Milky Way Galaxy, that form a small cluster.

Local Supercluster The supercluster of galaxies in which the Virgo Cluster, the Local Group, and other nearby clusters reside.

logarithmic A scale in which equal intervals stand for multiplying by ten or some other base, as opposed to linear, in which increases are additive.

longitude The angular distance around a body measured along the equator from some particular point; for a point not on the equator, it is the angular distance along the equator to a great circle that passes through the poles and through the point.

long-period variable A Mira variable.

lookback time The duration over which light from an object has been travelling to reach us.

luminosity The total amount of energy given off by an object per unit time; its power.

luminosity class Different regions of the H–R diagram separating objects of the same spectral type: supergiants (I), bright giants (II), giants (III), subgiants (IV), dwarfs (V).

lunar eclipse The passage of the Moon into the Earth's shadow.

lunar occultation An occultation by the Moon.

lunar soils Dust and other small fragments on the lunar surface.

Lyman-alpha The spectral line (1,216 Å) that corresponds to a transition between the two lowest major energy levels of a hydrogen atom.

Lyman-alpha forest The many Lyman-alpha lines, each differently Doppler-shifted, visible in the spectra of high-redshift quasars.

Lyman lines The spectral lines that correspond to transitions to or from the lowest major energy level of a hydrogen atom.

M$_\odot$ Solar mass; the mass of the Sun, used as a unit of measurement. For example, a star with 5 M$_\odot$ has 5 times the mass of the Sun.

MACHO Massive compact halo object – like dim stars, brown dwarfs, or black holes – that could account for some of the dark matter. The name was chosen to contrast with WIMPs.

Magellanic Clouds Two small irregular galaxies, satellites of the Milky Way Galaxy, visible in the southern sky.

magnetic-field lines Directions mapping out the direction of the force between magnetic poles; the packing of the lines shows the strength of the force.

magnetic lines of force See **magnetic-field lines**.

magnetic monopole A single magnetic charge of only one polarity; may or may not exist.

magnetosphere A region of magnetic field around a planet.

magnification An apparent increase in angular size.

magnitude A factor of $\sqrt[5]{100} = 2.511886\ldots$ in brightness. See **absolute magnitude** and **apparent magnitude**. An order of magnitude is a power of ten.

magnitude scale The scale of apparent magnitudes and absolute magnitudes used by astronomers, in which each factor of 100 in brightness corresponds to a difference of 5 magnitudes.

main sequence A band on a Hertzsprung–Russell diagram in which stars fall during the main, core-hydrogen-burning phase of their lifetimes.

major axis The longest diameter of an ellipse; the line from one side of an ellipse to the other that passes through the foci. Also, the length of that line.

mantle The shell of rock separating the core of a differentiated planet from its thin surface crust.

mare (plural: **maria**) One of the smooth areas of the Moon or on some of the other planets. From the Latin for sea.

mascon A concentration of mass under the surface of the Moon, discovered from its gravitational effect on spacecraft orbiting the Moon.

mass A measure of the inherent amount of matter in a body.

massive compact halo object See MACHO.

mass-luminosity relation A well-defined relation between the mass and luminosity for main-sequence stars.

mass number The total number of protons and neutrons in a nucleus; also known as **atomic weight**.

Maunder minimum The period 1645–1715, when there were very few sunspots, and no periodicity, visible.

mean solar day A solar day for the "mean Sun," which moves across the sky at a constant rate during the year.

merging The interaction of two galaxies in space with a single galaxy as the result; the merging of two spiral galaxies to form an elliptical galaxy, or of two galaxies interacting with a ring galaxy resulting, are examples.

meridian The great circle on the celestial sphere that passes through the celestial poles and the observer's zenith.

Messier numbers Numbers of fuzzy-looking objects in the 18th-century list of Charles Messier.

metal (a) For stellar abundances, any element higher in atomic number than 2, that is, more massive than helium. (b) In general, neutral matter that is a good conductor of electricity.

meteor A track of light in the sky from rock, ice, or dust burning up as it falls through the Earth's atmosphere.

meteor shower The occurrence at yearly intervals of meteors at a higher-than-average rate as the Earth goes through a comet's orbit and the debris left behind by that comet.

meteorite An interplanetary chunk of rock after it impacts on a planet or the Moon, especially on the Earth.

meteoroid An interplanetary chunk of rock smaller than an asteroid.

micrometeorite A tiny meteorite. The micrometeorites that hit the Earth's surface are sufficiently slowed down that they can reach the ground without being vaporized.

midnight sun The Sun seen around the clock from locations sufficiently far north or south during the suitable season.

Milky Way The band of light across the sky from the stars and gas in the plane of the Milky Way Galaxy.

Milky Way Galaxy The collection of gas, dust, and perhaps 400 billion stars in which we live; the Milky Way is our view of the plane of the Milky Way Galaxy.

mini black hole A black hole the size of a pinhead (and thus the mass of an asteroid) or less, thought by some to be left over from the big bang; Stephen Hawking deduced that such mini black holes should seem to emit radiation at a rate that allows a substantial temperature to be assigned to it.

minor axis The shortest diameter of an ellipse; the line from one side of an ellipse to the other that passes midway between the foci and is perpendicular to the major axis. Also, the length of that line.

minor planets Asteroids.

Mira variable A long-period variable star similar to Mira (omicron Ceti).

missing-mass problem The discrepancy between the mass visible and the mass derived from calculating the gravity acting on members of clusters of galaxies, or on stars in a galaxy.

model A physical or mathematical equivalent to a situation; the former is typified by a model airplane that scales down a real airplane, while the latter is typified by a set of equations or of tables that describes, for example, the interior of Jupiter or processes that work there.

molecule Bound atoms that make the smallest collection that exhibits a certain set of chemical properties.

momentum A measure of the tendency that a moving body has to keep moving. The momentum in a given direction (the "linear momentum") is equal to the mass of the body times its speed in that direction. See also **angular momentum**.

mountain ranges Sets of mountains on the Earth, Moon, etc.

multiverse The set of parallel universes that may exist, with our observable universe as only one part.

naked singularity A singularity that is not surrounded by an event horizon and is therefore potentially visible.

neap tides The tides when the gravitational pulls of the Sun and Moon are perpendicular, making the tides relatively low.

Near-Earth Object A comet asteroid whose orbit is sufficiently close to Earth's that a collision is possible or even likely in the long term. See **Apollo asteroids**.

nebula (plural: **nebulae**) Interstellar regions of dust or gas.

nebular hypothesis The particular nebular theory for the formation of the Solar System advanced by Laplace.

nebular theories The theories that the Sun and the planets formed out of a cloud of gas and dust.

neutrino A spinning, neutral elementary particle with little rest mass, formed in certain radioactive decays.

neutron A massive, neutral elementary particle, one of the fundamental constituents of an atom.

neutron star A star that has collapsed to the point where it is supported against gravity by neutron quantum mechanical degeneracy pressure.

New General Catalogue *A New General Catalogue of Nebulae and Clusters of Stars* by J. L. E. Dreyer, 1888.

new moon The phase when the side of the Moon facing the Earth is the side that is not illuminated by sunlight.

Newtonian (telescope) A reflecting telescope in which the beam from the primary mirror is reflected by a flat secondary mirror to the side.

NGC *New General Catalogue.*

nonthermal radiation Radiation that is *not* produced by the random motions of particles having a distribution of energies determined only by their temperature. For example, synchrotron radiation is nonthermal.

nova (plural: **novae**) A star that suddenly increases in brightness; an event in a binary system when matter from the giant component falls on the white dwarf component, or suddenly fuses on the surface of the white dwarf.

nuclear bulge The central region of spiral galaxies.

nuclear burning Nuclear fusion.

nuclear force The strong force, one of the fundamental forces.

nuclear fusion The amalgamation of lighter nuclei into heavier ones.

nucleosynthesis The formation of the elements.

nucleus (plural: **nuclei**) (a) Of an atom, the core, which has a positive charge, contains most of the mass, and takes up only a small part of the volume; (b) of a comet, the chunks of matter, no more than a few tens of km across, at the center of the head; (c) of a galaxy, the innermost region.

O and B association A group of O and B stars close together.

objective The principal lens or mirror of an optical system.

oblate With equatorial greater than polar diameter.

Occam's razor The principle of simplicity, from the medieval philosopher William of Occam (approximately 1285–1349): The simplest explanation for all the will be accepted.

occultation When a planet or asteroid passes in front of a star and blocks its light, or when a planet blocks our view of one of its moons.

Olbers's paradox The observation that the sky is dark at night contrasted to a simple argument that shows that the sky should be uniformly bright.

one atmosphere The air pressure at the Earth's surface.

one year The length of time the Earth takes to orbit the Sun.

Oort comet cloud The trillions of incipient comets surrounding the Solar System in a roughly 50,000 au (radius) sphere.

open cluster An irregular, relatively loosely bound group of up to a few thousand stars, generally in or near the plane of our Galaxy or other galaxies. Also sometimes called a galactic cluster, not to be confused with a cluster of galaxies.

open universe A big-bang cosmology in which the Universe has infinite volume and will expand forever.

opposition A planet outside Earth's orbit is said to be in opposition when it is in the direction of the sky opposite that of the Sun.

optical In the visible part of the spectrum, 4,000–7,000 Å, or having to do with reflecting or refracting that radiation.

optical double A pair of stars that appear extremely close together in the sky even though they are at different distances from us and are not physically linked.

organic Containing carbon in its molecular structure.

Orion Molecular Cloud The giant molecular cloud in Orion behind the Orion Nebula, containing many young objects.

Ozma A project that searched nearby stars for radio signals from extraterrestrial civilizations.

ozone layer A region in the Earth's upper stratosphere and lower mesosphere where O_3 absorbs solar ultraviolet radiation.

paraboloid A 3-dimensional surface formed by revolving a parabola around its axis.

parallax (a) Half the angle through which a star appears to be displaced when the Earth moves from one side of the Sun to the other (2 au); it is inversely proportional to the distance; (b) other ways of measuring distance, as in spectroscopic parallax.

parallel light Light that is neither converging nor diverging.

parsec The distance from which 1 au subtends one second of arc; approximately 3.26 lt-yr.

particle physics The study of elementary nuclear particles.

peculiar motion The motion of a galaxy with respect to the Hubble flow.

penumbra (a) For an eclipse, the part of the shadow from which the Sun's disk (photosphere) is only partially occulted; (b) of a sunspot, the outer region, not as dark as the umbra.

perfect cosmological principle The assumption that on a large scale the Universe is homogeneous and isotropic in space and unchanging in time.

perihelion The near point to the Sun of the orbit of a body orbiting the Sun.

period The interval over which something repeats.

phase (a) Of a planet, the varying shape of the lighted part of a planet or moon as seen from some vantage point; (b) the relation of the variations of a set of waves.

phase transition Change from one state of matter to another, as from solid to liquid or liquid to gas; phase transitions in the early Universe marked the separation of the fundamental forces.

photometry The electronic measurement of the amount of light.

photomultiplier An electronic device that through a series of internal stages multiplies a small current that is given off when light is incident on it; a large current results.

photon A packet of energy that can be thought of as a particle travelling at the speed of light.

photon sphere The sphere around a black hole, $3/2$ the size of the event horizon of a nonrotating black hole, within which exit cones open and in which light can orbit.

photosphere The region of a star from which most of its light is radiated. The disk or apparent surface of a star.

plage The part of a solar active region that appears bright when viewed in Hα.

Planck curve See **blackbody curve**.

Planck's constant The constant of a proportionality, h, between the frequency of an electromagnetic wave and the energy of an equivalent photon. $E = hf = hc/\lambda$.

Planck's law The formula that predicts, for an opaque object at a certain temperature, how much radiation is emitted per second at every wavelength.

Planck time The time very close to the big bang, 10^{-43} seconds, before which a quantum theory of gravity would be necessary to explain the Universe and which is therefore currently inaccessible to our computations.

planet Defined by a resolution of the 2006 General Assembly of the International Astronomical Union to be a celestial body of size large enough to be approximately round (technically, in hydrostatic equilibrium), sufficiently massive to be gravitationally dominant in its neighborhood (technically, dynamically dominant, said to "clear its neighborhood"), of insufficient mass for nuclear reactions ever to begin, and generally in orbit around a star. Those that have escaped from their parent star because of gravitational interactions are known as free-floating planets.

planetary nebulae Shells of matter ejected by low-mass stars after their main-sequence lifetime, ionized by ultraviolet radiation from the star's remaining core.

planetesimal One of the small bodies into which the primeval solar nebula condensed and from which the planets formed.

plasma An electrically neutral gas composed of exactly equal numbers of ions and electrons.

plates Large, flat structures making up a planet's crust.

plate tectonics The theory of the Earth's crust, explaining it as plates moving because of processes beneath.

plumes Thin structures in the solar corona near the poles.

plutoid A dwarf planet whose orbit is beyond that of Neptune. Only four plutoids (including Pluto) had been designated as of 2012, but there are surely many more.

point objects Objects in which no size is distinguishable.

pole star A star approximately at a celestial pole; Polaris is now the north pole star; there is no south pole star at this time.

poor cluster A cluster of stars or galaxies with few members.

positive ion An atom that has lost one or more electrons.

positron An electron's antiparticle (charge of $+e$), where e is the unit of electric charge.

precession The very slowly changing position of stars in the sky resulting from variations in the orientation of the Earth's axis; a complete cycle takes 26,000 years.

precession of the equinoxes The slow variation of the position of the equinoxes (intersections of the ecliptic and celestial equator) resulting from variations in the orientation of the Earth's axis.

pre-main-sequence star A ball of gas in the process of slowly contracting to become a star as it heats up before beginning nuclear fusion.

pressure Force per unit area.

primary cosmic rays The cosmic rays arriving at the top of the Earth's atmosphere.

primary distance indicators Ways of measuring distance directly, as in trigonometric parallax.

prime focus The location at which the main lens or mirror of a telescope focuses an image without being reflected or refocused by another mirror or other optical element.

primeval solar nebula The early stage of the Solar System in nebular theories.

primordial nucleosynthesis The formation of the nuclei of isotopes of hydrogen (such as deuterium), helium, and lithium in the first few minutes of the Universe.

principal quantum number The integer n that determines the main energy levels in an atom.

prograde motion The apparent motion of the planets when they appear to move forward (from west to east) with respect to the stars; see also **retrograde motion**.

prolate Having the diameter along the axis of rotation longer than the equatorial diameter.

prominence Solar gas protruding over the limb, visible to the naked eye only at eclipses but also observed outside the eclipses by its emission-line spectrum.

proper motion Angular motion across the sky with respect to a framework of galaxies or fixed stars.

proteins Long chains of amino acids; fundamental components of all cells of life as we know it.

proton Elementary particle with positive charge $+e$, where e is the unit of electric charge, one of the fundamental constituents of an atom.

proton-proton chain A set of nuclear reactions by which four hydrogen nuclei combine one after the other to form one helium nucleus, with a resulting release of energy.

protoplanets The loose collections of particles from which the planets formed.

protostar A nebula of gas and dust that is collapsing on its way to becoming a star.

protosun The Sun in formation.

pulsar A celestial object that gives off pulses of radio waves; thought to be a rotating neutron star with a very strong magnetic field.

QSO Quasi-stellar object, formally a radio-quiet version of a quasar, though now radio-loud quasars are also sometimes included.

quantized Divided into discrete parts.

quantum A bundle of energy.

quantum fluctuation The spontaneous formation and disappearance of virtual particles; it may be a quantum violation of the classical law of conservation of energy because of Heisenberg's Uncertainty Principle.

quantum mechanics The branch of 20th-century physics that describes atoms and radiation.

quantum theory The set of theories that evolved in the early 20th century to explain atoms and radiation as limited to discrete energy levels or quanta of energy, with quantum mechanics emerging as a particular version.

quark One of the subatomic particles of which modern theoreticians believe such elementary particles as protons and neutrons are composed. The various kinds of quarks have positive or negative charges of $1/3e$ or $2/3e$, where e is the unit of electric charge.

quasar A very-large-redshift object that is almost stellar (point-like) in appearance, but has a very non-stellar spectrum consisting of broad emission lines; thought to be the nucleus of a galaxy with an accreting supermassive black hole.

quiescent prominence A long-lived and relatively stationary prominence.

quiet Sun The collection of solar phenomena that do not vary with the solar activity cycle.

radar The acronym for radio detection and ranging, an active rather than passive radio technique in which radio signals are transmitted and their reflections received and studied.

radial velocity The velocity of an object along a line (the radius) joining the object and the observer; the component of velocity toward or away from the observer.

radial-velocity method See **Doppler-wobble method**.

radiant The point in the sky from which all meteors in a meteor shower appear to be coming.

radiation Electromagnetic radiation; sometimes also particles such as protons, electrons, and helium nuclei.

radiation belts Belts of charged particles surrounding planets.

radioactive Having the property of spontaneously changing into another isotope or element.

radio galaxy A galaxy that emits radio radiation orders of magnitude stronger than do normal galaxies.

radio telescope An antenna or set of antennas, often together with a focusing reflecting dish, that is used to detect radio radiation from space.

radio waves Electromagnetic radiation with wavelengths longer than about one millimeter.

red giant A post-main-sequence stage of the lifetime of a star; the star becomes relatively luminous and cool.

reddened See **reddening**.

reddening The phenomenon by which the extinction of blue light by interstellar matter is greater than the extinction of red light so that the redder part of the continuous spectrum is relatively enhanced.

redshift A shift of optical wavelengths toward the red, or in all cases toward longer wavelengths.

red supergiant Extremely luminous, cool, and large star; a post-main-sequence phase of evolution of stars of more than about 8 or 10 solar masses.

reflecting telescope A type of telescope that uses a mirror or mirrors to form the primary image.

reflection nebula Interstellar gas and dust that we see because it is reflecting light from a nearby star.

refracting telescope A type of telescope in which the primary image is formed by a lens or lenses.

refraction The bending of electromagnetic radiation as it passes from one medium to another or between parts of a medium that has varying properties.

refractory Having a high melting point.

relativistic Having a speed that is such a large fraction of the speed of light that the special theory of relativity must be applied.

resolution The ability of an optical system to distinguish detail; also called "angular resolution."

rest mass The mass an object would have if it were not moving with respect to the observer.

rest wavelength The wavelength radiation would have if its emitter were not moving with respect to the observer.

retrograde motion The apparent motion of the planets when they appear to move backward (from east to west) from the direction that they move ordinarily with respect to the stars.

retrograde rotation The rotation of a moon or planet opposite to the dominant direction in which the Sun rotates and the planets orbit and rotate.

revolution The orbiting of one body around another.

revolve To move in an orbit around another body.

rich cluster A cluster of many galaxies.

right ascension Celestial longitude, measured eastward along the celestial equator in hours of time from the vernal equinox.

rims The raised edges of craters.

Roche limit The sphere for each mass inside of which blobs of gas cannot agglomerate by gravitational interaction without being torn apart by tidal forces; normally about 2.5 times the radius of a planet.

Roche lobe In a binary-star system, the Roche lobe of each star is the region around it where its gravity dominates over that of the other star.

rotate To spin on one's own axis.

rotation Spin on an axis.

rotation curve A graph of the speed of rotation vs. distance from the center of a rotating object like a galaxy.

RR Lyrae variable A short-period variable star. All RR Lyrae stars have approximately equal luminosity and so are used to determine distances.

S0 A transition type of galaxy between ellipticals and spirals; has a disk but no arms.

scarps Lines of cliffs; found on Mercury, Earth, the Moon, and Mars.

scattered Light absorbed and then reemitted in all directions.

Schmidt telescope (Schmidt camera) A telescope that uses a spherical mirror and a thin lens to provide photographs of a wide field.

scattered disk A region beyond the orbit of Neptune that extends to a distance of about 100 au and is thicker than the Kuiper belt, populated with icy bodies that are thought to have been gravitationally flung outward by the migration of Neptune early in the history of the Solar System. Many comets near the ecliptic plane probably came from there.

Schmidt telescope (Schmidt camera) A telescope that uses a spherical mirror and a thin lens to provide photographs of a wide field.

Schwarzschild radius The radius that, according to Schwarzschild's solutions to Einstein's equations of the general theory of relativity, corresponds to the event horizon of a black hole.

scientific method No easy definition is possible, but it has to do with a way of testing and verifying hypotheses.

scientific notation Exponential notation.

secondary cosmic rays High-energy particles generated in the Earth's atmosphere by primary cosmic rays.

secondary distance indicators Ways of measuring distances that are calibrated by primary distance indicators.

sedimentary Formed from settling in a liquid or from deposited material, as for a type of rock.

seeing The steadiness of the Earth's atmosphere as it affects the angular resolution that can be obtained in astronomical observations.

seismic waves Waves travelling through a solid planetary body from an earthquake or impact.

seismology The study of waves propagating through a body and the resulting deduction of the internal properties of the body. "Seismo-" comes from the Greek for earthquake.

semimajor axis Half the major axis; that is, for an ellipse, half the longest diameter.

semiminor axis Half the minor axis; that is, for an ellipse, half the shortest diameter.

Seyfert galaxy A type of galaxy that has an unusually bright nucleus and whose spectrum shows emission lines that cover a wide range of ionization stages; in some cases the lines are very broad, indicating rapid gas motions.

Shapley-Curtis debate   The 1920 debate (and its written version) on the scale of our Galaxy and whether the "spiral nebulae" are in our Galaxy or are independent galaxies.

shear wave   A type of twisting seismic wave.

shock wave   A front marked by an abrupt change in pressure caused by an object moving faster than the speed of sound in the medium through which the object is travelling.

shooting stars   Meteors.

showers   A time of many meteors from a common cause.

sidereal   With respect to the stars.

sidereal day   A day with respect to the stars.

sidereal year   A circuit of the Sun with respect to the stars.

significant figure   A digit in a number that is meaningful (within the accuracy of the data).

singularity   A point in space where quantities become exactly zero or infinitely large; one is present in a black hole, according to classical general relativity.

slit   A long, thin gap through which light is allowed to pass.

small Solar System body   An object in the Solar System that does not satisfy the criteria for planets, dwarf planets, or their moons (satellites). This includes most asteroids and most Kuiper-belt objects, as well as meteoroids, comets, and centaurs.

snow line   See **frost line**.

SOHO   The Solar and Heliospheric Observatory, a joint NASA and European Space Agency mission to study the Sun.

solar-activity cycle   The 11-year cycle with which solar activity like sunspots, flares, and prominences varies.

Solar and Heliospheric Observatory   See **SOHO**.

solar atmosphere   The photosphere, chromosphere, and corona of the Sun.

solar constant   The total energy hitting each unit area at the top of Earth's atmosphere at Earth's average distance from the Sun; now Total Solar Irradiance.

Solar Dynamics Observatory (SDO)   A NASA spacecraft, launched in 2010.

solar eclipse   An eclipse of the Sun, when the Moon blocks part or all of the Sun's disk (photosphere).

solar flares   An explosive release of energy on the Sun.

solar mass   The mass of the Sun, 1.99×10^{33} grams, about 330,000 times the Earth's mass.

solar rotational period   The time for a complete rotation with respect to the Sun.

solar time   A system of timekeeping with respect to the Sun such that the Sun is overhead of a given location at noon.

solar wind   An outflow of particles from the Sun representing the expansion of the corona.

solar year (tropical year)   An object's complete circuit of the Sun; a tropical year in between vernal equinoxes.

solstice   The point on the celestial sphere of northernmost or southernmost declination of the Sun in the course of a year; colloquially, the time when the Sun reaches that point.

space velocity   The velocity of a star with respect to the Sun.

spallation   The break-up of heavy nuclei that undergo nuclear collisions.

special theory of relativity   Einstein's 1905 theory of relative motion.

speckle interferometry   A method obtaining higher resolution of an image by analysis of a rapid series of exposures that freeze atmospheric blurring.

spectral class   See **spectral type**.

spectral lines   Wavelengths at which the brightness is abruptly different from the brightness at neighboring wavelengths.

spectral type   One of the categories O, B, A, F, G, K, M, L, T, Y into which stars can be classified from study of their spectral lines, or extensions of this system. The above sequence of spectral types corresponds to a decreasing sequence of surface temperatures.

spectrograph   A device to make and record a spectrum.

spectrometer   A device to make and electronically measure a spectrum.

spectroscopic binary   A type of binary star that is known to have more than one component because of the changing Doppler shifts of the spectral lines that are observed.

spectroscopic parallax   The distance to a star derived by comparing its apparent brightness with its luminosity deduced from study of its position on a temperature-luminosity diagram (determined by observing its spectrum–spectral type and luminosity class).

spectroscopy   The use of spectrum analysis.

spectrum (plural: **spectra**)   A display of electromagnetic radiation spread out by wavelength or frequency.

speed of light   By Einstein's special theory of relativity, the speed at which all electromagnetic radiation travels in a vacuum, and the largest possible speed of an object moving through space.

spherical aberration   For an optical system, a deviation from perfect focusing by having a shape that is too close to that of a sphere.

spicule   A small jet of gas at the edge of the quiet Sun, approximately 1,000 km in diameter and 10,000 km high, with a lifetime of about 15 minutes.

spin-flip transition   Transition in the relative orientation of the spins of an electron and the nucleus it is orbiting.

spiral arms   Bright regions looking like a pinwheel.

spiral density wave   A wave that travels around a galaxy in the form of a spiral, compressing gas and dust to relatively high density and thus beginning star formation at those compressions.

spiral galaxy   A class of galaxy characterized by arms that appear as though they are unwinding like a pinwheel.

spiral nebula   The old name for a shape in the sky, seen with telescopes, with arms spiralling outward from the center; proved in the 1920s actually to be a galaxy.

sporadic   Not regular.

sporadic meteor   A meteor not associated with a shower.

spring tides   The tides at their highest, when the Earth, Moon, and Sun are in a line (from "to spring up").

Square Kilometre Array   An array of radio telescopes to be built in the southern hemisphere, partly in Namibia and partly in Western Australia, with outliers extending over a still larger region. The total collecting area will be the equivalent of about 1 square km. Construction should begin in 2016, and useful observations will be possible long before the anticipated completion date of 2024.

stable   Tending to remain in the same condition.

standard candle   A modern-day use of an old term for a source of known brightness.

star   A self-luminous ball of gas that shines or has shone because of nuclear reaction in its interior.

star clusters   Groupings of stars of common origin.

stationary limit   In a rotating black hole, the radius within which it is impossible for an object to remain stationary, no matter what it does.

steady-state theory   The cosmological theory based on the perfect cosmological principle, in which the average properties of the Universe are unchanging over time.

Stefan-Boltzmann law   The radiation law that states that the energy emitted by a black body varies with the fourth power of its surface temperature.

stellar atmosphere   The outer layers of stars not completely hidden from our view.

stellar chromosphere   The region above a photosphere that shows an increase in temperature.

stellar corona   The outermost region of a star characterized by temperatures of 10^6 K and high ionization.

stellar evolution   The changes of a star's properties with time.

stellar-mass black hole   A black hole that results from the collapse of the core of a massive star; they typically have about 5 to 15 solar masses.

stelliferous era   The current era, with lots of stars in the Universe.

stones   A stony type of meteorite.

stratosphere   An upper layer of a planet's atmosphere, above the weather, where the temperature begins to increase. The Earth's stratosphere is at 20–50 km.

streamers   Coronal structures at low solar latitudes.

string theories   See **superstring theories**.

strong force   The nuclear force, the strongest of the four fundamental forces of nature.

strong nuclear force   The strong force.

subtend   The angle that an object appears to take up in your field of view; for example, the full Moon subtends 1/2°.

sunspot   A region of the solar surface that is dark and relatively cool; it has an extremely high magnetic field.

sunspot cycle   The 11-year cycle of variation of the number of sunspots visible on the Sun.

supercluster   A cluster of clusters of galaxies.

supercooled   The condition in which a substance is cooled below the point at which it would normally make a phase change. For the Universe, the point in the early Universe at which it may have cooled below a certain temperature without breaking its symmetry; for example, the strong and electroweak forces may have remained unified.

supergiant   A post-main-sequence phase of evolution of stars of more than about 8 or 10 solar masses. They fall in the upper right of the temperature-luminosity diagram; luminosity class I.

supergravity   A theory that attempts to unify the four fundamental forces.

superluminal speed   An apparent speed greater than that of light.

supermassive black hole   A black hole of millions or billions of times the Sun's mass, as is found in the centers of galaxies and quasars.

supernova (plural: **supernovae**)   The explosion of a star with the resulting release of tremendous amounts of radiation.

supernova remnant   The gaseous remainder of the star destroyed in a supernova.

superstring (string) theories   A possible unification of quantum theory and general relativity in which fundamental particles are really different vibrating forms of a tiny, one-dimensional "string" instead of being localized at single points.

symmetric   A correspondence of shape so that rotating or reflecting an object gives you back an identical form. Symmetry of forces in the early Universe corresponds to forces acting identically that now act differently as the four fundamental forces: gravity, electromagnetism, weak nuclear force, and strong nuclear force.

synchronous orbit   An orbit of the same period; a satellite in geostationary orbit has the same period as the Earth's rotation and so appears to hover.

synchronous rotation   A rotation of the same period as an orbiting body.

synchrotron radiation   Nonthermal radiation emitted by electrons spiralling at relativistic velocities in a magnetic field.

tail   Gas and dust left behind as a comet orbits sufficiently close to the Sun, illuminated by sunlight.

temperature-luminosity diagram   A diagram of a group of stars with surface temperatures on the horizontal axis and intrinsic brightness (luminosity) on the vertical axis; also called a temperature-magnitude diagram or a Hertzsprung–Russell diagram.

terminator   The line between night and day on a moon or planet; the edge of the part that is lighted by the Sun.

terrestrial planets   Mercury, Venus, Earth, and Mars.

theory   A later stage of the traditional form of the scientific method, in which a hypothesis has passed enough of its tests that it is generally accepted.

thermal equilibrium   The condition achieved when the energy generation rate in the interior of a body is balanced by the energy loss rate to the outside world.

thermal pressure   Pressure generated by the random motion of particles that can be characterized by temperature.

thermal radiation   Radiation produced by the random motion of particles having a distribution of energies determined only by their temperature. Blackbody radiation, which follows Planck's law, is thermal radiation.

thermosphere   The uppermost layer of the atmosphere of the Earth and some other planets, the ionosphere, where absorption of high-energy radiation heats the gas.

3° background radiation   The isotropic blackbody radiation at 3 K, thought to be a remnant of the big bang.

tidal force   The difference in gravitational force exerted by one body on various parts of another body.

tide   A periodic variation of the force of gravity on a body, based on the difference in the strength of gravitational pull from one place on the body to another (the tidal force); on Earth, the tide shows most obviously where the ocean meets the shore as a periodic variation of sea level.

time dilation   According to relativity theory, the slowing of time perceived by an observer watching another object moving rapidly or located in a strong gravitational field.

Total Solar Irradiance   The former solar constant. It is now measured from several spacecraft, including SOHO, ACRIMSAT, and SORCE.

TRACE   The Transition Region and Coronal Explorer (1998–2010), a satellite that studied the solar corona and the chromosphere-corona transition region.

transit   The passage of a celestial body in front of another one that subtends a larger angle. When a planet is *in transit*, we understand that it is passing in front of the Sun. Also, *transit* is the moment when a celestial body crosses an observer's meridian, or the special type of telescope used to study such events.

transit method   A method for detecting exoplanets whose orbital plane is nearly along the line of sight. There is a very slight drop in a star's light as an exoplanet transits across the face of its parent star. Also known as the **blink method**.

transition zone   The thin region between a chromosphere and a corona.

Trans-Neptunian Object (TNO)   One of the sub-planetary objects in the Kuiper belt; Pluto, now classified as a dwarf planet, is one of the largest such objects.

transparency   Clarity of the sky.

transverse velocity   Velocity along the plane of the sky.

triangulation   The process of determining the distance of a relatively nearby star by measuring apparent changes in its position relative to more distant stars while the Earth orbits the Sun. Also called **trigonometric parallax**.

trigonometric parallax   See **triangulation**.

triple-alpha process   A chain of fusion processes by which three helium nuclei (alpha particles) combine to form a carbon nucleus.

tropical year   The length of time between two successive vernal equinoxes.

troposphere   The lowest level of the atmosphere of the Earth and some other planets, in which all weather takes place.

T Tauri star   A type of irregularly varying star, like T Tauri, whose spectrum shows broad and very intense emission lines. T Tauri stars have presumably not yet reached the main sequence and are thus very young.

tuning-fork diagram   Hubble's arrangement of types of elliptical, spiral, and barred spiral galaxies.

21-cm line   The 1420-MHz line from neutral hydrogen's spin-flip.

twinkle   A scintillation – rapid changing in brightness – and slight changing in position of stars as their light passes through the Earth's atmosphere.

Type Ia supernova   A supernova whose distribution in all types of galaxies, and the lack of hydrogen in its spectrum, make us think that it is an event in low-mass stars, probably resulting from the incineration of a white dwarf in a binary system.

Type II supernova   A supernova often associated with spiral arms, and that has hydrogen in its spectrum, making us think that it is the explosion of a massive star.

ultraviolet   The region of the spectrum 100–4,000 Å; also used in the restricted sense of ultraviolet radiation that reaches the ground, namely, 3,000–4,000 Å.

umbra (plural: **umbrae**)   (a) Of a sunspot, the dark central region; (b) of an eclipse shadow, the part from which the Sun's disk (photosphere) cannot be seen at all.

uncertainty principle   Heisenberg's statement that the product of uncertainties of position and momentum is greater than or roughly equal to Planck's constant divided by 2π. Consequently, both position and momentum cannot be known to infinite accuracy. Another version is that the product of uncertainties of energy and time is greater than or equal to Planck's constant divided by 2π.

universal gravitation constant   The constant G of Newton's law of gravity: force = Gm_1m_2/d^2.

valleys   Depressions in the landscapes of solid objects.

Van Allen belts   Regions of high-energy particles trapped by the magnetic field of the Earth. Also, analogous belts around other planets.

variable star   A star whose brightness changes over time.

vernal equinox   The equinox crossed by the Sun as it crosses the celestial equator and moves to northern declinations.

Very Large Array   The National Radio Astronomy Observatory's set of radio telescopes in New Mexico, used together for interferometry. Now officially called the Karl G. Jansky Very Large Array after its upgrade, completed in 2012, with new electronics, fiber optics, and a supercomputer.

Very Large Telescope   The set of instruments, including four 8.2-m reflectors and several smaller telescopes, erected by the European Southern Observatory on a mountaintop in Chile through 2001.

Very Long Baseline Array   The National Radio Astronomy Observatory's set of radio telescopes dedicated to very-long-baseline interferometry and spread over an 8,000-km baseline across the United States.

very-long-baseline interferometry   The technique using simultaneous measurements made with radio telescopes at widely separated locations to obtain extremely high resolution.

visible light   Light to which the eye is sensitive, about 4,000–7,000 Å.

visual binary   A binary star that can be seen through a telescope to be double.

VLA   See **Very Large Array**.

VLBA   See **Very Long Baseline Array**.

VLBI   See **very-long-baseline interferometry**.

VLT   See **Very Large Telescope**.

void   A giant region of the Universe in which few galaxies are found.

volatile   Evaporating (changing to a gas) readily; the opposite of refractory.

wave front   A plane in which parallel waves are in step.

wavelength   The distance over which a wave goes through a complete oscillation.

weakly interacting massive particle   See **WIMP**.

weak nuclear force   One of the four fundamental forces of nature, weaker than the strong force and the electromagnetic force. It is important only in the decay of certain elementary particles, such as neutrons.

weight   The force of the gravitational pull on a mass.

white dwarf   The final stage of the evolution of a star initially between 0.075 and about 8 or 10 solar masses but with a final mass no larger than 1.4 M$_\odot$. It is supported by electron degeneracy pressure. White dwarfs are found to the lower left of the main sequence of the temperature-luminosity diagram.

white-dwarf supernova   See **Type Ia supernova**.

white light   All the light of the visible spectrum together.

Wien's displacement law   The expression of the inverse relationship of the temperature of a black body and the wavelength of the peak of its emission.

WIMP   Weakly interacting massive particle, an as yet undiscovered massive particle, interacting with other elementary particles only through the weak nuclear force and gravity, that may be a major type of cold dark matter.

winter solstice   For northern-hemisphere observers, the southernmost declination of the Sun, and its date.

wormhole   The hypothetical connection or bridge between two black holes; they probably do not exist except in idealized black holes. The formal name is Einstein-Rosen bridge.

x-rays   Electromagnetic radiation between about 0.1 and 100 Å.

year   The period of revolution of a planet around its central star; more particularly, the Earth's period of revolution around the Sun.

zenith   The point in the sky directly overhead an observer.

zero-age main sequence   The curve of a temperature-luminosity diagram determined by the locations of stars at the time they begin nuclear fusion.

zodiac   The band of constellations through which the Sun, Moon, and planets move in the course of a year.

zodiacal light   A glow in the nighttime sky near the ecliptic from sunlight reflected by interplanetary dust.

INDEX

References to illustrations, either photographs or drawings, are in italics. Significant initial numbers followed by letters are alphabetized under their spellings; for example, 21 cm is alphabetized as *twenty one*. M4 appears at the beginning of the Ms. Greek letters are alphabetized under their English equivalents.

SPRING SKY

DATE	LOCAL TIME	D.S.T.
March 1	2 AM	3 AM
March 15	1 AM	2 AM
April 1	Midnight	1 AM
April 15	11 PM	Midnight
May 1	10 PM	11 PM
May 15	9 PM	10 PM
June 1	8 PM	9 PM
June 15	7 PM	8 PM

MAP BY WIL TIRION; FOR JAY M. PASACHOFF

SUMMER SKY

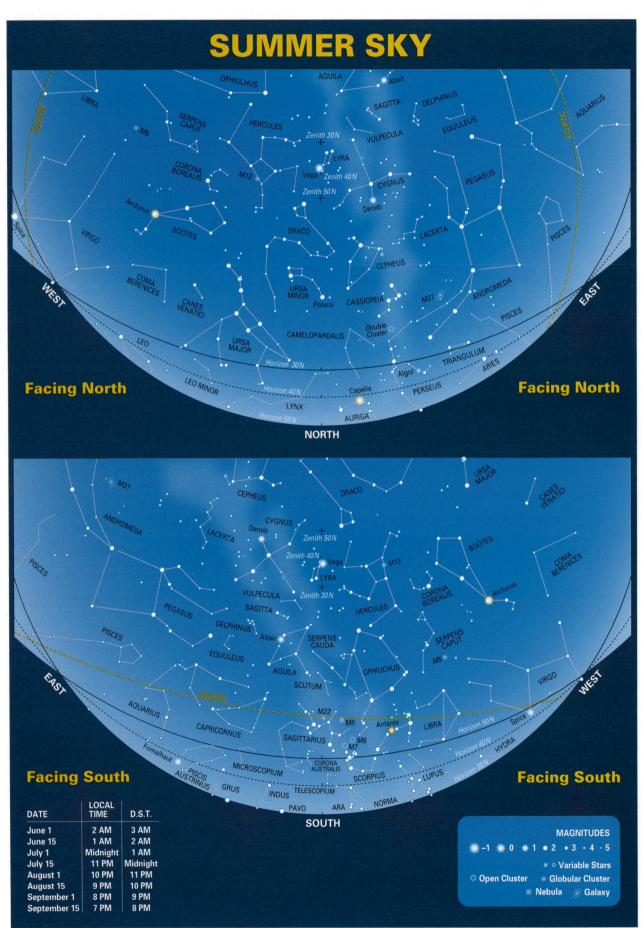

Facing North

Facing North

NORTH

Facing South

Facing South

SOUTH

DATE	LOCAL TIME	D.S.T.
June 1	2 AM	3 AM
June 15	1 AM	2 AM
July 1	Midnight	1 AM
July 15	11 PM	Midnight
August 1	10 PM	11 PM
August 15	9 PM	10 PM
September 1	8 PM	9 PM
September 15	7 PM	8 PM

MAGNITUDES
-1 0 1 2 · 3 · 4 · 5
Variable Stars
Open Cluster   Globular Cluster
Nebula   Galaxy

MAP BY WIL TIRION; FOR JAY M. PASACHOFF

AUTUMN SKY

Facing North

Facing North

Facing South

Facing South

DATE	LOCAL TIME	D.S.T.
September 1	2 AM	3 AM
September 15	1 AM	2 AM
October 1	Midnight	1 AM
October 15	11 PM	Midnight
November 1	10 PM	11 PM
November 15	9 PM	10 PM
December 1	8 PM	9 PM
December 15	7 PM	8 PM

MAGNITUDES
-1 0 1 2 · 3 · 4 · 5

Variable Stars
Open Cluster   Globular Cluster
Nebula   Galaxy

MAP BY WIL TIRION; FOR JAY M. PASACHOFF

WINTER SKY

Facing North

Facing North

NORTH

Facing South

Facing South

SOUTH

DATE	LOCAL TIME	D.S.T.
December 1	2 AM	3 AM
December 15	1 AM	2 AM
January 1	Midnight	1 AM
January 15	11 PM	Midnight
February 1	10 PM	11 PM
February 15	9 PM	10 PM
March 1	8 PM	9 PM
March 15	7 PM	8 PM

MAGNITUDES

-1　0　1　2　3　4　5

Variable Stars
Open Cluster　　Globular Cluster
Nebula　　Galaxy

MAP BY WIL TIRION; FOR JAY M. PASACHOFF